Ultrastructure of endocrine cells and tissues

ELECTRON MICROSCOPY IN BIOLOGY AND MEDICINE

Current Topics in Ultrastructural Research

SERIES EDITOR: P.M. MOTTA

Ultrastructure of Endocrine Cells and Tissues

Edited by
P.M. Motta, M.D., Ph.D.
Department of Anatomy, Faculty of Medicine
University of Rome, Italy

1984 **MARTINUS NIJHOFF PUBLISHERS**
a member of the KLUWER ACADEMIC PUBLISHERS GROUP
BOSTON / THE HAGUE / DORDRECHT / LANCASTER

Distributors

for the United States and Canada: Kluwer Boston, Inc., 190 Old Derby Street, Hingham, MA 02042, USA
for all other countries: Kluwer Academic Publishers Group, Distribution Center, P.O. Box 322, 3300 AH Dordrecht, The Netherlands

Library of Congress Cataloging in Publication Data

Main entry under title:

Ultrastructure of endocrine cells and tissues.

(Electron microscopy in biology and medicine ; v. 1)
Includes index.
1. Endocrine glands. 2. Ultrastructure (Biology) I. Motta, Pietro. II. Series.
QM576.U47 1983 599'.0873 83-2349
ISBN-13: 978-1-4613-3863-5 e-ISBN-13: 978-1-4613-3861-1
DOI: 10.1007/978-1-4613-3861-1

Preface

Innovative microscopic techniques, introduced during the last two decades, have contributed much to creating a new picture of the dynamic architecture of the cell, which can now be more exactly correlated with specific biochemical and physiopathological events.

These developments have led to significant advances in our understanding of the physiomorphological and pathological aspects of the secretory mechanism, as well as the pharmacologic methods used to control, experimentally, the function of exocrine and endocrine glands. The integration of new ultrastructural methods such as freeze-fracture/etching, immunocytochemistry, scanning and high-voltage electron microscopy, cytoautoradiography, etc., has proven to be of great value when applied to the study of endocrine cells and tissues. Because information on this topic has appeared in a variety of scientific and medical journals, this book:

(1) reviews the results of an integrative approach presenting a comprehensive ultrastructural account of the main aspects of the field;

(2) points out gaps or controversial topics in our knowledge; and

(3) outlines pertinent directions for future research.

The chapters, prepared by recognized authorities in the field, present traditional information on the topic in a concise manner and, with a valuable selection of original illustrations, show what the integration of new microscopic methods can contribute to the subject in terms of new concepts.

This volume will be useful to cell biologists, anatomists, embryologists, histologists, pharmacologists, pathologists, and, of course, endocrinologists. It will also be of interest to students, practitioners of medicine, and to all others dealing with clinical research and diagnosis.

I wish to express my sincere thanks to the contributors of the volume and to all the members of the advisory scientific committee for having responded to my numerous requests during the preparation of the volume; to H. Bell for reading selected chapters of the volume making helpful suggestions to improve the style; to J.K. Smith, medical publisher; and to the editorial and production staff of Martinus Nijhoff Publishers for their outstanding support during the entire preparation of this Series.

Finally, I have to express my deep gratitude to my wife, Silvia, whose incredible patience and love, behind the scenes, greatly helped me to realize this project.

Rome, June 1983

P.M. Motta

Contents

List of contributors

Amsterdam, Abraham. Department of Hormone Research, The Weizmann Institute of Science, Rehovot 76100, Israel

Barberini, Fabrizio. Department of Anatomy, Faculty of Medicine, University of Rome, viale R. Elena 289, I-00161 Rome, Italy

Chattarjee, Prasad. Department of Anatomy, School of Medicine, University of Leeds, Leeds LS2 9NL, United Kingdom

Correr, Silvia. Department of Anatomy, Faculty of Medicine, University of Rome, viale R. Elena 289, I-00161 Rome, Italy

Coupland Rex E. Department of Human Morphology, The Medical School University of Nottingham, Nottingham NG7 2UH, United Kingdom

Cutz, Ernest. Department of Pathology, The Hospital for Sick Children, 555 University Avenue, Toronto, Ont., Canada M5G 1X8

Ericson, Lars E. Department of Anatomy, University of Göteborg, Box 33031, S-400 33 Göteborg, Sweden

Fujita, Hisao. Department of Anatomy, Osaka University Medical School, Nakanoshima, Kitaku, Osaka 530, Japan

Girod, Christian. Department of Histology-Embryology-Cytogenetics, Fàculté de Médicine Alexis Carrel, rue Guillaume Paradin, F-69372 Lyon Cédex 2, France

Gulyas, Bela J. Pregnancy Research Branch, National Institute of Child Health and Human Development, National Institute of Health, Bethesda, MD 20205, U.S.A.

Hellström, Sten. Department of Anatomy, University of Umeå, S-901 87 Umeå, Sweden

Hervonen, Antti. Department of Biomedical Sciences, University of Tampere, Tampere 10, Finland

Horvath, Eva. Department of Pathology, St. Michael's Hospital, 30 Bond Street, University of Toronto, Toronto, Ont., Canada M5B 1W8

Kikuta, Akio. Department of Anatomy, Okayama University Medical School, 2-5-1 Shikata-cho, Okayama 700, Japan

Kjaergaard, Johan. Winslow Institute of Human Anatomy, University of Odense, DK-5230 Odense, Denmark

Kobayashi, Shigeru. Department of Anatomy, Yamanashi Medical School, Tamaho, Yamanashi 409-38 Japan

Kovacs, Kalman. Department of Pathology, St. Michael's Hospital, 30 Bond Street, University of Toronto, Toronto, Ont., Canada M5B 1W8

Kurosumi, Kazumasa. Department of Morphology, Institute of Endocrinology, Gunma University, Showa-Machi 3-39-15, Maebashi 371, Japan

Larsson, Lars-Inge. Unit of Histochemistry, University Institute of Pathology, Frederik dem V's vej 11, DK-2100 Copenhagen Ø, Denmark

Lindner, Hans R. Department of Hormone Research, The Weizman Institute of Science, Rehovot 76100, Israel

Lupulescu, Aurel. Wayne State University, 323 Medical Research Building School of Medicine, 550 East Canfield Avenue, Detroit, MI 48201, U.S.A.

Mascorro, Joe A. Department of Anatomy, Tulane University School of Medicine, New Orleans, LA 70112, U.S.A.

Mazzocchi, Giuseppina. Department of Anatomy, Laboratory of Electron Microscopy, University of Padua, via A. Gabelli 65, I-35100 Padua, Italy

McComb, Donna J. Department of Pathology, St. Michael's Hospital, 30 Bond Street, University of Toronto, Toronto, Ont., Canada M5B 1W8

Meneghelli, Virgilio. Department of Anatomy, Laboratory of Electron Microscopy, University of Padua, via A. Gabelli 65, I-35100 Padua, Italy

Mori, Hiroshi. Department of Pathology, Osaka University School of Medicine, Kitaku, Osaka 530, Japan

Motta, Pietro M. Department of Anatomy. Faculty of Medicine, University of Rome, viale R. Elena 289, I-00161 Rome, Italy

Murakami, Takuro. Department of Anatomy, Okayama University Medical School, 2-5-1 Shikata-cho, Okayama 700, Japan

Nussdorfer, Gastone G. Department of Anatomy, Laboratory of Electron Microscopy, University of Padua, via A. Gabelli 65, I-35100 Padua, Italy

Ohtani, Osamu. Department of Anatomy, Okayama University Medical School, 2-5-1 Shikata-cho, Okayama 700, Japan

Ohtsuka, Aiji. Department of Anatomy, Okayama University Medical School, 2-5-1 Shikata-cho, Okayama 700, Japan

Olivier, Léon. Laboratoire d'Histologie et d'Embryologie, Faculté de Médicine Pitié-Salpêtrière, 105 Boulevard de l'Hôpital, F-75634 Paris Cedex 13, France

Paull, Willis K. Department of Anatomy, University of Missouri-Columbia School of Medicine, Columbia, MO 65212, U.S.A.

Pearl, Gary S. Division of Neuropathology, Department of Pathology and Laboratory Medicine, Emory University School of Medicine, Atlanta, GA 30322, U.S.A.

Pelletier, Georges. MRC Group in Molecular Endocrinology, Le Centre Hospitalier de l'Université Laval, Québec, Canada G1V 4G2

Porte, Aimé. Laboratoire de Physiologie générale ULP, LA 309 CNRS Neuroendocrinologie comparée, 21, rue René Descartes, F-67084 Strasbourg Cedex, France

Rappay, György. Institute of Experimental Medicine, Hungarian Academy of Sciences, Szigony u. 43, H-1083 Budapest, Hungary

Scott, David, E. Department of Anatomy, University of Missouri-Columbia School of Medicine, Columbia, MO 65212, U.S.A.

Segi, Mitsuo†. Tohoku University, Segi Institute of Cancer Epidemiology, Yomeicho 2-5, Mizuho-ku Nagoya 457, Japan

Stoeckel, Marie-Elisabeth. Laboratoire de Physiologie Générale ULP LA 309 CNRS Neuroendocrinologie comparée, 21 rue René Descartes, F-67084 Strasbourg Cedex, France

Sundler, Frank. Department of Histology, University of Lund, Biskopsgatan 5, S-223 62 Lund, Sweden

Takey, Yoshio. Division of Neuropathology, Department of Pathology and Laboratory Medicine, Emory University School of Medicine, Atlanta, GA 30322, U.S.A.

Thiele, Jürgen. Institute of Pathology, Medical School of Hannover, Karl-Wiechert-Allee 9, D-3000 Hannover 61, F.R.G.

Unsicker, Klaus. Department of Anatomy and Cell Biology, Philipps-University, Robert-Koch-Str. 6, D-3550 Marburg, F.R.G.

Vila-Porcile, Evelyne. Laboratoire d'Histologie et d'Embryologie, Faculté de Médicine Pitié-Salpêtrière, 105 Boulevard de l'Hôpital, F-75634 Paris Cedex 13, France

Yates, Robert D. Department of Anatomy, Tulane University School of Medicine, New Orleans LA 70112, U.S.A.

CHAPTER 1

Mechanism of secretion in endocrine glands

KAZUMASA KUROSUMI

1. Introduction

The main difference between exocrine and endocrine glands is the absence of ducts in the endocrine glands. Therefore, the latter are often called ductless glands, 'glandulae sine ductibus'. Because they have no means of conveying secretory products away from the gland, the secretion enters the circulation via the blood vessels richly distributed throughout the gland. In a few cases lymphatic vessels receive the secretory products. As the action of secretion of endocrine glands transported either by blood or lymph is usually stimulation of other sensitive organs, the endocrine substance is called a hormone, meaning stimulator or accelerator. The sensitive organs which are stimulated by the hormones are termed target organs.

There are various types of endocrine glands, among which three main types, are the follicular type, the cell mass or cell cord, and the neuronic type (Fig. 1).

2. Endocrine glands of visceral type

A variety of visceral organs, for example the liver, pancreas and lung, are composed of lobes and lobules. The lobules, consist of units known as acini or alveoli. Endocrine glands such as the thyroid gland and anterior pituitary also consist of lobes and lobules. In the thyroid, the lobules contain follicles, while the lobules of the anterior pituitary consist of cell masses. Massive connective tissue septa bearing small blood vessels and peripheral nerves divide lobes into lobules. Within the lobule the interfollicular or interacinar spaces are filled with delicate strands of connective tissue stroma, within which are thin blood capillaries.

2.1. Follicular endocrine glands

Follicles are completely closed small sacs containing liquid or semiliquid material. Typical follicles are observed in the thyroid gland, while in the hypophysis and parathyroid gland follicular structures are rarely seen. In the latter organs these structures sometimes are termed cysts instead of follicles, but the two terms actually indicate the same structural entity. The follicles in the ovary are also liquid-containing sacs, but the ovarian follicles are slightly different in structure from those of other endocrine glands. The wall of the ovarian follicle is thick and, when developed, consists of a stratified cuboidal epithelium (stratum granulosum or membrana granulosa) which encloses a large oocyte.

The wall of the thyroid follicle consists of a single layer of secretory cells and its lumen contains a substance termed colloid. Cysts or follicles in the parathyroid and pars intermedia and pars distalis of the pituitary gland are similar in structure to the thyroid follicles, but their outlines are not so regularly spherical in shape as in the thyroid. Furthermore they lack the typical basal lamina which is always present in the thyroid and ovarian follicles.

In endocrine glands follicles are always surrounded by a complicated network of blood capillaries. They

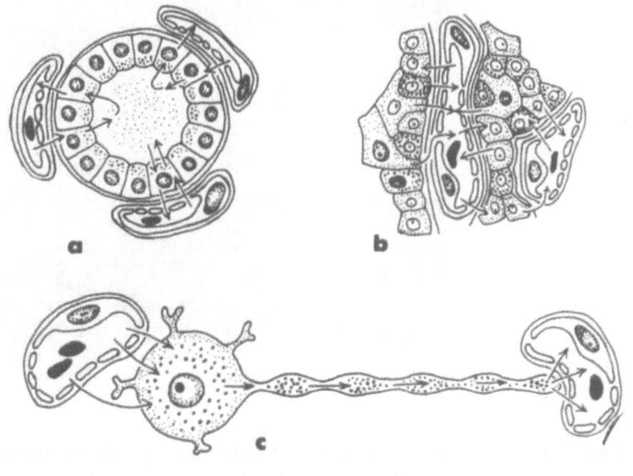

Fig. 1. Diagram of three main types of endocrine glands (a) follicular type; (b) type of cell mass or cell cord; (c) neuronic endocrine cell.

Motta, PM (ed): Ultrastructure of endocrine cells and tissues. ISBN-13: 978-1-4613-3863-5

are situated very close to the basal lamina, and often push into the follicular epithelium. The material for the production of hormones is brought into the neighborhood of endocrine cells by blood capillaries, which also transport the hormones. The follicular lumen is the storage site of incompletely synthesized hormone secreted from the follicular epithelium. Ultimately, the hormone is transported from the lumen to the bloodstream outside the follicle. Therefore, the luminal content must be reabsorbed into the follicular cells and modified in its chemical structure (usually decomposition), before it is released toward the blood vessels surrounding the outer surface of the follicle.

In this type of endocrine glands, the secretory substances (hormones and their carrier proteins)

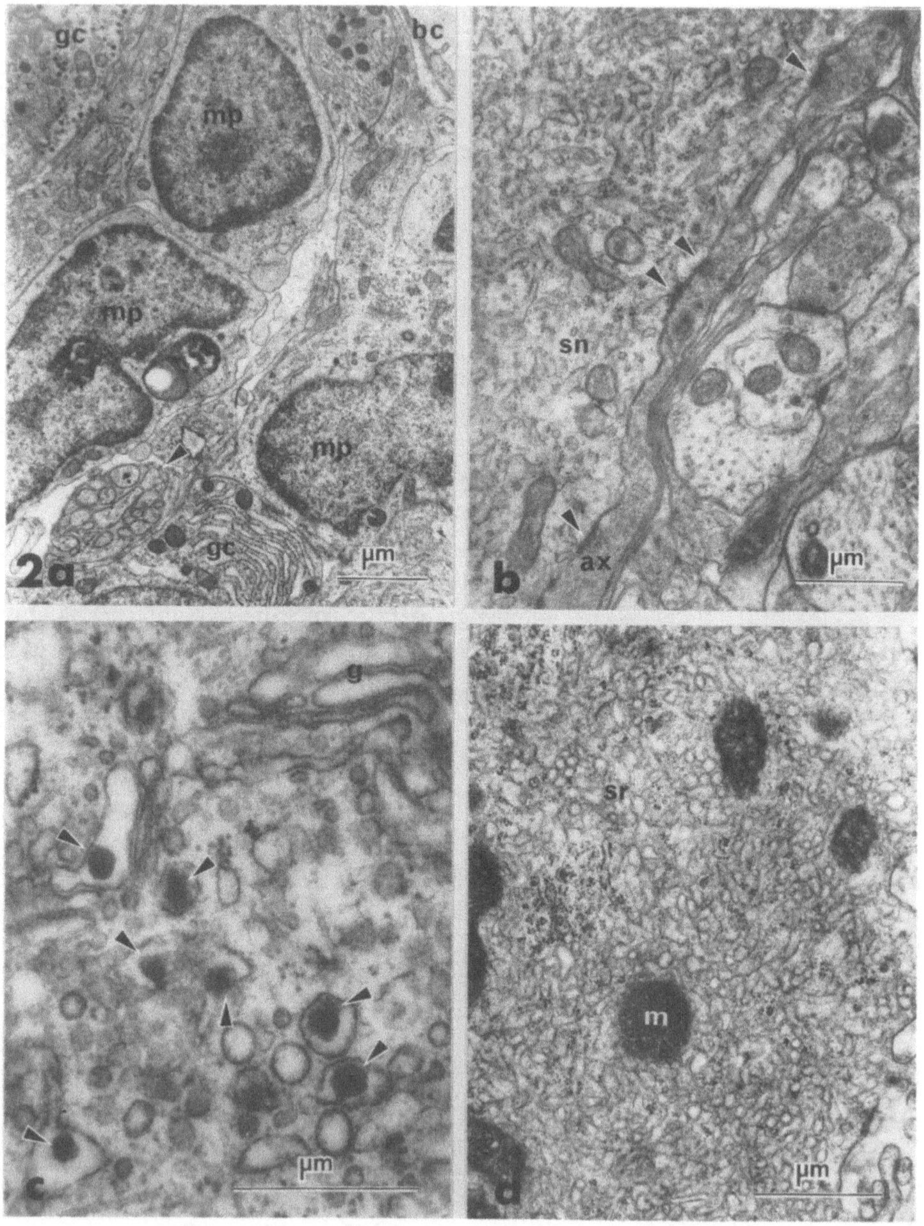

Fig. 2. (a) Pericapillary region of the anterior pituitary gland of the rat. Arrowhead: a bundle of unmyelinated nerve fibers. bc: blood capillary; mp: macrophages; gc: glandular cells; (b) A secretory neuron (sn) in the paraventricular nucleus of the rat. The cell surface is covered with an axon (ax) of another neuron, which makes axosomatic synapses (arrowheads); (c) The Golgi apparatus (g) of a neurosecretory cell in the supraoptic nucleus of the rat. Many neurosecretory granules (arrowheads) are formed within the sacs of Golgi apparatus; (d) A part of a steroid-secreting endocrine cell filled with smooth ER (sr) and mitochondria (m). Rat corpus luteum.

must be transported in two directions across the follicular wall, but in other types of endocrine tissue, such a complicated transportation of secretory substance never occurs.

2.2. Endocrine glands in the form of cell masses or cell cords

In this type of gland the secretory cells make a cluster which is either an irregular mass or an elongate cord. The cell mass is bound irregularly mostly by delicate bundles of collagen fibrils, and surrounded by networks of blood capillaries.

When the endocrine cells are arranged in cords, straight blood capillaries run parallel to the secretory cells. In such cases, the blood vessels are always provided with a basal lamina which incompletely surrounds the secretory cell masses or cords. When the basal lamina of the epithelial side is complete, the intervening tissue between the two parallel basal laminae is called the perivascular or pericapillary space. This space often contains a loose connective tissue with cells such as fibroblasts, mast cells, plasma cells and macrophages. Peripheral nerve fibers are also found in this space (1) (Fig. 2a).

2.3. Disseminated solitary endocrine cells

Endocrine cells occurring singly separated from a cell group are often observed in the mucous membrane of the digestive tract. There are many different types of enteroendocrine cells, but usually they appear singly dispersed among the exocrine or absorptive epithelial cells of the gastrointestinal mucosa. Most of the enteroendocrine cells are of neural origin, and therefore are classified as paraneurons (2). They also show characteristics of epithelial cells, such as well developed microvilli on the free surface facing the gut lumen and are provided with distinct junctional complexes. Occasionally such paraneuronic endocrine cells of the gastrointestinal tract are structurally similar to the adjacent secretory epithelial cells.

The islets of Langerhans in the pancreas are very similar to the gastrointestinal endocrine cells, and occur frequently as single separate islet cells among the exocrine pancreatic tissue. Many other cells with endocrine functions may occur as single or small group of cells. For example, the interrenal (adrenocortical) cells are found in places in the peritoneal cavity other than the usual position, and chromaffin cells which make up the adrenal medulla and some other groups of monoamine-secreting cells are found in a variety of tissues and often called paraganglia.

3. Endocrine glands of neuronal origin

The nervous system and endocrine glands are two major systems engaged in regulatory functions of the animal body. These two systems, however, are not clearly separated from each other. Very often the nerve cells also produce hormones, and act as endocrine tissues. Such a function of nerve cells is termed neurosecretion. In many invertebrate species, neurosecretion is predominant over the ordinary endocrine functions of non-neuronic cells. Typical neurosecretion is mediated by specially differentiated neurons which exert both impulse conduction and hormone production. However, some endocrine cells are presumed to be of neuroectodermal origin or to derive from some anlage related to the nervous tissue. These latter cells are usually called paraneurons, and the term indicates their intermediate position between neurons and epithelial endocrine cells.

3.1. Neuronic endocrine cells (Typical neurosecretory cells)

Neurons of the magnocellular neurosecretory system of the hypothalamus, which produce vasopressin and oxytocin, are one of the most characteristic types of neurosecretory cells. In these cells the structure of the perikaryon, axons and dendrites is that of typical neurons. On the surface of the cell body, a number of terminals from other neurons come in contact to form axo-somatic synapses (Fig. 2b) (3). Within the cytoplasm of neuronic endocrine cells is a great amount of rough endoplasmic reticulum (Nissl bodies) for the synthesis of peptide hormones and their carrier proteins. These substances are transported to the Golgi apparatus, where they are packed in vacuoles or vesicles and form secretory granules (Fig. 2c). These granules are transported through the axon toward its end situated in the posterior pituitary. Certain portions of the axon containing a great number of neurosecretory granules expand enormously and are called Herring bodies. Sometimes lysosomes are contained in the axons and Herring bodies, where they probably digest secretory granules by a mechanism termed 'crinophagy'. Physiological studies have shown that neurosecretory axons of the magnocellular system have the capability of conducting impulses (4).

The parvocellular hypothalamic neurosecretory cells also produce smaller neurosecretory granules which contain either releasing or inhibiting hormones. These, in turn, control the secretion of anterior pituitary hormones after transportation from the median eminence to the anterior lobe through the hypophyseal portal system.

The pineal body is also a site of neurosecretion. Synaptic ribbons are often observed in these cells (pinealocytes) and are known to be characteristic structures of sensory cells such as those in the retina and the inner ear. However, it is not known whether the pinealocytes have the ability to conduct impulses.

Production of neurotransmitter such as catecholamines or acetylcholine may be considered a sort of neurosecretory activity. However, these transmitters are not called hormones, because they act as stimulators of adjacent neurons or other excitable cells. Furthermore, the same substance may also be classified into different categories, such as neurohormones, neurotransmitters or neuromodulators. For example, noradrenalin produced by sympathetic neurons is a transmitter, but the same substance produced by the adrenal medulla is a hormone, because it may be released into the systemic circulation and transported to distant tissues.

3.2. Paraneuronic endocrine cells

Many endocrine cells have been assumed to be paraneurons, a designation introduced by Fujita (2). They are known to produce peptides, amines and ATP, which usually are contained in the same secretory granules. Most of these cells develop from the neural crest and are closely related to neurons. The APUD-cell system advocated by Pearse (5) occupies the main part of the realm of paraneurons. Most of the gastrointestinal endocrine cells, those of the pancreatic islets, chromaffin cells of the adrenal medulla and certain other organs, parafollicular cells or C-cells of the thyroid, ACTH-producing cells in the anterior pituitary have been presumed to be paraneurons.

Some other cells producing amines and/or peptides such as melanocytes, mast cells, Merkel cells and carotid body cells also belong to the paraneuron system. Some of these cells are known to produce bioactive substances, though these substances are not classified as hormones. Some paraneurons function as sensory receptors. Pinealocytes might be paraneurons, because these cells, in lower vertebrates, are typical photoreceptors. Secretory activities of circumventricular organs are well known, but it has yet to be determined whether their cells should be termed neurosecretory neurons or paraneurons.

4. Differences in ultrastructure between peptide hormone secreting cells and steroid hormone-secreting cells

The chemical structure of hormones is widely variable and can be closely related to the ultrastructure of secretory cells. There are a great number of peptide hormone-producing cells in different kinds of endocrine glands in which the secretions are either proteins, polypeptides or oligopeptides. If a hormone is rather small in molecular weight, the hormone is always bound to a carrier protein. For example, the oligopeptide hormone of the posterior pituitary (vasopressin or oxytocin) is bound to a carrier protein called neurophysin.

Amines are akin to amino acids which are the components of peptides, and the amine hormones are also bound to carrier proteins. For instance, the adrenal medullary hormones, that is catecholamines such as epinephrine and norepinephrine, are associated with the protein chromogranin.

Modified amino acids are also found among hormones; for example, thyroid hormones are iodinated amino acids; that is, T_4 (thyroxine) and T_3 (triiodothyronine). T_4 and T_3 are bound to a macromolecular substance, thyroglobulin that is a kind of glycoprotein stored in the follicular lumen. As these small molecular hormones are always bound to carrier proteins, the ultrastructure of the cells containing these hormones is quite similar to that of other protein-secreting cells, both in exocrine and endocrine glands. They have a well developed rough endoplasmic reticulum (ER), Golgi apparatus and secretory granules. Mitochondria and lysosomes are moderately developed and are variable in number depending upon the cell function.

Another large group of endocrine cells is that of the steroid hormone-producing cells. Steroids are derived from cholesterol and are soluble in water and lipid. Therefore, the steroid accumulations themselves cannot be recognized morphologically even under the electron microscope, but the cells producing steroids have common morphological characteristics. Functionally, steroid hormones are classified into corticoids and sex steroids. Both of these classes can be further subdivided. Corticoids are subdivided into mineralocorticoids and glucocorticoids, while the sex steroids are divided into male and female sex hormones called androgen and estrogens respectively. Progesterone is one of the female sex steroids, termed the corpus luteum hormone, and chemically the precursor for all other steroid hormones.

Steroidogenic enzymes reside either in the mitochondria or smooth endoplasmic reticulum. The first step of steroidogenesis is the formation of pregnenolone from cholesterol by the action of a side-chain cleavage enzyme which is localized in mitochondria. The enzyme 11 β-hydroxylase is also localized in mitochondria and aids in the completion of corticoids. Therefore, during the formation of cortical steroid hormones, intermediate metabolites must be transported from mitochondria to the smooth ER, and again into mitochondria. Ultrastructural features of steroid-producing cells are large and characteristic mitochondria and abundant smooth ER (Fig. 2d). Lipid droplets are also abundant in these cells, and are thought to be site of stored of cholesterol as the raw material for steroid production. Most mitochondria of steroid-secreting cells contain cristae in the form of vesicles or tubules. Mitochondria are often very large in these cells.

As the steroid hormones are watersoluble and small in molecular size, their release from secretory cells may occur by a mechanism of diacrine secretion, where apparently no morphological movement of membranes takes place.

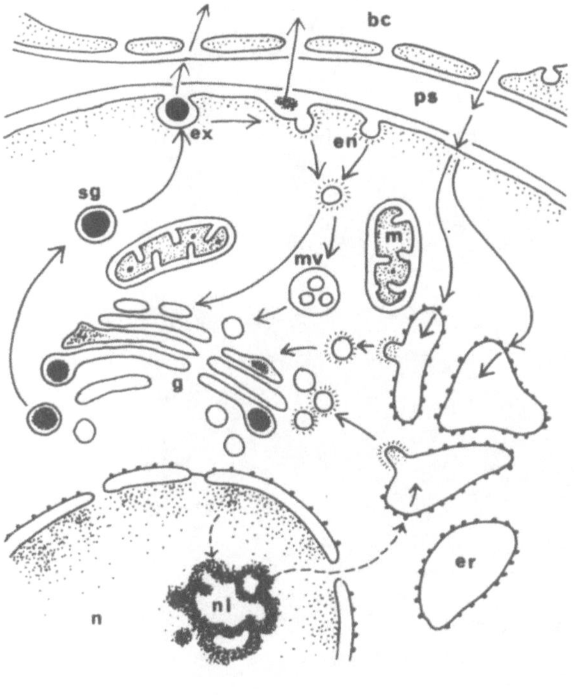

Fig. 3. Diagram of peptide-secreting endocrine cell. bc: blood capillary; ps: pericapillary space; n: nucleus; nl: nucleolus; er: rough ER; m: mitochondria; g: Golgi apparatus; sg: secretory granule; ex: exocytosis; en: endocytosis; mv: multivesicular body (lysosome). Arrows with solid lines indicate the transport of material and product of secretory biosynthesis. Arrows with broken lines indicate the transport of genetic information concerning the structure of secretory peptides.

5. Functional phases of secretory activity

5.1. Absorption of secretory material

The uptake of raw material for synthesis of secretory substances by gland cells is usually accompanied by no particular structural features, because these substances are mostly very low in molecular weight and easily transported through the plasma membrane, though some exocrine cells are provided with ample folds of plasma membrane which may serve as an ion pump for transportation of low molecular substance with water (6).

On the surface of endocrine cells of a variety of glands, many pits highly variable in size are observed. They are probably related to pinocytosis (endocytosis) of some materials outside the secretory cells. But they may also be concerned with the retrieval of surface membrane after exocytosis of secretory granules. This retrieval prevents the increase of cell surface area due to exocytosis on the one hand (7), but on the other hand it may also play a role in the uptake of material for the production of secretory substance. However, most of the absorption of secretory material may be accomplished without apparent membrane movement.

5.2. Synthesis of secretory substances

As mentioned, steroid hormones are synthesized in mitochondria and smooth ER, but it is well known that the synthesis of protein takes place in the rough ER. Both protein (peptide) hormones and carrier proteins of oligopeptide or amine hormones are synthesized on the ribosomes attached on the outer surface of rough ER. Biological events occurring in the gland cells which synthesize protein for secretion were morphologically and biochemically studied chiefly in pancreatic acinar cells (8) and also in anterior pituitary cells (9) (Fig. 3).

The rough ER consists of sacs of variable sizes and shapes, each of which is bounded by a membrane whose outer surface is studded with ribosomes, most of which are clustered to polysomes (Fig. 4a, b). The sacs of rough ER often called 'cisternae' are sometimes flattened and arranged parallel to one another (Fig. 4b). In other cases the cisternae are rather dilated and contain fluid in the cavity (Fig. 4c). If the protein hormone-secreting cells are stimulated and the biosynthesis of protein in the rough ER is accelerated, cisternae become dilated and filled with newly synthesized protein (Fig. 4c). Segregation of protein in the cisternal cavity may produce globular granules which were first observed in the pancreatic exocrine cells and called 'intracisternal granules' by Palade (10). Similar intracisternal granules are observed in the dilated cavities of rough ER of thyrotrophs of the anterior pituitary after thyroidectomy (Fig. 4d) (11). Typical gonadotrophs of the anterior pituitary are strongly hypertrophied after castration. They are also termed 'signet-ring cells', because the dilated cisternae of rough ER are fused to form a large cavity and the nucleus is pushed to the periphery of the cell (12). Despite an extraordinary dilatation of ER, the cisternae do not contain intracisternal granules in cells following castration.

The protein secretory granules are always formed in the Golgi apparatus. Therefore, the newly synthe-

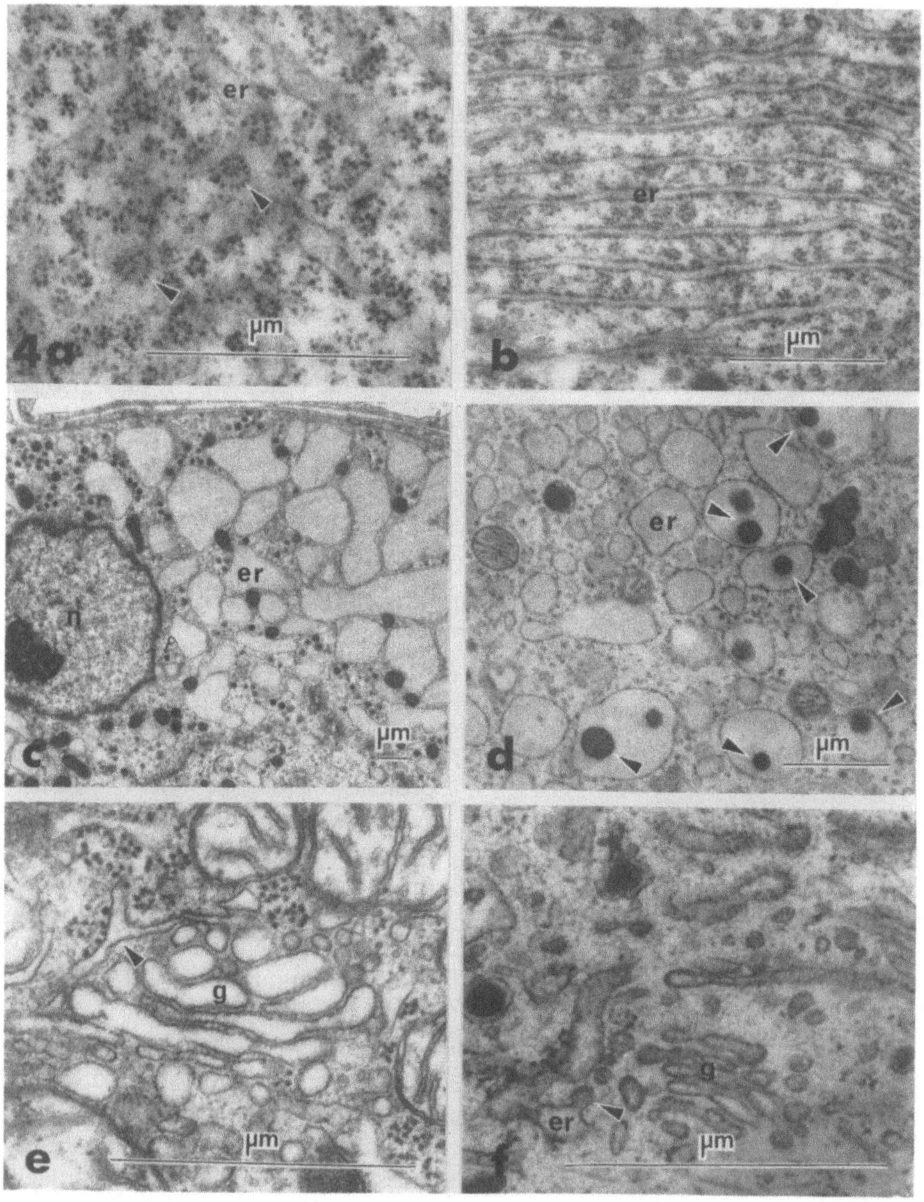

Fig. 4. (a) Polysomes either attached on the surface of ER (er) or free in the cytoplasm. Some polysomes are spiral-shaped (arrowheads). Neurosecretory cell of the rat supraoptic nucleus; (b) Parallel-arranged and flat cisternae of ER (er). Ribosomes appear as clusters (polysomes). Neurosecretory cell of the rat supraoptic nucleus; (c) Dilated ER filled with a homogeneous substance (er). A hypertrophied gonadotroph in the anterior pituitary of the castrated rat. n: nucleus; (d) Dilated ER (er) containing round granules (arrowheads). A hypertrophied thyrotroph in the anterior pituitary of the thyroidectomized rat; (e) The Golgi apparatus (g) of a secretory neuron of the rat supraoptic nucleus. Arrowhead indicates the continuity between the rough ER and Golgi cisternae; (f) The Golgi apparatus (g) of a pars intermedia cell of the rat pituitary gland. Arrowhead indicates the budding from the rough ER (er).

sized substance must be transported from the rough ER to the Golgi apparatus and this occurs in different ways in various types of secretory cells. For example a direct communication between rough ER and Golgi cisternae was found in the neurosecretory cells in the supraoptic nucleus (Fig. 4e) (13), while in many other protein-secreting cells there occurs a budding or blebbing from the cisternae of rough ER facing the Golgi apparatus, termed 'transitional elements' of the ER (14). Small vesicles formed by budding from the rough ER may transport half-made secretory substances toward the Golgi apparatus (Fig. 4f).

5.3. Condensation of secretory substances and granule formation

Condensation of the protein secretory substance, addition of polysaccharides, and packaging of this material into secretory granules take place within the Golgi apparatus. The Golgi apparatus consists of sacs bounded with smooth-surfaced membranes. Some large and more or less dilated sacs are called vacuoles; some are flattened and arranged parallel to one another forming a stack or heap of lamellae; others, round and very small, are termed vesicles. If rapid-freeze substitution fixation is applied to secretory cells, most of the Golgi sacs are flat and the so-called vacuoles are very rarely observed (Fig. 5a). Therefore, most Golgi vacuoles may be a result of fixation. The stacks of lamellae (flattened sacs) are usually curved; their convex side is the so-called 'forming face' or 'cis side', while the concave side represents the 'maturing face' or 'trans side'. The newly synthesized secretory substance in the rough ER may be transported by the vesicles of transitional elements and fuse to the Golgi sacs at the forming or cis side. The formation of secretory granules may occur at the other side of Golgi lamellae (maturing or trans side). Shift of the sacs from the cis side to the trans side is postulated, but it has not been proved.

Novikoff (15) proposed a kind of membrane-bound structure which he termed GERL. This is a part of the ER situated near the Golgi stack that produces lysosomes. The Golgi lamellae can be divided into three parts as observed by electron microscopic cytochemistry: a) the outermost (cis side) lamellae are strongly blackened after immersion in osmium tetroxide, b) the middle lamellae are positive to the detection of thiamine pyrophosphatase (TPPase) activity and, c) the innermost (trans side) lamellae positively react to the test of acid phosphatase (AcPase) activity (16). Novikoff's GERL is characterized by the positive reaction to AcPase test. Thus the innermost lamellae may correspond to GERL. Sometimes straight and flat sacs are found at the inner aspect of Golgi lamellae and react positively to AcPase activity. These flat sacs are called 'rigid lamellae' (17) and thought to be a part of GERL of Novikoff. Though Pelletier and Novikoff (18) argued that the secretory granules are formed at the GERL but not in the Golgi stack, we demonstrated granule formation both at the Golgi lamellae and the GERL (Fig. 5b) (16). Fujita and Okamoto (19) showed overlapping between Golgi lamellae positive to TPPase and those positive to AcPase. Therefore, the GERL is a part of the Golgi apparatus and not a part of the ER. In conclusion, secretory granules containing protein hormones are formed in any part of the Golgi lamellae including GERL in a variety of endocrine cells (Fig. 5a, b).

Some hormones or carrier substances are glycoprotein and the binding of protein secretory substance with saccharides and sulfates may occur at the TPPase positive cisternae in the middle portion of the Golgi lamellae (20, 21).

The size and shape of secretory granules are characteristic for each type of secretory cell. They depend upon the mechanism of formation of secretory granules by the Golgi apparatus, which might be determined genetically. The morphological characteristics of secretory granules are useful to identify the cell type in a certain endocrine gland. Small round granules are common in protein hormone-producing cells. They are usually smaller than the secretory granules of exocrine cells. Some endocrine cells have irregularly shaped secretory granules, as for example the prolactin-secreting cells of the anterior pituitary. Other cells contain crystallized secretory granules such as beta cells of pancreatic islets of some animals.

5.4. Storage and transport of secretory substance within the cell

In endocrine glands, there are three modes of processing of secretory products. The first is intracellular storage, which is observed most frequently. The secretory substance is packed in secretory granules, which accumulate in the cytoplasm. The second type is the extracellular storage and corresponds to the follicular type of endocrine glands. Here the secretory substance packed into granules at the Golgi apparatus is discharged into the follicular lumen where it is stored. When the follicular epithelial cells are stimulated, the follicular substance (colloid) is engulfed by the epithelial cells, where it is broken down by lysosomes with a consequent liberation and discharge of small molecular weight hormones into the blood stream. In the third type of processing the hormone is not stored but is released into the blood circulation immediately after its production. The steroid-producing cells belong to this type of endocrine glands.

The overproduction of hormone is controlled by the breaking down of excess secretory granules by multivesicular bodies, which are known to be a variety of lysosomes (Fig. 5c). This process of hydrolysis of secretory granules by lysosomes was first described by Smith and Farquhar (22) and called 'crinophagy'.

Once formed, secretory granules are transported

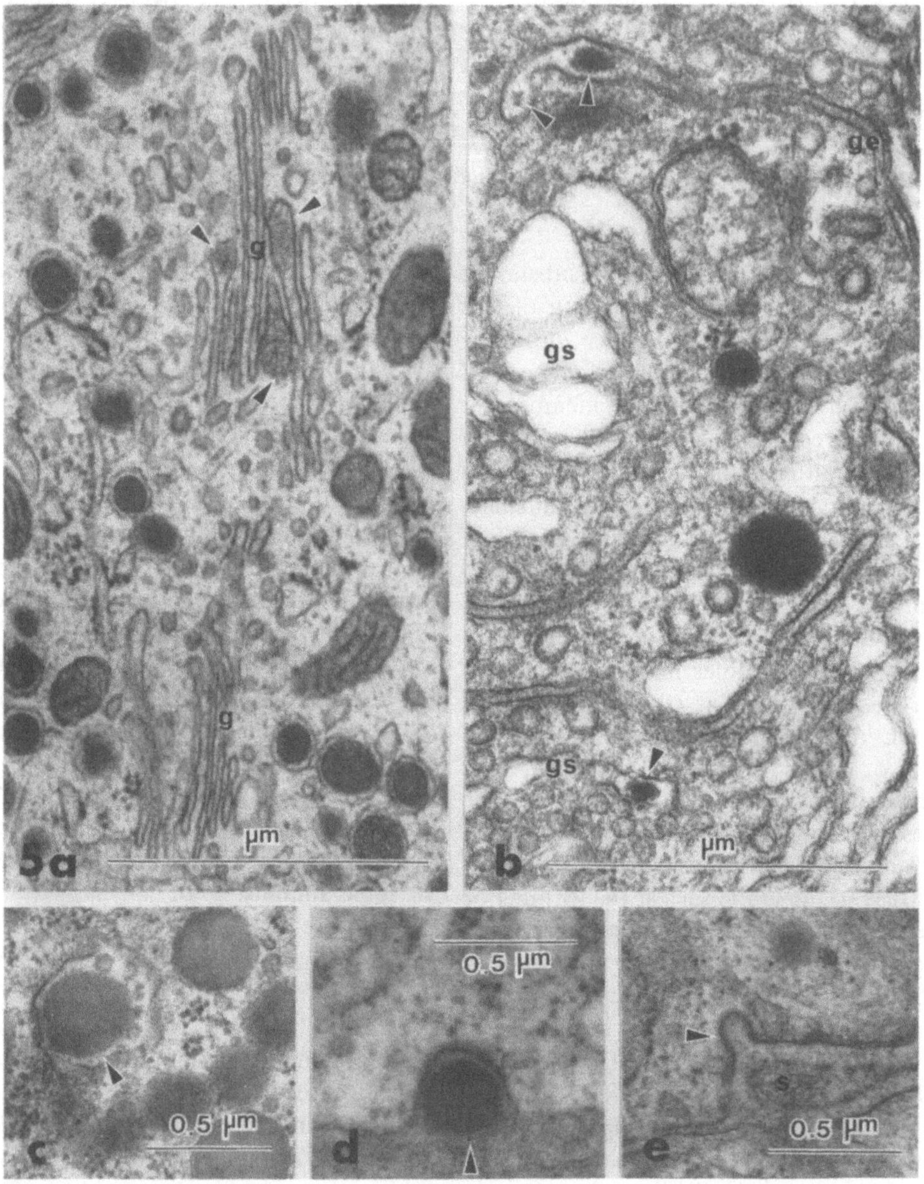

Fig. 5. (a) The Golgi apparatus (g) of a pars intermedia cell of the rat pituitary processed by the rapid-freeze substitution fixation. The Golgi cisternae arranged in parallel lamellae are not dilated, except for the parts condensing secretory substance (arrowheads); (b) Golgi apparatus and its neighborhood of an anterior pituitary cell of the rat fixed with glutaraldehyde and osmium tetroxide. The cisternae of the Golgi stack (gs) are highly dilated. GERL (ge) occurring inside the Golgi area is not dilated. Secretory granules arise both in the Golgi stack and GERL (arrowheads); (c) A multivesicular body containing a secretory granule (arrowhead), suggesting crinophagy. The rat anterior pituitary; (d) A secretory granule released by the mechanism of exocytosis (arrowhead). The rat anterior pituitary; (e) On the surface of an anterior pituitary cell is observed a pit for the endocytotic retrieval of the surface membrane (arrowhead). Secretory substance (s) is seen in the extracellular space.

from their site of production in the Golgi apparatus to the cell surface where their release takes place. In secretory neurons, the distance between the production and release site is very long, consisting of neurosecretory axons.

In the study of beta cells of pancreatic islets, Lacy and associates (23) suggested that microtubules and microfilaments are related to the transport of secretory granules, but detailed morphological evidence of this process is lacking.

5.5. Release of secretory substances and recycling of membrane material

The mechanism of release of secretory substances was first studied on skin glands by light microscopy and classified as three types, holocrine, apocrine and eccrine. The apocrine and eccrine secretions were named generically as merocrine secretion, but this term is often used as the synonym of eccrine secretion especially in English-speaking countries. The author (24) by electron microscopy, proposed a new classification of five types: type I (holocrine), type II (macroapocrine), type III (microapocrine), type IV (exocytosis or eruptocrine) and type V (diacrine).

There has been no report on the capacity of normal endocrine glands to release their secretion by a holocrine mechanism. Occasionally cell death may occur and if the dead cells contain active hormone, it may be released into the blood stream after the breakdown of the cell. This liberation may be classified as a type I (holocrine) mechanism, but it is not a physiological event. This type has been suggested in certain tumors, i.e., insulinoma (25). Such a holocrine-like cytolitic effect has been reported also in degenerating interstitial tissue such as in the ovary of some mammalian species (26).

Apocrine secretion (types II and III) is also very rare in endocrine glands, but processes of cytoplasm with the appearance of apocrine processes of exocrine glands may be seen in the surface epithelium of the pars intermedia of the rat pituitary, facing the residual cavity of Rathke's pouch (27). This cavity sometimes contains colloid-like substance, but it is not certain whether this substance contains hormone or not. Similar bulbous protrusions extending from the surface of marginal cells of the residual cavity of Rathke in the rat pituitary were observed by Correr and Motta (28, 29) with both transmission and scanning electron microscopy. The investigators thought these structures might be related to the endocytotic absorption of colloid or fluid in the residual cavity, but were not secretory by an apocrine mechanism.

In the follicular cells of the thyroid gland, tongue-like processes are often observed on the luminal surface. These processes were first thought to be apocrine secretory processes (30), but later were referred to as pseudopods for phagocytosis of colloid (31). Therefore the concept of apocrine secretion by thyroid follicular cells has been dismissed. Apocrine-like protrusions which actually are associated with pinocytosis (endocytosis) were not only observed in thyroid follicular cells, but also found in the endometrium and named 'pinopods' (32, 33). The processes of marginal cells lining the pituitary cavity (Rathke) are thought to be an example of pinopods (28, 29).

Most endocrine cells secreting protein hormone extrude the secretory granules by exocytosis. The membrane covering the granule is derived from Golgi sacs and comes in contact with the inner aspect of the surface plasma membrane. The two membranes eventually are fused and form a small pore through which the secretory substance may leak out into the extracellular space (Fig. 5d). The hormone then may be transported by the flow of extracellular fluid to the pericapillary space and transferred into the blood stream.

In the anterior pituitary, most types of gland cells release hormones by exocytosis with the exception of typical gonadotroph cells (type I gonadotroph cell) where a diacrine mechanism has been suggested (27). After stimulation by castration, however, exocytosis has been found in rat castration cells which were probably changed from the typical gonadotrophs (7).

Diacrine secretion is also called molecular permeation, because no morphological changes are found during the release of secretion. The secretory substance may be released in molecular form penetrating through the plasma membrane. This may occur in some gland cells which secrete a low molecular weight substance, such as steroid hormone. Negative data for exocytosis cannot be used for the evidence of diacrine secretion, because the exocytosis may be observed after strong stimulation, such as typical gonadotrophs after castration mentioned above, and after careful observations on the sections made by rapid-freeze substitution fixation. The mechanism of release of posterior pituitary hormones has been repeatedly debated. Though some investigations argued that neurosecretory granules might be extruded by exocytosis (34), its frequency is very low (35), and it is often very difficult to observe (36). Because the author (37) found flattened small sacs in the neurosecretory terminals after dehydration, he thought that the flat sacs may be the remnant of secretory granules, from which the interior substance leaked out by a diacrine mechanism. However, recent observations after rapid-freeze substitution fixation with liquid propane (38) demonstrated that exocytosis is not a rare event, when animals were dehydrated (39).

In the first step of exocytosis, granule and plasma membranes fuse at the cell surface. If the sample was processed by conventional fixation and sectioning, both membranes appear as trilaminar structures. In the areas of contact, the membranes display a pentalaminar structure, but later the middle dark layer disappears and the fused membrane becomes trilaminar (14). This fused membrane ruptures and the granule membrane is incorporated into the cell surface membrane. As a result of the exocytosis, the cell surface may increase in area. To maintain a constant cell surface area, the surface membrane must be taken into the cytoplasm by endocytosis (pinocytosis) (Fig. 5e). A great number of small vesicles are contained in the neurosecretory terminals and they increase after active release of the neurosecretory substance due to dehydration (37). Nagasawa et al. (34, 40) demonstrated the uptake of horseradish peroxidase (HRP) into these small vesicles and concluded that these vesicles may be formed by micropinocytosis. In the anterior pituitary, micropinocytosis by gonadotrophs becomes active after castration (7). In the same tissue, Pelletier (41) and Farquhar et al. (42) showed that the ingested HRP was localized in the innermost lamellae of the Golgi apparatus. Other authors (43) argue that the place where the ingested substance resides after exocytosis is nothing but GERL of Novikoff (15). Farquhar (44) also demonstrated the cationized ferritin in each lamella of the Golgi stack at the periphery of forming secretory granules, GERL and lysosomes. The retrieved membrane and ingested material in pinocytotic vesicles are incorporated into lysosomes, where they are digested. They subsequently appear in GERL, Golgi lamellae and forming secretory granules, suggesting that the same membrane material can be recycled and used to form the membrane covering new secretory granules.

Freeze-fracture images showed the place of membrane fusion to be smooth without membrane particles (45). At the bottom of a pit, which was a secretory granule before exocytotic opening, an accumulation of membrane particles was observed. Ishimura et al. (45) thought that such a gathering of membrane particles is the place where micropinocytosis for membrane retrieval may take place. On the contrary, other researchers (46, 47) who studied the pancreatic islets with freeze-fracture method found an accumultion of membrane-associated particles at the region of membrane junction during exocytosis. Such junctions between granule membrane and plasma membrane are permeable to ions and small molecules, and effective for solubilization of insulin prior to the extrusion. Chandler and Heuser (48) studied the exocytosis of cortical granules of sea urchin eggs immediately after fertilization with a quick-freezing and freeze-fracture technique and compared the results of this technique (49) with that of the conventional fixation and glycerination. The intramembrane particle free area occurring at the junction between the two membranes was ascribed to an artefact of glycerination.

In anterior pituitary cells, secretory granules are formed (50) and exocytosed during the course of mitosis (51). An old notion that fully differentiated cells such as actively secreting endocrine cells have lost the ability to divide by mitosis is no longer tenable for anterior pituitary cells. After castration gonadotrophs reactive to anti-LHβ serum increased enormously in number as well as in size, and such an increase in cell number is due to mitotic division (52).

References

1. Kurosumi K, Kobayashi Y: Nerve fibers and terminals in the rat anterior pituitary gland as revealed by electron microscopy. Arch Histol Jap 43: 141–155, 1980.
2. Fujita T: The gastro-enteric endocrine cell and its paraneuronic nature. In: Chromaffin, enterochromaffin and related cells. Coupland RE, Fujita T (eds), Amsterdam, Elsevier, 1976, pp 191–208.
3. Kurosumi K, Yukitake Y: Morphological and morphometric studies on the terminal boutons on the neurosecretory cells of the rat paraventricular nucleus. Arch Histol Jap 40: Suppl 293–302, 1977.
4. Yagi K, Azuma T, Matsuda K: Neurosecretory cell capable of conducting impulse in rats. Science 154: 778–779, 1966.
5. Pearse AGE: The cytochemistry and ultrastructure of polypeptide hormone-producing cells of the APUD series, and the embryologic, physiologic and pathologic implications of the concept. J Histochem Cytochem 17: 303–313, 1969.
6. Matsuzawa T, Kurosumi K: The ultrastructure, morphogenesis and histochemistry of the sweat glands in the rat foot pads as revealed by electron microscopy. J Electron Micr 12: 175–191, 1963.
7. Kurosumi K, Inoue K: Surface pits of typical gonadotrophs and castration cells of the rat anterior pituitary suggestive of exocytosis and micropinocytosis. Arch Histol Jap 43: 373–382, 1980.
8. Palade GE, Siekevitz P, Caro LG: Structure, chemistry and function of the pancreatic exocrine cell. In: The exocrine pancreas, de Reuck AVS, Cameron MP (eds), London, J and A Churchill Ltd, 1962, pp 23–49.
9. Farquhar MG, Wellings RS: Electron microscopic evidence suggesting secretory granule formation within the Golgi apparatus. J Biophys Biochem Cytol 3: 319–322, 1957.
10. Palade GE: Intracisternal granules in the exocrine cells of the pancreas. J Biophys Biochem Cytol 2: 417–422, 1956.
11. Farquhar MG, Rinehart J: Cytologic alterations in the anterior pituitary gland following thyroidectomy: an electron microscopic study. Endocrinology 55: 857–876, 1954.
12. Kurosumi K, Kawarai Y, Yukitake Y, Inoue, K: Electron microscopic morphometry of the rat castration cells. Gunma Symp Endocrinol 13: 221–236, 1976.

13. Kawabata I: Electron microscopy of the rat hypothalamic neurosecretory system. III The supraoptic nucleus after vital staining with trypan blue. Arch Histol Jap 26: 215–240, 1966.
14. Palade GE: Intracellular aspects of the process of protein synthesis. Science 189: 347–358, 1975.
15. Novikoff AB: GERL, its form and function in neurons of rat spinal ganglia. Biol Bull 127: 358A, 1964.
16. Inoue K, Kurosumi K: Cytochemical and three-dimensional studies on Golgi apparatus and GERL of rat anterior pituitary cells by transmission electron microscopy. Cell Str Func 2: 171–186, 1977.
17. Claude A: Growth and differentiation of cytoplasmic membranes in the course of lipoprotein granule synthesis in the hepatic cell. I Elaboration of elements of the Golgi complex. J Cell Biol 47: 745–766, 1970.
18. Pelletier G, Novikoff AB: Localization of phosphatase activities in the rat anterior pituitary gland. J Histochem Cytochem 20: 1–12, 1972.
19. Fujita H, Okamoto H: Fine structural localization of thiamine pyrophosphatase and acid phosphatase activities in the mouse pancreatic acinar cells. Histochemistry 64: 287–295, 1979.
20. Nakagami K, Warshawsky H, Leblond CP: The elaboration of protein and carbohydrate by rat parathyroid cells as revealed by electron microscope radioautography. J Cell Biol 51: 596–610, 1971.
21. Kurosumi K, Shibuichi I, Tosaka H: Ultrastructural studies on the secretory mechanism of goblet cells in the rat jejunal epithelium. Arch Hiatol Jap 44: 263–284, 1981.
22. Smith RE, Farquhar MG: Lysosome function in the regulation of the secretory process in cells of the anterior pituitary gland. J Cell Biol 31: 319–347, 1966.
23. Lacy PE, Howell SL, Young DA, Fink CJ: New hypothesis of insulin secretion. Nature 219: 1177–1179, 1968.
24. Kurosumi K: Electron microscopic analysis of the secretion mechanism. Internat Rev Cytol 11: 1–124, 1961.
25. Honjin R, Takahashi A, Maruyama H, Hanyu T: Electron microscopy of a case of insulinoma. J Electron Micr 14: 183–188, 1965.
26. Guraya SS, Motta PM: Interstitial cells and related structures. In: Biology of the ovary. Motta PM, Hafez ESE (eds), The Hague, Martinus Nijhoff, 1980, pp 68–85.
27. Kurosumi K, Fujita H: An atlas of electron micrographs: functional morphology of endocrine glands, Tokyo, Igaku-Shoin, 1974.
28. Correr S, Motta PM: Relationship between the marginal layer and parenchymal cells of the rat adenohypophysis as revealed by scanning electron microscopy. Biomed Res 2: Suppl 109–113, 1981.
29. Correr S, Motta PM: The rat pituitary cleft: a correlated study by scanning and transmission electron microscopy. Cell Tiss Res 215: 515–529, 1981.
30. Fujita H: Electron microscopic studies on the thyroid gland of domestic fowl, with special reference to the mode of secretion and the occurrence of central flagellum in the follicular cell. Z Zellforsch 60: 615–632, 1963.
31. Fujita H: Studies on the iodine metabolism of the thyroid gland as revealed by electron microscopic autoradiography of ^{125}I. Virchow Arch Abt B Zellpath 2: 265–279, 1969.
32. Enders AC, Nelson DM: Pinocytic activity of the uterus of the rat. Am J Anat 138: 277–300, 1973.
33. Parr MB, Parr EL: Endocytosis in the uterine epithelium of the mouse. J Reprod Fert 50: 151–153, 1977.
34. Nagasawa J, Douglas WW, Schulz RA: Ultrastructural evidence of secretion by exocytosis and of 'synaptic vesicle' formation in posterior pituitary glands. Nature 227: 407–409, 1970.
35. Krisch B, Becker K, Bargmann W: Exocytose im Hinterlappen der Hypophyse. Z Zellforsch 123: 47–54, 1972.
36. Kodama Y, Fujita H: Some findings on the fine structure of the neurohypophysis in dehydrated and pitressin-treated mice. Arch Histol Jap 38: 121–131, 1975.
37. Kurosumi K: Morphological and morphometric studies on the ultrastructural changes during the active release of neurosecretory substance from the neurohypophyseal nerve terminals in dehydrated rats. Arch Histol Jap 40: 225–242, 1977.
38. Inoue K, Kurosumi K, Deng Z-P: An improvement of the device for rapid freezing with use of liquid propane and an application of immunocytochemistry to resin section of rapid-frozen, substitution fixed anterior pituitary gland. J Electron Micr. 31: 93–97, 1982.
39. Kurosumi K, Inoue K, Yukitake Y: Ultrastructural studies on the mechanism of hormone release from the neurohypophysis by the method of rapid freezing and osmium substitution. Proc 5th Internat Taniguchi Symp for Brain Research 1983, in press.
40. Nagasawa, J, Douglas WW, Schulz RA: Micropinocytotic origin of coated and smooth microvesicles ('synaptic vesicles') in neurosecretory terminals of posterior pituitary glands demonstrated by incorporation of horseradish peroxidase. Nature 232: 341–342, 1971.
41. Pelletier G: Secretion and uptake of peroxidase by rat adenohypophyseal cells. J Ultrast Res 43: 445–459, 1973.
42. Farquhar MG, Skutelsky EH, Hopkins CR: Structure and function of the anterior pituitary and dispersed pituitary cells, in vitro studies. In: The anterior pituitary gland. Tixier-Vidal A, Farquhar MG (eds), New York, Academic Press, 1975, pp 83–135.
43. Gonatas NK, Kim SU, Stieber A, Avrameas S: Internalization of lectins in neuronal GERL. J Cell Biol 73: 1–13, 1977.
44. Farquhar MG: Recovery of surface membrane in anterior pituitary cells. Variations in traffic detected with anionic and cationic ferritin. J Cell Biol 77: R35–42, 1978.
45. Ishimura K, Egawa K, Fujita H: Freeze- fracture images of exocytosis and endocytosis in anterior pituitary cells of rabbits and mice. Cell Tiss Res 206: 233–241, 1980.
46. Berger W, Dahl G, Meissner H-P: Structural and functional alterations in fused membranes of secretory granules during exocytosis in pancreatic islet cells of the mouse. Cytobiologie 12: 119–139, 1975.
47. Dahl G, Berger W, Meissner H–P: Intracellular membrane junctions during the exocytosis of insulin. J Physiol Paris 72: 703–709, 1976.
48. Chandler DE, Heuser J: Membrane fusion during secretion. Cortical granule exocytosis in sea urchin eggs as studied by quick-freezing and freeze-fracture. J Cell Biol 83: 91–108, 1979.
49. Heuser JE, Reese TS, Dennis MJ, Jan Y, Jan L, Evans L: Synaptic vesicle exocytosis captured by quick freezing and correlated with quantal transmitter release. J Cell Biol 81: 275–300, 1979.
50. Kurosumi K: Mitosis of the rat anterior pituitary cells: an electron microscope study. Arch Histol Jap 33: 145–160, 1971.
51. Kurosumi K: Formation and release of secretory granules during mitosis in the anterior pituitary gland. Arch Histol Jap 42: 481–486, 1979.
52. Inoue K, Kurosumi K: Mode of proliferation of gonadotrophic cells of the anterior pituitary after castration – Immunocytochemical and autoradiographic studies. Arch Histol Jap 44: 71–85, 1981.

Author's address:
Department of Morphology
Institute of Endocrinology
(Gunma University)
Showa-Machi 3–39–15
Maebashi 371, Japan

CHAPTER 2

Fine structure of the pituitary pars distalis

CHRISTIAN GIROD

1. Introduction

The cytological study of the mammalian and human pituitary pars distalis has been undertaken either with cytological, cytochemical, and immunocytochemical methods or electron microscopy. Older results concerning cytological data have been collected in Romeis' monograph (1). More recently, Holmes and Ball (2) have presented an overview of the knowledge on the morphological aspects of the vertebrate pituitary gland. We have published several reviews on these subjects (3-6). The ultrastructural study of the pars distalis has been mainly investigated in laboratory mammals, with the goal of defining not only microscopic characteristics but also the physiological significance of different cell categories with respect to endocrine secretions. Since the preliminary report by Fernandez-Morán and Luft (7), and the initial description by Rinehart and Farquhar (8), a great number of reviews has been devoted to this aspect of pituitary cytology, the gland being studied *in situ* as well as *in vitro* (dispersed pituitary cells or cell cultures) (9-25). Ultrastructural immunocytochemistry also has been used effectively (26-29). After the first approaches by Foster (30-31), only a few descriptions have been reported regarding the fine structure of the normal adult human pars distalis (32-44). In this review, we will consider only the ultrastructural aspect of the secretory cells in adult mammalian and human pars distalis.

2. Structure of anterior pituitary cells in mammals

Among common laboratory animals, the albino rat has been repeatedly studied by numerous authors under normal and experimental conditions. However, since 1970, many results have been reported in other domestic and wild species. It is possible to underline common aspects whatever the species considered, but some differences appear in relation to certain species. With the electron microscope, two most helpful criteria were considered for distinguishing cell types: secretory granule size and organization of cytomembranes and organelles. We will summarize the ultrastructural characteristics of the pars distalis cell types. The pars distalis produces several hormones variable in nature (proteins, glycoproteins, peptides). In accordance with the recommendation of the International Committee for Nomenclature of the Adenohypophysis (45), cell types are named using terms reflecting their secretory function. Emphasis is placed on reporting general cytological characteristics of the different cell types independent of the species in which they are found.

2.1. Somatotropic cells

These cells (also named STH-cells, GH-cells, somatotrophs, somatotropin-producing cells) are easily identified with electron microscopy. They are the most prevalent cell type constituting about 40-50% of the antehypophyseal cell population. They are found throughout the pars distalis but especially in the lateral portions of the anterior lobe. The average diameter is slightly larger than that of other cell types; frequently, these cells are larger in males than in females. Far from capillaries somatotropic cells form cords or clusters of spherical, oval or, more rarely, polygonal cells. First identified by electron microscopy by Hedinger and Farquhar (46), the fine structure of this cell type was well defined by numerous authors in the following years (47-60) (Figs. 1 and 2).

The nucleus is usually round or oval, rather eccentric in position and contains a developed nucleolus. Various stages of mitosis can be seen in nuclei of somatotropic cells under normal conditions; in electron micrographs, chromosomes scattered throughout the nucleus during prophase or connected with spindle fibers by a kinetochore during metaphase and anaphase can be seen; mitoses are

Motta, PM (ed): Ultrastructure of endocrine cells and tissues. ISBN-13: 978-1-4613-3863-5

most frequently observed in this cell type (61).

In the cytoplasm are round or ovoid secretory granules; under normal conditions there are numerous osmiophilic, homogeneously dense granules dispersed throughout the cytoplasm. Because of the incidental section plane, their diameter appears variable, ranging between 150-400 nm, with an average diameter of 280-300 nm. In some but not all cases, mature secretory granules are surrounded by a distinct cytomembrane. In young animals, fewer granules frequently are observed. These granules have a smaller diameter and are less osmiophilic and are arranged along the plasma membrane. This 'margination phenomenon' is generally interpreted as evidence of active discharge. Occasionally, certain micrographs show discharging granules (exocytosis) in perivascular spaces and, exceptionally, in pericellular spaces ('misplaced exocytosis'). The Golgi complex is usually located near the nucleus; its development evidently depends upon the stage of secretion. In a

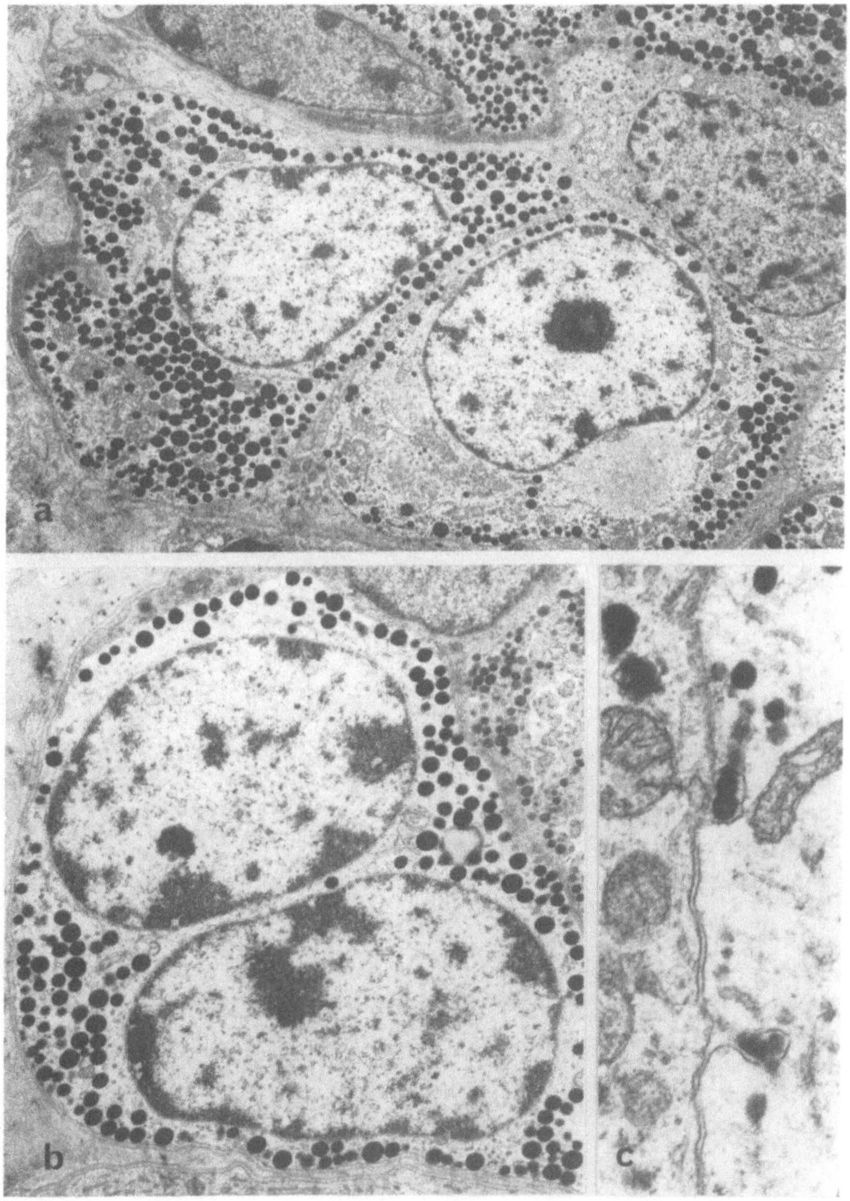

Fig. 1. Somatotropic cells in the monkey Macaca irus. (a) Various aspects of this cell type. On the right, a cell containing a globular fibrous body (× 7,840); (b) An unusual aspect of this cell-type: a binucleated cell (× 12,600); (c) Exocytosis in a somatotropic cell: numerous secretory granules are contained in folds of the plasmalemma (× 19,600).

cell engaged in granule formation, the Golgi complex consists of cisternae containing dense osmiophilic material and immature granules surrounded by a clear halo-like figure. In mature cells, the Golgi apparatus consists of flattened or slightly dilated saccules forming a rather large area near the nucleus. Mitochondria have a regular pattern of sparse internal cristae in a clear matrix which does not contain intramitochondrial granules. These organelles appear globular, filamentous or short and rodlike and they are especially numerous in the Golgi area. The rough endoplasmic reticulum appears either as randomly disposed flat cisternae or well developed parallel lamellae grouped in limited areas of the cytoplasm, often in a region opposite the Golgi apparatus. The importance of this cytomembranous system depends upon the degree of activity of the cell. Microtubules are dispersed throughout the cytoplasm or found in connection with Golgi cisternae or with centrioles (frequently organized as a diplosome). Lysosomes contain heterogeneous materials especially in large, hypertrophied cells. Single cilia with

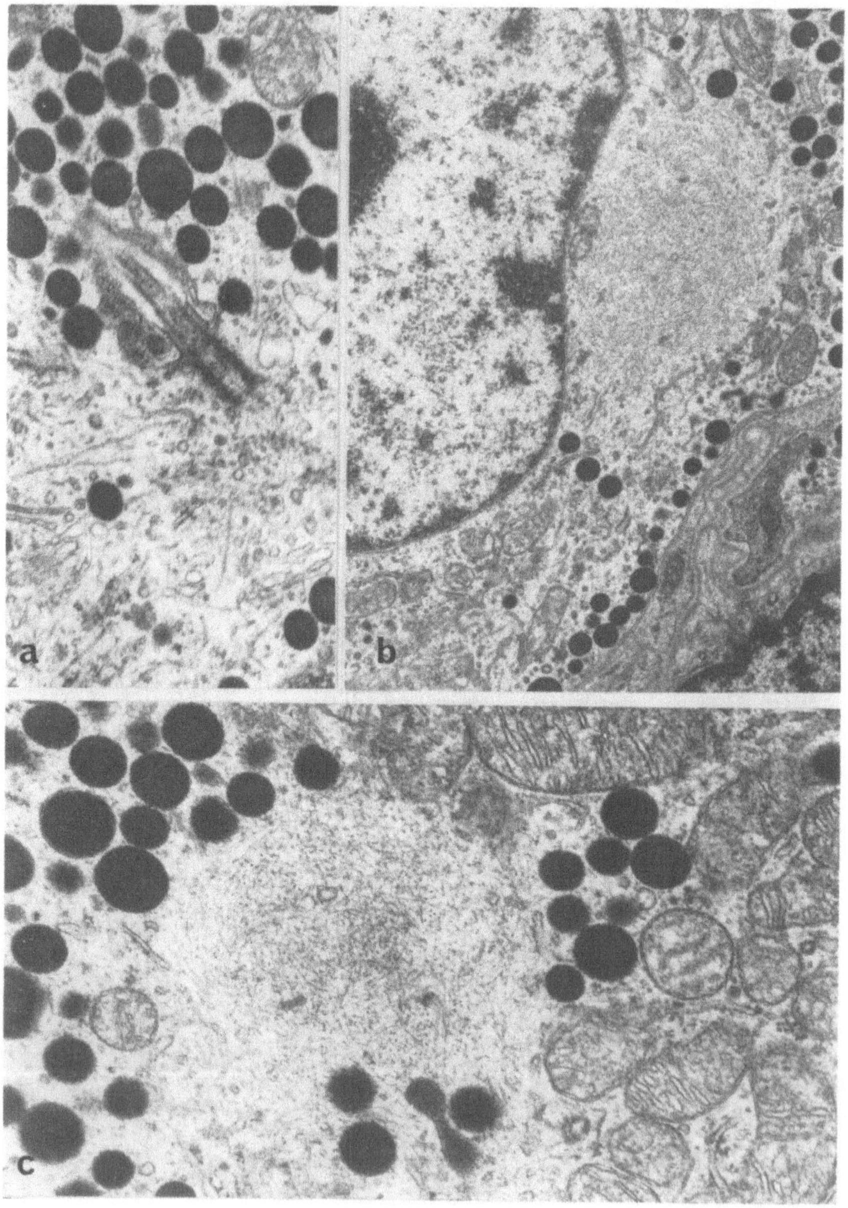

Fig. 2. Somatotropic cells in the monkey Macaca irus. (a) Single cilium and adjacent microtubules (× 20,000); (b, c) Two aspects of globular fibrous bodies (b: × 12,670; c: × 25,340).

a typical basal body, multivesicular bodies and, free ribosomes dispersed or grouped in 'rosettes' can also be seen.

In the same ultrathin section of the pars distalis, one can schematically observe two types of somatotropic cells. One type corresponds to a medium-sized cell (15-20 μm) with a large Golgi area and a few granules; the other type corresponds to a large cell (20-25 μm), the cytoplasm of which is filled with large secretory granules and contains a reduced Golgi area. The first type corresponds to an active cell engaged in hormone excretion, the second to a reserve stage in the cell cycle. These aspects correspond to the so-called functional alternation phenomenon.

2.2. *Prolactin cells*

These cells (also named PRL-cells, mammotrophs, prolactin-secreting cells, luteotrophs or LTH-cells [the latter two terms are applicable solely to rat hypophysis]) are easily recognizable with electron microscopy. The percentage of this cell type is highly variable according to the sex of the animal, predominating in the female. In addition, the percentage of prolactin cells in females depends upon the reproductive activity (nonpregnant, pregnant or lactating). It varies dramatically with experimental conditions e.g., injection of estrogens. Under normal conditions, prolactin cells appear similar in size and shape to somatotropic cells; during pregnancy and especially in lactation they are large and irregular in shape. This cell type predominates in the central portion of the lateral lobes of the pars distalis. First identified by electron microscopy by Hedinger and Farquhar (46), this cell type was well defined by numerous authors in the following years (47, 53, 56, 59, 62-68).

The Nuclei of prolactin cells have no specific characteristics and resemble those of somatotropic cells. Under normal conditions, mitoses can be seen, but more rarely than in somatotropic cells. By electron microscopy, it is possible to observe simultaneously mitoses, exocytosis and the formation of new secretory granules in the Golgi apparatus (69).

The cytoplasm contains secretory granules which are round or elliptical, or in rodents especially irregular and homogeneously dense. In most but not all species studied the prolactin cells have the largest secretory granules, being 600-800 nm in diameter. Immature granules, located in the Golgi area, are smaller (100 nm) and generally more variable in size and shape. In this area, within a golgian limiting membrane, the aggregation of smaller granules which gives a polymorphous aspect to these newly synthetized granules is observed frequently. Margination of granules and exocytosis are observed easily in this cell type notably in lactating females a few minutes after suckling. The development of the Golgi complex, located near the nucleus, depends upon the stage of secretion. In a cell engaged in granule formation, the Golgi complex contains immature secretory granules of small diameter. The Golgi apparatus is hypertrophied; halo-like figures around newly synthetized granules are evident. Mitochondria are numerous, rod-shaped, and distributed among the secretory granules. Parallel internal cristae are abundant. The rough endoplasmic reticulum, is moderately developed; it appears frequently as parallel flat cisternae at one side of the cell, often in the vicinity of the Golgi area. Microtubules, especially developed in this cell type and play a role in the displacement of secretory granules in the cytoplasm. Microtubules are also found in connection with centrioles and the Golgi apparatus. Lysosomes are especially abundant in animals in which the excretion is suppressed, e.g., lactating females after removal of suckling litters. The lysosomes move toward the secretory granules and fuse with them. This phenomenon, termed 'crinophagy', plays a role in regulating the secretory process as a result of the lysosomal degrading enzymes (proteases, peptidases) activated after the membranes coalescence. Other organelles such as multivesicular bodies and single cilia can be observed in this cell type.

In males and cycling females, prolactin cells do not show evidence of great activity. Well developed organelles appear especially during lactation or under experimental conditions (treatment with estrogens or autografts under the kidney capsule, for instance).

2.3. *Opiocortico-melanotropic cells*

One of the most controversial issues in antehypophyseal cytology has long been the identification of 'corticotropic' cells. A particular type of cell was initially described under the term 'corticotropic cells' (or ACTH-cells, corticotropin-secreting cells) because these cells were considered to be the cellular source of ACTH. However, as biochemical data established that ACTH originates from a precursor common to other peptides, the term 'corticotropic cells' could no longer be used. Since LPH, MSH and the endorphins originate from this common precursor, it seems more suitable, in a functional sense, to speak of 'opiocortico-melanotropic cells'. Furthermore, this term is in agreement with immunocytochemical data. Before the use of immunocytochemical techniques, the identification of this cell type in electron microscopy was presented first by Herlant and Klastersky in 1963 (70), and subsequently by others (71-79) (Fig. 5a).

This cell type consists of isolated or cluster-grouped cells, but not in microfollicular arrangements as emphasized by Farquhar (80). In many species, especially in rodents, these cells appear with an irregular outline and possess many thin processes intercalated between other cell types, some of which extend toward blood capillaries. The cells are not very abundant and correspond to approximately 2-5% of the antehypophyseal cell population. It seems that they are more abundant in females than in males. This cell type is located throughout the pars distalis; however, the cells are more numerous in the posteromedial part of the antehypophyseal lateral lobes.

The nucleus is usually eccentric in position and irregular in shape. As with other nuclei of the antehypophysis, mitoses can be observed but, at least under normal conditions they are exceptional in this cell type.

The cytoplasm contains irregular, pleomorphic and abundant secretory granules which present, even in the most mature granules, a halo-like figure, i.e., a clear space separating the dense granule core and the proper granular membrane. This cell type is the only one to possess such secretory granules resembling cored vesicles. This aspect depends upon the mode of fixation: it is marked when the hypophyseal fragment is fixed with osmium tetroxide alone, whereas it disappears with glutaraldehyde fixation or after successive fixation with glutaraldehyde and osmium tetroxide. In the case of double fixation the secretory granules have a high electron density. Secretory granules are dispersed throughout the cytoplasm; nevertheless, especially in thin extending processes, it is not exceptional to note a regular arrangement of the secretory granules along the plasmalemma. The granules are about 200 nm in diameter. Granular exocytosis can be seen, even under normal conditions, without any stimulation. Golgi complex is moderately developed depending upon the stage of secretion. In activated cells, e.g., 24-48 hours after adrenalectomy or metopirone treatment, the Golgi apparatus shows numerous dictyosomes encircling the nucleus. Each dictyosome consists of flattened saccules and vesicles. In the centers of a flattened saccule or in the dilated extremities of the Golgi lamellae, it is possible to see osmiophilic material corresponding to granular formation. Around the dense material of the immature granule a large clear space can be seen. Mitochondria are numerous and often regularly disposed around the Golgi area and the nucleus. Their size and shape are variable, most of them appearing as long vermicular organelles. Mitochondrial cristae are sparse and irregularly arranged in a clear matrix devoid of intramitochondrial granules. The rough endoplasmic reticulum is poorly developed under normal conditions but hypertrophied in stimulated cells and disposed in parallel lamellae or in 'Nebenkern', the smooth endoplasmic reticulum seems more abundant than in other antehypophyseal cell types. Microtubules are dispersed in the cytoplasm or in connection with the Golgi area. Lysosomes are more abundant in activated cells. Free ribosomes, dispersed or grouped in 'rosettes', can be seen.

According to Kurosumi (14), the electron microscope permits the distinction of two types of 'corticotropic' cells. In type I, the cytoplasmic processes are larger, small secretory granules are dispersed throughout the cytoplasm and appear as of cored vesicles. In type II, the processes are thinner, secretory granules are slightly larger in diameter and have the appearance of cored vesicles. These granules are usually located along the plasmalemma. In these secretory granules, the osmiophilic content reflects various degrees of electron density. Kurosumi has conjectured that type I corresponds to a storage type, and type II to a depletion type of cell.

2.4. Gonadotropic cells

Data acquired earlier by histochemical and electron microscopic methods led to the distinction of two types of 'gonadotropic' cells, classified as 'FSH' and 'LH' cells. However, the advent of immunocytochemistry resulted in the discovery that the same cell type can produce FSH and LH. This interpretation has been the subject of numerous discussions. In this presentation of morphological data obtained by electron microscopy, we will summarize the classical results (Fig. 4). The restrictions will be presented with regard to the discussion on functional data.

Gonadotropin-secreting cells are the most frequently observed type after somatotropic cells costituing 20-50% of the hypophyseal cell population. The separation of the two types, termed 'FSH-cells' or 'gonadotropes type I', and 'LH-cells' or 'gonadotropes type II' comes both from general aspects of the cells and of specific features of the cytoplasmic organelles. The existence of two kinds of gonadotropic cells was suggested as early as 1954 in the pioneer studies by Farquhar and Rinehart (81), and confirmed by different authors in following years (82-90). From a general viewpoint, FSH-cells have a large cell body while LH-cells are smaller. In many cases, the gonadotropic cells are located along blood capillaries. They are dispersed throughout the pars distalis but they are numerous either in the central portion of the lateral lobes ('central gonadotropic cells') or in a superficial area adjacent to the pars intermedia ('peripheral gonadotropic cells') the so-called 'gonadotropic zone' or 'sex-zone'.

The nucleus of the two types is not easily differentiable; the nucleus usually appears eccentric and occasionally is indented in type I and found about in the center of the cell body in type II. However, on ultrathin sections this criterion is unreliable. Mitoses can occur in gonadotropic cells, as in other cell types, but dividing gonadotropic cells are rarely encountered.

The appearance of the secretory granules has served as a basis for the distinction between the two types of gonadotropes. In type I, the secretory granules, regularly round in shape, are of two types. One type is large and less electron-dense; the other is small and dense. The diameter of the small granules varies not only with the species studied but also, in so

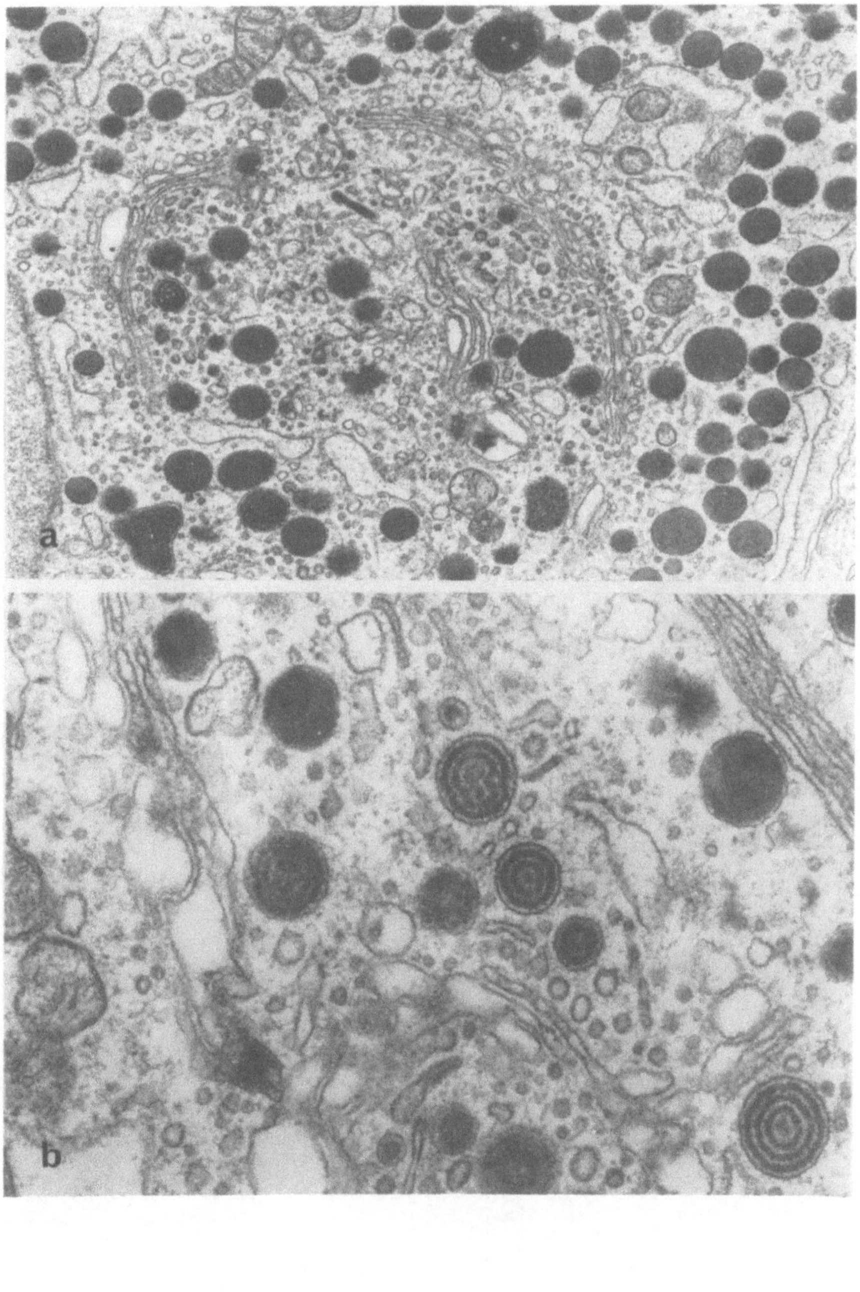

Fig. 3. Prolactin cells in the hedgehog. (a) Golgi complex with newly synthesized secretory granules (× 23,800); (b) Particular aspects of the newly synthesized secretory granules in this mammalian species (× 32,200).

far as studies of rat gonadotropes are concerned, with the findings of various authors. For example, Barnes (9) has reported that these granules may range in size from 150-200 nm, Costoff (16) from 50-200 nm and Kurosumi (19) from 200-250 nm. Large granules are about 500 nm or more in diameter. It must be emphasized that the ultrastructural aspect of secretory granules differs according to the mode of fixation. Osmium tetroxide alone, is frequently observed to rupture the membranes of large granules. Rupturing is uncommon in small granules. This phenomenon is interpreted as evidence for differing constitutions between the granular membranes of the two kinds of secretory granules. With double successive fixation (glutaraldehyde-osmium tetroxide), the ultrastructural aspect of large and small granules is identical: secretory granules appear with a fine granulated content. The ultrastructural aspect of secretory granules of gonadotropic cells differs from those of other cell types. The Golgi complex is well developed in the two kinds of gonadotropic cells and generally located in the center of the cell body. The

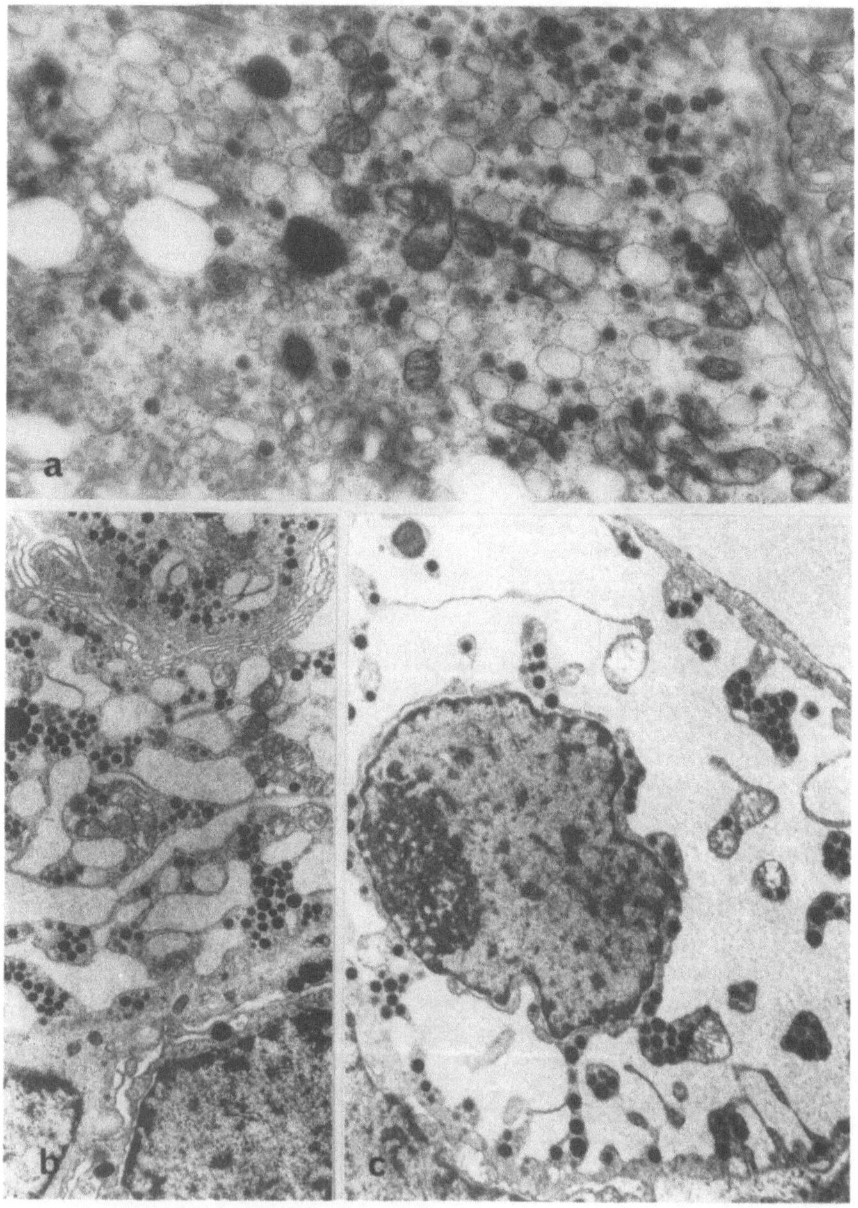

Fig. 4. Gonadotropic cells. (a) In the golden hamster (× 15,750); (b) In the rat, three weeks after orchidectomy (× 7,840); (c) In the rat, signet-ring cell, eight weeks after orchidectomy (× 7,840).

complex consists of saccules and vesicles arranged in a circular array. In the Golgi area, immature granules can be seen. The rough endoplasmic reticulum is largely developed but not arranged in parallel lamellae as seen in somatotropic, prolactin and opiocortico-melanotropic cells. The reticulum appears as irregular saccules, sometimes dilated. In activated cells, after gonadectomy for example, the rough endoplasmic reticulum is very hypertrophied. After some weeks, many saccules consist of large vacuoles filling the greater part of the cytoplasm (Fig. 4b). The maximal modification involves the coalescence of dilated saccules to form a unique vacuole. The cytoplasm and remaining organelles are pushed to one side of the cell. This castrative aspect of cell has resulted in the term 'signet-ring cell' (Fig. 4c). This image is encountered almost exclusively in type I gonadotropic cells. In type II, the modification affects the entire cell, which becomes enlarged. The rough endoplasmic reticulum is slightly hypertrophied with dilated saccules, but does not form large vacuoles and never a single vacuole. Therefore, the signet-ring cell does not appear in type II. Other organelles, such as free ribosomes, lysosomes, multivesicular bodies, centrioles and even single cilia are also observed in gonadotropic cells.

Undoubtedly, two kinds of presumed gonadotropic cells can be differentiated with the electron microscope. Is one type involved in the secretion of FSH, the other type in the secretion of LH? This exciting controversy will be discussed later; immunocytochemistry sheds some light on this subject (see section 4.1.4.).

2.5. Thyrotropic cells

The thyrotropic cell type was the first type to be welldefined with the electron microscope. As early as 1954, Farquhar and Rinehart (91), studying rat pituitary glands before and after thyroidectomy, attributed the source of thyrotropin to one cell type, irregular in shape and containing the smallest granules of any cell types. In following years, different authors have confirmed the existence of an ultrastructural type corresponding to thyrotropic cells (92-96) (Fig. 5b). These cells are few in number 2-3% of the hypophyseal cell population. For this reason they are rarely found in ultrathin sections; most of them are essentially located in the ventromedial area. Scattered cells are encountered in the midlateral area. Generally, thyrotropic cells are separated from each other, and usually arranged along blood capillaries. They almost always appear polygonal in shape but devoid of long extending processes.

The nucleus has no particularly unique characteristics. The cytoplasm contains spherical or, more rarely, spindle-shaped secretory granules. Relatively few in number, they are electron-dense but some are less dense than others. The granular membrane is poorly distinguished. Granule diameters range from 100-140 nm. Frequently, but not constantly, secretory granules tend to line up along the plasmalemma. Thus, this aspect of secretory granules is similar to that encountered in certain opiocortico-melanotropic cells. Exocytosis also appears in this cell type. The Golgi complex is located near the nucleus. Its apparent small proportion is due to the small size of the thyrotropic cell type. The complex consists of some flattened saccules. In the Golgi area, newly synthesized granules can be seen.

Mitochondria, ovoid or slender rod-shaped are very few in number. They contain internal cristae randomly distributed in a rather electron-dense matrix. The rough endoplasmic reticulum is vesicular in appearance and poorly developed under normal conditions. In activated cells, e.g., after thyroidectomy or after treatment with antithyroid drugs such as propylthiouracil, the endoplasmic reticulum is greatly dilated. This dilatation has resulted in the term 'thyroidectomy cell'. Only the activated thyrotropic cell type is marked by the appearance of intracisternal granules which become more abundant with increasing time after treatment. Other organelles, such as free ribosomes, lysosomes, centrioles are seldom seen.

In summary, electron microscopic examination of the mammalian pars distalis reveals the presence of at least five different cell types. This number perhaps may be extended to six if one considers that electron microscopy allows the identification of two kinds of gonadotropic cells. It is apparent from the data presented above that the diameter of secretory granules has been considered as a criterion for distinguishing one cell type from other. However, this criterion is not a valid reference because it is not suitable for all studied species. Consequently, the identification of antehypophyseal cell types in electron microscopy depends upon a series of characteristics and not especially upon the size of secretory granules.

2.6. Cytological peculiarities

The descriptions mentioned above summarize the general characteristics of glandular cells of the adult mammalian pars distalis. Nevertheless, some particular points should be mentioned and only some examples will be emphasized.

2.6.1. Cytological mechanism of secretion

In general secretory cells are round, ovoid or sometimes stellate in shape. However, other appearances may be encountered. For instance, in ultrastructural studies of antehypophyseal cells in monkeys, we have observed that the general shape of the glandular cells is very variable. Frequently the cells deform one another and their definite morphological outline is too variable for considering general shape as a suitable criterion for cell recognition. These apparent modifications of the cellular shape even involve inclusion-like appearances of one portion of a cell type into another (Fig. 5a).

Newly synthesized secretory granules appear in the Golgi area. Generally these granules are homogeneous in aspect and less electron dense than mature granules. We reported (97) that in the pars distalis of the hedgehog newly synthesized granules, originating in this area (Fig. 3a), are heterogeneous. The granules appear initially in the form of small, isolated masses

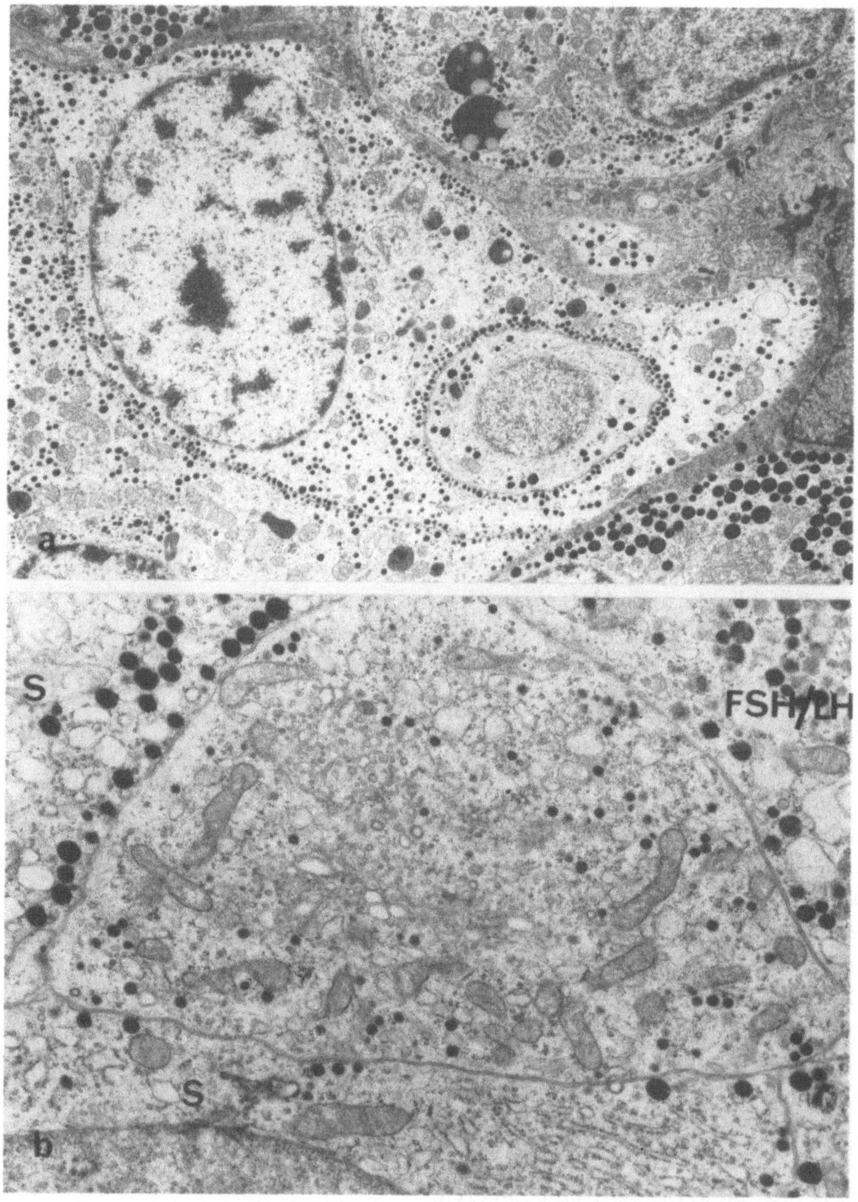

Fig. 5. (a) Opiocortico-melanotropin cell in the monkey Macaca irus. In the cytoplasm, an inclusion-like figure corresponds to a tangential section of a fragment of an other cell type (× 7,840); (b) Thyrotropic cell in the golden hamster (S = somatotropic cells; FSH/LH = gonadotropic cell) (× 10,360).

and subsequently with an appearance of concentric dense stripes (Fig. 3b). This feature disappears in mature granules.

Intranuclear inclusions can be seen especially in the prolactin cells. For instance, in the Mongolian gerbil such inclusions are not found in newborn animals of both sexes but develop soon after weaning. The inclusions increase in number in 45-50-day-old animals, or under the influence of low doses of estradiol benzoate (98).

Microtubules are especially developed in somatotropic and prolactin cells. These microtubules most likely play a role in the migration of secretory granules through the cytoplasm before exocytosis. After the injection of vinblastine sulfate (which disrupts microtubules and results in the formation of paracrystalline inclusions) into the pituitary gland of lactating rats, exocytosis is markedly depressed (99). In this study an accumulation of secretory granules became evident. Another type of morphological correlation between microtubules and cytoplasmic organelles of the glandular cells is demonstrated by the display of relations between microtubules, centrioles and Golgi apparatus which we recently have described (100).

Intercellular junctions, such as desmosomes, hemidesmosome, and zonula adherens, exist both in granular and folliculo-stellate cells. These intercellular junctions can join two glandular cells, two folliculo-stellate cells or a granular and a folliculo-stellate cell. All granular cell types can possess such membrane differentiations. Desmosomes are the most common type observed (101).

Globular fibrous bodies composed of cytoplasmic microfilaments have not been reported in normal pituitary cells. However, in all monkeys we have studied, fibrous bodies are encountered in certain somatotropic cells (Figs. 1a, 2b, 2c).

The presence of nerve fibers in the pars distalis remains open to discussion. Described initially by Théret and Tamboise (102), their existence has been denied by others. Nevertheless, nerve fibers and nerve terminals are found in intercellular spaces among the glandular cells, especially somatotropic cells and less often in contact with 'corticotropic' cells. It seems that they are not vasomotor nerves and that they are probably related to the endocrine activity of these cell types (103).

2.6.2. Cytological mechanisms of secretion

Since the first morphological observations by Farquhar (104) in somatotropic cells and by Sano (105) in prolactin cells, and the established correlations between increased prolactin secretion and increased exocytosis in prolactin cells demonstrated by Pasteels (63), exocytosis has been reported in various glandular cell types of the anterior pituitary gland. However, the recognition of this phenomenon is easier in somatotropic and prolactin cells. Under activated conditions, exocytosis undoubtedly appears in other cell types such as gonadotropic, thyrotropic and 'corticotropic' cells.

Exocytosis is the last phase of a sequence of events which is implicated in hormone secretion. The cytological mechanisms of secretion have been analyzed by Farquhar (106). She has described the following sequence of events: 1) synthesis on attached polyribosomes and segregation inside the rough endoplasmic reticulum 2) transport from the RER to the Golgi apparatus 3) concentration in the stacked Golgi cisternae and aggregation of small granules 4) storage in mature secretory granules 5) migration of these granules towards plasmalemma and 6) exocytosis. An alternative path exists in the final evolution of this process: mature secretory granules can fuse with protolysosomes. This 'crinophagic' phenomenon provides an adjustment mechanism for controlling secretory activity, especially in cases of granule overproduction. It seems probable that the degradation products can be reutilized by the glandular cell.

The phenomenon of exocytosis involves a succession of three stages. According to Vila-Porcile and Olivier (107), they are 1) contact between the granule membrane and the plasmalemma, followed by the fusion of the membranes and formation of a diaphragm; after dissolution or breaking up of the diaphragm, the granule appears 'opened' 2) during the opened phase ('omega phase') the granule content empties into the extracellular space 3) solubilization and disappearance of the granular content. After emiocytosis, the remaining membrane is internalized. Generally, exocytosis occurs towards basal laminae. Sometimes multiple exocytoses can occur, i.e., two or more granules can be seen in the same membrane pocket (Fig. 1c).

Exo- and endocytosis are well demonstrated using freeze-fracture techniques and scanning electron microscopic observation (108-110).

3. Anterior pituitary cells in human

The electron microscopic analysis of normal hypophyseal cell types in humans is more difficult than in animals cell types because the studies are generally based on the examination of specimens obtained from autopsy material or from patients hypophysectomized for different general pathological conditions. With respect to juxta-adenomatous tissue, the 'normal' character of antehypophyseal cells should be interpreted with caution. Nevertheless, in certain optimal conditions (hypophyses obtained from men and women in cases of accidental death with autopsy performed shortly after death, for instance), various investigations can be made.

3.1. General characteristics

From a general point of view, human anterior pituitary cells present numerous characteristics in common with mammalian anterior pituitary cells. A schematic description will be presented.

3.1.1. Somatotropic cells

This cell type is easily identified. As in mammals, it is the most prevalent cell type, comprising about 50% of the antehypophyseal cell population. It is dispersed throughout the pars distalis but predominates in the lateral lobes. The somatotropic cells are regularly formed and are spherical, ovoid or polygonal in shape. In humans, these cells frequently form the wall of pseudo-follicular structures, especially in the periphery of the pars distalis, under the glandular capsule. Their size is relatively larger than the average size of other cell types. The nucleus is usually spherical or oval, centrally located, with, in some sections, more than one large nucleolus.

The cytoplasm contains round or slightly ovoid secretory granules, evenly distributed in the cytoplasm. Nearly all the granules are homogeneously electron-dense. The granule membrane is not recognizable around these granules; in a few immature granules located in the Golgi area, the granule membrane occasionally can be seen. The diameter of the granules appears variable, ranging between 250-600 nm, with an average diameter of 320-420 nm. The variability probably is a function of the incidental section plane. Exceptional granular extrusions (exocytosis) can be observed. The Golgi complex is located near the nucleus. Numerous saccules and vesicles usually form a prominent area containing a few immature granules appearing with the halo-like figure described earlier. Mitochondria, show a regular pattern of internal cristae in a clear matrix. They often appear globular or slightly elongated and are disposed in small clusters in the vicinity of the Golgi apparatus. The rough endoplasmic reticulum generally is well-developed. The lamellae are located preferentially at the periphery of the cell. It is not uncommon to observe concentrically arranged lamellae forming the so-called 'Nebenkern'. As a general rule, there is an inverse relationship between the development of the RER and the size of the Golgi area and the number of secretory granules in densely granulated cells. Lamellae of RER are few in number with respect to RER in activated cells. Microtubules, centrioles, single cilia, lysosomes, and free ribosomes can be seen in this cell type.

In the normal adult pars distalis of both sexes, the greatest number of somatotropic cells consists of the densely granulated type. However, the mean diameter of secretory granules appears variable; in some cells, it varies between 250–400 nm. In others, the diameter is larger. In addition, there are larger secretory granules, ranging from 400–600 nm, which are more irregular in shape. It is not easy to speculate on the properties of these granules using the mean granular diameter as a criterion; nevertheless, it is not unlikely that the latter type corresponds, as in other mammalian species, to a reserve stage. Until the utilization of immunocytochemical methods, the cells containg the largest secretory granules were considered to be prolactin cells. At the present time, this affirmation, particularly without immunocytochemical controls, must be dismissed.

3.1.2. Prolactin cells

The recognition of this cell type in ultrathin sections is more difficult in the pars distalis of humans than laboratory animals (particularly rodents) because the size of secretory granules is not a valid morphological criterion for identification. Taking into consideration the results of immunocytochemical individualization in electron microscopy, it is certainly possible to describe a cell type corresponding to prolactin cells in the human pars distalis. The prolactin cells are relatively abundant, 15–20% of the antehypophyseal cell population. They are randomly distributed throughout the pars distalis but predominate in the posterolateral portions, in the vicinity of the neural lobe. They are dispersed or arranged in clusters. They are either oblong or pleomorphic in shape. Their size is variable and, schematically, two cell kinds are be encountered. One kind, with a large, polygonal or elongated cell body, bordering blood capillaries, is predominant in the lateral lobe. Another kind, smaller sparsely distributed or, more often, arranged in clusters, is seen more specifically in the posterolateral portions of the gland. The nucleus of prolactin cells is relatively large, centrally located, and contains one (or eventually two) prominent nucleoli.

The cytoplasm contains round or elliptical homogeneously dense secretory granules. The diameter of these granules is variable according to the type of cell. In large cells, it reaches 500–700 nm, but in small cells, it does not exceed 350 nm (in this latter type, numerous secretory granules have diameters ranging between 150–300 nm). In physiological conditions, such as lactation, exocytosis is relatively frequent. In the male, as well as the child or old age, exocytosis is exceptional. In patients treated with estrogens prior to obtaining the neurosurgical sample, secretory granules have a large diameter and frequently show irregular outlines; in this condition, misplaced exocytosis (i.e. extrusion of secretory granules away from the vascular pole of the cell) may be seen. The Golgi complex is located near the nucleus. Especially in small granulated cells, the Golgi area is prominent and newly synthesized granules are present in the dilated extremities of saccules. In the other type of prolactin cell, the Golgi apparatus is confined to a small juxta-nuclear area because of the abundance of secretory granules filling the cytoplasm. Mitochondria are small and randomly dispersed among secretory granules. The organelles do not possess unusual characteristics. The rough endoplasmic reticulum, generally well-developed, more particularly in the small cell type, appears as parallel rows of cytomembranes. In activated cells, for instance in the lactating woman or after estrogenic treatment, extensively developed RER consists of concentric whorls of cytomembranes (the 'Nebenkern'). Microtubules, centrioles and lysosomes are more frequently observed than in the somatotropic cells.

Prolactin cells exist in the human pars distalis not only in females but also in males. In the newborn human they are particularly abundant (owing to the stimulative effects of maternal estrogens); during childhood, they are small and scarce. As can be seen from the ultrastructural description, two cell types have been distinguished. The first type (large, densely granulated cells) probably corresponds to a storage stage; the second type (small cells with an enlarged Golgi area) to a stage of active hormone synthesis.

3.1.3. Opiocortico-melanotropic cells

The remarks presented for this type of cell in mammals could be repeated for the human antehypophyseal cell initially described under the name 'corticotropic' cell. This cell type reacts with various antibodies of the opiocortico-melanotropin group and now must be designated the 'opiocortico-melanotropic' cell. This cell type is relatively abundant comprising 15–20% of the antehypophyseal cell population. These cells are mainly located in the central part of the lateral lobes, but scattered cells are dispersed throughout the pars distalis. They are isolated or arranged in clusters. Pseudo-vesicular organizations containing this cell type in their wall can be seen. The cells are medium-sized. Their shape may be ovoid, or more rarely, irregular and stellate but they do not possess long, thin processess. The nucleus, centrally or often eccentrically located, is relatively large and contains one more or less prominent nucleolus.

The cytoplasm contains round or slightly irregular secretory granules. They appear with a variable electron density and with a wide range in size. If some secretory granules have a small diameter (250 nm), others have a much larger one (up to 800 nm). For the most part, the diameter ranges from 300–500 nm. Exocytosis can be seen occasionally. Golgi complex is located near the nucleus and it is generally prominent. It contains newly synthesized granules with the halo-like appearance. Mitochondria are sparsely distributed, and are round or slightly elongated. The rough endoplasmic reticulum is well-developed, and consists of widely dispersed flattened or slightly dilated saccules. Also in the cytoplasm are two organelles which seem quite specific: large perinuclear lysosomal bodies (so-called enigmatic bodies) and microfilaments of about 70 Å in width (type I microfilaments) which are dispersed in the cell body.

The apparent heterogeneity of the granule population in one cell raises the question of the functional significance of secretory granules showing such a variety in size and in electron density. This aspect may correspond either to a morphological reflection of various stages of the maturation process or to a population of granules containing different peptides. At this time, the problem has not been clearly resolved. This cell type is well-developed in young, adult and old subjects of both sexes. Under various treatments, for instance with corticoids, this cell type undergoes a morphological modification called 'Crooke's hyaline change', consisting of a localized accumulation of type I microfilaments.

3.1.4. Gonadotropic cells

In humans as well as mammals, the question of one or two types of gonadotropic cells has not yet been resolved by electron microscopy alone. Therefore, an ultrastructural description of two kinds of gonadotropes is unreliable. We will therefore only present a general description of the ultrastructural organization of presumed gonadotropic cells. This cell type is less abundant than the other previously mentioned cell types comprising 8–10% of the antehypophyseal cell population. Gonadotropic cells are medium-sized, and spherical or ovoid in shape. They are randomly distributed in the pars distalis with, however, a maximal concentration in the center of the lateral lobes. The nucleus of these cells is relatively small. The cytoplasm contains round secretory granules; they appear with a moderate electron density and with an appreciable variation in diameter (200–400 nm). The Golgi complex is located near the nucleus; generally, it is not prominent and consists of some flattened cisternae. Mitochondria are dispersed in the cytoplasm; they are small and can be round or sometimes slightly elongated. The rough endoplasmic reticulum is well-developed and consists of flattened or slightly dilated saccules. Microtubules, and lysosomes can be seen in this cell type.

Gonadotropic cells are poorly developed before puberty. In the gonadectomized adult, morphological changes resembling those described in laboratory animals may occur such as hypertrophied cells containing vacuolated cytoplasm. In some cells a large vacuole displaces the nucleus and the remaining cytoplasm to the cell periphery; this picture is quite similar to the 'signet-ring cell'.

3.1.5. Thyrotropic cells

This cell type of the human pars distalis is the least numerous comprising 2–5% of the antehypophyseal cell population. Thyrotropic cells take up a limited area of the pars distalis, located primarily in the ventromedial part of the gland. They are medium-sized, polygonal, and sometimes possess long cytoplasmic processess. The nucleus, centrally located, has no special characteristics.

The cytoplasm contains round secretory granules; they are the smallest in all glandular cells and their diameters are relatively homogeneous (125–200 nm). Usually, secretory granules are regularly disposed along the plasmalemma. However, such an arrangement is not specific to thyrotropic cells. It is observed in other cell types notably in opiocortico-melanotropic cells and less frequently in gonadotropic and somatotropic cells. The Golgi complex is poorly developed; it is not always encountered on ultrathin sections. Mitochondria are relatively large as compared with the other organelles. The rough endoplasmic reticulum is poorly developed and consists of small saccules. Other organelles are seldom seen.

Thyrotropic cells can be recognized in young, adult and old subjects. After thyroidectomy or treatments with antithyroid drugs, activated thyrotropic cells appear as hypertrophied and vacuolized cells, resembling the 'thyroidectomy cells' described in laboratory animals.

3.6. Some particular points

As can be seen from the morphological description, the fine structure of the pars distalis in humans is comparable in many aspects to that of laboratory or wild mammals. However, the results concerning granule size variation must be emphasized. The mean granule diameter is generally variable, not so much in thyrotropic cells (having the smallest secretory granules in all species), but more particularly in prolactin cells (which no longer possess the largest secretory granules, as was the case in rodents) and opiocortico-melanotropic cells (in which the secretory granules are larger than in most studied species). These results confirm the previous remark that the secretory granule size is not a suitable criterion for the identification of the antehypophyseal cell types.

Another particular point must be emphasized. We have pointed out repeatedly the particular distribution of cell type occurring in the human pars distalis. Some cell types are preferentially located in limited areas, notably the thyrotropic cells. On one hand, if a cell type is not randomly distributed, its non-appearance on an ultrathin section is hardly surprising. On the other hand if a cell type predominates in a more or less large area, it is risky, after examination of a small part of antehypophyseal glandular parenchyma, to conclude that there exists a hyperplasia of one cell type or another.

4. Functional data

4.1. Observations in laboratory animals

The functional significance of each cell type individualized by electron microscopy comes from observations under certain natural and experimental conditions. A more valuable approach, recently developed, is the immunocytochemical study at the ultrastructural level.

4.1.1. Somatotropic cells

The functional significance of this cell type is well established. In young animals, the somatotropic cells are numerous and appear with ultramicroscopic signs of a great activity. On the contrary, in the dwarf mouse, electron microscopic studies have confirmed the absence of STH-cells (25). Some treatments, known to enhance activation of this cell type, induce the exocytosis phenomenon. For instance, after starvation, it is easy to observe numerous exocytotic granules in somatotropic cells (55). Infusion of hypothalamic extracts also is accompanied by ultrastructural signs of cellular activation (54, 57).

Specific antibodies against GH react with the secretory granules of somatotropic cells. A positive reaction is exclusively located in the secretory granules.

4.1.2. *Prolactin cells*

The functional significance of this cell type is easy to demonstrate by electron microscopy. Confirming previous investigations using light microscopy, ultrastructural analyse have established that the percentage of this cell type increases during pregnancy. In lactating females, prolactin cells are especially developed (67, 68). Suckling enhances exocytosis (63). After removal of the suckling youngs, electron microscopic observations reveal activation of the lysosomes and the presence of numerous images of granules fusing with lysosomes (106). In estrogenized animals, a hyperplasia of this cell type is evident on electron micrographs (68). Specific antibodies against PRL stain positively the secretory granules.

4.1.3. *Opiocortico-melanotropic cells*

The functional significance of this cell type depends upon investigations using immunocytochemical techniques. Initially demonstrated by light microscopy, the localization of the immunoenzyme reaction in the secretory granules is now well established by electron microscopy. Pelletier et al. (111) have reported in the rat pituitary gland that antibodies against ACTH and β-LPH bound to the same granules. Weber et al. (112) have demonstrated, also in the rat pituitary gland, the existence of a concomitant storage of ACTH- and endorphin-like immunoreactivity in the secretory granules. The authors remarked that any granule was immunoreactive to only one of these antibodies. More recently, two experimental results supply further information concerning the functional significance of this cell type. On the one hand, Vaudry et al. (113) have confirmed these results and added that with γ-endorphin antibody immunoreactivity was restricted to the secretory granules which also contained ACTH and β-endorphin. On the other hand, Guy et al. (114), studying adjacent serial ultrathin sections of rat adenohypophysis, have demonstrated that the anti-16K fragment stains the same granules which react with the anti-ACTH antibody.

The same granules being labelled with anti-βMSH antibodies, it may be permissible to conclude that the pars distalis contains a cell type which is worth naming 'opiocortico-melanotropic cells'.

4.1.4. *Gonadotropic cells*

We have previously reported that examination of pituitary ultrathin sections reveals the existence of two kinds of gonadotropic cells. However, as early as 1970, Nakane emphasized FSH and LH cells are distributed in a identical manner, and in many instances both of these hormones are present within the same cell (115). More recently, Tougard et al. (116) have tried to identify rat pituitary gonadotropic cells by immunoelectron microscopy using anti-rat βFSH and anti-rat βLH antisera. They have concluded that on adjacent serial sections the same cells were stained by both antisera. Such an interpretation has been confirmed by several authors. Nevertheless, another group of results has been published. For some researchers, three types of gonadotropic cells may be distinguished by electron microscopy; one reacting simultaneously with βFSH and βLH, another either with βFSH or βLH, and yet another with βFSH only. According to Ellison et al. (117), between 62–88% of the gonadotropic cells in the rat pars distalis contain both hormones.

Thus, immunoelectron miscroscopy confirms the reality of gonadotropic cells in the pars distalis but, considering the functional significance, it seems that gonadotropic cells constitute a heterogeneous group. In addition, some results support the view that, in so-called gonadotropic cells some granules reacting with anti-βFSH and/or anti-βLH antibodies can also react with anti-$^{17-39}$ACTH antibody. The complete physiological significance of 'gonadotropic cells' remains to be determined.

4.1.5. *Thyrotropic cells*

By electron microscopy, this cell type is readily distinguishable. Because of specific changes occurring after thyroidectomy or treatment with antithyroid drugs, the recognition of this cell type does not entail serious problems. In immunoelectron microscopy, demonstration of specific labelling of secretory granules using anti-human βTSH permits the clear distinction between thyrotropic and gonadotropic cells. Applying immunocytochemical reactions on successive adjacent serial sections, one treated with anti-βLH antibody and the next with anti-βTSH, it is evident that labelling distinguishes two different cell categories. One condition for the application of immunocytochemistry at the ultrastructural level may be emphasized: relatively few fixatives preserve well the TSH molecules.

5. Observations in the human

It will be unnecessary to take up again the arguments previously presented, because the results obtained in the ultrastructural analysis of the functional significance of normal human antehypophyseal cytology are comparable in all respects to those reported in connection with the study of the antehypophysis of laboratory animals. Furthermore, the ultrastructural and the immunocytochemical data obtained in the identification of various types of pituitary adenomas confirm the individualization of the different cell types. This fascinating problem is explained in another chapter (8) of this book.

References

1. Romeis B: Hypophyse. In: Handbuch der mikroskopischen Anatomie des Menschen, von Möllendorf W (ed), Berlin, Springer, 1940, pp 1–625.
2. Holmes RL, Ball JN: The pituitary gland. A comparative account. Cambridge, The University Press, 1974.
3. Girod C: Etat actuel des connaissances sur la description morphologique et la signification fonctionnelle des cellules antéhypophysaires. Lyon Médical 236(16): 323–357, 1976.
4. Girod C: Histochemistry of the adenohypophysis. In: Handbuch der Histochemie, Graumann W, Neumann K (eds), Stuttgart, G. Fischer, 1976 (Vol VIII, suppl part 4), pp 1–325.
5. Girod C: Apport de l'immunohistochimie à l'étude cytologique de l'adénohypophyse, Bull Ass Anat (Nancy) 62(175): 417–603, 1977.
6. Girod C: Immunocytochemistry of the Vertebrate adenohypophysis. In: Handbuch der Histochemie. Graumann W, Neumann K (eds), Stuttgart, G. Fischer, 1983 (in press).
7. Fernandez-Morán H, Luft R: Submicroscopic cytoplasmic granules in the anterior lobe cells of the rat hypophysis as revealed by electron microscopy. Acta endocr (Kbh) 2(3): 199–211, 1949.
8. Rinehart JF, Farquhar MG: Electron microscopic studies of the anterior pituitary gland. J Histochem Cytochem 1(2): 93–113, 1953.
9. Barnes BG: The fine structure of the mouse adenohypophysis in various physiological states. In: Cytologie de l'adénohypophyse. Benoit J, Da Lage C (eds), Paris, CNRS, 1963, pp 91–109.
10. Herlant M: Apport de la microscopie électronique à l'étude du lobe antérieur de l'hypophyse. In: Cytologie de l'adénohypophyse. Benoit J, Da Lage C (eds), Paris, CNRS, 1963, pp 73–86.
11. Herlant M: The cells of the adenohypophysis and their functional sigificance. Int Rev Cytol 17: 299–382, 1963.
12. Green JD: Electron microscopy of the anterior pituitary. In: The pituitary gland. Harris GW, Donovan BT (eds), London, Butterworths, 1966, Vol 1, pp 233–241.
13. Yoshida Y: Electron microscopy of the anterior pituitary gland under normal and different experimental conditions. Meth Achiev exp Path 1: 439–454, 1966.
14. Kurosumi K: Functional classification of cell types of the anterior pituitary gland accomplished by electron microscopy. Arch histol jap 29(4), 329–362, 1968.
15. Foster CL: Relationship between ultrastructure and function in the adenohypophysis of the rabbit. Mem Soc Endocrinol 19: 125–146, 1971.
16. Costoff A: Ultrastructure of rat adenohypophysis. Correlation with function. New York-London, Academic Press, 1973.
17. Vila-Porcile E: La pars distalis de l'hypophyse chez le Rat. Contribution à son étude histologique et cytologique en microscopie électronique. Ann Sci natur, Zool 15(12e série): 61–138, 1973.
18. Farquhar MG, Skutelsky EH, Hopkins CR: Structure and function of the anterior pituitary and dispersed pituitary cells in vitro studies. In: The anterior pituitary. Tixier-Vidal A, Farquhar MG(eds), New York-San Francisco-London, Academic Press, 1975, pp 83–135.
19. Kurosumi K: Adenohypophysis. In: Functional morphology of endocrine glands. Kurosumi K, Fujita H (eds), Stuttgart-Tokyo, G Thieme-Igaku Shoin Ltd, 1975, pp 1–100.
20. Pantić VR: The specificity of pituitary cells and regulation of their activities. Int Rev Cytol 40: 153–195, 1975.
21. Tixier-Vidal A: Ultrastructure of anterior pituitary cells in culture. In: The anterior pituitary. Tixier-Vidal A, Farquhar MG (eds), New York-San Francisco-London, Academic Press, 1975, pp 181–229.
22. Tixier-Vidal A, Picart R, Moreau MF: Endocytose et sécrétion dans les cellules antéhypophysaires en culture. Action des hormones hypothalamiques. J Microscopie Biol Cell 25(2): 159–172, 1976.
23. Costoff A: Ultrastructure of the pituitary gland. In: The pituitary. A current review, Allen MB jr, Malesh VB (eds), New York-San Francisco-London, Academic Press, 1977, pp 59–76.
24. Farquhar MG: Secretion and crinophagy in prolactin cells. Adv exp Med Biol 80: 37–86, 1977.
25. Kurosumi K, Shimizu T, Takeda F: The pituitary gland. In: Electron microscopy in human medicine, Vol 10, Endocrine organs, New York, McGraw-Hill Intern Book Co, 1981, pp 1–26.
26. Sternberger LA: Electron microscopic immunochemistry: a review. J Histochem Cytochem 15(3): 139–159, 1967.
27. Moriarty GC: Adenohypophysis: ultrastructural cytochemistry. A review. J Histochem Cytochem 21(10): 855–894, 1973.
28. Nakane PK: Identification of anterior pituitary cells by immunoelectron microscopy. In: The anterior pituitary. Tixier-Vidal A, Farquhar MG (eds), New York-San Francisco-London, Academic Press, 1975, pp 45–61.
29. Kovacs K, Horvath E, Ryan N: Immunocytology of the human pituitary. In: Diagnostic immunohistochemistry. DeLellis RA (ed), New York-Paris, Masson Publ. USA, 1981, pp 17–35.
30. Foster CL: Some observations upon the cytology of the pars distalis of surgically-removed human pituitary. Quart J Micr Sci 97(3): 379–391, 1956.
31. Foster CL: Some observations upon the Golgi elements of the cells of the surgically-removed human anterior pituitary. Quart J Micr Sci 97(4): 481–486, 1956.
32. Foncin JF: Etudes sur l'hypophyse humaine au microscope électronique. Path-Biol 14(19–20): 893–902, 1966.
33. Foncin JF, LeBeau J: Cellules de castration et cellules FSH dans l'hypophyse humaine vue au microscope électronique. J Microscopie 5(4): 523–526, 1966.
34. Lederis K: Electron miscroscopic examination of the human pituitary obtained at hypophysectomy. In: La chirurgie endocrinienne majeure dans le traitement du cancer du sein en phase avancée, Dargent M, Romieu C (eds), Lyon, SIMEP Editions, 1967, pp 219–222.
35. Foncin JF: Morphologie ultra-structurale de l'hypophyse humaine. Neuro-chirurgie 17(suppl 1): 10–24, 1971.
36. Lawzewitsch I von, Dickmann GH, Amezúa L, Pardal C: Estudios citologicos y ultraestructurales de la adenohipofisis humana. Rev Soc argent Biol 46(5–8): 78–104, 1970.
37. Lawzewitsch I von, Dickmann GH, Amezúa L, Pardal C: Cytological and ultrastructural characterization of the human pituitary. Acta anat 81(2): 286–316, 1972.
38. Deaton PC, Dugger GS: The ultrastructure of the nonadenomatous anterior lobe of the pituitary gland in man. Surg Gynec Obst 135(6): 901–907, 1972.
39. Kovacs K, Horvath E, Stratman IE, Ezrin C: Cytoplasmic microfilaments in the anterior lobe of the human pituitary gland. Acta anat 87(3): 414–426, 1974.
40. Kovacs K, Horvath E: Gonadotrophs following removal of the ovaries: a fine structural study of human pituitary glands. Endokrinologie 66(1): 1–8, 1975.
41. Landolt AM: Ultrastructure of human sella tumors. Acta neurochir, suppl 23, 1975. See Chapter 3: Ultrastructure of the normal pituitary, pp 8–30.
42. Uei Y, Kanzaki M: Ultrastructure of the 'non-pathologic' human pituitary gland. Acta Path Jap 26(2): 191–203, 1976.
43. Gray AB: Analysis of diameters of human pituitary hormone secretory granules. Acta endocr (Kbh) 85(2): 249–255, 1977.
44. Ciocca DR, Rodriguez EH, Cuello CA: Comparative light- and electron microscopical study of the normal adenohypophysis in the human. Acta anat 103(1): 83–100, 1979.
45. Van Oordt PGWJ: Nomenclature of the hormone-producing cells in the adenohypophysis. A report of the international committee for nomenclature of the adenohypophysis. Gen Comp Endocrinol 5(1): 131–134, 1965.
46. Hedinger CE, Farquhar MG: Elektronenmikroskopische Untersuchungen von zwei Typen acidophiler Hypophysenvorderlappenzellen bei der Ratte. Schweiz Z allg Pathol Bakteriol 20(6): 766–768, 1957.
47. Azzali G: Studio citochimico ed al microscopo elettronico sulle cellule adenoipofisarie dei Primati (Comopithecus hamadryas). Biochim Biol sper 1(1): 67–91, 1961.
48. Barnes BG: Electron microscope studies on the secretory cytology of the mouse anterior pituitary. Endocrinology 71(4): 618–628, 1962.
49. Girod C, Dubois P. Curé M: Premières observations, au microscope

électronique, sur les cellules chromophiles de l'antéhypophyse du Hamster doré (Mesocricetus auratus Waterh). C R Soc Biol 158(8–9): 1641–1643, 1964.
50. Yoshimura F, Harumiya K: Electron microscopy of the anterior lobe of pituitary in normal and castrated rats. Endocrinol Japon 12(2): 119–152, 1965.
51. Girod C, Dubois P, Curé M: Etude cytologique et ultrastructurale de l'antéhypophyse du Hamster doré. C R Ass Anat 50(132): 460–468, 1966.
52. Yamashita K: Electron microscopic observations on the anterior pituitary of the crabeating monkey (Macacus irus). Okajimas Folia anat jap 43(6): 299–323, 1967.
53. Dekker A: Electron microscopic study of somatotropic and lactotropic pituitary cells of the syrian hamster. Anat Rec 162(2): 123–135, 1968.
54. De Virgiliis G, Meldolesi J, Clementi F: Ultrastructure of growth hormone-producing cells of rat pituitary after injection of hypothalamic extract. Endocrinology 83(6): 1278–1284, 1968.
55. Dubois P, Girod C: Influence du jeûne alimentaire sur les cellules antéhypophysaires du Hamster doré. Observations préliminaires en microscopie électronique. C R Soc Biol Paris 162(12): 2116–2118, 1968.
56. Rambourg A, Racadot J: Identification en microscopie électronique de six types cellulaires dans l'antéhypophyse du Rat à l'aide d'une technique de coloration par le mélange acide chromique-phosphotungstique. C R Acad Sc Paris 266(2): 153–155, 1968.
57. Couch EF, Arimura A, Schally AV, Saito M, Sawano S: Electron microscope studies of somatotrophs of rat pituitary after injection of purified growth hormone releasing factor (GRF). Endocrinology 85(6): 1084–1091, 1969.
58. Girod C, Dubois P: Recherches, en microscopie optique et en microscopie électronique, sur les cellules somatotropes antéhypophysaires du Hamster doré (Mesocricetus auratus Waterh). C R Acad Sc Paris 268(19): 2361–2363, 1969.
59. Amat P, Boya J: Estudio preliminar de la ultraestructura des lobulo anterior de la hipofisis del cobaya. An Anat 19(48): 437–454, 1970.
60. Mikami S: Light and electron microscopic investigations of six types of glandular cells of the bovine adenohypophysis. Z Zellforsch 105(4): 457–482, 1970.
61. Kurosumi K: Mitosis of the rat anterior pituitary cells: an electron microscope study. Arch histol jap 33(2): 145–160, 1971.
62. Garwood VF, Latta JS: Electron microscopic observations on the secretory processes in prolactin cells of the mouse anterior pituitary. Anat Rec 145(2): 231–232, 1963.
63. Pasteels JL: Recherches morphologiques et expérimentales sur la sécrétion de prolactine. Arch Biol (Liège) 74(4): 439–553, 1963.
64. Théret C, Renault H: Microscopie électronique des cellules adénohypophysaires dites à prolactine chez la Ratte Wistar dans divers états physiologiques et expérimentaux. C R Acad Sc Paris 259(20): 3618–3619, 1964.
65. Girod C, Dubois P, Curé M: Identification expérimentale, en microscopie optique et en microscopie électronique, des cellules à prolactine antéhypophysaires, chez le Hérisson (Erinaceus europaeus L). C R Acad Sc Paris 261(25): 5660–5663, 1965.
66. Pasteels JL: Les cellules à prolactine en microscopie électronique. Arch Anat micr Morphol exper 54(1): 635–636, 1965.
67. Young BA, Foster CL, Cameron E: Ultrastructural changes in the adenohypophysis of pregnant and lactating rabbits. J Endocr 39(3): 437–443, 1967.
68. Potvliege PR: Effects of estrogen on pituitary morphology in goitrogen treated rats. An electron microscopic study. Anat Rec 160(3): 595–605, 1968.
69. Kurosumi K: Formation and release of secretory granules during mitosis in the anterior pituitary gland. Arch histol jap 42(4): 481–486, 1979.
70. Herlant M, Klastersky J: Etude au microscope électronique des cellules corticotropes de l'hypophyse. C R Acad Sc Paris 256(12): 2709–2711, 1963.
71. Girod C, Dubois P, Curé M: Identification expérimentale des cellules corticotropes antéhypophysaires chez le Hamster doré (Mesocricetus auratus Waterh). Observations au microscope optique et au microscope électronique. C R Acad Sc Paris 259(5): 1229–1232, 1964.
72. Rennels EG: Electron microscopic alterations in the rat hypophysis after scalding. Am J Anat 114(1): 71–91, 1964.
73 Siperstein ER, Allison VF: Fine structure of the cells responsible for secretion of adrenocorticotropin in the adrenalectomized rat. Endocrinology 76(1): 70–79, 1965.
74. Kurosumi K, Kobayashi Y: Corticotrophs in the anterior pituitary glands of normal and adrenalectomized rats as revealed by electron microscopy. Endocrinology 78(4): 745–758, 1966.
75. Kurosumi K, Oota Y: Corticotrophs in the anterior pituitary gland of gonadectomized and thyroidectomized rats as revealed by electron microscopy. Endocrinology 79(4): 808–814, 1966.
76. Yamada K, Yamashita K: An electron microscopic study on the possible site of production of ACTH in the anterior pituitary of mice. Z Zellforsch 80(1): 29–43, 1967.
77. Nakayama I, Nickerson PA, Skelton FS: An ultrastructural study of the adrenocorticotropic hormone-secreting cell in the rat adenohypophysis during adrenal cortical regeneration. Lab Invest 21(2): 169–178, 1969.
78. Pelletier G: Identification en microscopie électronique des cellules corticotropes chez le rat intact; résultat de la surrénalectomie associée ou non à un traitement par la dexaméthasone. C R Acad Sc Paris 270(24): 2836–2838, 1970.
79. Siperstein ER, Miller KJ: Further cytophysiologic evidence for the identity of the cells that produce adrenocorticotrophic hormone. Endocrinology 86(3): 451–486, 1970.
80. Farquhar MG: 'Corticotrophs' of the rat adenohypophysis as revealed by electron microscopy. Anat Rec 127(2): 291 (Abstr.), 1957.
81. Farquhar MG, Rinehart JF: Electron microscopic studies of the anterior pituitary gland of castrate rats. Endocrinology 54(5): 516–541, 1954.
82. Girod C, Dubois P, Curé M: Etude au microscope électronique des cellules gonadotropes antéhypophysaires du Hamster doré (Mesocricetus auratus Waterh) intact ou castré. C R Acad Sc Paris 258(26): 6536–6538, 1964.
83. Dubois P, Girod C: Nouvelles recherches sur l'ultrastructure des cellules gonadotropes de l'hypophyse antérieure chez le Hamster doré (Mesocricetus auratus Waterh). C R Soc Biol Paris 158(11): 2102–2104, 1964.
84. Girod C, Dubois P: Etude ultrastructurale des cellules gonadotropes antéhypophysaires, chez le Hamster doré (Mesocricetus auratus Waterh). J Ultrastr Res 13(1–2): 212–232, 1965.
85. Young BA, Foster CL, Cameron E: Ultrastructural changes in the adenohypophysis of castrated rabbits. J. Endocr 35(1): 101–106, 1966.
86. Clementi F, De Virgiliis G: Ultrastructure de l'adénohypophyse après ovariectomie et traitement par les oestrogènes et la progestérone. Path Biol 15(3–4): 119–131, 1967.
87. Della Corte F, Angelini F: Dati ultrastruturali su alcune modificazioni da castrazione, nel coniglio. Arch zool ital 57(2): 221–235, 1967.
88. Girod C, Dubois P, Curé M: Recherches sur les corrélations hypophyso-génitales chez la femelle de Hérisson (Erinaceus europaeus L). Ann Endoc Paris 28(5): 581–610, 1967.
89. De Virgiliis G: Ultrastructure des cellules gonadotropes de l'adénohypophyse après ovariectomie. Ann Edoc Paris 29(5): 553–561, 1968.
90. Kurosumi K. Oota Y: Electron microscopy of two types of gonadotrophs in the anterior pituitary glands of persistent estrous and diestrous rats. Z Zellforsch. 85(1): 34–46, 1968.
91. Farquhar MG, Rinehart JF: Cytologic alterations in the anterior pituitary gland following thyroidectomy. An electron microscopic study. Endocrinology 55(6): 657–676, 1954.
92. Girod C, Dubois P, Curé M: Ultrastructure des cellules thyréotropes antéhypophysaires chez le Hamster doré (Mesocricetus auratus Waterh). C R Soc Biol Paris 159(8–9): 1694–1696, 1965.
93. Marescaux J, Rebel A: Modifications ultrastructurales de la pars distalis du Cobaye après thyroïdectomie. C R Soc Biol Paris 160(1): 190–193, 1966.
94. Dingemans KP: On the origin of thyroidectomy cells. J Ultrastr Res 26(5–6): 480–500, 1969.
95. Foster CL, Cameron E, Young BA: Ultrastructural changes in the adenohypophysis of hypothyroid rabbits. J Endocr 44(2): 273–277, 1969.
96. Kurosumi K, Baba N: Experimental and histochemical studies on the rat pituitary thyrotrophs by electron microscopy. Gunma Symp Endocrinol 9: 197–212, 1969.
97. Girod C, Lhéritier M, Guichard Y: Recherches sur la formation des granulations dans différents types cellulaires de l'antéhypophyse chez le Hérisson. Bull Ass Anat (Nancy) 58(162): 563–570, 1974.

98. Nickerson PA: Intranuclear inclusions in mammotrophs of the Mongolian gerbil: effect of low doses of estradiol benzoate and a study of females before weaning. Tissue & Cell 7(4): 773–776, 1975.
99. Shiino M, Rennels EG: Vinblastine-induced microtubular paracrystals in prolactin cells of anterior pituitary gland of lactating rats. Am J Anat 144(3): 399–405, 1975.
100. Girod C, Lhéritier M, Guichard Y: Relations cil-centriole-appareil de Golgi dans les cellules glandulaires de l'antéhypophyse du Hérisson (Erinaceus europaeus L.) C R Acad Sc Paris 290(5): 711–714, 1980.
101. Herbert DC: Intercellular junctions in the rhesus monkey pars distalis. Anat Rec 195(1): 1–6, 1979.
102. Théret C, Tamboise E: Etude ultrastructurale des rapports expérimentaux entre les cellules alpha et des fibres neurovégétatives dans l'adénohypophyse. Ann Endoc Paris 24(3): 421–440, 1963.
103. Kurosumi K, Kobayashi Y: Nerve fibers and terminals in the rat anterior pituitary gland as revealed by electron microscopy. Arch histol jap 43(2): 141–155, 1980.
104. Farquhar MG: Origin and fate of secretory granules in cells of the anterior pituitary gland. Trans New York Acad Sci 23(4): 346–351, 1961.
105. Sano M: Further studies on the theta cell of the mouse anterior pituitary as revealed by electron microscopy, with special reference to the mode of secretion. J Cell Biol 15(1): 85–97, 1962.
106. Farquhar MG: Secretion and crinophagy in prolactin cells. In: Comparative endocrinology of prolactin, Dellmann HD, Johnson JA, Klachko DM (eds), New York-London, Plenum Press, 1977, pp 37–86.
107. Vila-Porcile E, Olivier L: Exocytosis and related membrane events. In: Synthesis and release of adenohypophyseal hormones. Jutisz M, McKerns KW (eds), New York-London, Plenum Press, 1980, pp 67–103.
108. Daikoku S, Takahashi T, Kojimoto H, Watanabe YG: Secretory surface phenomena in freeze-etched preparations of the adenohypophysial cells and neurosecretory fibers. Z Zellforsch 136(2): 207–214, 1973.
109. Ishimura K, Egawa K, Fujita H: Freeze-fracture images of exocytosis and endocytosis in anterior pituitary cells of rabbits and mice. Cell Tissue Res 206(2): 233–241, 1980.
110. Krstić R: Secretory phenomenon of the adenohypophyseal cells viewed with the scanning electron microscope. Experientia (Basel) 36(5): 596–597, 1980.
111. Pelletier G, Leclerc R, Labrie F, Côté J, Chrétien M, Lis M: Immunohistochemical localization of β-lipotropic hormone in the pituitary gland. Endocrinology 100(3): 770–776, 1977.
112. Weber E, Voigt KH, Martin R: Concomitant storage of ACTH- and endorphin-like immunoreactivity in the secretory granules of anterior pituitary corticotrophs. Brain Res 157(2): 385–390, 1978.
113. Vaudry H, Pelletier G, Guy J, Leclerc R, Jegou S: Immunohistochemical localization of γ-endorphin in the rat pituitary gland and hypothalamus. Endocrinology 106(5): 1512–1520, 1980.
114. Guy J, Leclerc R, Pelletier G: Localization of a 16,000 daltons fragment of the common precursor of adrenocorticotropin and β-lipotropin in the rat and human pituitary gland. J Cell Biol 86(3), 825–830, 1980.
115. Nakane PK: Classifications of anterior pituitary cell types with immunoenzyme histochemistry. J Histochem Cytochem 18(1): 9–20, 1970.
116. Tougard C, Picart R, Tixier-Vidal A: Immunocytochemical localization of glycoprotein hormones in the rat anterior pituitary. A light and electron microscope study using antisera against rat β subunits: a comparison between preembedding and postembedding methods. J Histochem Cytochem 28(2): 101–114, 1980.
117. Ellison DG, Childs (Moriarty) GV, Lorenzon JR, Schwartz NB: Differential staining for LH and FSH on rough endoplasmic reticulum in castration cells. J Histochem Cytochem 28(6): 608, 1980.

Author's address:
Department of Histology-Embryology-Cytogenetics
Faculté de Médecine Alexis Carrel
Rue Guillaume Paradin
F – 69372 Lyon Cédex 2 (France)

CHAPTER 3

Fine structure and development of the pars tuberalis in mammals

MARIE-ELISABETH STOECKEL and AIMÉ PORTE

1. Introduction

The precise function of the tuberal lobe or pars tuberalis (PT) of the adenohypophysis identified anatomically by Tilney (1) and embryologically by Atwell (2) still poses an enigma (3). This lobe, consisting mainly of chromophobic cells in most species, was long considered an undifferentiated rostral extension of the pars distalis (PD). Electron microscopy, however, has revealed that the PT cells, except those lining the follicular cavities, clearly have the characteristics of peptide-producing glandular cells in mammals (4–13) as well as in other classes of vertebrates (3, 13, 17). The chromophobic cells of the rat PT containing small secretory vesicles were at first interpreted as corticotrophs (4, 6). Since typical gonadotrophic cells also proliferate in the PT after hypophysectomy, this lobe was thought to have mainly auxiliary or substitutional function in the case of high PD hormone demand (4, 8). This opinion has been corroborated by immunocytochemical data (18, 19). However, electron-microscope studies on the PT of various mammals under normal and experimental conditions (5, 10–13), and during its ontogenesis in the rat and the mouse (20, 21), and immunocytochemical data (22), indicate that PT glandular cells are in general peculiar to this lobe, i.e. are not found in the other adenohypophyseal lobes. So it seems highly probable that the PT has a specific but as yet undetermined function which should no longer be neglected in hypophyseal physiology.

2. General organization

The PT in mammals forms a cellular sheath running from the antero-ventral part of the PD along the median eminence (ME) and surrounding the hypophysial stalk, in close relationship with the primary vascular plexus of the hypophysial portal system. Lying along the external side of the pia-mater, it is also in direct contact with the cerebrospinal fluid of the subarachnoid space.

Covering the entire neuro-hemal contact zone, without any overlap, the PT is composed of anastomosing cell cords which, according to the species, are loosely or tightly packed and in more or less close contact with the ME and stalk external zones. The cell cords are always lined by a basal membrane and separated by a narrow connective space from fenestrated vessels and nerve endings of the neuro-hemal ME external zone. Even in cases of very close apposition of cell cords against the ME, for example in the garden dormouse (13), we never observed any direct contact between PT cell cords and elements of the palisadic zone of the ME, in particular with tanycytes which have been reported to make direct contact with PT cells in the rhesus monkey (7).

The PT merges with the anterior region of the PD at ventral or lateral level depending on the relative position of the lateral lobes of Rathke's pouch during embryonic development. It is separated from the pars intermedia (PI) and PD rostral junction zone ['rostrale Umschlagszone of Lothringer', (23)] by a conjunctivo-vascular space, a remnant of Atwell's fossa, through which the portal vasculature penetrates the PD. As clearly seen in muridae and in the cat, this vascular space extends between the PT and the rostroventral surface of the PD, the two lobes joining more caudally.

We exclude from this study the zona tuberalis which is the most anterior area of the PD. In some species, e.g. rabbit and man the pars tuberalis, differs from the rest of the PD by its cellular composition (in particular, the absence of acidophilic cells); but, as it clearly belongs to the PD, it should be distinghished from the tuberal lobe.

3. Cell composition

The PT cell cords contain both glandular and non-

Motta, PM (ed): Ultrastructure of endocrine cells and tissues. ISBN-13: 978-1-4613-3863-5

glandular cells. The latter, lining the cords and bordering follicular cavities, resemble the folliculo-stellate cells of the PD. In most species, the majority of glandular cells belong to a cell type peculiar to the PT and these cells are not found in other adenohypophysial lobes and are therefore called 'specific PT cells'. In the rodent PT, they are the sole glandular cell type present in the cephalic zone of the PT. Different quantities of PD cell types, mainly gonadotrophs, are found in the caudal PT of various mammals.

3.1. Glandular cells

3.1.1. Specific glandular cells

Viewed with the electron microscope, these cells show characteristic features of peptide-secreting cells. The moderately developed rough endoplasmic reticulum (rER) often forms clusters of parallel cisternae which sometimes appear as whorls (Fig. 2). Short rER cisternae are closely associated with mitochondria which are relatively large in this cell type. The Golgi ap-

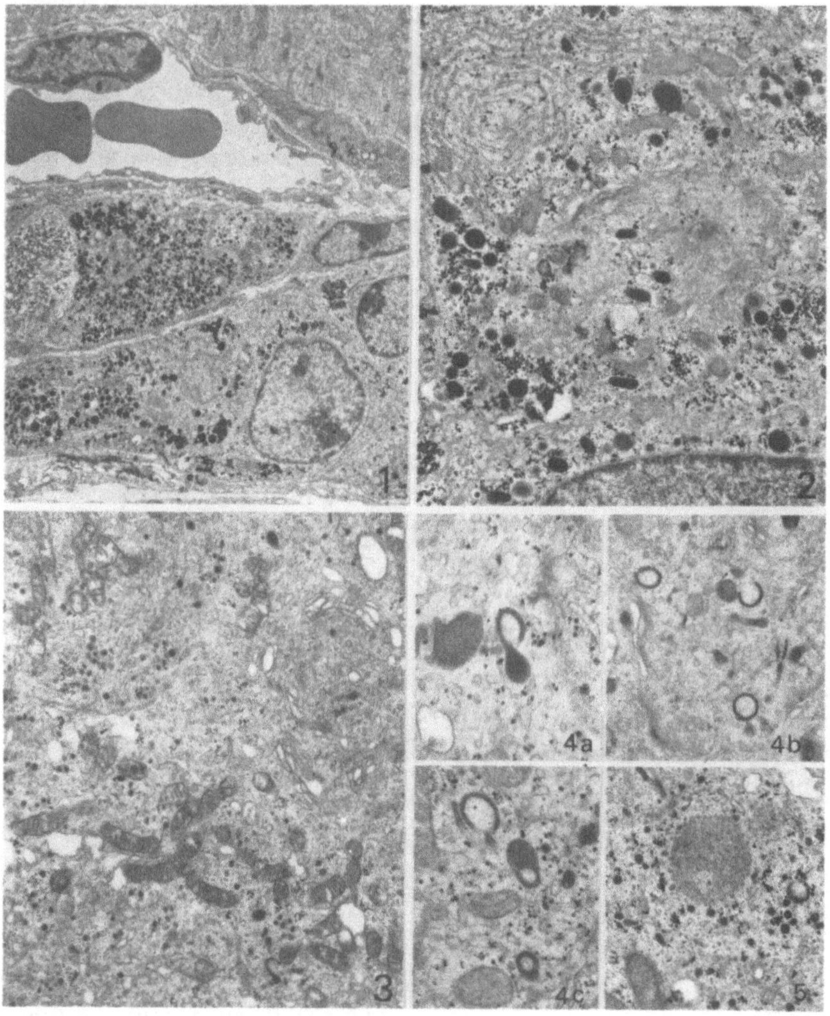

Figs. 1–5. (1) Mouse PT cell cords separated from the median eminence (ME) by a portal capillary and a fine connective tissue layer. Note numerous dense secretory vesicles (DSV) and glycogen particles in the specific glandular cells. Subarachnoid space at the bottom (× 3,920). (2) Mouse PT. Specific glandular cell containing large DSV, well developed Golgi apparatus and elongated cisternae of the rER, some forming a whorl. Numerous glycogen particles are intermingled with the DSV (× 8,000). (3) Rat (1 week old) PT. Specific glandular cells relatively rich in DSV in this young animal. Compared with the mouse, DSV are much smaller and glycogen is less abundant (× 8,000). (4) Rat PT. Characteristic lysosomes of the specific glandular cells seen in different section planes. In (a), dense body with cup-like expansion which accounts for the various ring shapes seen in (b) and (c) (a × 14,800; b × 8,750; c × 14,700). (5) Nucleolus-like body (nematosome) in a rat PT specific glandular cell (× 11,200).

paratus, generally conspicuous, occupies a large juxtanuclear area where several small sacculo-vesicular complexes are scattered. Dense secretory vesicles (DSV), abundant in the mouse (Figs. 1, 2), are rather scarce in other species (Figs. 3, 6) in which they accumulate only in a few cells. They are scattered, either isolated or in small clusters, at the periphery of the cytoplasm towards the vascular pole of the cells. These DSV originate from intravesicular condensations in the Golgi apparatus. Conspicuous multivesicular bodies are frequently observed in the Golgi area together with peculiar lysosomal formations. Mainly in rodents, these formations consist of a dense body with a cup-like expansion which accounts for the various, mostly ring-shaped figures observed on different sectional planes (Fig. 4). Glycogen particles, either aggregated in clusters or scattered throughout the cytoplasm, are an almost constant characteristic of these cells; their amount varies among species. They are abundant in the mouse and particularly in the European hamster where glycogen particles agglomerate in vast cytoplasmic areas (Fig. 8). Structures known as nucleolus-like bodies or nematosomes (Fig. 5) commonly observed in hypothalamic neurones can also be found in rat-specific PT cells as in MSH cells of the PI (24).

The diameter and number of DSV in specific glandular PT cells vary widely with the species. In the mouse PT, the DSV are especially abundant and much larger than in the rat PT where they are small and scarce; no such differences exist between other adenohypophysial cells in these two species. DSV diameter in the mouse can reach 300 nm (140–300); in most other species studied it does not exceed 200 nm. In the garden dormouse the diameter ranges from 120–180 nm, in the European hamster 120–200 nm. In the cells richest in DSV in the rat, guinea pig and rabbit the range is 100–120 nm, rarely reaching 150 nm. In the cat, DSV of 160–230 nm are only stored in a few cells; in most specific PT cells they are rare and often restricted to newly formed granules in Golgi vesicles.

The nature of the DSV of the specific secretory cells of the PT is unknown. In the rat, PT cells were first assimilated to PD cells containing small DSV, i.e., ACTH cells (4, 6), ACTH or TSH cells (8). However PT cells do not present any obvious ultrastructural changes after various experimental interventions on the endocrine system affecting the different adenohypophysial cell types or even after hypophysectomy (see section 6). They do not react with any antibody directed against the different hypophysial hormones, which has been shown by negative immunocytochemical reactions in the cephalic area of the muridae PT where the specific cells are the only glandular cell type present.

3.1.2. Gonadotrophic cells

Numerous cells with ultrastructural characteristics similar to gonadotrophic cells of the PD and often showing signs of secretory hyperstimulation have been observed by Stutinsky et al. (4) in the PT of hypophysectomized rats. These cells were interpreted as steuming from undifferentiated PT cells. Although less numerous, gonadotrophic cells also exist in normal rat PT but are restricted to its caudal part i.e. near the PD and along the infundibular stalk. These cells, well defined by their ultrastructural features (Fig. 6), were found in the PT of the different mammals we studied. Isolated or grouped in small clusters, they are more or less numerous in the PT areas near the PD and, except in man, decrease rostrally, being mostly absent from cell cords along the ME. These cells are distinguished easily from the specific glandular cells by their larger size, the extensive development of the rER in short, often dilated cisternae with amorphous weakly electron-dense content, and the abundance of DSV which, except in the mouse, are much larger than those of the specific PT cells. These cells clearly react after castration or hypophysectomy by an increase in size, and certain cells show dilatation and confluence of the rER elements resulting in large cavities filled with weakly electron-dense material, as observed in 'castration cells' of the PD.

Gonadotrophic cells have been detected in all mammalian PT investigated with immunohistochemical techniques (22, 25-35). For the rat PT, immunohistochemical data and electron microscopic findings concur on the distribution of these cells, i.e., their absence from the cephalic area and their progressive increase in number towards the PD. The same gradient appears to exist in other mammals studied, despite quantitative variations depending on the species (Fig. 7). After hypophysectomy, immunoreactive hyperplasic and hypertrophic gonadotropic cells have been found throughout the PT, even in rostral areas, in the rat (18, 19). In human PT, where gonadotrophic cells are particularly numerous (31, 32), they occur throughout the PT and, according to Baker (31), they form about 80% of the cell population.

Earlier immunohistological studies using antibodies against the whole LH molecule did not distinguish between FSH and LH immunoreactivity in gonadotrophic cells. Using antibodies against specific β chains, gonadotrophic PT cells reacting with anti-βLH and/or with anti-βFSH antibodies have been reported in the dog (36), monkey, macaca mulatta (34), rat (30) and man (31, 32). Gonadotrophic cells reacting with both immunosera appear to exist in different species, but cells reacting with only one antiserum also seem to occur in particular in the dog (33). In monkey, macaca irus, Girod et al. (22) only detected gonadotrophic cells reacting with anti-βLH antibodies in the PT, while cells staining for both βFSH and βLH, occurred in the zona tuberalis but not in the PT proper. It remains to be established if LH is also the main hormone secreted by PT gonadotrophic cells in other species.

3.1.3. Other glandular cells

Viewed with the electron microscope, other glandular PD cell types are occasionally found in the PT cords very close to the PD. This finding has been reported in the rabbit (12) and agrees with our observations in the rat and the mouse. In the PT proper of the latter species, we did not detect any immunoreactive ACTH or MSH cells. ACTH cells, moreover, remain absent from the PT of hypophysectomized rats (27). According to Baker and Yu (30), gonadotrophic cells are the only immunoreactive cell type normally present in the

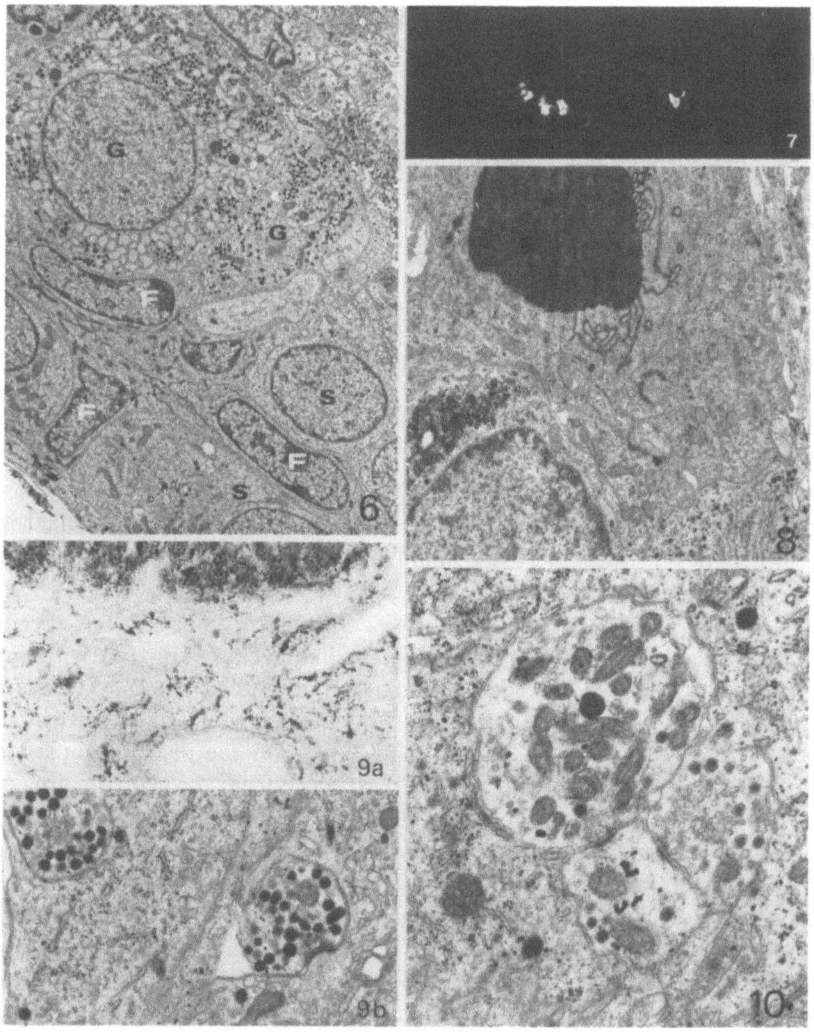

Figs. 6–10. (6) Guinea pig PT. Portion of cell cord containing two typical gonadotrophs (G), follicular cells (F) and specific glandular cells (S) containing cup-like lysosomes and scarce, small DSV only just visible at this magnification. M.E. external zone at the upper right (× 2,940). (7) Rabbit PT. Few immunofluorescent LH cells in the PT lining the hypophysial stalk (top) (× 161). (courtesy of J. Doerr-Schott). (8) European hamster PT. Follicular cavity with an amorphous electron-dense content separated from specific glandular cells (bottom), which contain small DSV (right) and glycogen aggregates (left), by non-glandular follicular cells (× 5,880). (9) Rabbit PT. (a) Light microscope immunoperoxidase characterization of oxytocinergic axons diffusely distributed throughout the PT cell cords; at the top, ME (× 175). (b) Electron microscopic aspect of the oxytocin immunoreactive axons containing numerous dense neurosecretory vesicles in synaptic contact (right) with specific glandular cells (× 11,760). (10) Garden dormouse PT. Axo-axonic contacts inside a cell cord: synaptic contacts between an axon and two presynaptic nerve endings, all containing small dense cored vesicles of undeterminated nature (× 21,000).

rat PT. Variable quantities of TSH immunoreactive cells have been reported in the monkeys, macaca mulatta (34), macaca rhesus (35) and in man (31, 32). Other scarce PD cell types, i.e., coticotrophs, somatotrophs, or prolactin cells have been found irregularly. Girod et al. (22), however, detected no TSH cells nor any other PD cell types in the PT of the monkey macaca irus. As pointed out by Baker (31), these cells, when they exist, are restricted to PT areas close to the PD. The frequent association of cells of different types suggests the existence of erratic cell clusters originating from the PD in the cell cords of the PT. These different antehypophysial cell types can be considered as 'invasive cells' (12) and incidental constituents of the normal PT.

In long-term hypophysectomized rats, the different PD cell types have been reported to occur more or less massively in the PT, except in the cephalic zone (19). These cells may have proliferated from cells incidentally or normally (i.e. gonadotrophs) present in caudal areas and/or from PD cell clusters remaining attached to the tip of the stalk after removal of the pituitary (10). The idea that the PT is a pool of stem cells capable of differentiating into the different antehypophysial cell types (4, 18, 19) seems rather improbable since the PT consists mainly of already differentiated, specific glandular cells which do not react to hypophysectomy (10, 11, 13).

3.2. Follicular cells

The PT cell cords enclose more or less conspicuous follicular cavities, the number and size of which vary with species and age. Cells lining the cavities, or follicular cells, have an irregular form and their extensions insinuate between the adjacent glandular cells towards the periphery of the cords. Comparable single cells can be seen inside the cords or lining them, apparently with no relation to the follicular cavities; however, their relationship with follicular structures outside the plane of section cannot be excluded. Viewed with the light microscope, these cells are seen to differ from chromophobic glandular cells by their smaller size and their irregularly shaped nucleus with dense chromatin. Viewed with the electron microscope, they are seen to be devoid of secretory characteristics and rarely contain glycogen particles, except in a few species, such as the garden dormouse where they are richer in glycogen than specific glandular cells, and the rabbit where glycogen, absent from specific glandular cells, is very abundant in the follicular cells. The rER is poorly developed. Numerous clear vesicles are concentrated near the Golgi area, which contains few saccular stacks, and are scattered throughout the apical cytoplasm, towards the follicular cavities. Whether such vesicles can store calcium, as has been shown in the folliculo-stellate cells of the PD and epithelium lining the hypophysial cleft with the pyroantimonate technique (36), has still to be determined for the follicular cells of the PT. In the rabbit, some cells reported as 'interstitial cells' by Cameron and Foster (12) have particularly long extensions surrounding the glandular cells and often lining the cell cords, and are packed with intermediate-type (10 nm in diameter) microfilaments. Such a fibrillar concentration, which can also occur in basal extensions of follicular cells obviously in contact with follicular lumina, seems to be peculiar to the rabbit PT.

Follicular cells are linked by tight junctions at their apical poles; they often display microvilli and some project tufts of cilia into the follicular lumen. A peculiar feature, obvious in the mouse (21, 37), is the occurrence of mitochondria opposite to an intercellular space and with their extremities connected to desmosomes by fine fibrillar material. Such an unusual mitochondria-desmosome complex is still not clearly understood but might indicate ionic control over the intercellular spaces near the follicular lumina. Follicular cavities are either flattened or dilated, and usually have an electron-lucent content. Sometimes they are filled with electron-dense material, either amorphous (Fig. 8) or heterogeneous, and in the latter case contain recognizable cell residue. Calcified masses can fill the lumen of small follicles, in particular in the guinea pig PT. The content of the follicular cavities varies with the individual but even more with the species.

The role of follicular cells is a problem not restricted to the PT; it has been extensively studied in the case of the folliculo-stellate cell system of the PD (see chapter 6). Follicular formations similar to those of the PT are observed in bird ultimobranchial bodies where cysts, often containing cell debris, proliferate in aging animals and coincide with a decreae in glandular activity (38). Large follicular cavities are also more frequent in the PT of aging animals where glandular cell activity is obviously lower.

PT cysts differ by their content but also by their permeability to the tracers from the hypophysial cleft. Thus, while horseradish peroxidase (HRP) diffuses rapidly in the hypophysial cleft, where it can be detected only 3 minutes after intravenous injection, it does not penetrate the PT follicles, even after 30 minutes, although it is present in the intercellular spaces. Similar observations by Aguado et al. (39) show that HRP diffusing from the subarachnoid cerebrospinal fluid (CSF) into the PT cell cords fills the intercellular spaces, interpreted as pre-existing channels, without penetrating the follicular lumina. If the hypophysial cleft seems able to protect the poorly vascularized PI against excessive accumulation of extracellular metabolites (40), the follicular cavities in the PT apparently have no such function.

3.3. Human pars tuberalis

Electron microscopic observations of human PT obtained a few hours after death reveal glycogen-containing follicular cells lining more or less conspicuous cystic cavities, and a heterogeneous glandular cell population where still clearly recognizable gonadotrophic cells predominate. These observations have been confirmed by immunofluorescence data (31, 32). In view of the poor preservation of cellular structures under these conditions, we cannot ascertain whether the regularly scattered cells containing small (100 nm) DSV are a cell type peculiar to the human PT. In the absence of antibodies against the peptide(s) synthetized by the specific glandular cells of the PT, one cannot be sure that such cells are present in the human pituitary. Certain chromophobic adenomas of the pituitary gland with cells containing small (100 nm) DSV but with no obvious endocrine function (41) may contain this type of glandular cells.

4. Innervation of the PT

The penetration of nerve bundles from the ME into the PT has been observed by light microscopy and in most cases has been related to the blood vessels rather than to the glandular innervation (42, 43). The electron microscope, however, has revealed more or less discrete innervation of the PT cell cords in most of the species studied. The axons entering the cell cords make synaptic contacts or, more often, synaptoid ones (i.e., with no clear-cut synaptic cleft and no post-synaptic differentation) with the specific glandular cells and also with follicular cells. In the rat and mouse, only scattered, inconspicuous axons, rarely containing small dense-cored vesicles (DCV), are observed in the cell cords. In the garden dormouse, where the PT is very closely applied against the ME palisade zone, especially numerous axons containing small DCV (about 100 nm diameter) penetrate the PT cell cords, and synaptic contacts with the glandular cells are frequent (Fig. 11). Moreover synaptic contacts between axons, also containing small DCV, and nerve endings in contact with glandular cells can be found (Fig. 10) (13). Such preterminal axo-axonic contacts may have an inhibitory significance. In the rabbit, a fine plexus of axons staining with Gomori's techniques for neurosecretion in seen extending throughout the PT (and also the PI) with the light microscope. Electron microscopy shows these axons contain abundant neurosecretory granules and make synaptoid contacts with the PT glandular cells (12). Although these neurosecretory granules are definitely smaller than those of the hypothalamo-hypophysial system, immunocytochemical data demonstrate that they are peptidergic, oxytocinergic (but not vasopressinergic (Fig. 9). Such oxytocinergic axons also innervate the rabbit intermediate lobe while peptidergic axons are generally absent or rare in the PI of other species. Immunofluorescence also shows that the PT and PI both have discrete serotoninergic innervation in the rabbit.

In the rat PT, peptidergic axons are only seen after severing the hypophysial stalk (11, 44) or after mechanical lesion of the floor of the 3rd ventricle. This provokes a sprouting of the neurosecretory axons of the hypothalamo-hypophysial tract which then invades both the PI and the PT (44). Serotoninergic innervation has recently been shown in the normal rat by means of ultrastructural autoradiography (45); ^{3}H serotonin is captured by slender axons which, on usual preparations, would sometimes be difficult to distinguish from the thin extensions of the follicular cells between the glandular cells (Fig. 12). The same type of serotoninergic innervation is also present in the PI. In contrast, dopaminergic innervation which is characteristic of the PI has not yet been reported in the PT of any species. We have never detected axons immunoreactive for neuropeptides such as vasopressin, oxytocin, enkephalins, somatostatin or substance P in rat and mouse PT under normal conditions. GABAergic innervation cannot be excluded, at least in the guinea pig, in which the PT cells exhibit strong, histochemically detectable GABA-T activity (46).

The functional significance of the innervation of the PT remains to be established. It probably plays an important role in regulating the activity of PT cells, although in view of the close relationships with the capillary network of the ME neurohumoral control of these cells may be more important.

5. Embryonic development and secretory differentiation

Embryonic development and secretory differentiation of the PT in rat and mouse (20, 21) underline the originality of this hypophysial lobe and also plead strongly in favour of its functional specificity.

The PT anlage originates from the anteroventral area of Rathke's pouch; classically it develops from two symmetrical buds, the lateral lobes, which grow rostrally and fuse medially along the ME and surround the hypophysial stalk. Lateral lobes are not always very obvious in mammals, in particular in muridae. In the rat and mouse, the PT anlage arises from a well defined antero-basal area of Rathke's pouch, including its pharyngeal pedicle. After closure of Rathke's pouch (day 11 in the mouse, day 14 in the rat), a massive cell bud or anterior process forms on the basal zone of the anterior wall (Fig. 13). This anterior process is clearly separated from the PD anlage by Atwell's recess or fossa where tiny blood vessels are already present and via which the portal vasculature penetrates the PD in the later stages. The anterior process is slightly bilobed, which is the sole remnant of the lateral lobes in rat and mouse. Correlating with the shifting and the flattening of the hypophysis, the PT anlage, growing forward, extends along the ME and surrounds the hypophysial stalk. At the same time as the PT anlage is extending, its primi-

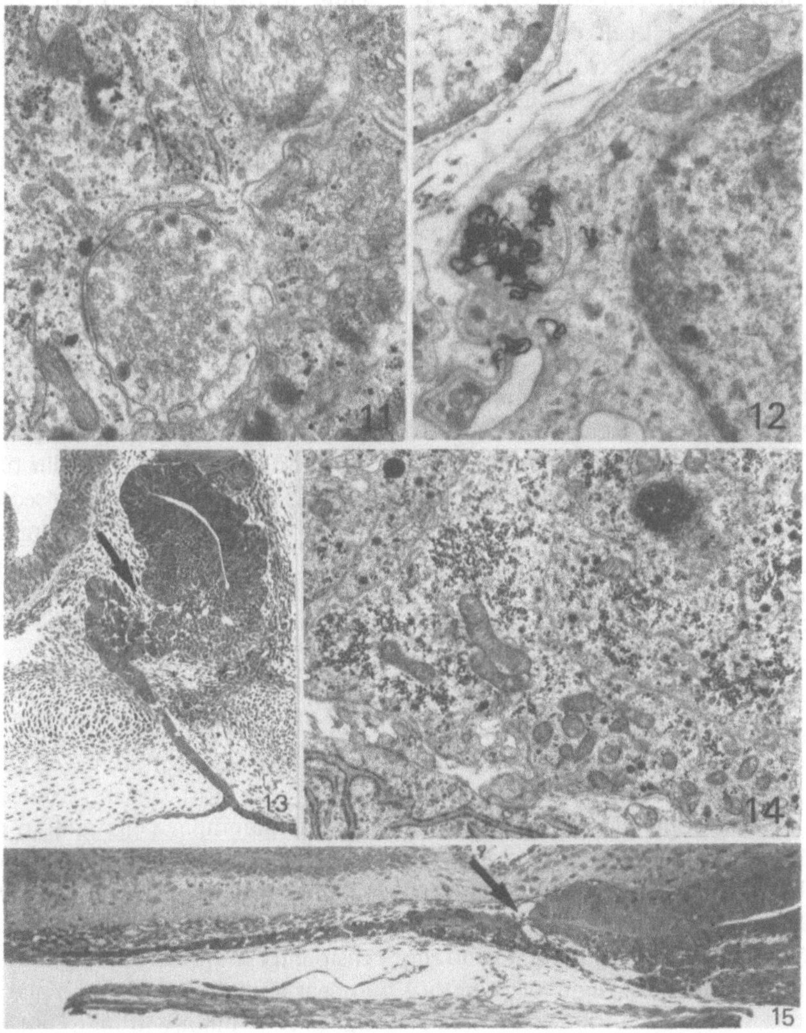

Figs. 11–15. (11) Garden dormouse PT. Axon containing small dense-cored vesicles, in synaptic contact with a specific glandular cell. Another similar contact is seen in tangential section (top) (× 17,500). (12) Rat PT. Autoradiographic identification of a serotoninergic nerve ending, in synaptoid contact with a specific glandular cell, after intravascular injection of ^{3}H serotonin (× 14,000) (courtesy of A. Calas). (13) Semithin sagittal section stained with PAS-hemalum of the hypophysial anlage in a rat foetus at 15 days gestation. The adenohypophysial lobes are clearly outlined; Atwell's recess draws the limit between the PD and PT anlagen (arrow). The PT anlage comprises the anterior process, already dividing into cell cords, and the pharyngeal pedicle (× 80). (14) Foetal mouse PT at 12 days gestation. Already well differentiated specific glandular cells containing abundant glycogen particles and numerous DSV albeit smaller than in the adult PT cells (× 10, 640). (15) Newborn rat. Semithin sagittal section stained with PAS-hemalum showing the extension of the PT along the ME, from which it remains separated by a loose conjunctivo-vascular layer. Caudally, the PT covers the rostro-ventral face of the PD from which it is separated by the remnant of Atwell's recess via which the large portal veins penetrate the PD (arrow) (× 88).

tive massive structure is being transformed into a loose arrangement of cell cords interwoven with capillaries. The foetal PT remains separated from the external palisadic zone of the ME by a loose vascular connective layer (Fig. 15). Close apposition of PT cords against the ME only occurs during the first week after birth, when capillary loops penetrate the palisadic zone of the ME.

Cell differentiation of the PT largely precedes its anatomical relationships with the ME. The presence of glycogen, which characterizes the adult tuberal lobe, is the first feature distinguishing the PT anlage from the other adenohypophysial anlagen. In the rat, the glycogen clearly labels the PT cells during development, allowing the entire PT anlage to be identified on PAS-stained semithin sections. In the mouse, cells of the PD (but not of the PI) anlage store glycogen temporarily but it appears at least 1 day later than in the PT, preceding the secretory differentiation of the cells.

Evidence of the secretory activity of the PT such as development of rER and Golgi apparatus and the formation of typical DSV in the Golgi area is obvious in PT cells as early as 12 days p.e. in the mouse (Fig. 14) and 15 days in the rat. In both species, cells of the pharyngeal pedicle of Rathke's pouch are among the first to differentiate. At 14 days in the mouse and 17 days in the rat, PT cells are already comparable with those of the adult by their richness in glycogen, rER and Golgi apparatus development and DSV content, although the latter remain smaller during foetal life than in the adult. The characteristic cup-like lysosomes are also seen from the earliest stage of secretory differentiation. At the time when the PT cells begin to differentiate, no differentiating cells are found in the PD anlage.

Even when the PT cells have an almost adult appearance, only a few cells exhibit signs of functional activity in the ventral area of the PD. Secretory differentiation of the PI occurs much later. Thus, it appears that adenohypophysial differentiation follows a ventro dorsal gradient, the ventral part, including the pedicle of Rathke's pouch, being the first (PT) and the dorsal part (PI) the last to show signs of secretory activity.

Follicular cells in the PT are only seen lining inconspicuous cavities after birth, when cell cords tend to abut the ME more closely. In contrast, the folliculo-stellate cell system develops early in the PD where it has been thought to derive from infoldings of the epithelium lining the hypophysial cleft (37). Since the primitive cavity of Rathke's pouch never extends into the PT anlage, such an origin for the follicular cells of the PT can be excluded. The frequency of mitoses in follicular cells after birth suggests that a few undifferentiated cells with a follicular potential exist in the foetal PT, but do not multiply before the close adhesion of the glandular cell cords to the ME.

The gonadotrophic cells which regularly occur in the caudal PT were not found in early post-natal stages (before 10 days) in the rat and mouse, and so are probably invasive elements from the PD. A few gonadotrophic cells appear to migrate also into the rostral PI. Depending upon the closeness of the apposition of the PD and the caudal part of the PT after birth, migration of gonadotrophic cells into the PT cords is greater or lesser according to the species. Thus, in the cat, where PT cords are very loosely arranged and are separated from the rostral PD by a continuous connective sheat, gonadotrophic cells are irregularly found through electron microscopy and immunocytochemistry, although they are seen scattered in rostral PI paraneural epithelium and even in the adjacent neural stalk. In the case of complete separation of the PT from the PD, as in the frog, only specific glandular cells are found by electron microscopy (15, 16) in the PT where no PD cell types are detected immunocytochemically (47).

From these observations, the PT emerges as a distinct hypophysial lobe, characterized by early differentiation of its specific glandular cells and early relation with the primary vascular plexus of the developing portal system. This early secretory differentiation does not seem restricted to mammalian PT, since it has also been reported in the developing lizard PT (17).

6. Functional significance of the pars tuberalis

These various data show that the PT is a distinct hypophyseal lobe, both in cell composition and ontogenesis. In most species studied it contains mainly specific glandular cells and so cannot be an undifferentiated lobe capable of providing PD cell types (mainly gonadotrophs) in the case of high hormonal demand. Among the PD cell types regularly or occasionally found in the PT, only gonadotrophic cells appear numerous enough to play an effective physiological role, mainly in man, but also in castrated animals where they are stimulated like the PD gonadotrophs. After hypophysectomy, they proliferate and may account for the residual gonadotrophic activity reported in the rat (48), monkey (49) and man (50). Other PD cell types incidentally present in the PT cannot appreciably contribute to or replace secretions of adenohypophysial hormones, in particular in rodents, in view of their small number. For example, in the rat, the PT cannot be responsible for the ACTH activity of the ME extracts, as suggested by Klein et al.

(8), but which might be explained by the high concentration of ACTH (and other adenohypophysial hormones) of PD origin in the portal circulation (51) and probably by the existence of pro-opiocortin producing neurones in the rat arcuate nucleus (52). The function(s) of the tuberal lobe will be elucidated only after further investigations have been made into the activities of its specific glandular cells.

Since the PT cannot be removed without seriously damaging the ME, target cells cannot be identified by this means. As pointed out by Dellmann et al. (13), experimental interventions which cause drastic modifications to most of the major endocrine systems do not obviously affect the specific secretory cells of the PT. These cells, as already mentioned, do not react with any antibody directed against the known hypophysial hormones and do not present any obvious ultrastructural changes after adrenalectomy, castration, thyroïdectomy or propylthiouracyl treatment, nor even after hypophysectomy (10, 11, 13). The same is true under other experimental conditions such as alloxane diabetes in the rat, chronic or acute ionic perturbations affecting natremia or calcemia in rat and mouse (13). We have not found changes in specific PT cells in the lactating rat and mouse, except for more abundant glycogen. In the hibernating garden dormouse, the number of DSV decreases considerably in these cells, coinciding with involutive aspects of the rER (13), but similar changes are known to occur under these conditions in other endocrine glands (53). Modifications of the PT thus do not prove a particular role of this lobe in the hibernating process.

The PT seems to be especially active in the young animal, as shown by the impressive accumulation of DSV in specific glandular cells in the first weeks after birth in rat and mouse (20, 21). This activity appears to decrease progressively with age. Signs of secretory activity are more or less pronounced according to species. This activity is especially marked in the mouse. In most species studied, the rather moderate development of the rER and, except in the mouse, the relative scarcity of DSV seem to indicate rather modest secretory activity. But the same is true for the parathyroid gland (53) which stores only a few DSV, and yet is known to ensure very rapid synthesis and release of parathormone.

From these data, it seems difficult to integrate the PT function into a known process of endocrine regulation. In view of its very early differentiation, this lobe might play a major role during foetal life.

As pointed out by Dellmann et al. (13), a striking feature of the PT is its privileged relationship with the primary capillary plexus of the portal system, already established in early stages of its development. Secretions of the PT can thus reach the PD directly, while the PT can also interact immediately with the ME. Data supporting the hypothesis of PT control over some PD cells, however, are lacking since conditions provoking the stimulation of the different PD cells, and even hypophysectomy, do not obviously affect the specific glandular PT cells. A modulatory effect on the PD, however, cannot be excluded under physiological conditions.

Little is known of the behaviour of specific glandular cells deprived of their vascular and nervous connections. No special attention was paid to these cells in a recent study of the guinea pig PT in organ culture (54). However, in intraocular homografts in the rat, the histological organization and the cytological characteristics of the specific glandular cells have been seen to persist when the median eminence component of the transplant was reduced to only tanycytes and glial cells (55).

As recently pointed out (39), relationships of the PT with the CSF should not be neglected. The PT cells are indeed close to the subarachnoid space, as well as the fenestrated capillaries which insinuate between the cell cords. Because the free diffusion of HRP tracer has been shown to occur throughout the PT cell cords the PT is not a diffusion barrier between the CSF and the ME. The CSF can influence the ME and the PT and in turn carry secretory products, including hormonal secretions of the PT, from the portal circulation to the brain. The possibility that the PT affects cerebral structures in this way is an interesting hypothesis.

In view of the present state of our knowledge of the PT, it seems unlikely that studies using only morphological criteria can lead to a better understanding of the specific function of the lobe. These data, however, should stimulate research to isolate a highly probable peptidic hormone specific to the PT.

References

1. Tilney F: An analysis of the juxtaneural portion of the hypophysis cerebri with an embryological and histological account of an hitherto undescribed part of the organ. Internat Monatschr Anat u Physiol 30: 258–293, 1913.
2. Atwell WJ: The development of the hypophysis cerebri of the rabbit. Amer J Anat 24: 271–337, 1918.
3. Fitzgerald KT: The structure and function of the pars tuberalis of the vertebrate adenohypophysis. Gen Comp Endocrinol 37: 383–399, 1979.
4. Stutinsky F, Porte A, Stoeckel ME: Sur les modifications ultrastructurales de la pars tuberalis du rat après hypophysectomie. C R Acad Sci (Paris) 259: 1765–1767, 1964.
5. Oota Y, Kurosumi K: Electron microscopic studies on the pars tuberalis of the rat hypophysis. Arch Histol Jap 27: 501–120, 1966.

6. Rinne UK: Ultrastructure of the median eminence of the rat. Z Zellforsch 74: 89–122, 1966.
7. Knowles F, Anand Kumar TC: Structural changes related to reproduction in the hypothalamus and in the pars tuberalis of the rhesus monkey. Phil Trans Roy Soc Ser B 256: 375–395, 1969.
8. Klein MJ, Stoeckel ME, Porte A, Stutinsky F: Arguments ultrastructuraux en faveur de l'existence de cellules corticotropes dans la pars intermedia et dans la pars tuberalis de l'hypophyse du rat. C R Acad Sci (Paris) 271: 2159–2162, 1970.
9. Clementi F, Cecarelli B: Fine structure of rat hypothalamic nuclei. In: The hypothalamus, Martini L, Motta M, Fraschini F (eds), New York – London, Academic press, 1970, pp 17–44.
10. Kotsu T: Studies on the pars tuberalis of the hypophysis. 2. Fine structural changes following hypophysectomy. Shikoku Acta med Jap 27: 483–493, 1971.
11. Kotsu T, Daikoku S: Ultrastructural changes of the pars tuberalis and median eminence of rats following hypophysectomy. Arch Histol Jap 34: 167–184, 1972.
12. Cameron E, Foster CL: Some light and electronmicroscopical observations on the pars tuberalis of the pituitary gland of the rabbit. J Endocrinol 54: 505–511, 1972.
13. Dellmann HD, Stoeckel ME, Hindelang-Gertner C, Porte A, Stutinsky F: A comparative ultrastructural study of the pars tuberalis of various mammals, the chicken and newt. Cell Tiss Res 148: 313–329, 1974.
14. Grignon G, Guedenet JC: Observations sur l'ultrastructure de la neurohypophyse et de la pars tuberalis chez le poulet au cours de la vie embryonnaire et après l'éclosion. Arch Anat Histol Embryol Norm Exp 51: 277–286, 1968.
15. Dierickx K, Lombaerts-Vandenberghe MP, Druyts A: The structure and vascularization of the pars tuberalis of the hypophysis of rana temporaria. Z Zellforsch 114: 135–150, 1971.
16. Doerr-Schott J: La pars tuberalis de rana temporaria L: Cytologie et ultrastructure. Gen Comp endocr 11: 516–523, 1971.
17. Pearson AK, Licht P: Embryology and cytodifferentiation of the pituitary gland in the lizard anolis carolinensis. J Morphol 144: 85–118, 1974.
18. Gross DS, Page RV: Luteinizing hormone and follicle-stimulating hormone production in the pars tuberalis of hypophysectomized rats. Am J Anat 156: 285–291, 1979.
19. Ordronneau P, Petrusz P: Immunocytochemical demonstration of anterior pituitary hormones in the pars tuberalis of long-term hypophysectomized rats. Amer J Anat 158: 491–506, 1980.
20. Stoeckel ME, Porte A, Hindelang-Gertner C, Dellmann HD: A light and electron microscopic study of the pre- and postnatal development and secretory differentiation of the pars tuberalis of the rat hypophysis. Z Zellforsch 142: 347–365, 1973.
21. Stoeckel ME, Hindelang-Gertner C, Porte A: Embryonic development and secretory differentiation in the pars tuberalis of the mouse hypophysis. Cell Tiss Res 198: 465–476, 1979.
22. Girod C, Dubois MP, Trouillas J: Immunohistochemical study of the pars tuberalis of the adenohypophysis in the monkey macaca irus. Cell Tiss Res 210: 191–203, 1980.
23. Lothringer S: Untersuchungen an der Hypophyse einiger Säugetiere und des Menschen. Arch Mikr Anat 28: 257–292, 1886.
24. Hindelang-Gertner C, Stoeckel ME, Porte A, Dellmann HD, Madarasz B: Nematosomes or nucleolus-like bodies in hypothalamic neurones, the subfornical organ and adenohypophysial cells of the rat. Cell Tiss Res 155: 211–219, 1974.
25. Midgely ARJr: Human pituitary luteinizing hormone. An immunohistochemical study. J Histochem Cytochem 14: 159–166, 1966.
26. Dubois MP: Cytologie de l'hypophyse des bovins: séparation des cellules somatotropes et des cellules à prolactine par immunofluorescence. Identification des cellules LH dans la pars tuberalis et la pars intermedia. C R Ass Anat 145: 139–146, 1970.
27. Dubois MP, De Reviers MM, Courot M: Activité gonadotrope de l'éminence médiane après hypophysectomie chez le rat. Etude en immunofluorescence. Exp Anim 4: 213–226, 1971.
28. Tramu G, Dubois MP: Identification par immunofluorescence des cellules à activité LH du cobaye mâle. C R Acad Sci (Paris) 275(D): 1159–1161, 1972.
29. Dubois MP: Localisation cytologique par immunofluorescence des secrétions corticotropes et mélanotropes au niveau de l'adenohypophyse des bovins, ovins, et porcins. Z Zellforsch 125: 200–209, 1972.
30. Baker, BL, Yu YY: Immunocytochemical analysis of cells in the pars tuberalis of the rat hypophysis with antisera to hormones of the pars distalis. Cell Tiss Res 156: 443–449, 1975.
31. Baker BL: Cellular composition of the human pituitary pars tuberalis as revealed by immunocytochemistry. Cell Tiss Res 182: 151–163, 1977.
32. Osamura RY, Watanabe K: An immunohistochemical study of epithelial cells in the posterior lobe and pars tuberalis of the human adult pituitary gland. Cell Tiss Res 194: 513–524, 1978.
33. El Etreby MF, Fath El Bab MR: Localization of gonadotropic hormones in the dog pituitary gland. A study using immunoenzyme histochemistry and chemical staining. Cell Tiss Res 183: 167–175, 1977.
34. Baker BL, Karsch FJ, Hoffman DL, Beckman WC: The presence of gonadotropic and thyrotropic cells in the pituitary pars tuberalis of the monkey (Macaca mulatta). Biol Reprod 17: 232–240, 1977.
35. Herbert DC: Identification of the LH and TSH-secreting cells in the pituitary gland of the rhesus monkey. Cell Tiss Res 190: 151–161, 1978.
36. Stoeckel ME, Hindelang-Gertner C, Dellmann HD, Porte A, Stutinsky F: Subcellular localization of calcium in the mouse hypophysis. I. Calcium distribution in the adeno and neurohypophysis under normal conditions. Cell Tiss Res 157: 307–322, 1975.
37. Dingemans KP, Feltkamp CA: Nongranulated cells in the mouse adenohypophysis. Z Zellforsch 124: 387–405, 1972.
38. Stoeckel ME, Porte A: Etude ultrastructurale de corps ultimobranchiaux du poulet. I. Aspect normal et developpement embryonnaire. Z Zellforsch 94: 495–512, 1969.
39. Aguado LI, Schoebitz K, Rodriguez EM: Intercellular channels in the pars tuberalis of the rat hypophysis and their relationship to the subarachnoid space. Cell Tiss Res 118: 345–354, 1981.
40. Stoeckel ME, Schmitt G, Porte A: Fine structure and cytochemistry of the mammalian pars intermedia. In: Peptides of the pars intermedia. Ciba foundation symposium 81, London, Pitman medical, 1981, pp 101–127.
41. Olivier L, Vila-Porcile E, Racadot O, Peillon F, Racadot, J: Ultrastructure of pituitary tumor cells: a critical study. In: The anterior pituitary. Tixier-Vidal A, Farquhar MG (eds), New York San Francisco London Academic Press, 1975, pp 231–276.
42. Stutinsky F: Sur l'innervation de la pars tuberalis de quelques mammifères. C R Ass Anat 55: 372–380, 1948.
43. Szentàgothai J, Flerko B, Mess B, Halasz B: Hypothalamic control of the anterior pituitary. Budapest, Akademiai Kiado, 1968.
44. Stutinsky F, Klein MJ, Stoeckel ME, Porte A: Réaction des fibres neurosécrétoires après irritation de la base du troisième ventricule chez le rat blanc. C R Ass Anat 57: 177–186, 1973.
45. Calas A: Personal communication.
46. Leonardelli J, Hermand E: Recherches histoenzymologiques hypothalamiques: la transaminase de l'acide gamma amino butyrique. C R Ass Anat 57: 141–148, 1973.
47. Doerr-Schott: Personal communication.
48. Lostroh AJ: Effect of follicle-stimulating hormone and interstitial cell-stimulating hormone on spermatogenesis of Long-Evans rats hypophysectomized for six month. Acta endocrinol (Kbh) 43: 592–600, 1963.
49. Niswender GD, Monroe SE, Peckham WD, Midgley AR, Knobil E, Reichert LE: Radioimmunoassay for rhesus monkey luteinizing hormone (LH) with anti-ovine LH serum and ovine LH-^{131}I. Endocrinology 88: 1327–1331, 1971.
50. Lachelin GD, Yen SSC, Alksne JF: Hormonal changes following hypophysectomy in humans. Obstet Gynecol 50: 333–338, 1977.
51. Oliver C, Mical RS, Porter JC: Hypothalamic-pituitary vasculature; evidence for retrograde blood flow in the pituitary stalk. Endocrinology 101: 598–604, 1977.
52. Bugnon C, Bloch B, Lenys D: Ultrastructural study of presumptive proopiocortin producing neurons in the rat hypothalamus. Neuroscience 6: 1299–1313, 1981.
53. Stoeckel ME, Porte A: Observations ultrastructurales sur la parathyroïde de mammifère et d'oiseau dans des conditions normales et expérimentales. Arch Anat Micr 62: 55–88, 1973.
54. Chatterjee P, Holmes RL: Some observations of the cell types of the guinea-pig pars tuberalis and their structural modifications in organ culture. J Endocrinol 85: 53P, 1980.
55. Dellmann HD: Personal communication.

Author's address: Laboratoire de Physiologie générale ULP, LA 309 CNRS
Neuroendocrinologie comparée, 21, rue René Descartes
F 67084 Strasbourg Cedex, France

CHAPTER 4

Ultrastructure of the pars intermedia: Development *in vivo* and in organ culture

PRASAD CHATTARJEE

1. Introduction

1.1. General plan

In this chapter the pars intermedia (PI) of the rabbit is taken as a model, but references to other species, including human, will be made where they are relevant. The ultrastructure of the cell types and the nerve supply in the adult will be described first. The gradual changes in the structural appearance of the PI that is related to the normal pre and post natal development *in situ*, will then be described. This will be followed by a description and discussion of the results of *in vitro* experiments designed to find out firstly if the cytodifferentiation of the mammalian PI is dependant on extrinsic factors, secondly if the two glandular cell types are really different and thirdly whether the non glandular cell has a role in the PI. Finally a conclusion is drawn from the facts presented in the chapter.

1.2. Nomenclature

The vertebrate pituitary gland (hypophysis) consists of two main parts, the adenohypophysis and the neurohypophysis. The adenohypophysis has three components (Fig. 1a, b): namely 1) the pars distalis (PD); 2) the pars intermedia, and 3) the pars tuberalis (PT). The neurohypophysis consists of three parts too, 1) the median eminence of tuber cinereum; 2) the infundibular stem, and 3) the infundibular process, which is also known as the neural lobe or pars nervosa (PN). The first two parts of the neurohypophysis are sometimes referred to collectively as the infundibulum or neural stalk. The infundibular stem together with the bulk of the PT which surrounds it, is called the hypophysial stalk. Owing to morphological diversity of the pituitary gland between different species or even at different ages in the same species, the above descriptive terms are not always interchangeable. In this article, the above terminology will be used as far as possible but as this is not suitable for the developing pituitary, terms used by other authors (1) will be used where necessary (Fig. 1c). In the adult rabbit the appropriate peripheral parts of the PI have sometimes been referred to as rostral, lateral and caudal zones (Fig. 1a, b).

2. Ultrastructure of the adult PI

2.1. Chief cells

The cell types found here are described in the order of their frequency in the PI. The most common cells in this tissue will be called here the chief cells (Fig. 2) following the terminology applied in the jird (2). Elsewhere, they have been named differently: in the rat, as pars intermedia cells (3), MSH cells (4), MSH – endorphin cells (5); in the rabbit, glandular cells (6) or PI-glandular cells (7). At a post natal immature stage the cell shape is variable and the nucleus is indented. In a mature cell the cytoplasm contains two types of membrane-bound granular inclusions: the cytoplasmic vesicles (6) to be referred to here as the vesicles, are the commonest. These are distinguished by their size upto 600 nm in diameter in the adult) and by their variable electron density. They are typically surrounded by a membrane which sometimes appears to be discontinuous. In contrast, dense secretory granules are less numerous, smaller (250–450 nm in diameter), always electron dense and each is enclosed in a continuous perigranular membrane.

2.2. Interstitial cells

The interstitial cell (6, 8) is non-granulated and its stellate form is noticeable as early as four weeks post coitum. (Fig. 2).

2.3. ACT-type cells

As early as the fifth week post coitum, a cell type

Motta, PM (ed): Ultrastructure of endocrine cells and tissues. ISBN-13: 978-1-4613-3863-5

appears which has some characteristics in common with immunocytochemically determined adrenocorticotropes (ACT) of the PD and PI of various mammals but differs in other respects to be discussed in section 3.4. In other mammalian PI is thas been called the ACTH cell in the past. In the rabbit it has not yet been labelled by antibody technique, hence it is named here as ACT-type cell and is found only infrequently, either localized in certain areas and/or sparsely scattered throughout the PI (Fig. 2). According to Stoeckel et al. (4) who worked on the mouse, rat and cat it is physiologically slightly different from an ACTH cell in the PD.

2.4. Dark cells

The so-called dark cells are a heterogeneous group of parenchymal cells whose only common distinguishing features are the general electron density in the cytoplasm and their frequent presence in the developing tissues (Fig. 3).

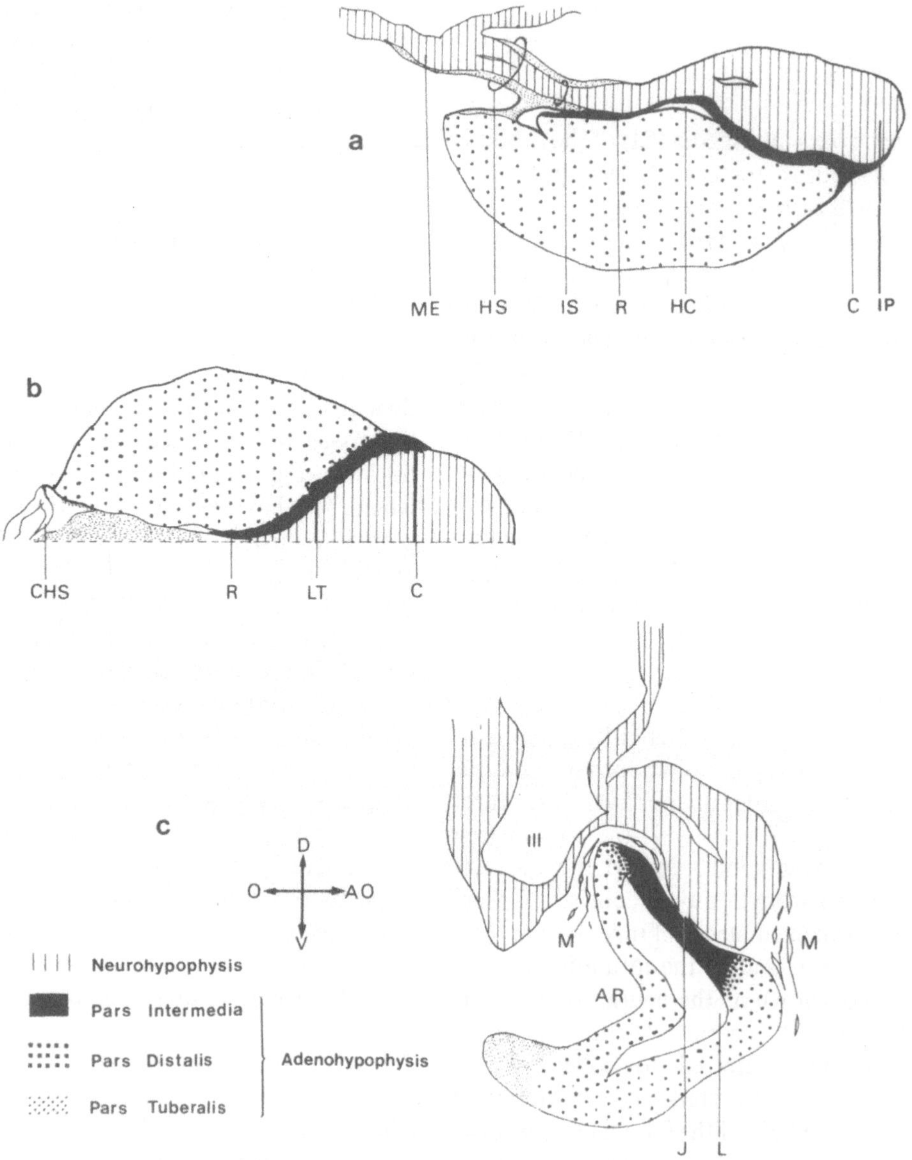

Fig. 1. (a) Sagittal section of the pituitary gland of a 16 months post partum rabbit; (b) Dorsal view of the right half of the above; (c) Sagittal section of the developing pituitary gland of a fetal rabbit 16 days post coitum. AO: aboral lobe; AR: Atwell's recess; C: caudal zone of the pars intermedia; CHS: cut hypophysial stalk; D: dorsal; HC: hypophysial cleft; HS: hypophysial stalk; IP: infundibular process; IS: infundibular stem; J: junction between the PI and neural lobe; L: lumen of the developing adenohypophysis; LT: lateral zone of the PI; M: mesenchyme; me: median eminence; O: oral; R: rostral zone of the PI; V: central; III: 3rd ventricle.

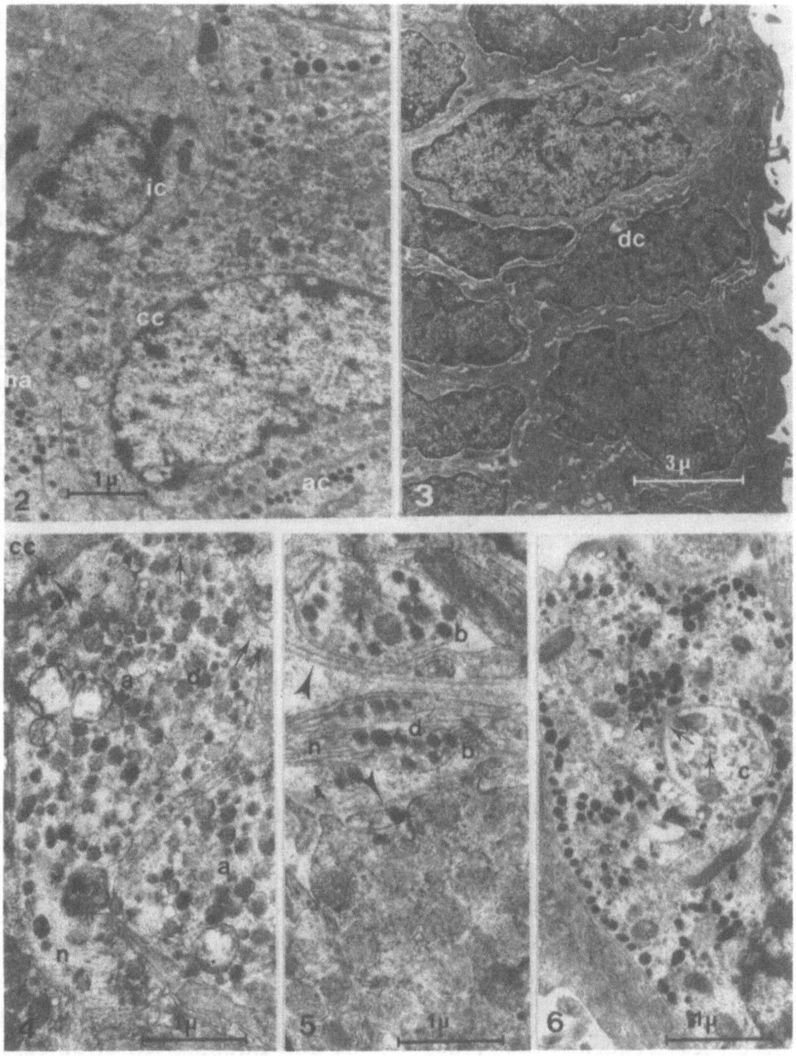

Figs. 2–6. (2) Pars intermedia tissue showing parts of a chief cell (cc) and ACT-type cell (ac) an interstitial cell (ic) and a nerve axon (na). 4 months post partum (× 10,500). (3) A group of dark cells (dc) bordering the hypophysial cleft. New born (30 days post coitum) (× 4,900). (4) Two type A nerve terminals (a) both showing dense granules (d), synaptic vesicles (arrow), neurotubules (n) and one with a synaptic junction (arrowhead) with a chief cell (cc). Double arrows indicate an apparent breakdown of the adjacent plasma membranes of the nerve and a chief cell. 5 months post partum (× 14,000). (5) Two type B nerve terminals (b) both showing dense granules (d) one with synaptic vesicles (arrow) and the other with neurotubules (n). The second terminal is in the perivascular space lined by basement membranes (arrowhead). 5 months post partum (× 14,000). (6) A type C nerve terminal (c) surrounded by an ACT-type cell and showing an apparent breakdown of the adjacent plasma membranes at one point (double arrow). Electron lucent vesicles (arrowhead) resembling those (arrow) of the nerve terminal, are seen in the ACT-type cell. 35 days post partum (× 14,000).

2.5. *Other glandular cells*

Other glandular cells resembling those of the PD are found only very infrequently in the PI. Cells of mesenchymal origin and nerve axons and terminals also occur in the PI. The neurosecretory nerve terminals are designated as peptidergic, aminergic and cholinergic (9) according to their resemblance to similar fibres elsewhere and in this chapter they are called types A, B and C respectively.

In recent years the ultrastructure of the adult PI has been described in a number of other mammalian species: rat (3, 10, 5), cat (9, 11), mouse (12, 13), ferret (14), rhesus monkey (15), fox (11), harp seal (16), experimentally treated hamster (17), jird (18) and dog (19).

2.6. *Nerve supply*

Light microscopy has revealed Gomori positive neurons penetrating the intermediate lobe in various groups of vertebrates (20). In the rabbit, Aldehyde fuchsin positive neurosecretory terminals first appear at the 7 day post partum stage (7). With fluorescent microscopy, adrenergic nerve fibres have been observed in the PI of the pig and rat (21). Electron microscopic studies have provided details of the relationship between the nerve fibres, neurosecretory elements and glandular cells of the PI of various vertebrates (22), rat (23), ferret (14) and rabbit (6, 7). Different forms of unmyelinated nerves and terminal varicosities in the PI of several species have been distinguished by the morphology of their contained granules and vesicles (24, 9, 23). The perikarya of these fibres are located in the hypothalamus. Their effect on the function of the PI is not fully understood and may differ in different vertebrates.

By electron microscopy adult nerve terminals can be divided into three different types based on the contained granules and electron-lucent vesicles. Type A terminals (Fig. 4) contain electron dense granules averaging 150 nm in diameter which show some variation in density and some electron lucent vesicles of 50 nm in diameter are seen.

Type B terminals (Fig. 5) contain electron dense granules ranging from 50 to 90 nm in diameter but fewer in number than type A and a large number of electron lucent vesicles averaging 50 nm in diameter. The dense granules usually appear darker than those in the type A nerve endings and sometimes granules with much larger perigranual membranes (haloed granules).

Type C (Fig. 6) occurs less frequently than types A and B and characteristically contains electron-lucent vesicles of about 50 nm in diameter but no dense granules. No intermediate forms between the type A and type B are seen but terminals with intermediate characters of type B and C do occur. All these terminals possess mitochondria; neurotubules are recognized in the type A and B only and type A contain large electron-lucent vesicles of up to 470 nm in diameter. Nerve axons – all containing mitochondria and neurotubules but no RER – are sometimes found. Unless the axons and their varicosities are connected in the same plane of section, structurally it is impossible to classify their types. It is possible that those axons which show junctions between axons, are either of the same functional type or not at all neurosecretory in nature.

Specialized synaptic contacts between chief cells and either type A or type B nerve endings are seen in the rabbit (7). These contacts resemble those in other mammals such as the cat (9) and the ferret (14). All these observations differ from the findings in the mouse (22, 25) where no specialized contact zone between type A endings and the glandular cells is encountered. Although some recent investigators (13) did not find cholinergic innervation in the mouse, rat and cat, an ultrastructurally distinguishable type C ending does exist in the rabbit (7). In this species the type C terminals maintain their exclusive relationship with the ACT-type cells and many form synaptic junctions, sometimes apparently penetrating into the cell body. In the mouse the ACT-type cells show synaptic contacts with not only type C terminals but also with type B terminals (25). Although in the rabbit actual synaptic contact between ACT-type cells and the type B terminals has not been encountered, their close association may indicate a dual control for these cells as may be the case in the mouse (25). The significance of this dual control may be that a predominant excitatory effect comes from the type C terminals [a supposition that is consistent with the findings (26) that if the nerve supply to the PI is disrupted it cannot maintain the adrenal control and function in the rat], and an inhibitory effect coming from the type B terminals (25).

The method of controlling the secretory function of the chief cells has been the subject of considerable speculation. Bargmann and Knoop (27) suggested that peptidergic fibres were responsible, whereas Enemar and Falck (28) speculated that only adrenergic nerves were involved. Knowles (24) suggested that both types of nerve endings were implicated. In the rabbit, the presence of identifiable type A, B and C terminals, however, supports the view of Bargmann et al. (9) and Baumgarten et al. (23) who suggested that peptidergic, adrenergic and cholinergic nerves were all involved. In addition, the distribution of various nerve endings in the rabbit (7) and mouse (25) is specific to the different types of glandular cells. Although peptidergic terminals are themselves localised at the juxta-PN zone their special association with blood vessels in the above two species (7, 25) indicates a far reaching effect. Hence it has been suggested that type A and B neurons are probably involved in controlling the activity of the chief cells whereas type B and C neurons appear to regulate the secretion of ACT-type cells (7). More recently other workers have reported that aminergic neurons (type B) in the rat directly control the secretion of MSH (29), beta endorphin and/or beta lipotropin (30).

3. Ultrastructure of development in vivo

3.1. First half of the prenatal development (1 to 13 days post coitum)

Rathke's pouch, the anlage of the adenohypophysis, is formed as a result of an active proliferation of an ectodermal evagination from the dorsal aspect of the stomodeum (see also Chapter 5). This evagination elongates and expands at its blind end, and part-way along its length a constriction forms which eventually separates off the characteristic pouch. Further modelling is dependent not only on local cellular multiplication but also on the continued flexure of the developing head and the differential growth of the surrounding tissues.

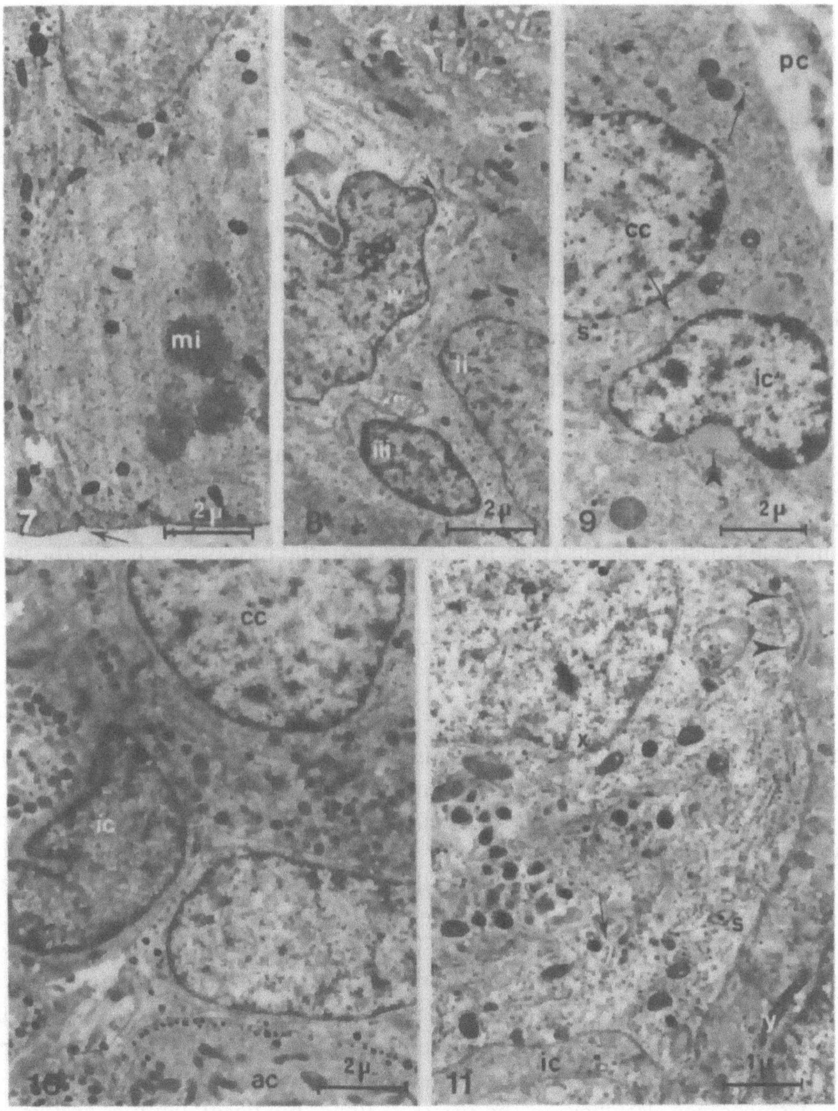

Figs. 7–11. (7) Undifferentiated cells of the pars intermedia bordering the lumen of the Rathke's pouch showing a mitotic figure (mi) and tight junctions (arrow). 16 days post coitum (× 5,600). (8) Examples of four immature forms of cells (i, ii, iii, iv) as recognised in an early fetus. Forms I and II contain secretory granules only but no vesicles. The endoplasmic reticulum in form I is dilated but not in form II. Form IV is small and non-granulated, form IV is associated with collagen (arrow) and separated from the other cell types by basement membranes. 21 days post coitum (× 5,600). (9) Partially differentiated chief cell (cc) showing no vesicles in its cytoplasm but secretory granules (s) two of which indicate incipient exocytosis (arrow). Also seen here is an interstitial cell (ic) with its perinuclear bundle of microfilaments (arrowhead); and part of a perivascular cell (pc). 24 days post coitum (× 5,600). (10) Pars intermedia showing advanced state of differentiation in a chief cell (cc) ACT-type cell (ac) and an interstitial cell (ic). new born (32 days post coitum) (× 5,600). (11) Differentiating chief cells in two stages of maturity designated as form X (x) and form Y (y) and an interstitial cell (ic). Vesicles are apparently budding off at two points (arrowhead) from a stretch of rough endoplasmic reticulum. Near the Golgi area the membranes of a mature and an immature secretory granule are apparently in continuity (arrow) with the RER. Elsewhere a secretory granule (s) can be seen in relation to the RER. 7 days post partum (× 10,500).

At about the 12th day post coitum in the rabbit (1) a constriction called Atwell's recess near the middle of the pouch separates the latter into oral and aboral lobes quite early in its development. According to some reports (31) two lateral lobes subsequently evaginate from the antero-ventral part of the oral lobe, which bend upwards and, together with the remnants of the oral lobe, eventually form the pars tuberalis, though in the rabbit Grant (32) demonstrated a single median origin of the pars tuberalis from the anteroventral region of the oral lobe by the 15th day post coitum.

Soon after Atwell's recess is formed at the 13th day post coitum in the rabbit (1), a bulge (the anlage of the neurohypophysis) grows down from the floor of the third ventricle; enlarging posteroventrally it comes into close contact with the posterodorsal surface of Rathke's pouch. The future median eminence will subsequently differentiate at the base of this neural protrusion and the neural stalk and pars nervosa are formed from the proximal and distal parts respectively.

The juxtaneural epithelial wall of the aboral lobe is the primordial pars intermedia and near this junctional zone neighbouring mesenchyme cells migrate in to form the primordial capillary plexus (Fig. 1c). The rest of this lobe of the pouch later forms the PD.

3.2. Second half of the prenatal development (14 to 26 days post coitum)

Differentiation is gradual. Not all cells of the same type begin to differentiate at once. The structural signs of cytodifferentiation become evident after two weeks of development. Upto 14 days post coitum, the presumptive PI is composed of 4 to 6 layers of ovoid cells which are swollen, elongated and contain oval nuclei.

The most striking feature at 15 days post coitum is the appearance of small (70 nm in diameter) secretory granules in the PI cell cytoplasm (33). The cell profiles become rhomboidal and keep close contact with each other through tight junctions at the luminal border of the cleft and by desmosomes elsewhere in the tissue. On the aboral face the junctional zone between the presumptive pars intermedia and the neural lobe is now invaded by the primitive blood vessels. A few as yet undifferentiated cells show slight irregularities in their nuclear profiles.

The Golgi complex is still immature at this stage. Frequently the rough endoplasmic reticulum (RER) is dilated and both tubular and dilated forms may contain electron dense secretory products.

At 16 days post coitum there is an increased nuclear: cytoplasmic ratio and an enlargement of the intercellular spaces. Mitotic figures are hitherto common (Fig. 7), but start to decline in number. More cells show secretory granules, commonly in groups and in the peripheral cytoplasm but there is still no vesicle formation. There is a great increase in the amount of RER and free ribosomes. Meanwhile nerve fibres with neurotubules are found in the neural lobe and there is an actual contact between the PI and PN for the first time. No blood vessels are found either in the PI or in the junctional zone between the PI and PD. After the third week of fetal life most of the PI cells assume an angular form. Dark cells appear in the PI during the fifth week of development, and at birth they are found in groups, especially at the hypophysial cleft border where they usually persist; cytoplasmic processes are first seen projecting into the cleft in the fifth week post coitum.

The PI and PD cells are distinguishable from each other by the third week post coitum. The PI tissue shows increased staining with Periodic acid-Schiff (PAS). On the other side of the PI, contact with the PN is complete by the time of birth. Mitotic figures are to be seen up to the perinatal stage, though in the rat they persist even in the adult (5).

By 21 days post coitum the PI is invaded by occasional blood vessels and the intercellular spaces are reduced. Most cells are irregular in outline and show a diversity of both size and number of cytoplasmic organelles and nuclear contents.

At this stage the PI consists of a variety of different cells that can be classified into forms I–V. Forms I and II represent two different stages in the development of one cell type – the chief cell. In Form I the RER is dilated, secretory granules are localized near the Golgi area and the polyribosomes are abundant; in Form II the tubular RER is tortuous, there are scattered granules, and ribosomes are reduced in number (Fig. 8). Form III at this time is recognized by the small size tendency towards stellate shape, absence of secretory granules and frequency of microfilaments of 5 nm in diameter (Fig. 8); later it develops into an interstitial cell, Form IV (Fig. 8) – always occurs in spaces lined by basal lamina and is probably of mesenchymal origin. Form V are undifferentiated cells and may give rise to any of the former and/or ACT-type and dark cells.

During the latter part of the 4th week post coitum a new population of glandular cells appears which shows further structural specialization of the Form II cells.

In the cytoplasm some of the cells contain well developed Golgi areas. Secretory granules are infrequent, small and ranging between 80 to 130 nm in diameter. The membrane bound vesicles that are characteristic of a mature cell are few, small and are never in the absence of secretory granules in the same cell.

Of the non-granulated cells, some appear similar to the adult forms while others still resemble the Form III or Form IV cells of the earlier stage. In all these cells the most prominent feature is an abundance of microfilaments which are usually scattered at random in the cytoplasm but are sometimes collected at the perinuclear areas (Fig. 9). So far microvilli are only seen rarely.

At this stage most specimens show the beginning of an internal vasculature. Both collagen and extracellular fine filaments are recognizable in the perivascular spaces. The perivascular cells are recognizable by their paucity of cytoplasmic organelles. The endothelial cells in developing vessels show characteristic indentations of the nucleus and occasionally a dark cytoplasm. Cytoplasmic projections towards the perivascular side, a characteristic of fetal endothelial cells are common in this stage.

3.3. Fifth week fetal (28 and 29 days post coitum)

At 29 days post coitum almost all cells, both granulated and non-granulated, show a distinct system of interdigitating cytoplasmic processes. The most striking cellular feature is to be found in the chief cells. The cells previously described as Form II now outnumber the Form I cells and the majority of these cells now show PI type vesicles. Most of the cells, however, still differ from the adult in the appearance of their secretory granules which are smaller and often paler than those of the adult cells. These vesicles contain a homogenous particulate material similar to that found in the adjacent tubular RER suggesting that the vesicles may be derived from the RER. Also the vesicles occur in groups in the vicinity of RER at this early stage. This is in keeping with the probability of the vesicles being formed by budding of the RER.

Interstitial cells are no longer to be found in groups but rather dispersed among the chief cells. Strikingly the cytoplasm is filled with microfilaments of 4 to 6 nm in diameter. Another feature of these cells is the general paucity of cytoplasmic organelles. ACT-type cells first appear at this stage but are very rare and not localized in distribution. Unlike the chief cells, these are small cells without cytoplasmic processes. The cytoplasm contains well developed RER with only a few secretory granules which are already arranged along the cell membrane.

At this stage, many cells bordering the hypophysial cleft appear dark or dense. These bordering cells do not resemble typical interstitial cells nor do they show any sign of granulation or vesicle formation like glandular cells. A number of other dark cells are also seen in a more central part of the PI tissue. Both varieties possess microvilli – a feature known to occur in the adult interstitial cells. These dark cells also contain an abundance of free ribosomes but this factor alone does not account for their general electron density.

3.4. Perinatal stage (30, 31 days unborn post coitum unborn, 30, 31, 32 days post coitum newborn)

A major feature of the perinatal PI is the localization of differentiated cells which in many cases show a general paucity of secretory granules. In isolated areas almost all cells resemble the adult form (Fig. 10). In contrast differentiation is far from complete in other areas where recognizable chief cells still show a wide range of differentiation. The branching and anastomosing RER is commonly dilated and islands of cytoplasm are trapped in this complex network. Form I cells with dilated RER but without vesicles are still to be found. Occasional mitotic figures are encountered, and microfilaments are very few.

Interstitial cells in contrast show a great abundance of microfilaments. ACT-type cells still remain but their number increases slightly and they are seen amongst other cell types (Fig. 10). The diameter of the secretory granules averages 135 nm, but they vary considerably in size, shape and electron density. An abundance of dilated RER is a conspicuous feature of these cells.

Non-granulated dark cells are mostly seen lining the hypophysial cleft. They characteristically contain bundles of filaments of 9 nm in diameter which are also seen in association with junctional complexes. These cells contain a large amount of tubular RER and free ribosomes but few other cytoplasmic organelles.

In the deeper region of the tissue a variety of dark cells already containing intracellular vesicles and free ribosomes now also show a large amount of dilated RER and a few granular inclusions.

3.5. The general pattern of post natal development

Though many cell types become distinguishable from one another during the 4th and 5th week post coitum, cytodifferentiation is far from complete even at parturition. During the first few weeks of neonatal life the difference in the distribution of specific cell types, neurosecretory fibres and capillaries begin to emerge; in the rostral zone blood vessels and intercellular spaces become more numerous than in the caudal zone, the parenchyma is segmented into small groups of cells. During the first 7 weeks after birth a connective tissue septum with occasional gaps becomes established between PI and PN. By five months post partum the PI consists of 6–10 layers of cells (as in the adult) and at the extreme rostral point very little contact is seen between the PI and PT. Intermingling of the cells with the PD cells at the rostral and caudal extremities of the PI becomes evident after the 7th week post partum. Gonadotropes (granules 160 nm in diameter) appear at the 2nd week post partum near the caudal PI. The granules in the chief cells still remain infrequent and are generally localized in the Golgi area which is either inconspicuous or is only moderately developed. Mitochondria are also localized at that zone of the cell and many cells show local elaboration of tubular cisternae of RER in parallel arrays. Free ribosomes occur in the cytoplasm both individually and as polyribosomes, and with increasing age the rest of the cytoplasm is gradually filled with vesicles.

Interstitial cells are initially few, scattered and small; they contain abundant microfilaments. The number of cells rapidly increases by the 3rd week post partum when elaborate junctional complexes between them are commonly seen. ACT-type cells show a scattered distribution, though common in the posterodorsal region. The hypophysial cleft remains.

After the cells are differentiated, no significant correlation between the age of the animal and the amounts of contained hormone is detectable in some species (13). Junctional complexes of several types (tight junctions, gap junctions and desmosomes) are seen throughout the development, and desomsomes in particular occur between both similar and dissimilar cells (34). Cilial are largely confined to the epithelial lining of the hypophysial cleft in post-partum animals, but cells of the ACT-type and interstitial cells occasionally possess them. The cilia are often surrounded by cytoplasm at the plane of the section.

3.6. Early neonatal (between 2nd and 4th week post partum)

At the beginning of the 2nd week two immature forms of chief cells are evident (Fig. 11); one has very few vesicles which are sometimes associated with the RER. The other, which initially appears at the posterodorsal region adjacent to PN, contains abundant vesicles. Both of these forms contain immature secretory granules, characterized by small size and a distinct membrane, occurring mainly at the Golgi area (Fig. 11). During the next two weeks the population of cells with few vesicles gradually disappears and is presumably transformed into the latter form which at the same time increases in number and spreads into different areas of the PI tissue. The changes in other cell types are obvious.

3.7. Late neonatal (5th and 6th week post partum)

During this period the PI attains an almost fully differentiated state. The chief cells, except near the cleft (anteroventral region), are well differentiated and the relative proportion and distribution of their cytoplasmic organelles is equivalent to those of an adult. In particular the secretory granules reach a size range of 300 to 600 nm.

The interstitial cells are well developed, often joined by desmosomes with associated microfilaments and for the first time abundant microvilli are seen on the surface of these cells. The ACT-type cells are well differentiated and become rather stellate (Fig. 12). The dark cells are now scattered among other cells and their cytoplasm is filled with microfilaments.

3.8. Juvenile (6 weeks to nearly 3 months of age)

By the 8th week all the PI chief cells take the adult form, except that the elaboration of the RER reaches its peak in most parts of the PI (Fig. 13). The RER may occur as long sinuous parallel stretches, sometimes encircling the nucleus and sometimes gathered near the Golgi, or it may be arranged in *Nebenkern* formation where a large amount of tubular RER is concentrically arranged in spiral or parallel arrays. By the end of this stage the chief cells acquire lipid droplets in their cytoplasm. Microtubules are rare and microfilaments rarer.

Most of the interstitial cells are well developed though they show few cytoplasmic organelles. The ACT-type cells now reach full maturity. Both granulated and non-granulated dark cells remain scarce.

3.9. Adult (3 months to 3 years)

After 6 months the adult male and non-pregnant female PI are essentially similar to what has been described before but there are minor differences. The interstitial cells of a 6 month animal differ from those of the 4 months old (Fig. 2) in the relative paucity of randomly scattered microfilaments of 5 nm in diameter in the latter. Cytoplasmic filaments of 9 nm in diameter are commoner than in the previous stage.

There is an increase in the number of ACT-type cells and well developed Golgi area. Their relationship with type C nerve endings (Fig. 15) becomes more intimate than the other two types of terminals although these cells amy also be associated with type b terminals. Non-granulated dark cells are only seen near the hypophysial cleft and the granulated variety become commoner than in the younger material.

In pregnancy the chief cells show a striking modification (34). Adjacent chief cells show interdigitations of the cell membranes. The majority of them show only a few scattered secretory granules but the vesicles are numerous. Sometimes groups of chief cells show accumulations of secretory granules along the periphery of the cytoplasm and expelled granules occur in the perivascular spaces and in vascular lumina (Fig. 14). This may indicate that transportation of the granules across the endothelial cells may take place without the former being broken up into molecules. A more consistent feature is the increased extent of the RER which is often continuous with the outer membrane of the nuclear envelope. Discontinuous patches of branching and anastomosing RER surrounding apparent pockets of cytoplasm are a characteristic feature except at some areas in the

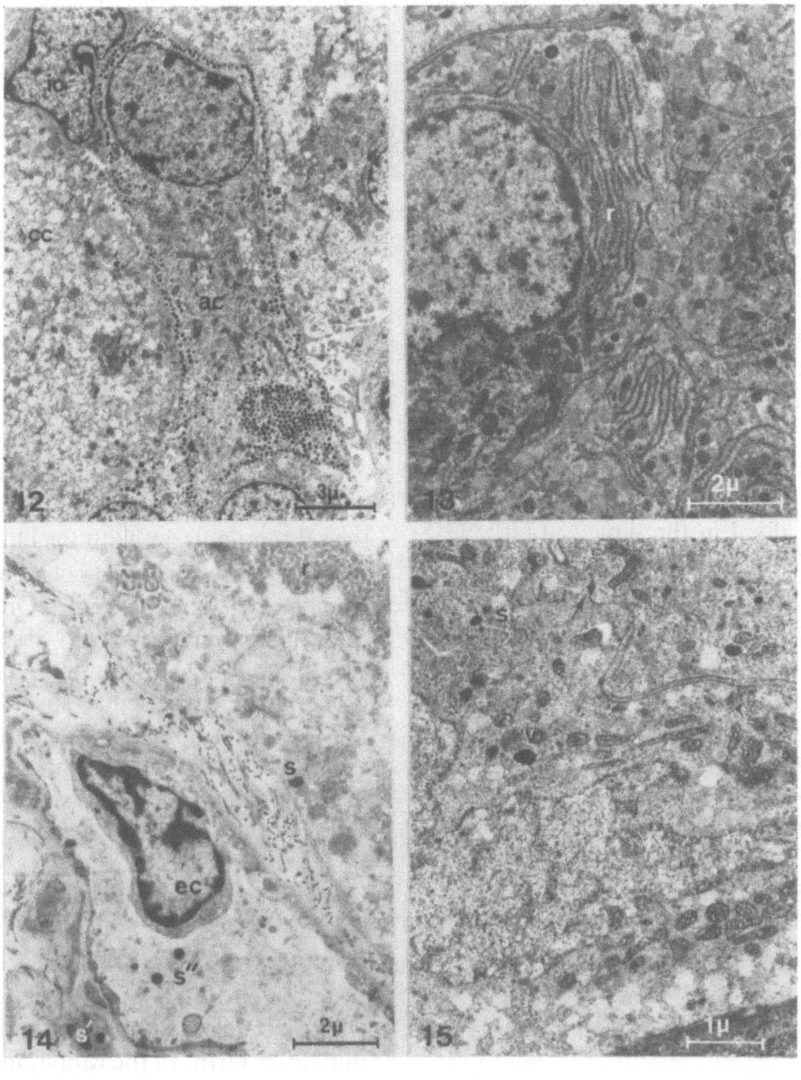

Figs. 12–15. (12) Profile of an ACT-type cell (ac) showing peripheral distribution of its secretory granules. Also seen are parts of chief cells (cc) and an interstitial cell (ic) with scanty cytoplasm containing no granular inclusion. 35 days post partum (× 3,500). (13) Parts of a group of chief cells, one showing pear shaped contour, rounded nucleus and large amount of rough endoplasmic reticulum (r). 53 days post partum (× 6,300). (14) Evidence of exocytosis and transportation of secretory granules from the chief cells to the capillary lumen, are shown. Dense granules are present in the chief cell at S, in the perivascular space at S′ and in the capillary lumen at S″ respectively. EC: endothelial cell; r: dilated and frequently anastomosed endoplasmic reticulum. 6½ months post partum (4 weeks pregnant) (× 5,600). (15) Parts of two chief cells showing the presence of vesicles (v) as well as secretory granules (s), one of the latter is in the process of exocytosis (arrow). 7 days post partum (38 days total development). (× 10,500).

caudal and lateral zones where cells of normal adult type are also observed. ACT-type and interstitial cells are fewer than in non-pregnant adult tissue.

Svalander (35) found that the contact between the PN and the rudimentary Rathke's pouch in the rat is first established on the 13th day post coitum but later mesenchyme cells partially separated the two parts. Structural evidence suggests that PI cells differentiate later than PD and PT cells *in vivo* (35) as is seen in other parallel investigations *in vitro* (36).

In the mouse also the PI cells are the last of the adenohypophysial tissue to differentiate (37). The differentiation starts on the 17th day post coitum and this is followed within the next two days by a sharp increase in the number of cells and in their granular content (37).

In the developing PI of the human, cells containing both MSH and ACTH were identified by immunocytochemical means in the 8 to 12 weeks old fetus. These cells were localized on the edge of the glandular cords in contact with vascular mesenchyme (38). Chatelain et al. (36) found cells of the rat fetus were immunoreactive to both anti hormones.

Other investigators employed immunocytochemical techniques for the ultrastructural localization of several PI hormones. While in the developing rat PI these same cells produce MSH, ACTH and LPH (39), it was demonstrated that in unborn human pituitaries, α- and β-endorphins, β-LPH, β-MSH and ACTH are all localised in the same cell (40).

In the human fetus MSH activity appears after 11 weeks of intra-uterine life (41). In addition to α- and β-MSH other smaller peptides such as CLIP (corticotropin-like intermediate lobe peptide) and β-endorphin are found to be predominant hormones secreted by human and other primate fetuses (42). By using the peroxidase-antiperoxidase complex, Chatelain et al. (36) found that ACTH, LPH and α- and β-MSH containing cells were present in a 17 day old rat fetus. Although the post natal PI of the rat also contains β-endorphin (43) while that of the monkey, ox and sheep contain LPH (44), their activities in these fetuses are not clear.

3.10. Development of nerve supple in vivo

Ultrastructurally axons without granules and vesicles first appear in the five week rabbit fetuses. Light microscopical recognition of the terminal varicosities of the neurosecretory fibres in the PI is possible at the beginning of the second week post partum and within two weeks after that they become frequent and axons are found deep in the tissue. During development, the first terminal to be ultrastructurally identified is type B which is already in intimate association with the chief cell in the second week post partum. For the following two weeks this type of terminal and unclassified nerve axons are the sole representatives seen occurring anywhere except at the oral and lateral edges.

At the 4th week post partum, type A terminals appear near the PN border. Type C terminals are first seen during the 5th week post partum but are usually infrequent compared to the other types. After 6 weeks post partum the numbers of the fibres and all types of terminals increase but their distribution and proportion remain unaltered. The occurrence of the type C terminals, either in close proximity to, or in intimate relation with, ACT-type cells becomes evident.

In the PI of specimens between 3 months and 2 years old, axons increase in number, and they may extend up to the cleft border. They are found in close proximity to all types of parenchymal cells and often in the perivascular spaces. The distribution of type A terminals never extends deeper than the posterodorsal zone (next to PN), but their location is no longer restricted to the intercellular spaces of parenchymal cells and they frequently appear in the perivascular spaces where type C terminals are never found and type B terminals occur infrequently (Fig. 5).

The abundance of the type B terminals remains unaltered from that of the previous stage but these endings are now extended to all parts of the PI. Although the only synaptic contacts shown by the type B terminals are with the chief cells, they are also seen in close proximity to, and sometimes surrounded by, ACT-type cells. On rare occasions chief cells may be simultaneously associated with a type A as well as a type B terminal. The latter may rarely occur in the perivascular space.

Type C terminals still remain the least common although occurring in all parts of the tissue.

Glandular cells with secretory granules are seen at a stage of development before any neuronal connection between the PI and PN can be found. This makes it unlikely that neurons are necessary for the synthesis of active substances and packaging into secretory granules at this early stage. A parallel situation is found in the rat (35) and mouse (25) where the nerve supply to the PI does not develop until after birth. In the rabbit, many axons appear by 28 days PC but neurosecretory granules are seldom seen at their terminals, nor do they show synaptic vesicles. Two explanations are possible; either the dense granules have not yet been synthesised in their perikaryon or they have not reached the PI.

In the rabbit the first neurosecretory granules ultrastructurally identified were in the type B nerve terminals (12) but the fluorescence microscopy in the rat PI (45) shows that neurons of type A are the first to develop in this species.

4. Ultrastructure of development in organ culture

4.1. Technique

The methods for culturing are essentially the same in all the expRiments discussed. Either the whole pituitary gland without its stalk, or fragments of it containing different areas of the PI, are cultured in HEPES buffered medium 199 on wire bridges as open culture, in 5% CO^2 in air. Viability and cell proliferation of such a culture are assessed by light microscopy using fluorescent stains and autoradiography respectively. Morphological characteristics of differentiation and function and studied by electron microscopy. The problems of using this technique have been discussed elsewhere (46).

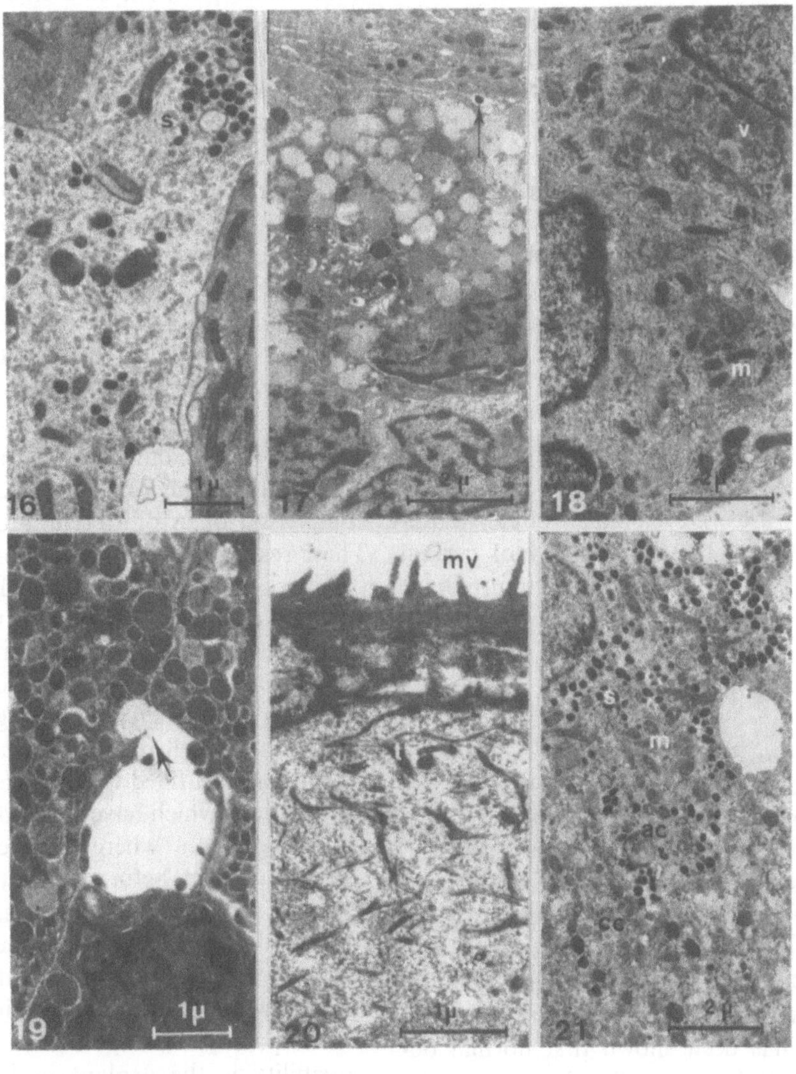

Figs. 16–21. (16) Part of a chief cell from a 38 day material that has undergone 24 days of intital development *in vivo*, followed by a further 14 days of organ culture. The cytoplasm shows secretory granules (s) only. (× 10,500) (17) Adult chief cells cultured for 4 weeks. Secretory granules and vesicles are present in the cytoplasm; also some incomplete granules are present in the Golgi area. One secretory granule is in the process of exocytosis (arrow) (× 7,00). (18) Chief cells from adult tissue, after 9 weeks in culture. Secretory granules are absent and vesicles (v) are few, but the mitochondria (m) are healthy (× 7,000). (19) Chief cells from adult tissue, cultured for 3 weeks. Vesicles are seen in the process of extrusion (arrow). (× 10,500). (20) Non-granulated cells at the peripheral zone of adult tissue cultured for 3 weeks. Note bundles of tonofilaments (t) scattered in the cytoplasm of one cell which is apparently an interstitial cell. The other cell is electron dense, and hence is a dark cell which shows microvilli (mv) (× 14,000). (21) ACT-type (ac) next to a chief cell (cc) from 2 week culture of adult tissue. Secretory granules (s) and healthy mitochondria (m) are clearly visible. (× 6,300).

Survival depends mainly on a combination of size and age of the tissue before explantation. For example, in the minute explant of Rathke's pouch of a 14 day fetus there is no evidence of cell death after 2 weeks of culture. Even when the entire pituitary gland of a 29 day fetus is cultured most PI cells remain healthy. If, however, the PI tissue is taken from an older animal, a high proportion of viable cells is achieved only if the fragments of the explants are less than 1 mm^3 in size. In cultures of 5th week fetal pituitaries, cells at the surface of the explant survive better than those in the centre. When ^3H thymidine is added to the culture 1 hour before fixation, nuclei of many cells near the periphery of the pars distalis together with at least a few PI cells are labelled indicating there is DNA synthesis associated with cell proliferation even after 2 weeks of culture (47, 48). In these explants, cells remain healthy and continue to proliferate when they are either near the periphery or close to the large intercellular cavities which develop during the first few weeks of culture. In prolonged cultures of 12 weeks the only cells which remain healthy are those at the superior surface of the explant and hence at the gas/liquid interphase zone of the culture. A detailed description of the chief cell is given first because it is abundant and easily recognized in many immature stages. Certain obvious aspects of other cell types will be discussed afterwards. Development of extracellular structures (49) and the morphology of certain nuclear inclusions (50) in prolonged culture have been described elsewhere.

4.2. Chief cells

4.2.1. Undifferentiated cells

In section 2 it was stated that before the differentiation *in vivo* starts at 15 days post coitum the cells are epithelioid and undifferentiated. When tissue from a 14 day post coitum fetus is cultured for up to two weeks, cells remain epithelioid in appearance and do not show the differentiation into any of the cell types which characterizes the PI tissue developing *in vivo* for a equivalent time (48). Perhaps this is to be expected since it has been shown that normal development of the PI *in vitro* will only occur if the neural lobe is present in the culture (51). It has been suggested that absence of the neural lobe in human anencephalics may have been the cause of a lack of PI development (52).

4.2.2. Partially differentiated cells

It has already been mentioned that secretory granules appear in the chief cells early in the 3rd week of fetal life – several days before the appearance of the vesicles late in the 4th week of fetal life. For instance, a 24 day post coitum PI shows secretory granules but no vesicles. This criterion can be used to study the process of differentiation in these cells. The fate and potency of this cell type has been analyzed; firstly by fixing and examining it in a particular stage of development; secondly by following the normal subsequent cytological development in an equivalent tissue *in vivo;* and finally both of these observations were separately compared with the modifications in an *in vitro* tissue either on a comparable common initial development or on their common ultimate age. Accordingly in one group, the PI of 24 day fetuses was allowed to differentiate undisturbed *in vivo* for a further two weeks before examination. In the second group the PI of 24 day fetuses were collected and maintained in organ culture for a further two weeks. In this way it was possible to compare material after 38 days of development, one *in vivo* for the whole time, the other cultured for the last 14 days. The PI of the first group contained not only secretory granules but also the vesicles (Fig. 15), the second group failed to develop vesicles although the secretory granules were seen in the cytoplasm (Fig. 16). It seems that after the appearance of granules the developing cell is not irrevocably committed to develop vesicles and it may be that some extrinsic factor is necessary for completion of cytodifferentiation of the chief cells. These results provide no support for the hypothesis that vesicles are formed by further differentiation of granules (3) and are consistent with the existence of these inclusions as two separate entities.

After the first appearance of the vesicles their number rapidly increases during the development *in vivo* until the cell reaches maturity, when the cytoplasm is almost filled with vesicles. From the isolated observations which have so far been made, it appears that in cultures where the vesicles have already started to form before explantation, the cell continues to form vesicles but the speed, degree and the nature of accumulation of these vesicles are not identical to *in vivo* development. Also compared to the *in vivo* tissue, the distribution of differentiating cells is patchy. Perhaps this is to be expected. One possibility is the explant itself did not contain a homogenous population of partially differentiated cells before the culture. Secondly, in the absence of a capillary network in culture there is a dependancy on the proximity to the gas/medium interphase, hence cellular activities are likely to be different in different areas of the same tissue.

4.2.3. Fully differentiated cells

The next aim was to find out whether the fully differentiated cell, that is a cell with both granules and vesicles, could maintain its specialized ultrastructure in organ culture, that is in the absence of its normal microenvironment. Adult material was, therefore, cultured for varying periods of time. The chief cells contain granules as well as vesicles after 4 weeks of culture. Moreover, the Golgi area is well developed and associated with it there are many immature granules suggesting they are being formed *in vitro*. There is also morphological evidence of exocytosis (Fig. 17), the normal process by which many glandular cells liberate their granules *in vivo*. These findings imply that up to four weeks *in vitro*, no outside factor is necessary either to maintain the state of full differentiation or to permit normal physiological activity such as formation and extrusion of granules. These experiments do not of course exclude the possibility of influences exerted by neighbouring cells also present in the culture itself.

When the cells are cultured for much longer periods their appearance alters dramatically. The granules gradually become fewer and eventually disappear altogether as can be seen in a 9 week culture (Fig. 18). The mechanism for this is suggested by the cell morphology, in that they only contain granules in the cytoplasm so long as the formation of the granules remains unimpaired. We have already seen that granules are lost from the cell by exocytosis as happens *in vivo*. It may well be that after four weeks of culture the formation ceases so that expelled granules are no longer replaced. Moreover, a chief cell after prolonged culture shows a general paucity of vesicles. This is explained by another process taking place in organ culture. Extrusion of vesicles also occurs *in vitro* (Fig. 19) but unlike exocytosis of granules this is not a normal physiological process as it was never encountered in any of over 1000 electron micrographs of the *in vivo* material examined. This abnormal process in culture is difficult to interpret. It is possible that the hypothalamic factor which is claimed to exert an inhibitory effect on the liberation of secretory granules (53) may in fact act to inhibit the release not of granules but of PI-vesicles. If this were so the absence of this factor *in vitro* might lead to the expulsion of vesicles which is seen here.

Unfortunately, little is known about the fine structure and function of the PI in other mammalian species in long term organ culture. In the frog the ultrastructural appearance of explants remained intact during 6 months in culture (54). The bioassay of samples of the explants revealed continuous secretion of MSH and this led to the inference that *in vivo* the hypothalamus exerts an inhibitory influence on MSH secretion in the frog (54). One report based on the immunocytochemical investigation of organ cultures of the rat fetus also suggested that adenohypophysial primordia were dependant on hypothalamic influence (39). In contrast, by administration of dopamine-receptor blocking agents, Tilders (55) found that a blockade of hypothalamic input to the rat pituitary gland did not produce a sustained hypersecretion of α-MSH.

It is not easy to assess the factors responsible for initiating and/or influencing the differentiation. We have seen that a more advanced stage of development is reached *in vitro* when the embryonic PI is explanted at a later stage of development. So far there is no evidence that the hypothalamus can induce cytodifferentiation of the chief cells though mesenchyme cells are thought to be responsible for inducing differentiation in various organs (56). In the uncultured PI of the fetal rabbit a thick layer of mesenchyme cells lies between the developing PI and PN (8, 33). It is also clear that the cells near the PN and hence also near the mesenchyme layer, are liable to achieve full differentiation sooner than others into fully differentiated cells. Further evidence of mesenchymal influence comes from the work on 8–12 week old human fetuses where cells with both ACTH and MSH functions appear next to the vascular mesenchyme (38).

The possibility of neurohormones being initiators is unlikely because the chief cells *in vitro* begin to differentiate before the development of a nerve supply to the tissue and appearance of neurosecretory granules in the nerve terminals (7). It is possible that the same sequence takes place in organ culture, with differentiation starting in cells that are adjacent to the PN and then spreading elsewhere in the explant. This point is, however, difficult to confirm due to the complicating factors such as the secondary position of the PI cells with respect to the vicinity to the gas-medium interphase (see section 4.1. Technique). In culture these cells continue to differentiate in the presence of only those axons that are severed from their perikaria. It is unlikely, therefore, that the controlling influence for differentiation comes from the hypothalamus.

The failure of the partially differentiated cells to continue differentiation is presumably due to the absence, in the conditions of culture, of necessary cellular or chemical inductive influences. Since cells of mesenchymal origin are frequently seen in explants containing partially differentiated cells that may show no sign of further development, it is assumed that after the cells reach a specific stage of development there must be a different factor other than that coming from the mesenchyme cells, needed for the completion of their development.

By examining *in vitro* cloned cells of the epithelium of Rathke's pouch, and by comparing this with similar cells grafted either under the kidney capsule or into the hypothalamus of hypophysectomised rats, Shino et al. (57) concluded that differentiation into all types of adenohypophysial cells is dependent upon hypothalamic hormones or some other factors conveyed by the blood. Using a different approach on fetal rat *in vitro* (58, 59, 36) or as a graft *in vivo* in normal and hypophysectomized rat (60, 61) or embryonic quail *in vitro* (62) several workers concluded that differentiation of the pituitary anlage is independent of hypothalamic neurosecretion. In the quail it seems to be independent of mesenchymal influence as well (62).

In addition, Chatelain et al. (36) found that pituitary glands of 21 day fetal rats, previously encephalectomized on day 16, contained as many immuno-reactive corticotropes and MSH-secreting cells as those of their littermate controls, and the cell types of the PD were revealed more than a day earlier than those in the PI. Ishikawa et al. (63), who used the clone culture technique concluded that corticotropes and other PD cells obtained from 11 to 13 days post coitum rats may develop from a single progenitor cell.

4.3. Interstitial cells

The non-granulated cells survive well in culture and retain their identifiable characteristics for a long as twelve weeks when all other cell types have disappeared. In a shorter period *in vitro* they increase in number, become more stellate and develop extreme attenuation of their cytoplasm, reaching up to the surface of the explant. After three weeks of culturing of a fully differentiated explant these cells accumulate at the superficial zone of the explant as is found in reaggregated cell culture (64). In organ cultures these cells develop bundles of tonofilaments (8–10 nm in diameter) and become progressively darker (Fig. 20). Since analogous morphological changes are found in the epithelial cells of the skin (65) it is possible that this is an adaptation by these cells to the culture conditions which might enable them to perform a protective role for the explant. A supportive function, equivalent to that of glial cells, is assumed by Stoeckel et al. (66), who found an intense reaction by these cells *in vivo* using labelled antibody to gliofibrillar acid protein in mouse, rat and cat.

4.4. ACT-type cells

Explants taken from adult material show that these cell types maintain their secretory granules for up to two weeks in culture (Fig. 21). The fact that both the chief cell and the ACT-type cell retain their separate identities in organ culture supports the view that they are in fact different cells and the difference in their granule measurements indicates that two separate hormones are possibly produced by them. A similar conclusion was reached by Naik (67) who determined immunocytochemically the existence of similar cells in the rat and mouse.

A good deal of information on the hormonal content of the rat PI has been accumulated during the last decade. Among other peptides ACTH has been found in the rat PI (68). It is known that more than one type of cell contains ACTH in the rat (69).

In the rabbit there is no ultrastructural evidence for or against the possibility that there is a decline in the rate of secretion in these cells in organ culture compared to the *in vivo* condition. Hadley (70), however, suggested that when both ACT and chief cells are maintained in tissue culture the ACT cells may show only low or basal secretory activity and Jackson et al. (71) found that proopiocortin (the precursor of ACTH, β-LPH and MSH) is synthesized in the PI and the release of these hormones from the ACT cells of the PI is positively controlled by a hypothalamic corticotropin releasing factor.

Some authors have described a type of cell in the mammalian PI which is functionally related to the adrenal gland (4, 72, 12, 67, 66). This cell type occurs and shows correlation with the adrenal gland not only in normal rats and mice but also in those specimens which have been either adrenalectomized to induce hyperactivity of the ACT cell (73) or been subjected to cortisol treatment to decrease ACTH secretion (74). Immunocytochemical techniques have been used to investigate their identity in mammals (67, 44). Using autoradiography, Gosbee et al. (74) deduced that ACTH is synthesized in the 'dark stellate type cells' of the PI of adrenalectomized rats.

Several striking differences between the ACT cells in the PD and ACT-type cells in the PI of the rat has been reported (72). Those in the PD are stellate and contain secretory granules which are usually round or ovoid, uniformly electron dense and arranged in a row near the plasma membrane. Those in the PI are generally polyhedral, with secretory granules scattered throughout the cytoplasm. These granules in the cells of the rabbit PI show considerable variation in size, shape and electron density, and are often mainly concentrated at one pole of the cell.

It has been shown (34) that the ACT-type cells of the PI in juvenile and young adult rabbits differ from those of the rat or mouse described by other authors, and from the ACT cells of the PD of the adult rabbits in experimental conditions (8). However, in the fetal and early neonatal rabbits, these cells of the PI have certain similarities (e.g. already peripherally distributed granules and dilated RER) to those in the PD (75); and in this species they seem to differentiate at the same age (33, 75).

It is interesting to find the special association between these cells and the nerve terminals in the rabbit *in vivo*. The fact that these cells in organ culture can survive for at least two weeks does not exclude the possibility that organized production and timely release of hormone(s) from these cells may be under direct nervouw control and indeed in the rat the secretion is stopped in the denervated PI (76). This direct relationship with the neurons makes them different from the ACT cells in the PD. This difference is further emphasized by Moriarty and Halmi (72) who, in spite of identifying *both* as ACT cells by radioimmunological techniques, were able to see ultrastructural differences between them. These differences, however, do not exactly correspond to the features described here. The situation is further complicated by the fact that in the human PD immunocytochemically determined ACT cells also contain other secretory granules not reacting to the antisera to ACTH and may have other hormonal functions (77). In the human and rat PI there is a massive reaction to the appropriate antisera indicating that the occurrence of ACTH and β-lipotrophin is very widespread and may not be restricted to ACT-cells alone. Since the ultrastructure of the ACT-type cells resembles that of immunohistochemically determined ACT cells as distinct from chief cells in the rat and mouse (67), it is considered likely that in the rabbit these cells may be a major source of ACTH in the PI which would justify its naming as an ACT-type cell.

4.5. Dark cells

The term 'dark cells' in the PI is given to the heterogeneous population of cells which are artificially grouped together by reason of their common electron dense appearance. The commonest is the type I dark cell, not seen *in vivo* but frequent in cultures taken from all stages after the second week of fetal development (Fig. 20). Few of them are recognizable at the onset of culture but the number increases with the time. Groups of these cells with many morphological characteristics in common with interstitial cells such as irregular contours, presence of cytoplasmic filaments and absence of granules are evident in prolonged cultures in areas such as the peripheral zone of the explant *in vitro*. In this zone the interstitial cells appear first and a decrease in their number coincides with the increase in number of the dark cells indicating a possible transformation from the former to the latter.

Type II dark cells are identical to the dark cells of the hypophysial cleft border as described earlier with their irregular stretches of RER scattered in dense cytoplasm containing filaments but no granules. The nucleus is elongated and always darker than the type I cells (47). Because they spread from their site of origin as the period of culturing is increased, tracing of their number and distribution is difficult. Since these cells are dark before culturing this may reflect some hitherto unexplainable characteristic of the cells themselves.

Type III cells are smaller but darker than type I cells and are the rarest variety of the dark cells. The absence of granules and filaments in the scanty cytoplasm and a very dark indented nucleus are easily identifiable features. They are likely to have been derived from perivascular cells since transitional forms are sometimes encountered.

Type IV cells are the commonest variety and randomly distributed in the explant. They are granulated and resemble chief cells in many respects but the cytoplasm is very electron dense (47). They show little structural difference from the equivalent cells *in vivo* in the rabbit as well as in other mammalian PI (8, 78, 67). Kobayashi (78) assumed that the dark appearance of the cells in the rat PI *in vivo* was due to the condensation of the cytoplasmic organelles. If this is the case, the excessive dark appearance of the nucleus and of some areas of cytoplasm without recognizable cytoplasmic organelles (including free ribosomes) remains unexplained. It has been suggested that the dark cells in the suprarenal cortex of rats and mice (79) and the dark granulated cells in the rat PI (80) are normal cells at a particular stage of functional activity. Secretory granules have been seen in the adjoining intercellular spaces and it can occasionally be seen that the contents of the vesicles are discharged from the granulated dark cells into the surrounding spaces (48). Consequently, if these cells are simply a stage in the normal life span of the chief cells, it must be an active stage, rather than a resting one. This is in agreement with the conclusion of Ziegler (80) who, after experimenting on salt loaded rats, thought that the increase of dark cells was an expression of a rise in the activity of the PI.

5. Conclusions

5.1. Cellular development

Observations on the ultrastructural changes that occur in one mammalian pars intermedia during development *in vivo* and in organ culture have been presented. Chief cells, interstitial cells, ACT-type cells and a heterogeneous group of dark cells were described; the time of onset and duration of differentiation varies between these groups of cells. The presence of different glandular cell types indicates production of different hormones. In the rabbit, secretory granules appear on the 15th day post coitum, one day before connection is made with the primitive neural lobe and more than one week before vascularisation. This raises the question of how the hormone, if liberated, could be transported to the target organ.

Within the PI both the time of onset and the duration of differentiation differ for each cell type. Chief cells can first be recognized in a 21 days post coitum animal and by 53 days post partum these cells have reached maturity. Immature ACT-type cells, on the other hand, first appear at 28 days post coitum, but it would seem that all have reached full differentiation by 35 days post partum. In the chief cells there are indications of granules being formed inside the RER at the 4th week of fetal development, and they probably start to form in the Golgi area after this stage of development. The role of the vesicles is uncertain.

Microfilaments occur in the undifferentiated parenchymal cell, especially after the 3rd week of intrauterine development. Among differentiated cells they are particularly numerous in intersitial cells and darker cells of the nongranulated variety. The chief cells of the pregnant rabbit show structural modifications thought to be indicative of hyperactivity.

5.2. Nerve supply

Three types of non-myelinated nerve endings resembling peptidergic, aminergic and cholinergic types are present in the PI. The first two make synaptic contacts only with the chief cells though the second type also invaginate into the ACT-type cells. Cholinergic-type neurons, presumably efferent, make synaptic contacts only with the ACT-type cells. Invasion of the PI by nerve fibres from the PN starts in the 5th week post coitum.

5.3. Organ culture

Mammalian PI can be successfully maintained in organ culture for up to 9 weeks, but small explants survive better than larger ones and explants from younger animals survive better than those from older ones. Cells at the surface survive better than those at the centre of the explant. The potential intercellular and perivascular spaces expand and increase in number with the duration of culture. Microvilli develop at the exposed surface of the explant.

In the chief cells, ultrastructural differentiation is arrested if the PI is explanted before the 4th week post coitum. Cells in an explant that has developed for 4 weeks *in vivo* may achieve partial differentiation *in vitro*. Full differentiation may only take place *in vitro* if the development *in vivo* is uninterrupted before the cells reach a specific stage of differentiation. Adult cells retain their differentiation for at least 5 weeks in organ culture. Evidence of extrusion of the vesicles in culture but not *in vivo* may mean that they are under inhibitory control *in vivo*.

Differentiated ACT-type cells can also be maintained in organ culture. During culture some of the interstitial cells lose their usual appearance and become peripheral dark cells. The number of dark cells increases during long term culture. All the main types of cells, with the exception of ACT-type cells, can become dark. Non-granulated cells of uncertain origin become progressively more common in older cultures and some of them resemble fibroblasts. Intracellular filaments increase in number and are often grouped in dense bundles in the interstitial cells.

Acknowledgements

I wish to thank Prof. R.L. Holmes, Dr. J.A. Sharp and Dr. J.W.B. Bradfield for critical reading of the manuscript. Part of the research was financed by the Standing Award Committee, St. Mary's Hospital, London. Research facilities were provided by the then Department of Cellular Biology and Histology (now Anatomy), St. Mary's Hospital Medical School, London. Electron micrographs were taken at the Anatomy Department, Middlesex Hospital Medical School, London. I am grateful to Mrs. L. Barrett and Miss H. Rogers for typing the manuscript and Mr. P.W. Hargreaves for printing the photographs.

Permission to reproduce the figure has been given as follows:

Figs. 2, 11, 13, 14 *Journal of Anatomy*, Cambridge University Press, Cambridge. 122: 415–433 (Figs. 9, 12, 15), 1975.

Figs. 4, 6, 7, 8, 9 *Cell and Tissue Research*, Springer-Verlag, Heidelberg. 152: 113–128 (Figs. 2, 10), 1974: and 1964: 481–501 (Figs. 9, 12, 15), 1975.

Figs. 15, 17, 20, 21 *Organ Culture in Biomedical Research* Cambridge University Press, Cambridge. pp 193–200 (Figs. 1d, 2b, 2e, 2f), 1976.

References

1. Atwell WJ: The development of the hypophysis cerebri of the rabbit (Lepus cuniculus), Amer J Anat 24: 271–337, 1981.
2. Bhattacharjee DK, Chatterjee P, Holmes RL: Follicles and related structures in the pars intermedia of the adenohypophysis of the jird (Meriones unguiculatus) J Anat 130(1): 63–67, 1980.
3. Howe A, Maxwell DS: Electron microscopy of the pars intermedia of the pituitary gland in the rat. Gen Comp Endocrinol 11: 169–185, 1968.
4. Stoeckel ME, Dellmann HD, Porte A, Klein MJ, Stutinsky F: Corticotrophic cells on the rostral zone of the pars intermedia and the adjacent neurohypophysis of the rat and mouse. Z Zellforsch 136: 97–110, 1973.
5. Saland LC: Mitosis in pituitary MSH/ endorphin cells of adult male rat pars intermedia: light and electron microscopic observations. Anat Rec 200(3): 315–319, 1981.
6. Cameron E, Foster CL: Some light and electron-microscopical observations on the pars intermedia of the pituitary gland of the rabbit. J Endocr 54: 479–485, 1971.
7. Chatterjee P: Ultrastructural studies on the development of the nerve supply in the rabbit pars intermedia. Cell Tiss Res 152: 113–128, 1974.
8. Chatterjee P: Observations on the histology and ultrastructure of the rabbit adenohypophysis with special reference to: (i) the development of the pars intermedia *in vivo*, (ii) the changes during organ culture of the developing and adult pars intermedia, (iii) the cell culture of the pars distalis. PhD Thesis, University of London, 1974.
9. Bargmann W, Lindner E, Andres KH: Über Synapsen an endokrinen Epithelzellen und die Definition sekretorischer Neurone. Untersuchungen am Zwischenlappen der Katzen-Hypophyse. Z Zellforsch 77: 282–298, 1967.
10. Alonso JL, Amat P, Vazquez R: Nerve fibres and neuroglandular synapses in the pars intermedia of the pituitary gland of rats and guinea pigs. Bull Assoc Anat (Nancy) 60(169): 373–380, 1976.
11. Bugnon C, Lenys D, Bloch D, Fellmann D: Les cellules corticotropes et melanotropes de l'adenohypophyse chez le chat, le renard, le rat et le foetus humain: étude avec diverses techniques de fluorescence induite, de cytoimmunologie et a l'hématoxyline phlombique. C R Soc Biol (Paris 169(5): 1271–1276, 1975.
12. Naik DV: Electron microscopic studies on the pars intermedia in normal and in mice with hereditary nephrogenic Diabetes Insipidus. Z Zellforsch 133: 415–434, 1972.
13. Stoeckel ME, Dellmann HD, Porte A, Gertner C: The rostral zone of the intermediate lobe of the mouse hypophysis, a zone of particular concentration of corticotrophic cells. A light and electron microscopic study. Z Zellforsch 122: 310–322, 1971.
14. Vincent DS, Anand kumar TC: Electron microscopic studies on the pars intermedia of the ferret. Z Zellforsch 99: 185–195, 1969.
15. Anand kumar TC, Vincent DS: Fine structure of the pars intermedia in the rhesus monkey. J Anat (Lond) 118: 155–169, 1974.
16. Leatherland JF, Ronald K: Structure of the adenohypophysis in parturient female and neonate harp seals (Pagophilus groenlandicus). Cell Tiss Res 192(2): 341–357, 1979.
17. Saluja PG, Hamilton JM, Thody AJ, Ismail AA, Knowles J: Ultrastructure of intermediate lobe of the pituitary and melanocyte stimulating hormone secretion in oestrogen-induced kidney tumours in male hamsters. Arch Toxicol (supp) (2): 41–45, 1979.
18. Bhattacharjee DK: A study of pars intermedia and pars tuberalis of the adenohypophysis of the jird. PhD Thesis, University of Leeds, 1979.
19. Halmi NS, Peterson ME, Colurso GJ, Liotta AS, Krieger DT: Pituitary intermediate lobe in dog: two cell types and high bioactive adrenocorticotropic content. Science 211(4477): 72–74, 1981.
20. Bargmann W: Weitere Untersuchungen am neurosekretorischen Zwischenhirn-Hypophysensystem. Z Zellforsch 42: 247–272, 1955.
21. Björklund A: Monoamine containing fibres in the pituitary neurointermediate lobe of the pig and the rat. Z Zellforsch 89: 573–589, 1968.
22. Belenky MA, Konstantinova MS, Polenov AL: On neurosecretory and adrenergic fibres in the intermediate lobe of the hypophysis in albino mice. Gen Comp Endocrinol 15: 185–197, 1970.
23. Baumgarten HG, Björklund A, Holstein AF, Nobin A: Organization and ultrastructural identification of the catecholamine nerve terminals in the neural lobe and pars intermedia of the rat pituitary. Z Zellforsch 126: 483–517, 1972.
24. Knowles F: Evidence for a dual control, by neurosecretion of hormone synthesis and hormone release in the pituitary of the dogfish. Phil Trans Royal Soc Lond Ser B 249: 435–456, 1965.
25. Jarskär R: Electron microscopical study on the development of the nerve supply of the pituitary pars intermedia of the mouse. Cell Tiss Res 184: 121–132, 1977.
26. Kraicer J, Morris AR: *In vitro* release of ACTH from dispersed rat pars intermedia cells II Effect of neuro transmitter substances. Neuroendocrinology 20(1): 79–96, 1976.
27. Bargmann W, Knoop A: Uber die morphologischen Beziehungen des neurosekretorischen Zwischenhirn-systems zum Zwischenlappen der Hypophyse (Licht und Electrononmikroskopische Untersuchungen). Z Zellforsch 52: 256–277, 1960.
28. Enemar A, Falck B: On the presence of adrenergic nerves in the pars intermedia of the frog Rana temporaria. Gen Comp Endocrinol 5: 577–583, 1965.
29. Tilders FJH, Smelik PG: Direct neural control of MSH secretion in mammals: the involvement of dopaminergic neurons. In: Melanocyte stimulating hormone: Control, chemistry and effects. Tilders FJH, Swaab DF, Wimersma V, Greidanus TJB van (eds). Frontiers of hormone research 4, Basel, Karger 1977, pp 140–152.
30. Vemes I, Mulder GH, Smelik PG, Tilders FJH: Differential control of beta-endorphin/beta-lipotropin secretion from anterior and intermediate lobes of rat pituitary gland *in vitro*. Life Sci 27(19): 1761–1768, 1980.
31. Harris GW, Donnovan BT (eds) The pituitary Gland 1, London, Butterworth, 1966.
32. Grant SCD: Some light and electron microscopic observations on the development of the pars tuberalis of the rabbit hypophysis. B.Sc. Thesis, Anatomy Dept., Leeds University, 1980.
33. Chatterjee P: Development and cytodifferentiation of the rabbit pars intermedia I Fetal and perinatal stages. Cell Tiss Res 164: 481–501, 1975.
34. Chatterjee P: Development and cytodifferentiation of the rabbit pars intermedia II Neonatal to adult. J Anat 122(2): 415–433, 1976.
35. Svalander C: Ultrastructure of the fetal rat adenohypophysis. Acta Endrocrinol (Suppl) (Kbh) 188: 1–113, 1974.
36. Chatelain A, Dupouy JP, Dubois MP: Ontogenesis of cells producing polypeptide hormones (ACTH, MSH, LPH, GH, prolactin) in the fetal hypophysis of the rat: influence of the hypothalamus. Cell Tiss Res 196(3): 409–427, 1979.
37. Euranius L, Jarskär R: Electron microscopy of neurosecretory nerve fibres in the neural lobe of the embryonic mouse. Cell Tiss Res 149: 333–347, 1974.
38. Bugnon C, Lenys D, Bloch B, Dessy C: Cytoimmunological detection of corticotropic and melanotropic cells in the human fetal adenohypophysis in early stages of development. Bull Assoc Anat (Nancy) 59(166): 571–582, 1975.
39. Bégeot M, Dubois MP, Dubois PM: Hypothalamic influence on differentiation of the immunoreactive ACTH, beta-LPH, alpha- and beta-endorphin containing cells in adenohypophysial primordia of rat fetus in organ culture (authors transl). J Physiol (Paris) 75(1): 27–31, 1979.
40. Dupouy JP: Differentiation of MSH-, ACTH-, endorphin- and LPH-containing cells in the hypophysis during embryonic and fetal development. Int Rev Cytol 68: 197–249, 1980.
41. Kastin AJ, Gennser G, Arimura A, Miller MC, Schally AV: Melanocytestimulating and corticotrophic activities in human foetal pituitary gland. Acta Endocrinol (Kbh) 58: 6–10, 1968.
42. Silman RE, Street C, Hollan D, Chard T, Falconer J, Robinson JS: The pars intermedia and the fetal pituitary-adrenal axis. In: *Peptides of the pars intermedia*. Evered D, Lawrenson G (eds) (Ciba Foundation Symp 81) London, Pitman medical, 1981, pp 180–195.
43. Martin R, Weber E, Voigt KH: Localisation of corticotropin- and

endorphin-related peptides in the intermediate lobe of the rat pituitary. Cell Tiss Res 196(2): 307–319, 1979.
44. Pelletier G, Leclerc R, Labrie F, Cote J, Chretien M, Lis M: Immunohistochemical localization of beta-lipotropic hormone in the pituitary gland. Endocrinology 100: 770–776, 1977.
45. Partanen S, Rechardt L: Histochemically demonstrable monoamines in the pituitary gland and median eminence of the female rat during the post natal development. Z Zellforsch 147: 41–57, 1973.
46. Chatterjee P: Histological and ultrastructural studies of the rabbit pars intermedia in organ culture II. Developing tissue. Cell Tiss Res 167: 387–405, 1976.
47. Chatterjee P: Histological and ultrastructural studies of the rabbit pars intermedia in organ culture. I. Adult and young adult tissue. Cell Tiss Res 169: 485–500, 1976.
48. Chatterjee P: Organ culture of the pars intermedia of the rabbit pituitary gland. In: Organ culture in biomedical research. Balls M, Monnickendam MA (eds), 1, London, Cambridge University Press, 1976, p 193–200.
49. Chatterjee P: Extracellular banded structures in cultures of the rabbit adenohypophysis. J Anat (Lond) 115: 414–420, 1973.
50. Chatterjee P: Ultrastructural studies on the intranuclear inclusions of the rabbit adenohypophysis *in vivo* and *in vitro*. Cell Tiss Res 160: 273–277, 1975.
51. Gaillard PJ: An experimental contribution to the origin of the pars intermedia of the hypophysis. Acta Neerl Morphol (1): 3–11, 1938.
52. Swaab DF: p52, discussion following paper by Mains RE, Eipper BA: Comparison of rat anterior and intermediate pituitary in tissue culture: corticotropin (ACTH) and β-endorphin. In: Peptides of the pars intermedia. Evered D, Lawrenson G (eds), (Ciba Foundation Symp 81), London, Pitman Medical 1981, pp 32–54.
53. Voitkevich AA, Ovchinnikova FA: The nature of the regulatory influences exerted by the hypothalamus on the anterior lobe and pars intermedia of the hypophysis. Biull Eksp Biol Med 55: 207–210, 1963.
54. Semoff S, Fuller BB, Hadley ME: Secretion of melanophore-stimulating hormone (MSH) in long-term cultures of pituitary neurointermediate lobes. Cell Tiss Res 194, 55–69, 1978.
55. Tilders, FJH: p49, discussion following paper by Mains RE, Eipper BA: Comparison of rat anterior and intermediate pituitary in tissue culture: corticotropin (ACTH) and β-endorphin. In: Peptides of the pars intermedia. Evered D, Lawrenson G (eds), (Ciba Foundation Symp 81), London, Pitman Medical, 1981, pp 32–54.
56. Golosow N, Grobstein C: Epithelio-mesenchymal interaction in pancreatic morphogenesis. Devel Biol 4: 242–255, 1962.
57. Shino M, Ishikawa H, Rennels EG: *In vitro* and *in vivo* studies on cytodifferentiation of pituitary clonal cells derived from the epithelium of Rathke's pouch. Cell Tiss Res 181(4): 473–485, 1977.
58. Watanabe YG, Matsumura H, Daikoku S: Electron microscopic study of rat pituitary primordium in organ culture. Z Zellforsch 146–453–461, 1973.
59. Watanabe YG, Daikoku S: An immunohistochemical study on the cytogenesis of adenohyophysial cells in fetal rats. Dev Biol 68(2): 557–67, 1979.
60. Gash D, Roos T, Chambers W: Development of Rathke's pouch tissue transplanted into adult hypophysectomized female rats. Neuroendocrinology 19: 214–226, 1975.
61. Gash D, Ahmed N, Schechter J: Cytodifferentiation of pituitary primordia transplanted to the kidney capsule of adult hosts. Anat Rec 189: 149–160, 1977.
62. Frémond PH, Ferrand R: *In vitro* studies on the self-differentiating capacities of quail adenohypophysis epithelium. Anat Embryol (Berl) 156(3): 255–267, 1979.
63. Ishikawa H, Shino M, Arimura A: Functional clones of pituitary cells developed from Rathke's pouch epithelium of fetal rats. Endocrinology 100(4): 1227–1230, 1977.
64. Chatterjee P: The effect of culture on ultrastructure of dissociated rabbit adenohypophysial cells. J Anat 121(2): 241–258, 1975.
65. Farbman AI: The dual pattern of keratinization in filiform papillae on rat tongue. J Anat 106: 233–42, 1970.
66. Stoeckel ME, Schmitt G, Porte A: Fine structure and cytochemistry of the mammalian pars intermedia. In: Peptides of the pars intermedia. Evered D, Lawrenson G (eds) (Ciba Foundation Symp 81), London, Pitman Medical, 1981, pp 101–127.
67. Naik DV: Electron microscopic-immunocytochemical localization of adrenocorticotropic and melanocyte stimulating hormone in the pars intermedia cells of the rats and mice. Z Zellforsch 142: 305–328, 1973.
68. Mains RE, Eipper BA: Synthesis and secretion of corticotropins, melanotropins and endorphins by rat intermediate pituitary cells. J. Biol Chem 254(16): 7885–94, 1979.
69. Moriarty GC: Heterogeneity of ACTH-containing cells in the rat pituitary, with emphasis on the structure and function of the intermediate lobe. Ann NY Acad Sci 197: 183–200, 1977.
70. Hadley ME: p48, discussion following paper by Mains RE, Eipper BA: Comparison of rat anterior and intermediate pituitary in tissue culture: corticotropin (ACTH) and β-endorphin. In: Peptides of the pars intermedia. Evered D, Lawrenson G (eds), (Ciba Foundation Symp 81), London, Pitman Medical, 1981, pp 32–54.
71. Jackson S, Hope J, Estivariz F, Lowry PJ: Nature and control of peptide release from the pars intermedia. In: Peptides of the pars intermedia. Evered D, Lawrenson G (eds), (Ciba Foundation Symp 81), London, Pitman Medical, 1981, pp 32–54.
72. Moriarty GC, Halmi NS: Electron microscopic study of the adrenocorticotropin-producing cell with the use of unlabelled antibody and the soluble peroxidase-antiperoxidase complex. J Histochem Cytochem 20: 590–603, 1972.
73. Kraicer J, Beraud G, Lywood DW: Pars Intermedia ACTH and MSH content: effect of adrenalectomy, gonadectomy and a neurotropic (noise) stress. Neuroendocrinology 23(6): 352–67, 1977.
74. Gosbee JL, Kraicer J, Kastin AJ, Schally AV: Functional relationship between the pars intermedia and ACTH section in the rat. Endocrinology 86: 560–567, 1970.
75. Miner D: Observation on the histology and ultrastructure of the pars distalis of the foetal and newborn rabbit. M. Phil. Thesis, University of London, 1973.
76. Kraicer J: Lack of release of ACTH from the denervated rat pars intermedia *in vivo*. Can J Physiol Pharmacol 54(6): 809–813, 1976.
77. Pelletier G, Robert F, Hardy J: Identification of human anterior pituitary cells by immunoelectron microscopy. J Clin Endocrinol Metab 46: 534–542, 1978.
78. Kobayashi Y: Functional morphology of the pars intermedia of the rat hypophysis as revealed with the electron microscope. IV Effects of corticosterone on the pars intermedia of the intact and adrenalectomized rats. In: Structure and function of adenohypophysis, Gunma Symp Endocr 6, 89. pp 107–122. Maebashi, Japan, Institute of Endocrinology, 1969.
79. Lever JD: Electron microscopic observations on the adrenal cortex. Amer J Anat 97: 409–429, 1955.
80. Ziegler B: Licht- und elektronenmikroskopische Untersuchungen am Pars intermedia und Neurohypophyse der Ratte. Z Zellforsch 59: 486–506, 1963.

Author's address:
Department of Anatomy
School of Medicine
University of Leeds
Leeds LS2 9NL United Kingdom

CHAPTER 5

Ultrastructure of Rathke's pituitary cleft

FABRIZIO BARBERINI and SILVIA CORRER

1. Introduction

Rathke's pituitary cleft represents the original recess of so-called Rathke's pouch derived in the embryo from the ectodermal evagination of the stomodeum (1,2). In several mammals, for example the rat, dog and cat, this recess persists during postnatal life as a wide fissure separating the pars distalis from the pars intermedia. It is lined by a continuous layer of cells termed 'marginal cells', and contains a fluid-like material (colloid) of uncertain function, origin and chemical composition (3).

The submicroscopic arrangement of the pituitary cleft has been described in ultrastructural studies with special emphasis upon 1) the role of the marginal cells in colloid formation (4–8) and 2) cleft cells as a possible 'renewal cell system' (9) due to the similarity between marginal cells and follicular cells of the hypophysis (10–12).

In this chapter will be summarized previous studies on the subject as well as the fine morphology of the pituitary cleft in the rat by correlated scanning and transmission electron microscopic technique (TEM and SEM). In order to evaluate better the three-dimensional architecture of associated glandular parenchyma, the marginal layer will be described with special reference to its relationship with the subjacent granular and agranular cells of the pars distalis and pars intermedia of the hypophysis.

2. Ultrastructure of pituitary cleft

The pituitary cleft is a wide fissure lined on the anterior (toward the pars distalis) side and the posterior (toward the pars intermedia) side by a continuous layer of cuboidal cells ('marginal cells'). The content of the cleft is composed of an amorphous material having a high electron opacity and probably corresponding to colloid. This material is rather dense and copious, especially in the lateral recesses where the marginal cells of both layers come in close contact. By contrast, the material filling the central part of the cleft often shows a fluid-like appearance and a very low electron density. In a few zones small regions of attachment between the pars intermedia and pars distalis appear to interrupt the continuity of the luminal surface of the cleft, forming smaller cavities which divide the pituitary fissure.

Some morphological differences occur between the anterior and the posterior sides of the pituitary cleft, mainly concerning the surface features of the marginal cells and their relations with the subjacent glandular parenchyma (Figs. 1 and 2).

2.1. The anterior marginal layer

The marginal cells of the distal part constitute a single layer of flattened epithelioid cells. Particularly, the anterior marginal layer of the cleft, observed by SEM, is made up of cells which have over their surface a varying number of microvilli ('microvillous cells'). Other apical surfaces possess long cilia ('ciliated cells'). Microvillous cells are more abundant than ciliated cells, the latter appearing singly or associated in small groups. Other cell types possessing a single short cilium or a reduced number of cilia, as well as cells showing predominantly smooth surfaces with rare microvilli, also are present. Other cell characteristics, such as a large size and a rounded shape are noted; in addition to the microvilli these consist of bulbous protrusions and laminar extensions.

Furthermore SEM reports have revealed that in some areas of the anterior as well as posterior marginal layers of the fissure are long finger-like evaginations, measuring 40–80 μm in lenght. The cells covering such villous projections are flattened in shape and appear continuous with the marginal cells lining the cleft. Their surface has numerous short microvilli and flattened microplicae.

By TEM, the anterior marginal cells have large

Motta, PM (ed): Ultrastructure of endocrine cells and tissues. ISBN-13: 978-1-4613-3863-5

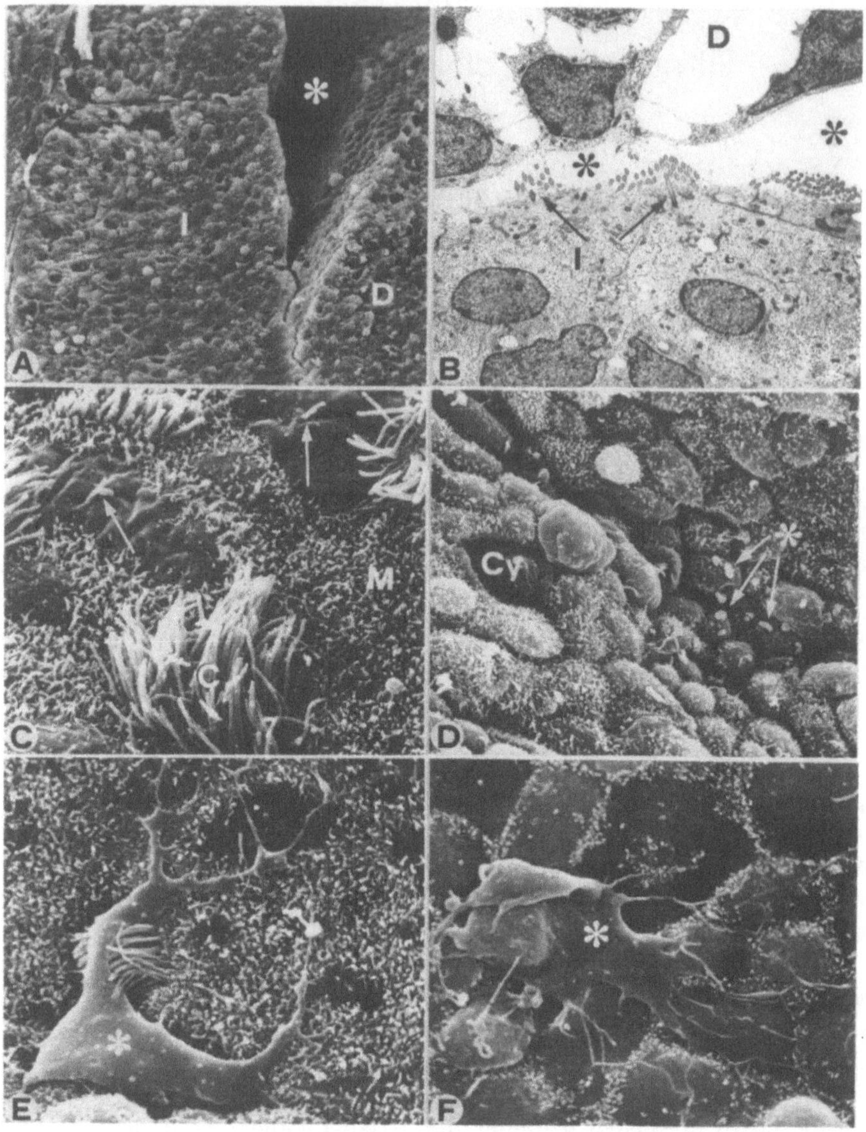

Fig. 1. (A, B) Pars distalis (D) and pars intermedia (I) with the interposed pituitary cleft (✱). This contains a fluid-like material and its lumen is lined by marginal cells. Some of these are ciliated cells (→). Rat (SEM: × 875; TEM: × 3416). (By permission from ref. 57); (C, D) The marginal layer lining the anterior side of the Rathke's cleft consists of ciliated (C) and microvillous cells (M). Some of these have a smooth surface with isolated cilia (→) and others possess bulbous protrusions (✱ →). In the fig. D the marginal layer appears infolded in a crypt-like invagination (Cy). Rat (SEM: × 4375, × 1750); (E, F) Irregularly shaped cells into the pituitary cleft (✱). They are located on the surface of the marginal cells lining respectively the pars intermedia (E) and the pars distalis (F). The exact nature of such elements (macrophagic and/or neuronal-like cells?) is still unknown. Rat (SEM: × 4375, × 3500).

irregularly shaped nuclei with dense chromatin and scant cytoplasm, and contain few membranes of rough endoplasmic reticulum (RER), polysomes, small mitochondria and a reduced apparatus of Golgi. Lysosomes, multivescicular bodies with large vacuoles are occasionally present in these cells, as well as discrete bundles of microfilaments. In some instance, the continuity of this layer is interrupted because of small irregular gaps and/or infoldings lined by marginal cells penetrating into the glandular parenchyma (crypts) and through which the smooth apical surfaces of subjacent cells are evident. The flattened elements lining the anterior part of the cleft can be reduced to very thin and irregular cytoplasmic extensions. These extensions, which may assume the features of filopodia when observed in SEM preparations, come in close contact with similar evaginations of adjacent cells and subjacent star shaped cells

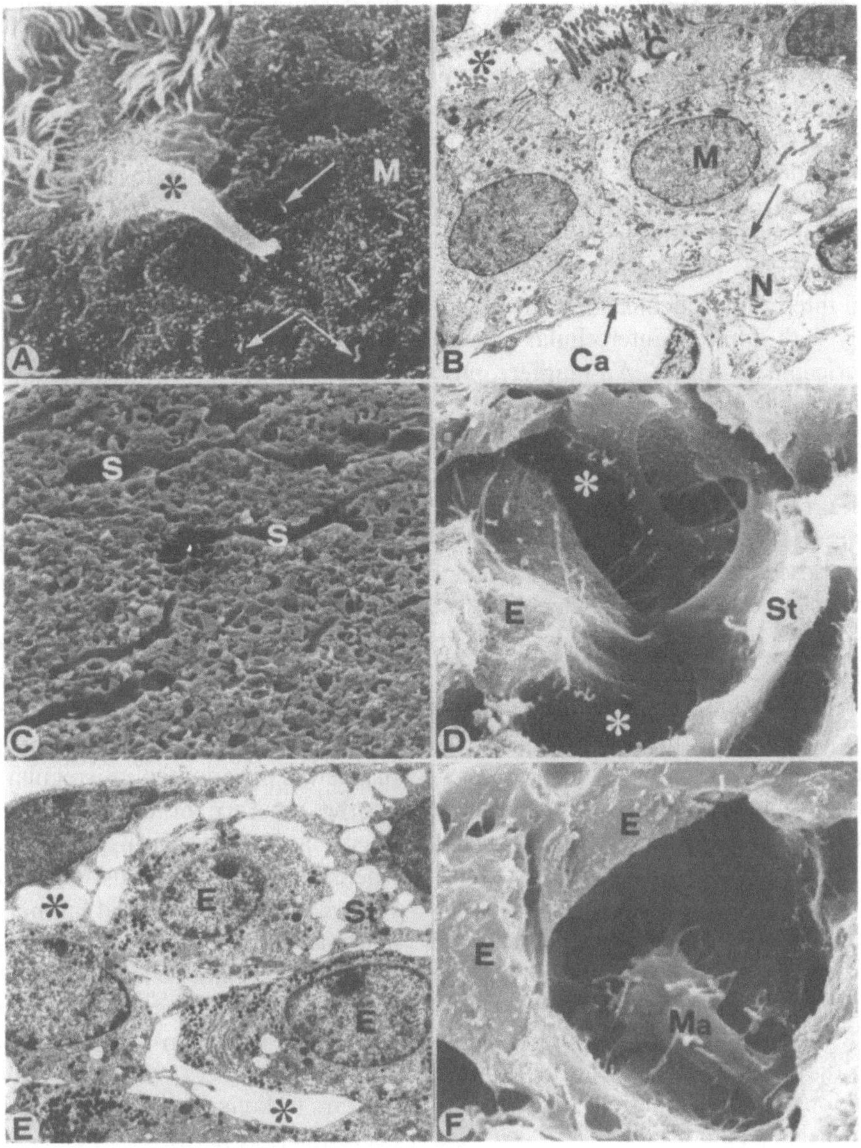

Fig. 2. (A) Villous-like protrusion (*) arising from the posterior marginal layer of the pituitary cleft. The marginal cells show abundant (C) or isolated cilia (→) and some have smooth surfaces and/or microvilli (M). Rat (SEM: × 2360); (B) Marginal cells (M) partially ciliated (C) cover the pars intermedia fronting the lumen of the Rathke's cleft (*). Nerve fibers (N) from the pars intermedia, intermingled with capillary vessels (Ca), are apparently in close contact with the marginal cells (→). Rat (TEM: × 3500) (By permission from ref. 57); (C, D, E, F) The pars distalis of the rat pituitary gland consists of sinusoids contained (S) in a three-dimensional network of cords and lacunar spaces. These, in correlated TEM and SEM preparations (C and D), appear as intercellular spaces delimited by endocrine glandular (E) and stellate cells (St) (D and F). Isolated macrophages (Ma), closely related to parenchymal cells and capillary vessels, are also suspended in these spaces. (SEM: × 350, × 4375; TEM: × 3590; SEM: × 4375). (By permission from ref. 48).

('stellate cells') (12) where they form typical junctional complexes. Such junctions are present only between marginal and/or 'stellate cells' and parenchymal elements. Thus, the cytoplasmic processes of the anterior marginal cells, together with the branching evaginations of 'stellate cells' of the pars distalis, form a series of large intercellular and irregular spaces containing granular cells (endocrine cells). These granular elements, as a rule, appear rounded and possess a relatively smooth surface. Occasionally, the lumen of the cleft is separated from the subjacent intercellular spaces only by very thin cellular evaginations, so that the granular cells appear bulging into the lumen or free on the cleft surface. In SEM preparations, the marginal, stellate, and granular cells form a continuous network of cord-like structures (Figs. 1 and 2).

A thin basal lamina covers the basal surfaces of marginal cells as well as the plasmalemma of many stellate and parenchymal cells, mainly where the latter cells come in close contact with the capillary walls (13). The endocrine cells of the adenohypophysis are separated from the capillary endothelium with wide perivascular spaces, which are continuous with intercellular lacunae. Observed in stereo-views, the submarginal intercellular lacunae appear in connection not only with adjacent intercellular spaces, but also with pericapillary spaces. An uninterrupted three-dimensional labyrinth of lacunar microchannels can be envisaged to be interposed among the parenchymal cells and capillary walls.

The capillaries of the pars distalis, showing a tortuous course, are regarded as sinusoids. The sinusoids are composed of endothelial cells expanded in very thin cytoplasmic extensions, which, as revealed by SEM, have a discrete number of gaps. The cytoplasmic extensions possess a few microvilli and fenestrations variable in shape and size and having a diameter of approximately 50–200 nm (15). Occasionally isolated cilia from parenchymal cells project into the perisinusoidal spaces or very close to the lumen of sinusoids and their fenestrated endothelial cells (15, 16). Furthermore, some large cells irregular in shape and most likely corresponding to macrophages have been observed in the intercellular pericapillary spaces, on the surfaces of the pituitary cleft, or interposed among the parenchymal cells. Occasionally similar cells are noted within the lumen of the sinusoids where they may assume, by means of their irregular projections, a close relationship with the fenestrated endothelial wall.

2.2. The posterior marginal layer

The posterior layer of the cleft appears generally similar to the anterior layer. Both ciliated and microvillous cells are present, but on the posterior the majority of the cells possess cilia. The cilia are long and arranged in regular bundles. Some cells with bulbous protrusions among microvilli or with microvilli concentrated on the border of the cellular body and a centrally located single cilium often are observed. In contrast to the surface of the anterior layer, the posterior side does not have interruptions or crypt-like invaginations of the marginal layer. As a rule the cells lining this side of the cleft have a typical epithelial arrangement. In SEM and TEM preparations, particularly in those obtained parallel to the major axis of the cleft, the posterior layer appears to be formed by polyhedral and/or columnar cells, with the features of a cylindrical epithelium. Generally these elements form a single thick cellular layer, but occasionally they may arranged in two or more layers rather like a stratified epithelium occurring mainly in the zones corresponding to the lateral extension of the cleft. Such cells always are separated from the glandular parenchyma of the pars intermedia by a thick basal lamina.

The cilia of these tall cells have the same arrangement of the common vibratile cilia (9 + 2 microtubular pattern). Both ciliated and microvillous cells possess abundant cytoplasm containing numerous elongated mitochondria, free polysomes and often conspicuous bundles of microfilaments. Dense bodies, lysosomes and a developed Golgi complex appear concentrated in the periapical areas of the cytoplasm. On many microvillous surface areas large vacuoles and irregular apical protrusions are also evident. The ciliated elements and the cells provided with microvilli are connected to each other on their luminal boundaries by typical junctional complexes. The intercellular spaces appear narrow. In some TEM observations, the marginal posterior layer of the cleft is invaginated in irregular and long recesses by the presence of elongated nerve fibers. These fibers, apparently arising from the subjacent areas where the connective tissue surrounds the blood vessels, fold into the basal lamina and penetrate the marginal cells. Typical of these zones of close contact, the marginal cells and nerve fibers form cytoneural (synaptical) neuroepithelial complexes similar to those revealed in the infundibulum in the pars intermedia proper (3). However, a thick basal lamina appears to be interposed between marginal cells and nerve fibers (Fig. 2).

On both side of the pituitary cleft, some cells apparently free on the surface of the anterior and posterior layers have irregularly distributed cilia and/or microvilli. Occasional erythrocytes, parenchymal (glandular) cells and other elements, irregular in shape (most likely macrophages) also are noted.

3. Origin and nature of pituitary cleft

During embryonic life the hypophysis of vertebrates arises from the interaction between ectodermal cells of the stomodeum (Rathke's pouch) and the neuroectodermal cells of the floor of the forebrain (infundibular process) (1). In the rat and other animals the pars distalis of the adenohypophysis is clearly separated from the pars intermedia by a cleft, commonly considered to be a remainder of the dorsal outpocketing of the roof of the primitive oral cavity

(9, 11). As above described by SEM, the cellular populations lining both sides of the pituitary cleft ('marginal cells') mainly include microvillous and ciliated cells. The latter are more abundant on the posterior side (toward the pars intermedia); the majority of the elements lining the anterior layer (toward the pars distalis) appears with microvillous features. Occasionally on the anterior side are small gaps and/or other infoldings (cript-like invaginations). Such surface patterns of the marginal elements covering the cleft resemble, in their topography and distribution, epithelial cells populating the upper parts of the respiratory tract (17, 18). According to these reports, the numerous ciliated cells found in the pituitary gland have been regarded as simple embrional remnants derived from the rhino-pharyngeal mucosa. This morphological similarity has suggested possible embryonic relationship between these epithelial layers (13).

Investigations on the development of the hypophysis in the chick embryos (19) have also suggested that the anterior pituitary gland might be of neuroectodermal rather than ectodermal (stomodeal) origin, so that the hypothalamo-hypophyseal complex can be considered as a single rather than a composite entity. Such alternative hypothesis on the origin and structure of the hypophysis seems also in agreement with the actual concepts concerning the embryology of the diffuse neuroendocrine system and its relationship to the so-called 'APUD cells', 'paraneurons' or cells producing 'common peptides' (20, 21, 22). Recent immunocytochemical reports at the microscopic level on the site of the brain-specific S-100 protein in both hypophyseal lobes of the rat (23), seem confirm the above theory. Thus, the marginal cells lining the cleft might be regarded as transformed ependymal cells ('tanycytes'). There are, in fact, striking surface similarities between the above cells (24), occurring in both groups of ciliated and non-ciliated cells, and possessing comparable general arrangement and surface morphology.

Moreover, tanycytes line the lateral recesses of the third ventricle and the infundibular process, with which the pituitary cleft has a very close topographical relationship (25–27). If the origin of the hypophysis is neuroectodermal (19), the pituitary cleft can be regarded as a closed cavity originally in relation to the hypothalamo-hypophyseal recess (infundibulum) and showing similar origin, nature and, possibly, function. Therefore the marginal cells of the cleft should be considered tanycytes. Cells occasionally free on the luminal surface of the cleft may correspond to both macrophages (Kolmer cells) (28, 29) and/or neuronal (supraependymal) cells (30, 31) such as those observed in the brain ventricles (32).

4. Marginal layer of the cleft and related pituitary parenchyma

Marginal cells of the pituitary cleft possess surfaces and cytoplasmic features very similar to those displayed by 'stellate cells', or follicular cells, located in the adenohypophysis. In both cell types similar thin brandhing projections are noted, often interconnected by means of junctional complexes (6, 12). Such features differ considerably from those of granulated cells, which probably correspond to active endocrine elements (3). The functional significance of 'stellate cells' is not known with certainty. It has been hypothesized that these cells may be involved in ACTH secretion (33, 34), but it also has been suggested that these cells may have a metabolic and supportive function in the pituitary tissue (6, 10, 12, 35–37). Considering the essential surface similarity of both cell types, which form a cellular continuum, it is probable, that they perform a similar function. Marginal cells and 'stellate cells', closely interposed between granulated elements and pericapillary spaces, contribute to the three-dimensional cellular network. Thus it may be surmised that they develop a metabolic as well as a supporting role similar to that which occurs for glial cells in their relationship to neurones (12).

Both SEM and TEM observations reveal that a close relationship occurs between parenchymal cells of the pars distalis, stellate cells and anterior marginal cells of the cleft. With their elongated and irregular extensions ('filopodia') they form a sort of three-dimensional labyrinthine system of narrow anastomotic cavities pervading the gland. Such a labyrinth, which appears to be continuous with the marginal spaces, has been assumed to be in continuity with the lumina of the pituitary follicles (7, 12) and seems to be in direct connection with the lumen of the cleft through the openings and/or infoldings often present on its anterior layer. This network of intercellular and perivascular spaces is filled by a fluid-like material similar to that frequently occupying the lumen of the cleft (Fig. 2).

The characteristic topography of this lacunar cavity around vessels and glandular cells appears similar to the pericapillary and intercellular spaces found in other endocrine glands and tissues such as corpus luteum (38), adrenal cortex (39), epiphysis (40, 41), Leydig cells (42) and in organs showing an endocrine-like arrangement, such as liver plates (43, 44). The presence of such spaces has a functional significance, in view of the fact that the above glands and tissues lack typical lymphatic capillaries (12, 45–47).

Thus the lacunar system, as in other endocrine glands, probably functions to collect blood and tissue filtrates. These intercellular and perivascular microchannels, interposed among the cellular surfaces and the wall of the vessels, can be considered as a means of storage and transport of secretory products from cells to the blood stream. Moreover, nutritious substances from the blood to the endocrine cells might temporarily occur in these lacunae, as a site where both cellular and blood filtrates are stored (48).

Finally, regarding the irregularly shaped cells found in the intercellular and pericapillary spaces and on the surface of the pituitary cleft and within the lumen of some capillaries, such cells are believed to be macrophages. These cells are responsible for the phagocytic activity in the pituitary gland tissues (15). The question of whether these macrophage-like elements are actually stellate cells transformed in certain experimental conditions (6, 49, 50), or whether these cells are simply macrophages derived from the circulating monocytes, is still unresolved.

5. Content of the pituitary cleft

Correlated SEM and TEM observations also have revealed that a number of microvillous cells lining both sides of the pituitary cleft possess some surface features, such as bulbous protrusions, laminar evaginations and large periapical vacuoles. These appearances are most likely the morphological expression of an active transport of fluids (13, 51, 52). Similar features have been described by SEM in the uterine epithelium (51–53), on the luminal surface of thyroid follicles during the resorption of colloid (54, 55) and in tanycytes lining some ventricular areas (26). Such characteristic aspects have been regarded as rapid morphodynamic changes related to active processes of endocytosis (56).

The marginal cells of the cleft generally do not present by TEM the characteristic of secretory elements. Thus, the fluids and/or colloid material, which often is contained in the fissure, cannot be considered to be a direct product of secretion of these cells (48). Although it is possible that degraded cell bodies arising from the glandular tissue may occasionally contribute to the constitution of the colloid (3, 7), it seems more likely that the cleft content arises mainly from the fluids occuring in the lacunar spaces of the pars distalis of the hypophysis (48). It may be more a product deriving from the parenchymal cells and blood filtrates than a secretory product arising from the elements lining the cleft. The ability of the marginal cells to actively transport fluids, however, quite likely is also related to the possibility that these cells, by absorbing fluids from the cleft by active endocytosis, may convey to the pars intermedia substances (including hormones) produced in the distal part of the pituitary gland and vice versa (14, 57). Moreover, the cilia present on many cellular surfaces, having a typical 9 + 2 microtubular pattern, possibly have a kinetic role, so that they may facilitate the circulation and subsequent absorption of the fluids (57). Curiously enough a similar function has been described with regard to ciliated ependymal cells lining the brain ventricles (27, 31).

References

1. Hamilton WJ, Boyd JD, Mossman HW: Human embryology, 4th edit., W Heffer & Sons Ltd, Cambridge, 1973.
2. Langman J: Medical Embryology. Human development, normal and abnormal. 3rd edit. The Williams & Wilkins Company, Baltimore, 1975.
3. Kurosumi K, Fujita H: Functional Morphology of endocrine glands. An atlas of electron micrographs. Igaku-Shoin Ltd, Tokyo, 1975, pp 1–377.
4. Dubois P, Girod C: Observation au microscope électronique d'un reliquat de la fente hypophysaire chez la hamster doré adulte. Cr Séanc Soc Biol (Paris) 164: 157–160, 1970.
5. Vanha-Perttula T, Arstila AV: On the epithelium of the rat pituitary residual lumen. Z Zellforsch 108: 487–500, 1970.
6. Dingemans KP, Feltkamp CA: Non granulated cells in the mouse adenohypophysis. Z Zellforsch 124: 387–405, 1972.
7. Ciocca DR, Gonzales CB: The pituitary cleft of the rat. An electron microscopic study. Tissue and Cell 10: 725–733, 1978.
8. Ciocca DR: Scanning electron microscopy of the cleft of the rat pituitary. Cell Tissue Res 206: 139–143, 1980.
9. Yoshimura F, Soji T, Kiguchi Y: Relationship between the follicular cells and marginal layer cells of the anterior pituitary. Endocrinol Jpn 24 (3): 301–305, 1977.
10. Kagayama M, Ando A, Yamamoto TY: On the epithelial lining of the cleft between pars distalis and pars intermedia in the mouse adenohypophysis. Gumna Symp Endocrinol 6: 125–136, 1969.
11. Sano M, Sasaki F: Embryonic development of the mouse anterior pituitary studied by light and electron microscopy. Z Anat Entw Gesch 129: 195–222, 1969.
12. Vila-Porcile E: Le réasou des cellules folliculo-stellaires et les follicules de l'adenohypophyse du rat (pars distalis). Z Zellforsch 129: 328–369, 1972.
13. Correr S, Caggiati A, Motta P: Scanning electron microscopy of the rat hypophysary cleft. Proc 5th Europ Anat Congr September 10–14 (abstract) p 70, Prague, 1979.
14. Correr S, Motta PM: Relationship between the marginal layer and parenchymal cells of the rat adenohypophysis as revealed by scanning electron microscopy. Biomedical Research 2, supplement: 109–113, 1981.
15. Fujita H, Kataoka K: Electron microscopic observation on blood capillaries of the anterior pituitary of normal and acid dye-injected animals. Gumna Symp Endocrinol 6: 137–149, 1969.
16. Correr S, Caggiati S, Motta PM: Scanning and transmission electron microscopic observations of the rat hypophysary cleft. Folia Morpho-

logica 28: 82–87, 1980.
17. Andrews PM: A scanning electron microscopic study of the extra-pulmonary respiratory tract. Am J Anat 139: 299–324, 1974.
18. Smolich JJ, Stratford BF, Maloney JE, Ritchie BC: Postnatal development of the epithelium of larynx and trachea in the rat: scanning electron microscopy. J Anat 124: 657–673, 1977.
19. Takor Takor T, Pearse AGE: Neuroectodermal origin of ovarian hypothalamo-hypophyseal complex. The role of the ventral neural ridge. J Embryol Exp Morphol 34: 311–325, 1975.
20. Pearse AGE: The cytochemistry and ultrastructure of polypeptide hormon-producing cells of the APUD series and the embryologic, physiologic and pathologic implications of the concept. J Histochem Cytochem 17: 303–313, 1969.
21. Fujita T: Concept of paraneurons. Arch Histol Jp 40, Suppl 1–12, 1977.
22. Pearse AGE, Takor Takor T: Embryology of the diffuse neuroendocrine system and its relationship to the common peptides. Federation Proc 38: 2288–2294, 1979.
23. Cocchia D, Miani N: Immunocytochemical localization of the brain specific S-100 protein in the pituitary gland of adult rat. J Neurocytol 9: 771–782, 1980.
24. Kessel RG, Kardon RH: Tissues and organs. A text-atlas of scanning electron microscopy. WH Freeman and Co, San Francisco, 1979.
25. Allen DJ, Persky B, Low FN: Some regional variations in ventricular lining material in laboratory mammals and man. SEM: Vol II, Becker RP, Johari D (eds), SEM Inc AMF O'Hare, IL USA, 1978, pp 45–52.
26. Flament-Durand J, Vienne G, Dustin P: Scanning electron microscopic study of the ependyma of the hypothalamic region in man. SEM: Vol II, Becker RP, Johari D (eds), SEM Inc AMF O'Hare, IL USA, 1978, pp 151–156.
27. Mestres P: Old and new concepts about circumventricular organs: an overview. SEM: Vol II, Becker RP, Johari D (eds), SEM Inc AMF O'Hare, IL USA, 1978, pp 137–142.
28. Hosoya Y, Fujita T: Scanning electron microscope observation of intraventricular macrophages (Kolmer cells) in the rat brain. Arch Histol Jpn 35: 133–140, 1973.
29. Bleier R: Surface fine structure of supraependymal elements and ependyma of hypothalamic third ventricle of mouse. J Comp Neurol 161: 555–568, 1975.
30. Mestres P, Breipohl W: Morphology and distribution of supraependymal cells in the third ventricle of the albino rat. Cell Tissue Res 168: 303–314, 1976.
31. Coates PW: The third ventricle of monkeys: scanning electron microscopy of surface features in mature males and females. Cell Tissue Res 177: 307–316, 1977.
32. Allen DJ: Scanning electron microscopy of epiplexus macrophages (Kolmer cells) in the dog. J Comp Neurol 161: 197–214, 1975.
33. Dent JN, Gupta BL: Ultrastructural observations on the developmental cytology of the pituitary gland in the spotted newt. Gen Comp Endocrinol 8: 273–288, 1967.
34. Schechter J: The ultrastructure of the stellate cell in the rabbit pars distalis. Anat Rec 160: 422, 1968.
35. Cardell RR: The ultrastructure of stellate cells in the pars distalis of the salamander pituitary gland. Amer J Anat 126: 429–456, 1969.
36. Rinehart JF, Farquhar MG: The fine vascular organization of the anterior pituitary gland. An electron microscopic study with histochemical correlations. Anat Rec 121: 207–240, 1955.
37. Yamada K, Yamashita K: An electron microscopic study on the possible site of production of ACTH in the anterior pituitary of mice. Z Zellforsch 80: 29–43, 1967.
38. Van Blerkom J, Motta P: A scanning electron microscopic study of the luteo-follicular complex. III. Formation of the corpus luteum and repair of the ovulated follicle. Cell Tissue Res 189: 131–154, 1978.
39. Motta P, Muto M, Fujita T: Three-dimensional organization of mammalian adrenal cortex. A scanning electron microscopic study. Cell Tissue Res 196: 23–38, 1979.
40. Quay WB: Pineal canaliculi: demonstration, twenty-four-hour rhythmicity and experimental modification. Amer J Anat 139: 81–93, 1974.
41. Krstic R: Scanning electron microscope observations of the canaliculi in the rat pineal gland. Experientia (Basel) 31: 1072–1074, 1975.
42. Motta PM, Calvieri S, Palermo D: On the occurrence of spaces similar to intercellular canaliculi in the Leydig cells of mice. Experientia (Basel) 29: 1120–1122, 1973.
43. Motta P, Porter KR: Structure of rat liver sinusoids and associated tissue spaces as revealed by scanning electron microscopy. Cell Tissue Res 148: 111–125, 1974.
44. Motta P: The three-dimensional fine structure of the liver as revealed by scanning electron microscopy. Int Rev Cytol (Suppl 6). Studies in Ultrastructure. Academic Press, New York, San Francisco, London, 1977, pp 347–401.
45. Ottaviani G: Linfatico Sistema. In: Enciclopedia medica italiana, USE, Firenze, 1953, Vol 5, pp 1877–1881.
46. Harrison RG: The adrenal circulation. Blackwell, Oxford, 1960.
47. Brauer RW: Liver circulation and function. Physiol Rev 43: 115–213, 1963.
48. Correr S, Motta PM: The pituitary cleft and associated parenchymal tissues in the rat adenohypophysis as revealed by scanning electron microscopy. In: Three-dimensional microanatomy of cells and tissue surfaces. Didio LJA, Motta PM, Allen DJ (eds), Elsevier North Holland, Inc, 1981, pp 167–181.
49. Farquhar MG: Processing of secretory products by cells of the anterior pituitary gland. In: Memoirs of the society for endocrinology. Heller H, Lederis K (eds), Cambridge University Press, N° 19 1971, pp 79–122.
50. Young BA, Foster CL, Cameron E: Some observations on the ultrastructure of the adenohypophysis of the rabbit. J Endocrinol 31: 279–287, 1965.
51. Barberini F, Sartori S, Van Blerkom J, Motta P: Changes in the surface morphology of the rabbit endometrium related to the estrous and progestational stages of the reproductive cycle. A scanning and transmission electron microscopic study. Cell Tissue Res 190: 207–222, 1978.
52. Enders AC, Nelson DM: Pinocytotic activity of the uterus of the rat. Am J Anat 138: 277–300, 1973.
53. Parr MB, Parr ELL: Endocytosis in the uterine epithelium of the mouse. J Reprod Fert 50: 151–153, 1977.
54. Ketelbant-Balasse P, Rodesch F, Neve P, Pasteels JM: Scanning electron microscope observations of apical surfaces of dog thyroid cells. Exp Cell Res 79: 111–119, 1973.
55. Nunez É, Wallis J, Gershon MD: Secretory processes in follicular cells of the bat thyroid. III. The occurrence of extracellular vesicles and colloid droplets during hibernation. Am J Anat 141: 179–202, 1974.
56. Motta P, Andrews PM, Porter KR: Microanatomy of cell and tissue surfaces. An atlas of scanning electron microscopy Lea & Febiger, Philadelphia, 1977.
57. Correr S, Motta PM: The rat pituitary cleft: a correlated study by scanning and transmission electron microscopy. Cell Tissue Res 215: 515–529, 1981.

Authors' address:
Department of Anatomy
Faculty of Medicine
University of Rome
Viale R. Elena 289
00161 Rome, Italy

CHAPTER 6

The problem of the folliculo-stellate cells in the pituitary gland

EVELYNE VILA-PORCILE and LÉON OLIVIER

1. Introduction

In 1953, in one of the first papers dealing with the ultrastructure of the rat anterior pituitary, Rinehart and Farquhar (1) noticed the presence of peculiar stellate cells. In a second report (2), they described 'certain agranular stellate cells with processes extending between the parenchymal cells' and they concluded: 'it may be that they represent an additional structural and functional unit'. Two years later, Farquhar (3) introduced the notion of 'follicular cell' in a now famous abstract from Anatomical Record, as 'a sixth cell type' located 'throughout the anterior lobe in groups around follicles or ductules which contain colloid of low density'. Following this short paper, many electron microscopy works have been published, describing either 'stellate' cells, or 'follicular' cells (i.e. cells lining a follicular cavity) in the pituitary of various species.

In 1965, Kagayama (4) proposed that in the dog pituitary 'the cells lining such a follicular cavity are exclusively of the stellate type... Therefore, the stellate cells may also be called the follicular cells', but the question remained debated, and in 1971, Farquhar (5) considered, due to works in which no follicular cavities had been described, that 'the relationship between these two elements – follicular cells and stellate cells – are not entirely clear'. This point might be still questioned, but only in the pituitary of some lower vertebrates, where 'pure' stellate cells possibly exist. However, in a large majority of species, follicles were progressively found, and between 1972 and 1974, the notion of 'folliculo-stellate cell' had been definitively accepted as involving in a single entity, two aspects of the same agranular cell (6, 7). The 'folliculo-stellate' concept had been extended (6) to the 'reticular framework' firstly described also by Kagayama (4) as formed by the cells in question 'to pack the other parenchymal cells in the meshes'.

At the present time, more than 180 papers have been published, either brief descriptions or complete reports on these cells, in the anterior pituitary of a large variety of Vertebrate species. We have no possibility in such a paper, mainly focussed on rat and human pituitary, to quote all of these articles (see for reviews 5 to 15). However, we will mention some reports on fish pituitaries, as their study had brought up numerous new informations.

In this report, we will successively look at the morphological characteristics of the folliculo-stellate cells, at the problem of their cytogenesis and at the functional questions raised by the folliculo-stellate cell system.

2. Morphology of the folliculo-stellate cells and follicles in the normal adult pituitary

The folliculo-stellate cells share some common morphological structures, valid for almost all vertebrates from the myxine to the human and for the different parts of the adenohypophysis, leading to the identification of this cell type. On the other hand, these cells can show particular features or variations, mostly related to species.

2.1. General features

The most conspicuous features of these cells are their stellate shape and follicular organization. They encircle well-defined cavities, located in the core of the pituitary cords. This cell association implies a specificity of each face of the cavity boundary cells: the 'apical' or 'follicular pole', i.e. the face of the cell lining the lumen, differentiates microvilli, occasional cilia and cytoplasmic 'blebs', while at the same level the lateral faces are sealed by junctional complexes, thus building a continuous barrier between the subjacent granular cells and the cavity.

The opposite pole of the follicular cells originates cytoplasmic processes or sheets which extend between the granular endocrine cells, and reach the

Motta, PM (ed): Ultrastructure of endocrine cells and tissues. ISBN-13: 978-1-4613-3863-5

basal lamina separating the epithelial cells from the perivascular spaces or from the connective septa. There, the processes widen to form 'end-feet' or long cytoplasmic sheets expanding between the granular cells and basal lamina, as to constitute a thin discontinuous basal covering.

The organization of these processes within the pituitary cords leads to the formation of a tri-dimensional network, in the meshes of which the granular endocrine cells reside. The considerable extension of the plasmalemmal surface, due to the development of the cytoplasmic processes, is to be emphasized.

Besides those two structural features, i.e. follicular association and stellate shape, the cytological features of the folliculo-stellate cells can be summarized as follows: a) they completely lack secretory granules, b) their nucleus is angular in shape with a typical pulverulent chromatin and encircled by a narrow ring of cytoplasm, c) they are rather poor as far as the other organelles are concerned, and d) they are loaded with filaments. The Golgi apparatus is re-

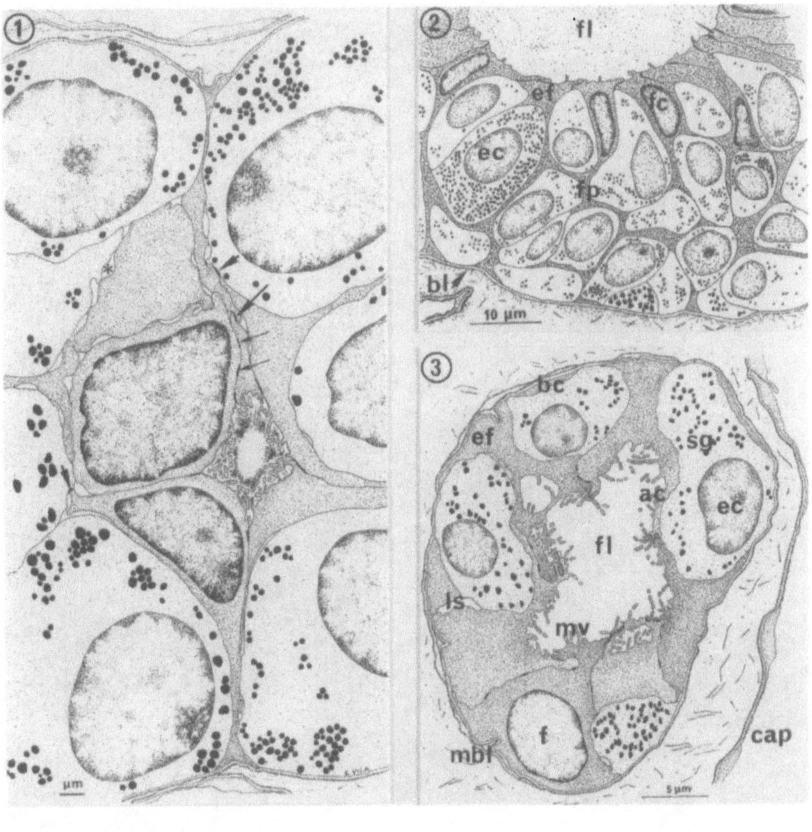

Diagrams 1–3. (1) Rat pituitary (Adult male). 'Réseau folliculo-stellaire', stressed by stippling. The follicular network rests on two perivascular spaces by the end-feet of the cell processes. One of them originates from a follicle boundary cell and directly abuts to the basal lamina (lower part of the diagram). The other processes reach the periphery through a relay (arrow). The follicle is equipped with long microvilli and is sealed off by junctions. Some junctions are present between follicular cells (arrows), and between follicular and endocrine cells (arrowheads). Presence of an intercellular space system (asterisks), more developed in rat than in other species pituitary. (2) An aspect of the folliculo-stellate network in fish pituitary. (Redrawn from a micrograph due to the courtesy of Dr. Lagios). fl: follicular lumen; fc: follicular cell; fp: follicular process; ec: endocrine cells; bl: basal lamina; ef: end-foot. The cavity is large-sized. The follicular cells form a complete barrier between the lumen and the endocrine cells, and most of the endocrine cells are isolated from each other by the slender cytoplasmic processes which form a tight network. The lumen contains colloid of low density. This structure is very similar to that of the human follicle as far as the apical barrier and the distribution of the endocrine cells into small clusters within small compartments are concerned. (3) Human pituitary. Structure of an isolated follicle, redrawn from a micrograph of the nontumoral tissue adjacent to a prolactinoma. The follicular cells form a typical complete barrier. At the apical pole, long microvilli project toward the lumen with interdigitations at the level of the junction complexes. At the basal pole, the covering is discontinuous. Some end-feet are underlined by multilayered basal laminae. f: follicular cell nucleus; fl: follicular lumen; ec: endocrine cell; mv: microvilli; sg: secretory granules; ef: end-foot; ac: apical covering; bc: basal covering; ls: lacunary space; mbl: multilayered lamina; cap: capillary with subjacent perivascular space.

In order to simplify the diagrams, the double membranes (nuclei) and the closely joined plasma membranes have been drawn as single lines.

duced and frequently located at the apical pole. The mitochondria are small-sized and curiously resemble those of the corticotrope cells. Lysosomal formations are sparse in the cell body and the processes as well. Microfibrils and microtubules, intermingled with coated and uncoated vesicles, are found in a terminal web at the apical pole.

2.2. Particular features and variations of the folliculo-stellate cells and follicles

2.2.1. Variations of the follicular cavity

The size, shape, content and the apical differentiations can vary not only from one species to another, but also from one area of a given pituitary to another one.

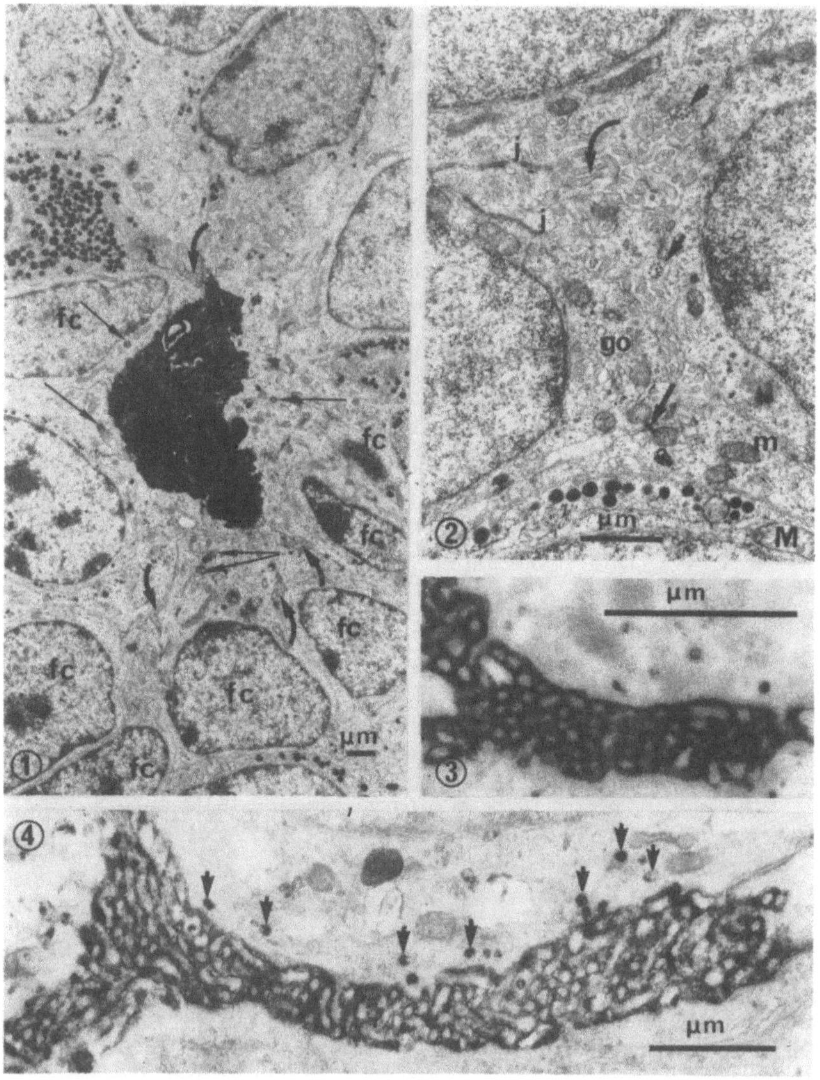

Figs. 1–4. (1) Young male rat (40 g) pituitary. Cluster of follicular cells with angular nuclei (fc) surrounding: i) a cavity filled with recognizable cell debris and fibrillar material, ii) several virtual cavities (curved arrows) with packed microvilli. In the cytoplasm of the follicular cells, centrioles (thin arrows) and glycogen particles. (2) Adult pituitary. Pregnant female. Typical submicroscopic follicle of the adult rat, with interdigitated microvilli (bented arrow) and cilia (arrow heads). Long junction complexes (j) sealing off the branched cavity. Association between 2 mitochondria and a desmosome (thick arrow). Mitochondria of follicular cells (m) are small and dense as compared to the clear ones of endocrine cells (M). A Golgi apparatus (go) located at a follicular pole. (3) Glycocalyx at the apical pole of follicular cells (castrated male rat). Thick cell coat, revealed over the microvilli by the Rambourg's method (PTA at low pH after GMA embedding). The presence of a thick cell coat is a general feature of the follicular cell apex. (4) Permeability of the follicle to exogenous horseradish peroxidase, 15 minutes after intravenous injection. The surface of the numerous microvilli and the narrow spaces between them are labelled. This indicates that the junctions sealing the cavity are not 'tight' and let the tracer free to flow into the follicular lumen. The numerous marked vesicles (arrowheads) show the high endocytotic activity of the follicular cells.

As far as the *size* of the follicular cavity is concerned, one can find submicroscopical, microscopical or 'macroscopical' cavities. These latter can be seen with the unaided eye. The three varieties can be observed in a given pituitary such as in man and fish (11, 16, 17) for instance. On the contrary, the submicroscopical follicle seems to be the normal form in the rat pituitary (Diag. 1). The cavity can even be virtual, entirely filled with interdigitated microvilli (Figs. 1, 2, 3, 4). Although animals show great individual differences, large follicles are few in the rat pituitary.

It is to be noted that the notion of 'follicles' belongs to the former light microscopists. However, the light microscope could not provide any information concerning the stellate cell processes or the

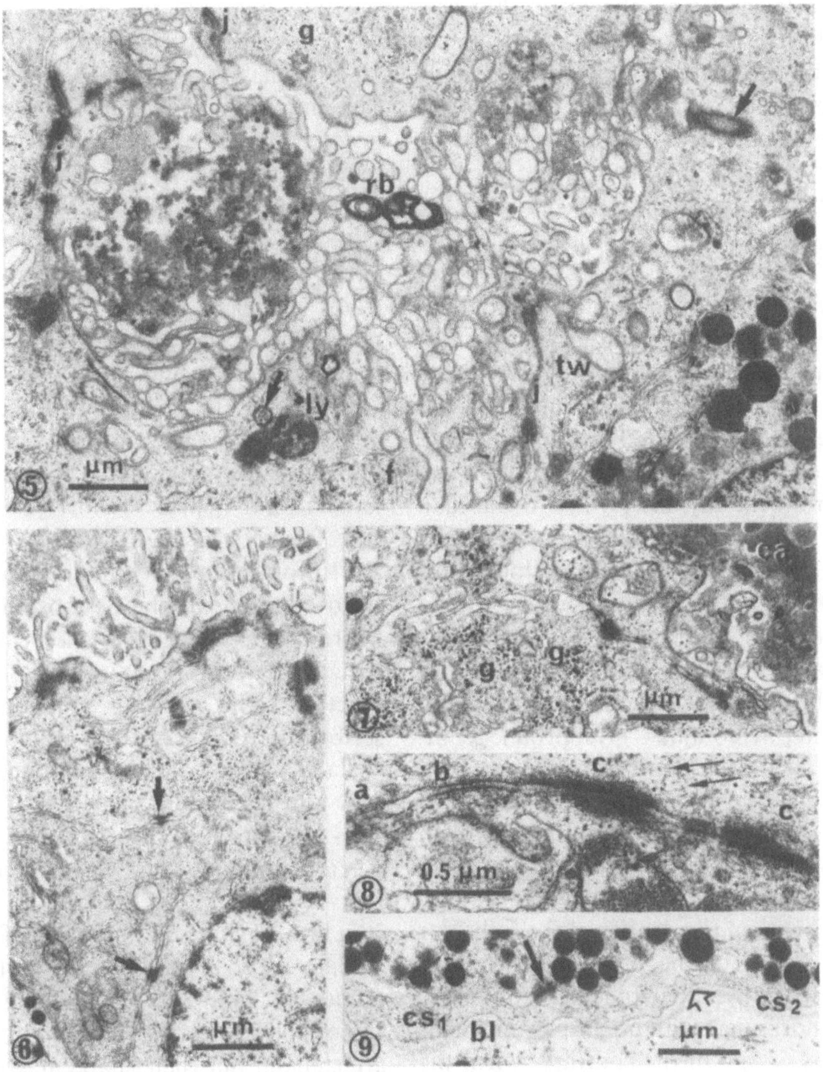

Figs. 5–9. (5) Non tumoral human pituitary (Diabetic patient with retinopathy). Large cavity filled with cell debris and residual bodies (rb). Tortuous infoldings and clear blebs of the apical plasmalemma. Cell junctions (j), cilia (arrows), terminal web (tw), filaments (f), lysosomes (ly) and glycogen particles (g). The endocrine cell (bottom right) does not reach the follicular lumen. (6) Non tumoral human pituitary (Diabetic retinopathy). Follicular pole with long microvilli. The interdigitations of the two adjacent follicular cells are typical of the human pituitary. The lateral densifications of the desmosomes are prominent. Macular junctions (arrows). The apical plasmalemma is contrasted as compared to the infolded lateral plasma membranes. (7) Non tumoral human pituitary (Breast cancer). Two interdigitated follicular cells with a heavy glycogen load (g). The cavity (ca) is filled up with dense debris, the microvilli are short and few. (8) Human pituitary. Portion of a long junction complex: (a) first segment; (b) intermediate segment; (c) desmosomes with particularly dense material in between (see text). Tiny network of filaments (arrows), originating from the lateral densifications of the desmosome. (9) Pituitary adenoma (Cushing's disease). Basal covering. Long cytoplasmic sheets (cs1, cs2) originating from two follicular cells (arrow), intercalated between a granular cell and the pericordonal basal lamina (bl). One macular junction between the endocrine cell and the follicular process (solid arrow).

submicroscopical cavities, which have been further revealed by the electron microscope, thus leading to the concepts of 'folliculo-stellate' cells and network (6).

The *shape* of the cavity varies according to its dimensions: circular when large, thus suggesting a globular volume for the follicle (Diag. 3), tubular and anfractuous when small (Diag. 1, Figs. 2, 3, 4), rising the question of the three-dimensional organization: whether the follicles are isolated units or intercommunicating channels has been debated (6), but new arguments for the existence of a branched system were brought up by different techniques and specially by the scanning electron microscope, which has shown the opening of narrow channels into large follicles (18) or even in the pituitary cleft (19).

Under the electron microscope, the *content* of the cavity may vary from a complete transparency to a high density. Generally, the small cavities appear with a light (Fig. 2), sometimes flocculous content, with some exceptions depending on the conditions of the animal (Fig. 1). In the normal adult rat, as well as in various other species (15, 20), the small follicles contain a low density material.

Although large cavities with a light content are possibly found from the myxine (9, 21) to the man pituitary (Diag. 3), the large follicles are generally cluttered with a dense material, either flocculous, granular (Figs. 6, 7) or fibrillar and often with a considerable amount of cell debris (Fig. 5), in the human pituitary as well as, for instance, in fishes (22), reptiles (23, 24), birds (25) or various mammals (26).

The dense material contains glycoproteic components, responsible for the positive reaction to PAS (light microscopy) and to Rambourg's method (electron microscopy) of the follicular colloid (6, 7) in addition to the strong positive reaction of the thick apical cell coat (Fig. 3).

A relationship between the density of the follicular content and the aspect of the *follicular cell apices* seems to show up. When the follicular content is dense, only few short microvilli or cytoplasmic blebs differentiate (Figs. 5, 7). On the contrary, when the content of the cavity appears to be clear or even flocculous, long microvilli may develop (Figs. 6, 13). However, under all circumstances, the high contrast of the apical plasmalemma is to be noticed. At the follicular pole, together with the microvilli, cilia are found, with reduced patterns such as 9 + 0, 8 + 1 (Figs. 2, 7) or even 6 + 1 (Fig. 5) (6, 27).

2.2.2. Variations of the junctional complexes

Contrarily to the granular endocrine cells, which seem to possess rather few junctional complexes (28), the follicular cells exhibit well-developed junctions, not only at the follicular pole as sealing differentiations for the cavity, but also along the plasmalemma of the processes, associating either follicular elements to each other, or follicular cells to glandular cells (Fig. 9, Diag. 1).

The junctional complexes binding the lateral faces of the follicular cells at their apical pole are composed of several elements, the organization of which depends on the species. From the lumen of the cavity, the first junctional segment is a very close association of the two opposite membranes. After having been regarded as a true 'tight junction' (zonula occludens), this junction is now considered as a punctate fusion point, due to experiments with horseradish peroxidase (see Fig. 4) (6, 7) and to the use of freeze-fracture (29). With this technique, Mira-Moser et al. (29) have demonstrated the presence of gap junctions at the follicular pole level, thus documenting both the tracer experiments and the electronic coupling measurements (30). Such gap junctions have not been found in Teleosteans (31).

Following this first, but not requisite segment, the most important elements are the intermediate junctions (zonulae adherentes). These seem to be characteristic of the species: short and straight in the rat (Fig. 2) or the dog (20) pituitary, very long, tortuous and infolded in the human pituitary (Fig. 5). The third component, the desmosomes, is rarely single. More or less developed, they cannot account for a specific pattern in any species (Fig. 8). Peculiar junctions, the 'spacing junctions', have been described by Abraham et al. (12) in a Teleost, between stellate cells and along the boundary between stellate cells and pericapillary space.

2.2.3. Variations of the folliculo-stellate cell body and processes

It is known that pituitary cells of other types can be stellate-shaped, for instance corticotrope or thyrotrope cells, but they *do not* belong to the network. On the opposite, the great majority of the follicular cells are stellate, but their shape may also depend on the location of their cell body, and on what kind of follicle they belong to.

Sometimes, the cell body is squat, with short and thick processes, as observed in the isolated globulous follicles of the human cystiform zone (type II cell of Ciocca et al. (32); see diag. 3). Most often, the folliculo-stellate cell takes a glial-like aspect (type I of Ciocca et al.; Fig. 4) with long and slender processes, sometimes not wider than 20 nm.

From data provided by thin sections, the question arises whether these processes are cytoplasmic sheets or finger-like or even thread-like extensions *in vivo*. It is known that the narrow ring of cytoplasm encircling the nucleus represents only a minor part of the whole cytoplasm, which spreads over a large area after the merging of the retracted cytoplasmic processes, as shown on dissociated (10, 33) or cultivated follicular cells (34, 35).

The possible existence of 'pure' stellate cells – i.e. cells belonging to the network but not directly related to a follicle – remains still debated, at least in Mammals (6). However, such cells certainly exist in fish pituitary (11, 12, 16, 36) and had been variously called (11).

2.2.4. Variations of the cavity wall

In section 2.1., the cavity wall has been described as constituting a complete barrier under normal conditions. This barrier can even be formed by a two-tiered arrangement of covering cells, as in a Teleost, the alewife (18). In some species, the situation is

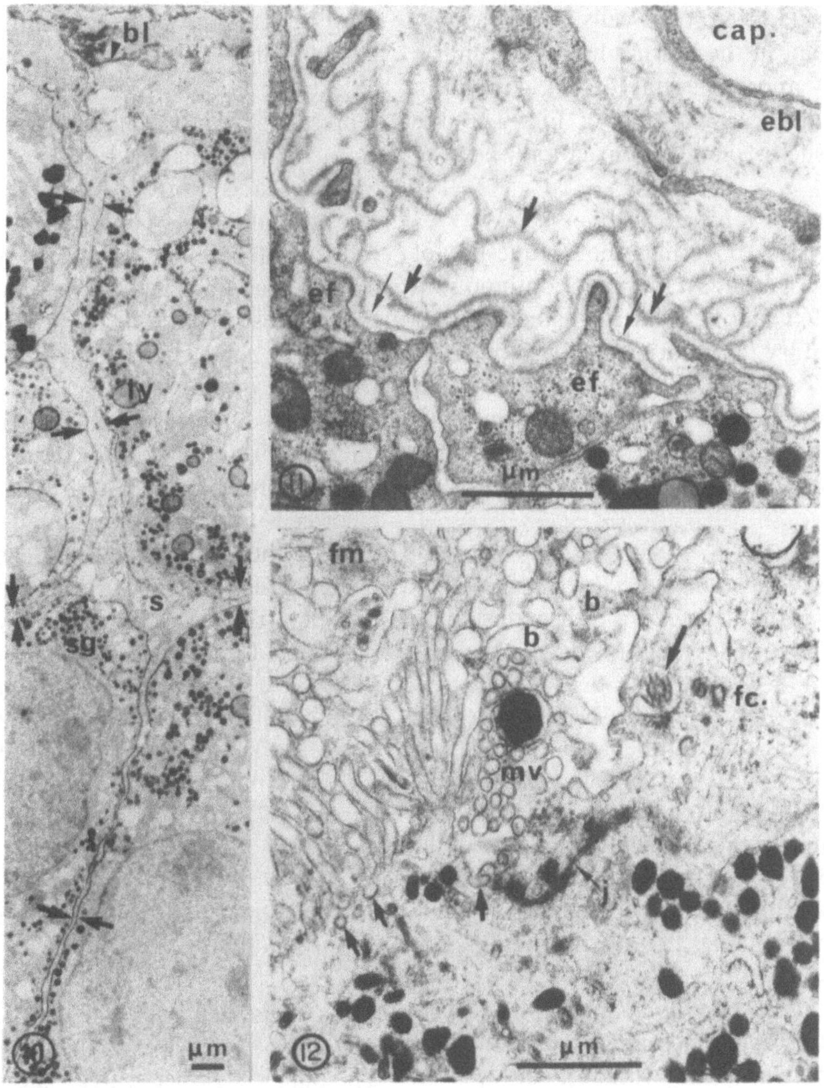

Figs. 10–12. (10) Non tumoral human pituitary (Breast cancer. Rambourg's technique). The staining procedure delineates the plasmalemma of both granular and stellate (s) cells. The slenderness of the cytoplasmic processes is obvious (arrows). Basal lamina (bl), lysosomal formations (ly) and secretory granules (sg) of endocrine cells positively react with the PTA at low pH. (11) Pituitary adenoma (Prolactinoma from a non diabetic female patient, 29 years old). Successive layers of basal lamina, lining two follicular end-feet (ef). The layer close to the cells is thinner (thin arrows) than the outer ones (thick arrows), suggesting their more recent synthesis. Upper right, fenestrated capillary (cap), underlined by a single endothelial basal lamina (ebl). (12) Pituitary adenoma (Cushing's disease). Disrupture of the apical covering: a granular cell projects long microvilli into the follicle cavity. At the basis of these microvilli, coated vesicles and pits (arrowheads) suggesting endocytotic phenomena. Junction complex (j) between this glanular tumor cell and an adjacent follicular cell (fc). Transverse sections of microvilli (mv), blebs (b) and a cilium (arrow). Flocculous material (fm) in the cavity.

unclear, due to the presence of 'cysts' or 'pseudofollicles' (22) in the wall of which are intermingled stellate cells and granular cells of various types. 'It is suggested the term cyst embraces a great variety of pituitary structures that are not necessarily similar in origin and significance in different animals' (22).

In the normal human adult pituitary, the observations are somewhat discrepant, some authors claiming that '... follicular cells are invariably interposed...' (40, 41) while others show the possibility of 'secretory cells in continuity with the colloid' (42). Although the presence of such follicles, including both follicular and granular cells in their cavity wall is still debated in the normal human pituitary, they indisputably appear in fetal (43, 44, 45) and in pathological pituitaries (40, 41, 46); Fig. 12). We propose to call such structures 'mixed-follicles'.

2.2.5. Variations of the basal covering

The presence of a basal covering is both a question of species and of topography. It is almost continuous in the human (Fig. 9, Diag. 3), in the salamander (47) or in the chick pituitary (25), but 'dotted' in numerous other species. In some fish pituitaries, a continuous sheet-like investment is provided by the cytoplasmic processes of stellate cells at the level of the penetration of neurohypophysial elements into the rostral pars distalis (39).

2.2.6. Variations of the follicular cell organelles

The morphology of the organelles is rather constant in the different species. This is valid for the nucleus, the mitochondria and the Golgi apparatus, all characterized by a poor development. On the opposite, the *filament* load appears quite different from one species to another. For instance, the follicular cells of the rabbit (15, 48, 49) of of the turtle (26) appear to contain a huge quantity of thin filaments, and those of the guinea pig (50, 51) exhibit numerous and larger ones (52). On the contrary, in the human pituitary, the filaments and microtubules are always present, but rather few.

The *glycogen load* in the adult pituitary is also a question of species and sometimes of physiological conditions. Prominent in man (Fig. 7), cat (53), fox (54) or dogfish (55) pituitary, the glycogen particles are not found in the follicular cells of adult rat, but only of young or very old animals.

Lysosomal formations as well as different populations of *vesicles* are always observable in the follicular cells, but their development depends on the physiological conditions of the subject (see section 4.5).

2.2.7. Variations of the follicular cells according to the topography of the gland

The distribution of the follicular elements throughout the pituitary gland, depends on the considered species. As early as 1967, Yamada and Yamashita (56) noticed they are 'numerous in the shoulder parts of the anterior lobe (of mice) but very small in number in the other regions'. Forbes (23) reported that 'in Anolis (Reptile) there is a substantial population of stellate cells in the caudal half of the gland, whereas fewer examples of such cells are found in the rostral half'. Opposite observations concern the fish, were follicles can be less prominent in the proximal pars distalis (11, 37), or vary in their aspect from one lobe to the other (55). In mice, the follicular cells represent 5% of the cell population in the pars distalis vs 50% in the pars tuberalis (8) (see chapter 3 by Stoeckel and Porte). Another difference between the pituitary lobes concerns the relationship of the folliculo-stellate network to the marginal cells of the cleft epithelium. The first histologists had considered the macroscopical follicles, for instance those of the cystiform zone in the human pituitary, as residues from the pituitary cleft. Electron microscopy has stated this 'kinship' and extended it to the submicroscopical follicles, since the epithelial cells lining the cleft share all their characteristics with the follicular cells and are part of the network (6, 8, 19, 57, 59). In the rat pituitary, the network of the pars distalis is continuous with the anterior epithelium of the cleft through contacts between processes and junctions. On the contrary, the posterior epithelium underlined by a thick basal lamina cannot be reached by the follicular processes of the intermediate lobe network (6), whereas in the mouse adenohypophysis a discontinuous basal lamina does not constitute a barrier (57).

3. Folliculo-stellate cell cytogenesis and evolution during the life

3.1. Cytogenesis

The data are rather scarce, since the majority of the authors had been more interested in the onset of the granular cells than in the study of the early follicular cells. Reports deal with man (43, 44, 45, 58), rat (59, 60, 61), mouse (62) chick (63) and quail (64), and have lead to contradictory analyses.

3.1.1. Onset of the follicular cells

The onset takes place very early in the proliferating buds originating from the Rathke's pouch epithelium

(59), but the mechanisms appear to be different according to the considered species. In the human embryo, the follicles get organized around canaliculi extending from the Rathke's pouch (58), whereas in the fetal rat, the follicles originate 'in situ' from the forming epithelial cords (59, 60).

3.1.2. Significance of the follicle lining cells

According to Yoshida (59) and to Svalander (60), the cells enclosing the first follicles are similar to the marginal cells of the residual cleft and constitute a specific cell line. This line results from the differentiation of the primordial cells of the Rathke's pouch and differs from the granular cell lines. The agranular cells are differentiated with their microvilli on the 14th day and with glycogen between days 17th and 20th for Yoshida and on day 16th for Svalander (60) and Yamashita (62).

On the opposite, for Yoshimura et al. (61), the follicular cells which appear from day 17th are not differentiated cells. They correspond to proliferating primordial cells tending to 'congregate' into clusters and to surround a pseudolumen equipped with microvilli, which 'must not be regarded as the specialized structure of the mature follicular cells'. 'Pituitary marginal cells resemble in fine structure the follicular cells. The former is a residue of the anterior wall of Rathke's pouch while the latter is a cluster of the cells differentiating for the specific purpose from the cell-cords ... The follicular cells located at the periphery of a cluster begin to granulate ... and finally transform into immature acidophils or basophils ... The follicular cells may be regarded as the stem cells of the pituitary'.

Andersen et al. (58) consider that the agranular cells lining the follicles have the same proliferative capacity as the lining cells of the cleft. They are undifferentiated and are the real progenitors of the granular adenohypophyseal cells. These forming granular cells can even participate in the lining of the follicles. A similar observation of 'mixed follicles' has been reported in the chick embryo (63).

Another interpretation is proposed in a study on the *in vitro* development of the quail Rathke's pouch (64). The follicular cells proliferating at the periphery of the explant could derive from undifferentiated epithelial cells, although other origins, for example dedifferentiation of glandular cells are not precluded.

3.2. Evolution of folliculo-stellate cells during the life

This problem has been considered only in rat and man pituitary, and three possibilities have been proposed.

3.2.1. First theory: early differentiated cells

In the first possibility, the folliculo-stellate cells remain as a differentiated cell line all along the life cycle. Their renewal is either nonexistent or inappreciable with present techniques. Some arguments can be brought up to support this idea:

a) the morphological characteristics of the follicles and folliculo-stellate cells are clear-cut, during growth and adulthood as well.

b) the ratio of follicular cells to granular cells decreases during the pituitary development from 25 to 30% in the neonate rat, to 17% in the young and 8% in adult (6). Evaluations in adult mouse, 5% (8), 6 to 10% (59) and in man pituitary, 11% (32) are similar. The decrease seems related to the proliferation of granular cells (6, 8, 59, 62) and not to the transformation of follicular cells into granular cells, since no cell with an intermediate structure was found (8, 65).

c) no mitose can be detectable in the follicular cells (personal observations), contrarily to the frequently dividing granular cells. This is supported by the failure in labelling the follicular cells with tritiated thymidine, thus demonstrating that they are not stem cells (66, 67).

d) recent immunocytochemistry works on mammal pituitary favor this idea of a definite and early differentiated cell line, through the labelling of the whole folliculo-stellate system with an anti-S-100 protein antibody in the pars distalis (68, 69) and in the pars intermedia as well (70). However, in the pars intermedia, another population of stellate cells exhibits a strong reaction with antiserum to GFA (gliofibrillar acid protein), thus suggesting their astrocytic nature (70). As the brain specific S-100 protein may be regarded as a molecular marker for cells of neuroectodermal origin, such an origin can be inferred for the positive folliculo-stellate cells of the pituitary. This brings up a new argument to consider these cells as an independent cell line (19). Moreover, since no cell contains both S-100 and secretory granules, the absence of transitional cells between stellate and granular cells suggests that the competence of the former to transform themselves into granular cells can be discarded (68).

3.2.2. Second theory: stem cells

A completely opposite theory is proposed by Yoshimura et al. (61, 71), who consider the follicular cell as a stem cell, keeping constant potentialities throughout the life cycle of the rat (see 2.1.2.). These cells may freely migrate and occasionally constitute a reticular framework. Their main argument rests upon observations of a mitotic activity of the follicular cell clusters, and relies on a paper from Shioda (72); but this latter work was performed in light microscopy and reported only on 'non specific mitoses' in chromophobic cells' after a tritiated-thymidin labelling. Nevertheless, Yoshimura et al. concluded that 'proliferation, due to mitotic division, and differentiation into the chromophils are two major activities inherent in the follicular cells'.

3.2.3. Third theory: non permanent structures
A third position has been to consider the follicles as non permanent structures in the pituitary, and the follicular cells as resulting from the dedifferentiation of impaired granulated cells (46). From their morphological data on human pituitaries, the authors have described a three-steps transformation and they argue from the presence of 'mixed-follicles' (see section 3.1.2.) to suggest a neoformation of follicles induced by the leaking of substances from injured cells, which trigger a mechanism leading to the formation of junctions.

However, the early differentiation of the complex folliculo-stellate cell system during the fetal life (see 3.1.1.) as well as its specific embryonic origin (see 3.2.1.d) do not fit with the hypothesis of non permanent structures. Moreover, the presence of 'mixed-follicles' is not the rule in the adult human normal pituitary (32, 40, 41) and, the cases of Cushing's disease excepted, most of the adenomatous pituitaries with a well-preserved tissue organization do not present 'mixed' but 'pure' follicles (41). We have also observed that some adenomas may acquire a 'follicular' arrangement: there, abnormal *granular* cells differentiate junctions and get organized around pseudo-follicles ('follicular adenomas', 41), but such cells do not belong to the folliculo-stellate network.

4. Possible functions of the folliculo-stellate cells

'The follicular cells ... represent a distinct, mature, pituitary cell type ... in search of a function' (20). Numerous roles have been proposed, inferred either from morphological, physiological or experimental data. They can be distributed under five headings: supporting role, contractile activity and motility, macrophagic and endocytotic functions, implication in the metabolism of pituitary fluids, and interrelations with endocrine cells.

The specific organization of the folliculo-stellate cells had suggested analogies with the organization of glial cells ('glial-like cells': 68, 70, 73, 74, 75) or of Sertoli cells as well (6, 10).

4.1. Supporting role

The possibility of a supporting role, evidently related to the network organization of the system, and also to the presence of junctional complexes and of a prominent cytoskeleton, has been proposed since 1965 (4, 6, 8, 16, 23, 24, 44, 47, 48, 49, 55, 60, 76, 77). In addition, the disposition of the end-feet upon the parenchymal basal laminae and the presence of thickenings over the plasmalemma, similar to hemidesmosomes (unpublished observations), at the contact areas, suggest such a function. However, the distinct inequality described in some species in the distribution of stellate cells (23) does not favor an exclusive supporting role.

4.2. Contractile activity and motility

Contractile properties of the folliculo-stellate cells are highly probable, as suggested by the prominence of filaments bundles (14, 37, 76, 78). It can be assumed that the agranular cells as a whole are motile and that they are capable of changes in form, especially the form and length of their cell protrusions. An influence on the shape of the follicle may also be considered (14, 18). The movements may occur during the phagocytic activity (10, 74, 78). They can also provide an explanation to the migration of follicular cells toward the periphery of cultivated explants (10, 33, 34, 35). Another argument accounting for possible movements of the end-feet, concerns the observation of multilayered basal laminae in the pituitary of elderly and diabetic patients (Fig. 11) suggesting the deposit of successive levels of material, due to retraction movements of the end-feet. This suggests an implication of the folliculo-stellate end-feet in the basal lamina synthesis (79, 80). We also suggest that a shift of some follicle lining cells may lead to the disrupture of the apical covering and thus to the formation of 'mixed-follicles'.

4.3. Endocytotic activity: phagocytosis and pinocytosis

4.3.1. Phagocytosis
The phagocytic activity of the follicular cells is now unanimously admitted. Reported for the first time in the folliculo-stellate cells of Mammals in vivo (65, 81), this phenomenon has been later evidenced in adult (5 to 8, 14, 15, 77, 80) as well as in developing pituitary (58, 60, 62) and in the stellate cells of non-mammalian species (17, 74, 82). The phagocytosis can be brought into light under certain physiological conditions, for instance after weaning in the female rat (83) or before spawning in the lamprey (78), as well as

after various experimental procedures such as endocrine stimulations through ablation of target organs (6, 8, 23, 84) or pituitary stalk section (73, 85; for review, see 74).

The phagocytosis has also been studied in vitro, either in incubated (8, 10, 13, 86) or transplanted (8) pituitary fragments, or trypsinized cells (8, 10).

The phagocytosis activity concerns either cell debris or complete cells. It occurs through a processus of encircling the injured elements by the follicular cells processes, followed by internalization into a large phagocytic vacuole within the follicular cell (10). As described in a short term incubation system, among the injured granular cells of the central necrotic portions of the explant, the follicular cells are hypertrophied, and a close correlation can be stated between the cytological response and the position of the cells within the explant (86). This may be due to differences in metabolic requirements between the follicular cells which are glycogen-loaded, and the granular cells affected by the lack of oxygen and energy reserves (10).

The fate of the phagocytosed material remains unclear, since one can frequently observe ghosts of organelles or cells within the follicular lumen. Two possibilities may be thought of: either a release of the phagocytic vesicle content, or a transfer of a follicular cell toward the lumen, after its transformation into a macrophage. As far as the presence of necrotic granular cells within the lumen is concerned, the disrupture of the follicular apical barrier may be involved.

4.3.2. Endocytosis

Besides those possible functions of phagocytosis and elimination of waste, the folliculo-stellate cells achieve an active pinocytosis. On routine samples, at the apical pole of the cells, coated and uncoated vesicles and pits located at the basis of the microvilli can be pointed out (Fig. 12). Such vesicles have been related to exchange phenomena between the lining cells and the content of the cavity, either from the colloid to the cell (25, 42, 74, 77) or in the reverse way (4). But the endocytosis is not restricted to the apical pole of follicular cells, and microvesicles have also been observed along the plasmalemma of end-feet (23, 24, 87) and of other sites of stellate cell membranes (25, 11, 87).

In the pituitary of mammals injected with horseradish peroxidase, the tracer immediately diffuses in the whole system of the intercellular spaces and into the follicular lumina (6, 7). The internalization rate of the tracer is much higher in the folliculo-stellate cells than in the granular cells, and the vesicles corresponding to those previously observed on routine samples are now labelled with the reaction product (Fig. 4). In rat pars distalis, the uptake starts earlier and is of longer duration (personal observations). In the neuro-intermediate lobe of a reptile, the phenomenon seems to be slower, and to increase after transection of the hypophysial stalk (85). Other experiments have been performed with peroxidase on nonmammal pituitaries (11, 23, 70, 88, 89), and have led Abraham et al. (88) to the conclusion of a permanent flow of molecules to and from the cells. However, if one knows where this flow comes from, we do not know where it flows to.

4.4. Implication in the exchanges and transport of pituitary fluids

Such a function has been suggested in numerous papers, and has been well documented by the tracer experiments (see 4.3.2.): 'a continuous pathway indeed exists between pericellular and pericapillary spaces' (88), and also between this first system and the system of follicular cavities and pituitary cleft, through the junctional complexes of the apical poles (6, 7). Another argument is brought up by the presence of blood proteins within the cleft. This accumulation of proteins is related to salt resistance or susceptibility of Dahl strain rats (90). Relationship between the behaviour of stellate cells and variations of the osmotic pressure of the medium have been stated in fish, either according to their environment (18, 91) or to the composition of culture media (92). The modifications can affect the number of the folliculo-stellate cells (12), their shape and size (92) or their luminal volume. This last point may reflect changes in the composition of the follicle fluid (18), and is in agreement with the results concerning the cleft, thus suggesting once more the 'kinship' between cleft and follicles, and between marginal cells of the cleft and folliculo-stellate cells.

Another possible intervention of the folliculo-stellate cells in the metabolism and circulation of pituitary fluids might be related to the ionic requirements of the secretory process, specially as far as calcium is concerned (30, 70). Moreover, the localization of the Na^+–K^+–ATPase on the plasma membranes of the pars intermedia stellate cells has suggested both their participation in the regulation of the extracellular ionic environment and their similarity to the glial cells of the central nervous system (75).

4.5. Interrelations with endocrine cells

The possibility of a ionic regulation raises the question of a functional association between the folliculo-stellate and the endocrine cells. Such a role has been frequently evoked, either as related to the providing of nutrients (2, 15) and/or neurohormones (31) to secretory cells, or to the transport of hormonal products toward the vascular network. The association of stellate cell processes with exocytosis sites (11, 38) has suggested on the one hand a facilitation of granule release due to changes induced in the properties of the granulated cell membranes, and on the other hand, an uptake of the extruded material (11; see 4.3.2.).

Approaches by stimulation of endocrine cells, either physiological (6, 8, 15) or experimental, through stress (93) and ablation of target organs (5, 6, 8, 15, 23, 65, 66, 72, 84), have led to the disappointing finding of a similar response, whatever the operating way (6, 8): the folliculo-stellate cells hypertrophy, develop prominent lysosomes and multiply their microvilli and filaments.

From this bulk of works, if a consensus has been found concerning the identity of the folliculo-stellate system in the pituitary of all Vertebrates, we only have a sketchy picture of the situation; the real role played by these cells still remains unclear and open to further investigation, all the more since similar structures exist in other endocrine glands (94).

Acknowledgements

The assistance of Mrs A. Combrier, Mrs D. Nève and Mrs B. Roncier in preparing the manuscript is gratefully acknowledged.

References

1. Rinehart JF, Farquhar MG: Electron microscopic studies of the anterior pituitary gland. J Histochem Cytochem 1: 93–113, 1953.
2. Rinehart JF, Farquhar MG: The fine vascular organization of the anterior pituitary gland. An electron microscopic study with histochemical correlations. Anat Rec 121: 207–240, 1955.
3. Farquhar MG: 'Corticotrophs' of the rat adenohypophysis as revealed by electron microscopy. Anat Rec 127: 291 (Abstr), 1957.
4. Kagayama M: The follicular cell in the pars distalis of the dog pituitary gland: an electron microscope study. Endocrinology 77: 1053–1060, 1965.
5. Farquhar MG: Processing of secretory products by cells of the anterior pituitary gland. Mem Soc Endocrinol 19: 79–124, 1971.
6. Vila-Porcile E: Le réseau des cellules folliculo-stellaires et les follicules de l'adénohypophyse du rat (Pars distalis). Z Zellforsch 129: 328–369, 1972.
7. Vila-Porcile E: La pars distalis de l'hypophyse chez le rat. Contribution à son étude histologique et cytologique en microscopie électronique. Ann Sc Nat (Zool) 15: 61–138, 1973.
8. Dingemans KP, Feltkamp CA: Nongranulated cells in the mouse adenohypophysis. Z Zellforsch 124: 387–405, 1972.
9. Holmes RL, Ball JN: The pituitary gland. A comparative account. In: Biological structure and function, 4 Cambridge, Mass, University Press 1974.
10. Farquhar MG, Stutelsky EH, Hopkins CR: Structure and function of the anterior pituitary and dispersed pituitary cells. In vitro studies. In: The anterior pituitary gland. Tixier-Vidal A, Farquhar MG (eds), New-York, Academic Press, 1975, pp 82–135.
11. Leatherland JF, Percy R: Structure of the nongranulated cells in the hypophyseal rostral pars distalis of cyclostomes and actinopterygians. Cell Tissue Res 166: 185–200, 1976.
12. Abraham M, Dinari-Lavie V, Lotan R: The pituitary of Aphanius dispar (Rüppell) from hypersaline marshes and freshwater. II Ultrastructure of the rostral pars distalis. Cell Tissue Res 179: 317–330, 1977.
13. Rawdon BB: Ultrastructure of the nongranulated hypophysial cells in the teleost Pseudocrenilabrus philander (Hemohaplochromis philander) with particular reference to cytological changes in culture. Acta Zool (Stockh) 59: 25–33, 1978.
14. Schultz HJ, Patzner RA, Adam H: Fine structure of the agranular adenohypophysial cells in the hagfish, Myxine glutinosa (Cyclostomata). Cell Tissue Res 204: 67–75, 1979.
15. Shiotani Y: An electron microscopic study on stellate cells in the rabbit adenohypophysis under various endocrine conditions. Cell Tissue Res 213: 237–246, 1980.
16. Lagios MD: Follicle boundary cells in the adenohypophysis of the chondrostean and holostean fishes: an ultrastructural study of their relationship to the follicular lumen, to endocrine cells, and to the hypophysial cleft. Gen Comp Endocrinol 20: 362–376, 1973.
17. Benjamin M: The origin of pituitary cysts in the rostral pars distalis of the nine-spined stickleback, Pungitius pungitius L. Cell Tissue Res 214: 417–430, 1981.
18. Betchaku T, Douglas WW: Cellular composition of the rostral pars distalis of the anterior pituitary gland of alewife, Alosa pseudoharengus, during the spawning run. Anat Rec 199: 403–421, 1981.
19. Correr S, Motta PM: The rat pituitary cleft: a correlated study by scanning and transmission electron microscopy. Cell Tissue Res 215: 515–529, 1981.
20. Nunez EA, Gershon MD: Specific paracristalline structures of rough endoplasmic reticulum in the follicular (stellate) cells of the dog adenohypophysis. Cell Tissue Res 215: 215–221, 1981.
21. Fernholm B: The ultrastructure of the adenohypophysis of Myxine glutinosa. Z Zellforsch 132: 451–472, 1972.
22. Benhamin M, Williams JG: Pituitary cysts in the nine-spined stickleback, Pungitius pungitius L. II. Light and electron microscopy. Acta Zool (Stockh) 60: 241–250, 1979.
23. Forbes MS: Fine structure of the stellate cell in the pars distalis of the lizard, Anolis carolinensis. J Morphol 136: 227–245, 1972.
24. Tseng MT, Yntema CL: Fine structure of the chromophobe in the pars distalis of the common snapping turtle, Chelydra serpentina. Cell Tissue Res 166: 235–240, 1976.
25. Jover Moyano A, Riviera Pomar JM: Ultrastructura de las cavidades y celulas foliculares en la hipofisis del pollo (Gallus domesticus). Anales de Anatomia (Zaragoza) 19: 61–73, 1970.
26. Girod C. Lhéritier M: Ultrastructure des cellules folliculo-stellaires de la pars distalis de l'hypophyse chez le spermophile (Citellus variegatus Erxleben), le graphiure (Graphiurus murinus Desmaret) et la hérisson (Erinaceus europaeus linnaeus). Gen Comp Endocrinol 43: 105–122, 1981.
27. Dubois P, Girod C: Les cellules ciliées de l'antéhypophyse. Etude au microscope électronique. Z Zellforsch 103: 502–517, 1970.
28. Herbert DC: Intercellular junctions in the Rhesus monkey pars distalis. Anat Rec 195: 1–6, 1979.
29. Mira-Moser F, Schofield JG, Orci L: Tight junctions between follicular cells of the anterior pituitary gland: a freeze-fracture study. J Microscopie 22: 117–120, 1975.
30. Fletcher WH: Intercellular communication in the rat anterior pituitary gland. An in vivo and in vitro study. J Cell Biol 67: 469–476, 1975.
31. Abraham M: The ultrastructure of the cell types and the neurosecretory innervation in the pituitary of Mugil Cephalus L from fresh water, the sea, and a hypersaline lagoon. I. The rostral pars distalis. Gen Comp Endocrinol 17: 334–350, 1971.
32. Ciocca DR, Rodriguez EM, Cuello CA: Comparative light and electron microscopical study of the normal adenohypophysis in the human. Acta Anatom 103: 83–99, 1979.

33. Hopkins CR, Farquhar MG: Hormone secretion by cells dissociated from rat anterior pituitaries. J Cell Biol 59: 276–303, 1973.
34. Tixier-Vidal A: Ultrastructure of anterior pituitary cells in culture. In: The anterior pituitary gland. Tixier-Vidal A, Farquhar MG (eds), New-York, Academic Press, 1975, pp 181–229.
35. Li JY: L'antéhypophyse foetale humaine in vitro. Etude morphologique et fonctionnelle. Mémoire DERBH, Univ Claude Bernard Lyon 1 France, 1980.
36. Follenius E: Personal communication, 1981.
37. Batten T, Ball JN, Benjamin M: Ultrastructure of the adenohypophysis in the teleost Poecilia latipinna. Cell Tissue Res 161: 239–261, 1975.
38. Batten T, Ball JN, Grier HJ: Circadian changes in prolactin cell activity in the pituitary of the teleost Poecilia latipinna in freshwater. Cell Tissue Res 165: 267–280, 1976.
39. Betchaku T, Douglas WW: Fine structure of the rostral pars distalis of the adenohypophysis of the killifish, Fundalus heteroclitus, in fresh and salt water. Anat Rec 198: 595–609, 1980.
40. Bergland RM, Torack R: An ultrastructural study of follicular cells in the human anterior pituitary. Am J Pathol 57: 273–297, 1969.
41. Olivier L, Vila-Porcile E, Racadot O, Peillon F, Racadot J: Ultrastructure of pituitary tumor cells: a critical study. In: The anterior pituitary gland. Tixier-Vidal A, Farquhar MG (eds), New-York, Academic Press, 1975, pp 231–276.
42. Paiz C, Hennigar GR: Electron microscopy and histochemical correlation of human anterior pituitary cells. Am J Pathol 59: 43–52, 1970.
43. Li JY, Dubois MP, Dubois PM: Ultrastructural localization of immunoreactive corticotropin, β-lipotropin, α and β-endorphin in cells of the human fetal anterior pituitary. Cell Tissue Res 204: 37–51, 1979.
44. Andersen H, von Bülow FA, Møllgård H: The histochemical and ultrastructural basis of the cellular function of the human foetal adenohypophysis. Progr Histochem Cytochem 1: 153–184, 1970.
45. Fukuda T: Agranular stellate cells (so-called follicular cells) in human fetal and adult adenohypophysis and in pituitary adenomas. Virchows Arch Abt A Path Anat 359: 19–30, 1973.
46. Horvath E, Kovacs K, Penz G, Ezrin C: Origin, possible function and fate of 'follicular cells' in the anterior lobe of the human pituitary. Am J Path 77: 199–205, 1974.
47. Cardell RR Jr: The ultrastructure of stellate cells in the pars distalis of the salamander pituitary gland. Am J Anat 126: 429–456, 1969.
48. Salazar H: The pars distalis of the female rabbit hypophysis; an electron microscopic study. Anat Rec 147: 469–497, 1963.
49. Schechter J: The ultrastructure of the stellate cell in the rabbit pars distalis. Am J Anat 126: 477–488, 1969.
50. Smith RE: An electron microscopic study of the adenohypophysis of the guinea pig. Anat Rec 145: 352 (Abstr), 1963.
51. Young BA: Some observations on the ultrastructure of the stellate cells of the pars distalis of the guinea pig. J Anat (London) 124: 153–156, 1977.
52. Nickerson PN: Filament containing chromophobe in anterior pituitary of the guinea pig. Tiss and Cell 6: 663–668, 1974.
53. Olivier L, Vila-Porcile E, de Brye C, Nouet JC: Les cellules folliculaires du lobe antérieur de l'adénohypophyse du chat adulte. Bull Ass Anat (Nancy) 152: 814–815, 1971.
54. Bugnon C: Personal communication, 1981.
55. Alluchon-Gérard MJ: Ultrastructure of the dogfish adenohypophysis. Cell Tissue Res 193: 139–154, 1978.
56. Yamada K, Yamashita K: An electron microscope study on the possible site of production of ACTH in the anterior pituitary of mice. Z Zellforsch 80: 29–43, 1967.
57. Kagayama M, Ando A, Yamamoto TY: On the epithelial lining of the cleft between pars distalis and pars intermedia in the mouse adenohypophysis. Gunma Symp Endocrinol 6: 125–136, 1969.
58. Andersen H, von Bülow FA, Møllgård K: The early development of the pars distalis of human foetal pituitary gland. Z Anat Entwickl Gesch 135: 117–138, 1971.
59. Yoshida Y: Electron microscopy of the anterior pituitary gland under normal and different experimental conditions. Meth Achievm exp Path 1: 439–454, 1966.
60. Svalander Ch: Ultrastructure of the fetal rat adenohypophysis. Acta Endocrinol (Kbh) 76: suppl 188, 1–113, 1974.
61. Yoshimura F, Soji T, Sato S, Yokoyama M: Development and differentiation of the rat pituitary follicular cells under normal and some experimental conditions with special reference to an interpretation of renewal cell system. Endocrinol Japon 24: 435–449, 1977.
62. Yamashita K: Electron microscopic observations on the postnatal development of the anterior pituitary of the mouse. Gumna Symp Endocrinol 6: 177–196, 1969.
63. Franco N, Guédenet JC, Grignon G: Etude ultrastructurale des cellules folliculaires de l'adénohypophyse du poulet. Bull Ass Anat (Nancy) 60: 515–526, 1976.
64. Frémont PH, Ferrand R: The differentiation of follicular-like cells from the epithelium of Rathke's pouch grown in vitro. Anat Embryol 160: 275–284, 1980.
65. Bhattacharjee DK, Chatterjee P, Holmes RL: Follicles and related structures in the pars intermedia of the jird (Meriones unguiculatus). J Anat (Lond) 130: 63–67, 1980.
66. Dingemans KP: Undifferentiated cells in the mouse adenohypophysis. In VIIth Congr Intern Micr Electr, Favard P (ed), Paris, Soc Fr Micr Electr, 1970, vol 3, pp 563–564.
67. Stratmann IE, Ezrin C, Sellers EA, Simon GT: The origin of thyroidectomy cells as revealed by high resolution radioautography. Endocrinology 90: 728–734, 1972.
68. Cocchia D, Miani N: Immunocytochemical localization of the brain specific S-100 protein in the pituitary gland of adult rat. J Neurocytology 9: 771–782, 1980.
69. Nakajima T, Yamaguchi H, Takamashi K: S-100 protein in folliculostellate cells of the rat pituitary anterior lobe. Brain Res 191: 523–531, 1980.
70. Ştoeckel ME, Schmitt G, Porte A: Fine structure and cytochemistry of the mammalian pars intermedia. In: Peptides of pars intermedia, Ciba Foundation Symposium 81, London, Pitman Medical, 1981, pp 101–127.
71. Yoshimura F, Soji T, Kiguchi Y: Relationship between the follicular cells and marginal layer cells of the anterior pituitary. Endocrinol Japon 24: 301–305, 1977.
72. Shioda T: Autoradiography of tritiated thymidine labelled anterior pituitary cells in thyrotrophin releasing hormone treated rats. In Proc 10th Intern Congr Anat, Yamada E (ed), Tokyo, Science Council of Japan, 1975 p 279.
73. Castel M: Ultrastructure of the anuran pars intermedia following severance of hypothalamic connection. Z Zellforsch 131: 545–557, 1972.
74. Perryman EK, De Vellis J, Bagnara JT: Phagocytic activity of the stellate cells in the anuran pars intermedia. Cell Tissue Res 208: 85–98, 1980.
75. Semoff S, Hadley ME: Localization of ATPase activity in the glial-like cells of the pars intermedia. Gen Comp Endocrinol 35: 329–341, 1978.
76. Yamashita K: Electron microscopic observations on the anterior pituitary of the crab-eating monkey. Okaj Fol Anat Jap 43: 299–323, 1967.
77. Foster CL: Relationship between ultrastructure and function in the adenohypophysis of the rabbit. Mem Soc Endocrinol 19: 125–146, 1971.
78. Båge G, Fernholm B: Ultrastructure of the pro-adenohypophysis of the river lamprey, Lampetra fluviatilis, during gonad maturation. Acta Zool (Stockh) 56: 95–118, 1975.
79. Olivier L, Vila-Porcile E, Racadot O, Lesobre B, Canivet J: Relations between the abnormal parenchymatous basal laminae and the follicular cells in the pituitary of diabetic patients with retinopathy. Diabetologia 11: 367 (abstr), 1975.
80. Vila-Porcile E: Morphological and functional relationships between the different compartments of the anterior pituitary. Proc of the Xth Intern Congr of Anat, Yamada E (ed), Tokyo, Science Council of Japan, 1975, p 24.
81. Young BA, Foster CL, Cameron E: Some observations on the ultrastructure of the adenohypophysis of the rabbit. J Endocrinol 31: 279–287, 1965.
82. Rawdon BB: Ultrastructural observations on nongranulated cells in the adenohypophysis of the cichlid, Hemihaplochromis philander. Gen Comp Endocrinol 29: 261–262, 1976.
83. Vila-Porcile E, Olivier L, Racadot O: Exocytose polarisée des corps résiduels lysosomiaux des cellules à prolactine dans l'adénohypophyse de la ratte en post-lactation. C R Acad Sci (Paris) 276: 355–357, 1973.
84. Surks MI, De Fesi ChR: Determination of the cell number of each cell type in the anterior pituitary of euthyroid and hypothyroid rat. Endocrinology 101: 946–958, 1977.
85. Larsson L: Control of the pars intermedia of the lizard, Anolis carolinensis V Extravascular transfer and cellular uptake of horseradish peroxidase in the neuro-intermediate lobe. Cell Tissue Res 214: 1–22, 1981.
86. Yamashita K: Fine structure of the mouse anterior pituitary maintained

in a short term incubation system. Z Zellforsch 124: 465–478, 1972.
87. Perryman EK: Ultrastructure of the stellate cell in the pars intermedia of the frog, Rana pipiens. Cell Tissue Res 164: 387–399, 1975.
88. Abraham M, Kieselstein M, Lisson-Begon S: The extravascular channel system in the rostral pituitary of Mugil cephalus (Teleostei) as revealed by use of horseradish peroxidase. Cell Tissue Res 167: 289–296, 1976.
89. Perryman EK, Bagnara JT: Extravascular transfer within the anuran pars intermedia. Cell Tissue Res 193: 297–313, 1978.
90. Rapp JP, Bergon L: Characteristics of pituitary colloid protein and their correlation with blood pressure in the rat. Endocrinology 101: 93–103, 1977.
91. Leatherland JF, Ball JN, Hyder M: Structure and fine structure of the hypophyseal pars distalis in indigenous african species of the genus Tilapia. Cell Tissue Res 149: 245–266, 1974.
92. Benjamin M, Baker BI: Ultrastructural studies on prolactin and growth hormone cells in anguilla pituitaries in vitro. Cell Tissue Res 174: 547–564, 1976.
93. Rennels EG: Electron microscopic alterations in the rat hypophysis after scalding. Amer J Anat 114: 71–91, 1964.
94. Boquist L: Follicles in human parathyroid glands. Lab Invest 28: 313–320, 1973.

Authors' address:
Laboratoire d'Histologie et d'Embryologie,
Faculté de Médecine Pitié-Salpêtrière
105 Boulevard de l'Hôpital
75634 PARIS Cedex 13 - France

CHAPTER 7

Ultrastructure of the human neurohypophysis

YOSHIO TAKEY and GARY S. PEARL

1. Introduction

The ultrastructure of the neurohypophysis has been extensively studied in non-human mammals as well as non-mammalian species in the past. By contrast, studies on human tissue have been only rarely conducted. This can largely be attributable to the difficulty in obtaining the normal human neurohypophysis suitable for electron microscopy. However, with the advent of hypophysectomy for the palliative treatment of patients with diabetic retinopathy or with metastasizing carcinomas of breast or prostate, the problem has been alleviated and adequate tissue has become available for ultrastructural study (1–6).

In the human pars nervosa, there are two parenchymal elements that are unique to this organ, nerve fibers and pituicytes. The nerve fibers are axons of hypothalamic neurons in the supraoptic and paraventricular nuclei. It is now understood that the neurohypophysial hormones and carrier proteins are produced in the perikarya of these neurons and transported through the axons to the pars nervosa in the form of neurosecretory granules (NSGs). It is still unclear, however, as to the exact mechanism(s) by which the hormones in NSGs are released into the extracellular space and then to the blood stream.

In this chapter, we summarize our recent studies on the ultrastructure of the human neurohypophysis (4–6) with a review of the literature in an attempt to synthesize these data into a concept of the functional morphology regarding transport, storage and release of the neurohypophysial hormones in the human pars nervosa.

2. Nerve fibers

The nerve fibers in the human neurohypophysis are unmyelinated, neurosecretory axons measuring 0.5 – 1,0 μ in diameter. They contain parallel rows of microtubules measuring 22–24 nm in diameter as well as scattered fine filaments. These axons, as well as those in non-human species, display focal enlargements or dilatations (1, 3, 4, 7–16), which may occur in the course of the axon or at the terminal. The largest of these dilatations correspond to the 'Herring bodies' described at the light microscopic level (17). The microfilaments and tubules are located at the periphery of the enlargements, suggesting that they are bulbous protrusions of the axolemmal surface rather than fusiform enlargement of the entire axis cylinder (4, 7). The dilatations contain NSGs, as well as mitochondria, endoplasmic reticulum, ribosomes and lysosomes in varying proportions.

The NSGs, which are also found in smaller numbers in the undilated portions of the axons, are divided into two groups in the human neurohypophysis (1, 4): Type A NSGs, measuring 100–300 nm in diameter, and type B NSGs, measuring 50–100 nm in diameter. Type A NSGs have a central core of varying electron density, whereas type B NSGs have a more uniform, denser core. NSGs resembling those of type A are present in the human fetus (18) but both types are seen in the fetal monkey (19). They are somewhat smaller than those in the adult but exhibit an increase in diameter with age, as in the rat (20, 21), reaching the adult size by approximately 19 weeks of gestation. Since the two types of granules never coexist in the same axon, neurosecretory fibers can be classified as type A or type B fibers depending on their NSG content (1, 4). Type A fibers are more abundant than type B fibers in the pars nervosa (4), whereas type B fibers predominate in the infundibular region (1).

2.1. *Type A fibers and their dilatations*

The NSGs of type A fibers exhibit a wide range of variability in the electron density of their core material and their seems to be an inverse relationship between the electron density of the core and the size of the NSG.

Motta, PM (ed): Ultrastructure of endocrine cells and tissues. ISBN-13: 978-1-4613-3863-5

Another important feature in type A fibers is the presence of microvesicles measuring 30–40 nm in diameter. They lack an electron dense core and, though generally spherical, exhibit some pleomorphism. They are scattered within the axoplasm in some axons, whereas in others, they cluster in 'synaptoid' aggregates adjacent to the plasmalemma. However, adjacent pituicytes or axons do not have post-synaptic specializations. Type A fiber dilatations which contain NSGs generally measure 1–5 μ in diameter, although some measure up to 50μ. These axonal dilatations have been classified in different ways, depending on their most prominent morphological feature (4, 10, 11, 13). In humans, they have been classified into six types, with various transitional forms, based on their internal fine structure (4).

Type I-dilatations (Fig. 1a) are characterized by abundant NSGs of type A exhibiting a wide range of size and content. Some of the dilatations are tightly packed with NSGs. A few mitochondria and occasional small dense bodies and/or lamellar bodies are present. Microvesicles are generally absent or few in number, except in rare foci with 'synaptoid' accumulations.

Type II-dilatations (Fig. 1b) characteristically contain numerous mitochondria. Pleomorphic dense or lamellar bodies are common. Most NSGs are 'empty' appearing because of extremely electron-lucent core material, although well-preserved NSGs are rarely present. There are membranous profiles, some of which are apparently tubular and approximately twice the size of neurotubules. Microvesicles, which

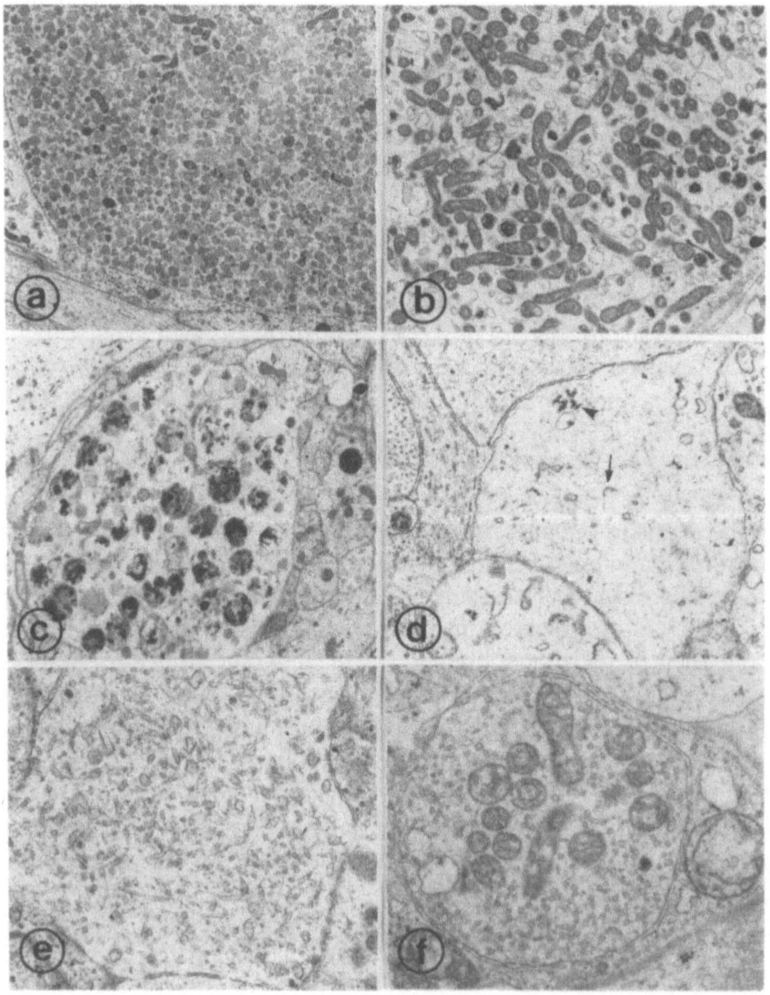

Fig. 1. Type A fiber dilatations. (a) Type I dilatation filled with NSGs (× 5,200); (b) Type II dilatation containing numerous mitochondria (× 6,160); (c) Type III dilatation with abundant lysosomal bodies (× 4,480); (d) Type IV dilatation with an electron-lucent, organelle-sparse axoplasm containing fragments of NSG membrane (arrow) and glycogen (arrowhead) (× 7,200); (e) Type V dilatation containing tubuloreticular profiles (× 14,400); (f) Type VI dilatation filled with microvesicles (× 12,000). Reprinted from reference 4.

occasionally form 'synaptoid' accumulations, are also present. In addition, microfibrils are occasionally scattered in this type of dilatation.

Type III-dilatations (Fig. 1c) are generally characterized by an abundance of lysosomal bodies of the cytosegresome type. Numerous autophagic vacuoles and dense lamellar bodies, some of which appear to contain multiple NSGs are present. NSGs, most of which are 'empty', are rare. A small number of mitochondria, tubular structures and glycogen particles are occasionally present.

Type IV-dilatations (Fig. 1d) exhibit an electron-lucent axoplasmic matrix without significant accumulation of cytoplasmic organelles. Scattered, irregular, crescent-shaped membranous profiles, some of which have slight condensation of granular material representing 'empty' type A-NSGs, are present. 'Synaptoid' accumulations of microvesicles are easily identified because of the relatively clear axoplasmic matrix. Mitochondria, tubular structures and clusters of glycogen particles are also occasionally observed.

Type V-dilatations (Fig. 1e) are occupied predominantly by tubuloreticular profiles. Other axoplasmic organelles, except for microtubules and smooth endoplasmic reticulum are scarce or absent. NSGs are absent in most dilatations of this type, thereby often obscuring the nature of the fiber type.

Type VI-dilatations (Fig. 1f) are generally smaller than the other types and are characterized by numerous microvesicles with 'synaptoid' accumulations. The number of NSGs is variable; when absent, the tubular microvesicles indicate that the dilatation originated from a type A-fiber. Although these dilatations are most commonly attached to the perivascular basal lamina, they are also found distant from vascular structures and are surrounded by other axons and pituicytes.

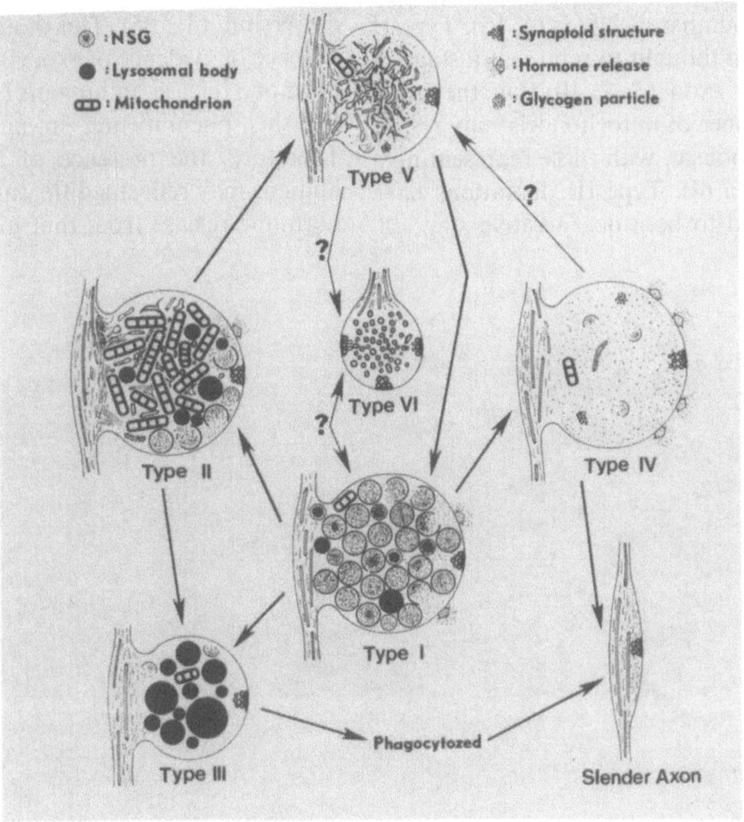

Fig. 2. Interrelation between type A-fiber dilatations. Type I dilatations are the storage site for hormone, which is released via molecular dispersion in type IV dilatations. Excess hormone is phagocytosed in type III dilatations. Types II and V may represent stages in the regeneration of type I dilatations. The significance of type VI dilatations is uncertain. Reprinted from reference 4.

2.2. Functional significance of type A fiber dilatations

While the type A NSGs have been shown to contain polypeptide hormones (22), there have been controversial opinions as to possible mechanisms whereby the neurohypophysial hormones are released from NSGs into the extracellular space (4). These opinions may be condensed to two theories, 'exocytosis' and 'molecular dispersion'. Although each theory has been advocated almost equally in many experimental settings (23–31), it must be noted that our recent ultrastructural studies on the human neurohypophysis have suggested that molecular dispersion is more likely to be the major physiologic process at the fiber dilatations (4).

Because of the ultrastructural heterogeneity, these dilatations have been considered by many investigators to be the site of intraaxonal hormone turnover (8, 10, 12, 14–16). Although the exact functional significance of these dilatations remains conjectural, we have recently proposed (4) a possible interrelationship between the six different types (Fig. 2).

In this schema, type I dilatations probably represent the storage site for the hormone because they contain the largest number of NSGs (4, 13). Type II dilatations have been thought to represent a stage of degeneration of the axon (3, 8, 10). On the other hand, the large number of mitochondria may reflect an increase in metabolism, with these representing a stage of regeneration (4). Type III dilatations have also been considered to be a degenerative stage (8, 10), since similar changes have been seen following stalk transection (32), lesioning of the paraventricular nuclei (33), or administration of cytotoxic agents (12, 34), as well as with aging (7). Similar changes have also been seen following stimulation of the neurohypophysis with water deprivation (35, 36), with a decreased need for hormones (12, 15, 16) or in the Brattleboro rat (37). Therefore, Type III dilatations appear to represent the site of degradation of unused or excess NSGs (Fig. 4). Type IV dilatations contain few organelles, including scattered empty NSGs and remnants of NSG membranes (4). Similar dilatations have been seen in the fetal rat and have been thought to remain as such into adult life (9). In humans, these dilatations are regarded as the site of 'readily releasable pools' of hormone (4). Such pools are thought to exist (38–40) and may explain why the hormonal content of the gland does not correlate with the electron density of the NSGs (41, 42). In support of this concept, immunocytochemical studies have demonstrated neurohypophysial hormone within (43–46) and outside (44) of NSGs, although such data have not been reported in humans. If such pools exist, the hormone would be released by 'molecular dispersion' (31, 35). This would be supported by the absence of evidence of exocytosis, the other means of hormone release, in humans (4, 18), despite evidence for this phenomenon in animals (23–25, 27–30). Therefore, the presence of type IV dilatations in humans may reflect a different predominant mode of hormone release from that in other animals.

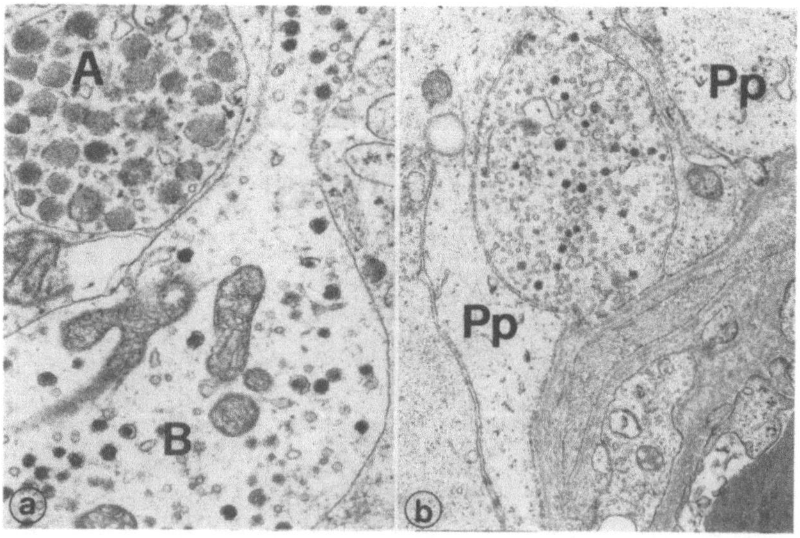

Fig. 3. Type B fibers. (a) Type B fibers (B) contain smaller granules than type A fibers (A) and the microvesicles in type B fibers do not assume the tubular form seen in type A fibers (× 16,800); (b) Type B fiber terminals are often seen attached to the abluminal basement membrane of a capillary. pp = pituicyte processes. The terminal contains numerous microvesicles and well-preserved NSGs (× 12,950). Reprinted in part from reference 4.

Type V dilatations are thought to be regenerative in nature (4, 8, 10). The tubuloreticular structures are of uncertain origin, but may relate to smooth endoplasmic reticulum or may be a stage in the reforming of NSG membrane.

Type VI dilatations contain abundant microvesicles, with the number of microvesicles present being in inverse proportion to the number of NSGs (1, 4). Some investigators have considered the microvesicles to reflect recaptured membrane after exocytosis (47, 48). The observation that these vesicles increase in number in states of increased hormone release could support this notion (3, 25, 42, 49). However, exocytosis has not been demonstrated in the human neurohypophysis (4). An alternative hypothesis has been that these may be cholinergic terminals (50), but acetylcholine is not located in hormone-containing terminals (51, 52) and these dilatations contain NSGs. Therefore, the significance of type VI dilatations remains quite uncertain. It is interesting to note, however, that axons and dilatations containing microvesicles represent a greater proportion in the human fetus than in the adult (18).

2.3. *Type B fibers*

In comparison to those of type A fibers, the NSGs of type B fibers are smaller and have a denser core with a sharper limiting membrane (Fig. 3). Most of these NSGs are located, along with microvesicles and other organelles, in terminal dilatations which are generally found in a perivascular location. The microvesicles in type B fibers are spherical and do not exhibit the pleomorphism seen in those of type A fibers. Although similar fibers have been observed in the median eminence and infundibulum of animals (53–55) and in the infundibulum of humans (1), Type B fibers are only rarely found in the human pars

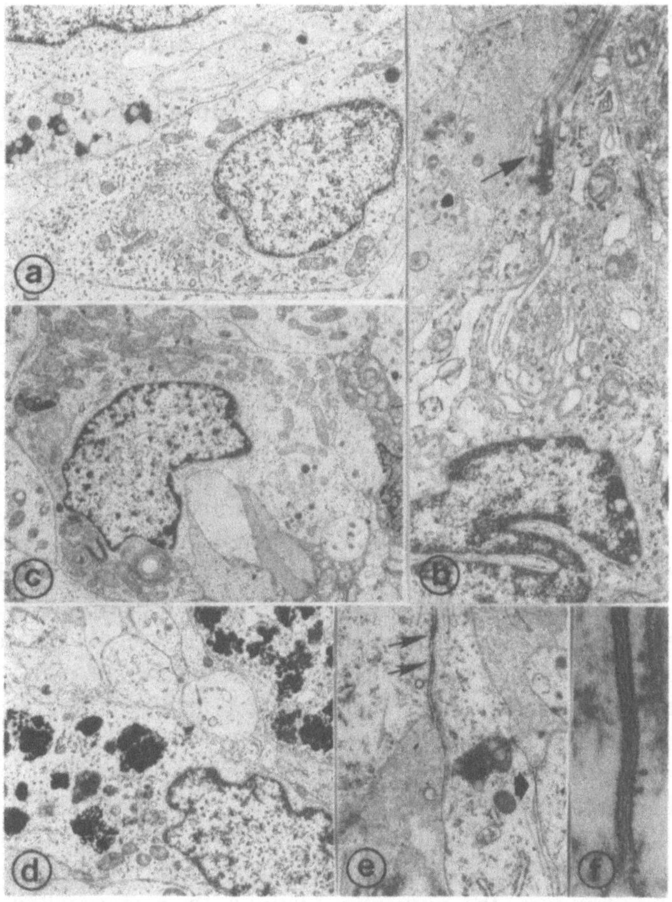

Fig. 4. Pituicytes. (a) Major pituicyte with an electron-lucent cytoplasm (× 4,000); (b) Ependymal pituicyte with a prominent cilium (arrow) (× 6,300); (c) Oncocytic pituicyte containing atypical mitochondria (× 2,100); (d) Granular pituicyte filled with amorphous, electron-dense bodies (× 4,000); (e) Pituicyte processes connected by desmosomes (thin arrows) and gap junctions (thick arrow) (× 6,440); (f) Higher magnification of an interpituicytic gap junction (× 35,900). Reprinted from reference 5.

nervosa (4). The perivascular terminal dilatations are quite reminiscent of those containing dopamine in the rat neurohypophysis (56) and may represent the catecholaminergic innervation of the neurohypophysis in humans as well. It is interesting to note that the number of NSGs and microvesicles present in type B fiber dilatations varies roughly in inverse proportion to one another (1, 3, 4).

3. Pituicytes

Pituicytes are the most predominant cellular elements in the human neurohypophysis (5). They have been well studied in different animals and have been shown to be composed of a heterogeneous cellular population, although there is still disagreement with regard to their classification (3, 26, 41, 57–60). In our recent studies on the human pars nervosa, five different types of pituicytes have been identified: major pituicytes; dark pituicytes; ependymal pituicytes; oncocytic pituicytes; and granular pituicytes (5).

3.1. Major pituicytes

The major pituicytes (Fig. 4a) are the most common type. They have abundant, electron lucent cytoplasm and an oval to slightly irregular nucleus with a thin chromatin rim and occasional nucleoli. The cytoplasm contains prominent Golgi complexes, randomly distributed polysomes, a moderate number of mitochondria and sparse rough endoplasmic reticulum. Lysosomes and lipid droplets are also present, often in association with one another. Occasional pinocytotic vesicles are seen along the plasmalemma. Cellular processes are prominent, often slender, containing bundles of fine fibrils measuring 5–7 nm in diameter. Some structures resembling Rosenthal fibers may be found in association with these fibrils (5). In the human fetus, pituicytes with similar cytoplasmic features are quite prominent (18). They differ from those in the adult since they have fewer processes, but these increase by 11 weeks of gestation. This morphology is closely related to that of astrocytes in the central nervous system (61). Indeed, the presence of S-100 protein (62) and glial fibrillary acidic protein (63, 64) also suggest their astroglial nature. Unlike astrocytes, however, the pituicytes have a close contact with nerve fibers: The neurosecretory axons of type A fibers are often embedded within the cytoplasm of the pituicytes or are surrounded by their processes (5). Although no definite synapses or junctions are found between pituicytic processes and neurosecretory axons in the human pars nervosa, some of the processes are closely apposed to axolemma in which subjacent 'synaptoid' aggregates of microvesicles are present (2, 5). In other mammals, some authors have observed true axopituicytic junctions (21, 60, 65), and a role for axopituicytic interactions in the control of hormone release has been suggested (21, 58, 66, 67). Such interactions have also been considered important for the growth or regeneration of axons in the neurohypophysis (20, 21, 32, 68).

Two types of junctions are found between pituicytic processes in the human neurohypophysis; one is the gap junction and the other is the desmosomal or intermediate junction (Fig. 4e, f). Similar junctions have also been described in pituicytes of other species (60, 69, 70) and may play a role in controlling extracellular hormone flow (vide infra). These axopituicytic or pituicyte-pituicytic contacts are generally not seen in the astrocytes of the central nervous system (71).

3.2. Dark pituicytes

Dark pituicytes have less abundant cytoplasm, which is of greater electron density than that seen in major pituicytes. Nuclei are more pleomorphic with denser chromatin. However, the organelle content is similar to that of major pituicytes.

Dark pituicytes have previously been described in animals (20, 26, 57, 59). Since they closely resemble major pituicytes, they probably represent a different functional state of the same cell. The dark cells predominate in the fetal rat, where they are thought to be more active and possibly secretory, whereas the lighter pituicytes are thought to be inactive and predominate in the adult rat (20, 21). In fact, an apparent progression from a predominance of dark pituicytes in the fetus to light cells in the adult has been demonstrated in the rat (21), but a similar progression was not seen in the human fetus (18).

3.3. Ependymal pituicytes

Ependymal pituicytes (Fig. 4b) have cilia, generally of the 9+0 type, and distinct intercellular junctions interconnect ependymal pituicytes. Some cells form aggregates surrounding an extracellular lumen containing their microvillous processes. Similar cells have also been reported in fetal and adult rats (20). These cells are of interest since the neurohypophysial cells in lower vertebrates are ependymal cells (72, 73). An inverse relationship has been noted between pituicytes and ependymal cells as one ascends the phylogenetic tree (74), suggesting that at least some pituicytes may be derived from ependymal cells. Further, the pituicytes in the fetal rat are seen to originate as a cluster in the subependymal region with contacts with the ependymal cells in the floor of the third ventricle (21). The finding of ependymal pituicytes suggests that human pituicytes may also be derived from ependymal cells.

3.4. *Oncocytic pituicytes*

Oncocytic pituicytes (Fig. 4c) contain numerous mitochondria, which often fill their cytoplasm. The mitochondria are often pleomorphic and atypical and some contain granular inclusions as well as crystalloid structures. These oncocytic pituicytes satisfy the criteria for oncocytes occurring anywhere (75), including the accumulation of large numbers of mitochondria, some of which are enlarged or pleomorphic, within a mature organ. Similar changes have been seen in the adenohypophysis (76), as well as in epithelial cells of other endocrine glands. Oncocytic change has not previously been described in cells of the neurohypophysis, but it occurs to a variable degree and the presence of transitional forms between oncocytic and major pituicytes suggests that this is a metaplastic change of major pituicytes. This hypothesis has been applied to oncocytic change in other organs (77), but the functional significance of this metaplasia is unknown.

3.5. *Granular pituicytes*

Granular pituicytes (Fig. 4d) contain abundant dense granules which mostly are of the cytosegresome type. These vary in shape and are usually surrounded by a limiting membrane, but some of them are amorphous aggregates of electron dense material without such a membrane. Lipid droplets and myelin figures are also present in some of these cells. The granular pituicytes appear to be the ultrastructural correlate of the iron- and melanin-positive cells observed at the light microscopic level. Since similar dense bodies in rat pituicytes contain acid phosphatase (78) they are presumably lysosomal in origin. Dense bodies (33, 58, 66), as wel as acid phosphatase (37) and intracellular lipid (37, 79), as seen in granular pituicytes, have been shown to increase with increased neurohypophysial secretion secondary to water deprivation or salt loading. Further, an inverse relationship between neurosecretory material in nerve terminals and dense bodies in pituicytes has been reported (80). These observations suggest the transport of macromolecules between axonal terminals and pituicytes. Such transport has been termed 'ultraphagocytosis'. Since pituicytes can phagocytose extracellular material by endocytosis with the phagocytosed material appearing in lysosomes (81), a mechanism is available for the granular pituicytes to endocytose and degrade excess neurosecretory material within cytosegresomes. In addition to ultraphagocytosis, pituicytes may be capable of phagocytosing foreign bodies, such as axonal debris after stalk transection (32, 82–84). This mechanism differs from ultraphagocytosis, however, which occurs at the molecular level. Since histiocytes are also present within the neurohypophysis, the significance of phagocytosis by pituicytes is uncertain.

Cells containing numerous cytosegresomes are also characteristic of granular cell tumor or choristoma of the neurohypophysis (85, 86), as well as granular cell tumors elsewhere in the body (87). Therefore, these tumor cells resemble granular pituicytes, especially since they may also be iron and melanin-positive. Thus, it appears that the neurohypophysial choristoma may, at least in some cases, be a tumor of granular pituicytes. The possible cause for such a change, especially in the elderly where choristoma is most common (88), is unknown.

4. Vascular and perivascular structures

Capillaries in the human neurohypophysis (Fig. 5) are lined by a fenestrated endothelium, with the fenestrae 30–50 nm in diameter. Tight junctions connect endothelial cells, which are similar to those seen elsewhere in the body. They contain prominent Golgi complexes, filaments, microtubules and pinocytotic vesicles, as well as membrane-bound microtubular bodies. The endothelial or luminal basement membrane is continuous and measures approximately 50 nm in thickness. An independent abluminal basement membrane of similar thickness lies external to the luminal basement membrane and is separated from it by a gap of 0.1 to several microns, representing the perivascular space. The abluminal basement membrane is complex, extending around cells and processes into the parenchyma, where it is focally discontinuous (1, 3, 6).

By contrast with the adult, the neurohypophysis of the 7.5 week fetus (18) contains few capillaries, which are lined by a thick endothelium, and has a poorly developed perivascular space. By 8.5 weeks, some vessels have a thinner endothelium, which is focally fenestrated, and the perivascular space has expanded but does not ramify in the neural parenchyma. Two distinct basement membranes are seen at 8.5 weeks.

By 11 weeks, the perivascular space extends from fenestrated vessels into the parenchyma, although some primitive vessels persist, even at 19 weeks of gestation.

Mesenchymal cells, including pericytes, histiocytes, mast cells, fibroblasts and rare lymphocytes, lie within the perivascular space. The pericytes are surrounded by basement membrane, which frequently fuses with the luminal basement membrane. The pericytes partially surround the capillaries with their prominent processes extending along the luminal basement membrane. Their cytoplasm contains abundant microfilaments which are oriented along the long axis of the processes and form focal condensations similar to those in smooth muscle. Abundant pinocytotic vesicles are located along the cytoplasmic membrane and lipid droplets are present within the cell soma. The pericyte nuclei are ovoid and have marginated chromatin. The histiocytes are characterized by abundant intracytoplasmic vacuoles and lysosomes as well as irregular filopodial projections of their cell surface. The filopodia occasionally enclose amorphous material of unknown type. The fibroblasts have well developed Golgi complexes and contain abundant saccular rough endoplasmic reticulum. Mast cells, which are often clustered, contain characteristic cytoplasmic granules and have microvillous surface projections.

Axons and pituicyte processes generally abut the outer surface of the abluminal basement but never contact the luminal basement membrane directly and rarely have direct contact with the perivascular space. Pituicyte processes and type VI dilatations occasionally contact the perivascular space in areas where the abluminal basement membrane is discontinuous. An important feature of the perivascular space is that sinusoidal channels often lined on each side by abluminal basement membrane occasionally extend from the perivascular space into the neural parenchyma (Fig. 5). In areas, this abluminal basement membrane becomes discontinuous, allowing the neural extracellular space to open directly into the perivascular space. This is the only site where NSG-containing axons, other than occasional type VI dilatations, contact the perivascular space.

5. Ultrastructural morphology of extracellular hormone pathyway and its regulation

Since no neurosecretory axons are present in direct contact with the capillary endothelium or its luminal basement membrane (6), it appears that the hormone is first released into the extracellular space, where superfluous material may be phagocytosed by pituicytes (5). In the neural parenchyma, there are intertwining channels of extracellular space which exhibit sinusoidal widenings seamed at both ends by either

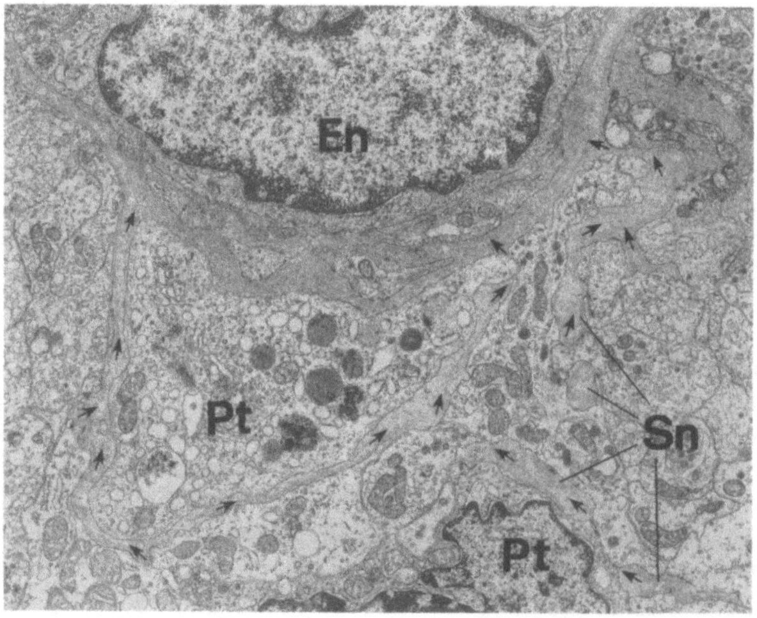

Fig. 5. Channels of extracellular space for hormone transport (arrows) lined with basement membrane extend from the neural parenchyma into the perivascular space. Interpituicytic junctions contribute to the formation of sinusoids (Sn). Pt: pituicyte; En: endothelial cell (× 7,000). Reprinted from reference 6.

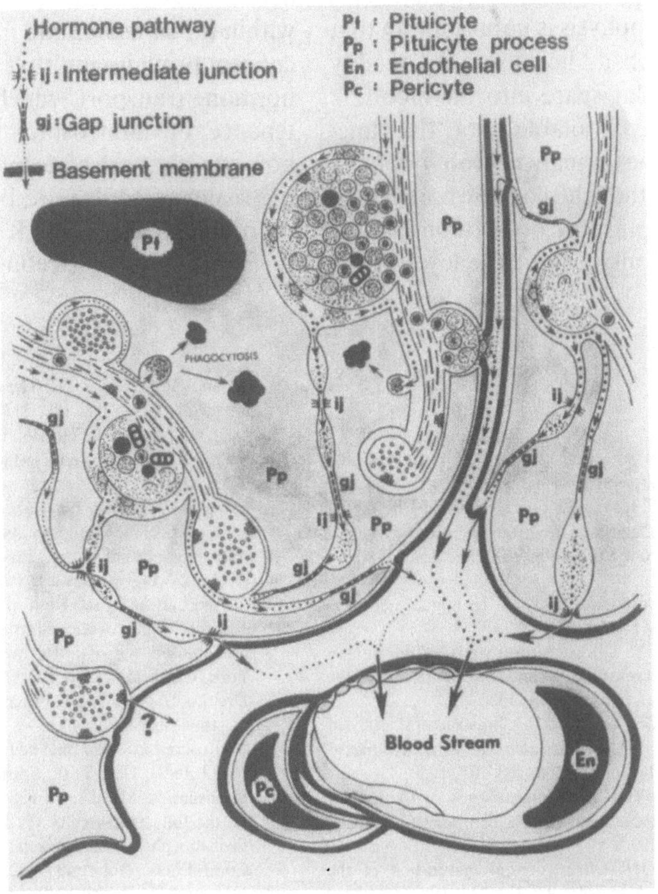

Fig. 6. Extracellular hormone pathway. Hormone is released via molecular dispersion (open arrows) and possibly by exocytosis into the extracellular or perivascular space. Hormone travels through the sinusoids, with flow controlled by intercellular junctions, to the perivascular space, while excess hormone is phagocytosed by pituicytes. Hormone then crosses the abluminal basement membrane, the perivascular space and the luminal basement membrane to enter the bloodstream through the fenestrated endothelium. Reprinted from reference 6.

gap junctions or intermediate junctions between pericytes (5, 6). These sinusoids occur more frequently near the perivascular space where they are often lined by abluminal basement membranes, thus connecting indirectly or directly to the perivascular space (Fig. 5). This complex network of extracellular channels has been described both in humans (1, 3, 6, 18) and in other animals (41, 89–91), and the sinusoids have been considered to be yet another site for storage of the readily releasable pool of hormone (41, 89, 90, 92) with the interpituicytic junctions serving as a flow regulatory device (5, 6).

The perivascular space itself appears to play an important role in the transport of hormone from neurosecretory axons to the blood stream. It is lined by a luminal and an abluminal basement membrane and contains such cells as pericytes, histiocytes, fibroblasts and mast cells. The pericytes, which resemble smooth muscle cells, are thought to be contractile (93–95). This is further suggested by the observation that their filaments resemble or are actin since they react with heavy meromyosin (97). If this is the case, they may play a role in modulating blood flow by altering capillary diameter. Since neurohypophysial blood flow increases under conditions of hormone release (90, 98, 99), the pericytes may play a role in the passage of hormone into the systemic circulation via alterations in the local circulation. Mast cells may also be involved in altering hormonal transport, since they may release substances that alter capillary permeability (100).

Capillaries in the pars nervosa of humans (1–3, 6) and other species (37, 41, 89) are fenestrated. This is one of the ultrastructural features that may correlate with the absence of the blood-brain barrier. This property in the neurohypophysis is important, in that it allows substances, such as hormones, to readily cross from the perivascular space into the blood.

Based on the currently available data, the functional morphology of the human neurohypophysis may be synthesized into the schema shown in Figure 6. In this schema, the transport of hormone to the systemic circulation is controlled at the level of the axon, based on the amount of hormone released, within the sinusoids and in the local vasculature. This concept provides multiple sites for the modulation of hormone transport, which is important in the maintenance of homeostasis. It must be emphasized, however, that the concept must partly remain within the realm of conjecture because of the limitation of morphological studies. It will require further studies for confirmation or refinement.

References

1. Bergland RM, Torack RM: An electron microscope study of the human infundibulum. Z Zellforsch 99: 1–12, 1969.
2. Bloodworth JMB, Horvath E, Kovacs K: Fine structural pathology of the endocrine system. In: Diagnostic electron microscopy, Vol 3. Trump BF, Jones RT (eds), New York, John Wiley and Sons, 1980, pp 359–527.
3. Lederis K: An electron microscopical study of the human neurohypophysis. Z Zellforsch 65: 847–868, 1965.
4. Seyama S, Pearl GS, Takei Y: Ultrastructural study of the human neurohypophysis. I. Neurosecretory axons and their dilatations in the pars nervosa. Cell Tissue Res 205: 253–271, 1980.
5. Takey Y, Seyama S, Pearl GS, Tindall GT: Ultrastructural study of the human neurohypophysis. II. Cellular elements of neural parenchyma, the pituicytes. Cell Tissue Res 205: 273–287, 1980.
6. Seyama S, Pearl GS, Takei Y: Ultrastructural study of the human neurohypophysis. III. Vascular and perivascular structures. Cell Tissue Res 206: 291–302, 1980.
7. Cannata MA, Tramezzani JH: Ultrastructural maturation of the neurohypophysis of the rat. Acta Anat 97: 213–223, 1977.
8. Dellmann HD: Degeneration and regeneration of neurosecretory systems. Int Rev Cytol 36: 215–315, 1973.
9. Dellmann HD, Castel M, Linner JG: Ultrastructure of peptidergic neurosecretory axons in the developing neural lobe of the rat. Gen Comp Endocrinol 36: 477–486, 1978.
10. Dellmann HD, Rodriquez EM: Herring bodies; an electron microscopic study of local degeneration and regeneration of neurosecretory axons. Z Zellforsch 111: 293–314, 1970.
11. Heap PF, Jones CW, Morris JF, Pickering BT: Movement of neurosecretory products through the anatomical compartments of the neural lobe of the pituitary gland: An electron microscopic autoradiographic study. Cell Tissue Res 156: 483–497, 1975.
12. Lullmann-Rauch R: Alteration in the neurohypophysis of rats treated with chlorphentermine or tricyclic antidepressants. Cell Tissue Res 169: 501–514, 1976.
13. Morris JF: Distribution of neurosecretory processes of the pituitary gland; A quantitative ultrastructural approach to hormone storage in the neural lobe. J Endocrinol 68: 225–234, 1976.
14. Polenov AL, Garlov PE: The hypothalamo-hypophysial system in Acipenseridae I. Ultrastructural organization of large neurosecretory terminals (Herring Bodies) and axoventricular contacts. Z Zellforsch 116: 349–374, 1971.
15. Polenov AL, Ugrumou MV, Belenky HA: On degeneration of peptidergic neurosecretory fibers in the albino rat. Cell Tissue Res 160: 113–123, 1975.
16. Rufener C: Autophagy of secretory granules in the rat neurohypophysis. Neuroendocrinology 13: 314–320, 1974.
17. Herring PT: The histological appearances in the mammalian pituitary body. Quart J Exp Physiol 1: 121–159, 1908.
18. Okado N, Yokota N: An electron microscopic study on the structural development of the neural lobe in the human fetus. Amer J Anat 159: 261–273, 1980.
19. Holmes RL: The neurohypophysis of the foetal monkey. Z Zellforsch 69: 288–295, 1966.
20. Dellmann HD, Sikora K: Pituicyte fine structure in the developing neural lobe of the rat. Dev Neurosci 4: 89–97, 1981.
21. Galabov P, Schiebler TH: The ultrastructure of the developing neural lobe. Cell Tissue Res 189: 313–329, 1978.
22. Dean CR, Hope DB: The isolation of purified neurosecretory granules from the bovine posterior pituitary lobe. Biochem J 104: 1082–1088, 1967.
23. Douglas WW: How do neurons secrete peptides. Exocytosis and its consequence, including 'synaptic vesicle' formation, in hypothalamo-neurohypophysial system. Prog Brain Res 39: 21–38, 1973.
24. Douglas WW, Nagasawa J, Schulz R: Electron microscopic studies on the mechanism of secretion of posterior pituitary hormones and significance of microvesicles (synaptic vesicles), In: subcellular organization and function of endocrine tissue. Heller H, Lederis K (eds). New York, Cambridge Univ Press, 1971, pp 353–378.
25. Dreifuss JJ, Normann K, Akert K, Sandri C, Moor H: Exoendocytosis in the neurohypophysis as revealed by freeze-fracturing. in: Neuroscretion – The final neuroendocrine pathway. Knowles F, Vollrath L (eds), New York, Springer Verlag, 1974, pp 31–37.
26. Kurosumi K, Matsuzawa T, Shibasaki S: Electron microscope studies on the fine structures of the pars nervosa and pars intermedia, and their morphological interrelation in the normal rat hypophysis. Gen Comp Endocrinol 1: 433–452, 1961.
27. Nagasawa J, Douglas WW, Schulz RA: Ultrastructural evidence of secretion by exocytosis and 'synaptic vesicle' formation in posterior pituitary glands. Nature 227: 407–409, 1970.
28. Nordmann JJ: Neurosecretion by exocytosis. Int Rev Cytol 46: 1–78, 1976.
29. Theodosis DT, Dreifuss JJ: Ultrastructural evidence for exo-endocytosis in the neurohypophysis. In: Neurohypophysis. Moses AM, Share L (eds), Basel, S Karger, 1977, pp 88–94.
30. Theodosis DT, Dreifuss JJ, Orci L: A freeze-fracture study of membrane events during neurohypophysis secretion. J Cell Biol 78: 542–553, 1978.
31. Vollrath L: New trends in vertebrate neurosecretion. In: Neurosecretion – The final neuroendocrine pathway. Knowles F, Vollrath L (eds), New York, Springer Verlag, 1974, pp 276–284.
32. Dellmann HD, Stoeckel ME, Porter A, Stutinsky F: Ultrastructure of the neurohypophysial glial cells following stalk transection in the rat. Experientia 30: 1220–1222, 1974.
33. Zambrano D, DeRobertis E: The ultrastructural changes in the neurohypophysis after destruction of the paraventricular nuclei in normal and castrated rats. Z Zellforsch 88: 496–510, 1968.
34. Vazques R, Amat P: The ultrastructure of Herring bodies in rats subjected to different experimental conditions. Cell Tissue Res 189: 41–51, 1978.
35. Krisch B, Becker K, Bargmann W: Exocytose im hinterlassen der Hypophyse. Z Zellforsch 123: 47–54, 1972.
36. Whitaker S, LaBella FS: Ultrastructural localization of acid phosphatase in the posterior pituitary of the dehydrated rat. Z Zellforsch 125: 1–15, 1972.
37. Tasso F, Rua S: Ultrastructural observations on the hypothalamo-posthypophysial complex of the Brattleboro rat. Cell Tissue Res 191: 267–286, 1978.
38. Thorn NA: In-vitro studies of the release mechanism for vasopressin in rats. Acta Endocrinol 53: 644–654, 1966.
39. Sachs H, Sare L, Osinchak J, Carpi A: Capacity of the neurohypophysis to release vasopressin. Endocrinology 81: 755–770, 1967.
40. Sachs H, Haller EW: Further studies on the capacity of the neurohypophysis to release vasopressin. Endocrinology 83: 251–262, 1968.

41. Barer R. Lederis K: Ultrastructure of the rabbit neurohypophysis with special reference to the release of hormone. Z Zellforsch 75: 201–239, 1966.
42. Daniel AR, Lederis K: Effect of ether anaesthesia and hemorrhage on hormone storage and ultrastructure of the rat neurohypophysis. J Endocrinol 34: 91–104, 1966.
43. Aspeslagh HR, Vandesande F, Dierickx K: Electron microscopic immunocytochemical demonstration of neurophysin-vasopressinergic and neurophysin-oxytocinergic fibers in the neural lobe of the rat hypophysis. Cell Tissue Res 171: 31–37, 1976.
44. Silverman AJ, Zimmerman EA: Ultrastructural immunocytochemical localization of neurophysin and vasopressin in the median eminence and posterior pituitary of the guinea pig. Cell Tissue Res 159: 291–301, 1975.
45. Vandesande F, Dierickx K: Indentification of the vasopressin producing and of the oxytocin producing neurons in the hypothalamic magnocellular neurosecretory system of the rat. Cell Tissue Res 164: 153–162, 1975.
46. Van Leeuwen FX, Swaab DF: Specific immunoelectronmicroscopic localization of vasopressin and oxytocin in the neurohypophysis of the rat. Cell tissue Res 177: 493–501, 1977.
47. Holmes RL, Knowles F: 'Synaptic' vesicles in the neurohypophysis. Nature 185: 710–711, 1960.
48. Morris JF, Nordmann JJ: Membrane recapture after hormone release from nerve endings in the neural lobe of the rat putuitary gland. Neuroscience 5: 639–649, 1980.
49. Santolaya RC, Bridges TE, Lederis K: Elementary granules, small vesicles and exocytosis in the rat neurohypophysis after acute hemorrhage. Z Zellforsch 125: 277–288, 1972.
50. DeRobertis F: Ultrastructure and function in some neurosecretory systems. In: Neurosecretion. Heller J, Clark RB (eds), New York, Academic Press, 1962, pp 3–20.
51. Lederis K, Livingston A: Neuronal and subcellular location of acetylcholine in the posterior pituitary of the rabbit. J Physiol (Lond) 210: 187–204, 1970.
52. Livingston A, Lederis K: Functional ultrastructure of the neurohypophysis. In: Subcellular organization and function in endocrine tissue. Heller J, Lederis K (eds), New York, Cambridge Univ Press, 1971, pp 233–262.
53. Campbell DJ, Holmes RC: Further studies on the neurohypophysis of the hedgehog (Erinaceus europaeus). Z Zellforsch 75: 35–46, 1966.
54. Duffy PE, Menefee M: Electron microscopic observation of neurosecretory granules, nerve and glial fibers, and blood vessels in the median eminence of the rabbit. Amer J Anat 117: 251–286, 1965.
55. Monroe BG: A comparative study of the ultrastructure of the median eminence, infundibular stem, and neural lobe of the hypophysis of the rat. Z Zellforsch 76: 405–432, 1967.
56. Baumgarten HG, Bjorklund A, Holstein AF, Nobin A: Organization and ultrastructural identification of the catecholamine nerve terminals in the neural lobe and pars intermedia of the rat pituitary. Z Zellforsch 126: 483–517, 1972.
57. Fujita H, Hartmann JF: Electron microscopy of neurohypophysis in normal, adrenaline-treated and pilocarpine-treated rabbits. Z Zellforsch 54: 734–736, 1961.
58. Krsulovic J, Bruckner G: Morphological characteristics of pituicytes in different functional stages. Z Zellforsch 99: 210–220, 1969.
59. Nakai Y: Electron microscopic observation on synapse-like contacts between pituicytes and different types of nerve fibers in the anuran pars nervosa. Z Zellforsch 110: 27–39, 1970.
60. Olivieri-Sangiacomo C: Ultrastructural features of pituicytes in the neural lobe of adult rats. Experientia 29: 1119–1120, 1973.
61. Wendell-Smith CP, Blunt MJ, Baldwin F: The ultrastructural characterization of macroglial cell types. J Comp Neurol 127: 219–239, 1966.
62. Cocchia D, Miani N: Immunocytochemical localization of the brain-specific S-100 protein in the pituitary gland of adult rat. J Neurocytol 9: 771–782, 1980.
63. Velasco ME, Roessmann U, Gambetti P: Immunohistochemical demonstration of glial fibrillary acidic (GFA) protein in human pituitary gland. J Neuropath Exp Neurol 38: 347, 1979.
64. Suess U, Pliska V: Identification of the pituicytes as astroglial cells by indirect immunofluorescence-staining for the glial fibrillary acidic protein. Brain Res 221: 27–33, 1981.
65. Olivieri-Sangiacomo C: Ultrastructural features of pituicytes in the neural lobe of adult rats. The peripheral processes. Experientia 29: 1536–1537, 1973.
66. Kurosumi K, Matsuzawa T, Kobayashi Y: On relation between the release of neurosecretory substance and lipid granules in the rat neurohypophysis. Gunma Symposia on Endocrinology 1: 87–118, 1964.
67. Tweedle CD, Hatton GI: Evidence for dynamic interactions between pituicytes and neurosecretory axons in the rat. Neuroscience 5: 661–667, 1980.
68. Kiernan JA: Pituicytes and the regenerative properties of neurosecretory and other axons in the rat. J Anat 109: 97–114, 1971.
69. Boudier JL, Boudier JA: Jonctions entre pituicytes dans la neurohypophyse du rat. J Microscopie 20: 27a, 1974.
70. Dreifuss JJ, Sandri C, Akert K, Moor H: Ultrastructural evidence for sinusoid spaces and coupling between pituicytes in the rat. Cell Tissue Res 161: 33–45, 1975.
71. Trombley IK, Mirra SS: Ultrastructure of tuberous sclerosis: Cortical tuber and subependymal tumor. Ann Neurol 9: 174–181, 1981.
72. Rodriquez EM: The comparative morphology of neural lobes of species with different neurohypophysial hormones. In: Subcellular organization and function in endocrine tissue. Heller J, Lederis K (eds), London, Cambridge Univ Press, 1971, pp 263–292.
73. Rodriguez EM, LaPointe J: Histology and ultrastructure of the neural lobe of the lizard, Klauberina riversiana. Z Zellforsch 95: 37–57, 1969.
74. Holmes RL, Ball JN: The pituitary gland, a comparative account, London, Cambridge Univ Press, 1974.
75. Tandler B, Hutter RVP, Erlandson EA: Ultrastructure of oncocytoma of parotid gland. Lab Invest 23: 567–580, 1970.
76. Kovacs K, Horvath E, Bilbao JM: Oncocytes in the anterior lobe of human pituitary gland. Acta Neuropath 27: 43–45, 1974.
77. Hamperl H: Benign and malignant oncocytoma. Cancer 15: 1019–1027, 1962.
78. Whitaker S, Labella FS, Anwal M: Electron microscopic histochemistry of lysosomes in neurosecretory nerve endings and pituicytes of rat posterior pituitary. Z Zellforsch 111: 493–504, 1970.
79. Boer GJ, Van Rheenen-Verberg CMF: Acid phosphatase in rat neurohypophysial dispersions and its fractions enriched for neurosecretosomes and pituicytes after water deprivation and lactation. Brain Res 114: 279–292, 1976.
80. Krsulovic J, Ermisch A, Sterba G: Electron microscopic and autoradiographic study with special consideration of the pituicyte problem. In: Aspects of Neuroendocrinology. Bargmann W, Scharrer B (eds), New York, Springer Verlag, 1970, pp 166–172.
81. Theodosis DT: Endocytosis in glial cells (pituicytes) of the rat neurohypophysis demonstrated by incorporation of horseradish peroxidase. Neuroscience 4: 417–425, 1979.
82. Dellmann HD, Owsley PA: Investigations on the hypothalamoneurohypophyseal neurosecretory system of the grass frog (Rana pipiens) after transection of the proximal neurohypophysis. II. Light and electron microscopic findings in the disconnected distal neurohypophysis with special emphasis on the pituicytes. Z Zellforsch 94: 325–336, 1969.
83. Rodriquez EM, Dellmann HD: Hormonal content and ultrastructure of the disconnected neural lobe of the grass frog (Rana pipiens). Gen Comp Endocrinol 15: 272–288, 1970.
84. Sterba G, Bruckner G: Elektronenmikroskopische Untersuchungen über die Reaktion der Pituizyten nach Hypophysenstieldurchternnung bei Rana esculenta. Z Zellforsch 93: 74–83, 1969.
85. Popovitch ER, Sutton CH, Besker NH, Zimmerman HM: Fine structure and histochemical studies of choristomas of the neurohypophysis. J Neuropathol Exp Neurol 29: 155–156, 1970.
86. Ulrich J, Landolt A, Benini A: Granularzelltumor im 3 Ventrikel des Grobhirns. Klinische Befunde, Licht- und Electronenmikroskopie. Acta Neuropathol 27: 215–223, 1974.
87. Sobel HJ, Marquet E, Avirin E, Schwarz R: Granular cell myoblastoma. An electron microscopic and cytochemical study illustrating the genesis of granules and aging of myoblastoma cells. Am J Pathol 65: 59–78, 1971.
88. Rubinstein LJ: Tumors of the central nervous system, Atlas of tumor pathology, second series, fascicle 6, Washington, Armed Forces Institute of Pathology, 1972.
89. Livingston A: Morphology of the perivascular regions of the rat neural lobe in relation to hormone release. Cell Tissue Res 159: 551–561, 1975.
90. Livingston A, Wilks PN: Perivascular regions of the rat neural lobe. Cell Tissue Res 174: 273–280, 1976.

91. Endemar A, Eurenius L: Organization and development of the perivascular space system in the neurohypophysis of the laboratory mouse. Cell Tissue Res 199: 99–116, 1979.
92. Norstrom A: The heterogeneity of the neurohypophysial pool of neurophysin. In: Neurosecretion – The final neuroendocrine pathway. Knowles F, Vollrath L (eds), New York, Springer Verlag, 1974, pp 86–93.
93. Rhodin JAG: Ultrastructure of mammalian venous capillaries, venules, and small collecting veins. J Ultrastruct Res 25: 452–500, 1968.
94. Stensaas LJ: Pericytes and perivascular microglial cells in the basal forebrain of the neonatal rabbit. Cell Tissue Res 158: 517–541, 1975.
95. Zimmermann KW: Der feinere Bau der Blutcapillaren. Z Anat Entwicklungsgesch 68: 29–109, 1923.
96. Weibel ER: On pericytes, particularly their existence on lung capillaries. Microvasc Res 8: 218–235, 1974.
97. Lebeux YJ, Willemot J: Identification of actin-like filaments in rat brain by means of HMM labeling. J Cell Biol 67: 236a, 1975.
98. Sooriyamoorthy T, Livingston A: Variations in blood volume of the neural and anterior lobes of the pituitary of the rat associated with neurohypophysial hormone-releasing stimuli. J Endocrinol 54: 407–415, 1972.
99. Sooriyamoorthy T, Livingston A: Blood flow studies in the neural lobe of the pituitary of rabbit associated with neurohypophysial hormone releasing stimuli. J Endocrinol 57: 75–85, 1973.
100. Bodian O: Cytological aspects of neurosecretion in opossum neurohypophysis. Bull John Hopk Hosp 113: 57–93, 1963.

Authors' address:
Division of Neuropathology
Department of Pathology and Laboratory Medicine
Emory University School of Medicine
Atlanta, GA 30322, USA

CHAPTER 8

The fine structure of pituitary tumors

KALMAN KOVACS, EVA HORVATH and DONNA J. McCOMB

'So little done, so much to do'
Cecil John Rhodes

1. Introduction

The advent of the electron microscope opened new avenues in biologic research. The spectacular progress achieved in the study of ultrastructural characteristics of different cell types, made possible the investigation of the fine structural features of endocrine glands clarifying the subcellular organization of hormone-producing cells in health and disease. In studies related to the pituitary gland, electron microscopy was widely used to define the various cell types, to shed light on the multiple steps of the secretory process, to elucidate morphologic responses of the cell to various injuries, and to classify tumors of the sella region.

Pituitary tumors can be epithelial or mesenchymal, benign or malignant, primary or secondary. In this review, only the fine structure of pituitary adenomas will be dealt with.

Pituitary adenomas are usually benign neoplasms arising from and consisting of adenohypophysial cells. In many animal species, they occur spontaneously or can be induced by various techniques and can serve as valuable models for the investigation of pathogenesis, endocrinology and therapeutic responsiveness of the human disease. Human pituitary tumors are commonly occurring lesions. The study of unselected human autopsy material reveals the presence of incidental pituitary adenomas in approximately 6–23% of the cases.

This review is divided into two parts: the first part concentrates on the ultrastructural features of pituitary adenomas in animals; the second part focuses on the electron microscopic findings of human pituitary adenomas. The concluding remarks attempt to outline the perspective for future research.

Although electron microscopy has substantially increased the knowledge on various aspects of pituitary adenomas, it must be emphasized that several problems remain unresolved. Basic differences exist in the fine structural features of adenohypophysial cells from species to species, and extrapolation of findings from one to another has lead to erroneous conclusions. It is also pertinent to note the difficulties of obtaining appropriate material for the fine structural study. Adenohypophysial cells, sensitive to hypoxia, show marked fine structural changes within a short period of time after cessation of circulation. Hence, human pituitary tissue obtained from autopsies cannot be used for detailed electron microscopic investigations, since most of the subcellular features are lost. Only surgically-removed human pituitaries are suitable for fine structural analysis. These tissues are not normal and lack adequate control material which makes comparison difficult, if not impossible, between the normal and pathological.

Furth and his associates (1), were doubtful of the value of electron microscopy in the study of pituitary adenomas harvested in animals. The main reason for their scepticism was the great variability in the size and shape of secretory granules found in adenomatous cells. Pleomorphism, already visible in hormone dependent tumors, was more extensive in autonomous neoplasms. Degenerative changes, focal hemorrhages and necrosis also obscured the fine structural characteristics of adenoma cells.

Despite all these difficulties, important details are being elucidated on the electron microscopic features of pituitary tumors in animals and in human subjects. Since space is limited to provide a comprehensive review of the literature, only the most important data will be summarized. Several types of rat pituitary tumors are being studied by electron microscopy in our laboratory. For the past 10 years, we have been studying more than 500 pituitary adenomas removed by surgery from human patients. The description of the fine structural appearance of human pituitary adenomas is based on our own classification.

Motta, PM (ed): Ultrastructure of endocrine cells and tissues. ISBN-13: 978-1-4613-3863-5

2. Pituitary tumors in animals

2.1. Spontaneous pituitary adenomas

The spontaneous occurrence of pituitary adenomas in rats was described in detail by Fischer (2) in 1926 and Wolfe et al. (3) presented the first comprehensive investigation on the gross and histologic features of these tumors in aging rats in 1938. These tumors varied in incidence between 1 and 68% in different strains of rats and were assumed to be endocrinologically inactive, chromophobic-acidophilic adenomas. However, manifestations of increased prolactin (PRL) secretion, such as lactation and mammary gland hyperplasia, were noted (4). Blood PRL concentrations were elevated (5) and the immunohistochemical technique revealed the presence of PRL in the cytoplasm of adenoma cells (6).

The fine structural features of spontaneous pituitary adenomas found in old female Long-Evans rats are described in detail (6, 7). By electron microscopy, these tumors are composed of oval or irregular cells with a large nucleus and well developed cytoplasm (Fig. 1). A large portion of the cytoplasm is occupied by extensively developed rough-surfaced endoplasmic reticulum (RER) networks; free ribosomes and polysomes are numerous. The Golgi complexes, consisting of crescent or ring-shaped sacculi occupy a considerable area of the cytoplasm. Mitochondria are rod shaped with lamellar cristae and moderately electron dense matrix. Centrioles, occurring singly, in diplosomal pairs or in multiple

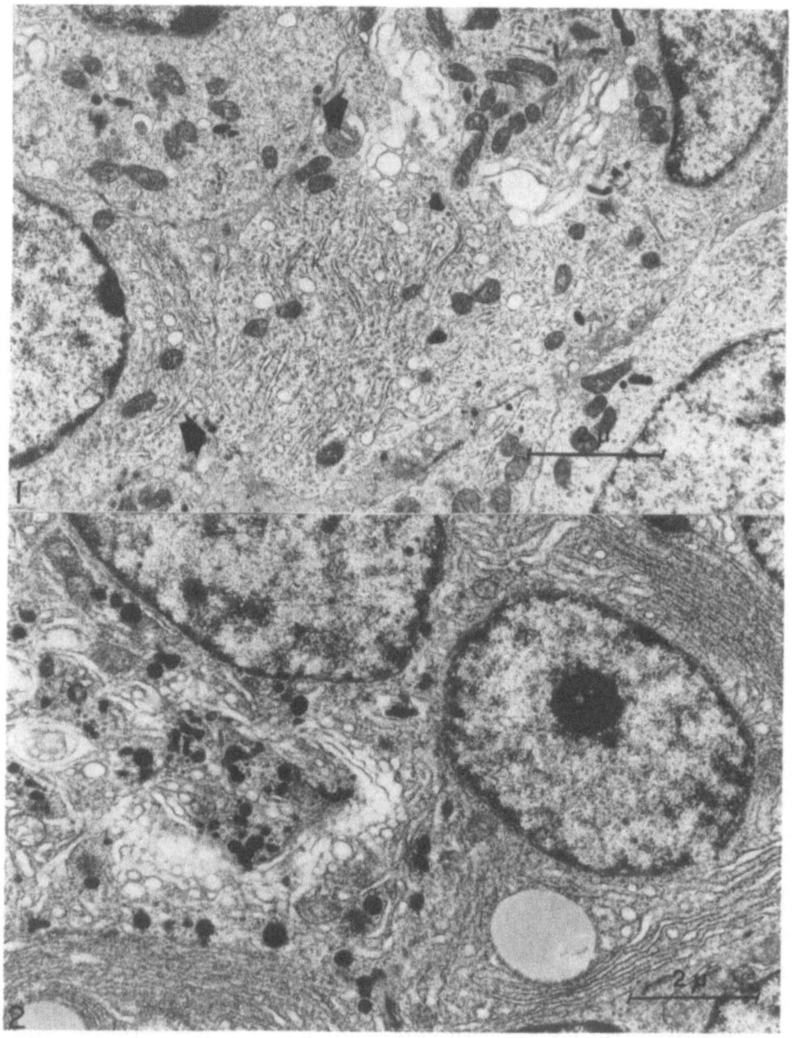

Figs. 1–2. (1) Sparsely granulated prolactin cell adenoma harvested from an aging female Long-Evans rat. Arrows show misplaced exocytosis (× 8,750). (2) Estrogen-induced pituitary adenoma in rat removed from the sella turcica showing the prolactin cell component (× 8,330).

groupings with rudimentary cilia, are frequently found. The adenoma cells are usually sparsely granulated, but there is often a marked variation in the number of secretory granules not only among various tumors, but also from cell to cell within the same tumor. The secretory granules are evenly electron dense and measure 150–500 nm. Misplaced exocytosis (8), the extrusion of secretory granules on the lateral cell surface into the extracellular space, distant from capillaries and intercellular extensions of the basement membrane, is a characteristic finding. Another common feature in many adenoma cells is the presence of pigment granules (6). Crinophagy, the fusion of secretory granules with lysosomes, seems to be the first step in pigment formation. Subsequently, lysosomes increase in size, number and electron density and may occupy large areas of the cytoplasm. Focal accumulation of pigment-laden macrophages is a conspicuous change in many adenomas. Occasionally, the formation of follicular structures with junctional complexes in their apical portions is also seen. The adenoma cells forming the follicles show no marked dedifferentiation and are similar to the follicular cells observed in the nontumorous human pituitary (9). Capillaries can be congested and erythrocytes can be found extravascularly in widened tissue spaces not lined by endothelium.

Based on the electron microscopic findings, it is evident that pituitary adenoma cells in aging female Long-Evans rats closely resemble estrogen-stimulated prolactin cells and human sparsely granulated adenomatous prolactin cells (10). These tumors are monomorphous and monohormonal, secreting only PRL. The development of adenomas in female Long-Evans rats is preceded and accompanied by prolactin cell hyperplasia. In this strain, prolactin cells seem to be more prone to undergo neoplastic transformation.

A recent histologic and immunocytologic study indicates that spontaneous pituitary adenomas, in Wistar rats, may contain not only prolactin but also GH, TSH and ACTH (11). A fine structural study of these tumors is awaited with great interest since electron microscopic investigation may unravel hitherto unknown tumor types.

2.2. Estrogen-induced pituitary adenomas

The pituitary tumors, induced by prolonged estrogen treatment (12, 13, 14), were widely used as disease models and their light microscopic and ultrastructural features were extensively investigated (1, 7, 15, 16, 17, 18).

Estrogen-induced pituitary tumors are chromophobic or slightly acidophilic, carmoisinophilic and PAS negative by light microscopy. The adenomas secrete primarily PRL, although production of GH has also been reported (16, 17). The immunoperoxidase technique demonstrates PRL in the majority of adenoma cells, while in a few cells GH can be noted (1, 7, 18).

By electron microscopy, estrogen-induced hypophysial adenomas resemble spontaneous pituitary tumors of old female Long-Evans rats (1, 7, 16). The predominating cell type is the stimulated, sparsely granulated prolactin cell (Fig. 2). The RER and Golgi complexes are prominent; the secretory granules are evenly electron dense and measure 150–400 nm. Misplaced exocytosis is a characteristic finding. Lysosomes are numerous; accumulation of pigment granules can also be encountered. Dilatation of capillaries and patchy congestion are conspicuous. Unlike spontaneous neoplasms, estrogen-induced tumors usually contain growth hormone cells. They occur singly or in small foci and seem to be small and unstimulated, containing numerous spherical secretory granules. The immunoperoxidase technique demonstrates that the two cell types contain either PRL or GH, but not both (7). Thus, estrogen-induced pituitary adenomas consist of two different cell populations – prolactin cells and growth hormone cells; they are bimorphous and bihormonal, resembling mixed growth hormone cell-prolactin cell adenomas of the human pituitary. These tumors are termed mammosomatotrophinomas by Furth et al. (1, 15).

2.3. Pituitary adenomas following irradiation

Total body, head and neck, ionizing or deuteron beam irradiation are often associated with the development of pituitary tumors in rats and mice (1, 19). Whether irradiation leads directly to tumor formation or only increases the incidence of spontaneous pituitary tumors is unresolved. Since the endocrine activity of radiation-induced pituitary adenomas differs from that occurring spontaneously in aging rats, it appears that irradiation is an inducer rather than a promoter.

Histologically, radiation-induced pituitary adenomas are chromophobic with pleomorphism, mitotic figures and a relatively rapid growth rate. The tumors are autonomous; their morphology, hormone production and growth are unaffected by changes in endocrine environment. They are transplantable, grow in the host animals and secrete PRL, GH or ACTH, differing from spontaneous pituitary adenomas of aging Long-Evans rats which produce PRL and no GH or ACTH. To date, this is the only way to induce an ACTH-secreting rat pituitary adenoma (1, 15). The radiation- induced multihormonal tumors are composed of undifferentiated, pleomorphic cells. Their derivation has yet to be elucidated (1, 15). It is not clear whether they consist of one single cell type, which accounts for multihormonal secretion, or two or three monohormonal cell types. Detailed ultrastructural and immunocytologic investigations may clarify the cytogenesis, hormone synthesis, storage and release of these neoplasms.

2.4. Pituitary adenomas in hypothyroid animals

Long-standing primary hypothyroidism may lead to the formation of pituitary tumors, indicating that stimulation of thyrotroph cells, due to a lack of negative feedback effect of thyroid hormones, may cause hyperplasia and, subsequently, neoplastic transformation (1, 15). Surgical, radioactive thyroidectomy, administration of various antithyroid compounds or a low iodine diet result in the development of thyrotrophic tumors in mice and rats (1, 15, 20, 21).

The histogenesis of thyrotrophic pituitary tumors is described in detail by Furth et al. (1, 22). Following radiothyroidectomy, small nodules form which are composed of a mixed population of hypertrophied thyrotrophs and of smaller cells. Transplantation of the enlarged pituitaries yields fully dependent tumors which grow only in hypothyroid hosts. By histology, these tumors are composed of large, chromophobic cells regarded as hypertrophied thyrotrophs. The immunoperoxidase technique reveals TSH in the adenoma cells. There is also positive immunostaining for gonadotrophic hormones due to the presence of immunologically similar subunits between TSH, FSH and LH. It may also be that the tumors secrete, in addition to TSH, gonadotroph hormones.

During subsequent transplantations the tumors become autonomous and change their biologic activity. They produce predominantly GH and not TSH. Furth et al. (1,22) interpret these tumors as monomorphous adenomas, composed of one single cell type capable of producing several hormones. The fine structural features of these tumors has not been studied in detail.

Astatine-211 which, in sufficient dosage, destroys the thyroid gland, induces pituitary adenomas in rats (23, 24). They grow well in athyroid hosts and poorly in euthyroid hosts. By electron microscopy, the adenoma cells contain a conspicuous Golgi apparatus, well developed endoplasmic reticulum and many mitochondria (23). The secretory granules are characteristic of those of the thyrotrophs, measuring up to 140 nm. Bioassays show that the tumors contain TSH, consistent with the assumption that they consist of thyrotrophs.

Transplantations of the tumors changes their behavior. The grafted tumors grow poorly in athyroid hosts and vigorously in normal hosts and their fine structure resembles mammosomatotrophic adenomas. According to Ueda and Mori (23), Astatine-211 induced pituitary tumors are originally mixed neoplasms consisting of thyrotrophic and mammosomatotrophic cells. During the passages, the thyrotrophic cell line is lost and the mammosomatotrophic cells survive and proliferate.

2.5. Pituitary adenomas in gonadectomized animals

Basophilic pituitary adenomas were reported in certain strains of neonatally gonadectomized mice (25) and rats (26). However, Furth and Clifton (15) found no pituitary tumors in gonadectomized mice and rats and expressed the view that induction of pituitary adenomas by gonadectomy had not yet been proven.

2.6. Miscellaneous pituitary tumors

Several workers reported the occurrence of pituitary tumors in animals under various conditions (1, 15). However, whether these tumors were causally related to the experimental procedures or were spontaneous pituitary neoplasms was not established. Conclusions, especially those in the older literature, must be viewed with caution. Information on immunocytologic and fine structural features of these neoplasms is scarce, and their cytogenesis and cellular composition are not known.

2.7. Adenomas in pituitary grafts

Transplantation of the pituitary distant from the hypothalamus disrupts the direct connection between the median eminence and the adenohypophysis and leads to an arrest of flow of hypothalamic hormones and factors via the portal circulation to the anterior lobe. Various hypothalamic substances regulate hormone release from adenohypophysial cells. In hypophysectomized animals bearing pituitary grafts GH, ACTH, FSH, LH as well as TSH secretion are markedly reduced, while blood PRL levels rise, indicating that PRL discharge is inhibited by the hypothalamus.

The fine structural features of the grafted pituitary are described (27, 28). Prolactin cells in the grafts show morphologic evidence of active secretion and undergo hyperplasia. The RER membranes are well developed in these prolactin cells; the Golgi complexes are conspicuous; secretory granules are sparse and small, measuring 100–350 nm. Exocytoses are also seen.

Pituitary grafts, in certain strains of rodents, may undergo neoplastic transformation (29, 30). Prolonged estrogen treatment can also lead to the development of pituitary adenomas in the grafts transplanted under the renal capsule or to the anterior chamber of the eye (31, 32). The fine structural features of these tumors are indistinguishable from those revealed in the sella turcica of estrogen-treated rats and are preceded by prolactin cell hyperplasia (32). Whether tumors in the grafts contain growth hormone cells has not been sufficiently explored.

2.8. Transplanted pituitary tumors

Successive transplantation of pituitary adenomas in animals made it possible to investigate their ultrastructural features, endocrine function, growth rate and biologic behavior (1, 15).

The ultrastructural characteristics of pituitary adenomas transplanted to hormonally normal or abnormal animals has been the subject of several reports (1, 7, 33). It is important to stress that pituitary adenomas may change their secretory pattern following transplantation and there is no satisfactory explanation for this change in secretory activity. Tumor cell heterogeneity may provide an answer: cells which produce GH, PRL and/or ACTH continue to grow; those producing other hormones are more sensitive

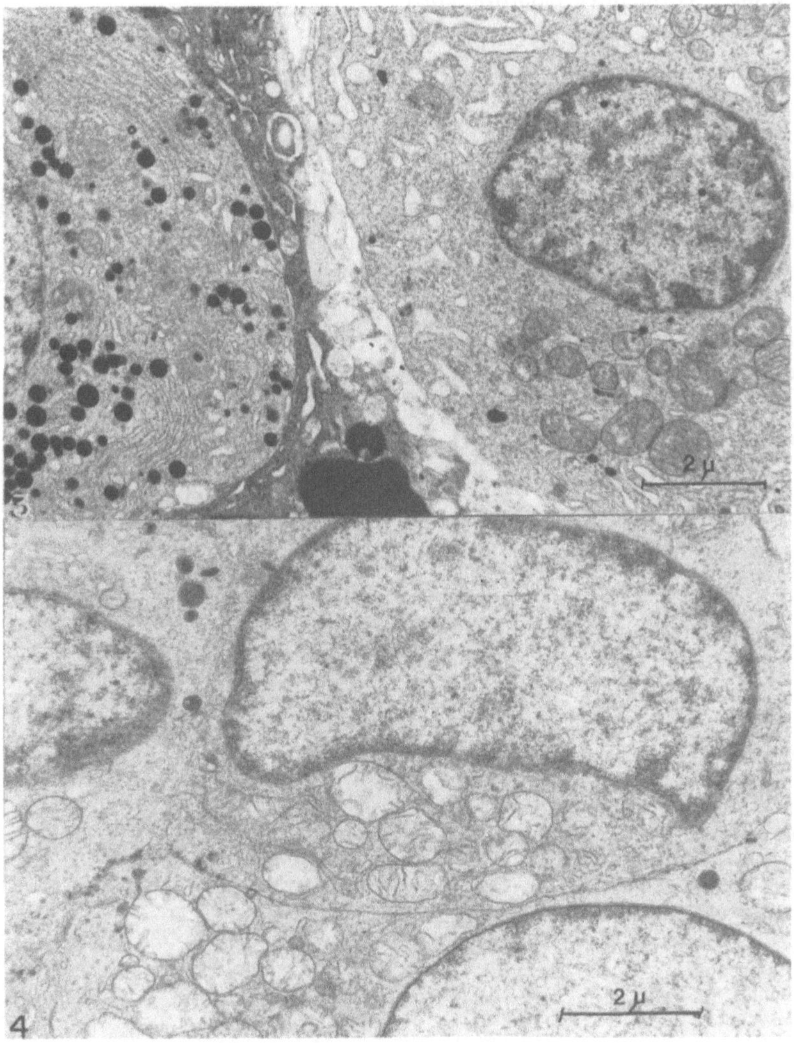

Figs. 3–4. (3) MtT.W 10 pituitary tumor transplant removed from the subcutaneous tissue of a Wistar-Furth rat (× 8,330). 4) MtT.F4 pituitary tumor transplant removed from the subcutaneous tissue of a Fisher rat (× 9,100).

to the new milieu and fail to survive with successive transplantation. Based on these considerations, the ultrastructural features of a transplanted tumor must be investigated in close correlation with its secretory activity.

Transplanted tumors may also change their fine structural appearance. They may be pleomorphic and may contain degenerative areas with hemorrhage and necrosis. In some cases, the cellular composition and cytogenesis of grafted pituitary adenomas can only be determined with immunoelectron microscopy. At the light microscopic level, lack of immunostaining has been encountered and can be interpreted in several different ways: the cells may be actively manufacturing hormones which are not stored, but immediately released; the secretory product may be immunologically different and not recognized by conventional antisera; or the cells are inactive and do not synthesize hormones. At the ultrastructural level, many cells possess few RER profiles, inconspicuous Golgi regions and small secretory granules. Occasionally even in these cases, immunoelectron microscopy may fail to reveal hormones known to be present in the tumor.

Three transplanted rat pituitary tumors have been studied recently in our laboratory by transmission electron microscopy (Fig. 3, 4) and immunoelectron microscopy (7). The MtT.W10 tumors contain immunoreactive PRL and GH in the same cell, as well as in separate cells.These neoplasms seem to be composed of three separate cell types. The MtT.F4 tumors which produce GH, PRL and ACTH are characterized by a single cell type with a lack of morphologic differentiation, few secretory granules and no definite immunopositivity. The MtT.W5 tumors secrete GH and contain one discernible cell type. Immunocytochemistry is inconclusive in revealing the hormone content of the tumor cells.

2.9. Pituitary tumors in culture

Pituitary adenoma cell cultures have great potential in the better understanding of structure-function correlation and hypophysial pathology. With appropriate methodology, animal pituitary adenoma cells survive and grow *in vitro* and may also maintain their endocrine activity. The cultures can be investigated by electron microscopy (34).

3. Pituitary tumors in human subjects

Based on their tinctorial properties, human pituitary adenomas were originally classified into three separate entities: chromophobic, acidophilic and basophilic adenomas (35, 36, 37, 38). Chromophobic adenomas were assumed to be endocrinologically inactive tumors. Acidophilic adenomas, often associated with acromegaly or gigantism, were thought to secrete GH. Basophilic adenomas accounting for the development of Cushing's disease were considered to be the source of ACTH. It became apparent that classifications based on the staining affinities of the cell cytoplasm were not reliable, did not distinguish clearly among the various cell types, and yielded no information on the cellular derivation and

Table 1. Classification and incidence of pituitary adenomas.

Type	Incidence	
	Number of cases	In percentage
Densely granulated growth hormone cell adenomas	40	8
Sparsely granulated growth hormone cell adenomas	47	10
Densely granulated prolactin cell adenomas	1	<1
Sparsely granulated prolactin cell adenomas	148	30
Mixed growth hormone cell-prolactin cell adenomas	22	4
Acidophil stem cell adenomas	18	4
Mammosomatotroph cell adenomas	6	1
Corticotroph cell adenomas associated with Cushing's disease	25	5
Corticotroph cell adenomas associated with Nelson's syndrome	17	3
Silent corticotroph cell adenomas	34	7
Gonadotroph cell adenomas	10	2
Thyrotroph cell adenomas	2	<1
Null cell adenomas	82	17
Oncocytomas	28	6
Unclassified adenomas	10	2
Total	490	100

biologic behavior of the tumors.

The application of electron microscopy resulted in marked progress. The fine structural investigation of adenohypophysial cells rendered it possible to define the various cell types in the nontumorous anterior lobe and reveal the cellular origin and composition of various neoplasms (10, 39, 40, 41, 42). With the application of the immunoperoxidase technique, it became possible to localize different hormones in specific cells in formalin-fixed and paraffin-embedded tissue (43, 44, 45). Autopsy material, kept in paraffin for many years, is also suitable for immunocytologic study (46, 47). The immunoperoxidase technique, also used at the electron microscopic level, demonstrates the subcellular localization of hormones (44, 45, 48, 49, 50). Radioimmunoassay is used to measure small quantities of various hormones in the blood and tissue (51, 52).

Based on electron microscopy and immunocytology, a new pituitary adenoma classification has been introduced by which pituitary adenomas can be separated into well defined entities (10, 40, 41, 42, 53). The various adenoma types are described as they appear in Table 1. This new classification relies on the fine structural features and hormone storage of adenoma cells and takes into account the clinical manifestations and biochemical results. The incidence of various adenoma types is also tabulated. The figures are based on our own material, comprising 490 surgically-removed pituitary adenomas studied by histology, immunocytology and electron microscopy.

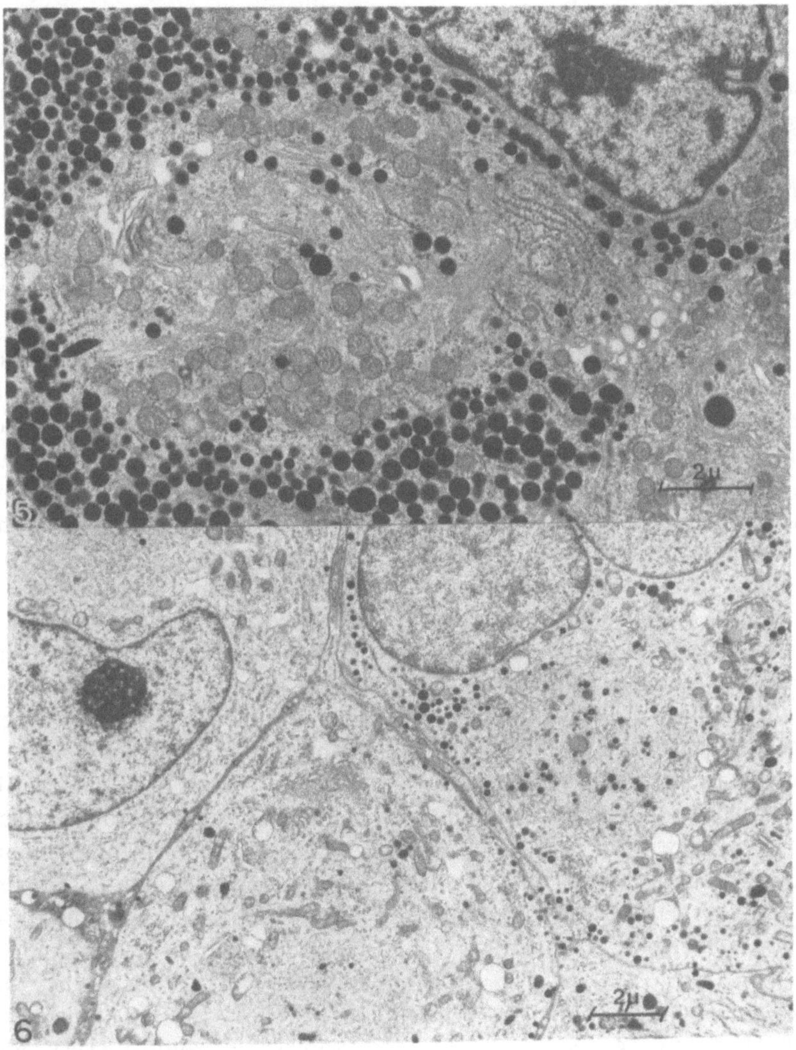

Figs. 5–6. (5) Densely granulated growth hormone cell adenoma of the human pituitary (× 6,300). (6) Sparsely granulated growth hormone cell adenoma of the human pituitary (× 4,340).

It can be seen from Table 1 that sparsely granulated prolactin cell adenomas occur most frequently in our material. If mixed growth hormone cell-prolactin cell adenomas, acidophil stem cell adenomas and mammosomatotroph cell adenomas are also included with these tumors, it becomes evident that 40% of hypophysial adenomas are capable of secreting PRL! It is also evident from Table 1 that tumors originating in the acidophilic cell line are more common than those deriving from the various basophilic cell lines, such as corticotroph, gonadotroph and thyrotroph cells. These findings are consistent with the assumption that the acidophilic cell line is more susceptible to neoplastic transformation than other cell types.

3.1. Densely granulated growth hormone cell adenomas (10, 39, 40, 41, 42, 53, 54, 55)

By light microscopy, these tumors are acidophilic adenomas. They secrete GH and are accompanied by acromegaly or gigantism.

By electron microscopy, the adenoma cells closely resemble growth hormone cells of the nontumorous pituitary, and the ultrastructural diagnosis of this tumor type is usually easy and creates no major problems. The adenoma cells are oval, similar to one another and show no marked pleomorphism (Fig. 5). The nuclei are spherical, usually centrally located, possess a light chromatin pattern and a prominent nucleolus. The cytoplasm is abundant and electron lucent. The RER membrane networks are well developed and consist of parallel arrays, studded with ribosomes on their external aspects and located chiefly in the periphery of the cell. Free ribosomes (polysomes) are numerous. The Golgi complex is conspicuous and crescent or ring shaped. The Golgi sacculi are slightly dilated at some places and contain a few immature secretory granules. Mitochondria are present in moderate number; they are oval or rod shaped with lamellar or cleft-like cristae and a moderately electron dense matrix. Secretory granules are numerous, mostly spherical with no variations in their electron density. They have closely apposed limiting membranes and the majority measures 300–600 nm. In a few cells, larger secretory granules are also encountered. Exocytosis is not seen.

3.2. Sparsely granulated growth hormone cell adenomas (10, 39, 40, 41, 42, 53, 54, 55)

These tumors are chromophobic adenomas by light microscopy. They produce GH and are associated with acromegaly or gigantism. There is no close correlation between GH release, size and number of secretory granules, and no conclusions can be drawn from the volume density of secretory granules about the endocrine activity of the tumor (53).

Unlike densely granulated growth hormone cell adenomas, cells of the sparsely granulated growth hormone-producing tumors do not resemble growth hormone cells seen in the nontumorous human pituitary (Fig. 6). The adenoma cells are irregular in shape, vary considerably in size and may show different degrees of pleomorphism. The nuclei are crescent shaped, and often exhibit deep indentations. Large cells with eccentrically located multiple nuclei may be noted. The RER membranes may be abundant in the cytoplasm of many adenoma cells. Nebenkern formations, concentric whorls of RER membranes, may also be found. In contrast to cells of the nontumorous pituitary or those of densely granulated growth hormone cell adenomas, sparsely granulated adenomatous growth hormone cells usually contain tubular, smooth andoplasmic reticulum membranes (SER) found in close association with the fibrous bodies. In some cells, free ribosomes and polysomes may also be numerous. The Golgi complex may be well developed. The Golgi sacculi are arranged in a flattened ring-like pattern, are often crescent shaped and usually contain a few immature secretory granules. Secretory granules are sparse, evenly electron dense, spherical, membrane bound and measure 100–250 nm. Cells containing a few larger secretory granules up to 500 nm may also occur. No exocytoses are encountered.

The most crucial distinguishing feature of sparsely granulated adenomatous growth hormone cells is the presence of fibrous bodies (56). The fibrous bodies are globular and can already be identified by light microscopy located at the concave side of the nucleus; they are PAS and lead hematoxylin negative, slightly acidophilic and exhibit no staining for GH by the immunoperoxidase technique. Electron microscopy reveals that they consist of type II microfilaments with an average width of 115 Å. The microfilaments show no periodicity; they are arranged in a circular manner forming globular aggregates and engulfing various cytoplasmic organelles. The SER is always associated with the fibrous bodies and may represent their principal component. Although the frequency of fibrous bodies varies from tumor to tumor, they are often numerous and detectable in the cytoplasm of almost every adenoma cell. Fibrous bodies are diagnostic for the acidophilic cell line, indicating that the adenoma originates in growth hormone cells or their precursors.

Another common feature of sparsely granulated

growth hormone cell adenomas is the occurrence of supernumerary centrioles usually situated adjacent to or within the fibrous bodies (57). In cells of the nonadenomatous adenohypophysis or in various other pituitary adenoma types, a pair of centrioles, with or without cilia, is a frequent finding. The mechanism accounting for their accumulation in sparsely granulated growth hormone cell adenomas is obscure.

Densely granulated growth hormone cell adenomas in tissue culture show a gradual reduction in the number and size of secretory granules and a dedifferentiation of the cell cytoplasm. Even the development of fibrous bodies can be followed in degranulating adenomatous growth hormone cells in tissue culture, providing evidence that densely granulated and sparsely granulated growth hormone cells belong to the same cell type. Sparsely granulated growth hormone cells are less differentiated than their densely granulated counterparts. The separation into densely or sparsely granulated tumors has practical significance, since sparsely granulated growth hormone cell adenomas are usually more aggressive and exhibit a faster growth rate than the densely granulated growth hormone cell tumors. Endocrinologically, as far as GH secretion and blood GH levels are concerned, there is no basic difference between the two variants of growth hormone cell adenomas.

3.3. Densely granulated prolactin cell adenomas (10, 40, 41, 42, 58, 59)

These tumors are rare. We examined only one case of densely granulated prolactin cell adenoma and few cases appear in the literature. Our patient, a young woman, had amenorrhea, galactorrhea and hyperprolactinemia, indicating that despite considerable hormone storage, the adenoma cells were capable of actively secreting PRL. More cases must be studied before correlations can be made between ultrastructural features and secretory activity.

By light microscopy, the tumors are acidophilic adenomas and are indistinguishable from densely granulated growth hormone cell adenomas. In contrast with these tumors, however, the immunoperoxidase technique reveals the presence of PRL in the cytoplasm of adenoma cells.

By electron microscopy, the adenoma cells are very similar to densely granulated prolactin cells found in the nontumorous pituitary. They are oval or oblong and possess a regular nucleus. The nucleoli are usually well developed. The cytoplasm is abundant and contains prominent rough-surfaced endoplasmic reticulum membranes which are chiefly in the periphery of the cytoplasm. The Golgi apparatus is conspicuous. The Golgi sacculi are frequently crescent shaped and contain dense, pleomorphic, immature secretory granules. The mitochondria are moderate in number, spherical or oval with lamellar cristae and moderately electron dense matrix. The secretory granules are oval or slightly irregular, evenly electron dense, membrane bound and measure 500–1200 nm. Exocytoses may be present. Some secretory granules, especially those adjacent to the Golgi region, are pleomorphic and have loosely-fitted limiting membranes. The plasma membranes are closely apposed with occasional intercellular junctions.

3.4. Sparsely granulated prolactin cell adenomas (10, 40, 41, 42, 59, 60, 61)

These tumors are the most common among pituitary neoplasms. They occur in both women and men and may be associated with amenorrhea, galactorrhea, infertility, decreased libido and impotence. However clinical symptoms may not be obvious, especially in men, and the morphologic examination of the tumor is useful in revealing its true nature. Sparsely granulated prolactin cell adenomas are always associated with various degrees of hyperprolactinemia.

In normal men and women and in patients with other pituitary tumor types, provided that functioning pituitary tissue is present, administration of TRH causes an increase in blood TSH levels and also a marked rise in blood PRL concentrations. An interesting abnormality in patients with prolactin-secreting tumors is the lack of blood PRL response to TRH stimulation (62). Administration of TRH induces an elevation of blood TSH levels, but no significant increase occurs in blood PRL levels. The mechanism accounting for this unresponsiveness is not clearly understood, but it is a useful laboratory test in distinguishing prolactinomas from other types of sellar tumors.

Larger prolactinomas are associated with higher elevations of blood PRL concentrations (63); thus there is a correlation between the size of the tumor and the degree of hyperprolactinemia. No correlations are evident between blood PRL levels and light or electron microscopic features of the adenoma cells, including volume density of rough-surfaced endoplasmic reticulum, Golgi complexes, mitochondria and secretory granules, or size of secretory granules and number of exocytoses (63).

It is noteworthy that hyperprolactinemia does not necessarily mean the presence of a prolactin-producing adenoma. Various diseases such as different tumors or inflammatory processes in the sella region may compress or damage the hypothalamus or pituitary stalk and interfere with the production or release of prolactin-inhibiting factor (which appears to be dopamine), or suppress its downward flow in the stalk via the portal vessels from the median eminence to the adenohypophysis (64). Consequently, adenohypophysial prolactin cells escape hypothalamic inhibition and PRL secretion increases. Only a careful morphologic investigation, including immunocytology and electron microscopy, can reveal the causes of the hyperprolactinemia. It must be emphasized that very high blood PRL levels and failure of a rise in blood PRL concentrations following TRH stimulation are strong indications that a prolactin-secreting tumor is directly responsible for the elevations and the endocrine changes. However, morphologic techniques can make the diagnosis of prolactin cell tumors with certainty.

By light microscopy, sparsely granulated prolactin cell adenomas are chromophobic and, using conventional staining procedures, are indistinguishable from other chromophobic adenoma types. The application of the immunoperoxidase technique is more useful in making the diagnosis, since it demonstrates the presence of immunoreactive PRL in the cytoplasm of adenoma cells (47, 60).

By electron microscopy, the adenoma cells are closely apposed, oval or irregular and possess large, indented, pleomorphic nuclei. The nucleoli are usually prominent. The cytoplasm is well developed and

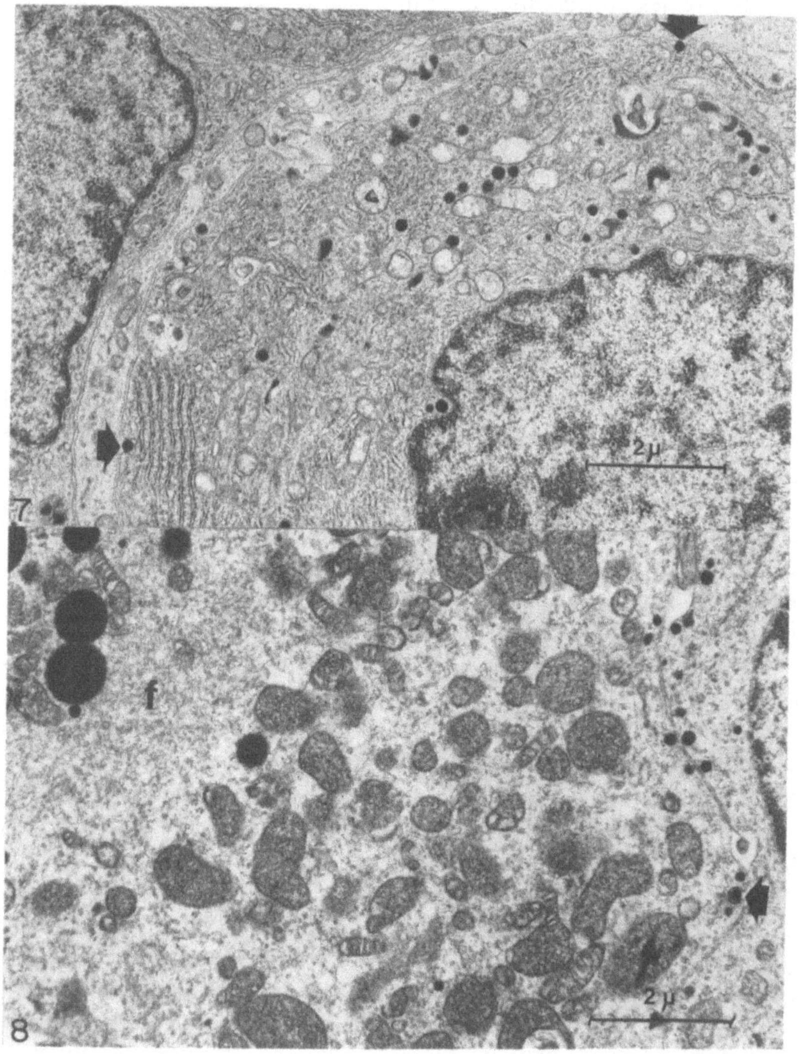

Figs. 7–8. (7) Sparsely granulated prolactin cell adenoma showing misplaced exocytosis (arrows) (× 9,520). (8) Acidophil stem cell adenoma showing a fibrous body (f) and misplaced exocytosis (arrows) in the same cell (× 9,100).

contains large amounts of RER membranes. Formation of Nebenkerns, concentric whorls of RER membranes, is a frequent finding. Free ribosomes and polysomes are numerous. The Golgi complexes are conspicuous, ring like, crescent shaped and may fill as much as a third of the entire cytoplasmic area. The Golgi sacculi are prominent and show focal dilatations containing pleomorphic immature secretory granules. The mitochondria are oval or rod shaped with lamellar cristae and fine granular matrix. The secretory granules are oval or irregular, membrane bound, evenly electron dense and measure 130–500 nm, the majority being 200–300 nm.

A distinctive ultrastructural feature of sparsely granulated prolacting cell adenomas is the extrusion of secretory granules on the lateral cell membranes into the extracellular space distant both from the perivascular area and the intercellular extensions of the basement membranes (Fig. 7). This type of secretory granule release is called 'misplaced exocytosis' (65) to stress that secretory granule discharge does not occur only on the capillary side of the cell, which is assumed to be the location of granule extrusion. The discharged secretory granules have fuzzy outlines and are situated in a membranous pit formed by the fusion of the plasmalemma and the limiting membrane of the secretory granules. Sparsely granulated adenomatous prolactin cells are similar to estrogen-stimulated nontumorous prolactin cells and those of spontaneous prolactin cell adenomas of the rat pituitary. These findings conclusively indicate that sparsely granulated adenomatous prolactin cells are in an active phase of secretion.

3.5. Mixed growth hormone cell-prolactin cell adenomas (10, 40, 41, 42, 66, 67, 68)

These unusual pituitary tumors are accompanied by acromegaly and, in some cases, by amenorrhea and galactorrhea. Blood GH and PRL concentrations are elevated. Morphologically, they represent true mixed adenomas and are composed of two different cell types: growth hormone cells and prolactin cells.

By light microscopy, mixed growth hormone cell-prolactin cell adenomas are acidophilic or chromophobic or partly acidophilic, partly chromophobic, and the immunoperoxidase technique reveals the presence of GH and PRL in the cytoplasm of adenoma cells. The two hormones are found in separate cell types.

By electron microscopy, the adenoma cells show the characteristics of either growth hormone cells or prolactin cells. The fine structural features of the tumor cells are identical to those seen in pure growth hormone cell or pure prolactin cell adenomas and may be densely or sparsely granulated. Every combination may occur; densely granulated growth hormone cells may mix with sparsely granulated prolactin cells or vice versa, or both growth hormone cells and prolactin cells may be densely or sparsely granulated. Adenoma cells belonging to the same cell type usually form smaller groups or individual growth hormone cells may be interspersed with single prolactin cells. A careful ultrastructural study is very important in diagnosing this tumor type. Most commonly, these tumors are composed of well differentiated growth hormone cells and prolactin cells although, in some cases, a few undifferentiated cells are also found. In our view, these tumors also arise from two distinct cell types. However, the possibility cannot be excluded that the source of these tumors is a precursor cell which can subsequently differentiate into two different mature cell types.

3.6. Acidophil stem cell adenomas (10, 41, 42, 69, 70)

Unlike mixed growth hormone cell-prolactin cell adenomas, acidophil stem cell adenomas are composed of one cell type, assumed to be the common committed precursor cell of growth hormone cells and prolactin cells. These immature tumors may appear as endocrinologically inactive neoplasms and are clinically often unaccompanied by acromegalic features. Blood GH concentrations are seldom elevated, but hyperprolactinemia is a common finding in the majority of patients.

By light microscopy, acidophil stem cell adenomas are chromophobic or slightly acidophilic; in most cases, chromophobic cells prevail. The immunoperoxidase technique reveals the presence of GH and PRL in varying numbers of cells. The two hormones are present in the cytoplasm of the same adenoma cell.

By electron microscopy, acidophil stem cell adenomas consist of closely-apposed, elongated or attenuated cells, possessing oval or indented nuclei. The nucleoli are usually not prominent. In the cytoplasm, the rough-surfaced endoplasmic reticulum networks are widely-dispersed. Free ribosomes and polysomes are numerous. The Golgi complexes are inconspicuous or moderately developed. The number and size of mitochondria may be considerably increased in many adenoma cells, indicating oncocytic transformation. The increase of mitochondrial volume density is associated with profound mitochondrial changes, such as loss of mitochondrial cristae, formation of cavities in the internal compartment, swelling, ballooning and gigantism. In the internal compartments of giant mitochondria, tubular structures are formed which are similar to those of the centriolar wall and of ciliary axonemes. Secretory granules are sparse, spherical or pleomorphic, evenly electron dense, membrane bound and measuring 150–300 nm. In some adenoma cells larger secretory granules, measuring up to 600 nm, may be observed. The presence of SER membranes and type II microfilaments in the Golgi region is a characteristic finding. Fully developed fibrous bodies, similar to those observed in sparsely granulated growth hormone cell adenomas and misplaced exocytosis as seen in sparsely granulated prolactin cell adenomas, are found in the same cell (Fig. 8).

The electron microscopic findings are consistent with the immunocytologic results supporting the view that these tumors may derive from the common progenitor of growth hormone cells and prolactin cells. The diagnosis of this tumor can be made conclusively by using electron microscopy and immunocytology. Acidophil stem cell adenomas are immature, rapidly growing tumors. Thus their diagnosis has definite practical significance.

3.7. Mammosomatotroph cell adenomas (71, 72)

These tumors have only recently been recognized. However, more work is required to establish the incidence of this tumor type, define its morphologic features, prove its cytogenesis and shed light on its biologic behavior. It appears that they have a slow growth rate and a good prognosis.

Mammosomatotroph cell adenomas are accompanied clinically by acromegaly. Blood GH concentrations are high and various degrees of hyperprolactinemia may also be evident. In our view, this adenoma is composed of one cell type showing the fine structural characteristics of mature growth hormone cells with the additional feature of extruded secretory material deposited in the intercellular space. It remains to be established whether mammosomatotroph cell adenomas are the more differentiated variants of acidophil stem cell adenomas.

By light microscopy, mammosomatotroph cell adenomas exhibit varying degrees of acidophilia and the immunoperoxidase technique reveals the presence of GH and PRL in the cytoplasm of the same adenoma cell. In some cases, immunoreactive PRL can only be detected at the electron microscopic level.

By electron microscopy, the adenoma cells exhibit the ultrastructural characteristics of well differentiated, densely granulated growth hormone cells. The RER is well developed. The Golgi complexes are prominent and are composed of several wide sacculi containing a few immature secretory granules. The secretory granules are usually numerous, oval or irregular, evenly electron dense, membrane bound and measure 200–1000 nm. The wide range in the size of secretory granules and the presence of very large secretory granules are characteristic features of this adenoma type. Misplaced exocytosis occurs. Large extracellular deposits of electron dense secretory material, distinguishing markers of this neoplasm, are revealed in several areas between neighboring adenoma cells (73).

3.8. Corticotroph cell adenomas associated with Cushing's disease (10, 40, 41, 42, 53, 74)

These tumors secrete ACTH and related peptides, such as β-LPH and endorphins and cause hypercorticism which becomes clinically manifest as Cushing's disease. They are often microadenomas, developing in the central mucoid wedge of the anterior lobe and can be removed by selective adenomectomy, usually via the transsphenoidal route.

Pituitary adenomas accounting for the development of Cushing's disease are composed of corticotroph cells. By light microscopy, they are most often basophilic adenomas and the cytoplasmic secretory granules yield a positive staining with PAS and lead hematoxylin. In rare cases, the tumors are chromophobic and only a few small PAS and lead hematoxylin positive cytoplasmic granules can be demonstrated. The immunoperoxidase technique reveals the presence of ACTH, β-LPH and endorphins in the cytoplasm of adenoma cells. The biologic and clinical significance of β-LPH and endorphin immunopositivity has yet to be elucidated.

By electron microscopy, corticotroph cell adenomas associated with Cushing's disease are composed of closely apposed cells which resemble corticotroph cells of the nontumorus adenohypophysis (Fig. 9). They are oval or polyhedral and possess spherical nuclei and conspicuous nucleoli. The cytoplasm is abundant and relatively electron dense. The RER is well developed and consists of scattered, short and focally dilated stacks. Free ribosomes and polysomes are numerous, partly accounting for the high electron opacity of the cytoplasm. Golgi complexes are conspicuous and the sacculi contain a few small, immature secretory granules. Mitochondria are present in moderate numbers. Secretory granules are spherical or slightly irregular, vary in electron density and measure 250–700 nm, the majority being 300–350 nm. Some cells are less densely granulated and possess smaller secretory granules. The secretory granules frequently line up along the plasmalemma, but exocytoses are not apparent. In some secretory granules a relatively wide electron lucent halo is seen between the electron dense core and the prominent limiting membrane, in others the granule content is electron lucent. Some secretory granules appear as basically empty shells; they consist only of the uninterrupted limiting membrane. This special form of secretion, in which the granule content is secreted directly into the hyaloplasm, is characteristic of corticotroph cells and is termed transmembrane effusion (75).

Similar to nontumorous corticotroph cells, adenomatous corticotroph cells contain a few bundles of type I microfilaments usually around the nucleus (76). These microfilaments have no periodicity and their width is about 70 Å. Under the influence of cortisol excess, type I microfilaments accumulate and can occupy large areas of the cytoplasm. They correspond to the Crooke's hyaline material by light microscopy. This change is found in corticotrophs of the nontumorous portions of the adenohypophysis in patients with Cushing's disease harboring ACTH-producing corticotroph cell adenomas (76, 77, 78, 79). In most of these cases, Crooke's hyalinization is extensive outside the adenoma, while no hyaline deposition occurs in the cytoplasm of adenoma cells. This difference between the adenomatous and non-adenomatous corticotroph cells indicates that the adenoma cells lose their responsiveness to the inhibitory effect of glucocorticoids; they become autonomous. Crooke's hyalinization also occurs in cortic-

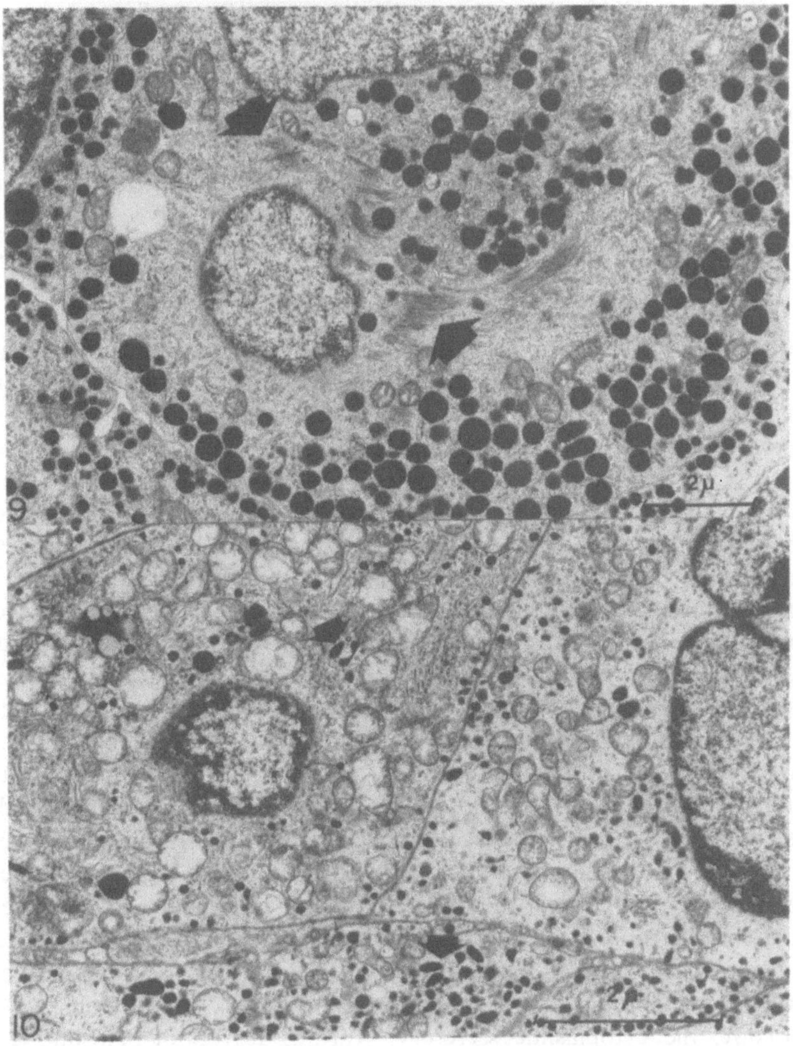

Figs. 9–10. (9) Densely granulated corticotroph cell adenoma removed from a patient with Cushing's disease. Arrows show type I microfilaments (× 7,560). (10) Silent corticotroph cell adenoma showing tear-drop-shaped secretory granules (arrows) (× 10,710).

otroph cells of the nontumorous pituitary of patients after protracted treatment with pharmacologic doses of cortisol or its derivatives, in ectopic ACTH syndrome and cortisol-producing adrenocortical tumors. We have seen three corticotroph cell adenomas with marked accumulation of type I microfilaments which was already evident by light microscopy as deposition of Crooke's hyaline material. It is conceivable that, in these cases, the adenoma cells did not lose their responsiveness to the inhibitory effect of cortisol. Hence, corticotroph cell tumors in the human pituitary may be autonomous or cortisol sensitive.

The chromophobic variant of corticotroph cell adenomas is less common than the basophilic tumor. By electron microscopy, the adenoma cells are sparsely granulated and appear to be less differentiated. The secretory granules are smaller than those of the densely granulated variant, measuring 150–350 nm. Type I microfilaments are usually inconspicuous or absent. Lysosomes and enigmatic bodies occur occasionally in both variants. Enigmatic bodies are large lysosomal formations which contain one or two large irregular vacuoles surrounded by electron dense amorphous material (80).

3.9. Corticotroph cell adenomas associated with Nelson's syndrome (10, 40, 41, 42, 53, 81, 82)

These tumors are usually large, basophilic neoplasms, composed of corticotroph cells and secreting ACTH and related peptides, such as β-LPH and endorphins. They occur in patients who had preexisting Cushing's disease and because of hypercorticism, underwent bilateral adrenalectomy. The patients show extensive skin pigmentation and blood ACTH concentrations are very high. Whether the pituitary tumor exists before adrenalectomy or whether it develops only after removal of the adrenals is an intriguing question. In the pituitaries of patients with Cushing's disease, dying without adrenalectomy or within a few months after adrenalectomy, the presence of adenoma can be demonstrated. Hence, there is little doubt that neoplastic transformation occurs before adrenalectomy and the tumor accounts for the clinical and biochemical manifestations of Cushing's disease. However, after removal of the adrenals, the tumor exhibits a more rapid growth rate and the microadenoma transforms to macroadenoma and causes local symptoms.

By light microscopy, the cytoplasmic secretory granules of the pituitary tumor cells in Nelson's syndrome stain positively with the PAS technique and lead hematoxylin. The immunoperoxidase technique reveals the presence of ACTH, β-LPH as well as endorphins in the cytoplasm of adenoma cells.

By electron microscopy, pituitary adenomas from patients with Nelson's syndrome are very similar to those with Cushing's disease. In patients without functioning adrenals, the pituitary tumors contain fewer type I microfilaments than seen in the tumor cells of patients who have not undergone adrenalectomy. The functional significance of microfilaments is not known, but it may be that they can suppress the intracellular migration of secretory granules from the Golgi complexes to the cell periphery. More work is required to shed light on the genesis and functional significance of these microfilaments.

Corticotroph cell adenomas of patients with Nelson's syndrome are usually densely granulated. In a few cases, however, sparsely granulated corticotroph cells predominate. These tumors show only slight PAS and lead hematoxylin positivity, are less differentiated, have a faster growth rate and correspond to chromophobic adenomas by light microscopy. There is no close correlation between volume density of secretory granules and endocrine activity of corticotroph cell adenomas. Actively-secreting tumors can also be densely granulated.

3.10. Silent corticotroph cell adenomas (41, 42, 83, 84)

These tumors are not uncommon and can pose difficult problems in the differential diagnosis. They contain immunoreactive ACTH, β-LPH and endorphins, but are clinically unassociated with signs of increased ACTH secretion. In some cases, blood ACTH may be elevated as measured by radioimmunoassay. Blood cortisol levels are usually within the normal range and manifest Cushing's disease is not evident. There are, however, border line cases and it is difficult to draw a definite line between adenomas with Cushing's disease and silent tumors. It is known that periodic hypersecretion of ACTH may occur in Cushing's disease and, for reasons obscure at present, spontaneous remission and recurrence may occur. The natural history of Cushing's disease is insufficiently explored, and more knowledge is needed to better understand the secretory process and the factors which regulate the extent and pace of ACTH synthesis and release.

By light microscopy, silent corticotroph cell adenomas are chromophobic or basophilic showing various degrees of PAS and lead hematoxylin positivity. The immunoperoxidase technique has an important role in the diagnosis for it demonstrates the presence of ACTH, β-LPH and endorphins in the cytoplasm of adenoma cells.

The fine structural features of silent corticotroph cell adenomas are not uniform. By electron microscopy they can be indistinguishable from those corticotroph cell adenomas which actively secrete ACTH and are accompanied by Cushing's disease, whereas in other cases, ultrastructural investigation reveals striking morphological differences between silent and functioning corticotroph cell adenomas. It is possible that more subtypes exist, but these have yet to be defined morphologically, histochemically, as well as clinically. Some silent corticotroph cell adenomas possess unusual ultrastructure and only the immunocytologic results can prove that these tumors consist of peptide hormone-producing cells. It may be that some of these tumors originate in the pars intermedia corticotrophs which spread to the neurohypophysis from the anterior lobe or pars intermedia. ACTH-containing cells are also evident in the lining epithelium of the pars intermedia cysts, but the functional significance of these cells is not well known. It may be that they have a different endocrinological function than adenohypophysial corticotroph cells.

The mechanisms accounting for the hormonal inactivity are not clear. In some silent tumors, lysosomes are increased in number and size, and crinophagy as well as autophagy are evident, suggesting that the hormonal inactivity is due to increased lysosomal function leading to intracellular degradation of the hormonal products. It may be that a biologically active hormone is synthesized, then taken up and digested by lysosomes and not discharged. In two of these tumors, a striking involution of Golgi complexes suggests that an intracellular defect in hormone production and processing for discharge might exist. Abnormalities may occur in the release mechanism. It is also possible that silent corticotroph cell adenomas synthesize molecules which lack ACTH bioactivity but, immunologically, are similar to and cross-react with ACTH when the immunoperoxidase technique is applied. In several cases, adenoma cells with tear-drop-shaped secretory granules are found by electron microscopy (Fig. 10). Some of the silent corticotroph cell adenomas may represent undifferentiated tumors which have lost their ability to produce biologically active ACTH. The cell types comprising silent corticotroph adenomas may have an endocrine function other than ACTH production. The study of more cases and the introduction of new methods will possibly result in further separation of silent corticotroph cell adenomas, clarification of the relationship between hormonally seemingly inactive adenoma cells and actively secreting corticotroph cells, and a better understanding of the mechanisms accounting for the lack of signs of hormone hypersecretion. It is important at present to know that silent corticotroph cell adenomas exist and that they must be considered in the differential diagnosis of pituitary tumors by the surgical pathologist.

3.11. Gonadotroph cell adenomas (41, 42, 85, 86)

These infrequent tumors may produce the two gonadotroph hormones, FSH and LH, simultaneously. In some cases, however, the discharge of the two hormones is not equal and FSH or LH is the major secretory product. Gonadotroph cell adenomas have only recently been recognized, thus their incidence is not yet established, and they may occur more frequently than previously thought. The tumors do not have a uniform morphology and may differ markedly in cellular differentiation. In some cases, the tumors are diagnosed in patients with long-standing primary hypogonadism. It is evident that a lack of gonadal steroids stimulates the pituitary gonadotrophs and blood FSH and LH levels are elevated, but whether or not protracted stimulation of gonadotroph cells causes neoplastic transformation is not clear.

By light microscopy, using conventional stains, gonadotroph cell adenomas are chromophobic, although fine cytoplasmic PAS positive granules may be noted. The immunoperoxidase technique reveals the presence of FSH and/or LH in the cytoplasm of adenoma cells. Immunocytologic techniques are very valuable in making the diagnosis, for some of these tumors are incompletely differentiated and their fine structural features are not sufficiently distinctive to permit their identification by electron microscopy. It is important to stress that, for the immunocytologic diagnosis, antisera raised against the β-subunits of FSH and LH must be used, since the α-subunits of FSH, LH and TSH are immunologically very similar and cross react with one another.

By electron microscopy, adenomatous gonadotroph cells may seem similar to nontumorous gonadotrophs, or appear undifferentiated and differ considerably from their nontumorous counterparts. In the majority of cases the adenoma cells are closely apposed and polyhedral with long cytoplasmic processes. The nuclei are eccentrically located, spherical or oval and the nucleoli are usually conspicuous. The cytoplasm is moderately developed. The RER membrane networks are randomly distributed and arranged in slim cisternae or unevenly dilated profiles containing a fine granular, electron lucent substance. Free ribosomes and polysomes are usually numerous. The Golgi profiles are prominent and exhibit cyst-like dilatations, endowing the complex with a honeycomb appearance. Secretory granules inside the sacculi are rarely found. Large, spherical membrane-bound bodies, measuring 500–800 nm in diameter are also revealed. They contain a fine granular substance and are usually close to the Golgi area. The origin of these structures remains to be established. Mitochondria, present in moderate numbers, are oval or slightly irregular with lamellar or tubular cristae and a fine granular matrix. Secretory granules vary in number. They are spherical, usually possess a prominent electron lucent halo between the dense core and the limiting membranes, vary somewhat in electron density and measure up to 200 nm, although the majority of them have a diameter of only about 100 nm. Granule extrusions are not encountered. Microtubules may be numerous and prominent.

In other tumors, the adenoma cells are elongated, possessing elongated or irregular nuclei and are so undifferentiated that their origin cannot be established. The cytoplasm is not abundant and contains relatively few scattered RER profiles in the form of short stacks. The Golgi complexes are usually not conspicuous. The secretory granules are sparse, spherical, vary in electron density and measure 100–200 nm. The lining up of secretory granules along the cell membrane is a frequent finding, however, exocytoses are not observed. Microtubules are numerous and conspicuous. The cellular derivation and composition of these tumors cannot be revealed conclusively by ultrastructural study. The demonstration of gonadotroph hormones in the tumors, by radioimmunoassay or immunocytology, is of fundamental importance in the diagnosis. To our knowledge, only a few cases of gonadotroph hormone-secreting adenomas have been studied by electron microscopy so far; more cases must be investigated to unravel the fine structural details of these interesting neoplasms, elucidate their biologic behavior and correlate their morphologic features with secretory activity. Long-term follow up studies are also needed. The defining of their clinical and biochemical presentation, will allow them to be diagnosed before surgical removal.

3.12. Thyrotroph cell adenomas (41, 42, 87, 88, 89, 90)

These tumors are rare and few have been studied so far by immunocytology and electron microscopy. Some patients with thyrotroph cell adenomas have long-standing primary hypothyroidism, suggesting that the tumor is due to protracted overstimulation of the pituitary thyrotroph cells, resulting from sustained thyroid deficiency. In the pituitaries of these patients, before the development of an adenoma, thyrotroph cells gradually enlarge, become vacuolated and transform to thyroidectomy cells. Further stimulation leads to focal thyrotroph cell hyperplasia which becomes more and more extensive. Subsequently, small nodules are formed composed of thyrotroph cells similar to those seen in the pituitary of rodents with protracted primary hypothyroidism. In many patients with primary hypothyroidism, the sella turcica is enlarged when seen in x-ray examination. Hence, the formation of adenomas is assumed to be preceded by hyperplasia.

In some cases, thyrotroph cell adenomas are not associated with pre-existing primary hypothyroidism. These patients show clinical and biochemical evidence of hyperthyroidism, accompanied by high or not suppressed blood TSH levels and pituitary tumor. Hyperthyroidism with elevated blood TSH concentrations and pituitary tumor is indicative of TSH-secreting pituitary adenoma. The hyperthyroidism in these patients is secondary to TSH oversecretion by adenomatous thyrotroph cells.

By light microscopy, thyrotroph cell adenomas are chromophobic. The cytoplasm of adenoma cells may possess a few small secretory granules which stain positively with the PAS technique, aldehyde fuchsin or aldehyde thionin. The immunoperoxidase technique may or may not reveal the presence of TSH in the cytoplasm of adenoma cells. The reason for negative results is not understood.

By electron microscopy, thyrotroph cell adenomas are composed of closely apposed, elongated or angular cells which have long cytoplasmic projections. The nuclei are irregular, contain deep indentations and are moderately rich in chromatin. The nucleoli are prominent. In the relatively small cytoplasm, the RER membranes are poorly to moderately developed. Free ribosomes are usually numerous. Golgi complexes are prominent and are composed of flat sacculi and numerous Golgi vesicles. Mitochondria are oblong or rod shaped with regular transverse lamellar cristae and moderately electron dense matrix. Oncocytic transformation may be apparent. Secretory granules are few, scattered, spherical, vary slightly in electron density, measure 100–200 nm and frequently line up along the cell membrane. Although the limiting membrane of the secretory granules may be attached to the plasma membrane, no granule extrusions are seen. The prominent electron lucent halo between the electron dense core of the secretory granule and its limiting membrane is a characteristic feature. The granule cores often show differences in electron density. Microtubules are numerous in the cytoplasm of some adenoma cells. It is interesting to note that adenomatous thyrotroph cells are not thyroidectomy cells, the stimulated form of TSH cells which can be seen in the rodent or human pituitary in cases of primary hypothyroidism. No interpretation of the absence of thyroidectomy cells in TSH cell adenomas has yet been proposed. More cases must be thoroughly studied before the fine structural details of this adenoma and its biologic behavior can be elucidated.

3.13. Null cell adenomas (10, 40, 41, 42, 53, 91, 92, 93)

These tumors are unassociated with clinical or biochemical manifestations of increased hormone secretion. They are called nonfunctioning adenomas, inactive adenomas or chromophobic adenomas. The former names are unrelated to their morphology; the latter name is misleading, since chromophobic adenomas can consist of various cell types and can secrete different hormones. The terms undifferentiated or precursor cell adenoma are also used, but it is not clear if these tumors consist of undifferentiated cells or precursor cells. The term null cell adenoma is more appropriate, for it emphasize that these tumors lack immunologic or ultrastructural markers. This name is very convenient because the cellular derivation and composition of these tumors is obscure at present. Later advancing knowledge and introduction of new techniques will, hopefully, permit further classification of these neoplasms. Currently they represent 17% of pituitary adenomas. Clinically, they cause local symptoms, such as visual disturbances and the consequences of increased intracranial pressure, but biochemical investigations fail to disclose endocrine abnormalities. However, null cell adenomas may be accompanied by hyperprolactinemia if the production and/or release of prolactin-inhibiting factors are impaired or their transport from the median eminence to the anterior lobe via the portal vessels is suppressed by the growing tumor. Once pituitary prolactin cells escape the hypothalamic dopaminergic inhibition, PRL secretion will increase. The growing tumor may press or damage the nontumorous portion of the pituitary, or may suppress adenohypophysial blood flow, resulting in infarction in the nontumorous pituitary. Various degrees of hypopituitarism may be manifest in these patients.

By light microscopy, null cell adenomas are chromophobic. In some cells the cytoplasm may have a granular acidophilic appearance related not to the presence of secretory granules, but to mitochondrial abundance. This phenomenon is not rare in null cell adenomas; its extent, however, ranges quite considerably from one tumor to another. The adenoma cells are PAS and lead hematoxylin negative and exhibit negative staining for the known pituitary hormones by applying the immunoperoxidase technique. In a few cases, scattered cells may be positive for one or more pituitary hormones, most commonly β-subunit and α-subunit of glycoprotein hormones. What substance is stored in the secretory granules is not yet known, and it may be a hormonal fragment, a precursor, an inactive substance, or a hormone not yet revealed.

By electron microscopy, null cell adenomas consist of closely apposed relatively small cells possessing irregular, deeply indented nuclei (Fig. 11). The nucleoli are not conspicuous. The electron lucent cytoplasm is poorly developed. The RER membranes are scanty and composed of a few short stacks. The Golgi complexes may be conspicuous but in most cases the Golgi sacculi are flat and contain only a few or no secretory granules. Mitochondria are oblong or rod shaped and have a few transverse lamellar cristae and a moderately electron dense matrix. In some adenoma cells, the number and volume density of mitochondria are low, in others they may be moderately or markedly increased. In some null cell adenomas, oncocytic transformation is pronounced and widespread; it may reach a stage in which almost the entire cytoplasmic area is filled by mitochondria while the other cytoplasmic organelles are obscured. Cells with marked mitochondrial abundance are known as oncocytes; tumors consisting of oncocytes are termed oncocytomas.

Secretory granules, always present, are spherical, vary somewhat in electron density, measure 100–250 nm and often line up along the plasmalemma. Granule extrusions are not apparent. Secretory granules possess a characteristic prominent halo between the relatively electron dense cores and their limiting membrane. Lysosomes are rare, whereas cytoplasmic microtubules are abundant. Intercellular junctions may be encountered between neighboring adenoma cells.

3.14. Oncocytomas (10, 40, 41, 42, 53, 94, 95, 96, 97, 98)

Since 1973, when this tumor was first described in the human pituitary, several cases have been reported, thus the fine structural features are well defined. Basically, pituitary oncocytomas belong to the null cell adenoma group, are endocrinologically nonfunctioning neoplasms and unassociated with clinical or biochemical manifestations of increased hormone secretion.

In our material, they are found to be relatively slow growing tumors, diagnosed only in patients over 40 years of age. Endocrine abnormalities are rare and the patients present with local symptoms, such as visual disturbances or headaches. In some patients, hypopituitarism or hyperprolactinemia may develop. The former abnormality is due to distruction of the nontumorous portion of the adenohypophysis as a result of compression or suppression of blood flow. The latter is usually due to the fact that the nontumorous pituitary escapes dopaminergic inhibition. Because of damage of the hypothalamus or pituitary stalk, the synthesis, discharge or adenohypophysial transport of prolactin-inhibiting factors are blocked and, consequently, hyperprolactinemia becomes evident as a result of increased secretory activity by nontumorous prolactin cells.

By light microscopy, oncocytomas may be chromophobic, although they are often diagnosed as acidophilic adenomas. The granular acidophilic staining is due to uptake of acid dye by the accumulating mitochondria. Oncocytes are not specific for the pituitary; they can be found in various other organs, such as thyroid, parathyroid, salivary gland or breast. Also, tumors composed of oncocytes can derive from various organs, such as the thyroid, parathyroid, salivary gland or kidney. Oncocytomas yield negative staining with the PAS technique and lead hematoxylin. In some cases, the mitochondria can be made visible by the phosphotungstic acid hematoxylin staining technique. The immunoperoxidase technique demonstrates the absence of the known adenohypophysial hormones, indicating a lack of hormone synthesis and storage. Similar to

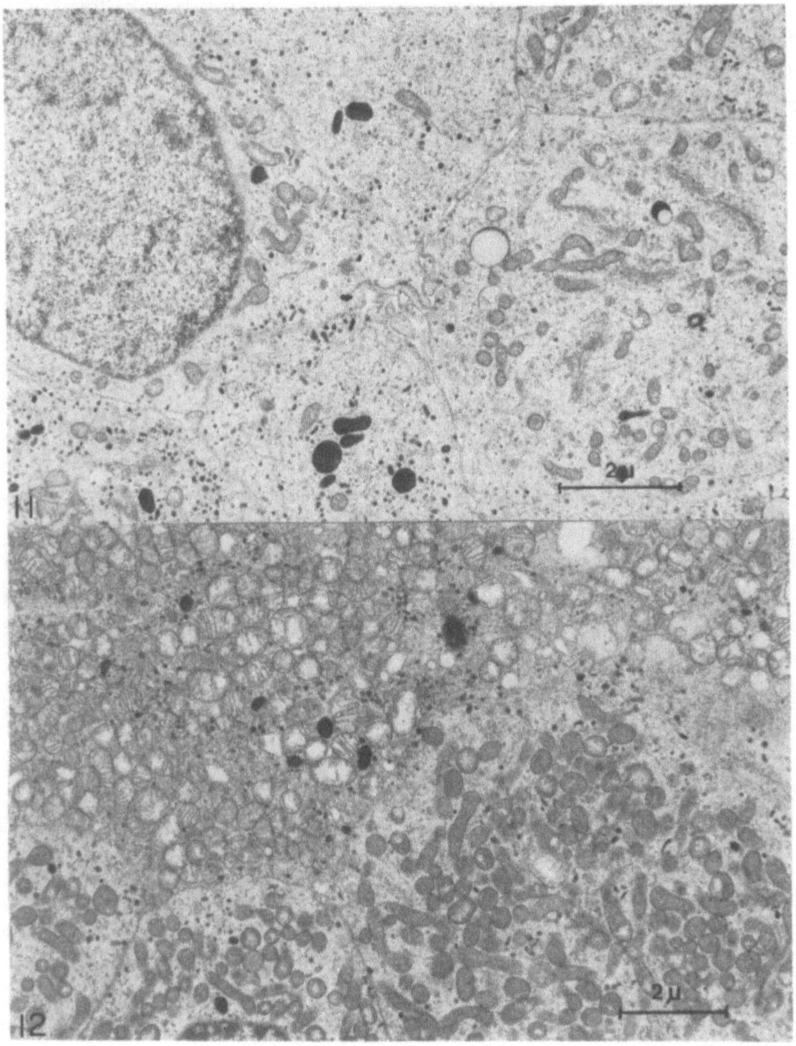

Figs. 11–12. (11) Null cell adenoma of the human pituitary (× 7,910). (12) Oncocytoma of the human pituitary (× 6,720).

non-oncocytic null cell adenomas, however, scattered cells of oncocytomas may show positive staining for one or more pituitary hormones without clinical and biochemical evidence of hypersecretion. This finding suggests that oncocytomas may arise in various hormone-producing cells which gradually transform to oncocytes and lose their hormone-producing abilities. Oncocytomas are described with ACTH, GH or PRL secretion. We use the term pituitary oncocytoma to designate only those tumors which show no evidence of endocrine activity. The underlying mechanism for oncocytic transformation is obscure at present. In our view, diagnosis of oncocytoma is justified only if all the cells contain abnormally large numbers of mitochondria. With such severe criteria, the occurrence of pituitary oncocytomas is about 6%.

By electron microscopy, the only characteristic feature of oncocytomas is the marked abundance of mitochondria (Fig. 12). This morphologic abnormality must be associated with some functional defect, but its cause and significance are not known. We believe oncocytic transformation is an irreversible process and, once mitochondrial abundance develops, the changes cannot regress. However, this is not proven, nor is it known whether oncocytic change precedes or simply follows neoplastic transformation. In the former case, it is evident that oncocytes are capable of dividing; in the latter case, oncocytoma develops by gradual oncocytic transformation of non-oncocytic cells. In our material, we have seen dividing oncocytes.

Oncocytomas are composed of relatively large, closely-apposed cells. The nuclei are oval or irregular with deep indentations. The nucleoli are not conspicuous. The cytoplasm of the adenoma cells is almost completely filled with mitochondria, and the remaining cytoplasmic organelles are obscured. The mitochondria are numerous and can be abnormally large. They usually have lamellar or irregular cristae and a rather electron lucent matrix. Signs, indicating mitochondrial fusion and/or division, are also apparent. In some mitochondria, cavitations are seen in the internal compartments with loss of cristae and the formation of whorl-like structures. The RER membranes consist of a few short, focally dilated stacks. Free ribosomes and polysomes are not numerous. The Golgi complexes are relatively prominent. Centrioles are frequently encountered in the Golgi regions. The secretory granules, noted in every cell, are few, spherical, vary slightly in electron density and measure 100–250 nm, averaging approximately 150 nm. Exocytoses are not noted. There is usually no lining up of secretory granules along the plasmalemma, although the secretory granules appear to be more numerous at the peripheral portion of the cytoplasm. Lysosomes are, in general, inconspicuous, although in a few cells they may be increased in size and number. In some cells, the abundance of mitochondria is somewhat less and transition forms can be detected between null cells with normal mitochondrial content and fully developed oncocytes. Oncocytic transformation is gradual and it progresses as the tumor proliferates.

Although the fine structural features of pituitary oncocytomas are known in detail, only limited information is available on their biologic behavior and endocrine activity. They have no immunologic markers but they can be conclusively recognized by electron microscopy. This fact underlines the significance of fine structural investigation in the differential diagnosis of pituitary adenomas.

3.15 Unclassified adenomas (41, 99, 100, 101, 102, 103)

Provided fixation, processing and embedding are appropriate, the diagnosis of unclassified pituitary adenoma is rare. The morphologic investigation of the tumors yields very valuable information, and by immunocytology and electron microscopy, the substantial majority of tumors can be classified and the cellular derivation and composition of adenomas can conclusively be determined. In our material, the incidence of unclassified pituitary adenomas is approximately 2%. The clinical data and results of biochemical investigation are diagnostic in many cases. It should be stated that, in our laboratory, the majority of pituitary adenomas are investigated 'blindly', without any previous knowledge of the clinical presentation and biochemical results.

Unclassified adenomas consist of cells which contain combinations of various hormones and cannot be identified even by detailed electron microscopic studies. In most cases, these adenoma cells appear to be rather well differentiated, but they do not resemble any known adenohypophysial cells. Obviously, these tumors are a heterogeneous group and do not represent a specific pituitary tumor category.

The functional activity of unclassified adenomas may be unusual. Some tumors may produce two or more hormones markedly different in chemical composition, immunoreactivity and biologic action. Adenomas producing GH and TSH – GH, PRL and ACTH, – GH, PRL and TSH, – PRL and FSH, – PRL and TSH, – PRL, α-subunit, FSH and TSH – may occasionally be found. In these cases, the application of the immunocytologic technique or radioimmunoassay of the tumor extract may provide answers regarding hormone content of the neoplasms. Tissue culture studies and detailed investigation of the patients, including stimulation tests and blood hormone measurements, are helpful in exploring what hormones are produced by these unusual and rather uncommon adenomas.

According to the one cell- one hormone theory, adenohypophysial cells secrete only one hormone but with two exceptions: gonadotroph cells produce the two gonadotroph hormones, FSH and LH simultaneously within the same cell; corticotroph cells contain ACTH and related peptides, such as β-LPH and endorphins. Based on the investigation of multiple hormone-producing adenomas, it appears that this theory is oversimplified and our current thinking and interpretation of the morphogenesis of pituitary adenomas requires substantial modification. Tumors secreting multiple hormones, as discussed in the first part of this review, are known to occur and to be frequent in the animal pituitary, and it appears they may also develop in the human adenohypophysis. It may be that further studies will reveal more unusual variants. Results of research in this field are just beginning to emerge and, with the application of sophisticated techniques, substantial progress can be expected in this exciting and hitherto unexplored field.

3.16. Invasive adenomas, adenohypophysial carcinomas (41, 42, 104, 105)

The vast majority of pituitary adenomas are benign tumors which grow slowly, do not invade surrounding tissue, are not pleomorphic and do not give rise to distant metastases. Invasion of surrounding tissue is evident in several pituitary adenomas. These tumors, called invasive adenomas, have a more rapid growth rate and spread to their neighborhood, they grow into the sphenoid bone, infiltrate the optic nerve and even invade the brain. They may produce GH, PRL or ACTH, or be endocrinologically inactive. Clinically, they may be associated with different signs and symptoms, indicating increased hormone secretion by the tumor. Accordingly, blood hormone levels may be elevated.

By light microscopy, invasive tumors are chromophobic, acidophilic or basophilic and the immunoperoxidase technique may demonstrate hormone storage in the cytoplasm.

By electron microscopy, invasive adenomas are similar to their noninvasive counterparts. Some tumor cells may show pleomorphism or mitotic figures, but the fine structural investigation contributes no answers on the invasiveness of the tumor. The question of malignancy is best assessed by light microscopy and the ultrastructural study provides no additional help in the differential diagnosis.

Primary adenohypophysial carcinomas arise in and consist of adenohypophysial cells. They are very rare, represent the malignant counterpart of pituitary adenomas, and may produce GH, PRL or ACTH, or may be unaccompanied by hormonal activity. Hormone production can be assessed by immunocytologic techniques. The diagnosis of pituitary carcinoma is justified only when distant metastases are demonstrated. Histologic examination may fail, in some cases, to reveal the biologic behavior of the tumors. Adenohypophysial carcinomas are usually very cellular tumors; cellular and nuclear pleomorphism may be conspicuous. Formation of multiple nuclei or giant cells may occur and mitotic figures may be numerous. Invasion of surrounding tissue is also evident. All these findings are important but not pathognomonic and the diagnosis solely depends on the disclosure of distant metastases.

Electron microscopy provides no answers to the diagnosis of carcinoma and it has no application in assessing malignant behavior of the tumor cells.

4. Concluding remarks

The objective of this review was to summarize the ultrastructural features of adenomas in the animal and human pituitary. Several electron microscopic studies on nontumorous and adenomatous pituitaries, in the last two decades, disclosed in detail the fine structural characteristics of many adenomas. No doubt exists that electron microscopy proved to be a valuable technique in distinguishing between various cell types and in reaching a meaningful adenoma classification.

Various pituitary adenoma models can be developed with induction of pituitary adenomas in animal experiments, and the identity of hormones secreted by the tumor cells can be clarified. Pituitary adenomas, cultured in vitro, continue to produce hormones, provided the conditions are suitable and their fine structural features can be investigated. When pituitary tumors are transplanted in animals, their growth rate, functional activity, morphology and hormone dependence can be examined. Adenoma cells can be cloned by special techniques and monoclonal cell lines produced. Detailed study of these pure cell lines may lead to a better understanding of the multiple phases of hormone secretion, the biologic behavior and growth rate of tumor cells and the various factors which affect and regulate these processes.

The introduction of a new human pituitary adenoma classification is regarded as an important step in endocrine pathology. Electron microscopy is fundamental in separating pituitary adenomas into distinct entities. The fine structural features of various adenoma types are now clearly defined and the ultrastructural adenoma classification can effectively be used in surgical pathology. Human pituitary adenoma cells can also be studied in vitro (106, 107, 108, 109); they grow in culture and their hormone production continues. They can be transplanted to athymic, immunologically incompetent nude mice and the responsiveness of adenoma cells to various hormones and drugs can be investigated (110, 111). With the more widespread application of electron microscopic immunocytology and ultrastructural morphometry substantial further advances are anticipated.

Despite spectacular achievements, many important questions remain unanswered. It is still obscure whether or not stem cells exist in animal and human adenohypophyses. The details regarding differentiation and dedifferentiation of various adenohypophysial cell types and the factors affecting the maturation process have yet to be elucidated. In the normal pituitary it is believed that: growth hormone cells produce GH; prolactin cells, PRL; corticotroph cells, ACTH, β-LPH and endorphins; thyrotroph cells, TSH; and gonadotroph cells, the two gonadotroph hormones FSH and LH. It is very difficult to understand, based on this theory, how those pituitary adenomas develop which are capable of producing two or more hormones. Which cell type is the origin of these multiple hormone-producing tumors? It may be that one or more stem cells exist which can synthesize more than one hormone. If these cells, during the maturation process, specialize and synthesize only one hormone, then perhaps they have the potential to dedifferentiate and in the course of neoplastic transformation again acquire the ability to produce more than one hormone.

Evidence is accumulating that various tumors do not necessarily consist of monoclonal cells. If this is so, tumor cell heterogeneity may provide an explanation for the formation of mixed adenomas and multiple hormone secretion. If one cell type overgrows the others, ultrastructural features as well as hormone secretion of the adenoma may undergo substantial changes in the course of proliferation. Certain cells may have a shorter cell cycle or may be less susceptible to changes in their environment than other forms. In tissue culture, we have seen by electron microscopy dedifferentiation of cells, the gradual loss of secretory granules and the transformation of densely granulated growth hormone cells to sparsely granulated growth hormone cells with the formation of fibrous bodies.

More work is required on the ultrastructural aspects of pituitary adenomas. Fine structural features must be correlated with immunocytologic findings at the light and electron microscopic levels in order to obtain a deeper insight into hormone storage and release. When the ultrastructure of the tumor is correlated with clinical findings and blood hormone levels, the biologic behavior of adenomas is unraveled. Criteria can be established to reveal those features which would predict the growth rate, endocrine function and biologic behavior of a given tumor.

The pituitary adenoma classification, as described in this review, provides important information to clinical endocrinologists, neurosurgeons, pathologists, as well as biomedical scientists. This classification is not final. Further electron microscopic studies will lead to modifications in adenoma classification, and hitherto undiscovered entities will emerge. Thus, it is imperative that more cases be studied by electron microscopy and the results correlated with biologic behavior, clinical and biochemical findings. Adenomas producing multiple hormones require further study for they do not fit into the classification presently in use.

Despite numerous efforts, the pathogenesis of pituitary adenomas is still not known. Obviously, electron microscopy alone will not resolve this complicated problem, but fine structural study can contribute important details which would help unravel the mystery of neoplastic transformation.

The role of the hypothalamus in pituitary tumorigenesis is not clearly understood. Chronic overstimulation of adenohypophysial cells by hypothalamic hormones may lead to adenoma formation, although evidence is lacking that protracted administration of hypothalamic hormones induces pituitary tumors. It is known that pituitary tumors may develop in the grafted adenohypophysis which possess no direct connections with the hypothalamus. The role of various hormones and the endocrine environment in the development of pituitary adenomas requires further study.

It is known that estrogen induces tumors in the rodent pituitary. This is very important from the practical point of view, since recently pituitary adenomas have been reported in women taking contraceptive pills (112). Contraceptive pills contain estrogens and may be causally or casually related to the formation of pituitary adenomas. It remains for future studies to establish whether the contraceptive pill acts as an inducing agent, accounting directly for the neoplastic transformation or only to promote the growth of an already developed neoplasm.

Prolactin cell adenomas may develop in patients with no previous estrogen treatment. Moreover, prolactin cell adenomas occur not only in women but also in men (113, 114). It is known that pituitary adenomas may develop in animals with long-standing hypofunction of the peripheral endocrine glands. The occurrence of pituitary adenomas have been reported in human subjects with long-standing primary hypogonadism and hypothyroidism. The presence of pituitary adenomas in patients with Addison's disease suggests that protracted hypofunction of adrenal cortices may lead to adenoma formation (115, 116). However, gonadotroph cell, thyrotroph cell adenomas may develop in patients with no preexisting hypogonadism, hypothyroidism, indicating that factors other than decreased functional activity of target glands may be responsible for adenoma formation. In the overwhelming majority of cases, corticotroph cell adenomas are not preceded by manifest hypocorticism. It also remains to be seen whether loss of receptor sites or the development of abnormal receptors play a role in the formation of pituitary adenomas. Recent studies indicate that dopamine receptors are reduced in prolactin-producing adenomas compared with the nontumorous adenohypophysis (117).

Conventional transmission electron microscopy gives information of the functional state of the cell, yet it yields only qualitative images. It is important to obtain quantitative data. For this purpose, ultrastructural morphometry has great potential, for it permits quantitative expression of the findings and allows for comparison of diameters, volume densities, surface areas and number of exocytoses (53, 63, 118, 119, 120). We believe that ultrastructural morphometry should be applied much more frequently in assessing the subcellular features of adenoma cells. Electron microscopic autoradiography, if used effectively, can also yield quantitative results to label hormone precursors, to study the cell cycle or to investigate various steps of the secretory process (121, 122). However, it has little significance in service pathology. Immunoelectron microscopy is time consuming, difficult and expensive, but it appears to be a very valuable technique, revealing hormone storage and cellular identity. This technique has a far reaching future and its use should be much more widespread than it is at present.

Scanning electron microscopy will probably not prove useful in the study of pituitary adenomas (107). By scanning electron microscopy, the various cell types cannot be distinguished conclusively, and adenomatous cells cannot be separated from the nontumorous adenohypophysial cells. Certainly more scanning electron microscopic works is required to establish the value of this approach in pituitary adenoma studies.

The last words of Sir Cecil John Rhodes: 'So little done – so much to do', are the motto of this review. Although this quote is not related to electron microscopy, it is applicable. Future work in this exciting field is called for.

Acknowledgement

This work was supported in part by Grant MA-6349 of the Medical Research Council of Canada and by Grant 1R01 CA 21905-01 awarded by the National Cancer Institute, DHEW. The authors wish to thank Mrs. Gezina Ilse, Mr. Gerhard Penz, Mrs. Nancy Ryan and Mrs. Nora Wood for their valuable contribution throughout the undertaking of this project. The authors wish to express their appreciation and gratitude to Mrs. Wanda Wlodarski who performed the secretarial work. We are indebted to many colleagues who provided material for our study.

References

1. Furth J, Ueda G, Clifton KH: The pathophysiology of pituitaries and their tumors. Methodological advances. In: Methods in cancer research. Busch H (ed), Academic Press, 1973, pp 201–277.
2. Fischer O: Uber Hypophysengeschwülste der weissen Ratten. Arch Pathol Anat 259: 9–29, 1926.
3. Wolfe JM, Bryan WR, Wright AW: Histologic observations on the anterior pituitaries of old rats with particular reference to the spontaneous appearance of pituitary adenomata. Am J Cancer 34: 352–372, 1938.
4. Kim U, Clifton KH, Furth J: A highly inbred line of Wistar rats yielding spontaneous mammo-somatotrophic pituitary and other tumors. JNCI 24: 1031–1055, 1960.
5. Kwa HG, van der Gugten AA, Verhofstad F: Radioimmunoassay of rat prolactin. Europ J Cancer 5: 571–579, 1969.
6. Kovacs K, Horvath E, Ilse RG, Ezrin C, Ilse D: Spontaneous pituitary adenomas in aging rats. A light microscopic, immunocytochemical and fine structural study. Beitr Pathol 161: 1–16, 1977.
7. McComb DJ, Ryan N, Horvath E, Kovacs K, Nagy E, Berczi I, Domokos I, Laszlo FA: Five different adenomas derived from the rat adenohypophysis: Immunocytochemical and ultrastructural study. JNCI 66: 1103–1111, 1981.
8. Kovacs K, Horvath E: Misplaced exocytosis. A distinct form of secretion in the anterior pituitary. Indian J Pathol Micr 19: 85–89, 1976.
9. Horvath E, Kovacs K, Penz G, Ezrin C: Origin, possible function and fate of 'follicular cells' in the anterior lobe of the human pituitary. An electron microscopic study. Am J Pathol 77: 199–212, 1974.
10. Kovacs K, Horvath E, Ezrin C: Pituitary adenomas. Pathol Annu part 2) 12: 341–382, 1977.

11. Berkvens JM, van Nesselrooy JHJ, Kroes R: Spontaneous tumors in the pituitary gland of old Wistar rats. A morphological and immunocytochemical study. J Pathol 130: 179–191, 1980.
12. Cramer W, Horning ES: Experimental production by oestrin of pituitary tumors with hypopituitarism. Lancet 9: 247–249, 1936.
13. McEuren CS, Selye H, Collip JB: Some effects of prolonged administration of oestrin in rats. Lancet 9: 775–776, 1936.
14. Zondek B: Tumors of the pituitary induced with follicular hormone. Lancet i: 776–778, 1936.
15. Furth J, Clifton KH: Experimental pituitary tumors. In: The pituitary gland, Vol. 2. Harris GW, Donovan BT (eds), Buttersworth, London, 1966, pp 460–497.
16. Zambrano D, Dies RP: The adenohypophysis of female rats after hypothalamic oestradiol implants. An electron microscopic study. J Endocrinol 47: 101–110, 1970.
17. Ito A, Martin JM, Grindland RE, Takizawa S, Furth J: Mammotropic and somatotropic hormones in sera of normal rats and in rats bearing primary and grafted pituitary tumors. Int J Cancer 7: 416–429, 1971.
18. Ueda G, Tanizawa O, Hamanaka N, Nishiura H: Changes of growth hormone-containing cells during tumorigenesis and subpassages of estrogen induced pituitary tumors in rat. Endocrinol Jap 17: 447–452, 1970.
19. Yokoro K, Furth J, Haran Ghera N: Induction of mammotropic pituitary tumors by x-ray in rats and mice. The role of mammotropes in development of mammary tumors. Cancer Res 21: 178–186, 1961.
20. Gorbman A: Pituitary tumors in rodents following changes in thyroid function: A review. Cancer Res 16: 99–105, 1956.
21. Doniach I, Williams ED: The development of thyroid and pituitary tumours in the rat two years after partial thyroidectomy. Br J Cancer 16: 222–231, 1962.
22. Furth J, Moy P, Hershman JM, Ueda G: Thyrotropic tumor syndrome. Arch Pathol 96: 217–226, 1973.
23. Ueda G, Mori T: Astatine-211 induced transplantable pituitary tumor in rat with a brief analytic review of cell types of pituitary tumors. Am J Pathol 51: 601–619, 1967.
24. Yokoro K, Kunii A, Furth J, Durben PW: Tumor induction with astatine-211 in rats: Characterization of pituitary tumors. Cancer Res 24: 683–688, 1964.
25. Dickie MM, Woolley GW: Spontaneous basophilic tumors of the pituitary glands in gonadectomized mice. Cancer Res 96: 372–384, 1949.
26. Houssay BA, Houssay AB, Cardeza AF, Foglia VG, Pinto RM: Adrenal tumors in gonadectomized rats. Acta Physiol Lat Am 3: 128–130, 1953.
27. Rennels EG: An electron microscope study of pituitary autograft cells in the rat. Endocrinology 71: 713–722, 1962.
28. Viragh S, Tiboldi T, Kovacs K, Julesz M, Toro I: Electron microscopic structure of anterior pituitaries transplanted to the anterior chamber of the eye in rats. Acta Biol Acad Sci Hung 18: 77–89, 1967.
29. Muhlbock O, Boot LM: Induction of mammary cancer in mice without mammary tumor agent by isographs of hypophyses. Cancer Res 19: 402–412, 1959.
30. Gardner WU: Tumors in transplanted pituitary glands in mice. Proc Amer Assoc Cancer Res 3: 113, 1960.
31. Clifton KH, Furth J: Changes in hormone sensitivity of pituitary mammotropes during progression from normal to autonomous. Cancer Res 21: 913–920, 1961.
32. Tiboldi T, Viragh S, Kovacs K, Hodi M, Julesz M: Electron-microscopic studies of estrogen induced adenomata developing in intracellar and transplanted rat anterior pituitaries. J Microscopie (Paris) 6: 677–689, 1967.
33. Nakayama I, Nickerson PA: Ultrastructure of a transplantable mammosomatotropic pituitary tumor (MtT.W10). Acta Pathol Jap 23: 237–248, 1973.
34. Tixier-Vidal A, Gourdji D, Tougard C: A cell culture approach to the study of anterior pituitary cells. Int Rev Cytol 41: 173–239, 1975.
35. Benda C: Beitrage zur normalen und pathologischen Histologie der menschlichen Hypophysis cerebri. Berl Klin Wochenschr 36: 1205–1210, 1900.
36. Erdheim J: Zur normalen und pathologischen Histologie der Glandula thyreoidea, parathyreoidea und Hypophysis. Beitr Pathol Anat 33: 158–236, 1903.
37. Cushing H: The pituitary body and its disorders. JB Lippincott, Philadelphia, PA, 1912.
38. Kernohan JW, Sayre GP: Tumors of the pituitary gland and infundibulum. Armed Forces Institute of Pathology, Washington DC, 1956.
39. Schelin U: Chromophobe and acidophil adenomas of the human pituitary gland. A light and electron microscopic study. Acta Pathol Microbiol Scand (Suppl) 158: 1–80, 1962.
40. Horvath E, Kovacs K: Ultrastructural classification of pituitary adenomas. Can J Neurol Sci 3: 9–21, 1976.
41. Kovacs K, Horvath E: Pituitary adenomas: Pathologic aspects. In: Clinical neuroendocrinology. A patho-physiological approach. Tolis G, Labrie F, Martin JB, Naftalin F (eds), Raven Press, NY, NY 1979, pp 367–384.
42. Horvath E, Kovacs K: Pathology of the pituitary gland. In: Pituitary diseases. Ezrin C, Horvath E, Kaufman B, Kovacs K, Weiss MH (eds), CRC Press Inc, Boca Raton, Florida, 1980, pp 1–83.
43. Nakane PK: Application of peroxidase-labelled antibodies to the intracellular localization of hormones. Acta Endocrinol (Kbh), Suppl, 153: 190–204, 1971.
44. Sternberger LA: Immunocytochemistry (Basic and Clinical Immunology Series), J Wiley and Sons, NY, NY, 1979.
45. Kovacs K, Horvath E, Ryan N: Immunocytology of the human pituitary. In: Diagnostic immunohistochemistry. DeLellis RA (ed), Masson Publ USA Inc NY, NY, 1981, pp 17–35.
46. Halmi NS, Parsons JA, Erlandsen SL, Duello T: Prolactin and growth hormone cells in the human hypophysis: A study with immunoenzyme histochemistry and differential staining. Cell Tiss Res 158: 497–507, 1975.
47. Kovacs K, Corenblum B, Sirek AMT, Penz G, Ezrin C: Localization of prolactin in chromophobe pituitary adenomas: Study of human necropsy material by immunoperoxidase technique. J Clin Pathol 29: 250–258, 1976.
48. Moriarty CM: Adenohypophysis: Ultrastructural cytochemistry. A review. J Histochem Cytochem 21: 855–894, 1973.
49. Moriarty CM, Sternberger LA: Ultrastructural immunocytochemistry with unlabelled antibodies and the peroxidase-antiperoxidase complex. J Histochem Cytochem 21: 825–833, 1973.
50. Pelletier G, Robert F, Hardy J: Identification of human anterior pituitary cells by immunoelectron microscopy. J Clin Endocrinol Metabol 46: 534–542, 1978.
51. Conway LW, Schalch DS, Utiger RD, Reichlin S: Hormones in human pituitary sinusoid blood: Concentration of LH, GH and TSH. J Clin Endocrinol 29: 446–456, 1969.
52. Hwang P, Guyda H, Friesen H: A radioimmunoassay for human prolactin. Proc Natl Acad Sci USA 68: 1902–1906, 1971.
53. Landolt AM: Ultrastructure of human sella tumors. Correlations of clinical findings and morphology. Acta Neurochir (Suppl 22): 1–167, 1975.
54. Halmi NS, Duello T: 'Acidophilic' pituitary tumors: A reappraisal with different staining and immunocytochemical techniques. Arch Pathol Lab Med 100: 346–351, 1976.
55. Trouillas J, Girod C, L'Héritier M, Claustrat B, Dubois MP: Morphological and biochemical relationships in 31 human pituitary adenomas with acromegaly. Virchows Arch (Path Anat) 389: 127–142, 1980.
56. Horvath E, Kovacs K: Morphogenesis and significance of fibrous bodies in human pituitary adenomas. Virchows Arch (Cell Pathol) 27: 69–78, 1978.
57. Horvath E, Kovacs K, Ezrin C: Centrioles and cilia in non-tumorous anterior lobes and adenomas of the human pituitary. Pathol Eur 11: 81–86, 1976.
58. Peake GT, McKeel DW, Jerett L, Daughaday WH: Ultrastructural, histologic and humoral characterization of a prolactin-rich human pituitary tumor. J Clin Endocrinol Metabol 29: 1383–1393, 1969.
59. Kovacs K, Horvath E, Corenblum B, Sirek AMT, Penz G, Ezrin C: Endocrinol (Oxf) Suppl 6: 71s–79s, 1977.
60. Kovacs K, Horvath E, Corenblum B, Sirek AMT, Penz G, Ezrin C: Pituitary chromophobe adenomas consisting of prolactin cells. A histologic, immunocytological and electron microscopic study. Virchows Arch (Path Anat) 366: 113–123, 1975.
61. Robert F, Hardy J: Prolactin-secreting adenomas. A light and electron microscopical study. Arch Pathol 99: 625–633, 1975.
62. Biller BJ, Boyd A, Molitch ME, Post KD, Wolput SM, Reichlin S: Galactorrhea syndrome. In: The Pituitary adenoma. Post KD, Jackson IMD, Reichlin S (eds), Plenum Publ Corp NY, NY, 1980, pp

65–118.
63. McComb DJ, Kovacs K, Horvath E, Singer W, Killinger DW, Smyth HS, Ezrin C, Weiss MH: Correlative ultrastructural morphometry of human prolactin-producing adenomas. Acta Neurochir 53: 217–225, 1980.
64. Frantz AG: The assay and regulation of prolactin in humans. Adv Exp Med Biol 80: 95–133, 1977.
65. Horvath E, Kovacs K: Misplaced exocytosis. Distinct ultrastructural feature in some pituitary adenomas. Arch Pathol 97: 221–224, 1974.
66. Guyda H, Robert F, Colle E, Hardy J: Histologic, ultrastructural and hormonal characterization of a pituitary tumor secreting both hGH and prolactin. J Clin Endocrinol Metab 36: 531–547, 1973.
67. Corenblum B, Sirek AMT, Horvath E, Kovacs K, Ezrin C: Human mixed somatotrophic and lactotrophic pituitary adenomas. J Clin Endocrinol Metab 42: 857–863, 1976.
68. Kameya T, Tsumuraya M, Adachi I, Abe K, Ichikizaki K, Toya S, Demura R: Ultrastructure, immunohistochemistry and hormone release of pituitary adenomas in relation to pituitary production. Virchows Arch (Path Anat) 387: 31–46, 1980.
69. Horvath E, Kovacs K, Singer W, Ezrin C, Kerenyi NA: Acidophil stem cell adenoma of the human pituitary. Arch Pathol Lab Med 101: 594–599, 1977.
70. Horvath E, Kovacs K, Singer W, Smyth HS, Killinger DW, Ezrin C, Weiss MH: Acidophil stem cell adenoma of the human pituitary: clinicopathologic analysis of 15 cases. Cancer 47: 761–771, 1981.
71. Horvath E, Kovacs K, Killinger DW, Smyth HS, Platts ME, Weiss MH, Ezrin C: Mammosomatotroph cell adenomas of the human pituitary. EMSA Proc 38: 726–727, 1980.
72. Horvath E, Kovacs K., Killinger DW, Smyth HS, Weiss MH, Ezrin C: Mammosomatotroph cell adenoma of the human pituitary. A morphologic entity. Virchows Arch 398: 277–289, 1983.
73. Landolt AM, Rothenbuhler V: Extracellular growth hormone deposits in pituitary adenoma. Virchows Arch (Path Anat) 378: 55–65, 1978.
74. Robert F, Pelletier G, Hardy J: Pituitary adenomas in Cushing's disease. Arch Pathol Lab Med 102: 448–445, 1978.
75. Ryder DR, Horvath E, Kovacs K: Fine structural features of secretion in adenomas of the human pituitary gland. Arch Pathol Lab Med 104: 518–522, 1980.
76. Kovacs K, Horvath E, Stratmann IE, Ezrin C: Cytoplasmic microfilaments in the anterior lobe of the human pituitary gland. Acta Anat 87: 414–426, 1974.
77. Crooke AC: A change in the basophil cells of the pituitary gland common to conditions which exhibit the syndrome attributed to basophil adenoma. J Pathol Bacteriol 41: 339–349, 1935.
78. Halmi NS, McCormick WF, Decker DA: The natural history of hyalinization of ACTH-MSH cells in man. Arch Pathol 91: 318–326, 1971.
79. De Cicco FA, Dekker A, Yunis EJ: Fine structure of Crooke's hyaline change in the human pituitary gland. Arch Pathol 94: 65–70, 1972.
80. Horvath E, Ilse G, Kovacs K: Enigmatic bodies in human corticotroph cells. Acta Anat 98: 427–433, 1976.
81. Garcia JH, Kalimo H, Givens JR: Human adenohypophysis in Nelson's syndrome. Ultrastructural and clinical study. Arch Pathol Lab Med 100: 253–258, 1976.
82. Kovacs K., Horvath E., Kerenyi NA, Sheppard RH: Light and electron microscopic features of a pituitary adenoma in Nelson's syndrome. Am J Clin Pathol 65: 337–343, 1976.
83. Kovacs K, Horvath E, Bayley TA, Hassaram ST, Ezrin C: Silent corticotroph cell adenoma with lysosomal accumulation and crinophagy: A distinct clinicopathological entity. Am J Med 64: 492–499, 1978.
84. Horvath E, Kovacs K, Killinger DW, Smyth HS, Platts ME, Singer W: Silent corticotrophic adenomas of the human pituitary gland. A histologic, immunocytologic and ultrastructural study. Am J Path 98: 617–638, 1980.
85. Kovacs K, Horvath E, Van Loon GR, Rewcastle NB, Ezrin C, Rosenbloom AA: Pituitary adenomas associated with elevated blood follicle-stimulating hormone levels. A histologic, immunocytologic and electron microscopic study of two cases. Fertil Steril 29: 622–628, 1978.
86. Kovacs K, Horvath E., Rewcastle NB, Ezrin C: Gonadotroph cell adenoma of the pituitary in a woman with long-standing hypogonadism. Arch Gynecol 229: 57–65, 1980.
87. Leong AS, Chawla JC, Teh EC: Pituitary thyrotropic tumour secondary to long-standing primary hypothyroidism. Pathol Eur 11: 49–55, 1976.
88. Samaan NA, Osborne BM, McKay B, Leavens MF, Duello TM, Halmi NS: Endocrine and morphologic studies of pituitary adenomas secondary to primary hypothyroidism. J Clin Endocrinol Metab 45: 903–911, 1977.
89. Afrasiabi A, Valenta L, Gwinup G: A TSH-secreting pituitary tumor causing hyperthyroidism: Presentation of a case and review of the literature. Acta Endocrinol (Kbh) 92: 448–454, 1979.
90. Katz MS, Gregerman RI, Horvath E, Kovacs K, Ezrin C: Thyrotroph cell adenoma of the human pituitary gland associated with primary hypothyroidism: Clinical and morphological features. Acta Endocrinol (Kbh) 95: 41–48, 1980.
91. Schechter J: Electron microscopic studies of human pituitary tumors. I. Chromophobic adenomas. Am J Anat 138: 371–386, 1973.
92. Racadot, J, Vila-Porcile E, Olivier L, Peillon F: Electron microscopy of pituitary tumours. Prog Neurol Surg 6: 95–141, 1975.
93. Kovacs K, Horvath E, Ryan N, Ezrin C: Null cell adenoma of the human pituitary. Virchows Arch (Path Anat) 387: 165–174, 1980.
94. Kovàcs K, Horvath E: Pituitary 'chromophobe' adenoma composed of oncocytes. A light and electron microscopic study. Arch Pathol 95: 235–239, 1973.
95. Landolt AM, Oswald UW: Histology and ultrastructure of an oncocytic adenoma of the human pituitary. Cancer 31: 1099–1105, 1973.
96. Kovacs K, Horvath E, Bilbao JM: Oncocytes in the anterior lobe of the human pituitary gland. A light and electron microscopic study. Acta Neuropathol (Berl) 27: 43–53, 1974.
97. Saeger W: Vergleichende licht- und elektronenmikroskopische Untersuchungen an onkocytaren Hypophysenadenomen. Virchows Arch (Path Anat) 369: 29–44, 1975.
98. Bauserman SC, Hardman JM, Schochet SS, Earle KM: Pituitary oncocytoma. Indispensable role of electron microscopy in its identification. Arch Pathol Lab Med 102: 456–459, 1978.
99. Horn K, Erhardt F, Falbusch R, Pickardt CR, Werder K, Scriba PC: Recurrent goiter, hyperthyroidism galactorrhea and amenorrhea due to a thyrotropin and prolactin-producing pituitary tumor. J Clin Endocrinol Metab 43: 137–143, 1976.
100. Duello TM, Halmi NS: Pituitary adenoma producing thyrotropin and prolactin. An immunocytochemical and electron microscopic study. Virchows Arch (Pathol Anat) 376: 255–265, 1977.
101. Tolis G, Bird C, Bertrand G, McKenzie JM, Ezrin C: Pituitary hyperthyroidism. Case report and review of the literature. Am J Med. 64: 177–181, 1978.
102. Heitz PU: Multihormonal pituitary adenomas. Hormone Res 10: 1–13, 1979.
103. Waldhäusl W, Bratusch-Marrain P, Nowotny P, Buchler M, Forssmann W-G, Lujf A, Schuster H: Secondary hyperthyroidism due to thyrotropin hypersecretion: Study of pituitary tumor morphology and thyrotropin chemistry and release. J Clin Endocrinol Metab 49: 879–887, 1979.
104. D'Abrera VSE, Burke WJ, Bleasel KF, Bader L: Carcinoma of the pituitary gland. J Pathol 109: 335–343, 1973.
105. Queiroz L, de S, Facure NO, Facure JJ, Modesto NP, de Faria JL: Pituitary carcinoma with liver metastases and Cushing syndrome. Report of a case. Arch Pathol 99: 32–35, 1975.
106. Kohler PO. Bredson WE, Rayford PL, Kohler SE: Hormone production by human pituitary adenomas in culture. Metabolism 18: 782–788, 1969.
107. Harris RD, Seljeskog EL, Murray KJ, Chou SN, Cunningham WP, Douglas SD: Surface topography of normal and neoplastic human anterior pituitary cells maintained in vitro. J Neurosurg 49: 169–178, 1978.
108. Tixier-Vidal A, Brunet N, Tougard C, Gourdji D: Morphological and molecular aspects of prolactin and growth hormone secretion by normal and tumor pituitary cells in culture. In: Pituitary microadenomas. Faglia G, Giovanelli MA, MacLeod RM (eds), Proc Serono Symposia, Vol 29, Academic Press, London, 1980, pp 73–90.
109. Bethea CL, Weiner RI: Human prolactin secreting adenoma cells maintained on extracellular matrix. Endocrinology 108: 357–360, 1981.
110. O'Sullivan JP, Alexander KM, Jerkins JS: Maintenance of functioning human pituitary tumors in nude athymic mice. J Endocrinol 79:

139–140, 1978.

111. Usadel KH, Bastert G, Schwedes U, Althoff PH, Eichholz H, Steinau U, Klempa I: Xenotransplantation: A new model in experimental endocrinology. Acta Endocrinol Suppl 234: 153–154, 1980.
112. Sherman BM, Harris CE, Schlechte J, Duello TM, Halmi NS, Van Gilder J, Chapler FK, Granner DK: Pathogenesis of prolactin-secreting pituitary adenomas. Lancet 82: 1019–1021, 1978.
113. Carter JN, Tyson JE, Tolis G, van Vliet S, Faiman C, Friesen HG: Prolactin-secreting tumors and hypogonadism in 22 men. New Engl J Med 299: 847–852, 1978.
114. Pont A, Sheltion R, Odel WD, Wilson CB: Prolactin-secreting tumors in men: Surgical cure. Ann Int Med 91: 211–213, 1979.
115. Himsworth RL, Lewis JG, Rees LH: A possible ACTH secreting tumor of the pituitary developing in a conventionally treated case of Addison's disease. Clin Endocrinol 9: 131–139, 1978.
116. Jara-Albarra A, Bayort J, Caballero A, Portillo J, Laborda L, Sampedro M, Cure C, Palacios Mateos JM: Probable pituitary adenoma with adrenocorticotropin hypersecretion (corticotropinoma) secondary to Addison's disease. J Clin Endocrinol Metab 49: 236–241, 1979.
117. Bression D, Brandi AM, Martres MP, Nousbaum A, Cesselin F, Racadot J, Peillon F: Dopaminergic receptors in human prolactin-secreting adenomas: A quantitative study. J Clin Endocrinol Metab 51: 1037–1043, 1980.
118. Gray AB, Doniach I, Leigh PN: Correlation of diameters of secretory granules in clinically non-functioning chromophobe adenomas of the pituitary with those of normal thyrotrophs. Acta Endocrinol (Kbh) 79: 417–420, 1975.
119. Landolt AM, Rothenbuhler V: The size of growth hormone granules in pituitary adenomas producing acromegaly. Acta Endocrinol (Kbh) 84: 461–469, 1977.
120. McComb DJ, Kovacs K: Ultrastructural morphometry of sparsely granulated prolactin cell adenomas of the human pituitary. Acta Endocrinol (Kbh) 89: 521–529, 1978.
121. Farquhar MG, Reid JJ, Daniell LW: Intracellular transportant and packaging of prolactin: A quantitative electron microscope autoradiographic study of mammotrophs dissociated from rat pituitaries. Endocrinology 102: 296–311, 1978.
122. Walker AM, Farquhar MG, Preferential release of newly synthesized prolactin granules is the result of functional heterogeneity among mammotrophs. Endocrinology 107: 1095–1104, 1980.

Authors' address:
Department of Pathology,
St. Michael's Hospital
30 Bond Street
University of Toronto,
Toronto, Ontario, M5B 1W8 Canada

CHAPTER 9

The ventricular system in neuroendocrine mechanisms

DAVID E. SCOTT and WILLIS K. PAULL

1. Introduction

Basic precepts used to build our knowledge of neuroendocrine biology have historically identified the presence of neurosecretory neurons (1) in the endocrine hypothalamus as the 'sine qua non' of neuroendocrinology. A considerable body of circumstantial evidence suggests that these neuronal elements in the basomedial hypothalamus and possibly elsewhere in the central nervous system of vertebrates may terminate in large numbers upon the perivascular space that surrounds fenestrated portal capillaries of the hypophyseal portal circulation. This unique vascular network in the median eminence of the neurohypophysis apparently serves to functionally interconnect the ventral diencephalon with the adenohypophysis and was first described by Popa and Fielding in human post-mortem material over half a century ago (2). This vascular organization, apparently unique to the central nervous system, appears to be the sole and exclusive blood supply of the adenohypophysis since selective destruction of regions of the hypophyseal portal bed leads to focal necrosis in discrete regions of the pars distalis (3). Over 17 years elapsed before some functional significance was attached to the original observations of Popa and Fielding. In 1947, Green and Harris (60) hypothesized that the hypophyseal portal system was indeed a mechanism for delivery of 'neural principles' from the brain to the anterior pituitary. A number of morphological investigations further demonstrated that close proximity of the adenohypophysis to the median eminence and portal bed was necessary in maintaining its viability and the production of trophic pituitary hormones (4, 5). It was further demonstrated that the basomedial hypothalamus exerted an intrinsic hypophysiotrophic influence upon explants of the adenohypophysis transplanted into it, whereas other regions of the brain (e.g., cortex) failed to maintain the viability and normal histological appearance of such transplants (6). More recently it has been demonstrated that explants of normal fetal anterior hypothalamus when stereotaxically implanted into the third cerebral ventricle of adult rats with homozygous autosomal diabetes insipidus will trigger remarkable alterations in the physiological parameters of the host recipients (7, 8). The polydipsia and polyuria characteristic of this syndrome is ameliorated and hence strongly suggest alternative routes for the delivery of magnocellular neuropeptides.

With the historic characterization and synthesis of the releasing factors, TRF and LRF, in the laboratories of Guillemin and Schally (9, 10, 11) from millions of ovine and porcine hypothalamic fragments, it was apparent that the elegant hypothesis enunciated by Green and Harris, had stood the test of time. It is now quite clear that the mammalian median eminence is a bona fide neuroendocrine transducer wherein bioelectric signals from putative pools of neurons in the endocrine hypothalamus (and probably other integrative regions of the central nervous system) are translated into blood-borne signals of the releasing factors. Thus, the mammalian median eminence has been analogized as the 'final common neuroendocrine pathway' (12) and the first documented neuroendocrine transducer and serves as a template to investigate other circumventricular organs that are in anatomical juxtaposition to the third and fourth cerebral ventricles.

Conceptually, the idea of transduction with respect to the function of the median eminence (and perhaps other regions of the vertebrate brain the so-called circumventricular organs (13), is undoubtedly correct and has been supported by a wealth of morphological and physiological data. The crucial issue now at hand is the *precise location* and *distribution* of *cellular* (neuronal) pools responsible for the *synthesis*, *storage*, and *transport* of the various relevant *neuro peptides*, *carrier proteins*, as well as the *biogenic amines* which reputedly modulate the way in which releasing factors are ultimately delivered to the aden-

Motta, PM (ed): Ultrastructure of endocrine cells and tissues. ISBN-13: 978-1-4613-3863-5

ohypophysis. A further important question is the overall role of specialized ependyma, the so-called tanycyte which is exclusive to mammalian circumventricular organs.

2. The parvocellular neuron

From the ultrastructural standpoint, parvocellular neurons in the arcuate-periventricular region fulfilled the morphologic criteria used to identify true neurosecretory neurons (14, 15, 16, 17, 18). These neurons also demonstrate populations of dense core vesicles which, in terms of their size and disposition, were identical to those seen in axons which terminated upon the portal bed in the adjacent ventral median eminence. Thus, it was understandable for electron microscopists to conclude (at first based chiefly upon morphologic results) that the arcuate nucleus was contributing substantially in a neurovascular (neurohemal) way to the hypophyseal portal circulation. Further support for this supposition evolved from the fact that arcuate and periventricular neurons undergo remarkable morphologic alterations following a spectrum of physiological manipulations. Castration elicits ultrastructural changes wherein neuronal perikarya enlarge, numerous whorled bodies appear, Golgi cisternae are measurably widened, and the population of dense core vesicles increases by as much as 50% (15, 19). Fine structural fluctuations have also been noted during the estrus cycle as well, wherein peak values of dense core vesicles in the perikarya of arcuate neurons were recorded on the morning of proestrus (20). Similar alterations have also been observed in the contact zone of the median eminence of the female North American mink, *Mustela vision*, during late proestrus (21). This monestrus mammal ovulates in early spring for a period of one week only. It has also been described that statistically significant increases in populations of dense core vesicles occur in certain tuberoinfundibular axon terminals in the median eminence of adrenalectomized-dexamethasone treated rats (22). These investigators surmised that the apparent increase in this species of vesicle may in some fashion be related to CRF metabolism and was in agreement with investigations by others (23). Previous data support the notion that inclusions of this nature may arise from neuronal elements within the endocrine hypothalamus, for if one surgically isolates this region from surrounding brain structures, few if any signs of focal anterograde degeneration can be noted in tuberoinfundibular axon terminals. However, arcuate-periventricular region has been described to results in anterograde degeneration in numerous parvocellular axons in the palisade contact zones of the mammalian median eminence (24). The ultrastructural organization of the median eminence remains relatively normal following surgical isolation. However, there are some conflicting reports of scattered anterograde degeneration (25), coupled with the structural correlates of heightened synthetic activity (26).

Recent autoradiographic investigations employing radiolabeled estrogen, testosterone or corticosteroids have demonstrated selective nuclear uptake by a number of hypothalamic neurons and others in the limbic continuum (27, 28, 29). These investigations discuss a potential steroid receptor role for such neurons. Recent investigations in this laboratory employing other radiolabeled compounds such as tritiated releasing factors and/or biogenic amines have consistently demonstrated selective uptake of high specificity in sub-populations of the arcuate-periventricular complex (30, 31, 32). The neuroanatomical patterns of uptake and localization vary significantly depending upon the releasing hormones or neural transmitter (biogenic amine) that is introduced. Thus, a considerable body of evidence has been marshalled to support the concept that the arcuate-periventricular region may serve as a chief neuronal bed nucleus with primary afferent (neurohemal) input to the hypophyseal portal system. These parvocellular axon terminals, which harbor both dense core and clear microviscles (33), may be chiefly derived from this functionally heterogeneous complex (34). These data are consistent with the autoradiographic findings cited above that certain subpopulations of neurons in the endocrine hypothalamus selectively sequester a variety of exogeneous radiolabeled hormones or transmitters and that this may be the structural correlate of a receptor role for such cells (35, 36). Selective uptake as observed with autoradiography could be interpreted as short loop-autoregulatory feedback mechanism for specified neurons of the arcuate-periventricular pool. The compartmentalization of different functions in the heterogenous vertebrate hypothalamus is an attractive hypothesis and has been enunciated in earlier investigations (34). Thus, the idea has persisted that tuberoinfundibular elements represent the major contribution to the portal bed, and based on the model of magnocellular neurosecretory neuronal function of the supraoptic and paraventricular nuclei (35, 36, 37), these smaller parvocellular elements are hypothesized to be a major focus for the active synthesis, sequestration, storage and ultimate release of hypophysiotropins and other as yet unidentified neuropeptides.

3. Alternative pathways

Despite the wealth of morphologic and physiologic data that supports the relatively austere theory of the arcuate-periventricular complex, recent data now challenges this rather simplistic concept and necessitate a careful re-examination of what may be a far more complex model of neuroendocrine control. The first serious doubts were raised from the elegant investigations of Olney (37) who demonstrated a dose-dependent selective destruction of the arcuate nucleus of mice with chemical lesions induced by monosodium glutamate injection. Using Olney's model, simultaneous and independent investigations in acute and chronic preparations, (38, 39) demonstrated the same phenomenon at the ultrastructural level. These investigations reported extensive and selective lesions involving as much as 80% of the total population of arcuate neurons. Furthermore, they examined the subjacent median eminence. Despite alterations in the endocrine status of experimental animals (e.g., failure to reproduce, alterations in body weight, etc.), paradoxically, there were no notable ultrastructural changes in the organization of the median eminence proper. Clearly, if the arcuate-periventricular complex represents the chief bed nucleus that projects axons to the median eminence, then following extensive lesioning of this putative bed nucleus, one would expect to observe evidence of substantial anterograde degeneration of parvocellular axon terminals in the adjacent contact and palisade zones of the median eminence. These studies were repeated (40), and again following extensive destruction of the arcuate nucleus with relatively few neurons remaining intact, after careful sampling, the ultrastructural picture of the rostral mid-central and caudal portions of the median eminence remained intact and normal appearing with no overt signs of anterograde degeneration. The obvious complexity of this region of the brain has been further reinforced by a series of superb immunohistochemical investigations (41, 42, 43). These investigations have clearly demonstrated the presence of immunoreactive products in the contact zone of the median eminence using antisera against neurophysins(s) and vasopressin. At the light microscopic level, the reaction product (the PAP molecule) was noted over dense core vesicles that exhibit a broad size range including that for both small parvocellular dense core vesicles (800–1220 Å) and magnocellular neurosecretory inclusions (2000–3000 Å) typically associated with neurohypophyseal hormone (oxytocin/vasopressin) sequestration (41). Although the presence of immunoreactive material against LRF in the median eminence (44, 45, 46, 47) was an expected finding, the presence of immunopositive material for magnocellular hormones such as vasopressin and oxytocin and their carrier proteins (the neurophysins) was an intriguing and unexpected finding in the contact zone (zona externa) of the median eminence. Further substantiation of these data has been the assessment of unusually high levels of vasopressin and neurophysin found in portal blood (48). Using a sensitive radioimmunoassay technique, these investigators determined the presence of vasopressin in portal plasma collected via cannulation of large portal veins from female rhesus monkeys. Values for vasopressin exceeded 13,800 pg/ml in contrast to normal peripheral serum values of 30–50 pg/ml. Despite the fact that numerous ultrastructural investigations have failed to detect the presence of magnocellular (supraoptic-paraventricul-lular) axons in the palisade contact zone, recent exhaustive collaborative investigations have established the presence of Herring bodies and what can be identified ultrastructurally as contact with the portal perivascular space (49). Supporting data for this hypothesis continue to accumulate. In light of these newer findings, a re-examination of the presence and potential role and distribution of magnocellular hormones and their carriers in the zona externa of the median eminence and the obvious implications of this with respect to adenohypophyseal function is clearly required.

4. Specializations of the ventricular system

Ever since the original findings of Höfer (50) attention has been directed toward other regions of the diencephalon which harbor unique midline structures that share common structural similarities in their vascular, neuronal, and glial organization. These have become known as circumventricular organs (CVO's) and have steadily gained the attention of neuroendocrine morphologists. These primitive midline structures include the median eminence, the neural lobe, the organum vasculosum of the lamina terminalis, the subfornical organ, the subcommissural organ, the pineal gland and the area postrema. These seven regions of the brain have been hypothesized to play an intregative role between blood, brain and cerebrospinal fluid (CSF) and have been analogized as 'Windows of the Brain'. Unlike surrounding neural structures, CVOs do not have a blood-brain barrier per se but possess instead fenestrated capillaries which were earlier characterized in the median eminence and neural lobe in a number of ultra-structural investigations (50, 51, 36). The barrier properties of many CVOs have been tested, and it has been well established that intravascularly injected tracers such as colloidal protein dyes (HRP, trypan blue, etc.) easily traverse fenestrated capillaries of CVOs and penetrate into the adjacent parenchyma of these regions (52, 53, 54, 55). However, unlike the rest of the cerebral ventricular system, ependymal cells that line CVO's possess tight junctions (zonulae occludentes) which prevent the passive insinuation of proteinaceous colloidal dyes from the vascular compartment into the CSF of the ventricular lumen. Thus, a barrier to blood-born substances has been preserved in these regions. However, in cases of CVOs the blood-brain (blood-CSF) barrier has been shifted from the vascular side to the ventricular side (52, 53). Recent immunohistochemical investigations of CVOs including the median eminence, neural lobe as well as the organum vasculosum have also demonstrated immunoreactive material against LRF (43, 56, 57, 58, 59, 61). Localization of catecholamines in other CVOs such as the area postrema as well as angiotensin-II mediated drinking behavior and osmotic and dipsogenic control mechanisms in the subfornical organ and possibly the organum vasculosum as well have also been established (62, 63, 64). Thus, it is becoming evident that the median eminence and the neural lobe may not be the sole neuroendocrine transducers of the brain, but that other circumventricular organs mentioned above may also participate in neuroendocrine events and await a more clear and precise characterization of their functional capacities and interrelationships.

5. The tanycyte

The fundamental anatomical observations of Horstmann (1954) and Löfgren (1959–1961) directed attention to a specialized population of ependymal cells that constitute a significant proportion of the parenchyma of the median eminence, and all other CVOs as well. These specialized metaglia were given the name tanycytes or 'stretch cells' by Horstmann and appropriately so, for they are observed to literally stretch from the lower walls and floor of the third ventricular lumen through all zones of the median eminence to the perivascular space of the hypophyseal portal bed below, (Fig. 18). In their 500–700 μ course through the parenchyma of the median eminence, they branch and arborize extensively and, as observed with low voltage electron microscopy (LVEM) and high voltage electron microscopy (HVEM), and represent a significant proportion of surface area that terminates upon the portal bed (63, 64, 65, 66, 67, 68). In all zones of the median eminence, substantial numbers of apparent parvocellular axon terminals appear to make contact with the plasmalemma of vertically oriented processes of ependymal cells (66, 69). The unique ultrastructural configuration of tanycytes and their 'synaptoid' relationship with parvocellular axon terminals has stimulated a number of working hypotheses with respect to their potential role. It has been postulated that tancycytic ependymal cells of the median eminence may serve as a structural link between the CSF of the third ventricle and the plasma of the hypophyseal portal system (70, 71) and extensive axonal termination upon their numerous processes may represent a potential mechanism for a unique type of 'neurogenic' control. A potential absorptive role has been postulated for tanycytes wherein biologically active molecules in the CSF might be absorbed, transported and released into the portal circulation. This putative absorptive capacity has been further reinforced by the fact that both SEM and TEM revealed that the apical (ventricular) surfaces of mature tanycytes exhibit a profuse nap of microvilli which is virtually identical with that of the choroid plexus (Figs. 1A, 1B and 1C). A number of investigations have established both a secretory and absorptive role for the choroid plexus (71, 72, 73) and based upon this ultrastructural parallelism, it has been speculated that tanycytes might function in a similar fashion. Other morphologic criteria began to accumulate with respect to a potential role for absorption and transport. It was demonstrated that tanycytes possess a rich assortment of organelles including an extensive array of microtubules, well suited for transport activity (64, 65) and that certain drugs such as colchicine or vinblastine significantly alter the ultrastructural organization of such subsystems of tanycytes (74). It has been amply demonstrated that colchicine inhibits the transport of axonal proteins generally and virtually inhibits movement of neurosecretory material in nearby axons of the neurohypophyseal system (75). One further piece of morphological evidence with respect to the potential role that tanycytes might play (other than maintaining structural integrity) is the observation that this cell type commonly sequesters secretion product(s) in their terminal cytoplasmic processes that abut upon the perivascular space of the hypophyseal portal bed (64, 65, 36). These secretory inclusions are strikingly similar to the population of dense core vesicles sequestered in nearby parvocellular axon terminals. It has been demonstrated that axons which terminate upon the plasmalemma of adjacent tanycytes will sequester radiolabeled dopamine (76, 36, 77). It is possible that dopamine may function as a transmitter in the control of tanycyte transport. Preliminary histochemical investigations in this laboratory employing ETPA and BIUL have failed to demonstrate the presence of true synapses in the median eminence. Despite this, the 'synaptoid' relationship between parvicellular axon terminals and tanycytic ependyma may serve as a mechanism for the neurological control of tanycyte activity. Recently del Cerro (78) demonstrated the presence of gap junctions between developing arcuate neurons and adjacent tanycytes in the perinatal hypothalamus. He has speculated that these may serve as points of functional con-

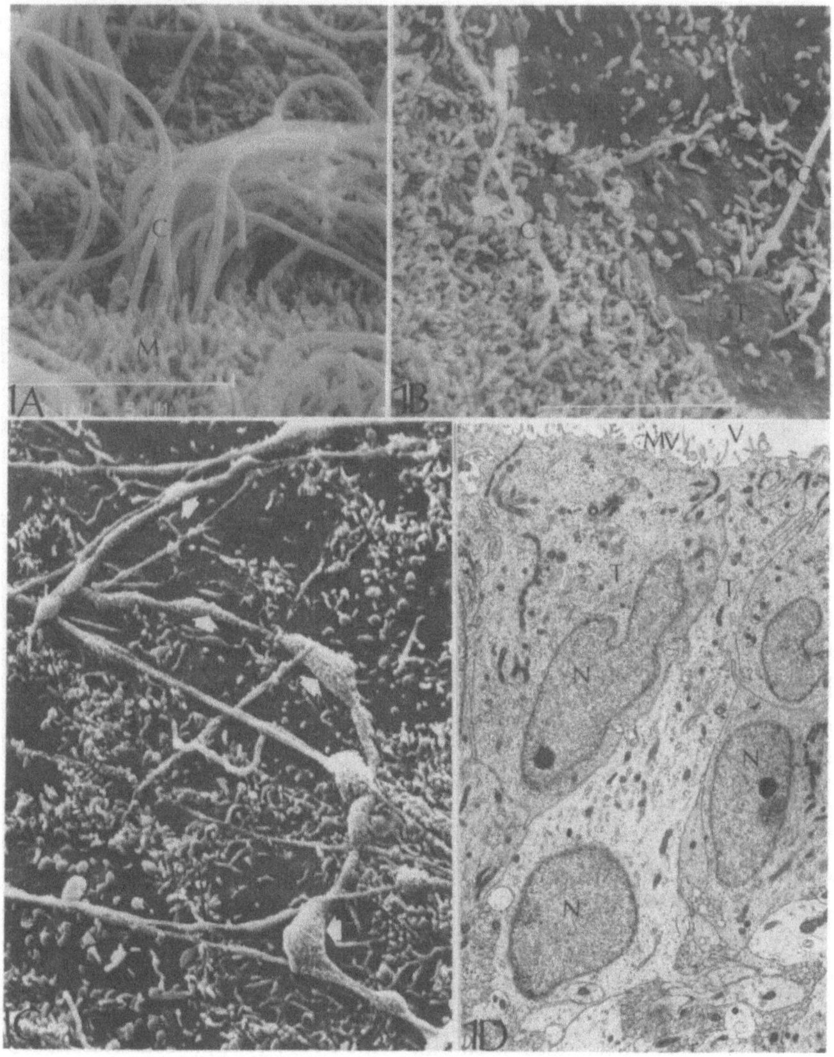

Fig. 1. (A) SEM of dorsal thalamic wall of third ventricle of normal rat. Noteable here are clusters of cilia (C) which punctuate this area of the third cerebral ventricle. A thick nap of microvilli (M) can be observed to underline the cilia (× 8,530); (B) SEM of floor of infundibular recess of normal male rat. Profiles of individual tanycytes (T) can be observed to be separated by microvilli (arrows) which anatomically represents the interface between adjacent cells. This cell type, which is exclusive to circumventricular organs, exhibits individual excentric cilia (C) and varying densities of microvilli. The variation in the density of microvilli from cell to cell has led a number of investigators to postulate that there either be several subclasses of tanycytes or conversely this cell line may undergo functional changes that mirror alternations in the physiological status of the animal (× 7,950); (C) SEM of floor infundibular recess of male rat with autosomal homozygous diabetes insipidus. Consistent with the stereotaxic placement of a fragment of normal fetal hypothalamus into the lumen of the third cerebral ventricle, one can note a remarkable increase in the number of neurites (axons-dendrites) (arrows) with numerous varicosities upon the ependymal surface. This increase is consistently observed in close anatomical juxtaposition with the fetal graft (× 4,340); (D) TEM of zona interna (ependymal-hypendymal zone) of the median eminence of a normal male rat. Tanycytes (T) constitute the anatomical boundary between the cerebrospinal fluid compartment of the third central ventricle (V) and the hypophyseal portal plexus below in contact zone of the median eminence. MV: microvilli; N: nucleus (× 4,760).

tinuity between these two diverse cell compartments. These observations were reenforced by Scott and Paull (79) who demonstrated an intimate structural relationship between presynaptic axon terminals and adjacent tanycytes. In an attempt to resolve the issue of a potential tanycytic transport capability, *in vitro* investigations utilizing radiolabeled non-metabolizable amino acids and thyroxine have been employed. Such studies have demonstrated rapid uptake by ependyma in explants of the median eminence (80). Tracer studies employing horseradish peroxidase have also demonstrated transependymal uptake and transport following intraventricular infusion (81, 82). Recent autoradiographic investigations at the light and electron microscopic level in this and other laboratories have demonstrated extraordinarily rapid uptake and transport of radiolabeled releasing hormones, catecholamines or catecholamine precursors such as ^{3}H-L-dopa following their infusion into the ventricular system. The labeled hormone or its metabolite appeared in the portal perivascular space within minutes following its infusion (83, 84, 36, 85). The elegant tracer studies by Brightman and Reese (*vide supra*) have unquestionably established the presence of tight junctions between ependyma (tanycytes) that constitute the floor of the third ventricle, as mentioned before. These tight junctions (zonulae occludentes) serve to prevent passive interstitial migration of substances into the CSF from the fenestrated portal vessels of the median eminence and vice versa. Therefore, a plausible mechanism wherein radiolabeled releasing hormones or biogenic amines could possibly reach the portal bed from the ventricular lumen with such rapidity would be through the medium of ependymal transport. However, in view of the fact that both non-metabolizable amino acids as well as horseradish peroxidase can apparently be transported by tanycytes, one is forced to conclude that a tanycyte transport mechanism may simply be a general phenomenon which may not necessarily be linked to a specific neuroendocrine phenomenon or event. However, recent investigations (86) have argued against this and convincingly demonstrated that following selective and complete destruction of tanycytes of the lower walls and floor of the third cerebral ventricle, without attendant damage to surrounding magnocellular and parvocellular elements, remarkable alternations occur in the endocrine status of the animals. These animals (rats), exhibit a classical panhypopituitarism coupled with hyperprolactinemia. This depression of hypothalamic releasing and inhibitory hormone actively following destruction of tanycytes of the median eminence underscores their potential functional capacity and putative integrative role.

Basic to any concept of a transport system for the movement of neurotransmitter peptides, catecholamines, etc. to or form the ventricular CSF to the portal bed is the presence of neural factors and trophic hormones in the CSF in significant titers. A number of biologically active substances such as vasopressin, neurophysin, prostaglandins, growth hormone, ACTH, arginine vasotocin and other physiologically active molecules normally exist in the CSF and have been characterized in numerous investigations (87, 88, 89, 90, 91, 92, 93). The intriguing question is how do such neuropeptides or trophic factors reach the CSF. Studies by Reichlin (94, 95, 96) reveal that TRF is found in high concentrations in non-hypothalamic regions such as the cerebral cortex and forebrain. The broad distribution of this tripeptide-amine throughout the mammalian forebrain argues for a multiple pharmacologic role for TRF. Furthermore, the pattern of distribution of extrahypothalamic TRF and LRF suggests the possibility that TRF and LRF (as well as other hormones) could enter the CSF from adjacent brain parenchyma by simple diffusion or bulk tissue flow through ependymal gap junctions in non-circumventricular regions. However, due to the presence of high concentrations of deactivating enzymes for TRF, LRF, and other trophic hormones in brain parenchyma which parenthetically are not found in CSF (96), a more likely and reasonable explanation for high concentrations of biologically active molecules in the CSF may be due to abundant nerve terminals and neuronal elements that insinuate between lining ependymal cells and project into the lumen of the third, lateral and forth ventricles as observed with scanning-transmission electron microscopy (SEM/TEM) and electron microscopic autoradiography (vide infra).

6. The CSF contacting neuron: neurocisternal secretion-absorption

In examining the fine structural correlates of neuroendocrine function, one must address a new model system that has recently become evident from extensive SEM/TEM investigations of the mammalian circumventricular organ system.

New data originating in laboratories from around the world have identified the presence of supraependymal and periventricular cell networks which are seen in large numbers to make direct contact with the CSF (97). These are especially common elements in rodent and primate brains. These supraependymal networks are most commonly observed upon the ventricular surfaces of CVOs themselves, although not exclusively confined to these regions. These networks are neuronal in nature. Bona fide neurons in direct contact with CSF are seen to project dendritic and axonal ramifications over the surface of the ventricle for substantial distances (Figs. 1C, 2B). Collaterals or secondary branches then penetrate into the underlying neuropil. The CSF (liquor) contacting neuronal and axonal populations, first described by Vigh and Vigh-Teichman (98) in lower vertebrates and more recently in rat and human (99, 100) have been identified and verified with simultaneous and correlative (TEM) analysis of the same sample of tissue of perinatal and adult hypothalamus (101, 102, 103). These supraependymal cells possess all of the fine structural characteristics and criteria that are used to identify true neurosecretory neurons. Autoradiographic data, both at light and electron microscopic levels, demonstrate selective sequestration of various radiolabeled biogenic amines by such supraependymal neurons and axons and suggest that they, like adjacent neurons in the arcuate periventricular complex, may serve as either receptors or

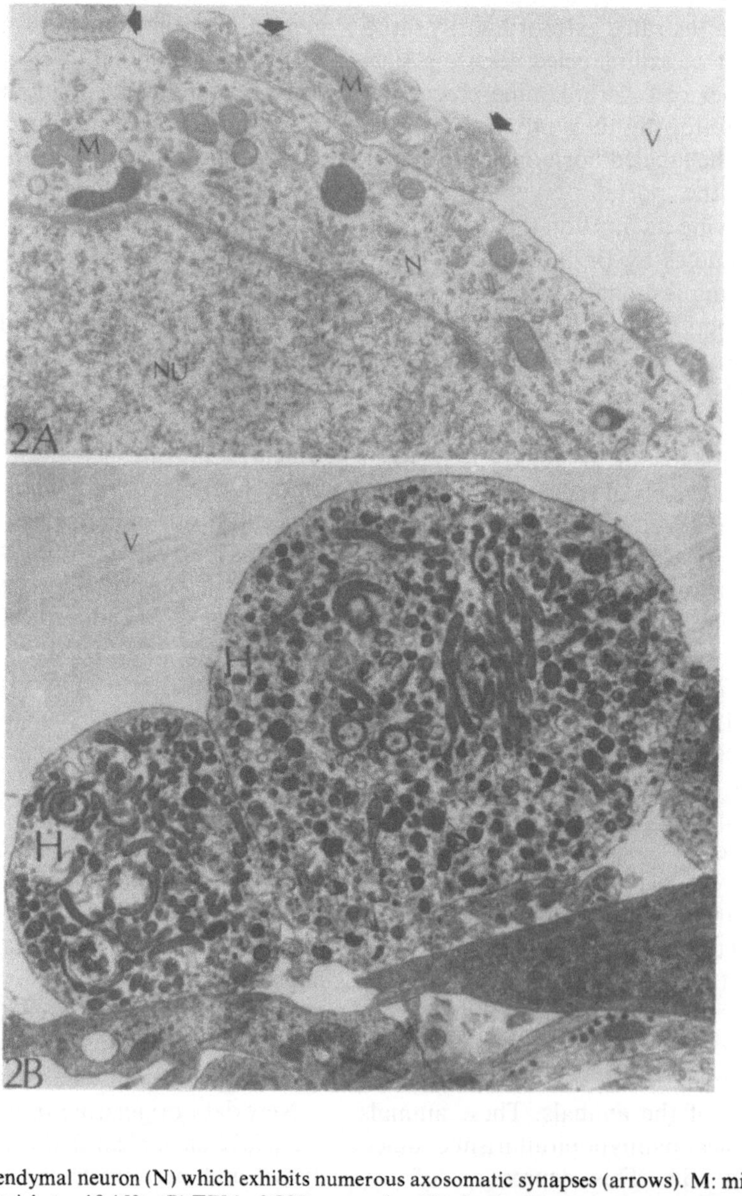

Fig. 2. (A) TEM of supraependymal neuron (N) which exhibits numerous axosomatic synapses (arrows). M: mitochondria; NU: nucleus; V: lumen, third cerebral ventricle (× 10,150); (B) TEM of CSF-contacting Herring bodies (neurosecretory axons) (H) on the floor of the infundibular recess of male rat. The size and disposition of these axon profiles and the dense core vesicles they harbor argue strongly that they may be magnocellular (supraoptic-paraventricular) in origin and hence may either sequester oxytocin, vasopression or both. V: lumen, third cerebral ventricle (× 9,100).

integrators of neuroendocrine function. SEM/TEM investigations in primates and rats also serve to suggest that these supraependymal neuronal networks in various CVOs may be interconnected or integrated with those that are observed upon the ventricular surfaces of other adjacent CVOs and that there may exist a supraependymal network of receptor cells that serves to assess ambient changes in the milieu of the CSF (103, 104). This hypothesis, first postulated by McKenna and Rosenbluth (105) was based upon their findings of catecholaminergic CSF-contacting neurons in the hypothalamus of male and female toads and the concomitant changes that were observed in catecholamine fluorescence in these cells during ovulatory cycle. Numerous ultrastructural investigations have established the presence of axon terminals which literally 'terminate' in the CSF of the third lateral and fourth ventricles (106, 107, 108, 109, 110, 64). Autoradiographic and histocytochemical investigations ultimately determined that a significant proportion of these terminals were either catecholaminergic or indoleaminergic (109) however, their origin remained a mystery. Not until recent investigations (110) was it established that a significant number of these axons arose from raphe nuclei of the brain stem. Conceptually, the

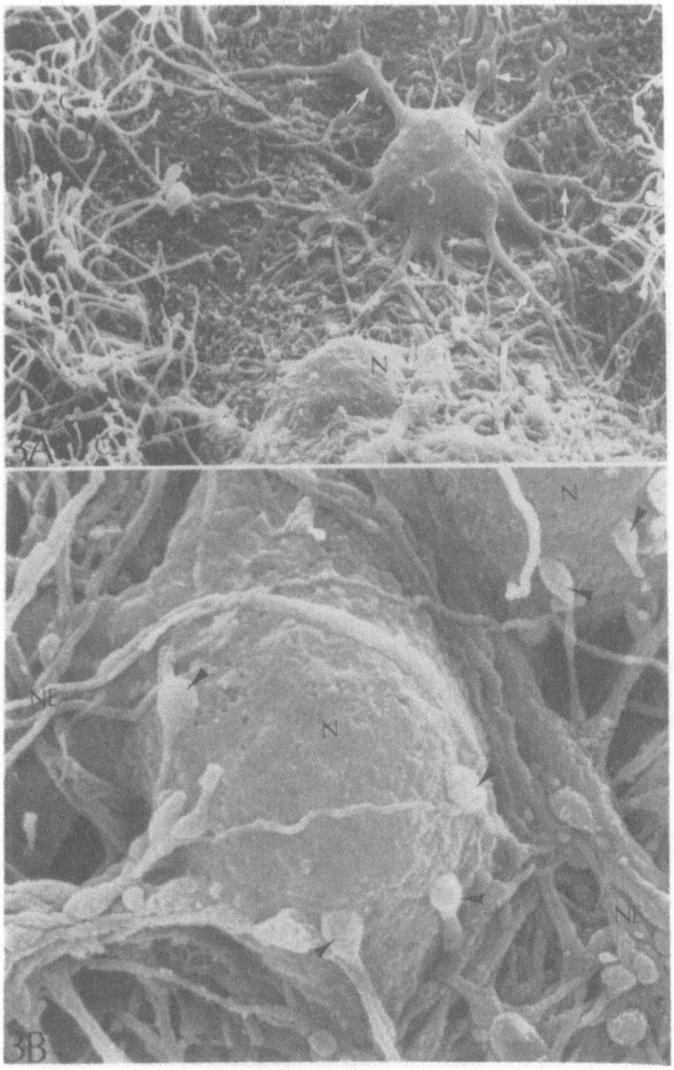

Fig. 3. (A) SEM of supraependymal CSF-contacting neurons (N) upon the third ventricular wall of male mouse. These cells exhibit multiple processes (arrows) which extend away from the cell body in a radial fashion. C: cilia (× 2,660); (B) SEM of supraependymal neuron (N) in a male DI rat which received a sterotaxic implant of normal fetal hypothalamus into the lumen of the third cerebral ventricle. Noteable here is the presence of boutons terminaux (arrows) which abut upon the somata of supraependymal CSF-contacting neurons (N) and constitute the SEM correlate of axosomatic synapses. NE: neurites (× 5,950).

notion of axons literally 'terminating' in the fluid medium of the cerebral ventricular system is not dissimilar to the finding of free noradrenergic endings in the cerebral cortex, (111, 112). Thus, rather than exerting a restricted influence via trans synaptic excitation or inhibition, free axon terminals that terminate in the CSF may have a broad modulatory influence of periventricular neurons which are juxtaposed to the CSF of the adjacent cerabral ventricles. This close relationship of neurons with hormone-laden CSF has fostered the idea that the CSF may subsume the role of trophic mediator.

These intriguing data may represent the ultrastructural correlate of a fundamental mechanism of *neurocisternal* control of various endocrine processes. This becomes especially relevant with respect to the presence of a putative sleep factor in the CSF (113).

7. The perinatal supraependymal neuron and sexual differentiation

Despite an extensive literature that has described in detail the fine structural organization of the median eminence-arcuate region in adult mammals, relatively little is known about the organization of the perinatal hypothalamus and the ultrastructural correlates that are attendant to its growth, development and plasticity. The basic anatomical organization of the perinatal median eminence has been described in virtually one class of mammals, namely the rat and until recently has been restricted to descriptive analyses with transmission electron microscopy (TEM) alone (114, 115, 116). Restrictive sample size of tissue analyzed has been the major limitation of these studies. More recent investigations (103) have, using

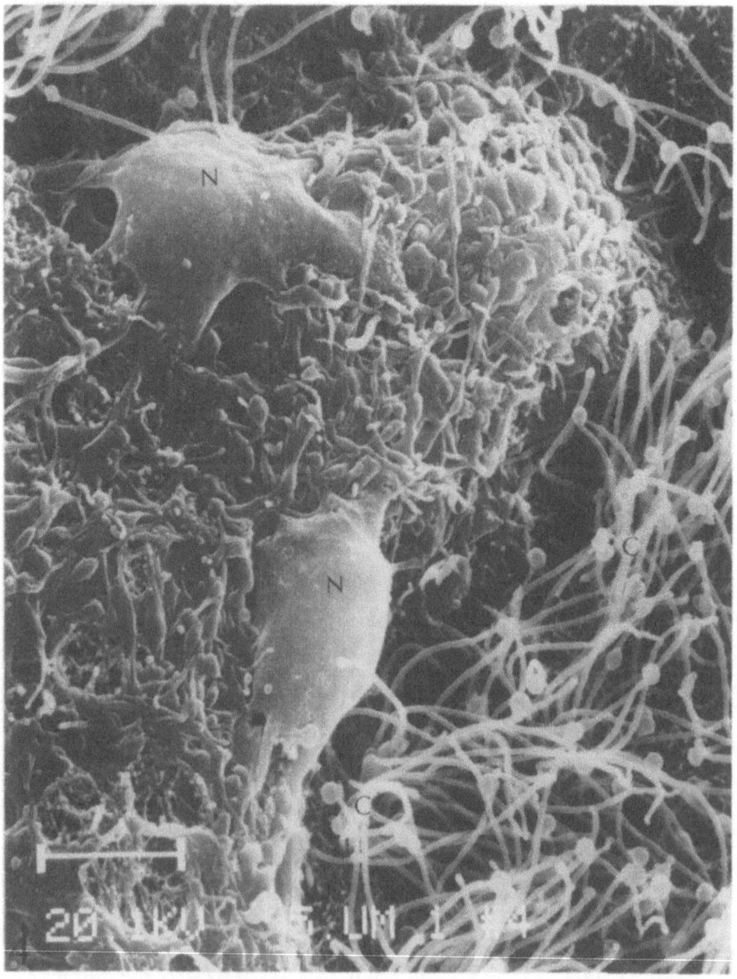

Fig. 4. SEM of the dorsal hypothalamic wall of an aged male mouse. Of note here are two supraependymal neurons which do not appear to possess numerous axosomatic synapses. C: cilia (× 3,150).

the combined-correlative SEM/TEM approach on the same sample of fetal tissue, demonstrated a number of intriguing findings that revealed remarkable fine structural alterations in the median eminence of the day 17–20 prenatal (fetal) rat and the day 1–16 postnatal (perinatal) pup. By day 17 *in utero* the ventricular (ependymal) surfaces of brain regions other than circumventricular organs (e.g., the thalamic wall, cerebral vesicle and rhomboid fossa) were well differentiated with microvilli and cilia. The histocyte (Kolmer cell) was the first supraependymal cell evident upon the ventricular lumen and appeared by day 17 *in utero*. In contrast, the apical surfaces of tanycytes of the infundibular recess as well as those of most circumventricular organs were poorly differentiated and unremarkable. A vivid example of that state of undifferentiation was the presence of tanycytes in the infundibular recess of the hypothalamus that exhibited a simple hexagonal mosaic pattern of apposed plasmalemmata and even 2 days postpartum few cilia or microvilli were evident. By day 5–6 postpartum, the classical supraependymal neurons and axons began to make their appearance upon the ventricular surface with many of them emerging from the underlying parenchyma of the endocrine hypothalamus and subjacent median eminence. By day 16 postpartum, with the exception of a significant increase of number of supraependymal neurons, the ventricular surface, and substructural organization of the median eminence, as well as other circumventricular organs of the hypothalamus assume a picture comparable to that of the adult. It is important to note that the population of perinatal supraependymal neurons peak in numbers at 3 to 7 days postpartum. This fact alone suggests that their presence and hence potential influence may be felt most markably during the critical period of sexual differentiation.

These data serve to suggest that supraependymal neurons in the perinatal and adult rat (virtually identical with those observed in other species) are remarkably similar in their ultrastructural organization with parvocellular neurosecretory neurons elsewhere in the endocrine hypothalamus which have been implicated in the synthesis of centrally acting peptides (Figs. 2, 3 and 4). Their consistent emergence at day 5–6 postpartum (in the rat) suggests a possible correlation with the critical period of sexual differentiation and also a potential receptor role for this cell line. The importance of this unique population of supraependymal neurons of CVOs and extensive axonal networks cannot be minimized and additionally may serve as an anatomical substrate for the functional integration of the circumventricular organ system. In addition, an equally exciting impact of the emergence of such neuronal elements in the cerebral ventricle of the neonate underscores a remarkable degree of neuronal plasticity in the postpartum perinatal brain. The consistency of the observations of neuritic elements in direct contact with the lumen of the mammalian cerebral ventricular system and CSF strongly suggest a bonafide physiological role for this poorly understood, yet intriguing neurological system.

References

1. Scharrer EA: Secretory cells in the midbrain of the european minnow, *Phoxinus paevis*. J Comp Neurol 55: 573–591, 1932.
2. Popa CT, Fielding V: A portal circulation from the pituitary to the hypothalamic region. J Anat 65: 88–101, 1930.
3. Daniel PM, Prichard MML: Anterior pituitary necrosis. Infarction of the pars distalis produced experimentally in the rat. OJ Exp Physiol 41: 215–229, 1956.
4. Everett JW, Nikitovich-Winer M: Physiology of the pituitary gland as affected by transplantation or stalk section. In: Advances in Neuroendocrinology. Nalbandov A V (ed) University of Illinois Press, Urbana, 1963, p 289.
5. Smith PE: Postponed pituitary homotransplants into the region of the hypophyseal portal circulation in hypophysectomized female rats. Endocrinology 73: 793–801, 1963.
6. Halasz B, and Pupp L: Hormone secretion of the anterior pituitary gland after physical interruption of all nervous pathways to the hypophysiotrophic area. Endocrinology 77: 553–562, 1965.
7. Gash DM and Sladek JR, Jr: Vasopressin neurons grafted into Brattleboro rats. Viability and Activity. Peptides, 1–14, 1980.
8. Gash DM, Scott DE: Fetal hypothalamic transplants in the third ventricle of the adult rat brain. Cell Tiss Res 211: 191–206, 1980.
9. Burgus R, Dunn TF, Desiderio D, Guillemin R: Structure moleculaire du facteur hypothalamique hypophysiotrope TRF d'origine ovine: mise en evidence per spectrométrie de masse de la séquence PCA-His-Pro-NH_2. CR Acad Sc Paris 169: 1879–1883, 1969.
10. Folkers K, Enzmann F, Boler J, Bowers CY, Schally AV: Discovery of modification of synthetic tripeptide sequence of thyrotropin releasing hormone activity. Biochem Biophys Res Commun 37: 123–126, 1969.
11. Schally AV, Arimura A, Baba Y, Nair RMG, Matsudo H, Redding TW, Develjuk L, White FF: Purification and properties of LH and FSH releasing hormone from porcine hypothalami. Proceedings of the 53rd Meeting of the Endocrine Society, 1971, pp 70A.
12. Scharrer E: The final common path in neuroendocrine integration. Arch Ant Micr Morphol Exp 54: 359–370, 1965.
13. Afifi A, Adel K, Bergman RA: Basic Neuroscience. Urban and Schwarzenberg, Baltimore, 1980.
14. Etcheverry GJ, Pellegrino de Iraldi A: Ultrastructure of neurons in the arcuate nucleus of the rat. Anat Rec 160: 239–254, 1968.
15. Brawer JR: The fine structure of the arcuate nucleus in castrated male rats. Anat Rec 169: 283–296, 1971.
16. Adamo NJ: Ultrastructural features of the lateral preoptic area, median eminence and arcuate nucleus of the rat. Z Zellforsch 27: 483–491, 1972.
17. Ratner A, Adamo NJ: Arcuate nucleus region in androgen sterilized female rats. Ultrastructural observations. Neuroendocrinology 8: 26–35, 1971.
18. Scott DE, Sladek JR, Jr, Kozlowski GP, McNeill TH, Paull WK, Krobisch-Dudley G: The median eminence as a neuroendocrine transducer. In: The Neuroendocrine Control of Fertility, A. Kumar (ed) Karger, Basel, 1976A, pp 57–70.
19. Zambrano E, Robertis E: Ultrastructure of peptidergic and mono-

aminergic neurons in the hypothalamic neurosecreting system of anuran bacitracians. Z Zellforsch 90: 200–244, 1968.
20. Zambrano E: On the presence of neurons with granulated vesicles in the median eminence of the rat and the dog. Neuroendocrinology 3: 144–155, 1968.
21. Scott DE, Krobisch-Dudley G: Ultrastructural analysis of the median eminence and third ventricle of the mink (Mustela vison). Anat Rec 172: 2, 1972.
22. Gibbs FP, Scott DE: Influence of glucocorticoid levels on fine structure of the rat median eminence. Endocrinology 94: 303–308, 1974.
23. Wittkowski W, Bock R: Electronmicroscopial studies of the median eminence following interference with the feedback system anterior pituitary adrenal cortex. In: Brain-Endocrine Interaction. I. Median Eminence: Structure and Function. Knigge KM, Scott DE, Weindl A (eds) Karger, Basel, 1972, pp 171–180.
24. Raisman G: A second look at the parvicellular neurosecretory system. In: Brain-Endocrine Interaction. Median Eminence: Structure and Function, International Symposium, Munich, 1971. Knigge KM, Scott DE, Weindl A (eds) Karger, Basel, 1972, pp 306–318.
25. Rethelyi M, Halasz B: Origin of the nerve endings in the surface zone of the median eminence of the rat hypothalamus. Exp Brain Res 11: 145–158, 1970.
26. Gross JH, Knigge KM, Sheridan MN: The fine structure of the arcuate neurons and median eminence of the golden hamster following immobilization. Cell Tissue Res 156: 1–17, 1976.
27. Stumpf WE: Estrogen-neurons and estrogen-neuron systems in the periventricular brain. Am J Anat 129: 207–218, 1970.
28. Pfaff DW: Autoradiographic localization of radioactivity in the rat brain after injection of tritiated sex hormones. Science 161: 1355–1356, 1968.
29. Morrell JI, Kelley DB, Pfaff DW: Autoradiographic localization of hormone-concentrating cells in the brain of an amphibian, Zenopik laevis. J Comp Neurol 164: 63–78, 1975.
30. Scott DE: The ultrastructural correlations of circumventricular organ function. The median eminence as a neuroendocrine transducer. Bis Conf Rep 40: 22–28, 1975.
31. Scott DE, Krobisch-Dudley G, Knigge KM: The ventricular system in neuroendocrine mechanisms. II. In vivo monamine transport by ependyma of the median eminence. Cell Tiss Res 154: 1–16, 1974.
32. Scott DE, Krobisch-Dudley G, Paull WK, Kozlowski GP: The ventricular system in neuroendocrine mechanisms. III. Supraependymal neuronal networks in the primate brain. Cell Tiss Res 197: 235–254, 1977.
33. Kobayashi H, Matusi T, Ishii S: Functional electronmicroscopy of the hypothalamic median eminence. Int Rev Cytol 29: 281–381, 1970.
34. Oksche A, Oehmke HJ, Hartwig HC: A concept of neuroendocrine cell complexes. In: Neurosecretion. The Final Common Pathway, Sixth International Congress of Neurosecretion. Knowles F, Vollrath L (eds) Springer-Verlag, Heidelberg, 1974, pp 154–164.
35. Scott DE, Knigge KM, Krobisch-Dudley G: The ventricular system in neuroendocrine mechanisms. II. In vivo monoamine transport by ependyma of the median eminance. Cell Tiss Res 154: 1–16, 1974.
36. Monroe BG, Scott DE: Ultrastructural changes in the neural lobe of the hypophysis of the rat during lactation and suckling. J Ultrastruct Res 14: 405–431, 1966.
37. Scott DE, Krobisch-Dudley G, Weindl A, Joynt R: An electron microscopic autoradiographic analysis of hypothalamic magnocellular neurons. Z Zellforsch 138: 421–438, 1973.
38. Holzwarth MA, Hurst EM: Manifestations of monosodium glutamate (MSG) induced lesions in the arcuate nucleus of the mouse. Anat Rec 178: 378, 1974.
39. Paull WK, R Lechan: The median eminence of mice with an MSG induced lesion. Anat Rec 178: 346, 1974.
40. Holzwarth-McBride MA, Hurst EM, Knigge KM: Monosodium glutamate induced lesions of the arcuate nucleus. L. Endocrine deficiencies and ultrastructure of the median eminence. Anat Rec 186: 185–196, 1976.
41. Silverman A, Zimmerman EA: Ultrastructural immunocytochemical localization of neurophysin and vasopressin in the median eminence and posterior pituitary of the guinea pig. Cell Tissue Res 159: 291–301, 1975.
42. Silverman AJ, Knigge KM, Zimmerman EA: Ultrastructural immunocytochemical localization of neurophysin in freeze-substituted neurohypophysis. Am J Anat 142: 265–271, 1975.
43. Kozlowski GP, Scott DE, Krobisch-Dudley G, Frenk S, Paull WK: The primate median eminence. II. Correlative high-voltage electron microscopy. Cell Tissue Res 175: 265–277, 1976.
44. Flerko B: Oestrogen-sensitive neurons and their role in the control of ovulation. In: Neuroendocrine Regulation of Fertility. Kumar A (ed) Karger, Basel, pp 114–123, 1976.
45. Kordon C: New data on hormone-neurotransmitter interaction in gonadotropic regulation. In: Neuroendocrine Regulation of Fertility. Kumar A (ed) Karger, Basel, 1976, pp 95–105.
46. Hoffman GE, Moynihan JA, Knigge KM: Immunocytochemical localization of luteinizing hormone-releasing hormone (LHRH) differences with different antisera. Neurosci Abstr 962, 1976.
47. Scott DE, Sladek JR, Jr, Kozlowski GP, McNeill TH, Paull WK, Krobisch-Dudley G: Ultrastructural correlates of circumventricular organ function. I. The median eminence as a neuroendocrine transducer. In: Proceedings of the Symposium of Neural Control of Human Reproduction and Fertility. Int Symp Simla, India. Kumar A (ed) Karger, Basel, 1976, pp 29–35.
48. Zimmerman EA, Carmel PW, Husain MK, Ferin M, Tannenbaum M, Franz AG, Robinson AG: Vasopressin and neurophysin: High concentrations in monkey hypophyseal portal blood. Science 182: 925–927, 1973.
49. Zimmerman EA, Krupp L, Hoffman DL, Mathew E, Nilaver G: Exploration of peptidergic pathways in brain by immunocytochemistry: A ten year perspective. pp 3–10. In: Brain-Endocrine Interaction Symposium IV. Neuropeptides, development and aging. Scott DE, Sladek JR Jr (eds) Vol. I, Supp. I. Peptides Ankho International Inc. NY, USA, 1980.
50. Höfer H: Zur Morphologie der circumventricularen Organe des Zwischenhirnes der Saugetiere. Vorh Dtsch Zool Ges Frankfurt/M 8: 202–251, 1958.
51. Bargmann W, Knoop A: Electronenmikroskopische Beobachtungen and der Neurohypophyse. Z Zellforsch 46: 242–258, 1957.
52. Weindl A, Joynt RJ: The median eminence as a circumventricular organ. In: Brain-Endocrine Interaction. Median Eminence: Structure and Function, Knigge JM, Scott DE, Weindl A (eds) Karger, Basel, 1972, pp 280–297.
53. Weindl, A: Neuroendocrine aspects of circumventricular organs. In: Frontiers in Neuroendocrinology, Ganang WF, Martini L (eds) Oxford University Press, New York, 1973, pp 3–33.
54. Brightman MW, Reese TS: Junctions between intimately opposed cell membranes in the vertebrate brain. J Cell Biol 40: 648–677, 1968.
55. Brightman MW, Prescott L, Reese TS: Intercellular junctions of special ependyma. In: Brain Endocrine Interaction. II. Knigge KM, Scott DE, Kobayashi H, Ishii S (eds). Karger, Basel, 1975, pp 146–165.
56. Weindl A, Shinko I: Vasculur and ventricular neurosecretion in the organum vasculosum of the lamina terminalis of the golden hamster. In: Brain-Endocrine Interaction. II. The Ventricular System in Neuroendocrine Mechanisms. Knigge KM, Scott DE, Kobayashi H, Ishii S (eds) Karger, Basel, 1975, pp 190–203.
57. Flerko B, Setalo S, Vigh A, Arimura Schally AV: The luteinizing hormone releasing hormones (LH-RH) neuron system in rat and rabbit. In: Brain-Endocrine Interaction. III. Scott DE, Kozlowski JP, Weindl W (eds) Karger, Basel, 1977, pp 108–116.
58. Bugnon C, Block B, Lenys D, Fellman D: Cytoimmunological study of the LH-RH neurons in humans during fetal life. In: Brain-Endocrine Interaction. III. Neural Hormone and Reproduction, 1978, pp 183–196.
59. Palkovits M, Mezey E, Ambach G, Kovovics P: Neural and vascular connections between the organum vasculosum of the lamina terminalis and preoptic nuclei. In: Brain-Endocrine Interaction. III. Neural Hormones and Reproduction. Scott DE, Kozlowski GP, Weindl A (eds) Karger, Basel, 1978, pp 303–312.
60. Green JD, Harris GW: The neurovascular link between the neurohypophysis and the adenohypophysis. J Endocrinol 5: 136–146, 1977.
61. McNeill TH, Scott DE, Sladek JR Jr: Simultaneous meoamine histofluorescence and neuropeptide immunocytochemistry: I. Localization of catecholamines and gonadotropin-releasing hormone in the rat median eminence. Peptides. Vol. 1, Ankho International, USA, 1980, pp 59–68.
62. Torack RM: The role of epinephrine in the function of the area postrema. Participation of nerve endings in uptake and release of

tritiated norepinephrine. I. Brain-Endocrine Interaction. II. The ventricular system in neuroendocrine mechanisms. Knigge KM, Scott DE, Kobayashi H, Ishii S (eds). Karger, Basel, 1975, pp 190–203.
63. Monroe BG: A comparative study of the ultrastructure of the median eminence, infundibular stem and neural lobe of the hypophysis of the rat. Z Zellforsch 76: 405–415, 1967.
64. Rodriguez E: Ultrastructure of the neurohemal region of the toad median eminence. Z Zellforsch 93: 183–212, 1969.
65. Rodriguez E, LaPointe J: Histology and ultrastructure of the neural lobe of the lizard, Kauberina riversiana. Z Zellforsch 95: 35–37, 1969.
66. Knowles, F, Kumar ATC: Structural changes related to reproduction in the hypothalamus and in the pars tuberalis of the rhesus monkey. Part I: The Hypothalamus. Part II: The pars tuberalis. Phil. Trans R Soc 256: 356–375, 1969.
67. Akmayev IG, and Fidelina OV: Morphological aspects of the hypothalamohypophyseal system. VI. The tanycytes. Their relationship to sexual differentiation of the hypothalmus. Cell Tiss Res 173: 407–416, 1976.
68. Kozlowski GP, Frenk SE, McNeill TH, Scott DE, HSU KC, Zimmerman EA: Light and electron microscopic localization of gonadotropin-releasing hormone in rat median eminence by immunoperoxidase technique. First International Symposium of Immunoenzymatic Techniques (INSERM), April 2–4, 1975, Auremeas S, Bigon J, Dwet P, Feldman G (Organizer), Clichy, Paris, 1976.
69. Kobayashi H, Matsui T, Ishii S: Functional electronmicroscopy of the hypothalamic median eminence. Int Rev Cytol 29: 281–381, 1970.
70. Löfgren F: The glial-vascular apparatus in the floor of the infundibular cavity. Lunds Univ Arsskr NF 57: 1–18, 1961.
71. Cserr HF: Physiology of the choroid plexus. Physiol Rev 51: 273–311, 1971.
72. Cserr HF, Van Dyke DH: 5-Hydroxyindoleacetic acid accumulation by isolated choroid plexus. Am J Physiol 220: 718–723, 1971.
73. Milhorat TH: Choroid plexus and cerebrospinal fluid production. Science 166: 1514–1516, 1969.
74. Schechter J, Yancy B, Weiner R: Response of tanycytes of rat median eminence to intraventricular administration of colchicine and vinblastine. Anat Rec 184: 233–249, 1976.
75. Norstrom A, Hansson HA, Sjostrand J: Effects of colchicine on axonal transport and ultrastructure of the hypothalamo-neurohypophyseal system on the rat. Z Zellforsch 113: 271–293, 1971.
76. Scott DE, Krobisch-Dudley G: Ultrastructural analysis of mammalian median eminence. I. Morphologic correlates of transependymal transport. In: Brain-Endocrine Interaction. II. The ventricular system in neuroendocrine mechanisms. Knigge KM, Scott DE, Kobayashi H, Ishii S (eds) Karger, Basel, 1975.
77. Scott DE, Paull WK: Correlative scanning-transmission electron microscopic examination of the perinatal rat brain. I. The third cerebral ventricle. Cell Tiss Res 190: 317–336, 1978.
78. delCerro M: Personal communication (1981).
79. Scott DE, Paull WK: The tanycyte of the rat median eminence. I. Synaptoid contacts. Cell Tiss Res 200: 329–334, 1979.
80. Silverman AJ, Knigge KM, Peck WA: Median eminence: *In vitro* transport of amino acids, thyroxin and thyreotroph releasing hormone (TRF) Anat Rec 169: 429–438, 1972.
81. Wagner HJ, Pilgrim CH: Extracellular and transcellular transport of horseradish peroxidase (HRP) through the hypothalamic tanycytic ependyma. Cell Tissue Res 152: 477–491, 1974.
82. Nakai Y, Naito N: Uptake and bidirectional transport of peroxidase injected into the blood and cerebrospinal fluid by ependymal cells of the median eminence. In: Brain-Endocrine Interaction. II. The ventricular system in neuroendocrine mechanisms. Knigge KM, Scott DE, Krobisch H, Ishii S (eds) Karger, Basel, 1975, pp 94–108.
83. Calas A, Kerdelhue B, Assenmacher I, Jutisz M: Etude ultrastructurale chez le canard par une technique immunocytochimique. CR Acad Sci Paris 278: 2557–2560, 1974.
84. Kobayashi H: Absorption of cerebrospinal fluid epenolymal cell of median eminence. In: Brain-Endocrine Interaction. II. The ventricular system in neuroendocrine mechanisms. Knigge KM, Scott DE, Kobayashi H, Ishii S (eds) Karger, Basel, 1975, pp 190–222.
85. Scott DE, Scott PM, Krobisch-Dudley G: Ultrastructural localization of radiolabelled L-dopa in the endocrine hypothalamus of the rat. Cell Tiss Res 195: 29–43, 1978.
86. Rodriguez E: The cerebrospinal fluid as a neuroendocrine integration. In: The cerebrospinal fluid and peptide hormones. Frontiers in Hormone Research. Rodriguez E, von Wimersma Greidanus T (eds) Karger, Basel, 1980.
87. Heller H, Hasa SH, Saifi AQ: Antidiurectic activity in the cerebrospinal fluid. J Endocrinol 41: 273–280, 1980.
88. Vorherr H, Bradbury NWB, Hoghougri N, Kleeman CR: Antidiuretic hormone in cerebrospinal fluid during endogenous and exogenous changes in blood levels. Endocrinol 83: 246–250, 1968.
89. Linfoot JA, Garcia GF, Wei W, Fink R, Sanin R, Born JL, Lawrence JH: Human growth hormone levels in cerebrospinal fluid. J Clin Endocrinol 31: 230–232, 1970.
90. Horton EW: The hypothesis on the physiological role of prostaglandin. Physiol Rev 49: 122–161, 1969.
91. Kendall JW, Jacobs J, Kromer RM: Studies on transport of hormones from the cerebrospinal fluid to the hypothalamus. In: Brain-Endocrine Interaction. I. Median Eminence Structure and Function, Knigge KM, Scott DE, Weindl A (eds) Karger, Basel, 1972, pp 343–349.
92. Kendall JW, Sedich JL, Allen JP: Pituitary-CSF relationships in man. In: Brain-Endocrine Interaction. II. The ventricular system in neuroendocrine mechanisms. Knigge KM, Scott DE, Kobayashi H, Ishii S, (eds) Karger, Basel, 1975, pp 313–323.
93. Pavel S: Tentative identification of arginine vasotocin in human cerebrospinal fluid. J Clin Endocrinol Metab 31: 369–371, 1973.
94. Reichlin S, Jackson IMD, Seyler LE, Grimm-Jorgenson Y: Regulation of the secretion of thyrotropin releasing hormone (TRH) and luteinizing hormone releasing hormone (LRH). In: Frontiers in Neurobiology and Neuroscience Research. Seeman P, Brown GM (eds) University of Toronto, 1974, pp 48–59.
95. Oliver C, Eskay L, Ben-Jonathan N, Porter JC: Distribution and concentration of TRH in the rat brain. Endocrinology 96: 540–546, 1974.
96. Kubek M, Schalch D, Knigge JM: Quantitation of TRH deamidation activity in vitro using a charcoal absorption procedure. Proceeding of the 56th Meeting of the Endocrine Society, IIC-21, 1974.
97. Coates PW: Supraependymal cells: light and transmission electron microscopy extends scanning electron microscopic demonstration. Brain Res 57: 502–507, 1973.
98. Vigh-Teichmann I, Vigh B: The infundibular cerebrospinal fluid contacting neurons. Adv Anat Embryol Cell Biol 50: 7–89, 1974.
99. Paull WK, Scott DE, Boldossen WG: A cluster of supraependymal neurons located in the infundibular recess of the rat third ventricle. Amer J Anat 140: 129–132, 1974.
100. Paull WK, Scott DE: Cerebral ventricular surfaces. In: Principles and Techniques of Scanning Electron Microscopy, Hyat MA (ed) 1973 (In press).
101. Scott DE, Krobisch-Dudley G, Paull WK, Kozlowski GP, Ribas J: The primate median eminence. I. Correlative scanning/transmission electron microscopy. Cell Tiss Res 162: 61–73, 1975.
102. Scott DE, Krobisch-Dudley G, Paull WK, Kozlowski GP: The ventricular system in neuroendocrine mechanisms. III. Supraependymal neuronal networks in the primate brain. Cell Tiss Res 197: 235–254, 1977.
103. Scott DE, Paull WK: Correlative scanning-transmission electron microscopic examination of the perinatal rat brain. I. The third cerebral ventricle. Cell Tiss Res 190: 317–336, 1978.
104. Card JP, Mitchell JA: Electron microscopic demonstration of a supraependymal cluster of neuronal cells and processes in the hamster third ventricle. J Comp Neurol 180: 43–58, 1978.
105. McKenna O, Rosenbluth J: Cytological evidence for catecholamine – containing sensory cells bordering the ventricle of the toad hypothalamus. J Comp Neurol 54: 133–148, 1974.
106. Leonhardt H, Lindner T: Norklose Nervenfaseren im III and IV ventrikel des kaninchen und katzengehirns. Z Zellforsch 78: 1–18, 1967.
107. Westergaand E: The lateral cerebral ventricles and the ventricular walls. An anatomical, histological and electron microscopic investigations on mice, rats, hamsters, guinea pigs and rabbits. Thesis Andelsbogtrykkeriet; Odense, 1970.
108. Richards JG, Lorez HP, Tranzer JP: Indolealkylamine nerve terminals in cerebral ventricles. Identification by electron microscopy and fluorescence histochemistry. Brain Res 57: 277–288, 1973.
109. Richards JG, Tranzer JP: Ultrastructural evidence for the localization of an indoleamine in the supraependymal nerves from combined cytochemistry and pharmacology. Experientia 30: 287–289, 1974.
110. Aghajanian GK, Gallager DW: Raphe origin of serotonergic nerves

terminating in the cerebral ventricles. Brain Res 88: 221–231, 1975.
111. Lapierre YJ, Beaudet A, Demianczuk N, Descarries L: Noradrenergic axon terminals in the cerebral cortex of rat. II Quantitative data revealed by light and electron microscopic radioautography of the frontal cortex. Z Zellforsch 114: 404–440, 1973.
112. Beaudet A, Descarries L: Quantitative data on serotonin nerve terminals in the adult rat neocortex. Anat Rec 184: 355, 1976.
113. Pappenheimer JR, Miller TB, Goodrich CA: Sleep promoting effects of cerebrospinal fluid from sleep deprived goats. Proc Nat Acad Sci (Wash.) 58: 513–517, 1967.
114. Daikoku S, Kotsu T, Hashimoto M: Electron microscopic observations on the development of the median eminence in perinatal rats. Z Anat Entwichi-Gesch 134: 311–327, 1971.
115. Eurenius L, Jaskar R: Electron microscopic studies on the development of the external zone of the mouse median eminence. Z Zellforsch 122: 488–502, 1971.
116. Monroe BG, Paull WK: Ultrastructural changes in the hypothalamus during development and hypothalamic activity: The median eminence. In: Progress in Brain Research. Swab DF, Shade TP (eds) 41: 185–208 Amsterdam, Elsevier, 1974.

Authors' address:
Department of Anatomy
University of Missouri-Columbia
School of Medicine
Columbia, MO 65212, USA

CHAPTER 10

Gut paraneurons and Segi's cap

SHIGERU KOBAYASHI and MITSUO SEGI†

1. Gut paraneurons

1.1. Paraneuron concept

The concept of paraneurons includes the idea that neurons are not as specialized a cell type as formerly believed. There are transitional forms between neurons and endocrine/sensory cells with regard to all their cell biological features. Some of them are differentiated more toward the nervous side, whereas others are differentiated more toward the endocrine/sensory side.

The 10th International Congress of Anatomists was held in Tokyo, Japan, in August 1975. As a satellite meeting of this international congress, a symposium on the 'chromaffin cells of the adrenal medulla, enterochromaffin cells of the gastro-intestinal tract and related cells' was held at Kawashima, a nearby town Gifu, Japan. The term 'paraneurons' was introduced for the first time at this symposium by Professor T. Fujita of Niigata (1). Fujita proposed the following criteria for considering a cell to be a paraneuron:

1. Chemical. A paraneuron is able to produce substance(s) identical with or related to the neurotransmitter or the suspected transmitter, and protein/polypeptide substance(s) which may possess hormonic actions.

2. Structural. A paraneuron possesses synaptic vesicle-like and/or neurosecretion-like granules.

3. Physiological. A paraneuron is recepto-secretory in function. It releases secretions in response to adequate stimuli acting upon the receptor site of its cell membrane.

4. Ontogenic. A paraneuron is of common origin with the neuron, i.e. neuroectodermal.

At present the chemical, structural and physiological criteria mentioned in the proposal article are still well-grounded. However, the ontogenic criterion is almost groundless because the results of many experimental and embryological studies have indicated that several paraneurons such as those in the gastro-entero-pancreatic (GEP) endocrine system originate from non-neuroectodermal tissues (2). The recently amended criteria of paraneurons are summarized as follows (2):

1. Possessing neurosecretion-like and/or synaptic vesicle-like granules.

2. Producing substances identical with, or related to, neuro-secretions or neurotransmitters.

3. Being recepto-secretory in function. Receiving adequate stimuli and releasing its secretory substances in response to them.

Paraneurons contain the amine/peptide-producing cells of the pineal gland, pituitary gland, thyroid gland, parathyroid gland, pancreatic islet, adrenal medulla, autonomic ganglion, carotid body, lung, skin and gastro-intestinal tract.

1.2. What are gut paraneurons?

It should be made clear, here, that gut paraneurons are basal-granulated cells which are distributed throughout the gastro-intestinal epithelial tissues. In this review the cytological features of basal-granulated cells as gut paraneurons will be explained by particular reference to the Segi's cap, i.e. a large aggregation of basal-granulated cells on the top of villi in the human fetus. From the viewpoint of the paraneuron concept the basal-granulated cells in the human gut represent one of the most primitive and essential types of neuro-endocrine-sensory cells (2).

Until the late 1960s, the cellular origin of gut hormones had not been demonstrated. It was generally accepted that basal-granulated cells with different appearances represent various functional and/or developmental stages of a single kind of serotonin secreting cell (monism) (3). The idea that the basal-granulated cells in the pyloric antrum

† Deceased on May 8th, 1982.

Motta, PM (ed): Ultrastructure of endocrine cells and tissues. ISBN-13: 978-1-4613-3863-5

secrete gastrin was proposed by Solcia and Sampietro (3, 4). Later McGuigan demonstrated gastrin in the G cells in the pyloric antrum by an immunocytological technique (3). Furthermore, recent advances in peptide chemistry resulted in the establishment of the amino-acid sequence of secretin (1966), cholecystokinin-pancreozymin (CCK-PZ, 1967), gastric inhibitory peptide (GIP, 1971), motilin (1972) and vasoactive intestinal polypeptide (VIP, 1972) (3). Antisera against these hormones were raised and the cellular origin of gut hormones was identified by immunocytological techniques.

Since later half of the 1960s, electron microscopic and immunocytochemical investigations of the basal-granulated cells have been performed in many laboratories such as those at Pavia (5), Geneva (6), London (7) and Niigata (8).

The basal-granulated cells are classified into ten or more types according to their morphology and function, especially by the ultrastructure of their secretory granules and by their immunocytochemical properties, and are thought to secrete different gut hormone types. International meetings on the ultrastructural classification and functional characterization of the basal-granulated cells as well as pancreatic endocrine cells were first held at Wiesbaden (9), and later at Bologna (10), Lausanne (11) and Santa Monica (12).

1.3. Features of gut paraneurons

As the name implies, the basal-granulated cells are provided with many secretory granules in the basal-cytoplasm. In the pyloric antrum of the stomach and duodenum, basal-granulated cells regularly reach the intestinal lumen. They are basal-granulated cell of an open type (3, 13). In the fundic mucosa of the stomach and in the lower large intestine, the basal-granulated cells possess no connection with the intestinal lumen, thus they are basal-granulated cells of a closed type.

The fine structure of the basal-granulated cells in the gut well corresponds to that in the typical peptide hormone-producing cells of the pancreatic islet and other endocrine glands. In the basal cytoplasm membrane-bound secretory granules are gathered (Fig. 11). Granular endoplasmic reticulum is seen in both the supra- and paranuclear areas. Its cisterns are narrow and arranged in parallel. The Golgi complex is seen in the supranuclear cytoplasm. Coated vesicles are frequently attached to the secretory granules in the Golgi area. Mitochondria are scattered among the secretory granules.

The open type basal-granulated cells are provided with a luminal process with a tuft of microvilli. These microvilli are thicker and longer than those of the neighboring cells. A single cilium is occasionally seen. There are many pinocytotic vesicles in the apical cytoplasm. Microtubules and fine filaments are contained in the apical cytoplasm. Junctional complexes are formed between the basal-granulated cell and the neighbouring epithelial cells. On the other hand, the closed type of basal-granulated cells are devoid of an apical process and are connected to neighbouring cells with occasional desmosomes. These cells frequently extend long cytoplasmic processes in the space between the neighbouring cell and the basement membrane.

The ultrastructure of secretory granules differs

Table 1. A classification of human GEP paraneurons.

Cell type	Pancreas	Stomach		Small intestine		Large intestine	Secretion	
		Oxyntic	Pyloric	Upper	Lower		Peptide	Amine
A	+	[(+)]	–	–	–	–	Glucagon	
B	+	–	–	–	–	–	Insulin	
D	+	+	+	+	+	+	Somatostatin	
D_1	+	+	+	+	+	+		
EC	[+]	+	+	+	+	+		Serotonin
ECL	–	+	–	–	–	–		Histamine
G	(+)	–	+	+	–	–	Gastrin	
M or I	–	–	–	+	+	–	CCK-PZ	
K	–	–	–	+	+	–	GIP	
L	–	–	–	+	+	+	GLI	
Mo	–	–	–	+	+	–	Motilin	
N	–	–	–	–	+	–	Neurotensin	
PP	+	–	[+]	–	–	–	PP	
S	–	–	–	+	+	–	Secretin	

( ) In animals, exceptionally in man
[ ] In fetus or newborns, exceptionally in adult

from cell type to cell type concerning the size, shape and internal structure. Table 1 shows a classification of GEP paraneurons. This table is mainly based on the Santa Monica 1980 classification of the human GEP endocrine-paracrine cells (12) with authors' own opinion included.

Gastrin secreting G cells are concentrated in the basal half of the gastric glands in the pyloric antrum. Gastrin secreting cells are also seen in the upper portion of the duodenum. It is difficult to fix the content of the secretory granules. Thus they are flocculent in appearance in the electron micrograph. The electron density of the secretory granules differs from granule to granule with a single G cell.

EC cells are characterized by their serotonin content, their argentaffinity and their characteristic ultrastructural appearance. Their secretory granules are of high electron density, variable in shape and irregular in outline. It has been suggested that the EC cells include plural subtypes in regard to their peptide secretion products. Some EC cells contain substance P immunoreactivity, but other do not. Enkephalin immunoreactivity was also demonstrated in a proportion of the EC cells. Although Pearse et al. stated that all the enterochromaffin cells contain motilin, it is becoming more evident that EC cells and motilin-containing cells are different (12, 14). In the Santa Monica classification of the GEP-endocrine-paracrine cells, the term Mo or D_1 (Mo) cell was proposed to designate motilin-containing cells in the human intestine (12).

ECL(enterochromaffin-like) cells show enterochromaffin cell-like properties when monoamine precursors are administered. However, the amine precursor uptake property (APUD) is not specific to ECL cells. Secretory granules of the ECL cell are variable in appearance. There are vacuolar secretory granules with a small core. Some of the secretory granules possess a compact matrix.

D_1 and PP cells possess small secretory granules (100–190nm). It is frequently difficult to distinguish these cells on the basis of their electron microscopic appearance. The term D_1 cells was introduced because, at the beginning of the study, these cells were thought to represent a subtype of D cells. The D_1 cells contain plural subtypes. The PP cells were termed on the basis of immunocytological studies. In the canine pancreas, tissues in the processus uncinatus contain abundant PP immunoreactive basal-granulated cells.

S, M and L cells are those basal-granulated cells which contain secretory granules with a small (200nm), medium sized (250nm) and large (260nm) diameter respectively. M(medium sized and microvilli-rich) cells are also called as I(intermediate) cells. N cells are distinguishable from the L cell in the immunocytochemical study. With regard to the existence of A cells in the gastric mucosa, a species variation is conspicuous. In the adult human beings, no A cell is demonstrated in the stomach, whereas in the dog, cat and monkey, fundic glands contain a considerable number of A cells. L cells contain GLI (glucagon-like-immunoreactivity, enteroglucagon, glycentin). K cells are characterized by the secretory granules with a dense core and with a halo consisting of fine granular matrix. K cells secrete GIP(gastric inhibitory peptide).

It should be pointed out here that the outline of the ultrastructural and immunocytochemical classification of the basal-granulated cells was constructed in the period between the late 1960's and early 1970's. At that time it was generally accepted that 'the one cell-one hormone relationship' in the pancreas and in other endocrine glands such as adenohypophysis could safely be applied to the GEP endocrine system (13). However, recent methodological advances have made it clear that some basal-granulated cells store more than one secretion product. Thus a dilemma occurred about how such cells should be classified. Studies on this problem could result in the clarification of the family-tree of the basal-granulated cells and the mechanism of their differentiation.

1.4. A putative secretory machinery

A cellular mechanism of the gut hormone secretion was reviewed previously (15, 16). The secretory process of the basal-granulated cells as a paraneuron consists of the following stages: synthesis of the products, intracellular transport, storage and release. Basal-granulated cells keep gut hormones in the secretory granules. The latter contain at least four kinds of substances. They are peptides or low molecular proteins including gut hormones, high molecular compound proteins, amines such as catecholamines and serotonin, and ATP and other adenine nucleotides. The raw materials of these substances are released from the blood capillary into the tissue space, and then incorporated in the basal-granulated cells from their basal and lateral surfaces. Peptide hormones and macromolecular proteins are synthesized in the granular endoplasmic reticulum, transported to the Golgi complex and packaged into secretory granules. Carrier proteins are supposed to be converted into glucoproteins in and near the Golgi complex.

All the gut hormones are synthesized as macromolecular prohormones and then cut into smaller

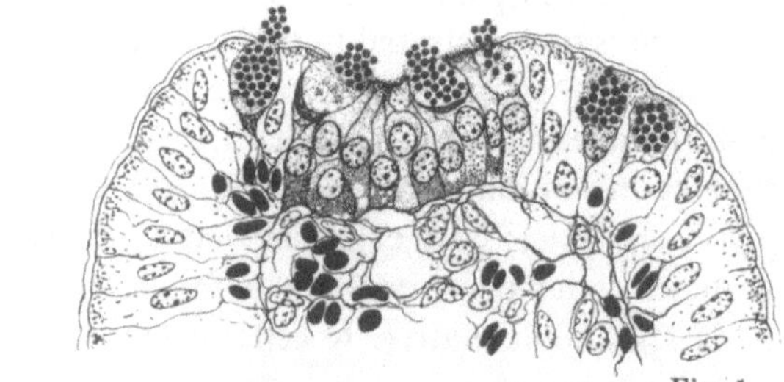

Fig. 1

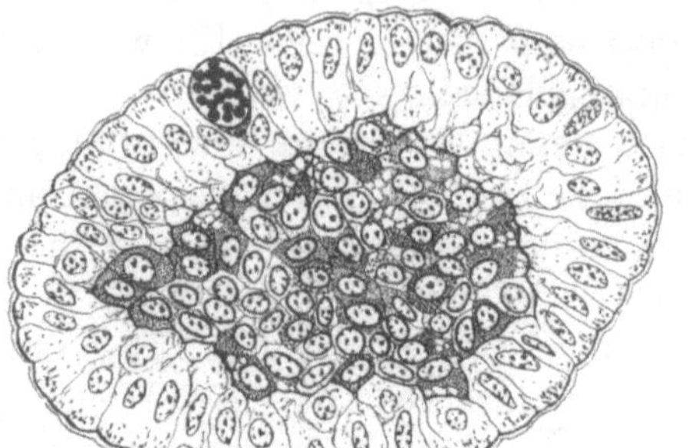

Fig. 2

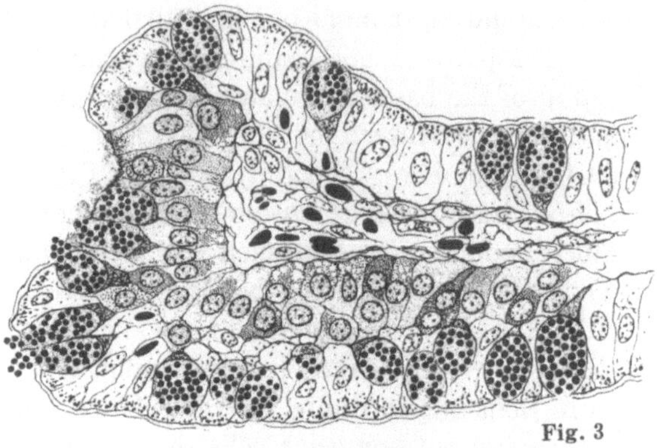

Fig. 3

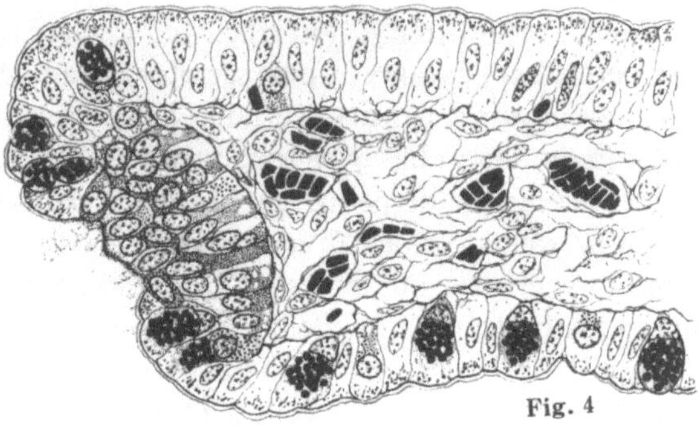

Fig. 4

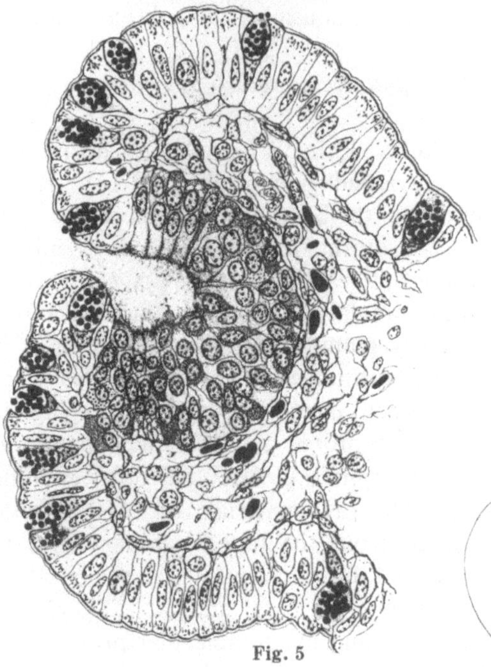

Fig. 5

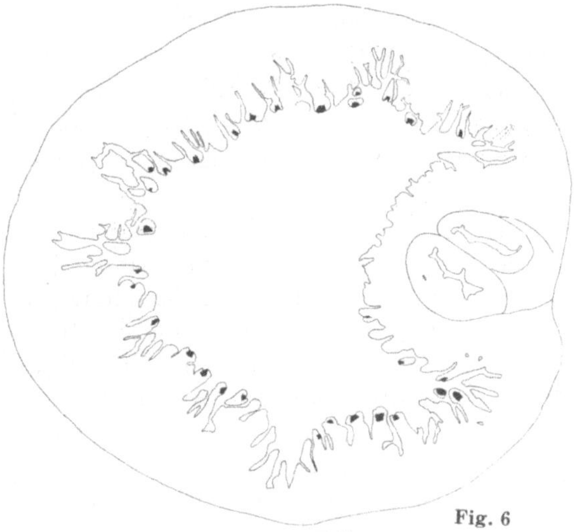

Fig. 6

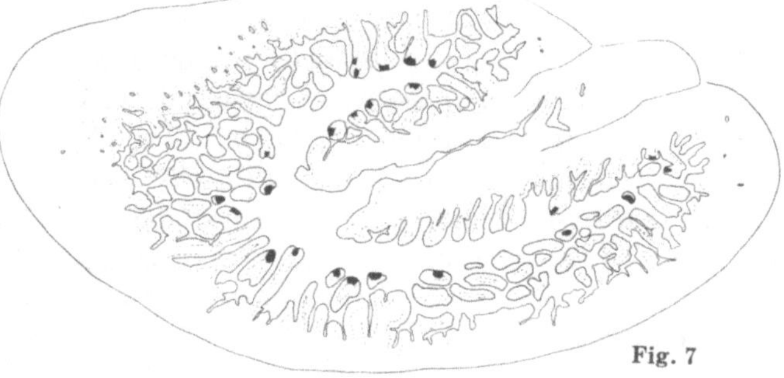

Fig. 7

molecules by proteases in the cytoplasm. In the case of insulin, pre-proinsulin is synthesized on the ribosomes and cut into proinsulin in the cisternae of the granular endoplasmic reticulum. Most of proinsulin is converted into insulin and C-peptide within the secretory granules by proteases (2). Furthermore, gastrin is first synthesized as big gastrin and then converted into gastrin in the Golgi complex (2).

ATP and amines directly enter the secretory granules from the cytosol without passing through the Golgi complex. Thus, the secretory granules become mature and are stored in the basal cytoplasm until they release their content in response to the secretory stimuli (3, 15, 16).

It is supposed that, when an adequate secretory stimulus reaches the basal-granulated cells, they incorporate Ca^{++} from the tissue fluid. The limiting membrane and cell membrane fuse, and exocytotic granule release takes place. The limiting membrane of the secretory granules is recaptured by the cell in the form of coated vesicles (16) and reutilized for the formation of new secretory granules.

Paraneurons are characterized as recepto-secretory cells (1, 2). Gut hormones are secreted by the basal-granulated cells in response to the secretory stimulus in the gut lumen. Fujita et al. proposed a hypothesis that the open type basal-granulated cells receive secretory stimulus by the cell membrane facing the gut lumen (1, 3). By this recepto-secretory hypothesis of the basal-granulated cells, the release mechanism of gut hormones can be explained including the feedback mechanisms (3).

The secretory mechanism of closed type basal-granulated cells is unknown. These cells are mainly located in the fundic gland of the stomach and in the caudal portion of the large intestine. It seems possible that they release gut hormones and other granule contents in response to physical changes of the gut lumen such as pressure and temperature (3). On the other hand, there is a possibility that the closed type basal-granulated cells secrete hormones in response to stimuli from the blood.

2. Segi's cap

2.1. Rediscovery of the Segi's cap

It has long been a general argument that the basal-granulated cells in the gut are disseminated among other epithelial cells such as chief and parietal cells of the gastric gland and enterocytes and goblet cells of the small intestine. Thus, a cluster formation of the basal-granulated cells has been regarded as a quite unusual phenomenon. However, it has recently been rediscovered that a striking cellular group occurs in a blind point of researchers, i.e. the top of the intestinal villus.

Segi (17, 18) reported a large aggregation of basal-granulated cells on the top of villi in the duodenum and upper jejunum of the human fetus 5 or more months of gestation. This characteristic structure as well as Segi's papers have long been forgotten ever since. Kobayashi, Iwanaga and Fujita (1980) found the Segi's old papers and called the aggregation of the basal-granulated cells as Segi's cap, after the discoverer (19).

Recent light microscopic studies in the human fetus have shown that the greater part of the grouped basal-granulated cells in the Segi's cap were argyrophil (20). About half the argyrophil cells were argentaffin, indicating that they contained serotonin. By the immunocytochemical technique using specific antisera against various gut hormones, somatostatin-, gastrin- and motilin-immunoreactive cells were identified in Segi's cap (14, 19, 20). It was also demonstrated that CCK-PZ, secretin- and pancreatic peptide-immunoreactive cells were dispersed in the intestinal epithelium, but they did not show a tendency to gather on the top of the villus. VIP-, substance P-

←

Figs. 1–7. Semidiagrammatic drawing of Segi's cap. These figures were reproduced from Segi's original paper (18). (1) Segi's cap as seen in the duodenum of a human fetus (male, crown-rump length: 205 mm). Eleven basal-granulated cells and 3 goblet cells form a cellular cap. (2) Cross section of an apical portion of a duodenal villus of a human fetus, 18 weeks and 4 days after the first day of the last mensus (male, crown-rump length: 146 mm). Sixty-five basal-granulated cells of a Segi's cap fill the central portion of a circular layer of enterocytes containing a goblet cell. (3) Segi's cap in the duodenum of a human fetus (male, crown-rump length: 205 mm). Many profiles of basal-granulated cells are seen in the space between the intestinal epithelium and the basement membrane. (4) Segi's cap in the upper jejunum of a human fetus, 31 weeks and 45 days after the first day of the last mensus (female, crown-rump length: 260 mm). Twenty-six nuclei of basal-granulated cells are seen. (5) Segi's cap containing 64 nuclei of basal-granulated cells. This cap has a diameter of 72 × 26 μm. From the duodenum of a human fetus, 31 weeks and 4 days after the first day of the last mensus (female, crown-rump length: 230 mm). (6) Semidiagrammatic drawing of a cross section of the duodenum of a human fetus, 31 weeks and 4 days after the first day of the last mensus (female, crown-rump length: 230 mm) showing the distribution of Segi's cap. Segi's cap is located on the top of almost all the intestinal villi. Papilla duodeni major is seen in the right portion of the specimen. In total 31 Segi's cap are seen. (7) Distribution of the Segi's cap in the cross section of the duodenum of a human fetus, 18 weeks and 4 days after the first day of the last mensus (male, crown-rump length: 146 mm). End segments of the ductus choledochus and ductus pancreaticus are seen in the right-upper portion of the specimen. There are 23 Segi's caps.

and insulin-immunoreactive cells were not found in the intestinal mucosa of the human fetus (20).

The ultrastructure of the grouped basal-granulated cells in the Segi's cap in the human fetus was similar to that of the disseminated basal-granulated cells of the adult (19). EC, D and S cells were identified by the ultrastructure of their secretory granules. However, many basal-granulated cells in Segi's cap were unclassifiable because of an atypical appearance of their secretory granules.

Ultrastructural features characteristic to those basal-granulated cells in Segi's cap are summarized as follows:

1. The apical pole of the basal-granulated cells in Segi's cap were provided with various cytoplasmic projections. Regularly arranged microvilli were rare, whereas club- or finger-shaped protrusions of a thicker diameter were frequently seen.

2. A large number of polymorphous lysosomes were dispersed in the cytoplasm of some of the basal-granulated cells.

3. Bundles of cytoplasmic microfilaments were abundantly interspersed among the secretory granules and mitochondria.

4. In addition to the secretory granules, large and rounded bodies of moderate electron opacity measur-

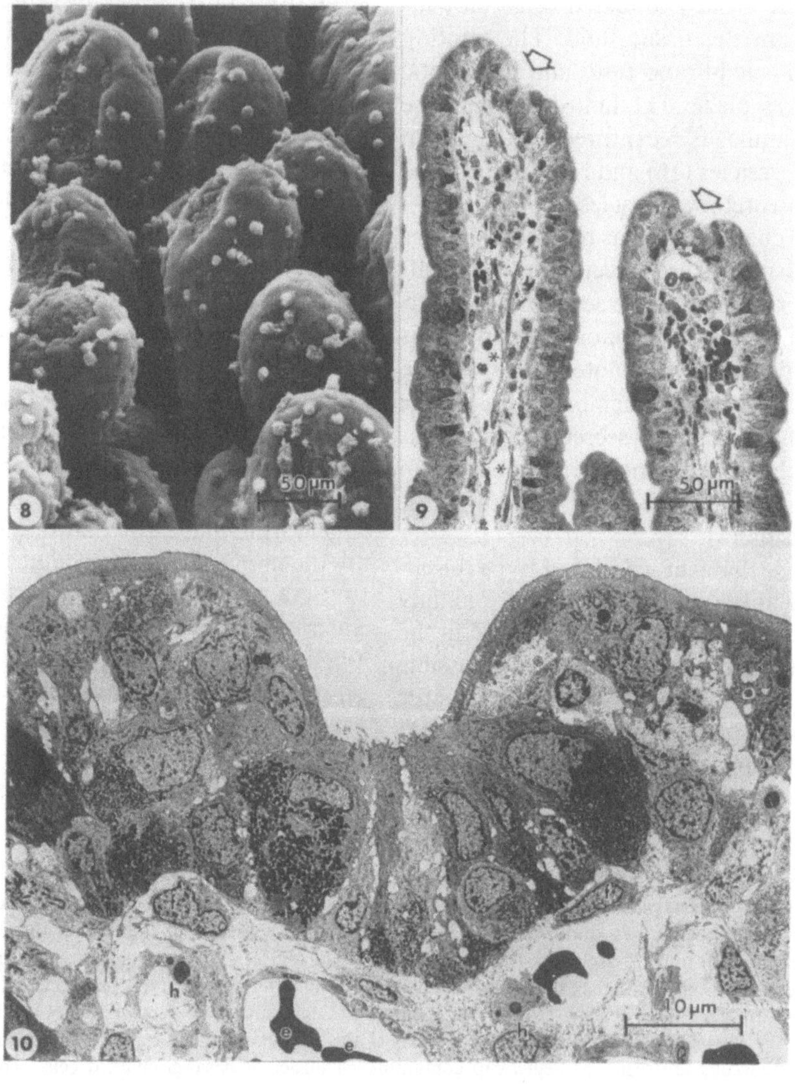

Figs. 8–10. (8) Scanning electron micrograph of the duodenum of a 7-month old human fetus. Characteristic dent on the top of the villus shows the concave surface of Segi's cap. (9) Transmission electron micrograph showing apical portions of the villi in the duodenum of a 7-month old human fetus. Segi's cap is indicated by the arrows. A central lacteal (asterisk) is seen in the center of the villus. (10) Apical portion of villus in the upper jejunum of a 6-month old human fetus. At least 19 basal-granulated cells are seen in Segi's cap shown in this micrograph. In the lamina propria mucosae, there are profiles of fibroblasts (f), histiocytes (h), and blood capillary containing erythrocytes (e).

ing up to 1μm in diameter were contained in the basal cytoplasm. These vacuoles resemble the 'precursor' type secretory granules reported by Moxey and Trier in the precursor and transitional cells in the intestine of human fetus (21). Further studies are needed to clarify the significance of these vacuoles.

Segi in his original light microscopic study, clearly illustrated the difference between types of basal-granulated cells concerning the size, shape and staining property of their secretory granules, though he did not notice that this difference in light microscopic features was due to different gut hormones (17, 18). The cells which showed strong chromaffin reaction in Segi's preparations probably correspond to EC cells in the electron microscopy, whereas those with rounded blue granules that he recognized in his dichromate-fixed and azan-stained sections are probably D cells. The other basal-granulated cells illustrated by Segi are still difficult to correlate with the ultrastructural and immunocytological classification due to the present lack of knowledge regarding the light microscopic appearance and ultrastructure of the basal-granulated cells in the human fetus. It may be worth mentioning here that those basal-granulated cells illustrated in yellow or brown color in the Segi's original drawing do not necessarily correspond to EC cells, because the yellow to brown colour of the azan-stained specimen used by Segi is not due to dichromate but due to orange G of the staining solution.

2.2. Segi's cap in experimental animals

Segi (17, 18) performed his embryological studies in the human fetus only. Thus it has remained unknown whether or not Segi's cap was present in the fetuses of other mammals.

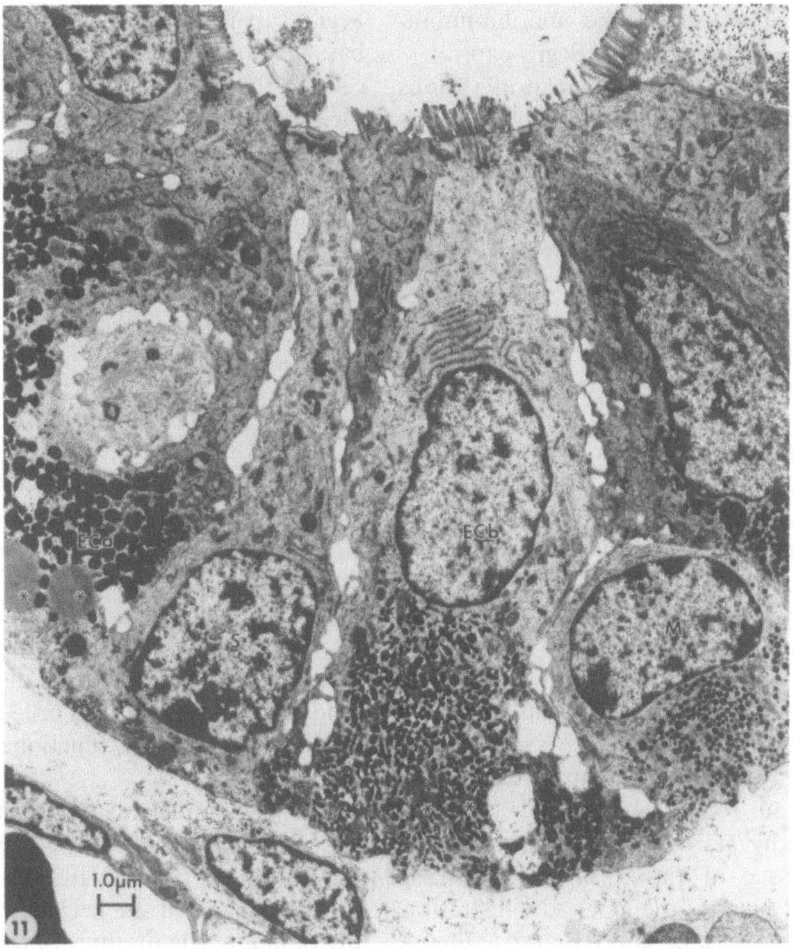

Fig. 11. Transmission electron micrograph illustrating the fine structure of the basal-granulated cells in Segi's cap. EC, M (M) and S (S) cells can be distinguished. There are at least two subtypes (ECa, ECb) of EC cells. Large rounded bodies indicated by asterisks correspond to the 'vacuoles' which were reported by Segi in his original papers (17, 18). Their nature and function is unknown.

Yamada et al. investigated the occurrence of Segi's cap in cattle and pig fetuses (22). They showed that the peculiar aggregation of the basal-granulated cells also occurs in both species of animal fetuses. Segi's cap in these domestic animals consists mainly of argyrophil basal-granulated cells. It is interesting that Yamada et al. found no argentaffin cells in Segi's cap of cattle and pig fetuses. By immunocytochemistry, gastrin-, somatostatin-, motilin- and secretin-containing cells were demonstrated in the Segi's cap of these animal species.

Kobayashi, Segi et al. (1981) examined monkey fetuses (unpublished). Since the animal fetuses examined are limited in number (2 *Macaca fuscata* fetuses and 1 *Macaca mulatta* fetus), it is still premature to make a decisive conclusion. However, it seems certain that the Segi's cap is present in both species of monkey fetuses at a later stage of pregnancy, that the aggregation of basal-granulated cells mixed with goblet cells occurs not only in the duodenum and jejunum but also in the ileum and colon, and that gastrin-, CCK-, somatostatin and motilin-immunoreactive cells are included in the Segi's cap.

Comparative studies of the Segi's cap in various experimental animals may facilitate the analysis of its function and significance in the human fetus.

2.3. The embryological mechanism of the formation of the Segi's cap

How does the formation of the Segi's cap take place in the human fetus? Segi discussed the possibilities that the top of intestinal villi was a generative spot for the basal-granulated cells (17, 18). He suggested that the basal-granulated cells are produced in the top of the villi. However, it is now generally accepted that the lower portion of crypts represents the cell proliferation site and that the epithelial cells with the exception of the Paneth cells move from the crypt toward the villus top. Therefore, it hardly seems possible that the Segi's cap is formed by the division of the basal-granulated cells in the villus and plays the role of supply-center of the basal-granulated cells during the prenatal life.

Isn't it possible to make Segi's cap by experimentation? Clarke and Kobayashi infused isosmotic (16.9% w/v) polyethylene glycol 4000 solution (PEG) into an isolated sac of rat upper jejunum (23). Between 6 and 72 hr after the PEG administration, the villi in the sac became shorter and a unique cellular cap was formed at their top. This cellular cap was composed of many goblet cells and a few basal-granulated cells. Therefore it was suggested that PEG infusion into the isolated intestinal sac in the rat resulted in severe damage to the enterocytes, whereas goblet cells and basal-granulated cells remained relatively intact (23). Since the epithelial cells that line the intestinal lumen undergo rapid and constant renewal, Clarke and Kobayashi (23) accounted for the mechanism of the PEG induced formation of the peculiar cellular cap as the disturbance of the healthy dynamics of the epithelial cells of the intestine. If the life span of the enterocytes is greatly reduced as compared with that of the basal-granulated and goblet cells, only goblet and basal-granulated cells might be able to reach the top of the villi, where the continuous 'escalator movements' of the three kinds of intestinal epithelial cells terminate, and the cell loss takes place under normal condition. As a result, the goblet and basal-granulated cells are assembled at the villus head and form a peculiar cellular cap.

It seems reasonable to speculate that an analogy between Segi's cap in the human fetus and the PEG-induced cellular cap in the isolated sac of rat small intestine may be possible. It is probable that the Segi's cap is closely related to the development of basal-granulated cells and other types of epithelial cells in the gut, and especially to the peculiarity of their cell cycle in the prenatal period. Studies on the kinetics of the intestinal epithelial cells of the human fetus are necessary to obtain a correct view concerning the embryological mechanism of the formation of Segi's cap.

2.4. Functional significance of the Segi's cap

What is the functional meaning of the occurrence of Segi's cap in the human fetus?

The reason why not all kinds of the basal-granulated cells seen in the duodenal and jejunal mucosa are present in Segi's cap remains unsolved. However, it seems worthwhile to correlate the characteristic distribution of different kinds of basal-granulated cells with their secretory, motor and/or trophic actions in the fetal tissues. Further and more detailed studies on the topographical distribution of different kinds of basal-granulated cells may be useful for the clarification of their functional significance in the fetal life.

It is fascinating to speculate that the basal-granulated cells in Segi's cap play significant role in the physiology of the human fetus in the later half of pregnancy and in the perinatal period. In this sense the timing of the disappearance of Segi's cap should be studied. It may be possible that the timing of the disappearance of the Segi's cap is related to the lactation. During the intra-uterine life, fetal intestine produces and stores meconium. On the other hand, after the birth the intestine starts to digest and absorb the mother's milk. Segi's cap may produce a set of hormones needed for the fetal intestine to be adjusted to the sudden and great change of the luminal content which must occur in the perinatal period.

The localization of Segi's cap at the very top of the villi seems most favorable for the perception of information in the intestinal lumen. It is tempting to suppose that Segi's cap may be a recepto-endocrine organ specific to the later half of pregnancy and perinatal life which functions as a monitor of the fetal environment. The fact that the basal-granulated cells in Segi's cap are open in type and possess an apical process which reaches the epithelial surface may provide a support for this idea. Segi's cap, as fetal sensory organ, receives information from the external environment (i.e. chemical and physical changes of the intestinal lumen and in turn the amniotic fluid), and transmits it to the target organs in the form of gut hormones. Studies on the functional significance of the fetal intestine in the sanitation of the amniotic environment are desired. The localization of the Segi's cap on the very top of the villus seems most convenient for it to control the intestinal epithelium in the absorption of amniotic fluid, if this function really takes place.

Segi's cap is comparable in size and number to the islet of Langerhans in which many cytological and electrophysiological studies have been performed. Thus the Segi's cap may provide useful material for the cell biological studies of the basal-granulated cells as gut paraneurons.

Acknowledgement

Partly supported by a grant from the Ministry of Education, Science and Culture of Japan 56440020.

References

1. Fujita T: Gastro-enteric endocrine cell and its paraneuronic nature. In: Chromaffin, enterochromaffin and related cells. Coupland RE, Fujita T (eds), Elsevier, Amsterdam, 1976, pp 191–208.
2. Fujita T, Kobayashi S: Current views on the paraneurone concept. Trend Neurosci 2: 27–30, 1979.
3. Fujita T, Kobayashi S: Structure and function of gut endocrine cells. Internat Rev Cytol Suppl 6: 187–233, 1977.
4. Solcia E, Sampietro R: Cytologic observations on the pancreatic islets with reference to some endocrine-like cells of the gastrointestinal mucosa. Z Zellforsch 68: 689–698, 1965.
5. Solcia E, Vassallo G, Sampietro R: Endocrine cells in the antro-pyloric mucosa of the stomach. Z Zellforsch 81: 474–486, 1967.
6. Forssmann WG, Orci L, Pictet R, Renold AE, Rouiller C: The endocrine cells in the epithelium of the gastrointestinal mucosa of the rat. J Cell Biol 40: 692–715, 1969.
7. Pearse AGE, Coulling I, Weavers B, Friesen S: The endocrine polypeptide cells of the human stomach, duodenum, and jejunum. Gut 11: 649–658, 1970.
8. Kobayashi S, Fujita T, Sasagawa T: The endocrine cells of human duodenal mucosa. An electron microscope study. Arch Histol Jap 31: 477–494, 1970.
9. Solcia E, Forssmann WG, Pearse AGE: Table 1. In: Origin, chemistry, physiology and pathophysiology of the gastrointestinal hormones. Creutzfeldt W (ed), Schattauer, 1970, p 95 (Abstract).
10. Solcia E, Pearse AGE, Grube D, Kobayashi S, Bussolati G, Creutzfeldt W, Gept W: Revised Wiesbaden classification of gut endocrine cells. Rendic Gastroenterol 5: 13–16, 1973.
11. Solcia E, Polak JM, Pearse AGE, Forssmann WG, Larsson L-I, Sundler F, Lechago J, Grimelius L, Fujita T, Creutzfeldt W, Gept W, Falkmer S, Lefranc G, Heitz PH, Hage E, Buchan AMJ, Bloom SR, Grossman MI: Lausanne 1977 classification of gastroenteropancreatic endocrine cells. In: Gut hormones. Bloom SR (ed), Churchill Livingstone, Edinburgh, 1978, pp 40–48.
12. Solcia E, Creutzfeldt W, Falkmer S, Fujita T, Greider MH, Grossman MI, Grube D, Håkanson R, Larsson L-I, Lechago J, Levin K, Polak JM, Rubin W: Human gastroenteropancreatic endocrine-paracrine cells: Santa Monica 1980 classification. In: Cellular basis of chemical messengers in the digestive system. Grossman MI, Brazier MAB, Lechago J (eds), Academic Press, London, 1981, pp 159–165.
13. Fujita T, Kobayashi S: The cells and hormones of the GEP endocrine system. The current of studies. In: Gastro-entero-pancreatic endocrine system. A cell-biological approach. Fujita T (ed), Igaku-Shoin, Tokyo 1973, pp 1–16.
14. Kobayashi S, Iwanaga T, Fujita T, Yanaihara N: Do enterochromaffin (EC) cells contain motilin? Arch Histol Jap 43: 85–98, 1980.
15. Kobayashi S, Sasagawa T: Morphological aspects of the secretion of gastroenteric hormones. In: Endocrine gut and pancreas. Fujita T (ed), Elsevier, Amsterdam, 1976, pp 255–271.
16. Kobayashi S: Cellular background in gut hormone secretion. In: Gut peptides. Secretion, function and clinical aspects. Miyoshi A (ed), Kodansha, Tokyo, 1979, pp 53–58.
17. Segi M: Über eine aus Chromaffine-Darmzellen bestehende Struktur auf den Zotten von Menschenembryo. (In Japanese with German summary). Acta anat nippon (Kaibogaku-Zasshi, Tokyo) 8: 276–280, 1935.
18. Segi M: Über die Entwicklung der verschiedenen Granularzellen im Darmepithel des Menschenembryo. (In Japanese with German summary). Acta anat nippon (Kaibogaku-Zasshi, Tokyo) 9: 850–937, 1936.
19. Kobayashi S, Iwanaga T, Fujita T: Segi's cap. Huge aggregation of basal-granulated cells discovered by Segi (1935) on the intestinal villi of the human fetus. Arch Histol Jap 43: 79–83, 1980.
20. Iwanaga T, Kobayashi S, Fujita T, Yanaihara N: Immunocytochemistry and ultrastructure of Segi's cap. Biomed Res 1: 117–129, 1980.
21. Moxey PC, Trier JS: Endocrine cells in the human fetal small intestine. Cell Tiss Res 183: 33–50, 1977.
22. Yamada J, Kuramoto H, Iwanaga T, Yamashita T, Misu M, Yanaihara N: Segi's cap, a large aggregation of endocrine cells on the intestinal villi, occurs also in cattle and pig fetuses. Arch Histol Jap 44: 193–197, 1981.
23. Clarke RM, Kobayashi S: The cytological effects of infusion of luminal polyethylene glycol on the rat small intestinal mucosa. Arch Histol Jap 38: 133–150, 1975.

Author's address:
S. Kobayashi
Yamanashi Medical School
Tamaho
Yamanashi, 409–38, Japan

CHAPTER 11

Cell types of endocrine pancreas by immunoelectron microscopy

GEORGES PELLETIER

1. Introduction

The first light microscopic studies using differential staining have established that the endocrine pancreas (also named islets of Langerhans) of a large number of mammals contained at least three different cell types (1, 2). In the early days of electron microscopy, two cell types, the alpha and the beta cell, were identified by using ultrathin sections adjacent to thick sections specifically stained for light microscopic investigation (3–5). Later on, the delta cell was identified on the basis of its location in the islet and on the ultrastructural characteristics of its secretory granules which appeared different from those of alpha and beta cells (6–10). In the seventies, the rapid development of immunocytochemical techniques, especially those applied to the ultrastructural localization of antigens (11), as well as the discovery of new pancreatic hormones (12, 13), has largely contributed to an easier identification of the cell types and organelles containing the different hormones of the endocrine pancreas. In this chapter, we have summarized what is presently known about the identification and classification of the different cell types of the mammalian endocrine pancreas.

2. Alpha cell

The alpha cell is generally located at the periphery of the islet of Langerhans in the rat and several other species. However, in man, the alpha cells are scattered throughout the islet. This cell type is characterized by the appearance of its secretory granules which are round or slightly ovoid and have a very high density. In man, the granules have a diameter of about 250–300 nm and have an excentric core of high-electron density separated from the limiting membranes by a halo of medium dense granular material.

Immunoelectronmicroscopic localization of glucagon performed in both rat and man showed that only the alpha cells could be labelled (14, 15). The reaction product indicating the presence of immunoreactive glucagon was mainly observed over the secretory granules and also to a lesser degree over the cytoplasm. In man, immunostaining was mostly detected over the dense core of the granules, indicating a high concentration of glucagon in this portion of the secretory granules (Fig. 1).

Recently, Ranazzola et al. (16) have shown that antibodies against the gut glucagon-like immunoreactants (GLI or glicentin) could react with the alpha cells. These studies support the hypothesis that glucagon and glicentin are coming from the same precursor.

3. Beta cell

The beta cells are by far the most numerous cells in the endocrine pancreas. They do not have any special localization within the islet. At the electron microscopic level, these cells can easily be identified by their secretory granules which are composed of an electron dense core surrounded by an empty space. The granule membrane is generally well preserved by glutaraldehyde fixation followed by osmication. In various species, such as dog (4) and man (5) the beta secretory granules have a dense core which have a barlike appearance.

The immunoelectron microscopic detection of insulin has confirmed that immunoreactive insulin was concentrated in the dense core of the secretory granules (Figs. 2 and 3). Only typical beta cells were labelled in several species (14, 15). When localization of C-peptide was performed, it was found that, in the rat, immunostaining was restricted to immature secretory granules located on the Golgi area (Pelletier, unpublished data) (Fig. 4). These recent results indicate that there are two populations of secretory granules in the beta cell and that, at least in the rat,

Motta, PM (ed): Ultrastructure of endocrine cells and tissues. ISBN-13: 978-1-4613-3863-5

the mature granules do not contain C-peptide or proinsulin or have a very low content of these two peptides.

4. Delta cell

The delta cells have usually the same distribution as the alpha cells. They have been long considered as a sub-population of alpha cells. Electron microscopy has contributed to establish that the delta cell is a distinct cell type. The delta cell can be easily identified by its secretory granules which are larger and less dense than the alpha granules and also more uniform than the beta-granules (6–10).

The function of this cell type has been long controversial. Early works using immunofluorescence had suggested that the delta cell could contain gastrin (17–19). These findings were not confirmed by other laboratories (14, 15, 20–22).

When somatostatin, a polypeptide first discovered in the hypothalamus (12), was shown to inhibit both insulin and glucagon secretion, investigators looked for a possible source of somatostatin in the endocrine pancreas. They found that somatostatin could only be detected in cells having the ultrastructural characteristics of delta cells (14, 15, 20–22). Immunostaining for somatostatin was mostly detected in the

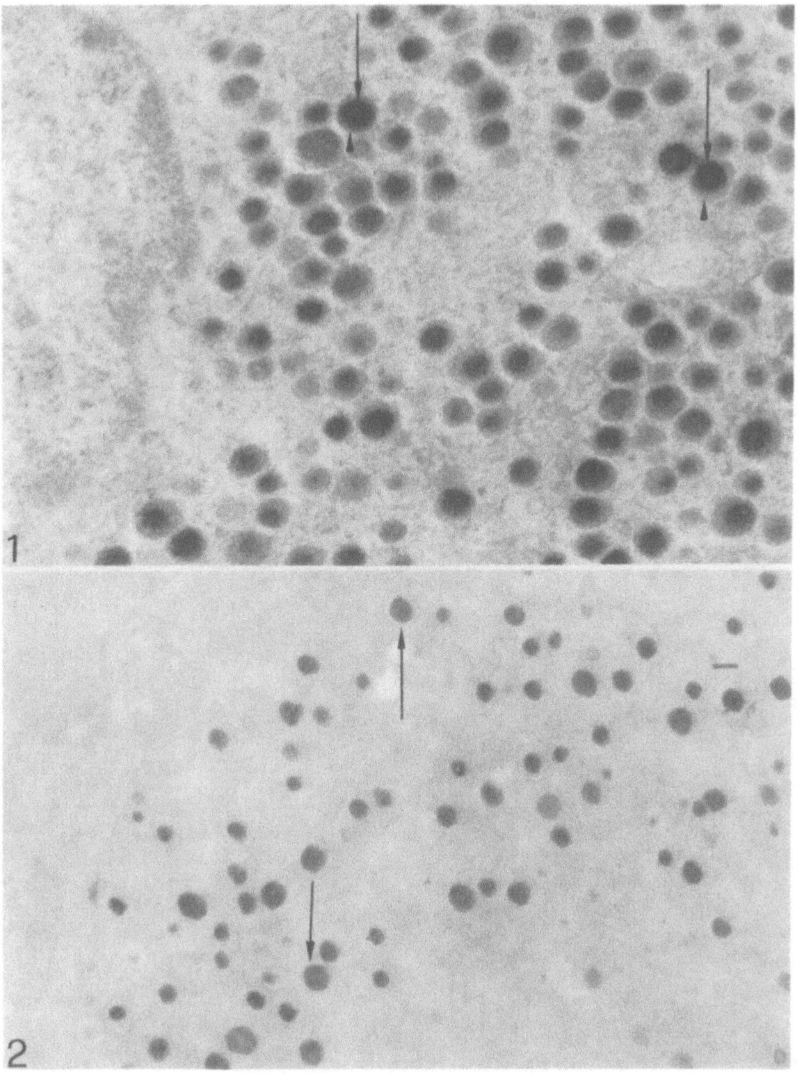

Figs. 1–2. (1) Localization of glucagon in an alpha cell of the human endocrine pancreas. The staining is mostly detected over the dense cores of the secretory granules (arrows). The pale halo (arrowheads) surrounding the dense cores remains unstained. Glutaraldehyde-osmium fixation (× 21,100). (2) Localization of insulin in a beta cell of the rat pancreas. The reaction product can only be detected over the secretory granules (arrows). Note the space between the core and the membrane of the granule. Glutaraldehyde fixation (× 15,400).

secretory granules (Fig. 5), thus indicating that this cell type was responsible for the synthesis and storage of somatostatin (14, 15, 23). The proximity of delta cells with alpha cells and some beta cells suggests that somatostatin could act as a local hormone to influence the secretion of glucagon and insulin.

5. Pancreatic polypeptide (PP) cell

Besides the alpha, beta and delta cells, other cells types named X, F or E cells have been described several years ago (24 and Chapter 12). The discovery of the pancreatic polypeptide (PP) by Lin and Chance (13) has prompted us to search for the identification of the cell types producing PP in the rat and human pancreas. With the help of immunocytochemistry, positively stained cells were found in both endocrine and exocrine pancreas of rat and man (14, 25). In the islets, only very few cells containing PP could be generally observed. With immunoelectron microscopy, the PP containing cell type has an elongated shape and is characterized by the presence of small secretory granules of about 100–150 nm in diameter (Fig. 6). As reported for the other cell types, the reaction product appears to be concentrated in the secretory granules. In the dog endocrine pancreas, Greider et al. (26) have identified the cell type

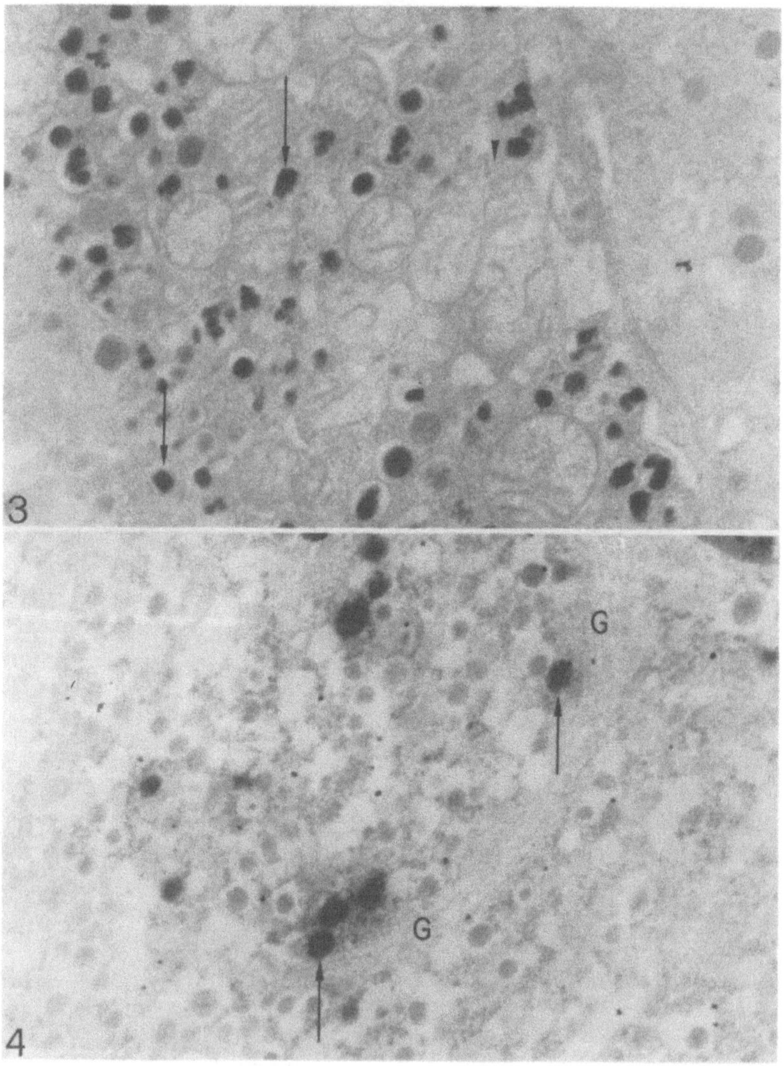

Figs. 3–4. (3) Localization of insulin in a beta cell of the human pancreas. The staining is restricted to the secretory granules (arrows). Note the barlike appearance of these granules. Glutaraldehyde-osmium fixation (× 21,700). (4) Detection of the C-peptide in a beta cell of the rat pancreas. The staining is present only in the immature granules (arrows) in the Golgi (G) area. The mature secretory granules (arrowheads) are unlabelled. Paraformaldehyde fixation (× 16,800).

which contains PP. The PP cell which seems to correspond to the previously described F cell contains granules larger (300 to 400 nm in diameter) than those found in the PP cells of the rat and man.

6. Conclusion

The recent development of immunoelectron microscopic techniques has largely contributed to a better knowledge of cells and organelles responsible for the secretion of a wide variety of hormones. In the endocrine pancreas, this new technology has permitted to confirm the identification of alpha and beta cells and to establish the identification of organelles containing not only glucagon, insulin, but also C-peptide immunoreactive material. The major contribution of immunoelectron microscopy has probably been the accurate identification of cells which are responsible of the production of newly discovered hormones, somatostatin and pancreatic polypeptide. The close association of delta cells which secrete somatostatin with alpha and beta cells clearly suggested that somatostatin is released locally to influence the secretion of both insulin and glucagon. The localization of pancreatic polypeptide in the secretory granules of the fourth cell type which is well distinct from the three other ones also suggests that

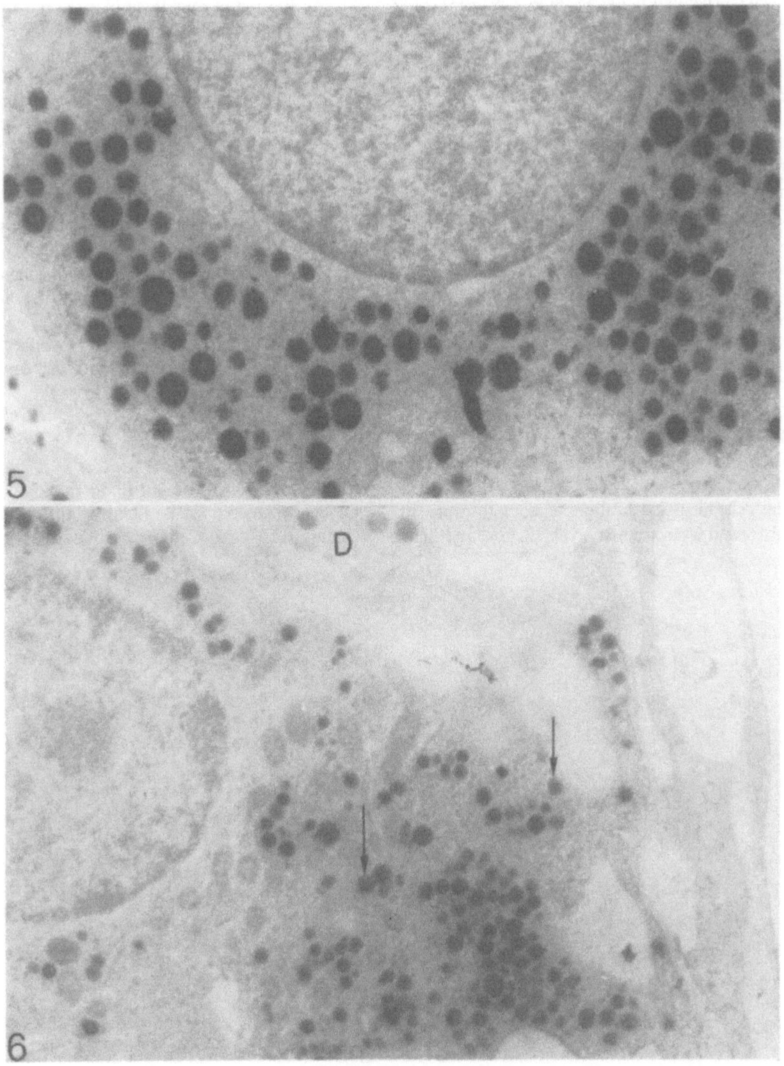

Figs. 5–6. (5) Localization of somatostatin in a delta cell of the human endocrine pancreas. A strong positive reaction can be detected in the secretory granules. A weak diffuse reaction is also observed outside the granules. Glutaraldehyde fixation (× 13,300). (6) Localization of the pancreatic polypeptide (PP) in the human pancreas. All the secretory granules (arrows) of a fourth cell type are strongly stained. An adjacent delta cell (D) is unstained. Glutaraldehyde-osmium fixation (× 14,000).

this peptide could be secreted into the islet itself to modulate the function of some other cells. However, the role of this peptide remains to be clarified.

It is likely that in a near future immunoelectron microscopic techniques will contribute to the identification of other cell types in the endocrine pancreas.

References

1. Hartroft WS, Wrenshall GA: Correlation of beta cell granulation with extractable insulin of the pancreas. Diabetes 4: 1–7, 1955.
2. Hellerström C, Hellman B: Some aspects of silver impregnation of the islets of Langerhans in the rat. Acta Endocrinol 35: 518–532, 1960.
3. Bencosme SA, Peace DC: Electron microscopy of the pancreatic islets. Endocrinology 63: 1–13, 1958.
4. Lacy PE: Electron microscopy of the normal islets of Langerhans: studies in the dog, rabbit, guinea pig and rat. Diabetes 6: 498–507, 1957.
5. Lacy PE: Electron microscopy of the beta cell of the pancreas. Am J Med 31: 851–859, 1961.
6. Sato TL, Herman, Fitzgerald PJ: The comparative ultrastructure of the pancreatic islet of Langerhans. Gen Comp Endocrinol 7: 132–157, 1966.
7. Björkman N, Hellerström C, Hellman B, Petersson B: The cell types in the endocrine pancreas of the human fetus. Z Zellforsch 72: 425–445, 1966.
8. Caramia F: Electron microscopic description of a third cell type in the islets of the rat pancreas. Am J Anat 112: 53–64, 1963.
9. Caramia F, Munger BL, Lacy PE: The ultrastructural basis for the identification of cell types in the pancreatic islets. I. Guinea pig. Z Zellforsch 67: 533–546, 1965.
10. Greider MH, Bencosme SA, Lechago J: The human pancreatic islet cells and their tumors. I. The normal pancreatic islets. Lab Invest 22: 344–354, 1970.
11. Sternberger LA: Immunocytochemistry. Englewood Cliffs, NJ, Prentice-Hall, 1974.
12. Brazeau P, Vale W, Burgus R, Ling N, Butcher M, Rivier J, Guillemin R: Hypothalamic polypeptide that inhibits the secretion of immunoreactive pituitary growth hormone. Science 179: 77–79, 1973.
13. Lin TM, Chance RE: The biology of secretory tumors of the pancreatic islets. In: Endocrinology of the Gut Cheng WY, Brooks FP (eds), Charles B Slack Inc 1974, pp 143–145.
14. Pelletier G: Identification of four cell types in the human endocrine pancreas by immunoelectron microscopy. Diabetes 26: 749–756, 1977.
15. Pelletier G: Immunohistochemical localization of somatostatin. In: Progress in Histochemistry and Cytochemistry, vol. 12, Gustav Fischer (ed) Verlag, Stuttgart, New York, 1979, pp 1–40.
16. Ranazzola M. Siperstein A, Moody AJ, Sundy F, Jacobsen H, Orci L: Glicentin immunoreactive cells: their relationship to glucagon-producing cells. Endocrinology 105: 499–508, 1979.
17. Lomsky R, Longr F, Vortel V: Immunohistochemical demonstration of gastrin in mammalian islets of Langerhans. Nature 223: 618–619, 1969.
18. Greider MH, McGuigan JE: A comparison of ulcerogenic tumors of the pancreas with the gastrin cell of normal human pancreas and hog antrum. Am J Pathol 59: 76–77, 1970.
19. Creutzfeldt W, Creutzfeldt C, Arnold R: Gastrin-producing cells. In: Endocrinology of the Gut. Cheng WY, Brooks FP, Thorofare, NJ (eds), Charles B Slack Inc, 1975, pp 35–62.
20. Orci L, Unger RH: Functional subdivision of islets of Langerhans and possible role of D-cells. Lancet 2: 1243–1244, 1975.
21. Pelletier G, Leclerc R, Arimura A, Schally AV: Immunohistochemical localization of somatostatin in the rat pancreas. J Histochem Cytochem 23: 699–701, 1975.
22. Dubois MP: Immunoreactive somatostatin is present in discrete cells of the endocrine pancreas. Proc Natl Acad Sci (USA) 72: 1340–1343, 1975.
23. Goldsmith PC, Rose JC, Arimura A, Ganong WF: Ultrastructural localization of somatostatin in pancreatic islets of the rat. Endocrinology 97: 1061–64, 1975.
24. Lacy PE, Greider MH: Ultrastructural organization of mammalian pancreatic islets. In: Handbook of Physiology, vol. 1, Endocrine Pancreas, Greep RO, Astwood ER (eds). Am Phys Soc, Washington DC, 1972, pp 77–89.
25. Pelletier G, Leclerc R: Immunohistochemical localization of human pancreatic polypeptide (HPP) in the human endocrine pancreas. Gastroenterology 72: 569–571, 1977.
26. Greider MA, Gersell DJ, Gingerish RL: Ultrastructural localization of pancreatic polypeptide in the F cell of the dog pancreas. J Histochem Cytochem 26, 1103–1108, 1978.

Author's address:
MRC Group in Molecular Endocrinology
Le Centre Hospitalier de l'Université Laval
Québec G1V 4G2, Canada

CHAPTER 12

Pancreatic polypeptide cells

LARS-INGE LARSSON

1. Introduction

Unlike most established hormones, the pancreatic polypeptides were not isolated due to distinctive physiological activities, but were detected as contaminants during purification of insulin and glucagon from the avian and mammalian pancreas (1, 2). Avian pancreatic polypeptide (APP) differs from the mammalian PP's in 20 of the 36 amino acid positions (2). Therefore, antibodies raised to APP usually do not cross-react with mammalian PP's and vice versa. Availability of highly specific PP antibodies permitted radioimmunoassay studies showing that APP circulated in blood and occurred in pancreatic extracts of many avian and reptilian species (3). Subsequently, immunocytochemical studies demonstrated the presence of APP in pancreatic endocrine-like cells, different from the B, A and D cells (4). Somewhat later, also mammalian PP's were localized to secretory granules in distinct endocrine-like cell types (5, 6).

These findings gave further support to the notion that the PP's represented candidate hormones of the pancreas and stimulated many investigators to study the secretion and physiological role of PP. Radioimmunoassay studies in man documented that human PP (HPP) is released very rapidly after ingestion of a meal and that this release is vagally mediated (7, 8). Physiological studies have documented a wealth of effects of PP on pancreatic secretion, gastric secretion and gastrointestinal motility (9). So far, however, convincing evidence that these actions represent important physiological functions of the PP's are lacking (9). Recently, Schwartz et al. while studying the biosynthesis of canine PP, found that a large molecular weight precursor was converted into PP and into another peptide fragment (10). The physiological role of the cosynthesized product is unknown. It is possible that this peptide may prove to possess more striking biological effects than PP.

2. Distribution of PP cells

In mammals, PP cells have been shown to occur in both the pancreas and in the gastrointestinal tract. There are, however, pronounced species differences both with respect to their distribution and their ultrastructure. A detailed species discussion can be found elsewhere (5, 6) and the following short description will only emphasize certain fundamental aspects. With few exceptions, PP cells are most concentrated in the juxtaduodenal portion of the pancreas (6). In the dog, this portion is well developed and has been referred to as the duodenal lobe (uncinate process) (11). We found that PP cells were particularly well represented here, being almost as numerous as the insulin cells, whereas in the tail of the pancreas, the insulin cells greatly outnumbered the PP cells (6). Ronald Chance (6) simultaneously showed by radioimmunoassay that the content of PP and insulin was about equal in the duodenal lobe (55 vs. 51 μg per g), but that the insulin concentration in the tail was much higher (13 vs. 61 μg per g). Similar findings were made in many other mammals including e.g. cat, rabbit, opossum, hamster, rat, mouse and man (6, 12). Subsequent studies revealed that a correspondingly PP-rich lobe of the human pancreas could be macroscopically dissected out (13). It has been conjectured that the uneven distribution of PP cells reflects a different derivation of the pancreatic sublobes from the ventral and dorsal 'Anlagen' (13). Interestingly, in teleosts, PP-like immunoreactivity has been localized in endocrine cells of the juxtapyloric Brockmann body (14).

Although, in most species examined, PP cells are most numerous in the duodenal lobe, they may still be plentiful in the splenic part of the pancreas (6). In the majority of species, PP cells occur both within islets and scattered between acinar cells in the exocrine parenchyma. Also other islet cells like e.g. the somatostatin cells may show a similar dual distribution. Whether this reflects a dual function is not

Motta, PM (ed): Ultrastructure of endocrine cells and tissues. ISBN-13: 978-1-4613-3863-5

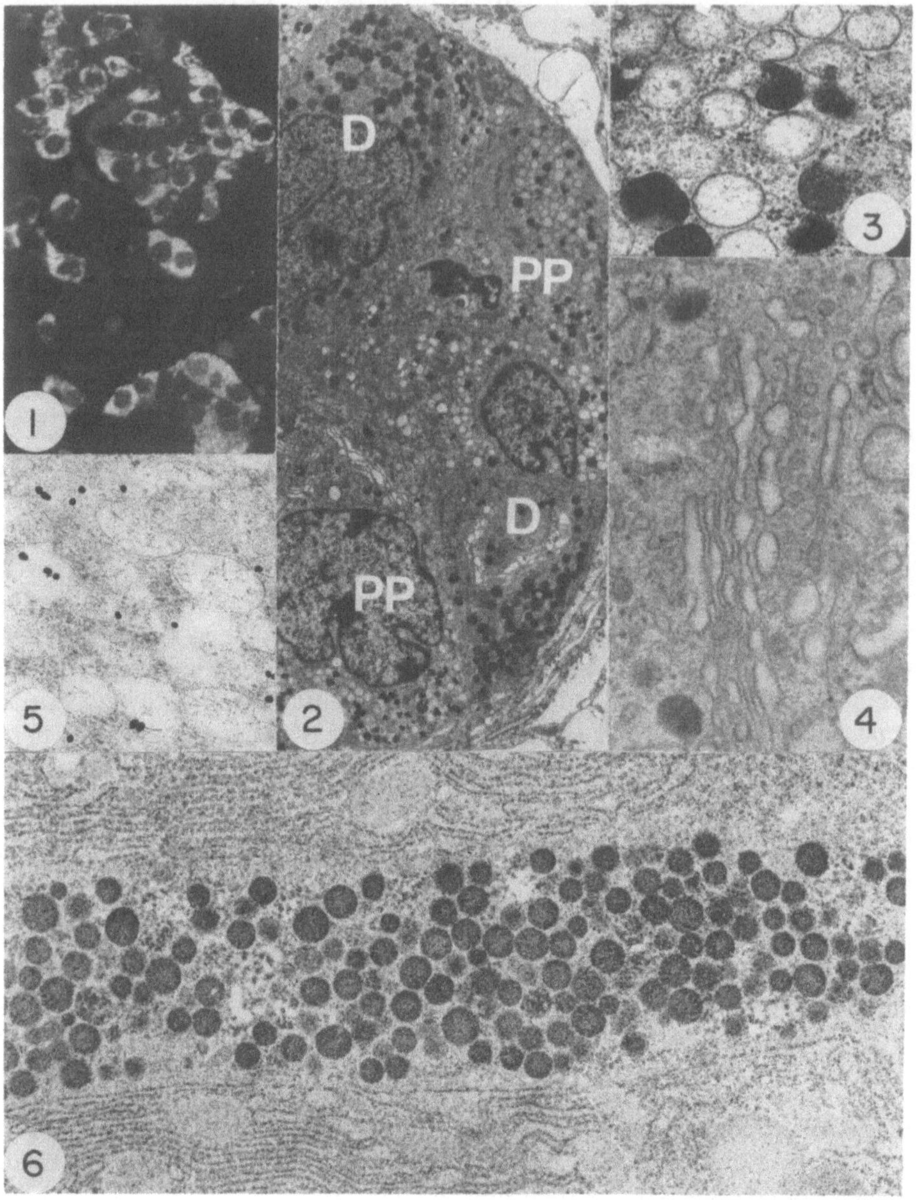

Figs. 1–6. Cat pancreas, duodenal lobe, fixed in a formaldehyde-glutaraldehyde mixture, postfixed in osmic acid and embedded in Epon 812. (1) Semithin plastic section stained with an HPP antiserum (indirect immunofluorescence (× 320). Note that the plentiful PP cells occur both in small clusters and scattered among non-immunoreactive acinar cells. (2) Electron micrograph of a cluster of endocrine cells consisting of a few D cells (D) and many PP cells (× 5,120). (3) Close up view of cat PP (F) cell granules. Note the presence of both dark and light granules (× 38,400). (4) Close up view of part of the Golgi complex of a cat PP cell (× 35,200). (5) Ultrathin section of cat pancreas (osmicated) stained with HPP antiserum, using the gold-labelled antigen detection (GLAD) method. Immunocytochemical staining of osmicated material requires removal of osmium by hydrogen peroxide treatment; hence the absence of discrimination between light and dark granules. Note that gold particles, indicating the site of PP immunoreactivity, occur as small black dots over the periphery of the PP cell granules. The advantages with the colloidal gold technique is that it allows simultaneous studies of membrane structures and also contrasting of the specimen (× 25,600). (6) Ultrathin section from a non-osmicated specimen otherwise treated as in Figure 1 and stained immunocytochemically with HPP antiserum using the peroxidase-antiperoxidase method. Note the presence of a long cytoplasmic process filled with PP immunoreactive (blackened) granules. The process extends between two exocrine pancreatic cells, the rich endoplasmic reticulum of which is readily apparent. Note that all granules of the process are PP immunoreactive and that most of the immunoreactivity occurs peripherally in the granules. From numerous staining experiments such as this, always revealing that all granules are stained, we have concluded that both the dark and light granule varieties contain PP immunoreactivity (× 16,000).

known. It is noteworthy, however, that in some species PP cells are almost exclusively insular (e.g. rat) whereas in others approximately equal numbers of PP cells occur in both insular and extrainsular locations (e.g. guinea pig). Man is intermediate between these two extremes (6).

Reports on PP cells in the gastrointestinal tract have been scanty. In healthy man, true PP cells have not been reproducibly observed in the gut and pancreatectomy has been reported to result in disappearance of PP from serum (15). In the dog and opossum, however, HPP and bovine PP (BPP) immunoreactive cells have been observed in the stomach (6). These observations may have a bearing on studies using dogs as models for studies on PP physiology. Additionally, a population of endocrine-like cells immunoreactive to an antiserum to BPP, but not to HPP, has been detected in the colon of dog, cat and man (17). Since in these species pancreatic PP cells react with both types of antisera, this immunoreactivity pattern may suggest some differences between the PP's stored in colonic cells and in the pancreas. The colonic PP immunoreactive cells have been shown to contain also glicentin and to be identical with intestinal L cells (17).

3. Ultrastructural heterogeneity of PP cells

In dogs and cats, PP cells have been shown to be identical with the previously identified F (or X) cells (6, 18, 19). These cells are characterized by cytoplasmic granules (200–300 nm in diameter) of greatly variable electron density (Fig. 1–4). Thus, empty-looking granules coexist with granules of medium to very high electron-density in the same cell. Whereas the light granules frequently are round and contain some filamentous or flocculent material, the electron-dense granules are often angular, containing a small impression at one pole. Both types of granules are seen in the immediate vicinity of the often well-developed Golgi complex of the F (PP) cell. Ultrastructural immunocytochemistry has revealed that both dark, medium and light granules contain PP immunoreactivity (Figs. 5–6). Interestingly, the PP immunoreactivity occurs concentrated at the periphery of the granular content, leaving a central unstained or poorly stained area. This appearance is often missed in immunocytochemically overstained specimens, using insufficiently diluted antisera in combination with the PAP method. These results may indicate the presence of molecules that are not PP immunoreactive in the center of the granules. By comparison to other endocrine cell types, it is unlikely that this appearance represents a fixation artifact (20).

Antropyloric gastrin (G) cells have also been shown to contain an admixture of secretory granules of greatly variable electron-densities. In these cells, the frequency of poorly electron-dense granules increases upon stimulation (e.g. feeding) – a phenomenon which has suggested that the light granules are empty of hormone (21). Ultrastructural immunocytochemistry, however, has shown that both in PP cells (6) and in G cells (22), light granules contain immunoreactive hormone. It has been suggested that the ultrastructural properties of the G cell granules vary with the pH of the fixative used, indicating that fixatives of neutral or high pH may dissolve out some component of the granules (23). This is a quite likely explanation. However, the constancy of occurrence of light and dark granules suggests that this possible artifact is a reproducible one, which not only may help to identify the cells but may also tell us something of granule heterogeneity and turnover in the PP cell.

In man and in most other species examined, including the rat, guinea pig and chinchilla, F cells are not found (6). In the past, claims to the opposite were made but close scrutiny of the photographic evidence presented revealed that the alleged 'F cells' most likely represented poorly fixed D cells. Accordingly, such false 'F cells' were most frequently claimed to be present in pathology specimens from man. It is easy to distinguish them from true F cells by the fact that well-fixed F cells (at pH 7.3) possess granules of highly variable electron density, whereas damaged D cells exclusively seem to contain electron lucent sacs, representing damaged granules, as well as to exhibit other signs of damage.

Immunocytochemical studies of human pancreatic specimens, fixed in formaldehyde-glutaraldehyde immediately upon their surgical removal, on human foetal pancreas, and on perfusion-fixed rats, guinea pigs and chinchillas revealed that, in these species, PP cells are quite distinct from F cells (6). Thus, their cytoplasmic granules are much smaller, more electron-dense and often possess a tiny halo between the granular dense core and the surrounding membrane. An immunocytochemically identified human (foetal) PP cell is illustrated in Figure 7, along with a picture of another small-granulated endocrine-like islet cell type of unknown function. When identifying the human PP cells, we noted the occurrence of at least two distinct small-granulated islet cell types (6). Such cells had been noted in previous investigations and had, in want of a functional classification, been labelled 'atypical cells', 'type IV, V and VI cells' etc.

(6, 24). The properties of the non-immunoreactive small-granulated cell type fit well with the descriptions of the D_1 cell by Solcia et al. (25). Subsequently, we were astonished to find a paper identifying the PP cells as D_1 cells (26). Soon afterwards, however, these authors revised their data and labelled these cells D_2 cells, instead. After a short period of total confusion, it was decided at an endocrine cell classification meeting in Lausanne to call all PP immunoreactive cells PP cells according to their original classification and to entirely delete the terms D_2 or X cells (the latter term leading to confusion with an endocrine cell type in the gut) and, when necessary, use the term F cell for PP cells of cat and dog pancreas (25). This classification seems to have stood the test of time and has been universally used since then (25). In the islets,

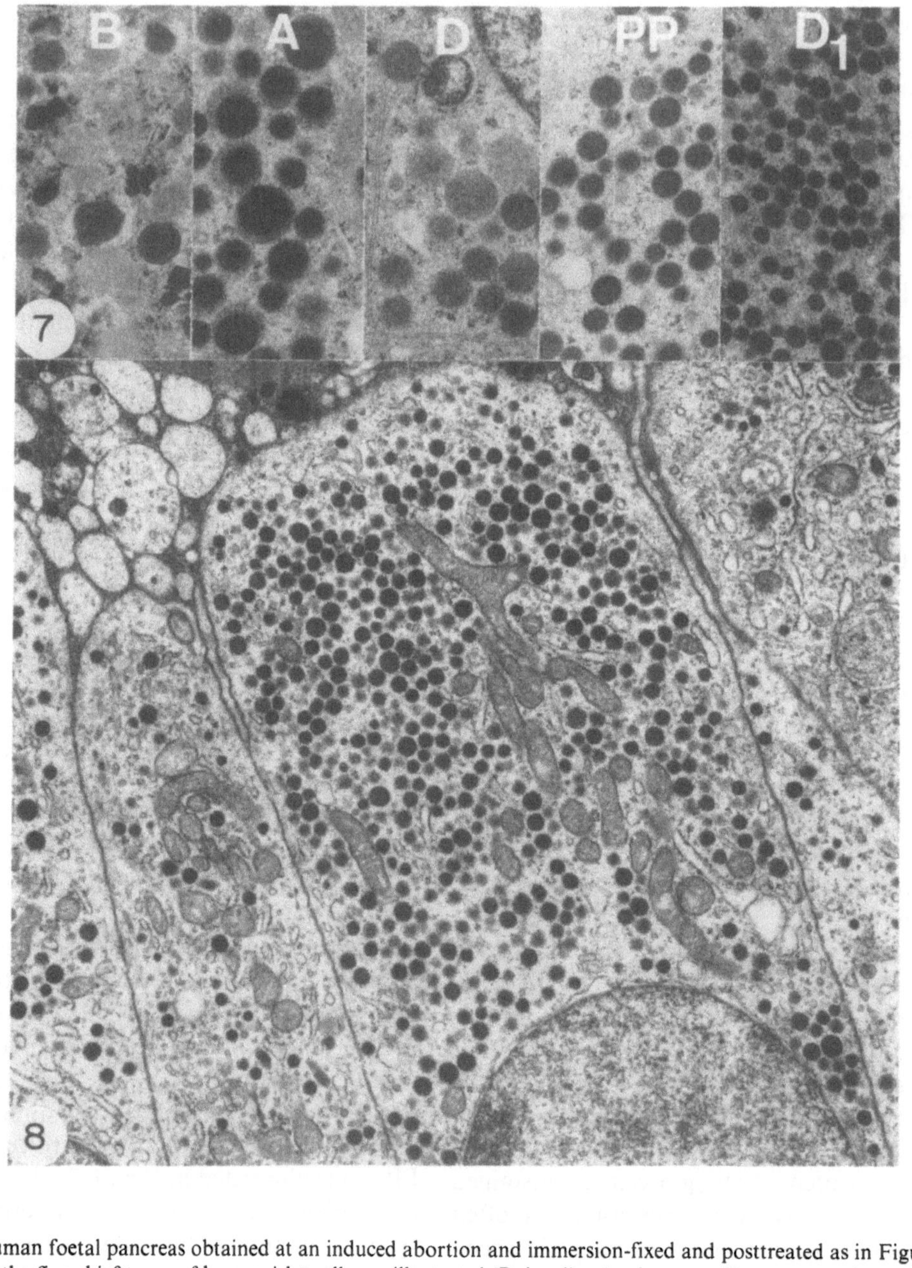

Figs. 7–8. (7) Human foetal pancreas obtained at an induced abortion and immersion-fixed and posttreated as in Figure 1. The granule morphologies of the five chief types of human islet cells are illustrated (B: insulin; A: glucagon; D: somatostatin and PP and D_1 cells). Note that the PP and D_1 cell types possess much smaller cytoplasmic granules than the B, A and D cells. The D_1 cell granules vary somewhat in size and electron density between cells, possibly indicating that D_1 cells may be divided into several subtypes. Such variation is not noted with PP cells, which, in contrast to D_1 cells, can be shown to. constitute a homogeneous cell population by immunocytochemistry for PP. (8) Ultrathin section of a human pancreatic tumour shown to consist exclusively of PP cells. This tumour was diagnosed and described by Bordi and colleagues (ref. 37) who kindly supplied this picture (× 9,600).

the term D_1 cell is reserved for the remaining group of small-granulated islet cells, which yet have to be functionally classified. It may well turn out that this group is heterogeneous, when appropriately studied, since both small- and large-granulated D_1 cell variants can be found.

4. Structure and function of PP cells

4.1. Paracrine/endocrine cells

There is much morphological, cytochemical and chemical evidence that PP is released into the bloodstream in the fashion of classical peptide hormones. PP is localized to the cytoplasmic (secretory) granules of cells which exhibit the morphological and ultrastructural characteristics of endocrine cells (6). Moreover, PP cells display the property of amine precursor uptake and decarboxylation (APUD) (4–6) – a cytochemical hallmark for peptide-secreting cells (27). Radioimmunoassay studies document release of PP to the blood-stream, following appropriate stimuli (7, 8). A possible hormonal role of PP in regulating pancreatic enzyme secretion has recently been postulated (28).

The morphology of the PP cell, however, is not always that of a classical pancreatic endocrine cell type. In man and cat, PP cells give off long cytoplasmic processes which come into close contact with numerous exocrine (and endocrine) pancreatic cells (29 and unpublished data) (Fig. 2). This observation, paired with the elegant studies by Lin et al. on the effects of PP on pancreatic enzyme secretion (2, 9) may suggest that exocrine cells directly respond to PP released in their immediate surroundings. Such a local regulatory action was named paracrine by the German histopathologist Friedrich Feyrter, who suggested that the disseminated endocrine cells of the gut and other organs exerted both distant (endocrine) and local (paracrine) actions (32). Recent studies have demonstrated that the potent inhibitory peptide, somatostatin, may be delivered to its target cells via long cytoplasmic processes (paraxons) (30). Many other secretory cell types of the body are now known to give off similar paraxonal processes (31). In the pancreas, both insular and extrainsular PP cells give off such processes, suggesting, but not proving, a potential paracrine function.

4.2. PP-like peptides in neurons

Recently, the occurrence of APP immunoreactivity in neurons of the rat central and peripheral nervous system has been reported (33). These findings are surprising, since previous studies have established that APP is antigenically dissimilar to mammalian PP's. Antisera to HPP did not stain the neurons, although they were potent in revealing pancreatic PP cells in the rat. Very recently, other workers have documented that an antiserum to BPP stains numerous neurons of the rat brain (34). Furthermore, an antiserum to the synthetic C-terminal hexapeptide of PP (PP-6) reveals nerves and endocrine cells in the rat (35). These data seem to bring some order to chaos and suggest that some parts of the PP sequence may occur in the stained neurons, although other parts (like those recognized by the HPP antiserum) apparently are lacking. Thus, whether the neuronal PP immunoreactivity reflects peptides related to true PP or represents cross-reacting peptides remains to be determined.

5. Pathology of PP cells

5.1. PP cell tumours

Although PP cells sometimes are found as components of mixed endocrine tumours of the pancreas, pure PP cell tumours are rare. No clear symptoms are associated with these tumours although, in one case, the watery diarrhoea, hypokalemia and achlorhydria (WDHA) syndrome was noted (36). This syndrome is most often associated with hyperproduction of the vasoactive intestinal polypeptide (VIP). VIP hypersecretion could, however, be excluded in the abovementioned case (36). Other PP cell tumours have not been associated with the WDHA syndrome (37). However, since pancreatic tumours frequently are multihormonal, with only one of the hormones causing clinically recognizable symptoms, these data are not sufficient to exclude a possible causative role of PP in a minority of WDHA patients (38). Only one PP cell tumour has been adequately preserved for ultrastructural studies (37). The tumour cells displayed the features of well-differentiated, typical HPP cells (37) (Fig. 8).

At the ultrastructural level, many endocrine pancreatic tumours are heterogeneous and frequently contain small-granulated endocrine-like cells. Immunocytochemistry has shown that some of these cells are PP cells, whereas others, for the time being, must be classified as D_1 cells or 'atypical cells' with no known peptide product (36–38). Studies on large materials have proven that, contrary to previous claims, PP is no better and no worse a marker for endocrine pancreatic tumours than are insulin, glucagon, somatostatin, gastrin and VIP (38). In view of the documented high frequency of multihormonal tumours, it is advisable to screen suspected patients for all of these peptides even though the clinical symptoms may only implicate one of them as a causative agent (38).

5.2. PP cells and obesity

In classical endocrinology studies of the effects of extirpation of a hormone-producing gland gave useful hints of its normal functions. This approach is not possible with the disseminated endocrine cells of the gut and pancreas. Occasionally, however, nature provides models which are impossible to construct in the laboratory. Thus in obese, but not in lean, Zucker rats the number of demonstrable PP cells is much reduced (39). The same seems also to apply to obese (ob/ob C57 BL/6J) mice, but not to their lean litter mates (40, 41). Thus, both obese Zucker rats and obese C57BL/6J mice seem to be relatively deficient in PP cells. This is of interest, since in such animals parabiosis experiments as well as implanted Millipore chambers with islets from lean animals, reverse the weight gain (39–41). Therefore, much speculation has been devoted to a possible role of an unknown islet hormone in preventing obesity through interaction with satiety centers. Indeed, in both obese C57BL/6J mice and in obese NZO mice (the latter shown to have reduced circulating levels of immunoreactive PP) injections of BPP or APP were found to reverse the weight gain whereas control injections of insulin A chain were without effect (40, 41).

These data are suggestive of a role for PP as a satiety signal, rapidly elicited by the ingestion of food. There is, however, data to suggest that the explanation to genetical obesity is not that simple. Thus, young obese Zucker rats have normal numbers of PP cells and develop a relative PP cell deficiency only with age (39). This is good evidence against a primary causative role of PP in the obesity syndrome of these animals. Unfortunately, similar age studies have not been done with the C57BL/6J or NZO mice although discrepant results with the former mice might well have such a basis (40–43). Additionally, upon diet restriction the islet morphology and hyperinsulinism of obese Zucker rats reversed (39). This reversal is associated with a reappearance of PP cells in their islets. Thus, the relative lack of PP cells in these animals may more likely be yet another symptom of the genetical obesity syndrome than a true primary defect (39). It should be noted that in all of the above-mentioned studies PP cells are defined immunocytochemically by their content of PP. It may therefore be that the obese animals do contain true PP cells, the hormone stores of which are depleted. Ultrastructural studies will be needed to clarify this.

Acknowledgement

Grant support was from the Danish MRC and Cancer Society.

References

1. Kimmel JR, Hayden J, Pollock HG: Isolation and characterization of a new pancreatic polypeptide hormone. J Biol Chem 250: 9369–9376, 1975.
2. Lin T-M, Chance RE: Bovine pancreatic polypeptide (BPP) and avian pancreatic peptide (APP): Candidate hormones of the gut. Gastroenterology 67: 737–738, 1974.
3. Langslow DR, Kimmel JR, Pollock HG: Studies on the distribution of a new avian pancreatic polypeptide and insulin among birds, reptiles, amphibians and mammals. Endocrinology 93: 558–565, 1973.
4. Larsson L-I, Sundler F, Håkanson R, Pollock HG, Kimmel JR: Localization of APP, a postulated new hormone, to a pancreatic endocrine cell type. Histochemistry 42: 377–382, 1974.
5. Larsson L-I, Sundler F, Håkanson R: Immunohistochemical localization of human pancreatic polypeptide (HPP) to a population of islet cells. Cell Tissue Res 156: 167–171, 1975.
6. Larsson L-I, Sundler F, Håkanson R: Pancreatic polypeptide – A postulated new hormone: Identification of its cellular storage site by light and electron microscopic immunocytochemistry. Diabetologia 12: 211–226, 1976.
7. Schwartz TW, Rehfeld JF, Stadil F, Larsson L-I, Chance RE, Moon NE: Pancreatic polypeptide response to food in duodenal ulcer patients before and after vagotomy. Lancet I: 1102–1105, 1976.
8. Schwartz TW, Holst JJ, Fahrenkrug J, Lindkaer-Jensen S, Rehfeld JF, Schaffalitzky de Muckadell OB, Stadil F: Vagal cholinergic regulation of pancreatic polypeptide secretion. J Clin Invest 61: 781–789, 1978.
9. Lin T-M: Pancreatic polypeptide: Isolation, chemistry and biological function. In: Gastrointestinal Hormones, GBJ Glass (ed). New York, Raven Press, 1980, pp 275–306.
10. Schwartz TW, Gingerich RL, Tager HS: Biosynthesis of pancreatic polypeptide. J Biol Chem 255: 11494–11498, 1980.
11. Bencosme SA, Liepa E: Regional differences of the pancreatic islet. Endocrinology 57: 558–593, 1955.
12. Orci L, Baetens D, Ravazzola M, Stefan Y, Malaisse-Lagae F: Pancreatic polypeptide and glucagon: non-random distribution in pancreatic islets. Life Sci 19: 1811–1815, 1976.
13. Malaisse-Lagae F, Stefan Y, Cox J, Perrelet A, Orci L: Identification of a lobe in the adult human pancreas rich in pancreatic polypeptide. Diabetologia 17: 361–365, 1979.
14. Stefan Y, Dufour C, Falkmer S: Mise en évidence par immunofluorescence de cellules à polypeptide pancréatique dans le pancréas et le tube digestif de poissons osseux et cartilagineux. C R Acad Sci 286: 1073–1075, 1978.
15. Floyd JC Jr: Pancreatic Polypeptide. Clinics in Gastroenterology 9: 657–678, 1980.
16. Buffa R, Capella C, Fontana P, Usellini L, Solcia E: Types of endocrine cells in the human colon and rectum. Cell Tissue Res 192: 227–240, 1978.
17. Fiocca R, Capella C, Buffa R, Fontana R, Solcia E, Hage E, Chance RE, Moody AJ: Glucagon-, glicentin- and pancreatic polypeptide-like immunoreactivity in rectal carcinoids and related colorectal cells. Am J Pathol 100: 81–92, 1980.

18. Lazarus SS, Shapiro SH: The dog pancreatic X cell: A light and electron microscopic study. Anat Rec 169: 487–499, 1971.
19. Munger BL, Caramia F, Lacy PE: The ultrastructural basis for the identification of cell types in the pancreatic islets. II Rabbit, dog and opossum. Z Zellforsch 67: 776–798, 1965.
20. Ravazzola M, Orci L: Glucagon and glicentin immunoreactivity are topographically segregated in the α granule of the human pancreatic A cell. Nature 284: 66–68, 1980.
21. Forssmann WG, Orci L: Ultrastructure and secretory cycle of the gastrin-producing cell. Z Zellforsch 101: 419–432, 1969.
22. Larsson L-I: Gastrointestinal cells producing endocrine, neurocrine and paracrine messengers. Clinics in Gastroenterology 9: 485–516, 1980.
23. Mortensen NJ Mc, Morris JF: The effect of fixation conditions on the ultrastructural appearance of gastrin cell granules in the rat gastric pyloric antrum. Cell Tissue Res 176: 251–263, 1977.
24. Deconinck JF, van Assche FA, Potvliege PR, Gepts W: The ultrastructure of the human pancreatic islets. II. The islets of neonates. Diabetologia 8: 326–333, 1972.
25. Solcia E, Polak JM, Larsson L-I, Buchan AMJ, Capella C: Update on Lausanne classification of endocrine cells. In: Gut Hormones, 2nd edition, SR Bloom, JM Polak (eds). Churchill-Livingstone, Edinburgh, 1981, pp 96–100.
26. Heitz PH, Polak JM, Bloom SR, Pearse AGE: Identification of the D_1 cell as the source of human pancreatic polypeptide (HPP). Gut 17: 755–765, 1976.
27. Pearse AGE: Common cytochemical and ultrastructural characteristics of cells producing polypeptide hormones (the APUD series) and their relevance to thyroid and ultimobranchial C cells and calcitonin. Proc Roy Soc B 170: 71–80, 1968.
28. Taylor IL, Solomon TE, Walsh JH, Grossman MI: Pancreatic polypeptide: Metabolism and effect on pancreatic secretion in dogs. Gastroenterology 76: 524–528, 1979.
29. Rahier J, Wallon J: Particular morphological aspect of pancreatic polypeptide cells in the human pancreas. Regulatory Peptides Suppl 1: S 89, 1980.
30. Larsson L-I, Goltermann N, de Magistris L, Rehfeld JF, Schwartz TW: Somatostatin cell processes as pathways for paracrine secretion. Science 205: 1393–1394, 1979.
31. Larsson L-I: Immunocytochemistry of secretory peptides: Immunochemical dissections of peptide diversity. J Histochem Cytochem 29: 1032–1042, 1981.
32. Feyrter F: Über die peripheren endokrinen (parakrinen) Drüsen des Menschen. Wilhelm Maudrich, Wien & Düsseldorf, 1954.
33. Lorén I, Alumets I, Håkanson R, Sundler F: Immunoreactive pancreatic polypeptide (PP) occurs in the central and peripheral nervous system: Preliminary immunocytochemical observations. Histochemistry 78: 179–186, 1979.
34. Olschawska JA, O'Donahue TL, Jacobowitz DM: The distribution of bovine pancreatic polypeptide-like immunoreactive neurons in the rat brain. Peptides (in press).
35. Vaillant C, Taylor IL: The distribution of PP-like peptides in endocrine cells and nerves revealed by a region-specific antiserum. Peptides (in press).
36. Larsson L-I, Schwartz TW, Lundquist G, Change RE, Sundler F, Rehfeld JF, Grimelius L, Fahrenkrug J, Schaffalitzky de Muckadell OB, Moon N: Occurrence of human pancreatic polypeptide in pancreatic endocrine tumours; possible implication in the watery diarrhoea syndrome. Am J Pathol 85: 675–684, 1976.
37. Bordi C, Togni R, Baetens D, Ravazzola M, Malaisse-Lagae F, Orci L: Human islet cell tumor storing pancreatic polypeptide: A light and electron microscopic study. J Clin Endocrinol Metab 46: 215–219, 1978.
38. Larsson L-I: Endocrine pancreatic tumors. Human Pathology 9: 401–416, 1978.
39. Larsson L-I, Boder GB, Shaw WN: Changes in the islets of Langerhans in the Zucker obese rat. Laboratory Invest 36: 593–598, 1977.
40. Malaisse-Lagae F, Carpentier J-L, Patel YC, Malaisse WJ, Orci L: Pancreatic polypeptide: A possible role in the regulation of food intake in the mouse. Hypothesis. Experientia 33: 915–917, 1977.
41. Gates RJ, Lazarus NR: The ability of pancreatic polypeptides (APP and BPP) to return to normal the hyperglycaemia, hyperinsulinaemia and weight gain of New Zealand obese mice. Hormone Res 8: 189–202, 1977.
42. Gingerich RL, Gersell DJ, Greider MH, Finke EH, Lacy PE: Elevated levels of pancreatic polypeptide in obese-hyperglycemic mice. Metabolism 27: 1526–1532, 1978.

Author's address:
Unit of Histochemistry
University Institute of Pathology
Frederik den V's vej 11
DK-2100 Copenhagen Ø
Denmark

CHAPTER 13

Neuro-endocrine (APUD-type) cells of the lung

ERNEST CUTZ

1. Introduction

The cells with neuro-endocrine (NE) characteristics, identified in pulmonary epithelium, represent a relatively recent addition to the system of diffuse neuroendocrine cells. Although these cells have been described almost 40 years ago, their potential functional significance became appreciated only during the last decade.

The first ultrastructural study on pulmonary NE cells was reported by Bensch et al. who described cells with cytoplasmic neuro-secretory granules in bronchial mucosa of human adult lung and noted similarities with endocrine cells of the gastrointestinal tract (1). Subsequent electron microscopic and histochemical studies revealed that pulmonary NE cells exhibit many of the features of APUD-type cells (2). These studies have also shown that NE cells occur as single cells, scattered throughout the tracheobronchial mucosa, and in groups forming innervated intra-epithelial corpuscles, referred to as neuro-epithelial bodies (NEB) (3). Both single NE cells and NEB have been identified in lungs of human as well as in various animal species, and were found to be particularly prominent during fetal development and in neonates.

The terminology of these cells is still confusing. In earlier literature these cells were referred to as clear cells, argyrophilic cells, Feyrter cells, Kultschitzky cells, small granulated cells and APUD cells. Currently, a term neuro-endocrine (NE) cells is being used to emphasize the possible relationship to neural and endocrine regulatory systems. This term, also adopted here, includes both the single NE cells and NEB when discussed collectively.

Although there has been a great deal of speculation about the possible function and significance of NE cells in the lung, the precise role of these cells remains largely unknown. Increased interest in these cells has been stimulated by a number of recent observations. Neuro-secretory type granules, found in certain pulmonary neoplasms (bronchial carcinoid and oat cell carcinoma) suggested that these tumours originate from lung NE cells (4). Studies of Lauweryns et al. have shown that NEB in newborn rabbit respond to airway hypoxia by increased exocytosis of its secretory granules, suggesting a possible role as intrapulmonary hypoxia sensitive chemoreceptors (5, 6). More recently, several peptide hormones have been identified in NE cells of human lung (7, 8). This chapter will review the current state of knowledge pertaining to the morphology and development of pulmonary NE cells with emphasis on the ultrastructural aspects. Experimental studies and functional considerations will be also discussed.

2. Morphology

2.1. *Light microscopic and cytochemical features*

The demonstration of pulmonary NE cells at light microscopy level requires special staining procedures. The basic techniques are similar to those used for demonstration of APUD-type cells in other tissues (2). The staining methods, cytochemical and molecular markers for pulmonary NE cells have been recently reviewed (9).

The reactivities of single NE cells and NEB are generally similar, however the results with individual methods may vary depending on the age and the species of animals studied. Argyrophilic stain according to Grimelius, is the most commonly used method. The single argyrophilic NE cells appear as pyramidal, flask-shaped or oval cells, which may form apical cytoplasmic processes extending to the airway lumen or others may show a basal process in contact with the basement membrane (Fig. 1, inset). Cells with lateral dendrite-like processes extending for some distance have also been described. Argyrophilic cells constituting an NEB are closely packed together, and form a distinct ovoid corpuscle within the bronchial epithelium.

Motta, PM (ed): Ultrastructure of endocrine cells and tissues. ISBN-13: 978-1-4613-3863-5

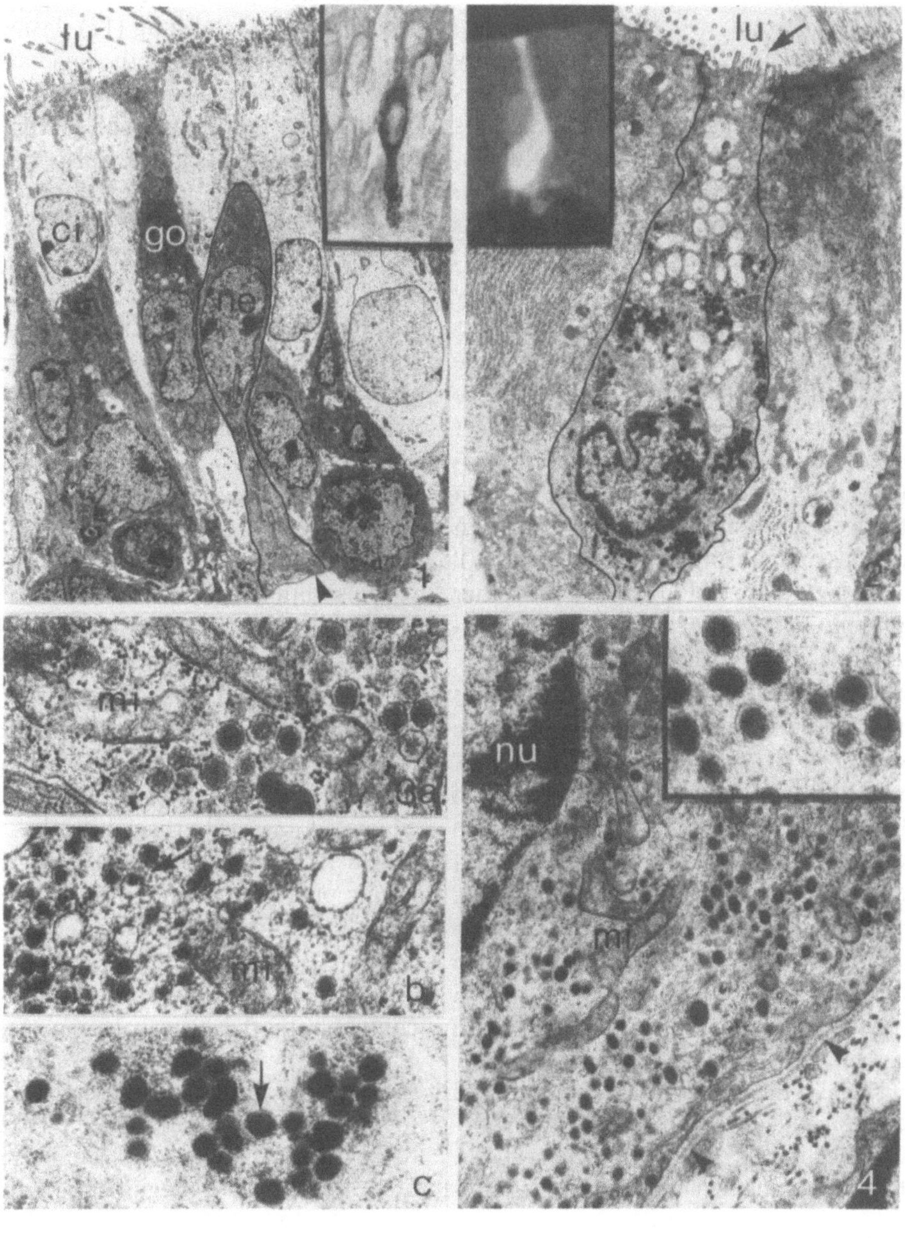

Figs. 1–4. (1) Low magnification electron micrograph of fetal lamb tracheal epithelium showing the position and shape of single NE cell (cytoplasm outlined in black) in relation to adjacent goblet (go) and ciliated (ci) cells. The base of NE cell is in contact with basement membrane (arrowhead), but apical portion does not extend to the lumen (lu) (TEM × 2,320). Inset: Argyrophilic cell in paraffin section appears in identical position as NE cell in Figure 1. LM, Grimelius stain (× 560). (2) Apical surface of a NE cell (outlined in black) in tracheal epithelium of adult rabbit shows short microvilli (arrow) facing the lumen (lu). Clusters of DCV are mainly located around the nucleus (TEM × 5,760) Inset: Intensly fluorescent cell in tracheal epithelium of adult rabbit shows apical process in contact with the lumen. (FIF × 560). (3) (a) Higher magnification of type 1 NE cell in lamb tracheal epithelium shows variation in the density of DCV matrix and a few mitochondria (mi) (TEM × 26,400); (b) Cytoplasmic detail of type II NE cell in fetal lamb trachea. DCV are smaller, have denser central core; a few show eccentrically placed central core (arrow), mi: mitochondria (TEM × 26,400); (3) Pleiomorphic and uniformly electron dense DCV in NE cell of adult rabbit trachea. The limiting membrane of DCV (arrow) is ill defined (TEM × 26,400). (Figs. 1–3 reproduced from Cell Tiss. Res. 158: 425–437, 1975 by permission of publisher Springer-Verlag, Heidelberg). (4) Basal aspect of NE cell in human child bronchiole. Numerous cytoplasmic DCV are concentrated in the region between the nucleus (nu) and basement membrane (arrowheads). Small mitochondria (mi) and accasional profiles of rough endoplasmic reticulum are dispersed in the cytoplasm (TEM × 16,000) Inset: Higher magnification of DCV's from Figure 4 showing the single limiting membrane and a homogeneously electron-dense matrix (TEM × 32,000). (Reproduced from 'Asthma–Physiology, Immunopharmacology and Treatment'. Lichtenstein LM, Austen F (eds), Academic Press, New York, San Francisco, London, 1977, with permission from publisher Academic Press Inc. New York, San Francisco, London.)

2.2. Amine and peptide content

The presence of endogenous amine and/or capacity for amine precursor uptake and decarboxylation has been demonstrated in NE cells of human and various animal species by the formaldehyde induced fluorescence (FIF) technique (Fig. 2, inset). From the FIF studies it appears that the amount of endogenous amine in NE cells varies between species. For example in human lungs, amine can be demonstrated by FIF, only after pretreatment with amine precursors (5-HTP or L-DOPA) (10). The predominant amine in pulmonary NE cells of most species appears to be serotonin. Using microspectrography, Lauweryns et al. demonstrated serotonin in NEB of rabbit neonatal lung (11). More recently, we have shown immunoreactivity to serotonin in NE cells of human lung and in various animal species (Fig. 5, inset) (9).

Based on analogy with APUD cells in other tissues, the presence of peptide hormones in pulmonary NE cells has been suspected for some time. However it was only recently that immunoreactivities to bombesin, calcitonin and leu-enkephalin have been demonstrated in NE cells of human lung (7, 8). Initially, bombesin was found in NE cells of human fetal lung (7), but subsequently this peptide was also demonstrated in post-natal and adult lungs (8). Of the three peptides so far identified in the lung, bombesin appears to be the most abundant. Bombesin-like immunoreactivity was localized in both single NE cells and NEB. Bombesin-immunoreactive NE cells appear particularly prominent in fetal and neonatal lungs. Calcitonin-like immunoreactivity also occurs in single NE cells and NEB, but is found mainly in neonatal and child lungs (8). Leu-enkephalin has been localized in single NE cells only (8).

Of interest is the recent demonstration of immunoreactivity for neuron-specific enolase (NSE) in NE cells of human fetal lung (12). Using serial sections, NSE was localized in both single and groups of NE cells, which were also immunoreactivities for bombesin and/or serotonin. The peribronchial nerve fibres and ganglion cells were also immunoreactive for NSE.

2.3. Ultrastructure

At ultrastructural level, single NE cells and the cells forming NEB share many features in common. However, NEB have a more complex structural organization and therefore will be considered separately.

Single NE cells have been identified in the tracheobronchial epithelium of human fetal (10, 13), newborn (14) and adult lung (15), as well as in various animal species (16).

According to the position within the tracheobronchial epithelium and the relationship to the airway lumen, two main categories of NE cells can be recognized. NE cells in the first category (closed type) are more frequent and make no direct contact with the lumen (Fig. 1). The shape of these cells is elongated or oval, their base lies directly on the basement membrane and often forms lateral cytoplasmic processes interdigiting between adjacent epithelial cells. In the second category, (open type) NE cells are usually pyramidal or flask shaped with an apical process, furnished by short microvilli in direct contact with the airway lumen (Fig. 2). The most characteristic ultrastructural feature of NE cells is the presence of cytoplasmic neurosecretory granules (dense core vesicles, DCV). The DCV are considered to be the storage site for amine and peptide hormones. At higher magnification the DCV show a

Table 1. Ultrastructural characteristics of DCV in single NE cells in tracheobronchial epithelium of human and various animal species.

Species	Age	Reference	DCV		
			Shape	Average size (nm)	Density of matrix
Human*	(NB, A)	(14, 15)	Spherical	120	Moderate to high
Lamb	(F, NB)	(16)	Type 1 Spherical	170	Moderate
			Type 2 Vesicular	120	High
Rabbit	(F, NB, A)	(10, 16)	Pleiomorphic	140	High
Rat	(F, NB)	(20)	Spherical	100	Moderate to low
Mouse	(NB, A)	(10)	Spherical	100	Moderate to low
Armadillo	(A)	(16)	Type 1 Spherical	170	High
			Type 2 Spherical	120	Moderate
Chicken	(A)	(17)	Pleiomorphic	140	Moderate to high

Age: F = fetal; NB = new born; A = adult.
* For human fetal lung see Table 3.

Table 2. Ultrastructural features of NEB in various animal species.

Species	Age	References	DCV			Innervation
			Shape	Average size (nm)	Density of matrix	
Rabbit	(F, NB)	3, 18)	Type 1 Pleiomorphic	140	High	Afferent and
			Type 2 Spherical	120	Moderate	efferent
Rat	(F, NB)	(20)	Spherical	100	Moderate to low	Afferent
Mouse	(NB, A)	(21)	Spherical	100	Moderate to low	Afferent
Hamster	(NB, A)	(22)	Pleiomorphic	110	Moderate to high	Afferent
Toad (Bufo marinus)		(19)	Vesicular and elongated	80	Moderate to high	Afferent and efferent
Tree Frog (Hyla arborea L.)	(A)	(25)	Spherical	80	Moderate to high	?Afferent and efferent

Age: F = fetal; NB = new born; A = adult.

single limiting membrane enclosing an electron dense matrix with a clear halo around it. The size, electron density and morphology of DCV may vary in individual NE cells as well as between species (Table 1). Based on the differences in DCV size and appearance, several types of pulmonary NE cells have been identified. By analogy with similar cells in the GI tract, it is assumed, that each of NE cell subtypes represents cells producing different hormonal substances.

Within the tracheobronchial epithelium of the lamb, two distinct NE cell types have been described (16). Type 1 contains spherical DCV with moderately electron dense matrix (Fig. 3a). In contrast, Type 2 cells contain smaller vesicular DCV with a dense, sometimes eccentrically placed core (Fig. 3b). In the rabbit tracheal epithelium, only one type of NE cells containing large pleiomorphic DCV with a homogeneously electron dense matrix have been identified (Fig. 3c).

In the human fetal lung, three distinct types of NE cells have been identified (10, 13) (see also section on development). In contrast, EM studies of human newborn and adult lung revealed only one NE cell type containing spherical DCV with a homogeneous electron dense matrix and a well defined limiting membrane (Fig. 4, inset) (15).

The other cytoplasmic components of NE cells include small mitochondria, perinuclear Golgi zone, ribosomes and small stacks of rough endoplasmic reticulum (Fig. 4). Varying numbers of microtubules and intermediate sized filaments are also present, the latter often arranged in bundles around the nucleus. The single NE cells do not seem to be innervated. Although in mammalian lungs only occasional nerve endings in the proximity of NE cells were observed; in chicken lung, close association of nerve endings with these cells has been reported (17). There is no evidence for constant association of single NE cells with submucosal capillaries.

Neuroepithelial bodies (NEB), consist of a group of three or more NE cells arranged in a distinctive corpuscular body within the bronchial epithelium. By electron microscopy, NEB were first described in the rabbit fetal and neonatal lungs, where these bodies are particularly prominent (3, 5, 6). Ultrastructural studies on NEB have been also reported in other animal species, but detailed EM studies in human are not available. The main ultrastructural features of NEB in various animal species are listed in Table 2.

In the near-term rabbit fetus or neonate, an average NEB is composed of 10 to 15 cells. However, the size and shape of NEB varies with its location along the airway mucosa or plane of sectioning. In a perpendicular section, NEB cells appear as tall columnar cells extending from the basement membrane to the airway lumen. In a tangential cut, they usually appear as distinct ovoid corpuscles situated within the bronchial mucosa. Some NEB, particularly those located close to or directly on airway bifurcations, are sometimes composed of up to 80 closely packed cells (Fig. 5). The cell membranes form slight interdigitations with occasional gaps and desmosomes. In the apical region, tight junctions (zonula occludens) are present and the plasma membrane forms short microvilli covered by a thin glycocalyx (Fig. 6a). The basal aspect is in close contact with the basement membrane and this region also contains the

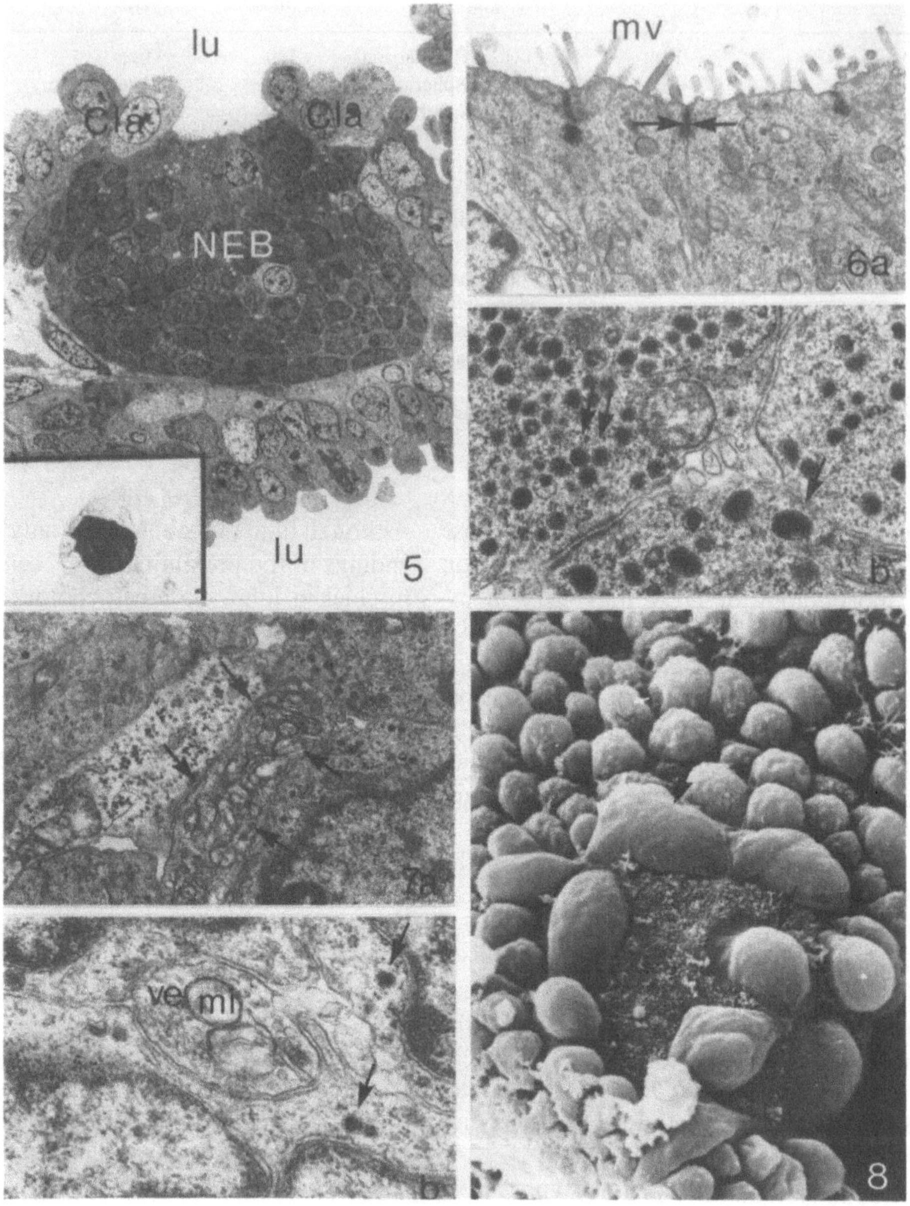

Figs. 5–8. NEB in rabbit fetal lung. (5) Low-magnification electron micrograph of an obliquely cut NEB near bifurcation, forming an 'organoid' structure within bronchiolar epithelium. The surface of NEB (at top) is exposed to the airway lumen (lu). Clare-like cells (Cla), adjacent to NEB, protrude above its surface (TEM × 800). Inset: Serotonin immunoreactivity in NEB of fetal rabbit lung. LM, Immunoperoxidase method (× 150). (6) (a) Apical surface of NEB shows surface microvilli (mv) and intercellular junctions (TEM × 4,800); (b) Close-up of basal portion of NEB cells with numerous cytoplasmic DCV. Type I DCV (single arrow) are elongated and have highly electron dense matrix, whereas type II granules (double arrow) appear uniformly spherical and the density of their matrix vary (TEM × 26,400). (7) (a) Afferent-like nerve fibre amongst NEB cells (arrows) containing numerous mitochondria (TEM × 4,800); (b) Small agranular vesicles (ve) and few mitochondria (mi) in an efferent-like nerve ending within NEB. Adjacent NEB cells contain few DCV's (arrows) (TEM × 20,000). (8) Surface view of an NEB at the bronchiolar bifurcation, showing fine microvilli. Flattened Clara-like cells (arrows) encircle NEB periphery and partially straddle the bifurcation. (SEM × 2,000) (Figs. 5, 7a, b, 8 reproduced from Anat. Rec. 193: 459–466, 1978 by permission of publisher. The Wistar Press, Philadelphia, Pa.).

highest concentration of DCV. In the rabbit, two distinct DCV types have been described (3). The granules of the first type are pleiomorphic with a homogeneous electron dense matrix, whereas the second type are spherical and their matrix is less electron dense (Fig. 6b). This variation in the appearance of DCV has been interpreted as possibly representing granules with different contents (amine vs peptide).

A unique feature of NEB is the presence of non-myelinated nerve fibres terminating between the granulated cells. In rabbit fetus and neonate, two types of nerve endings have been described (18). The afferent-like variety which contains numerous small mitochondria, microtubules and occasional agranular vesicles (Fig. 7a); and efferent-like nerve endings characterized by more numerous agranular vesicles, microtubules and only occasional mitochondria (Fig. 7b). Synaptical contacts between each type of nerve ending and granulated cells as well as between the two different types of nerve endings have been observed. Dual innervation, similar to the rabbit has been reported for NEB in toad lung (Bufo marinus) (19), but in the rat (20), mouse (21) and hamster (22) only one type of nerve ending (thought to be sensory) has been identified. The relationships between NEB and vascular structures in the submucosa have not been studied in detail. Lauweryns et al. (3) illustrated the presence of fenestrated capillaries immediately beneath NEB base, but this is not a constant observation.

Scanning electron microscopy (SEM) has been used only recently for investigation of the surface morphology of NEB. We have studied the surface features of NEB in rabbit fetal lung using a combination of light microscopy, transmission and scanning EM (23). In this species, NEB surfaces appear as crater-like depressions surrounded by palisading non-ciliated cells. The surfaces of NEB are covered by fine microvilli and are in direct contact with the airway lumen. When located directly on the bronchial bifurcations, NEB surfaces may be partially covered by flattened non-ciliated (Clara-like) cells (Fig. 8). In neonatal mouse lung, NEB surfaces were found to be almost entirely covered by attenuated Clara-like cells with apices of some NEB cells protruding through openings between the flattened cells (24). Similar findings have been recently reported for NEB in the lung of a tree frog (Hyla arborea) (25). It is not clear whether variations in the surface appearances of NEB are related to species differences or reflect different stages of development.

3. Distribution and frequency

The distribution of single NE cells and NEB differ in that the single cells are scattered throughout the entire length of the tracheobronchial mucosa, whereas NEB are located only within the epithelium of intrapulmonary airways (16). Within the lung, both single NE cells and NEB appear concentrated in the peripheral airways, where they extend up to the alveolar ducts (14). The NEB appear particularly frequent in close proximity to bronchial bifurcations and at the sites of bronchial branchings. In neonatal mouse lung, 86% of NEB were located at the branching points of the intrapulmonary airways (24). Precise quantitative data on the frequency of NE cells in normal lung are incomplete. A rough estimate of the total number of NE cells in the lung is between 1 to 2% of all bronchial epithelial cells. NE cells appear relatively more numerous in fetal lungs, because they differentiate early in gestation and the fetal lung consists mainly of airways. During postnatal growth of the lung, NE cells may become 'diluted' and dispersed over a larger surface area, because of elongation of airways and development of alveoli.

Table 3. Ultrastructural features of various types of NE cells in human fetal lung*

Cell type	DCV				Intermediate sized filaments
	Shape	Average size (nm)	Range (nm)	Density of matrix	
P_0 (?Precursor cells)	Spherical	90	70–100	Moderate	+
P_1	Vesicular Spherical	110	70–170	High to Moderate	+
P_2	Spherical	130	80–200	Moderate	++
P_3	Spherical	180	125–230	High	+

* Modified from Hage (10)

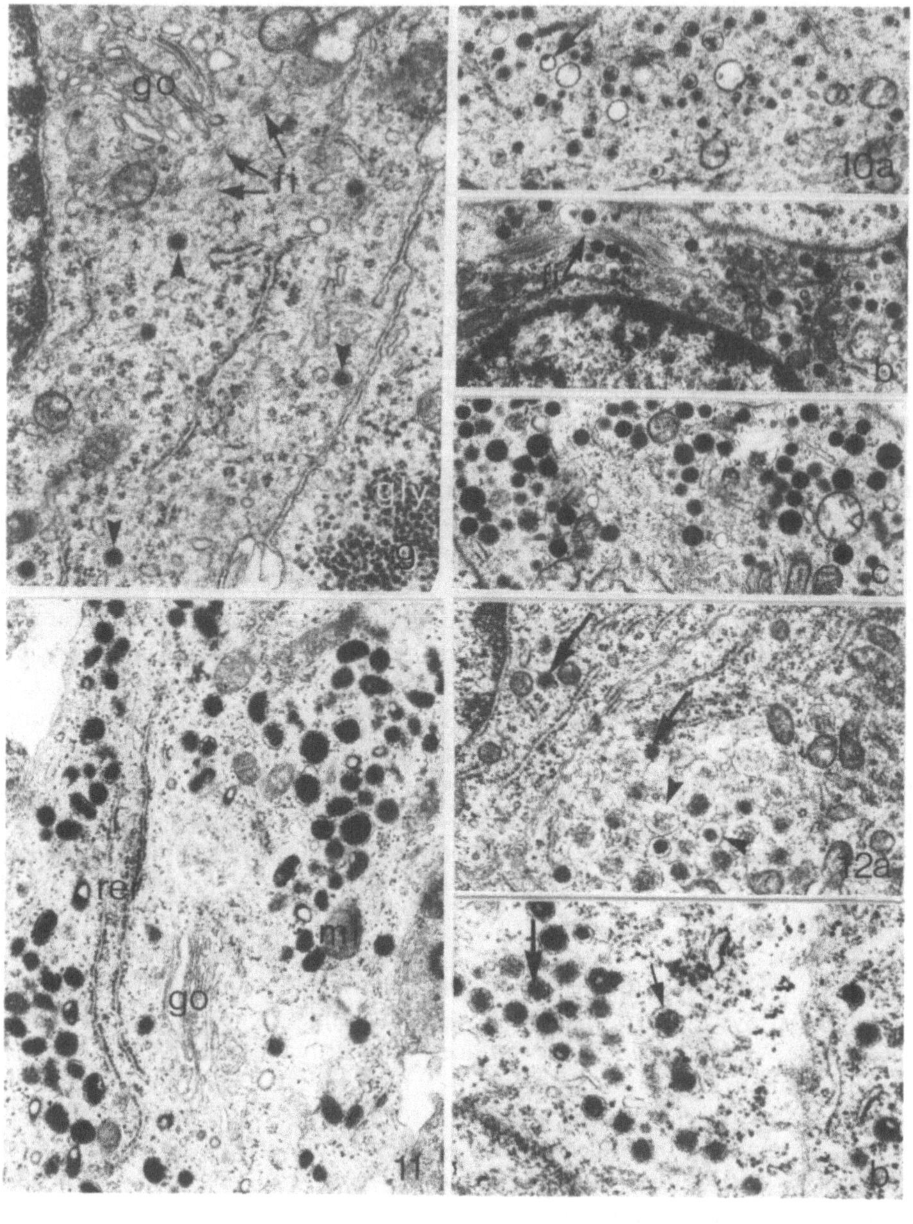

Figs. 9–12. (9) Portion of NE cell cytoplasm in 8 week gestation human fetal lung shows prominent Golgi complex (go), intermediate-sized filaments (fi) and scanty DCV (arrowheads). Numerous free ribosomes and a few cisternae of rough endoplasmic reticulum are also seen. The cytoplasm of an adjacent epithelial cell contains glycogen (gly). (TEM × 24,000). (10) (a) Cytoplasmic DCV in P1 type of NE cell in human fetal lung (16 wks gestation) shows a distinct limiting membrane and a few vesicular granules with eccentrically placed core (arrow) (TEM × 18,400); (b) Portion of P2 cell in human fetal lung (18 wks gestation) shows DCV with moderate to high electron density and a bundle of perinuclear filaments (fi) (TEM × 20,000); (c) P3 cell in human fetal lung (18 wks gestation) contains large, homogeneously electron dense granules (TEM × 20,000). (11) Portion of NEB cell cytoplasm from control explant culture of rabbit fetal lung maintained for 7 days *in vitro* with prominent Golgi zone (go), elongated rough endoplasmic reticulum (rer), mitochondria (mi) and numerous pleiomorphic, electron dense DCV (TEM × 28,000). (12) (a) Cores of DCV's from NEB in lung explant 7 days *in vitro*, then treated with reserpine (10 μg/ml) for 45 min, showing greatly reduced electron density and fine granularity (arrowheads). Few DCV appear unaltered by reserpine treatment (arrows) (TWM × 28,000); (b) Cores of DCV from NEB in lung explant 7 days *in vitro*, then treated with Ca^{++} ionophore (1 μg/ml) for 30 min, showing slightly decreased electron density (large arrow) and more prominent halo around some granules (small arrow) (TEM × 28,000). (Figs. 11, 12a, b reproduced from Cell Tissue Res. 199: 159–170, 1979 by permission of the publisher Springer-Verlag, Heidelberg).

4. Development and origin

In human fetal lung, NE cells differentiate in a centrifugal pattern and follow the general development of the airways. The first NE cells can be identified around 8 weeks gestation (early glandular period) (13). At EM level, they are characterized by lack of cytoplasmic glycogen and presence of abundant ribosomes, small cisternae of rough endoplasmic reticulum, single or multiple Golgi complexes, small mitochondria, microtubules and intermediate sized filaments, usually concentrated in perinuclear region (Fig. 9). The cytoplasmic DCV are scanty and are often seen in the proximity of the Golgi complex. The shape of DCV is mostly spherical and their mean diameter (90 nm) is smaller compared to later gestation.

At the end of glandular period (about 16 weeks), the differentiation of NE cells is characterized by formation of apical or lateral cytoplasmic processes and by emergence of three distinct NE cell types, each having different population of DCV (Table 3). Hage designated these cells as P_1, P_2, and P_3 based on the different size and morphology of granules (10). In P_1 cells, two different populations of DCV can be distinguished; the first is of spherical shape and has a moderately electron dense matrix with a distinct halo, whereas the second shows larger vesicular membrane with an eccentrically placed dence core (Fig. 10a). P_2 cells contain spherical DCV which vary in size and show a thin halo around a moderately electron dense core (Fig. 10b). The type P_3 cells are the least frequent, but contain the largest DCV with a homogeneous electron dense core and a distinct limiting membrane (Fig. 10c). Whereas P_1 and P_2 cells can be found in different regions of the developing bronchial tree, P_3 cells are usually located in the epithelium of larger, proximal airways. The different types of NE cells are distributed as single cells or in groups of 3 to 5 cells, consisting of the same or of different NE cell types. These agglomerates are reminiscent of basal granulated cells described in human fetal duodenum (26). These structures appear to be unique for human fetal lung, since they are not seen in the adult, nor in the fetal or adult animal lungs. Distinctive ovoid corpuscles (similar to NEB in the rabbit) can be identified between 17 to 20 weeks gestation. These corpuscles are usually composed of the same NE cell type (mainly P_1 type) (9).

The ultrastructure of NE cells during saccular period of lung development (18–25 weeks) has been reported by McDougall (27). NE cells with long cytoplasmic processes and DCV, similar to those of earlier stages of development, have been identified in terminal bronchioles and primitive saccules. These cells or their processes appeared in close association with capillaries and smooth muscle cells beneath the basement membrane. Occasional non-myelinated nerve fibres were seen in close proximity, but not in direct contact with granulated cells.

In late fetal or neonatal lungs, the ultrastructural distinction between different NE cell types is less apparent.

The lungs develop from foregut endoderm which also give rise to the GI tract and pancreas. Therefore it is reasonable to assume that the embryological origin of endocrine cells in these tissues would be similar. An earlier suggestion by Pearse, that all cells with APUD characteristics are derived from the neural crest (2), has been recently modified. The revised APUD system is now referred to as diffuse neuro-endocrine system (28). It consists of central and peripheral divisions, the latter includes pulmonary NE cells as well as GI and pancreatic endocrine cells. A possible origin for these cells from 'neuro-endocrine programmed' cells of embryonal epiblast (definitive endoderm precursor) has been proposed.

5. Experimental studies

The scattered distribution and relative paucity of NE cells in the lung complicates direct *in vivo* experimentation. Consequently, there are only a few experimental studies available. Thus far, the major emphasis of these studies has been on the influence of various gaseous environments on the reactivity of pulmonary NEB. Efforts have also been made to develop suitable *in vitro* systems to study the secretory and metabolic activity of NE cells.

5.1. Effects of hypoxia

Lauweryns et al. studied the effects of short term (acute) hypoxia on NEB in neonatal rabbit lung (5). When one or two day old neonatal rabbits were exposed to profound hypoxia (10% oxygen in nitrogen) for 20 min., increased exocytosis of DCV and decreased amine fluorescence of NEB was observed. Similar changes occurred if animals were exposed to hypercapnia, whereas none were observed with hyperoxia. In subsequent cross-circulation experiments it has been further shown that NEB react specifically to airway hypoxia, but not to hypoxaemia (6). These findings led to a hypothesis suggesting that NEB represent intrapulmonary hypoxia sensitive chemoreceptors (3, 5, 6). According to this hypothesis, NEB are directly responsive to changes in intraluminal composition of air; serotonin and/or peptide released from NEB cells upon hypoxic stimulus induces local vasoconstriction, altering the ventilation perfusion ratio in the lung by shunting the blood from poorly to better aerated areas. Since NEB possess dual innervation (at least in the rabbit), this reflex mechanism could be modulated by CNS.

NE cells have been also studied in animals exposed to chronic hypoxia by maintaining them in hypobaric chambers, simulating conditions at high altitude. NE cells of neonatal rats maintained in this environment for one month, exhibited ultrastructural changes in DCV indicating increased secretory activity (29). In the same animals, the carotid body cells showed similar changes, suggesting parallel responses and possible functional relationship between pulmonary NE cells and the carotid body. Taylor reported significant increase in the number of argyrophilic NE cells in neonatal rabbits born to does living at high altitude, compared to those born at sea level (30). Decreased intensity of amine fluoresence, without change in the number of argyrophilic cells has been reported in rabbit neonates exposed to hypobaric hypoxia for 2 to 2½ hours (31). The same authors recently examined the relationship between NE cell numbers, amine fluorescence and the medial thickness of pulmonary arteries in normoxic, acutely and chronically hypoxic neonatal rabbits (32). In hypoxic animals, the number of argyrophilic NE cells and the medial thickness of pulmonary arteries were significantly increased compared with normoxic controls. There was an inverse correlation between the intensity of amine (serotonin) fluorescence, NE cell numbers and medial thickness. This study concluded that pulmonary NE cells respond to changes in airway oxygen levels; hypoxia results in decreased cellular content of serotonin, increased number of argyrophilic NE cells and increased medial thickness of pulmonary arteries. These changes were reversible when hypoxic animals were returned to normoxic conditions.

5.2. *In vitro characteristics*

An *in vitro* system offers several advantages over an *in vivo* approach, particularly for the investigation of secretory and metabolic activities of NE cells. We have studied pulmonary NEB in organ cultures of rabbit fetal lung maintained up to 22 days *in vitro* (33). In this system, NEB retained their corpuscular shape, cytoplasmic argyrophilia and amine fluorescence, but several changes were observed at ultrastructural level. In cultured NEB, the cytoplasmic DCV became larger, more pleiomorphic and their matrix became more electron dense (Fig. 11). In addition, DCV became re-distributed throughout the cytoplasm rather than being confined chiefly to the basal regions. The Golgi zones were hypertrophied, which together with the increased size and density of DCV suggested increased synthesis and storage of amine and/or peptide *in vitro*. NEB in lung explants also responded to pharmacologic stimuli; incubation with reserpine greatly reduced the electron density of DCV matrix, indicating amine release (Fig. 12a). Whereas treatment with Ca^{++} ionophore produced slight decrease in electron density and coarse granularity of the matrix (Fig. 12b), suggesting that Ca^{++} is involved in the secretory process. This and other *in vitro* model systems (such as separated and enriched preparations of NE cells), coupled with quantitative biochemical studies, could provide an ideal research tool to further explore the responses of NE cells to various gas mixtures, pharmacologic and hormonal agents.

6. Functional considerations

As already indicated, the precise function of NE cells in the lung is presently unknown. Currently available morphological and experimental data suggest that the functional significance of NE cells may be closely related to lung development and neonatal adaptation. The role of NE cells may be particularly important during intrauterine life (when lung does not serve for gas exchange) and at the time of birth, when respiration and gas exchange becomes established. The primary physiological role of pulmonary NE cells and their principal secretory products (bombesin-like peptide and serotonin) may involve the control and modulation of pulmonary circulation *in utero* and at the time of birth. During intrauterine life, the release of amine and peptide mediators (in response to relative pulmonary hypoxia in the fetus) may contribute to the maintenance of active pulmonary vasoconstriction. At the time of birth, sudden exposure of fluid filled lung to higher oxygen concentrations in inspired air may terminate the release of vasopressor substances and in concert with mechanical and other humoral factors contribute to dilatation of the pulmonary vascular bed. Alternate functions of NE cells in fetal lungs may include local hormonal production exerting trophic effects on adjacent epithelial cells (paracrine function) or an endocrine activity with effects on distant organs.

In post-natal lungs, once the air breathing has been established, NE cells and NEB in particular may function as intrapulmonary hypoxia sensitive chemoreceptors, regulating local ventilation/perfusion ratio in the lung lobule (6). The functional relationships between NEB and single NE cells is not clear. NEB show many morphologic similarities with known chemoreceptors (e.g. carotid body, taste buds), whereas single NE cells appear to be analogous to endocrine cells in the GI tract.

Another area of interest, pertaining to the possible role of NE cells in adult lung, relates to pulmonary neoplasia. It has been shown that bronchial carcinoids and oat cell carcinomas originate from pulmonary NE cells (4). The neoplastic NE cells retain many ultrastructural and histochemical characteristics of their normal counterparts. An important question, whether these neoplasms develop from preexisting, differentiated NE cells, or develop from precursor stem cells is presently not known.

7. Concluding remarks

Pulmonary NE cells, initially thought to be rare and even 'vestigial' structures, have emerged as potentially important elements in the neuro-hormonal regulation of lung function. In spite of recent advances in morphologic characterization of pulmonary NE cells, it is apparent that our understanding of the role and function of these cells, under normal or pathologic conditions is manifestly incomplete. Further morphological and physiological studies are needed to better define the innervation of these cells, their relationship to other pulmonary structures such as the vascular supply, bronchial smooth muscle and others. Precise quantitative data are lacking on the distribution and frequency of NE cells during development and in post-natal lung. Studies are required to define in more detail the physiological stimuli, responses of NE cells and the effects of their amine and/or peptide mediators on target cells. The role of amine and peptide hormones, elaborated by NE cells, needs to be investigated with particular reference to neuro-hormonal regulation of pulmonary circulation and airway tonus. Undoubtedly, the identification of NE cells in the lung has opened a new area of investigation with potentially far reaching implications for pulmonary physiology and pathophysiology. Recent discovery of a number of new cell markers for NE cells, amenable to both morphologic and biochemical analysis should facilitate the many studies which lie ahead.

Acknowledgement

This work was supported by grants from Ontario Thoracic Society, Canadian Lung Association and Medical Research Council of Canada.

References

1. Bensch KG, Gordon GB, Miller LR: Studies on the bronchial counterpart of the Kultschitzky (argentaffin) cell and innervation of bronchial glands. J Ultrastruct Res 12: 668–686, 1965.
2. Pearse AGE: The cytochemistry and ultrastructure of polypeptide hormone-producing cells of the APUD series and the embryologic, physiologic and pathologic implications of the concept. J Histochem Cytochem 17: 303–313, 1969.
3. Lauweryns JM, Cokelaere M, Theunyck P: Neuroepithelial bodies in the respiratory mucosa of various mammals. A light optical, histochemical and ultrastructural investigation. Z Zellforsch 135: 569–592, 1972.
4. Bensch KG, Corrin B, Pariente R, Spencer H: Oat-cell carcinoma of the lungs. (Its origin and relationship to bronchial carcinoid). Cancer 22: 1163–1172, 1968.
5. Lauweryns JM, Cokelaere M, Deleersnyder M, Liebens M: Intrapulmonary neuroepithelial bodies in newborn rabbits. Influence of hypoxia, hyperoxia, hypercapnia, nicotine, reserpine, L-DOPA and 5-HTP. Cell and Tiss Res 182: 425–440, 1977.
6. Lauweryns JM, Cokelaere M, Lerut T, Theunyck P: Cross circulation studies on the influence of hypoxia and hypoxaemia on neuroepithelial bodies in young rabbits. Cell Tiss Res 193: 373–386, 1978.
7. Wharton J, Polak JM, Bloom SR, Ghatei MA, Solcia E, Brown MR, Pearse AGE: Bombesin-like immunoreactivity in the lung. Nature 273: 769–770, 1978.
8. Cutz E, Chan W, Track NS: Bombesin, calcitonin and leu-enkephalin immunoreactivity in endocrine cells of human lung. Experientia 37: 765–767, 1981.
9. Cutz, E: Neuro-endocrine cells in the lung – morphologic characteristics and development. (Overview). Exp Lung Res 3: 185–208, 1983.
10. Hage E: Endocrine-like cells of the pulmonary epithelium. In: Chromaffin, Enterochromaffin and Related Cells. Coupland RE, Fujita T (eds), Amsterdam, Oxford, New York, Elsevier Scientific Publishing Company, 1976, pp 317–332.
11. Lauweryns JM, Cokelaere M, Theunyck P: Serotonin producing neuroepithelial bodies in rabbit respiratory mucosa. Science 180: 410–413, 1973.
12. Wharton J, Polak JM, Cole GA, Marangos PJ, Pearse AGE: Neuron-specific enolase as an immunocytochemical marker for the diffuse neuroendocrine system in human fetal lung. J Histochem Cytochem 29: 1359–1364, 1981.
13. Cutz E, Conen PE: Endocrine-like cells in human fetal lung. An electron microscopy study. Anat Rec 173: 115–122, 1972.
14. Lauweryns JM, Peuskens JC, Cokelaere M: Argyrophil, fluorescent and granulated (peptide and amine producing?) AFG cells in human infant bronchial epithelium. Light and electron microscopic studies. Life Sci (I) 9: 1417–1429, 1970.
15. Hage E, Hage J, Juel G: Endocrine-like cells of the pulmonary epithelium of the humen adult lung. Cell Tiss Res 178: 39–48, 1977.
16. Cutz E, Chan W, Wong V, Conen PE: Ultrastructure and fluorescence histochemistry of endocrine (APUD type) cells in tracheal mucosa of human and various animal species. Cell Tiss Res 158: 425–437, 1975.
17. Wasano K, Yamamoto T: APUD-type recepto-secretory cells in the chicken lung. Cell Tiss Res 201: 197–205, 1979.
18. Lauweryns JM, Cokelaere M: Hypoxia sensitive neuro-epithelial bodies intrapulmonary secretory neuroreceptors, modulated by the CNS. Z Zellforsch 145: 521–540, 1973.
19. Rogers DC, Haller CJ: Innervation and cytochemistry of the neuroepithelial bodies in the ciliated epithelium of the toad lung (Bufo marinus). Cell and Tiss Res 195: 395–410, 1978.

20. Cutz E, Chan W, Wong V, Conen PE: Endocrine cells in rat fetal lungs: Ultrastructural and histochemical study. Lab Invest 30: 458–464, 1974.
21. Hung KS, Loosli CG: Bronchiolar neuro-epithelial bodies in neonatal mouse lungs. Amer J Anat 140: 191–200, 1974.
22. Edmondson NA, Lewis DJ: Distribution and ultrastructural characteristics of Feyrter cells in the rat and hamster airway epithelium. Thorax 35: 371–374, 1980.
23. Cutz E, Chan W, Sonstegard K: Identification of neuroepithelial bodies in rabbit fetal lungs by scanning electron microscopy: A correlative light, transmission and scanning electron microscopic study. Anat Rec 459–466, 1978.
24. Wasano K, Yamamoto T: A scanning and transmission electron-microscopic study on neuro-epithelial bodies in the neonatal mouse lung. Cell Tiss Res 216: 481–490, 1981.
25. Goniakowska-Witalinska L: Neuroepithelial bodies in the lung of the tree frog. Hyla arborea L. A Scanning and transmission electron microscopic study. Cell Tiss Res 217: 435–441, 1981.
26. Osaka M, Kobayashi S: Duodenal basal-granulated cells in the human fetus with special reference to their relationship to nervous elements. In: Endocrine Gut and Pancreas. Fujita T (ed). Amsterdam, New York, Elsevier Scientific Publishing Company, 1976, pp 145–158.
27. McDougall J: Endocrine-like cells in the terminal bronchioles and saccules of human fetal lung: An ultrastructural study. Thorax 33: 43–53, 1978.
28. Pearse AGE, Takor Takor T: Embryology of the diffuse neuroendocrine system and its relationship to the common peptides. Fed Proc 38: 2288–2294, 1979.
29. Moosavi H, Smith P, Heath D: The Feyrter cell in hypoxia. Thorax 28: 729–741, 1973.
30. Taylor W: Pulmonary argyrophil cell at high altitude. J Path 122: 137–144, 1976.
31. Keith IM, Wiley LA, Will JA: Pulmonary neuroendocrine cells: Decreased serotonin fluorescence and stable argyrophil-cell numbers in acute hypoxia. Cell Tiss Res 214: 201–205, 1981.
32. Keith IM, Will JA: Hypoxia and the neonatal rabbit lung: neuroendocrine cell numbers, 5-HT fluorescence intensity and the relationship to arterial thickness. Thorax 36: 767–773, 1981.
33. Sonstegard K, Wong V, Cutz E: Neuro-epithelial bodies in organ cultures of fetal rabbit lungs. Ultrastructural characteristics and effects of drugs. Cell Tiss Res 199: 159–170, 1979.

Author's address:
Department of Pathology
The Hospital for Sick Children
555 University Avenue
Toronto, Ontario, Canada
M5G 1X8

CHAPTER 14

Recent advances in the ultrastructure of the carotid body

STEN HELLSTRÖM and JOHAN KJAERGAARD

1. Introduction

Even after almost 250 years of research the morphological and physiological interrelationship of the cellular elements of the carotid body are still poorly understood (1). The only function which has been confirmed is the capacity of the carotid body to act as a specific chemoreceptor.

At the present stage of our knowledge of the intimate relationship between the nervous and the endocrine cell systems the question of whether the carotid body is a gland or part of the nervous system is of lesser interest. It is now more necessary to establish exactly how the various elements of the complex structure of the organ are involved in the mechanisms responsible for the onset of chemosensory discharge. Excellent reviews have appeared recently (2, 3) and several international workshops have summarized progress during the last decade (4, 7, 10). The present chapter will summarize the most recent advances in the study of the ultrastructure of the carotid body.

2. Methodological progress

The complexity and structural variations in the carotid body demonstrated within species as well as between species have clearly revealed the necessity of employing control materials and quantitative methods in this field. It appears that much research has been carried out neglecting these basic principles.

2.1. Three-dimensional reconstructions

Random single sections have been employed in the majority of electron and light microscopic studies of the carotid body. These are inadequate for tracing topographical relationships between the cellular elements. Analysis of serial sections, although somewhat time-consuming, has been used with considerable success on the vascular network of the carotid body (4) and on the organization of the nerve fibres (5).

2.2. Freeze-etching

This is another approach for studying the topographical relationship within the structures of the carotid body. The method provides information regarding the cell surface as well as other membranes (6, 7).

2.3. Morphometry

This technique has permitted the study of cellular events during non-stimulated and stimulated conditions. It has allowed the volumes and volume ratios of the cellular tissue of the carotid body as well as the subcellular elements to be determined (8, 9). Furthermore, it would appear useful for the correlation of biochemical data with structural changes during experimental conditions (10).

2.4. Analytical electron microscopy

Technical innovations in electron microscopy now permit atomic constituents to be analyzed from small structural elements. Employing this technique it is possible to study the catecholamines *in situ* in dense cored vesicles of the glomus Type I cells (11).

2.5. Immunohistochemistry

Immunohistochemistry has revolutionized our understanding of fine structural features of the nervous system. This tool has recently been employed on the carotid body and data regarding the localization of several neuropeptide transmitters to the glomus cells and nerve fibres may greatly influence the view of the chemosensory circuits (12, 13).

Motta, PM (ed): Ultrastructure of endocrine cells and tissues. ISBN-13: 978-1-4613-3863-5

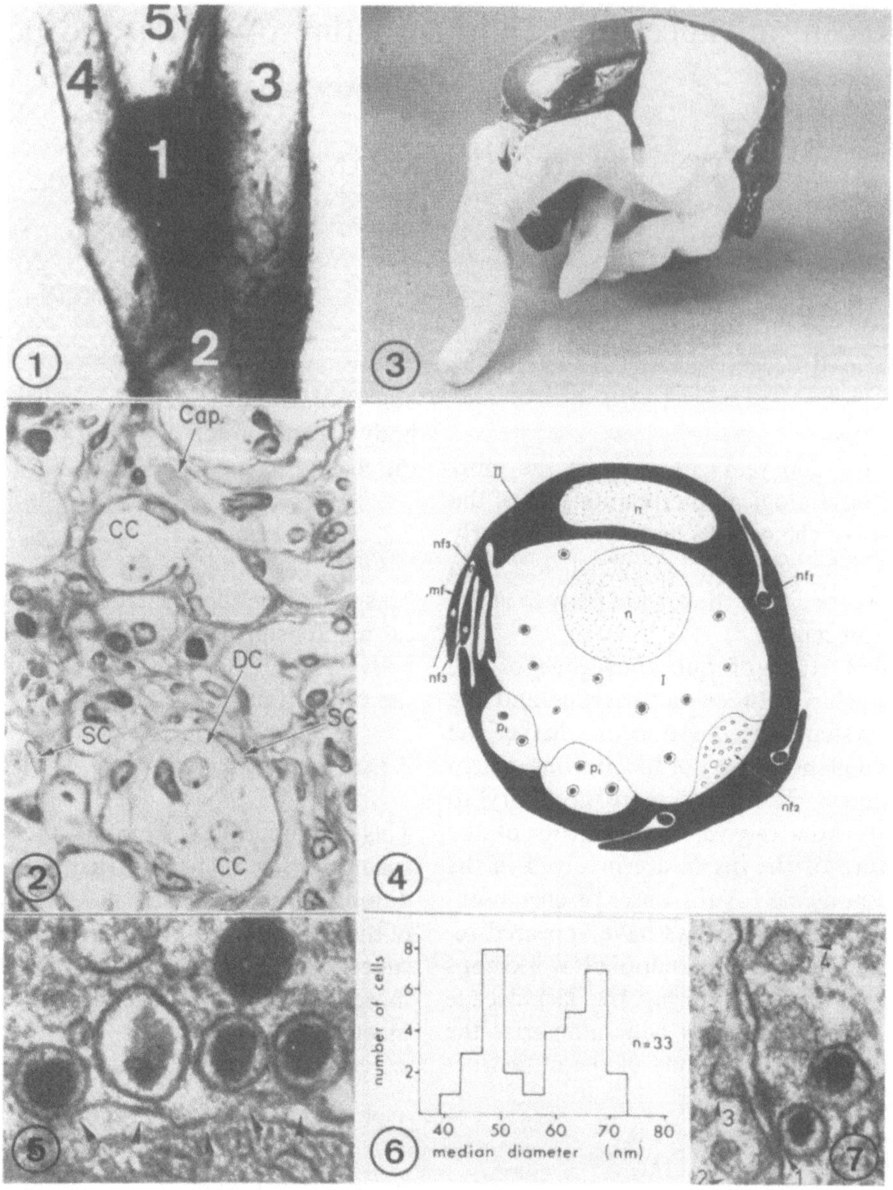

Figs. 1–7. (1) A preparation of the carotid bifurcation from a human foetus (crown-rump length 55 millimeters). The carotid body is ovoid in shape and lies medially in the carotid bifurcation. The photograph was taken by transillumination of the resin-embedded specimen before polymerization (× 25). 1: Carotid body; 2: Common carotid artery; 3: Internal carotid artery; 4: External carotid artery; 5: Sinus nerve; After Kjaergaard (1). (2) Light microscopic high power view of the human carotid body. Type I cells with round, pale nuclei. Type II cells with semi-lunar, darker nuclei. DC and CC: Type I cells; SC: Type II cell; CAP: Capillary. Epoxy section stained with toluidine blue and malachite green (× 560). After Glenner, Grimley (16). (3) Reconstruction of uninterrupted serial slices of a carotid body unit of the rabbit. A single vessel passes tortuously through the unit. Inflow and outflow vessels run parallel. An aggregation of cells contact the vessel on one side. Diameter of the unit, approximately 70 micrometers. After Seidl et al. (17). (4) Schematic drawing of Type I and Type II cells and their nerve supply. I: Type I cell, II: Type II cell, p1: Finger-like projections of Type I cells, nf_1, nf_2, nf_3: nerve fibers, infolding for a Type I cell, mf: Mesaxonic folding for a nerve fiber, n: nucleus. After de Kock, Dunn (18). (5) DCV containing an ovalar, homogeneous, electron-opaque core, which varies in shape and size. A clear space of varying width is present between the core and vesicle. The material in the concentric pale ring is less electron transparent than the cytoplasmic ground substance and contains small granules reminiscent of core material. Other structures represent peripheral cut pieces of DCV, where the cut has not hit the core but just the pale ring and the vesicles (× 70,000). After Böck, Gorgas (19). (6) Histogram showing the median diameters of vesicle profiles of individual Type I cells in the rat carotid body. Two sub-classes of Type I cells are recognized: Cells containing small and cells containing large DCV. After Hellström (8). (7) A series of profiles which may correspond the secretion of the granule content by exocytosis. It is difficult to demonstrate the fusion of the vesicle with the cell membrane convincingly. Most of the apparently fused vesicles would be found to be free of the cell membrane if a tilting device was used in the electron microscope. 1: Possible fusion of cell membrane and vesicule membrane, 2: Extrusion of matrix material, 3: Formation of a coated vesicle, 4: Pinching of the coated vesicle (× 70,000). After Böck, Gorgas (19).

2.6. *Chemical analysis*

Despite the minute amount of tissue available in a single carotid body, the introduction of mass fragmentographic assay of catecholamines (14) allows specific measurement of its catecholamines. Thus the distribution of the biogenic amines of the carotid body and their role in the chemosensory function have been studied (15).

3. Cellular elements

The carotid body (Fig. 1) is divided into approximately one-hundred lobules, consisting of clusters (16) of Type I and Type II cells (Fig. 2). A cell cluster together with a blood vessel and nerve fibers constitute a functional unit. The diameter of a unit is 70–100 μm. The size of human carotid bodies (mean values of the maximum diameter) when employing micrometry on serial sections is 3.1 × 2.2 × 1.7 mm.

The vessels are tortuous with a varying diameter and are in contact on the one side with an aggregation of specific cells (Fig. 3). The endothelial cells of the blood vessels are fenestrated (6). These endothelial cells are normally separated from the Type I cells by the perivascular space, in which there are pericytes, vascular basement membranes, strands of connective tissue and Type II cells, the latter being surrounded by a basement membrane.

In the cat the specific tissue of the carotid body constitutes approximately 20%, the vessel 20% and the nerves and connective tissue the remaining 60%. 60% of the specific tissue is made up of Type I cells and 15% of Type II cells, 5% by nerve fibres and 20% by connective tissue (17).

4. Ultrastructure of the Type I and Type II cells

The relationship between Type I and Type II cells and the nerves is very intricate. Type I cells have long dendritic processes and Type II cells flat projections and deep folds in the plasma membrane. The mesaxonic infoldings of the latter cells surround both the bodies and processes of the Type I cells as well as the nerve fibres (Fig. 4). It has been demonstrated (18) in the sole ultrastructural investigation based on serial sections focused on Type II cells, that Type I cells are almost always surrounded by thin flattened cytoplasmic projections from one or several Type II cells.

The interior structure of these cells does not exhibit characteristic structures. It has been emphasized that ultrastructurally they are indistinguishable from Schwann cells by their ultrastructure. A passive supporting function has been ascribed to the cells due to their morphology, and this is reflected in the various synonyms employed: supporting cells, sustentacular cells, sheet cells and enclosing cells. However, experiments have not been carried out specifically designed to demonstrate the exact function of these cells.

There is surprising conformity between the results of the large number of investigations which have been published and which are summarized by among others Kjaergaard (1) and Verna (3) with regard to the general fine structural characteristics of Type I cells. The cytoplasm comprises from one half to three quarters of the whole cell volume (16). In general the cytoplasm is considered rich in organelles. However, mitochondria comprise, at the most, 7% of the cytoplasmic volume (16) and the dense cored vesicles (DCV) 0.4 to 0.9% (8).

The basic structure of DCV (Fig. 5) has been described in detail (19). The profile diameter of the vesicles has been reported to be about 120 nm (range: 30 to 200 nm). Two investigations have been reported which comply with the theoretical requirements, random sampling and large number of samples (8, 20). Two cell populations were found (Fig. 6) in the rat, one containing small DCV (about 40% of the cells) and the other with large DCV (about 60% of the cells).

No corresponding division into small and large cored vesicles could be demonstrated in a similar study of the carotid body of rabbits (3). The size of the DCV shows a considerable variation from cell to cell and within a given cell.

In Type I cells the distribution of the various organelles in the cytoplasm is generally reported to be uniform in all areas, the cell thus showing no polarity. On the other hand, these reports are based on semiquantitative studies only and should thus be viewed with reservation. Quantitative measurements have not been carried out of the density of organelles as a function of the region within the cell. Some investigators state that DCV are accumulated below the plasma membrane in certain animals.

Among the large number of electron microscopic descriptions of carotid bodies published since 1957, there are only sporadic reports regarding the observations of so called omega figures (OF) in Type I cell plasma membranes, which could be interpreted as exocytosis of the content of DCV. The best of the published micrographs (Fig. 7) are not convincing, and may represent folded plasma membranes.

The fact that OF have not, or have only rarely, been observed in random sections does not exclude the possibility of exocytosis occurring under physiological conditions. The probability of observing an OF when the secretory rate is low and the OF only exists for a very short time is very slight.

A possible way of increasing the changes of seeing an OF is to enlarge the area of the plasma membrane under observation and by instantaneous fixation. These two prerequisites are complied with in freeze cleavage studies. No definite morphological exocytotic profiles (fractured neck as vesiculated stomata, membrane-associated particles as neck collars) could be observed in the large exposed plasma membrane surfaces of unstimulated and stimulated carotid bodies (6, 7).

The DCV can be fractured between the two la-

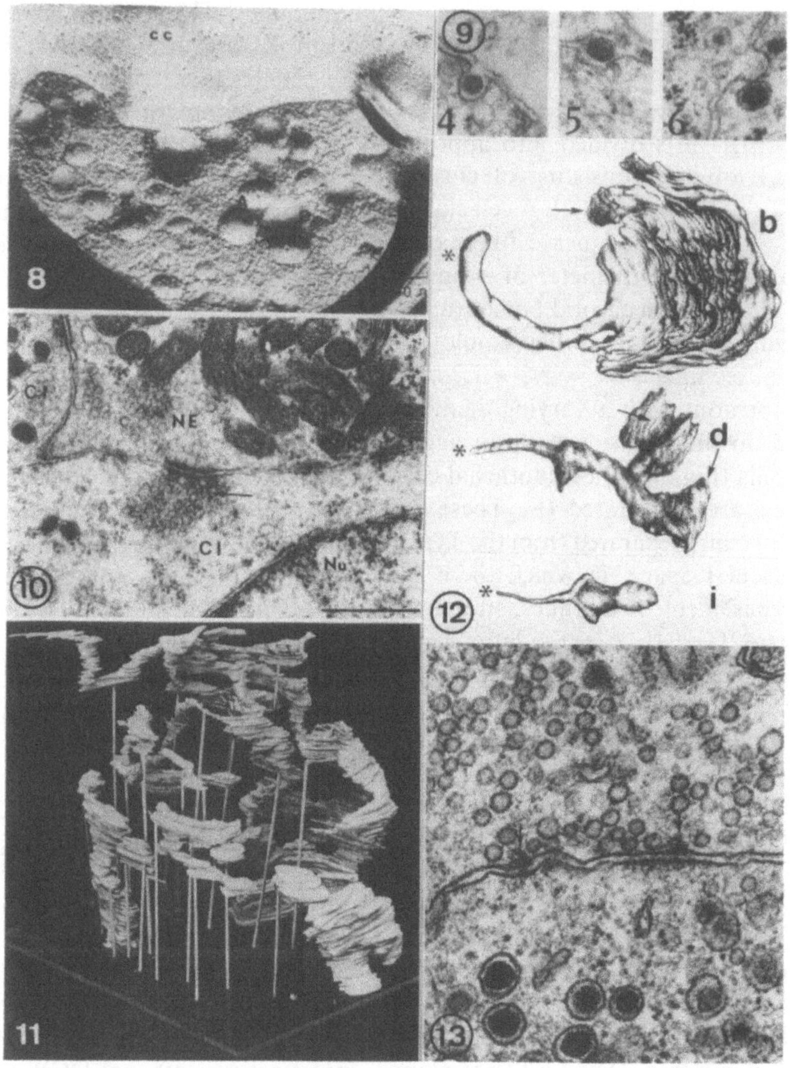

Figs. 8–13. (8) Freeze cleavage of a rat carotid body Type I cell. The membrane-associated particles of the P face on the plasma membrane (CC) seem randomly distributed, no neck collar-like arrangements are noted. The DCV present as split surfaces of the vesicle: basin-shaped depression (A) in the P face, and dome-like protrusions (B) on the E face of the vesicle membrane (× 56,000). After Kjaergaard, Hellström (6). (9) Exocytotic profiles from a rat carotid body following incubation in potassium-rich, calcium-rich medium. 4: Omega figure; 5: Dense core release; 6: Dense core almost dissolved, a coated pit remains (× 35,000). After Grönblad et al. (21). (10) Junctions between Type I cell (CI) and nerve ending (NE). Synaptic-like clear vesicles (arrow) accumulated on the Type I cell side. The junctional material is asymmetrically arranged. Nu: nucleus. Bar = 0.5 μm. After Verna (3). (11) Three-dimensional reconstruction of one large nerve fiber and four small aborizations in the central region of the rat carotid body. After Kondo (5). (12) Three-dimensional reconstruction of nerve endings in the cat carotid body. Large calyciform ending (b), small calyciform endings (d, i). The nerve enlargement arise from an unmyelinated nerve stem fiber which passes out of the plane of section (*); some nerve endings send a short process to a neighbouring Type I cell (arrow). Bar = 5 μm. After Nishi, Stensaas (25). (13) Junction between Type I cell and nerve ending (above). Synaptic-like clear vesicles accumulated on the nerve end side. The junctional dense material is asymmetrically arranged with the dense projection (arrows) in the neuroplasma (× 49,000). After McDonald, Mitchell (20).

minae of a unit membrane by means of freeze cleavage (Fig. 8). The pattern and density of the membrane-associated particles on both the P and E faces are equal to the density and distribution of the membrane-associated particles in the plasma membrane.

The probability of observing OF can also be increased by raising the rate of secretion. Exocytotic profiles (Fig. 9) have been demonstrated in rat Type I cells from carotid bodies incubated in potassium and calcium rich, oxygenated media (21). The study was carried out blind. The number of profiles of Type I cell plasma membrane increased following incubation of the carotid body in potassium/calcium rich media. However, the absolute number of OF was very low, of a magnitude of zero to one half OF per complete Type I cell profile.

An abundance of small clear vesicles (mean profile diameter approximately 60 nm, has been found in Type I cells. The morphology of these vesicles cannot be distinguished from that of ordinary synaptic vesicles. These small vesicles frequently occur in clusters near the plasma membrane (Fig. 10). In morphometric investigations of strongly stimulated carotid bodies of cats, an inverse correlation has been found between the number of DCV and small clear vesicles (22). This may indicate that the membrane, following release of the content of the DCV, is retrieved into the plasma as small clear vesicles.

5. Ultrastructure of nerves and nerve contacts

There was still considerable uncertainty as to exactly how many of the nerve fibers of the carotid body are afferent and how many are efferent when Kondo (2) published his review in 1977. It may definitely be assumed (3) from classical transection experiments under controlled conditions, electron microscopy and quantitation of the degeneration changes, later confirmed by autoradiographic studies, that the nerves in contact with Type I cells have their trophic centre in the petrosal ganglion of the glossopharyngeal nerve. Less than 5% of the nerve fibers to the carotid body are efferent and belong to the sympathetic and parasympathetic nervous systems. These fibers probably innervate the ganglion cells and blood vessels of the carotid body, although contact with Type I cells cannot be excluded.

Whether Type I cells or the afferent nerve terminals were the essential chemoreceptor, even at the time when Verna (3) published his review in 1979 is still controverted. Experiments have since been reported indicating that the afferent nerves themselves do not react to adequate stimuli (23, 24). In the experiments the carotid body was destroyed and the sinus nerve permitted to regenerate into the bifurcation of the carotid arteries. No nerve activity could be demonstrated after ventilation of the experimental animals with hypoxic gas mixtures. The endings of the regenerated sinus nerve could not be distinguished from the genuine nerve endings by their fine structure. Chemoreception is thus not an intrinsic property of sinus nerve fibers. Non-nervous factors, which are present in the carotid body must, therefore, be essential for the transduction of adequate stimuli to the nerve impulse.

Thus after underlining that the nerves of the carotid body are mainly afferent and without the property of intrinsic chemoreception, one may turn to the actual subject matter of the present section: the ultrastructure of the fibres and their contacts.

Nerve fibres and nerve endings may be classified according to several criteria, when evaluated from random section studies only. According to their course in relation to Type I cells they may thus be divided into 'en passant' contacts and terminal contacts. According to their two-dimensional form into basket-formed, bulbous, button-like and calyciform. According to their size into small and large terminals and finally after their organelle content into nerve endings dominated by small clear vesicles and those dominated by small mitochondria. Three-dimensional reconstructions have indicated that these classifications are all somewhat misguiding. The above mentioned subdivisions can be observed in sections of the same nerve fibres at different levels.

The longest series of ultra-thin sections of nerve branches analyzed consisted of 420 consecutive sections from the central portion of a lobule of the carotid body of a rat (5). The series contained 19 Type I cells and 5 Type II cells. The reconstruction contained 4 small nerve branches and one large branch (Fig. 11). The nerve had contact with 12 of the 19 Type I cells. The sites of contact were both of the 'en passant' and terminal types. The one and same nerve branch could have several contacts with the same Type I cell or several Type I cells.

Nerve terminals were reconstructed and analyzed in the next longest ultrastructural serial sections published (25). These consisted of 200 consecutive sections through 15 Type I cells. Twenty nerve contacts were found, representing a continuum from small terminals to large calyciform enlargements, covering up to 40 per cent of the surface of one Type I cell (Fig. 12).

The intracellular space between Type I cells and nerve endings is approximately 15 nm wide. Symmetrical and asymmetrical plasma membrane specializations are formed on the opposed membranes at limited point-shaped sites. Symmetrical densities on the membranes are the most common finding (desmosomes).

At other sites the densities form projections on one side of the membranes only. These dense projections are conical in shape and extend approximately 30 nm down into the cytoplasm. The membrane may be slightly thickened and fuzzy on the opposing side. Asymmetrical membrane specializations are found with the dense projections on both the Type I cell and nerve fibre sides. Clusters of small clear vesicles often occur in relation to dense projections (Fig. 13).

Dense projections on Type I cell plasma membranes plus small clear vesicles comprise 95 per cent of asymetrical membrane specializations. The remaining 5% are on the nerve fibre plasma membrane (2). Both types may be found in the same junction with a mutual distance of less than 200 nm (20).

It is presumed, based on knowledge of synapses in the central nervous system with known impulse pathways, that these junctions must be chemical synapses. Thus small clear vesicles contain a transmitter substance and are positioned on the presynaptic side together with the dense projections.

If this functional scheme is employed in respect of the carotid body, then it means that 95% of the asymmetrical junctions are afferent and that they release a transmitter from Type I cells which acts on the nerve endings. However, in the central nervous system, chemical synapses are also found with postsynaptic accumulations of small clear vesicles. Structures have also been found with the classical morphology of chemical synapses, but having no actual transmission. Thus conclusions cannot be drawn from structure to function in respect of chemical synapses.

It should be emphasized that the chemical composition of the content of small clear vesicles in Type I cells is unknown. As has been described in the section on the ultrastructure of Type I cells, an explanation can be provided for the presence of small clear vesicles, in that they participate in exocytosis-vesiculation process. The small clear vesicles represent, in this theory, a membrane retrieval mechanism, the object of which is to recirculate the surplus plasma membrane after exocytosis.

According to the classical theory, 5% of the synapses are efferent, i.e. transfer a transmitter substance from the nerve terminals to Type I cells. If this interpretation is *not* accepted, then an explanation must be found for the function of the membrane specializations with their dense projections on the nerve side and small clear vesicles in the nerve terminals.

It has been shown, ontogenetically, (2) with regard to the dense projections, that the relative ratio of dense projections on the nerve fibres side to Type I cell side is a dynamic process; with maturation an increase occurs in the number of dense projections on the Type I cell side.

Reverting to the presence of small clear vesicles in the afferent nerves, similar vesicles are present in nerves, the afferent function of which is well-documented, for example in nerves in muscle tendons and hair follicles. The mean concentration of small clear vesicles in the nerve terminals of the carotid body falls 27% after hypoxia or hypercapnia (26). This was interpreted by the author as indicating a release from the afferent nerves of a vesicular-bound transmitter with an excitory action on Type I cells. However, this assumption is purely hypothetical.

The chemical content of small clear vesicles is unknown. It has recently been rendered probable that the nerves contain vasoactive intestinal polypeptide (VIP) and substance P (12, 13). The localization and function within the carotid body of these neuropeptides is unknown.

6. Putative neurotransmitters

There is no doubt that catecholamines, dopamine and noradrenaline are located in the carotid body (27). Mass fragmentographic studies have shown that dopamine is the major amine stored, at least in the rat (14). The catecholamines are stored in the DCV. Denervation experiments, however, have shown that about 50% of the noradrenaline disappeared after sympathetic ganglionectomy (28). Thus a considerable amount of the noradrenaline must be within the sympathetic nerves. Short-term hypoxia releases dopamine selectively from the carotid body (15). Long-term hypoxia produces a large increase in dopamine and noradrenaline (29). This increase is augmented in ganglionectomized rats thus indicating a local regulating control of the noradrenergic nerves on dopaminergic as well as noradrenergic glomus cells.

The location of acetylcholine is less well-defined. This chemical is not affected by sectioning of the carotid sinus nerve, showing that acetylcholine is not contained in the nerves reaching the organ (30). Thus acetylcholine may be located to the glomus cells, coexisting with the catecholamines, and/or in the sustentacular cells.

The location of some neuropeptides has recently been elucidated by immunohistochemistry (12, 13).

Studies have shown that the Type I cell can be stained by enkephalin antisera, but not with VIP or substance P antisera. The occurrence of enkephalin and catecholamine within the same glomus cell is a likelihood, as neuropeptides and other transmitters have been shown to co-exist in other cell systems.

7. Physiomorphological correlations under stimulated and unstimulated conditions

The carotid body enlarges in man and in animals under chronic hypoxic conditions (29, 31). Morphometric studies (rats) have shown this enlargement to be due to a dilatation as well as new growth of capillaries and an increase in the total volume of the glomus cells.

Attempts have been made to determine, at the ultrastructural level, the specific structures which respond to the stimulus of hypoxia. Table 1 shows the conflicting results in respect of the fate of DCV during acute hypoxia. The only study, however, where stereological measurements were consequently used, is that by Hellström (10), and it indicated a reduction in the DCV. This is interesting when compared to the observed decrease in the catecholamine content under similar hypoxic conditions.

Table 2 shows the results of ultrastructural studies during chronic hypoxia; there is no agreement here either. One morphometric study on randomized sections (9) indicated an increase in the mean diameter of the DCV. As the catecholamines of the rat carotid body have been observed to multiply under chronic hypoxia (29), it is tempting to assume that these phenomena are reflected in the carotid body morphology by an increase in the mean diameter of the DCV, thus indicating an increased storage capacity for catecholamines.

8. Concluding remarks

Several questions remain to be answered within the sphere of carotid body research. The ultrastructure of individual cellular elements is well established, but their mutual relationship, e.g., the nerve terminals versus Type I, versus Type II cells, still has to be studied. This could be achieved by three-dimensional analysis.

Quantitation is essential in future morphological studies, where random samples taken under experimental conditions should be compared to adequate controls from the same species. Morphometry may provide an answer to the question whether dense cored vesicles in the chemosensory process decrease or increase in number during hypoxia. No conclusive evidence is available as yet with regard to a release mechanism for dense cored vesicles via exocytosis. Exocytotic profiles have been observed during stimulation, but only under extreme, non-physiological conditions. Up to the present no investigation has unequivocally shown the site of extrusion, should exocytosis occur. Thus the question remains as to whether it takes place at the perikaryon, cell processes, close to the blood vessels or close to the nerve endings.

Although various putative transmitters have been identified and quantitated in the carotid body, no metabolizing enzymes have been detected which could be responsible for the breakdown of catecholamines. Another unanswered question is which neurochemical is stored in the small clear vesicles of the afferent nerve endings.

The involvement of Type II cells in the chemosensorytransduction is obscure. It is still unknown whether these cells have properties other than simple support. A recent study has indicated that they may govern the specificity of the nerve contacts between

Table 1. Effects of acute hypoxia on the ultrastructure of Type I cells.

Species	Exposure to hypoxia	Effects	Reference
Rat	200 mm Hg	Disappearance of DCV	(32)
Rat	2.5% O_2 in N_2 10 min	Increased number of exocytotic profiles swelling of mitochondria and nuclei	(33)
Cat	9% O_2 in N_2 45 min	No significant change or a possible increase of the DCV; less opaque mitochondria with widened cristae	(34)
Syrian hamsters	5% O_2 in N_2 20 min-6h	No significant change of the DCV	(35)
Cat	5% O_2 in N_2 3 hrs	*In vivo* no effects on ultrastructure; *in vitro* a possible depletion of the DCV	(36)
Rat	9% O_2 in N_2 30 min	No effect on the DCV; enlargement of the mitochondria	(20)
Rat	5% O_2 in N_2 30 min	Decrease in volume density for DCV swelling of mitochondria	(10)

Table 2. Effects of chronic hypoxia on the ultrastructure of Type I cells.

Species	Exposure to hypoxia	Effects	Reference
Guinea pigs	High altitude about 4750 m	Vacuolation of the DCV	(37)
Rabbit	High altitude about 4000 m and simulated high altitude about 6000 m 7 days	For both groups increased DCV and of mitochondria	(38)
Rat	Simulated high altitude about 7000 m 103 days	Decreased number of DCV swelling of mitochondria	(39)
Rat	Simulated high altitude about 4300 m 27–35 days	Mean diameter of DCV increased, number of mitochondria increased	(9)

the afferent nerve endings and Type I cells (39). Perhaps they undergo dynamic changes, like a shutter, sometimes protecting Type I cells, at others exposing the Type I cell surfaces to agents emanating from the surrounding tissue.

Future studies on the carotid body must combine various methods, including structural analysis and controlled quantitation in order to correlate structure to function.

References

1. Kjaergaard J: Anatomy of the Carotid Glomus and Carotid Glomus-like Bodies (Non-chromaffin Paraganglia), Copenhagen, FADL's Forlag, 1973.
2. Kondo H; Innervation of the chief cells of the carotid body: An ultrastructural review. Arch Histol Jap 40: Suppl 221–230, 1977.
3. Verna A: Ultrastructure of the carotid body in the mammals. Int Rev Cytol 60: 271–330, 1979.
4. Seidl E: On the morphology of the vascular system of the carotid body of cat and rabbit and its relation to the glomus Type I cells. In: The Peripheral Arterial Chemoreceptors, Purves MJ (ed), Cambridge, Cambridge University Press, 1975, pp 293–299.
5. Kondo H: Innervation of the carotid body of the adult rat. A serial ultrathin section analysis. Cell Tiss Res 173: 1–15, 1976.
6. Kjaergaard J, Hellström S: Freeze cleaving of the unstimulated carotid glomus. Proceedings of the 9th Congress of the Nordic Society for Cell Biology, Odense, Odense University Press, 1976.
7. Kondo H: Evidences for the secretion of chief cells in the rat carotid body. In: Arterial Chemoreceptors, Belmonte C, Pallot DJ, Acker H, Fidone S (eds), Leicester, Leicester University Press, 1981, pp 45–53.
8. Hellström S: Morphometric studies of dense-cored vesicles in Type I cells of rat carotid body. J Neurocytol 4: 77–86, 1975.
9. Laidler P, Kay JM: A quantitative study of some ultrastructural features of the type I cells in the carotid bodies of rats living at a stimulated altitude of 4300 metres. J Neurocytol 7: 183–192, 1978.
10. Hellström S: Effects of hypoxia on carotid body Type I cells and their catecholamines. A biochemical and morphologic study. In: Chemoreception in the Carotid Body. Acker H, Fidone S, Pallot D, Eyzaguirre C, Lübbers DW, Torrance RW (eds), Berlin, Springer-Verlag, 1977, pp 122–127.
11. Stensaas LJ, Stensaas SS, Gonzalez Ç, Fidone SJ: Analytical electron microscopy of granular vesicles in the carotid body of the normal and reserpinized cat. In: Arterial Chemoreceptors, Belmonte C, Pallot DJ, Acker H, Fidone S (eds). Leicester, Leicester University Press, 1981, pp 176–186.
12. Wharton J, Polak JM, Pearse AGE, McGregor GP, Bryant MG, Bloom SR, Emson PC, Bisgard GE, Will JA: Enkephalin-, VIP-, and substance P-like immunoreactivity in the carotid body. Nature 284: 269–271, 1980.
13. Lundberg J, Hökfelt T, Fahrenkrug J, Nilsson G, Terenius L: Peptides in the cat carotid body (glomus caroticum): VIP-, Enkephalin-, and substance P-like immunoreactivity. Acta Physiol Scand 107: 279–281, 1979.
14. Hellström S, Koslow SH: Biogenic amines in carotid body of adult and infant rats, a gas chromatographic-mass spectrometric assay. Acta Physiol Scand 93: 540–547, 1975.
15. Hanbauer I, Hellström S: The regulation of dopamine and noradrenaline in the rat carotid body and its modification by denervation and by hypoxia. J Physiol Lond 282: 21–34, 1978.
16. Glenner GG, Grimley PM: Tumors of the Extra-Adrenal Paraganglion System (Including Chemoreceptors), Washington D.C., Armed Forces Institute of Pathology, 1974.
17. Seidl E, Schäfer D, Zierold K, Acker H, Lübbers DW: Light-microscopic and electron-microscopic studies on the morphology of cat carotid body. In: Chemoreception in the Carotid Body, Acker H, Fidone S, Pallot D, Eyzaguirre C, Lübbers DW, Torrance RW (eds), Berlin, Springer-Verlag, 1977, pp 1–6.
18. de Kock L, Dunn AEG: Electron-microscopic investigation of the nerve endings in carotid body. In: Arterial Chemoreceptors. Torrance RW (ed), Oxford, Blackwell Scientific Publications, 1968, pp 197–187.
19. Böck P, Gorgas K: Catecholamines and granule content of carotid body Type I-cells. In: Chromaffin, Enterochromaffin and Related Cells, Coupland RE, Fujita T (eds), Amsterdam, Elsevier Scientific Publishing Company, 1976, pp 355–374.
20. McDonald DM, Mitchell RA: The innervation of glomus cells, ganglion cells and blood vessels in the rat carotid body: A quantitative ultrastructural analysis. J Neutrocyt 4: 177–230, 1975.
21. Grönblad M, Åkerman KE, Eränkö O: Ultrastructural evidence of exocytosis from glomus cells after incubation of adult rat carotid bodies in potassium-rich calcium-containing media. Brain Res 189: 576–581, 1980.
22. Echeverria OM, Vázquez-Nin GH, Aguilar R: Cytochemical study on the mechanism of secretion of catecholamines. Acta Anat 100: 51–60, 1978.
23. Smith PG, Mills E: Physiological and ultrastructural observations on regenerated carotid sinus nerves after removal of the carotid bodies in cats. Neuroscience 4: 2009–2020, 1979.
24. Leitner L-M, Roumy M, Verna A: Further studies on the cryodestruction of the rabbit carotid body. In: Arterial Chemoreceptors, Belmonte C, Pallot DJ, Acker H, Fidone S (eds), Leicester, Leicester University Press, 1981, pp 85–89.
25. Nishi K, Stensaas KJ: The ultrastructure and source of nerve endings in the carotid body. Cell Tiss Res 154: 303–319, 1974.
26. McDonald D: Ultrastructural changes in sensory nerve endings accompanying increased chemoreceptor activity: A morphometric study of the rat carotid body. In: Chemoreception in the Carotid Body, Acker H, Fidone S, Pallot D, Eyzaguirre C, Lübbers DW, Torrance RW (eds), Berlin, Springer-Verlag, 1977.
27. Alfes H, Kindler J, Knoche H, Matthiessen D, Möllmann H, Pagnucco R: The biogenic amines in the carotid body. Prog Histochem Cytochem

10: 1–69, 1977.

28. Hellström S, Hanbauer I: Role of dopamine and norepinephrine in carotid body. In: Catecholamines: Basic and Clinical Frontiers, Usdin E, Kopin IJ, Barchas JD (eds), Pergamon Press, New York, 1979, pp 1539–1541.
29. Hanbauer I, Karoum F, Hellström S, Lahiri S: Effects of hypoxia lasting up to one month on the catecholamine content in rat carotid body. Neuroscience 6: 81–86, 1981.
30. Hellström S: Putative neurotransmitters in the carotid body: Mass fragmentographic studies. Adv Biochem Psychopharmacol 16: 257–263, 1977.
31. Arias-Stella J: Human carotid body at high altitude. Amer J Path 55: 82a–83a, 1969.
32. Hoffman H, Birrell JHW: The carotid body in normal and anoxic states: An electron microscopic study. Acta Anat 32: 297–315, 1958.
33. Blümcke S, Rode J, Niedort HR: The carotid body after oxygen deficiency. Z Zellforsch 80: 52–77, 1967.
34. Al-Lami F, Murray RG: Fine structure of the carotid body of normal and anoxic cats. Anat Rec 160: Suppl 4, 697–718, 1968.
35. Chen I-Li, Yates RD, Duncan D: The effects of reserpine and hypoxia on the amine-storing granules of the hamster carotid body. J Cell Biol 42: 804–816, 1969.
36. Zapata P, Hess A, Bliss EL, Eyzaguirre C: Chemical, electron microscopic and physiological observations on the role of catecholamines in the carotid body. Brain Res 14: 473–496, 1969.
37. Edwards C, Heath D, Harris P: Ultrastructure of the carotid body in high-altitude guinea-pigs. J Path 107: 131-152, 1972.
38. Møller M, Møllgård K, Sørensen SC: The ultrastructure of the carotid body in chronically hypoxic rabbits. J Physiol 238: 447–453, 1974.
39. Blessing MH, Kaldeweide J: Light and electron microscopic observations on the carotid bodies of rats following adaptation to high altitude. Virchows Archiv B: Cell Pathol 18: 315–329, 1975.
40. Kondo H, Pappas GD: The effect of postsynaptic deprivation on the presynaptic chief cell in the rat carotid body. Brain Res 215: 125–133, 1981.

Authors' addresses:
Sten Hellström
Department of Anatomy
University of Umeå
S-901 87 Umeå, Sweden

Johan Kjaergaard
Winslow Institute of Human Anatomy
University of Odense
DK-5230 Odense, Denmark

CHAPTER 15

Ultrastructural features of the mammalian adrenal medulla

REX E. COUPLAND

1. Introduction

The term adrenal medulla can only be applied with semantic verisimility to chromaffin cells aggregated to form a distinct mass within the adrenal glands of mammals. In lower forms the relationship of chromaffin and cortical (interrenal) elements is often less intimate, as in fishes and amphibia, or the chromaffin elements form multiple small islets scattered throughout the adrenal gland (interrenal tissue) in birds or exist as scattered islets or as a continuous envelope of cells associated with adrenal cortical tissue in reptiles (1–5).

The present account will relate to eutherian mammals only since scant attention appears to have been paid to monotremes and marsupials. Inevitably the majority of work has been performed on common laboratory animals with more recently an extension to primates, but to date no marked order or species difference has been noted with eutheria. It should be stressed that the adrenal chromaffin cells form only one part of the chromaffin system – the remainder being extra-adrenal where cells may form large discrete encapsulated bodies (paraganglia, (6) para-aortic bodies, (1, 7)) or exist as isolated cells or small groups within or adjacent to pre or paravertebral sympathetic ganglia or along the course of peripheral sympathetic nerve fibres (see Chapter 16).

The term chromaffin cell is used to designate an element that synthesizes and stores catecholamines in sufficient concentration to give a positive chromaffin reaction following fixation in aqueous solutions containing a fixative aldehyde and potassium dichromate. Adrenal medullary chromaffin cells and some extra-adrenal elements are innervated by preganglionic sympathetic nerve fibres.

In post-natal mammals the predominant catecholamine in the adrenal medulla is adrenaline although a variable but smaller amount of noradrenaline is always present. Some workers (8) have reported the presence of dopamine in the adrenal medulla but the amount recorded was relatively small (less than 1% of total catecholamines) and in our recent experience using high performance liquid chromatography with electrochemical detection it is less than 0.5% in normal mouse, rat, guinea-pig and marmoset adrenal medulla; in consequence it probably exists only in the role of a precursor of later stages of noradrenaline and adrenaline synthesis and not as a stored amine. Hence chromaffin cells of the post-natal mammal are concerned with the synthesis and storage and secretion of adrenaline and noradrenaline and the ultrastructure of the adrenal medulla reflects these facts as well as its neuroectodermal origin.

2. Historic background and development

Lever (9) published the first electron micrographs of the adrenal medulla and these demonstrated the presence of chromaffin granules, some with a peripheral limiting membrane, together with mitochondria within the cell cytoplasm. Following improvements in methods of fixation and in the design and resolution of electron miroscopes detailed accounts of the ultrastructure of the adrenal medulla of the rat were published by Coupland (1, 10, 11) and Elfvin (12) while the ultrastructure of the adrenal medulla and extra-adrenal chromaffin bodies was described by Coupland and Weakley (13); later Coupland (2, 14) commented on the ultrastructure of the adrenal chromaffin cells of a variety of other species.

Observations on the development of chromaffin cells in the rat adrenal medulla were published by Elfvin (12), Ratzenhofer and Müller (15) Daikoku et al. (16) El-Maghraby and Lever (17), Millar and Unsicker (18) and Diner (19) while the development of both adrenal medullary and extra-adrenal chromaffin cells was described by Coupland and Weakley (20) and Hervonen (21) in the rabbit and man respectively. From these various descriptions it is apparent that all stages in the development of chrom-

Motta, PM (ed): Ultrastructure of endocrine cells and tissues. ISBN-13: 978-1-4613-3863-5

affin cells from primitive sympathetic cells though phaeochromoblasts to typical chromaffin cells can be identified using electron microscopy. The small rounded primitive sympathetic cells have a high nuclear-cytoplasmic ratio, many nucleoli and no, or few amine storage granules. Phaeochromoblasts in some species and particularly in extra-adrenal situations may be elongated with respect to both cytoplasm and nucleus. In other forms and particularly in the adrenal medulla they are more polyhedral with a rounded nucleus that loses chromatin as it differentiates and there is a general increase in number and size of cytoplasmic amine storage granules which show ultrastructural features of primary amine or adrenaline storage before birth (22). Primitive sympathetic cells (sympathochromaffin anlage), phaeochromoblasts and phaeochromocytes (chromaffin cells) are closely associated with nerve fibres and supporting cells. During the pre-natal development of the chromaffin cell various intermediate forms between typical primitive cells, phaeochromoblasts and chromaffin cells exist and typical noradrenaline (NA) and adrenaline (A) storing cells were not observed in the rat adrenal medulla before the 21st day of gestation by Millar and Unsicker (18) and El-Maghraby and Lever (17) reported that all chromaffin cells contained a mixed population of adrenaline (A) and noradrenaline (NA) storing granules up to

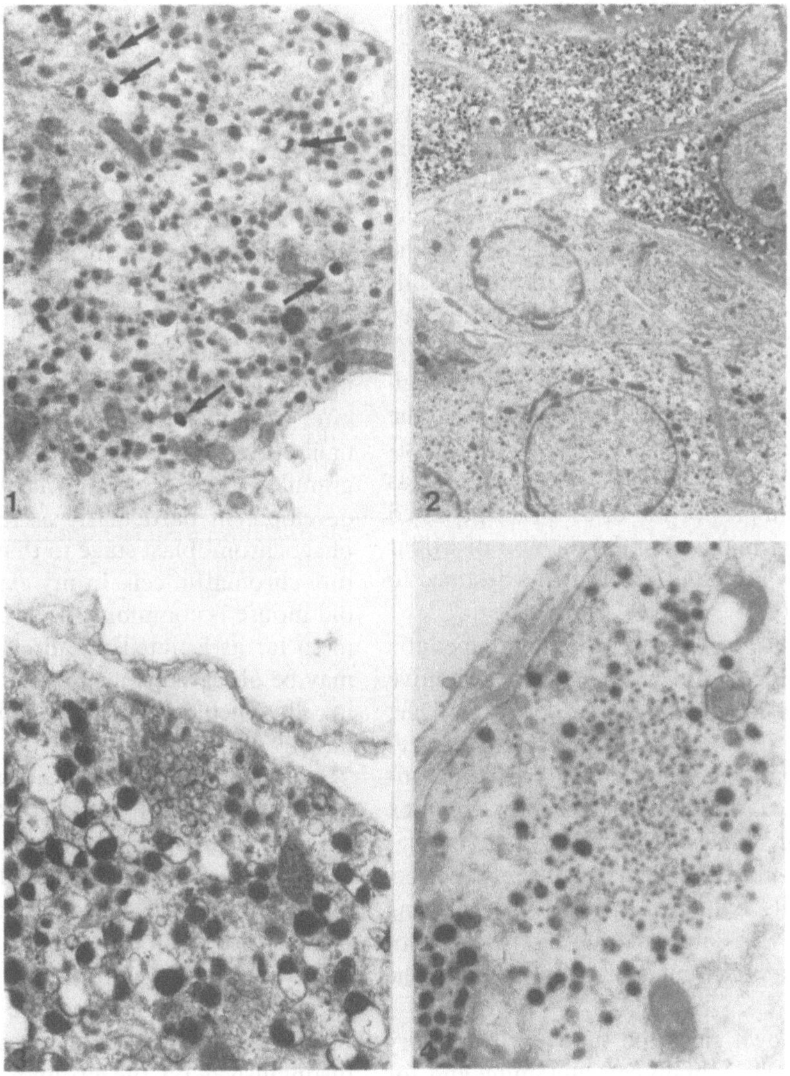

Figs. 1–4. (1) A and NA (arrow) chromaffin granules in MGA cell in mouse adrenal (× 3,175). (2) NA (top), SGC (centre) and A (bottom) cells in rat adrenal medulla (× 12,880) (3) Mouse SGC cell, fixed by glutaraldehyde/osmium sequence, showing synaptic-type vesicles adjacent to plasma membrane (× 25,900). (4) Mouse SGC cell, fixed in dichromate/chromate showing electron dense contents of synaptic-type vesicles (× 23,800).

birth and only developed into typical A cells and NA cells in the postnatal phase. In the rabbit, mixed populations of A and NA granules were noted in a few chromaffin cells late in foetal life though the majority of cells possessed only A granules and a minority stored only NA – as evidenced by granule structure and electron density (14, 22). Non-myelinated nerve fibres and associated Schwann cells were observed with evidence of synaptic contacts with developing chromaffin cells at 18 days gestation in the rabbit (20) and rat (17) while Millar and Unsicker (18) observed primitive synaptic contacts between nerve fibres and developing chromaffin cells in the rat at 15 days gestation. In man synapses on adrenomedullary cells were noted by Hervonen (21) from the 12th week of intrauterine life.

In the recent study of Millar and Unsicker (18) acetycholinesterase (AChE) histochemistry was applied to the developing and adult adrenal gland of the rat and revealed a strong reaction deposit, indicative of high AChE activity, throughout the endoplasmic reticulum of small granule containing chromaffin (SGC) cells and in sympathetic neurones. Furthermore, the same workers reported an absence of the AChE reaction product in developing adrenal medullary cells and nerve fibres at 15 days of gestation but noted that by 17 days some cells (less than 10%) showed a relatively strong AChE reaction. Although a weak AChE reaction developed in cisternae of the endoplasmic reticulum of typical phaeochromoblasts at 19 days and later – the intense reaction was confined to a small proportion of cells that persisted up to birth and beyond and in the opinion of Millar and Unsicker they represent the elements destined to become SGC cells.

Thus the pattern of the future adrenal medulla emerges during foetal life as groups of primitive chromaffin cell precursors invade the cortical anlage, together with nerve fibres, Schwann cells and precursors of sympathetic neurons and become intimately associated with an increasingly complex vascular bed and associated connective tissue elements including fibroblasts, macrophages and occasional mast cells. Scattered islands of developing chromaffin cells with supporting elements join together to form a more circumscribed tissue mass, the adrenal medulla.

It has been known since the turn of the century that chromaffin cells contain catecholamine and for many years it was assumed that the extractable pressor substance was adrenaline. The demonstration by Euler in the 1940s (23) that post-ganglionic sympathetic nerve fibres contain noradrenaline and that primary amine is also present in the adrenal medulla of some mammals (24) was rapidly followed by an attempt to localize adrenaline and noradrenaline within the adrenal medulla. Eränkö (25) noted discrete islets of cells in the rat adrenal medulla which showed fluorescence after fixation in formaldehyde, a negative acid phosphatase and a strongly positive argentaffin reaction and suggested that they represented a different cell type to that of the majority of the parenchyma. Hillarp and Hökfelt (26) introduced the iodate reaction which revealed stained islets of NA cells lying in association with, usually, a much larger mass of A cells. Subsequent work has demonstrated that the relative proportions of A and NA cells within the adrenal medulla approximates reasonably well with the relative proportions of the two amines as determined by assay of adrenal extracts and ranges from 60% to in excess of 90% adrenaline (for review see Coupland (3)).

3. Cell types in the post-natal adrenal medulla

With the development of better instruments and fixation techniques it became possible to distinguish between adrenaline and noradrenaline at the ultrastructural level (22, 27, 28) and it became apparent that in the majority of vertebrate adrenal glands ultrastructurally distinct A cells and NA cells exist after birth though cells containing mixtures of adrenaline-storing and noradrenaline-storing chromaffin granules often occur during the pre-natal phase of development particularly as the cells pass from the phaeochromoblast stage to that of a relatively immature chromaffin cell. In my experience the 3–6 week old mouse is commonly an exception to this pattern in so far as isolated noradrenaline storage granules may be observed (Fig. 1), after appropriate fixation, in cells which otherwise would have been classified as having A storage granules.

In tissues fixed initially in osmium tetroxide it is not possible to differentiate between A and NA cells with any degree of certainty and hence early accounts of the ultrastructure of the adrenal medulla (1, 10) made only passing reference to these two types of cells. However, following the use of the technique involving initial fixation in glutaraldehyde followed by osmication (27) or metal binding with silver (28) or chromium (22) the two types of cells can be readily distinguished on the grounds of differences in appearance and electron density of adrenaline and noradrenaline storage granules. A-storage granules show a limiting peripheral membrane within which granular moderately electron dense material is scattered or aggregated symmetrically – often leaving a distinct peripheral halo adjacent to the membrane. The contents of an A granule represent amine-binding protein but not fixed adrenaline – since this is lost by diffusion during the fixation and dehydration process. NA storage granules contain a substance that exhibits intense homogenous electron density and is usually asymmetrically located within the limiting membrane – probably as a result of the rapid *in situ* formation of a polymer of noradrenaline and glutaraldehyde – and hence the contents represent bound amine plus carrier and other proteins (see below). The chromate/dichromate fixation technique (29) also allows differentiation of A and NA storing granules on grounds of electron density, but aqueous potassium permanganate (30, 31) is of no value in differentiating the two cell types since amines and soluble proteins are lost from the cells on immersion in the fixative. In the writer's experience initial fixation in freshly prepared 3% purified glutaraldehyde buffered with 0.2% cacodylate to pH 7.2 at room temperature (20° C) followed by 1% osmium tetroxide is the most useful fixative; for the best differentiation between the various types of storage granule only lead citrate should be used for staining thin sections since uranyl acetate binds strongly to many cellular proteins, including those contained within the chromaffin granules, and by increasing the overall electron density of both A and NA granules makes differentiation more difficult. Although within a particular species A and NA storage granules may show different size distributions (14). The overlap, in typical chromaffin cells, is so great that granule size is of little use in determining cell type. Recently it has become apparent that in addition to typical A and NA cells a third type of amine storage cell exists (32, 33, 34) in the adrenal medulla in some species, the small granule chromaffin (SGC) cell, and this is characterized by the presence of granules having smaller profile diameters e.g. in the mouse 100–230 nm as compared with 50–500 nm in typical A and NA cells. These are usually situated in the peripheral part (outer zone) of the adrenal medulla. Similar cells had been observed in the adrenal glands of birds and reptiles (35, 36) and in extra-adrenal chromaffin cells of frogs (37). In the writer's experience they are most numerous in the mouse adrenal medulla accounting for some 3–5% of the total of chromaffin cells and present but less common in other forms such as guinea-pig, rat (Fig. 2) and rabbit: they have also been described in the dog adrenal medulla (38).

4. Cell types in the adult adrenal medulla

In the context of the present article adult means sexually mature animals since a gradual transition from the foetal to the adult state occurs during the first few days or weeks after birth in all species. The cellular secretory unit of the adrenal medulla is the chromaffin cell together with its preganglionic sympathetic nerve fibre that ends in synaptic contact with one face of the cell. Additional cellular elements include Schwann cells, sympathetic neurons, a species variable number of islands or cords of cortical cells of the juxta-medullary zone, vascular and connective tissue elements.

In the adult gland after appropriate fixation (as above) the majority of chromaffin cells can be clearly categorized as A or NA cells while a minority conform to typical SGC cells. SGC cells are characterized by a relatively high nuclear cytoplasmic ratio and are often elongated and sometimes exhibit very attenuated processes (34). Not only do they contain chromaffin granules that are smaller in average size and range than those of typical A and NA cells, but the granule contents often show an electron density intermediate between that of typical A and NA granules, a homogeneity of content more in keeping with that of NA granules and with a symmetry within the granule membrane more in keeping with that of an A than NA granule. In the mouse typical SGC cells also contain another characteristic organelle: synaptic-type vesicles with a mean diameter of 50 nm that may lack electron dense contents in glutaraldehyde/osmium tetroxide fixed specimens. These synaptic-type vesicles may lie centrally or peripherally in the cytoplasm and when in the latter situation often suggest the likelihood of discharge into adjacent tissue spaces (Fig. 3). After fixation by the method of Tranzer and Richards (29) electron dense contents typical of noradenergic synaptic vesicles can be identified (Fig. 4), indicating that they contain catecholamine. To date, we have not observed small synaptic-type vesicles in adrenal SGC cells of species ofther than mice.

The differences in the morphology of the granules of SGC cells and those of typical A and NA cells raised the possibility that SGC cells may store a different catecholamine, i.e. dopamine. However, recent application in these laboratories of high performance liquid chromatography followed by electrochemical detection in the mouse, in which SGC cells account for 3–5% of medullary volume during the first three months of post-natal life, failed to detect dopamine in amounts greater than 0.5% of total catecholamine. This level is considered consistent with the role of dopamine as an intermediate in adrenaline and noradrenaline synthesis rather than as an amine stored in a particular cell type. This conclusion is also in keeping with observations on amine synthesis in the mouse following labelling with ^{3}H-dopa (39, 40).

More recent observations by the writer on the mouse adrenal medulla has revealed the presence of cells that contain granules intermediate in size distribution between A cells and NA cells on the one hand and SGC cells on the other. Some contain granules exhibiting moderate electron density and internal granularity typical of A storage (Figs. 5 and 6), while others appear like smaller-than-average NA-storage granules (Fig. 6). The former granules are often elongated and show a higher density distribution in the cytoplasm than those of the typical SGC or A cell (Fig. 3). Due to the elongated shape meaningful measurement of granule size is difficult though in the shorter dimension they range from *c* 50–300μm, and could justifiably be called medium granule A (MGA) cells or intermediate granule A (IGA) cells. Unlike the typical elongated mouse SGC cell referred to above they are polyhedral in shape and less evidently peripheral in distribution in the adrenal medulla; they do not contain synaptic-type vesicles. The cells possessing smaller than average NA storage granules are also polyhedral in shape and occur singly or in groups adjacent to typical A or NA cells; in all respects they justify a designated of medium granule NA cells (MGNA) or intermediate granule NA (IGNA) cells and they do not contain synaptic-type vesicles (Fig. 4).

In guinea-pig SGC cells (Fig. 5) with a mean

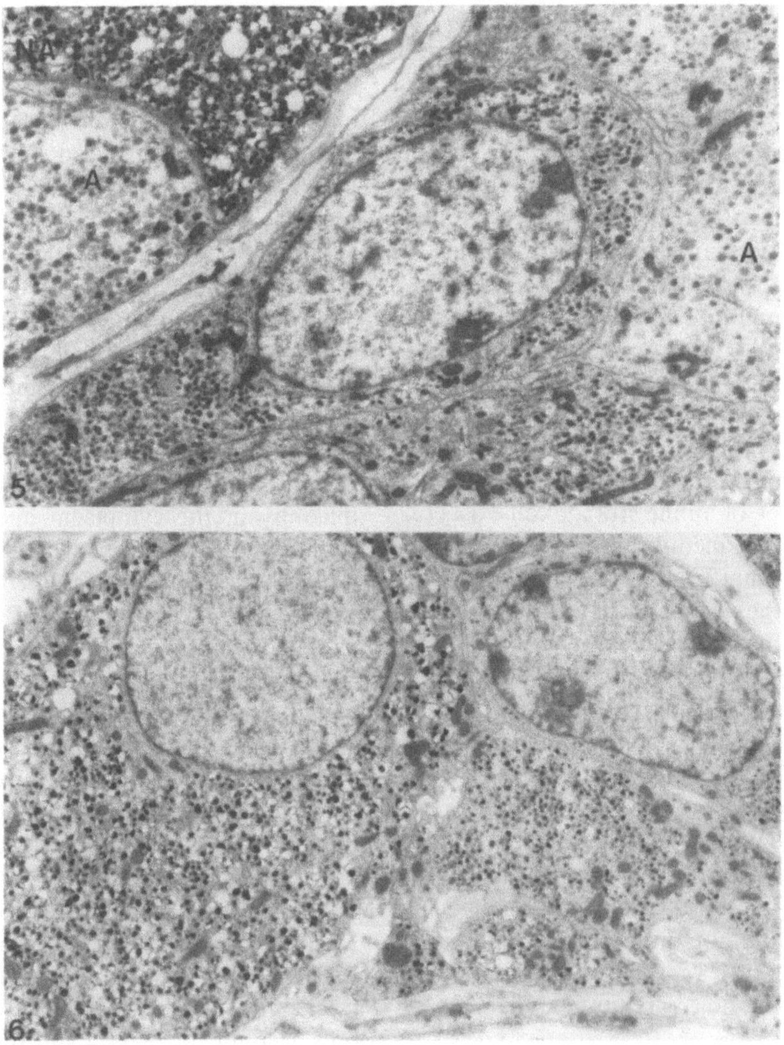

Figs. 5–6. (5) Mouse adrenal medulla showing MGA cells lying adjacent to A and NA cells (× 7,195). (6) Mouse adrenal medulla SGC cell lying adjacent to NA cell (× 4,650).

diameter of the granules approximately half that of the granules of typical chromaffin cells have been observed and while the majority contain symmetrically distributed contents of moderate electron density and internal granularity typical of A storage, occasional ones contain eccentric highly electron dense inclusion typical of NA storage. Small granule cells in the rat sometimes contain granules with characteristics of NA storage while others appear to be A storage granules similar to the MGA cells of the mouse.

The above observations taken together with those of Gorgas and Böck (33) on types of NA cells in the mouse adrenal medulla and of Kajihara et al. (38) on intermediate cell types between small granule cells and A and NA cells in the dog suggest that MGCA and MGCNA cells represent secretory phases of A and NA cells and that the typical mouse SGC cells, which differ from typical chromaffin cells in various nucleocytoplasmic features but especially in possessing two distinct types of cytoplasmic storage granules or vesicles, may represent an intermediate cell type between a typical chromaffin cell and a sympathetic neuron. Cells of this type do not appear to have been observed within the adrenal medulla of other species. However, although the mouse SGC cell may be an intermediate form it would appear that functionally it behaves more like an NA than an A cell as

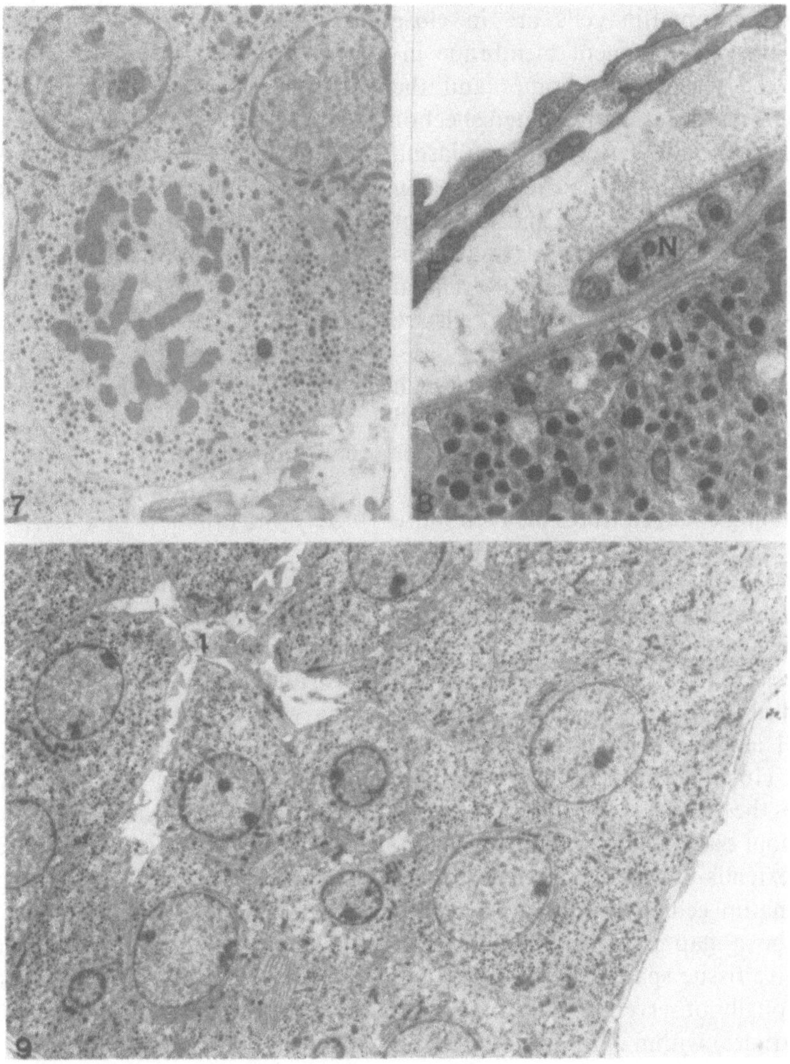

Figs. 7–9. (7) Mitotic figure in A cell of adrenal medulla in 9-week-old mouse (× 4,045). (8) Chromaffin cell in marmoset adrenal medulla lying adjacent to blood capillary (above), note basement membranes, process of fibroblast (F), collagen fibres and unmyelinated nerve fibres (N) associated with Schwann cell cytoplasm (× 10,950). (9) Collection of A cells in mouse adrenal medulla adjacent to blood capillary. Note intercellular spaces, occasional stacks of endoplasmic reticulum (× 2,620).

evidenced by its being unaffected in insulin hypoglycaemia and with respect to amine turnover (41). A further possibility would be that the SGC and MGCA and MGCNA cells may represent elements that have differentiated from primitive precursors only recently. Against this conclusion is the fact that mitotic figures occur only rarely and that when they do they are observed in fully granulated clls such as the A cell illustrated in Figure 7.

5. Chromaffin and supporting cells

In all eutherian mammals chromaffin cells together with intimately associated Schwann cells form a relatively compact cell mass associated with the blood vessels and connective tissue elements. Where Schwann cells and chromaffin cells are in close apposition (c. 20 nm) no basement membrane intervenes between the plasma membranes and the Schwann cell cytoplasm forms an attenuated sheet or tongue-like process partially covering the plasma membrane of the chromaffin cell and interposed in some situations between the latter and adjacent connective tissue elements. Where the plasma membranes of chromaffin cells directly abut general connective tissue spaces and where Schwann cells are related to connective tissue elements a distinct basement membrane intervenes (Fig. 8). In primates such as marmosets and in man, covering Schwann cell cytoplasm, although present, appears to be less abundant than in rodents and the perivascular connective tissue including collagen fibres is much more abundant.

A particular feature of the arrangement of chromaffin cells in the adrenal medulla is the presence of intercellular canaliculi (Figs. 6 and 9) free from connective tissue elements and containing microvillous processes and cilia derived from adjacent chromaffin cells as well as bare nerve fibres leading to synaptic endings (10, 11). As they border these intercellular spaces the plasma membranes of chromaffin cells are without basement membrane but basement membrane extends between the external faces of adjacent chromaffin cells to bridge across what could otherwise be a gap opening into the conventional connective tissue spaces.

Cilia are occasionally observed arising from basal bodies (distal centrioles) within chromaffin cells and extending for a variable distance into an adjacent intercellular space. Fibrils within them often show an irregular arrangement and usually show pairs in an 8 + 2 pattern (Fig. 10) although the range has varied from 9 + 2 to 6 + 2 (2, 10). In consequence it has been concluded that they are more probably atavistic motor cilia rather than sensory cilia. Their relative infrequency suggests that they are of little functional importance.

Nerve fibres are commonly and intimately associated with both Schwann cell elements and chromaffin cells and nerve endings are observed in synaptic contact with chromaffin cells. Both myelinated and non-myelinated nerve fibres are observed and the latter are associated with Schwann cell cytoplasm except when they lie in intercellular spaces or in contiguity with adjacent chromaffin cells; non-myelinated elements are by far the most numerous. The nerve fibre net is more evident in relation to NA cells than A cells. Although a nerve ending may appear to synapse with a single chromaffin cell some have been observed in synaptic contact with up to three adjacent cells. Characteristically the endings contain small synaptic vesicles with clear centres and larger vesicles with moderately electron dense contents typical of peptide storage vesicles. The author has never observed nerve endings adjacent to blood vessels within the adrenal medulla, including medullary arteries, although they are often associated with cortical vessels. Small groups of ganglion cells, intimately related to Schwann cells and chromaffin cells are a constant but occasional feature in all species examined by the writer.

6. Chromaffin cells

6.1. Cytoplasmic organelles

As reported previously(1, 10, 11, 14) chromaffin cells possess the usual admixture of cell organelles including a rounded or ovoid nucleus, with fine diffusely distributed chromatin and 1 to 3 nucleoli, mitochondria with well defined cristae and a well developed and usually paranuclear Golgi apparatus (Fig. 11 and 12) which may be associated with centrioles or cilia (Fig. 10): in some instances the distal centriole can be seen to form the basal body of a cilium. Isolated stacks of rough endoplasmic reticulum are observed and may be paranuclear in position or lie in the more peripheral cytoplasm (Fig. 9), though some cisternae associated with ribosomes can usually be demonstrated in the vicinity of the Golgi complex. Dark bodies (lysosomes) are usually evident in profiles of chromaffin cells and are variable in size and number; they are usually between 400–800 nm in diameter and contain acid phosphatase (42, 43) as do the contents of some of the more peripheral Golgi tubules and vesicles from which they appear to arise.

According to some authors (44) peroxisomes also occur. Multivesicular bodies are a common feature in chromaffin cells. Although not numerous and usually identifiable only in longitudinal section small numbers of microtubules or filaments (Fig. 12) with a diameter of 16–18 nm are a constant feature in both peripheral and central cytoplasm including the Golgi zone and Redburn et al. (45) have demonstrated the presence of tubulin in the adrenal medulla that has many of the characteristics of that present in the brain.

6.2. The chromaffin granule

The characteristic feature of the normal chromaffin cell is of course the chromaffin granule and in consequence it is of particular interest that under conditions of extreme stress (46) or a combination of the antiandrogen cyproterone acetate, in combination with the stress of handling and administration (47) (and Coupland, unpublished) it is possible to induce almost total loss of storage granules.

Chromaffin granules are membrane bound and

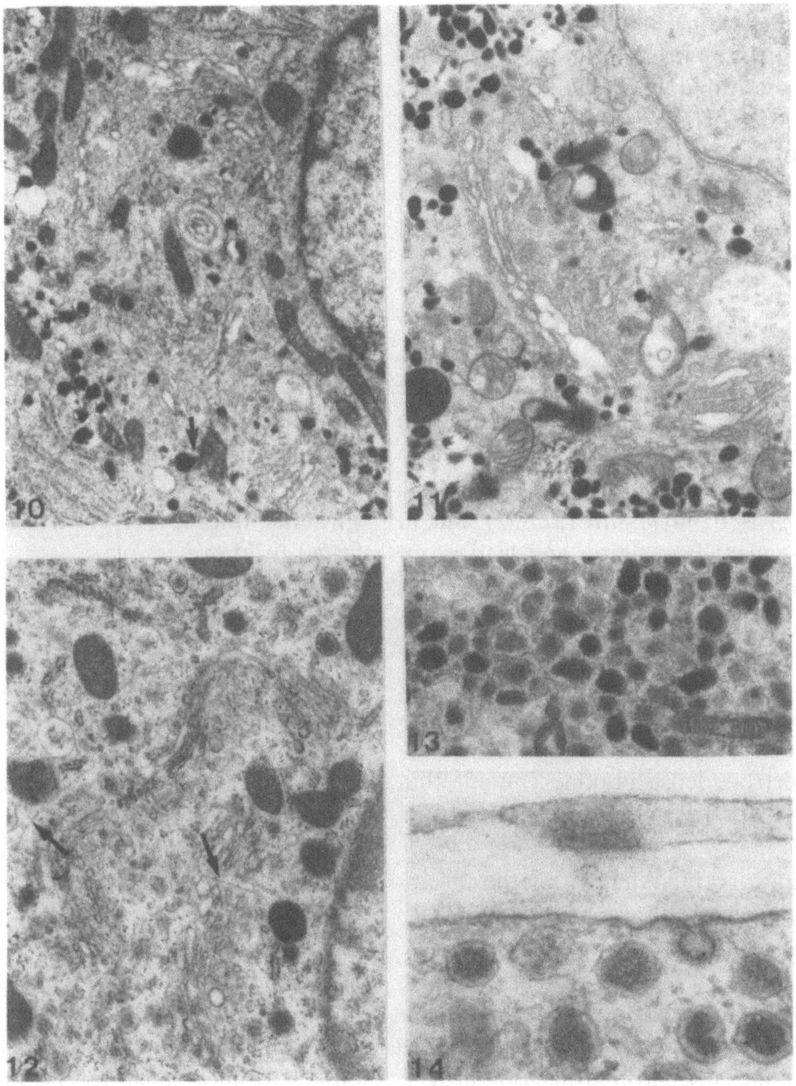

Figs. 10–14. (10) Section through Golgi apparatus of mouse NA cell fixed by glutaraldehyde/osmium sequence. Note 8 + 2 cilium and highly electron dense granules of various sizes in Golgi vesicles and tubules also possible fusion of granules (arrow (× 11,200). (11) Golgi apparatus of mouse NA cell fixed in dichromate/chromate. Note highly electron dense and less electron dense deposits in vesicles. L = lysosome. Some granule profiles in the periphery suggest the possibility of granule fusion (× 18,900). (12) Golgi apparatus of rat A cell showing absence of highly electron dense contents in vesicles etc. L = lysosome. Microtubule = arrow (× 18,900). (13) Chromaffin granules in marmoset adrenal chromaffin cell showing symmetry of contents but variable electron density (× 12,600). (14) Chromaffin cell below shows coated vesicle arising from plasma membrane of mouse A cell. Fenestrated endothelium of capillary is seen above with basement membranes interposed between it and the chromaffin cell (× 63,000).

possess contents that are electron dense – varying in degree from moderate to high – and may be homogenous or granular. By appropriate treatment (4, 10, 11, 14) the contents can be disorganized or extruded and following fixation show a complex substructure that suggests aggregated subunits. During the past twenty years chromaffin granules and their content have been analyzed in great detail and much of the molecular organization determined. This work has been extensively reviewed relatively recently (48, 49, 50) with reference to membranes, contents, including protein, lipid and carbohydrate together with nucleotide, and biogenesis of the chromaffin granule. The most abundant proteins in chromaffin granules are chromogranin A, met-and leu-encephalin and dopamine-β-hydroxylase. Although dopamine-β-hydroxylase forms a major insoluble component of the membrane this same enzyme and other proteins listed occur in soluble form within the granule matrix and will be fixed, in whole, or in part by appropriate fixatives especially glutaraldehyde, during normal fixation for electron microscopy. This denatured protein accounts or the granular content of A cell granules after fixation in glutaraldehyde since over 90 percent of adrenaline diffuses out during the fixation and dehydration procedure (22). By adding potassium dichromate to the initial fixative at least some of the adrenaline (probably 20–30%) is fixed *in situ* (51, 52) and in consequence subsequent osmication and lead staining results in a more electron dense and homogenous granule content though this never approaches the electron density or homogeneity of the NA storage granule, in which almost instantaneous polymer formation, between glutaraldehyde and noradrenaline, results in a usually eccentrically placed homogenous highly electron dense granule content that may appear to have fused with the peripheral limiting membrane.

It is now generally accepted that within the chromaffin granule catecholamines are bound mainly to ATP and chromogranin. By using a single intravenous pulse of ^{3}H-leucine it has been possible to follow the movement of labelled peptides or proteins through the endoplasmic reticulum to the Golgi membranes and finally to chromaffin granules (39, 53): these observations together with evidence reviewed by Winkler (1977) demonstrates the pathway taken by the protein component of the granule during biogenesis. However, evidence derived from the localization of recently synthesized catecholamines within the cytoplasm of chromaffin cells following pulse labelling with ^{3}H-dopa (39, 40) is in keeping with the synthesis of catecholamines in the cytoplasm unrelated to endoplasmic reticulum or Golgi complex (54, 55, 56). Thus it would appear that the moderately electron dense and highly electron dense material present in some Golgi tubules or vesicles represents the protein component of the chromaffin granule that has, in common with more peripheral chromaffin granules, already taken up a significant concentration of catecholamine. Some of the less electron dense granular contents of the Golgi tubular and vesicular profiles represent lysosomal proteins. In NA cells fixed by the conventional glutaraldehyde/osmium tetroxide technique and by chromate/dichromate method (29) and stained with lead citrate only, vesicles (Figs. 10 and 11) containing highly electron dense material can be seen apparently pinching off from Golgi tubules, and other granules of varying size and homogenous moderate to high electron density are seen in the immediate vicinity of the Golgi tubules or cisternae in line with the membrane bound spaces and on the convex and concave faces of the complex. The electron density of these vesicles indicates that they are recently formed chromaffin granules containing noradrenaline. In A-cells (Fig. 12) no highly electron dense material is seen in the Golgi complex indicating that, in fully differentiated A cells, the recently formed chromaffin granules do not pass through a stage in which only noradrenaline is stored. This is entirely in keeping with the findings of electron autoradiography referred to above.

Within an individual chromaffin cell a histogram of the diameters of chromaffin granules (14) together with their appearance and electron density suggests that one is dealing with a sized population ranging from small to large granules. Furthermore within an individual cell there is no zonal distribution according to size and granules of all sizes may lie either in the peripheral zone of cytoplasm or more centrally i.e. in a paranuclear position or adjacent to the Golgi complex. Whether the larger granules orginate directly from the Golgi complex or as a result of fusion of smaller granules within the cytoplasm is unknown. However, the irregular shape of some granules as illustrated in Figures 10 and 11, together with the presence of large granules within or adjacent to the Golgi complex suggests that both mechanisms may operate.

Pulse labelling experiments with ^{3}H-dopa (40) and uptake studies with ^{3}H-catecholamines (57, 58) suggest that chromaffin granules throughout the cell take up both endogenous and exogenous amines as soon as they are presented and that from the viewpoint of catecholamine storage there are no such things as old and new chromaffin granules – hence age with respect to chromaffin granules relates only to limiting membrane, protein and possibly nucleotide(ATP).

Although typical A and NA cells and often small granule chromaffin cells have been identified in the adrenal medulla of many mammals, in others, such as rabbit, guinea-pig and marmoset, in which noradrenaline accounts for only *c* 5% of total catecholamines (the remainder being adrenaline) no NA cells have been identified within the central mass of the medulla although they may occasionally be seen on the surface of the medial pole of the gland in situations where, during foetal life, continuity of intra- and extra-adrenal chromaffin tissue is often apparent. This suggests that the *c* 5% of noradrenaline is stored in the predominantly adrenaline storage granules in these forms. If so, it is in contrast with the amount of dopamine stored during the earlier part of the synthetic pathway. This may in part reflect the differences in the kinetics of the methylation process involving phenylethanolamine n-methyl transferase in adrenaline synthesis and dopamine-β-hydroxylase in noradrenaline synthesis. Otherwise it may be due to the greater loss from the cytosol, or less efficient binding, of endogenous dopamine than noradrenaline by chromaffin granules. However, since following intravenous injection of ^3H labelled dopamine and noradrenaline, both amines are taken up and stored in A and NA cells (57, 58) it would seem likely that the difference is more likely to reflect enzyme kinetics. In both guinea-pig and marmoset there is considerable variation in the electron density of what are otherwise typical A-cell granules (Fig. 13) and apart from in an occasional cell in the guinea-pig, referred to above, typical NA granules have not been observed. This variation may at least in part reflect a low level of NA binding by otherwise A-storage granules, and it may reflect variations in adrenaline concentration in individual granules with some fixation in situ of adrenaline. It should be noted that even in rodents some but usually a less marked variation in electron density in A cell granules does occur and this could be accounted for in either or both of the above ways.

6.2. Secretion

It is now generally accepted, from both morphological and biochemical evidence, that secretion from the adrenal medulla occurs by the process exocytosis (49). Morphological evidence of the exocytotic discharge of granule contents was presented by Diner (59) and Grynszpan-Winograd (60) in the hamster – a particularly suitable species for observing this phenomenon – and in the rat by Coupland (2). Although exocytotic profiles are only rarely observed coated pits or vesicles adjacent to the plasma membrane are commonly seen and are indicative of membrane retrieval (Fig. 14). More recently the phenomenon of exocytosis and formation of coated pits has been demonstrated by freeze-etching (61). It is now generally accepted that the secretion of granule contents including catecholamine is Ca^+ dependent (62) but the factors controlling the movement of storage granules within the cell as well as the mechanism of granule fusion are still incompletely understood even though morphologically identifiable elements including microtubules (63) and actin filaments (64) may be involved in granule movement and discharge. However, work aimed at the demonstration of the possible role of microtubules within the chromaffin cell has received a setback by the recent observation that the alkaloid vinblastine has a complex effect on chromaffin cells *in vivo* (6). Exocytosis may be observed on any face of a chromaffin cell and is not limited to the plasma membrane adjacent to a capillary, which may or may not show fenestrations. In consequence of the fact that granule discharge may occur into intercellular spaces or into the narrow gap of *c* 20 nm between adjacent chromaffin cells it is not surprising that cell adhesions between chromaffin cells are not numerous and are usually confined to simple gap junctional complexes. Recently the presence of gap junctions and both gap junctions and occasional focal tight junctions have been demonstrated in the hamster and guinea-pig respectively by a freeze-fracture study (65). In our experience gap junctions are the normal type of cell adhesion seen by transmission electron microscopy between adjacent chromaffin cells in all species studied: tight junctions, when present (and they are excessively rare in rodents) are confined to very short stretches of what would otherwise be a gap junction.

References

1. Coupland RE: The Natural History of the Chromaffin Cell. Longmans, London, 1965.
2. Coupland RE: The Chromaffin System. Handb exp Pharmak 33: 16–45, 1972.
3. Coupland RE: The adrenal medulla. In: The Cell in Medical Science, Vol. 3 (F. Beck and J.B. Lloyd Eds.) Academic Press, London, 1975, pp 193–242.

4. Coupland RE: Catecholamines. In: Hormones and Evolution Vol. 1 (E.J.W. Barrington Ed.). Academic Press, London, 1979, pp 310–340.
5. Varano L: Comparative aspects of the adrenal chromaffin cells in vertebrates. In: Biogenic Amines in Development (Parvez H and Parvez S eds.) Elsevier, Amsterdam, 1980, pp 213–240.
6. Kohn A: Die Paraganglien. Arch mikr Anat 62: 263–365, 1903.
7. Coupland RE: The prenatal development of the abdominal para-aortic bodies in man. J Anat, Lond, 86: 357–372, 1952.
8. Cession-Fossion A, Vandermullen R: La dopamine, constituant normal de la glande médullo-surrénale. Arch intern Physiol Biochem 77: 670–672, 1969.
9. Lever JD: Electron microscopic observations on the normal and denervated adrenal medulla of the rat. Endocrinology 57: 621–635, 1955.
10. Coupland RE: Electron microscopic observations on the structure of the rat adrenal medulla. I. The ultrastructure and organization of chromaffin cells in the normal rat adrenal medulla. J Anat (Lond) 99: 231–254, 1965.
11. Coupland RE: Electron microscopic observations on the structure of the rat adrenal medulla. II. Normal innervation. J Anat (Lond) 99: 255–272, 1965.
12. Elfvin LG: The development of the secretory granules in the rat adrenal medulla. J Ultrastruct Res 17: 45–62, 1967.
13. Coupland RE, Weakley BS: Electron microscopic observations on the adrenal medulla and extra-adrenal chromaffin tissue of the postnatal rabbit. J Anat (Lond) 106: 213–31, 1970.
14. Coupland RE: Observations on the form and size distribution of chromaffin granules and on the identity of adrenaline- and noradrenaline-storing cells in vertebrates and man. Mem Soc Endocr 19: 611–35, 1971.
15. Ratzenhofer M, Müller O: Ultrastructure of adrenal medulla of the prenatal rat. J Embryol exp Morph 18: 13–25, 1967.
16. Daikoku S, Kinutani M, Sako M: Development of the adrenal medullary cells in rats with special reference to synaptogenesis. Cell Tiss Res 179: 77–86, 1977.
17. El-Maghraby M, Lever JD: Typification and differentiation of medullary cells in the developing rat adrenal. A histochemical and elecron microscopic study. J Anat (Lond) 131: 103–120, 1980.
18. Millar TJ, Unsicker K: Catecholamine storing cells in the adrenal medulla of the pre- and postnatal rat. Cell Tiss Res 217: 155–170, 1981.
19. Diner O: Observations sur le dévelopment de la médullo-surrénale du rat: l'évolution de la partie non chromaffine. Arch Anat microsc 54: 671–718, 1965.
20. Coupland RE, Weakley BS: Developing chromaffin tissue in the rabbit: an electron microscopic study. J Anat (Lond) 102: 425–455, 1968.
21. Hervonen A: Development of catecholamine-storing cells in human fetal paraganglia and adrenal medulla. Acta physiol Scand Suppl 368, pp 1–94, 1971.
22. Coupland RE, Hopwood D: The mechanism of the differential staining reaction for adrenaline and noradrenaline storing granules in tissues fixed in glutaraldehyde. J Anat (Lond) 100: 227–243, 1966.
23. Euler US v: Noradrenaline (arterenol), adrenal medullary hormone and chemical transmitter of adrenergic nerves. Expt Physiol exp Pharmak 46: 261–307, 1950.
24. Euler US v, Hamberg U: L-nor-adrenaline in the suprarenal medulla. Nature (Lond) 163: 642–3, 1949.
25. Eränkö O: On the histochemistry of the adrenal medulla of the rat, with special reference to acid phosphatase. Acta Anat 16, Suppl 17 pp 1–60, 1952.
26. Hillarp N-Å, Hökfelt B: Evidence of adrenaline and noradrenaline in separate adrenal medullary cells. Acta physiol Scand 30: 55–68, 1953.
27. Coupland RE, Pyper AS, Hopwood D: A method of differentiating between noradrenaline- and adrenaline-storing cells in the light and electron microscope. Nature (Lond) 201: 1240–42, 1964.
28. Tremezzani JH, Chiocchio S, Wassermann GF: A technique for light and electron microscopic identification of adrenaline- and noradrenaline-storing cells. J Histochem Cytochem 12: 890–899, 1964.
29. Tranzer JP, Richards JG: Ultrastructural cytochemistry of biogenic amines in nervous tissue: methodologic improvements. J Histochem Cytochem 24: 1178–1193, 1976.
30. Hökfelt T, Jonsson G: Studies on reaction and binding of monoamines after fixation and processing for electron microscopy with special reference to fixation with potassium permangate. Histochemie 16: 45–67, 1968.
31. Kanerva L, Hervonen A, Rechardt L: Permanganate fixation demonstrates the monoamine-containing granular vesicles in SIF cells but not in the adrenal medulla or mast cells. Histochemistry 52: 61–72, 1977.
32. Lassermann H: Ein dritter Zelltyp im Nebennierenmark der Maus? Verh Anat Ges 68: 641–644, 1974.
33. Gorgas K, Böck P: Morphology and histochemistry of the adrenal medulla. I. Various types of primary catecholamine-storing cells in the mouse adrenal medulla. Histochemistry, 50: 17–31, 1976.
34. Kobayashi S, Coupland RE: Two populations of microvesicles in the SGC (small granule chromaffin) cells of the mouse adrenal medulla. Arch histol jap 40: 251–259, 1977.
35. Unsicker K: Fine structure and innervation of the avian adrenal gland. I. Fine structure of adrenal chromaffin cells and ganglion cell.s Z Zellforsch 145: 389–416, 1973
36. Unsicker K: Chromaffin, small-granule-containing and ganglion cells in the adrenal gland of reptiles. Cell Tiss Res 165: 477–508, 1976.
37. Hill CE, Watanabe H, Burnstock G: Distribution and morphology of amphibian extra-adrenal chromaffin tissue. Cell Tiss Res 160: 371–387, 1975.
38. Kajihara H, Akimoto T, Iijima S: On the chromaffin cells in dog adrenal medulla; with special reference to the small granule chromaffin cells (SGC cells). Cell Tiss Res 191: 1–14, 1978.
39. Coupland RE, Kobayashi S: Recent studies on the fixation of adrenaline and noradrenaline, and on amine synthesis and storage in chromaffin cells. In: Chromaffin, Enterochromaffin and Related Cells. Coupland RE, Fujita T (eds). Elsevier, Amsterdam, 1976, pp 59–81.
40. Coupland RE, Kent C, Kobayashi S: Observations on the localization of recently synthesized catecholamines in chromaffin cells after the injection of L(2, 5, 6-^{3}H) dopa. J Endocr 69: 139–148, 1976.
41. Coupland RE, Kent C, Kobayashi S: Amine turnover and the effects of insulin hypoglycaemia on small-granule chromaffin (SGC) cells of the mouse adrenal medulla. In: Peripheral Neuroendocrine Interaction. Coupland RE, Forssmann WG (eds). Springer-Verlag, Berlin, pp 86–96.
42. Bradbury S, Smith AD, Winkler H: The demonstration of lysosomes in the bovine adrenal medulla. Experientia 22: 142, 1966.
43. Coupland RE, Mastrolia L, Weakley BS: Localization of acid phosphatase in the adrenal medulla of the albino rat. In: Histochemistry of Nervous Transmission. Eränkö O (ed). Elsevier, Amsterdam, pp 456–464.
44. Gail A, Holtzman E: Peroxisomes in rat sympathetic ganglia and adrenal medulla. Brain Res 83: 509–515, 1975.
45. Redburn DA, Poisner AM, Sampson FE: Comparison of microtubule protein (tubulin) from adrenal medulla and brain. Brain Res 44: 615–624, 1972.
46. Kobayashi S, Serizawa Y: Adrenal chromaffin cells in the stressed mouse. In: Histochemistry and Cell Biology of Autonomic Neurons, SIF Cells and Paraneurons. Eränkö O, Soinila S, Päivärinta H (eds). Raven, New York, 1980, pp 195–200.
47. Coupland RE, Bustami F: Influence of drugs affecting the pituitary adrenal axis on chromaffin cells. In: Histochemistry and Cell Biology of Autonomic Neurons, SIF Cells and Paraneurones. Eränkö O, Soinila S, Päivärinta H (eds). Raven, New York, 1980, pp 175–183.
48. Winkler H: The composition of adrenal chromaffin granules: an assessment of controversial results. Neuroscience 1: 65–80, 1976.
49. Winkler H: The biogenesis of adrenal chromaffin granules. Neuroscience 2: 657–683, 1977.
50. Winkler H, Westhead E: The molecular organization of adrenal chromaffin granules. Neuroscience 5: 1803–1823, 1980.
51. Coupland RE, Kent C: Fixation of adrenaline and dopamine in tissue sections. J Anat (Lond.) 123: 247–248, 1977.
52. Coupland RE, Kobayashi S, Crowe J: On the fixation of catecholamines including adrenaline in tissue sections. J Anat (Lond), 122: 403–413, 1976.
53. Kobayashi S, Kent C, Coupland RE: Observations on the localization of labelled amino acid in mouse adrenal chromaffin cells after the injection of L-(4, 5-^{3}H) leucine. J Endocr 78: 21–29, 1978.
54. Blaschko H, Hagen P, Welch AD: Observations on the intracellular granules of the adrenal medulla. J Physiol 129: 27–49, 1955.
55. Kirschner N: Biosynthesis of adrenaline and noradrenaline. Pharmacol Rev 11, 350–357, 1959.
56. Kirschner N, Goodall McC: Formation of adrenaline from noradrenaline. Biochem Biophys Acta, 24: 658–659, 1957.
57. Kent C, Coupland RE: On the uptake of exogenous catecholamines by

adrenal chromaffin cells and nerve endings. Cell Tiss Res 221: 371–383, 1981.
58. Kent C, Monkhouse WS, Coupland RE: Effect of hydrocortisone, reserpine, propanolol and phentolamine on *in vivo* uptake of exogenous amines by adrenal chromaffin cells. Cell Tiss Res 221, 385–393, 1981.
59. Diner O: L'expulsion des granules de la médullo-surrénale chez le Hamster. CR Acad Sci Paris, 265: 616–619, 1967.
60. Grynszpan-Winograd O: Morphological aspects of exocytosis in the adrenal medulla. Phil Trans R Soc B 261: 291–292.
61. Smith U, Smith DS, Winkler H, Ryan JW: Exocytosis in the adrenal medulla demonstrated by freeze-etching. Science 179: 79–82, 1973.
62. Douglas WW, Poisner AM: On the mode of action of acetylcholine in evoking adrenal medullary secretion: increased uptake of calcium during the secretory response. J Physiol 162: 385–392, 1962.
63. Poisner AM, Bernstein J: A possible role of microtubules in catecholamine release from the adrenal medulla: effect of colehicine, vinca alkaloids and deuterium oxide. J Pharmacol exp Ther 171: 102–108, 1971.
64. Friedman JE, Lelkes PI, Rosenheck K, Oplatka A: The possible implication of membrane-associated actin in stimulus secretion coupling in adrenal chromaffin cells. Biochem Biophys Res Comm 96: 1717–1723, 1980.
65. Grynszpan-Winograd O, Nicolas G: Intercellular junctions in the adrenal medulla: a comparative freeze-fracture study. Tissue and Cell 12: 661–672, 1980.

Author's address:
Department of Human Morphology
The Medical School
University of Nottingham
Nottingham NG7 2UH
England

CHAPTER 16

Extraadrenal chromaffin organs (abdominal paraganglia): Distribution, histology and fine structure

JOE A. MASCORRO, ANTTI HERVONEN and ROBERT D. YATES

1. Introduction

The term paraganglia refers to microscopic aggregations of epithelioid cells which are encapsulated by connective tissue and positioned directly next to abdominal sympathetic ganglia and plexuses. These small cells exhibit chromaffinity, contain catecholamines and generally posses morphological and embryological characteristics like those of intraadrenal chromaffin cells. Consequently, paraganglion cells are homologous to the cells which comprise the endocrine adrenal medulla (1, 2, 3). Vincent (4) and later Wislocki (5) located paraaortic chromaffin organs which were significantly larger than paraganglia in the abdomen of various species. Recent studies by the present authors utilized anatomical and fluorescence histochemical methods to map the pattern of distribution and persistence of extraadrenal chromaffin tissue in dogs, rabbits and humans (6, 7, 8). In addition, the so-called 'paraganglia' also have been located in relation to the vagus and glossopharyngeal nerves (9, 10). A significant volume of similar chromaffin cells also resides within the sympathetic ganglia (11, 12). Chromaffin cells found in the superior cervical, celiac and abdominal mesenteric ganglia are known as small intensely fluorescent (SIF) or small granule-containing (SGC) cells because of their catecholamine granules which fluoresce very brightly. Thus when considering that paraganglia actually occur in paraganglionic, intraganglionic as well as paraaortic locations, and along major nerves, it is reasonable to assume that this large and widespread system of aminergic cells performs a function very necessary to the organism.

The current work applies histological, ultrastructural as well as anatomical mapping techniques and presents a timely review concerning the structure, distribution and persistence of extraadrenal chromaffin organs. Furthermore, it introduces novel information concerning the innervation of this important tissue in the ultimate organism, the human.

2. Chromaffin organs

2.1. Distribution and histology

Mongrel dogs of various ages and young New Zealand albino rabbits were perfused routinely with glutaraldehyde and their retroperitoneal tissue block (RTB) subsequently was treated with a combination of glutaraldehyde/potassium dichromate in phosphate buffer. The dual action of aldehyde and dichromate preserved tissue morphology and additionally produced a brown coloration (the chromaffin reaction) which identified chromaffin organs easily (6). The RTB from 3 week old dogs displayed a dichromate-positive organ which measured 33 mm in length and 2–3 mm at its widest margin (Fig. 1). This organ was ventral to the abdominal aorta and displayed a brown coloration identical to that of the classical chromaffin reaction produced by the adrenal medulla (7). The organ was widest in the midretroperitoneum and then gradually tapered into slender strands toward the rostral and caudal extremities of the abdominal tissue block. Smaller chromaffin organs were visible near the medial aspect of the left adrenal gland, but a direct continuity between intra- and extraadrenal chromaffin tissue at this point was not evident (Fig. 1). Light microscopic study of the dichromate-reacted organs from dogs revealed a conspicuous vascular pattern of branching and anastomosing blood vessels (Fig. 2). The endothelial lining of these sinusoidal-like vessels was quite attenuated and contained nuclei which protruded into the lumen. Also equally conspicuous were cells which contained large nuclei in addition to a pronounced cytoplasmic basophilia. The density of the cytoplasm was particularly obvious when contrasted against the large, pale nuclei. The individual cells were arranged as cords or clusters and were distinctly separated by the tortous blood vessels (Fig. 2).

Motta, PM (ed): Ultrastructure of endocrine cells and tissues. ISBN-13: 978-1-4613-3863-5

2.2. *Ultrastructure of paraaortic chromaffin organs*

Ultrastructural evaluation of the organs removed from the dog RTB following dichromate treatment revealed that they contained cells replete with darkly stained granules (Fig. 3). The occurrence of many distinct granules which occupied large areas of the cytoplasm correlated well with the cellular density initially observed at the light microscope level (compare arrows, Figs. 2 and 3). The granules were closely packed and often completely filled the perinuclear cytoplasm. However, it was not unusual to find granules disposed along the limiting plasma membrane. The specific granules predominantly were round or oval, but elongate and pleomorphic forms also were encountered. The granule-containing cells displayed the usual array of cytoplasmic organelles and inclusions, such as mitochondria, endoplasmic reticulum, Golgi membranes and glycogen-like particles. However, these structures were relatively unnoticed because of the tremendous volume of the cell granules (Fig. 3).

2.3. *Ultrastructural comparisons of paraganglion and intraganglion chromaffin cells in rabbits*

Extraadrenal chromaffin cells aggregated into discrete structures (paraganglia) which were easily visible following dichromate staining (6). On the other hand, the intraganglion small, granule-containing

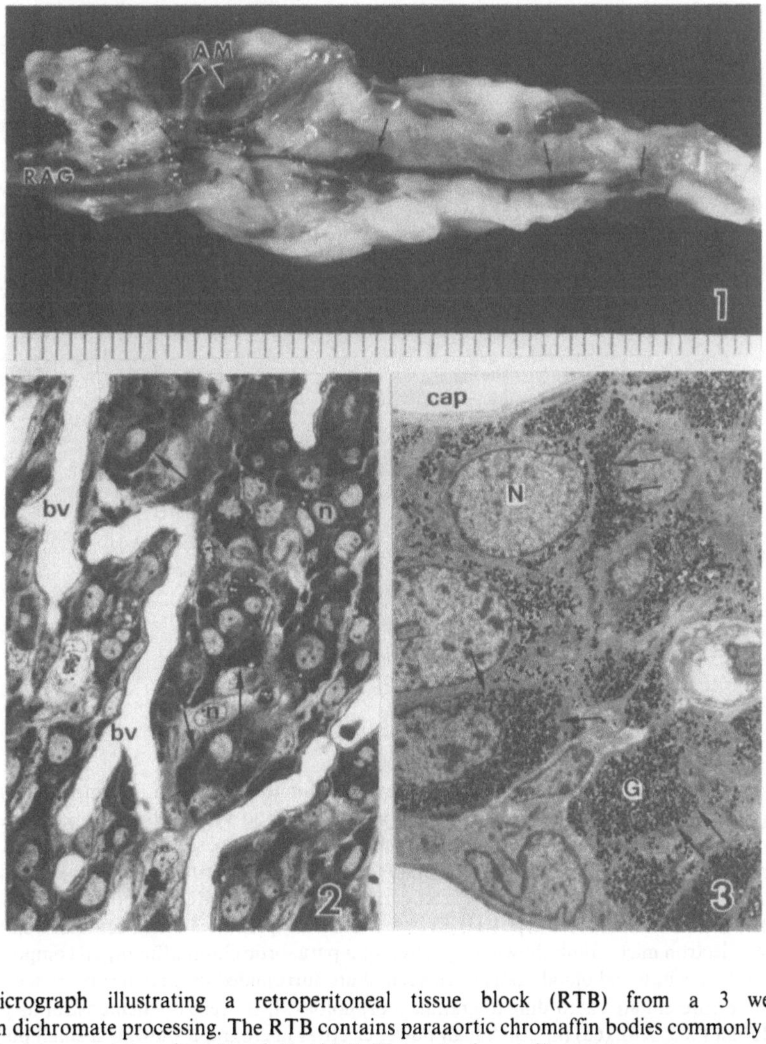

Figs. 1–3. (1) Photomicrograph illustrating a retroperitoneal tissue block (RTB) from a 3 week old dog following glutaraldehyde/potassium dichromate processing. The RTB contains paraaortic chromaffin bodies commonly known as the paraganglia (arrows). Aldehyde staining also maps the adrenal medulla (AM). The main chromaffin organ measures approximately 33 mm in length. RAG = right adrenal gland. Scale represents millimeters. (2) A light micrograph which illustrates the histological characteristics of paraaortic chromaffin organs. The organs contain many sinusoidal blood vessels with thin walls (bv), prominent nuclei (n), and a markedly basophilic cytoplasm (arrows). Dog (× 560). (3) Electron micrograph showing chromaffin cells with many dense granules (G). The rounded vesicles account for the cytoplasmic density observed with the light microscope (arrows). The chromaffin cells have large nuclei (N) and are closely associated with capillaries (cap). Dog (× 2,870).

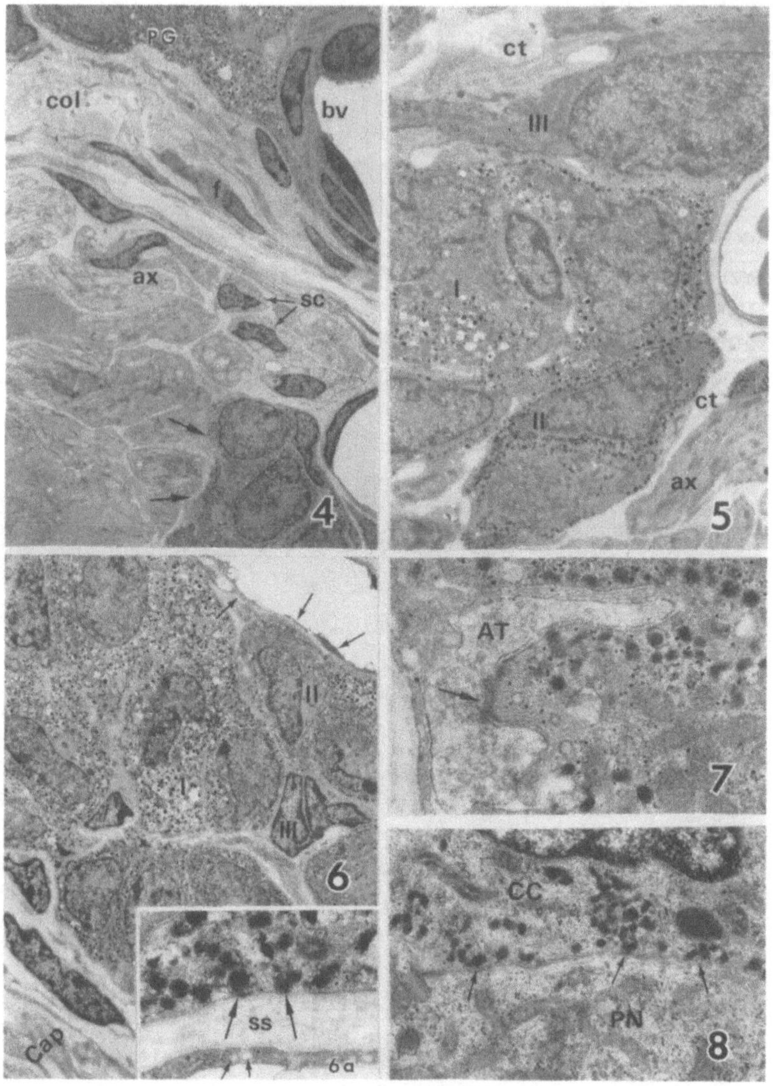

Figs. 4–8. (4) This illustration shows a small portion of a true paraganglion (PG) adjoining a sympathetic ganglion. The two structures are spearated by collagen fibers (col), fibroblasts (f), and large arteriolar blood vessels (bv). The ganglion contains a small cluster of chromaffin cells (arrows). Several unmyelinated axons (ax) and schwann cells (sc) are seen in the ganglion. Rabbit (× 3,000).(5) A cluster of intraganglion chromaffin cells similar to those seen in Figure 4. The group is surrounded by axons (ax) and fine connective tissue components (ct). Certain cells contain granules with off-center cores (I), while the granules in other cells have dense cores that completely fill the granule area (II). A third chromaffin cell type (III) gives rise to slencer processes which contain small granules along the periphery (× 3,710). (6) Low power electron micrograph showing a portion of a paraaortic chromaffin organ composed of many granule-containg cells. The cells closely border a sinusoidal blood vessel (arrows) and are surrounded by a connective tissue capsule (Cap) at the opposite pole. Two types of cells are differentiated according to granule morphology: I, II. Cells not demonstrating granules represent fibroblasts or satellite cells (III). Rabbit (× 2,520). (6a) (inset). A high power electron micrograph showing a small portion of chromaffin cell lying next to a blood vessel. The endothelial lining of the vessel is uniformly attenuated and shows pinocytotic activity (arrows). The subendothelial space (SS) contains fine collagen fibrils. The granules appear merged with the plasma membrane in several places (large arrows). Rabbit (× 19,320). (7) A high power electron micrograph showing an axon terminal (AT) next to an intraganglion chromaffin cell. The terminal contains synaptic-like vesicles with clear cores. Membrane densities are evident at the area of apposition between the cell and terminal (arrow). Rabbit (× 23,520). (8) A ganglion chromaffin cell (CC) in proximity to a postganglionic neuron (PN). A small space separates the two cells and granules approximate the chromaffin cell membrane (arrows). Rabbit (× 21,000).

cells (SGC) were microscopic groups rarely more than a few cells in diameter (Fig. 4). A plane of connective tissue and blood vessels intervened between a paraganglion proper and the adjacent sympathetic ganglion which contained the chromaffin cells. This connective tissue presumably was an extension of the distinct capsule which encased the paraganglion. Conversely, the intraganglion SGCs were surrounded only by fine endoneurial collagenous fibers (Figs. 4, 5). The SGCs often were close to blood vessels, but otherwise were surrounded by unmyelinated axons and endoneurium at their nonvascular poles (Fig. 5).

The fine structural characteristics of paraganglion and intraganglion chromaffin cells generally were similar (Fig. 4). All of the ganglion chromaffin cells displayed rounded cytoplasmic granules which contained homogeneously dense cores (Fig. 5). In one cell type, the core substance was separated from a surrounding membrane by a narrow electron lucid space (Fig. 7). However, other cells contained granules whose dark cores were located in an eccentric position within the membrane space. Both cell types occurred together in the form of small intraganglion groups (Fig. 5). Also associated with the cell clusters was a third cell type whose soma gave rise to slender processes. These cytoplasmic elongations were characterized by the presence of smaller and lighter granules arranged along the cell margin (Fig. 5). The paraganglia contained similar cells and granules. In addition, these organs contained a stromal framework rich in collagen fibers and fibroblasts (Fig. 6).

The chromaffin cells from all locations usually presented an intimate association with thin-walled blood vessels which represented capillaries with attenuated endothelium, or possibly small postcapillary venules (Figs. 5, 6, 6a). In many areas, the barrier between chromaffin cells and the blood was minimal and the subendothelial space contained only basal laminae and fine collagen components (Figs 5, 6a). Many of the cytoplasmic granules rested along the cell membrane and often contacted the membrane (Fig. 6a). Not all chromaffin cells approximated the vasculature this closely, since distinct barriers such as connective tissue and pericytes occasionally filled the wider subendothelial spaces.

Chromaffin cells in the sympathetic ganglia, but not those of the paraganglia, received nerve endings (Fig. 7). The axons ending upon the chromaffin cells were large structures, displayed synaptic membrane densities and were filled with vesicles showing clear cores (Fig. 7). Also, the chromaffin cells made somatic contacts upon principal postganglionic neurons (Fig. 8). At the area of somatosomatic proximity, the cell bodies were separated by a very small intercellular space devoid of connective tissue or satellite cell cytoplasm. Many of the chromaffin granules were aligned at the periphery of the cell (Fig. 8).

2.4. Distribution and fine structure of human paraganglia

Formaldehyde induced fluorescence revealed paraganglia in all human specimens studied (8). The paraganglia exhibited strong fluorescence and were distributed as groups of varying size throughout the paraaortic retroperitoneal connective tissue (Fig. 9). The fine structure of human paraganglia was similar to that described in other species (7, 9, 10). Following standard fixation in glutaraldehyde and osmium tetroxide, large dense cored granules (100–400 nm in diameter) dominated the cytoplasm. The granules usually possessed very electron dense cores, and only occasionally were granules with pale cores observed (Figs. 10, 11).

An important feature of the human paraganglia was the presence of cholinergic-like nerve endings upon the paraganglion cell soma or its processes (Fig. 12). The axon endings were large structures which were filled with small vesicles exhibiting clear cores. In addition, the endings also contained a sparse population of larger vesicles which contained cores of moderate density. The membranes at the axon ending displayed the usual densities associated with typical synapses (Fig. 12).

3. Persistence of extraadrenal chromaffin tissue

The anatomical distribution of extraadrenal chromaffin tissue in the 3 week old dog was accurately defined by using a glutaraldehyde/potassium dichromate fixing and staining method reported earlier (6). This method produced tissues which were morphologically preserved by the initial action of glutaraldehyde. Post-Treatment of the fixed tissues by the aldehyde/dichromate combination produced a gross chromaffin reaction based upon the interaction between dichromate ions and chromaffin cell catecholamines. The occurrence of a paraaortic chromaffin organ in the 3 week dog differed substantially from the distribution of similar tissue in mice and rats, species where the main extraadrenal chromaffin body regresses very soon in the postnatal specimen (13). The main bulk of extraadrenal chromaffin tissue in humans (*the Organ of Zuckerkandl*) is located around the level of the inferior mesenteric artery (14). This organ is rich in pressor amines and presumably can regulate blood pressure in the fetus and infant (15). The body fibroses in the young child and finally is not recognized as a distinct entity beyond the 7th year. Nevertheless, scattered chromaffin cells have been detected histologically in 49 year old humans in the areas of the celiac and superior mesenteric arteries (16). Furthermore, recent studies by Hervonen et al. applied fluorescence histochemistry, electron microscopy as well as immunocytochemical techniques to demonstrate categorically that the human paraganglia indeed form a catecholamine endocrine system which persists from fetal through adult ages (17, 18).

The largest mass of extraadrenal chromaffin tissue in the 3 week dog did not occur near the inferior mesenteric artery, as does the human *Organ of Zuckerkandl (OZ)*, but instead was located in the

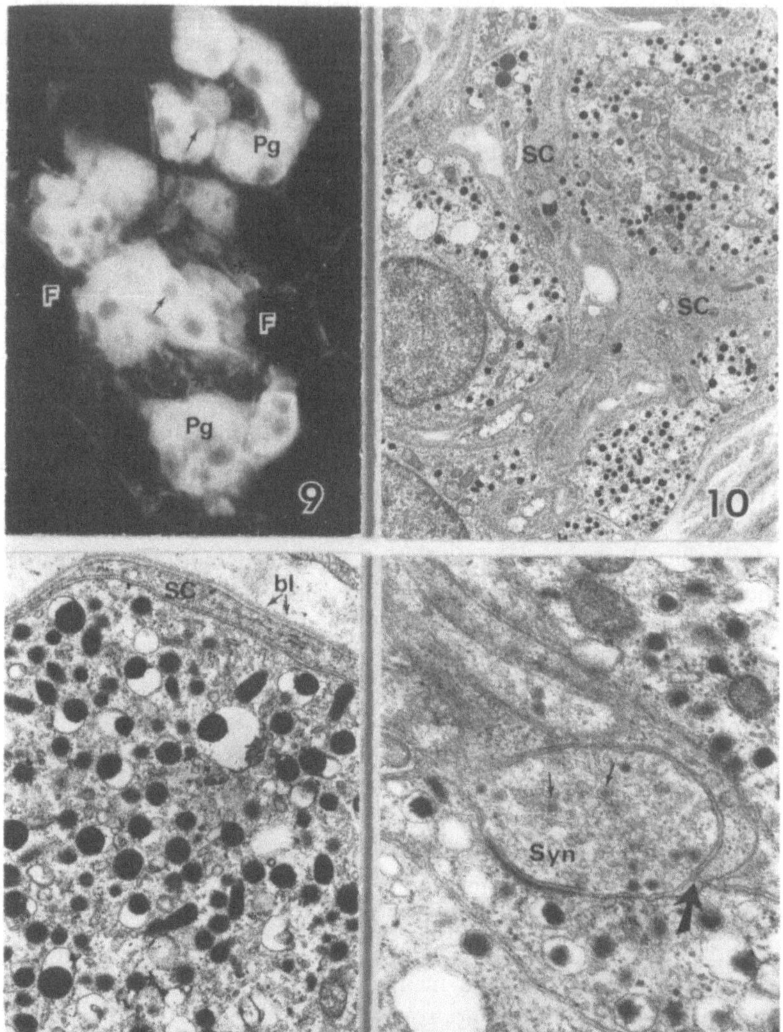

Figs. 9–12. (9) Photomicrograph illustrating a high degree of formaldehyde-induced fluorescence in the human retroperitoneal paraganglia. The paraganglia (Pg) occur in small groups and are surrounded by connective tissue (*) and islands of fat cells (F). Nonfluorescent nuclei appear as dark structures (arrows) (× 455). (10) Electron micrograph showing the fine structure of human paraganglia. The constituent chromaffin cells contain very dense granules and are surrounded by satellite cell cytoplasm (SC) (× 5,265). (11) A portion of a human chromaffin cell with a cytoplasm very rich in catecholamine granules. The granules are highly dense and many show a space between the core and surrounding membrane. The cell is bordered by a thin rim of satellite cell (SC) and a basal lamina (bl) (× 19,890). (12) Synaptic contact (Syn) upon a granule-containing process in human paraganglia. The ending predominantly contains small vesicles, but several larger ones with dense cores are present (small arrows). The surrounding satellite cell stops at the area of the synapse and does not intervene between the axon ending and cell process (large arrow) (× 13,300).

midretroperitoneum. This was a consistent observation in several age groups studied previously, and only in the 2 and 8 week dogs were organs noted in the lower retroperitoneum that could qualify as anatomical correlates of the OZ (7). From these data, it was surmised that the OZ in dogs was inconstant and certainly did not represent the largest deposit of abdominal extraadrenal chromaffin tissue.

4. Parenchymal cells and granule content

The histology and fine structure of paraaortic chromaffin tissue is well documented. The earliest descriptions of this tissue firmly established its chromaffinity and general likeness to the endocrine adrenal medulla (1, 2). Brundin (19) presented fluorimetric data which showed that paraganglia in fetal and newborn rabbits contained almost exclusively noradrenaline, and Brundin and Nilsson (20) reported that this amine existed at the subcellular level in the form of osmiophilic granules. The fortuitous advent of glutaraldehyde as a differential stain for catecholamines by Coupland et al. enabled them to differentiate lighter-staining (epinephrine) from darker-staining (norepinephrine) granules in tissues initially fixed in glutaraldehyde and treated secondarily with osmium tetroxide (21). Using this fixation method, the present authors defined the granules in various species, including the human. Studies with rabbit paraganglia showed that in addition to granules with dark centers, many others possessed granular cores with less density (22). Observations in dogs similarly indicated that many granules definitely were not as electron dense when compared to others in neighboring cells (7). The morphological appearance of granules cannot be used as a sole criterion in catecholamine identification, nevertheless it must be emphasized that many granules exhibiting the appearance usually attributed to epinephrine storage resided in paraganglion cells reported to contain predominantly norepinephrine. The idea that paraganglion cells can synthesize epinephrine appears reasonable, since immunohistochemical analyses have shown that similar chromaffin cells found within sympathetic ganglia (SIF cells) contain the enzyme necessary for converting norepinephrine to epinephrine (23).

Most granules in the human paraganglion cells displayed electron dense cores which often were placed eccentrically within the limiting granule membrane (17). The bright formaldehyde fluorescence from the granules and the microspectrofluorimetric analysis of their spectra indicated that the cytoplasmic granules were indeed noradrenergic in nature (24). In addition to catecholamines, the human paraganglia apparently also contained neuropeptides which coexisted within the granule structures (18). The release of catecholamines from paraganglion cells is associated with end organ effects, but the possible function of protein components that are exocytosed in tandem with the amines raises a question that surely requires attention.

5. Vascularity

All extraadrenal chromaffin tissue, whether in the form of small paraganglia or much larger paraaortic organs, possessed a very abundant blood supply. The blood vessels which most closely surrounded the chromaffin cells had very thin walls and represented capillary sinusoids or post-capillary venules. The most significant observation was that the subendothelial space between the cells and sinusoidal vessels often only contained collagen fibrils and basal laminae. Alterations in granule morphology have been reported in mouse paraganglion cells following reserpine treatment, a sympathomimetic stimulus responsible for catecholamine release from chromaffin cells (25). Since chromaffin granules represent storage sites for catecholamines, a deviation from normal form could indicate a release of cellular amines. Thus, the vascular relationship noted in extraadrenal chromaffin tissues would appear opportune for the release of cell products into the circulation. Caution must be exercised with this interpretation, however, since there is no experimental evidence that secretions from extramedullary chromaffin cells coincide with increased levels of blood catecholamines following stimulation. It was noted that many granules assumed a peripheral location close to the cell membrane. In this position frequent contacts between the granule and cell membranes occurred which were very suggestive of exocytosis, the mechanism primarily implicated in catecholamine extrusion (26).

6. Innervation

The innervation of sympathetic extraadrenal chromaffin organs which are comparable with the adrenal medulla in structure and endocrine function remains questionable in certain respects. Initial light microscope studies of the main paraaortic chromaffin organs in cats and dogs showed that they received a preganglionic cholinergic innervation like that of the adrenal medulla (27). However, Thompson and Gosling (28) did not detect appreciable quantities of acetylcholinesterase (AChE) in the human pelvic paraganglia or adjoining nerves. They concluded that pelvic chromaffin cells were not autonomically innervated, and that the absence of AChE disagreed with the concept that paraganglia received a rich network of cholinergic fibers, eventhough Jacobowitz (29) and Hervonen (30) previously had detected AChE in paraganglia of cats and human fetuses. Thompson and Gosling further noted that Hervonen (30) as well as Coupland and Weakley (31) were unable to locate nerve terminals at the ultrastructural level. The innervation of extraadrenal chromaffin tissue became better defined when Mascorro and Yates (32) located axon terminals in Syrian hamsters and squirrel monkeys at the fine structural level. Furthermore, the novel information presented in this paper by Hervonen proves that extraadrenal chromaffin cells in the human receive cholinergic-like endings. The paucity of nerve endings upon sympathetic paraganglion cells in dogs is unclear, particularly since synapses are found with regularity in the adrenal medulla (33) and in association with vagal paraganglion cells where complex axon terminals exist (9, 34). If synapses indeed are present on these chromaffin cells one would expect to locate them, even when examining small amounts of tissue as must be done in the electron microscope. The absence of a functional innervation to the large paraaortic chromaffin organs in this species is enigmatic, nevertheless this possibility must be considered.

7. Correlation between intraganglion and extraadrenal chromaffin cells

The existence of chromaffin cells within various sympathetic ganglia is a well-documented fact (11, 12). These small cells have been studied extensively and are known to contain catecholamines (23, 24). Some ganglion chromaffin cells are known as small intensely fluorescent (SIF) cells because of their brilliant catecholamine fluorescence. SIF cells are interposed within neural circuits and synaptically inhibit the functioning of postganglionic neurons (35). Single SIF cells situated in this manner are regarded as true interneurons. Others SIF cells occur in small clusters and bear a striking morphological resemblance to extramedullary chromaffin cells found throughout the retroperitoneum, particularly in cats (36). Ganglion chromaffin cells of this type are not innervated, do not synapse with postganglionic neurons, display an abundant vascularity and generally are regarded as endocrine rather than interneuronal in purpose (36, 37). The importance of intraganglion SIF cells is not totally understood. Although some certainly function as inhibitory interneurons, the infrequency of these small cells in human ganglia implies that any effect(s) on total ganglion transmission mechanisms would be quite localized (24). Conversely, SIF cells in the ganglia of rats, cats, rabbits and guinea-pigs greatly outnumber those in human ganglia (38). The clusters of chromaffin cells residing within the ganglia, since they are morphologically identical to those cells which collectively comprise the endocrine paraganglia, actually may represent 'miniparaganglia' with neurosecretory capabilities (36). The release of catecholamines into the ganglion circulation by the supposed miniparaganglia could negate the synaptic requirements for neuronal inhibition and would allow more neurons to come under the influence of catecholamines. This hypothesis, however, must be calculated against the ability of the chromaffin cells to release their hormonal contents, but there is no evidence as yet that this can occur following proper stimuli.

8. Concluding remarks

Extraadrenal chromaffin tissue classically known as the paraganglia is distributed extensively throughout the retroperitoneum and persists into adulthood. This tissue was localized in dogs and rabbits by utilizing a solution of glutaraldehyde in combination with potassium dichromate. The dichromate ions, upon reacting with catecholamines in chromaffin tissue, produced a gross chromaffin reaction which 'mapped' extraadrenal chromaffin organs, while glutaraldehyde preserved the morphological integrity of the dichromated organs for subsequent microscopic verification. All animals contained abundant chromaffin tissue in the form of large, discrete organs in a paraaortic position. Extraadrenal chromaffin tissue in rabbits showed a clear continuity with the adrenal medulla, but such a connection was not evident in dogs. In addition, smaller chromaffin bodies occurred next to sympathetic ganglia, a position where they could be classified as the true paraganglia. Groups of small chromaffin-like cells occurred within the celiac/superior mesenteric ganglion complex in rabbits. Ultrastructural study subsequently confirmed that these elements in fact represented intraganglionic 'miniparaganglia' which possessed the usual characteristics of chromaffin organs, including the presence of catecholamine granules. All paraaortic, paraganglion and intraganglion cells were extensively vascularized and showed intimate apposition with blood vessels which displayed an attenuated endothelium. Synapses were noted upon the intraganglion chromaffin cells in rabbits, but similar terminations could not be located on the chromaffin cells of the dog paraaortic organs.

The findings in man indicate that the paraganglia represent an endocrine system which likewise persists throughout life. Histochemical studies have shown that these paraganglia fluoresce intensely for catecholamines, and ultrastructural observations have confirmed the presence of cytoplasmic catecholamine granules. These data are convincing and strongly suggest that paraaortic and paraganglion catecholamine-containing organs in higher organisms, including humans, are permanent structures of physiological importance. It is probable that all sympathetically-related extraadrenal chromaffin cells are endocrine by nature, except possibly for the interneuronal SIF cells. However, producing irrefutable proof for this hypothesis remains a question of paramount importance. This matter assumes added significance when considering the latest discoveries that the human 'paraganglia' are cholinergically innervated structures whose granules contain neuropeptides in addition to catecholamines (18).

References

1. Stilling H: A propos de quelques experiences nouvelles sur la maladie d'Addison. Rev Medecine 10: 808–831, 1890.
2. Köhn A: Die paraganglien. Arch Mikroskop Anat 62: 263–365, 1902.
3. Mascorro JA, Yates RD: Fine structural comparisons between paraganglion and adrenal medullary cells in the Syrian hamster. Tex Rep Biol Med 31: 519–535, 1973.
4. Vincent S: The chromophil tissues and the adrenal medulla. Proc Soc Lond 82: 502–515, 1910.
5. Wislocki GB: Note on a modification of the chromaffin reaction, with observations on the occurrence of abdominal chromaffin bodies in mammals. J Hopkins Hosp Bull 33: 359–361, 1922.
6. Mascorro JA, Yates RD, Chen I: A glutaraldehyde/potassium dichromate tracing method for the localization and preservation of abdominal extraadrenal chromaffin tissues. Stain Technol 50: 391–396, 1976.
7. Mascorro JA, Yates RD: The anatomical distribution and morphology of extraadrenal chromaffin tissue (abdominal paraganglia) in the dog. Tissue Cell 9: 447–460, 1977.
8. Hervonen A, Partanen S, Vaalasti A, Partanen M, Kanerva L, Alho H: The distribution and endocrine nature of the abdominal paraganglia of adult man. Am J Anat 153: 563–572, 1978.
9. Chen I, Yates RD: Ultrastructural studies of vagal paraganglia in Syrian hamsters. Z Zellforsch 108: 309–323, 1970.
10. McDonald DM, Blewett RW: Location and size of carotid body-like organs (paraganglia) revealed in rats by the permeability of blood vessels to Evans blue dye. Neurocytol 10: 607–643, 1981.
11. Eränkö O, Harkonen M: Monoamine-containing small cells in the superior cervical ganglion of the rat and an organ composed of them. Acta Physiol Scand 63: 511–512, 1965.
12. Williams TH, Palay SL: Ultrastructure of the small neurons in the superior cervical ganglion. Brain Res 15: 17–34, 1979.
13. Coupland RE: The post-natal distribution of the abdominal chromaffin tissue in the guinea pig, mouse and white rat. J Anat 94: 244–256, 1960.
14. Zuckerkandl E: Uber nebenorgane des Sympaticus in Retroperitonealraum des Menschen. Anat Anz 15: 97–107, 1901.
15. West GB, Shepherd DM, Hunter RB, Macgregor AR: The function of the Organs of Zuckerkandl. Clin Sci 12: 317–324, 1953.
16. Coupland RE: The post-natal fate of the abdominal para-aortic bodies in man. J Anat 88: 455–464, 1954.
17. Hervonen A, Kanerva L, Partanen S, Vaalasti A: Histochemistry and fine structure of the paraganglia of man. In: Peripheral neuroendocrine interaction. Coupland RE, Forssmann WB (eds), Berlin, Springer-Verlag, 1978, pp 48–59.
18. Hervonen A, Pickel VM, Joh TH, Reis DJ, Linnoila I, Kanerva L, Miller RJ: Immunocytochemical demonstration of the catecholamine-synthesizing enzymes and neuropeptides in the catecholamine-storing cells of human fetal sympathetic nervous system. In: Histochemistry and cell biology of autonomic neurons, SIF cells, and paraneurons, Eränkö O, Soinila S, Pälvärenta H (eds), Raven Press, New York, 1980, pp 373–378.
19. Brundin T: Catecholamines in the preaortal paraganglia of fetal rabbits. Acta Physiol Scand 64: 287–288, 1965.
20. Brundin T, Nilsson SE: Osmiophilic granules in preaortal paraganglia from newborn rabbits. Acta Physiol Scand 65: 287–288, 1965.
21. Coupland RE, Pyper AS, Hopwood D: A method for differentiating between noradrenaline- and adrenaline-storing cells in the light and electron microscope. Nature 201: 1240–1243, 1964.
22. Mascorro JA, Yates RD: Morphological observations on the granule types in rabbit extraadrenal chromaffin cells. EMSA Proc 33: 318–319, 1975.
23. Elfvin LG, Hökfelt T, Goldstein M: Fluorescence microscopical, immunohistochemical and ultrastructural studies on sympathetic ganglia of the guinea pig, with special reference to the SIF cells and their catecholamine content. J Ultrastruct Res 51: 337–396, 1975.
24. Hervonen A, Hannu A, Helen P, Kanerva L: Small, intensely fluorescent cells of human sympathetic ganglia. Neurosci Lett 12: 97–101, 1979.
25. Mascorro JA, Yates RD: Ultrastructural studies of the effects of reserpine on abdominal sympathetic paraganglia. Anat Rec 170: 269–280, 1971.
26. Nagasawa J: Exocytosis: The common release mechanism of secretory granules in glandular cells, neurosecretory cells, neurons and paraneurons. Arch Histol Jpn 40 (Suppl): 31–47, 1977.
27. Hollinshead WH: The innervation of abdominal chromaffin tissue. J Comp Neurol 67: 133–143, 1937.
28. Thompson SA, Gosling JA: Histochemical light microscopic study of catecholamine containing paraganglia in the human pelvis. Cell Tiss Res 170: 539–548, 1976.
29. Jacobowitz D: Catecholamine fluorescence studies of adrenergic neurons and chromaffin cells in sympathetic ganglia. Fed Proc 29: 1929–1940, 1970.
30. Hervonen A: Development of catecholamine-storing cells in human fetal paraganglia and adrenal medulla. A histochemical and electron microscopical study. Acta Physiol Scand 368 (Suppl): 1–92, 1971.
31. Coupland RE, Weakley BS: Electron microscopic observation on the adrenal medulla and extra-adrenal chromaffin tissue of the postnatal rabbit. J Anat 106: 213–231, 1970.
32. Mascorro JA, Yates RD: Innervation of abdominal paraganglia: An ultrastructural study. J Morphol 142: 153–164, 1974.
33. Prentice FD, Wood JG: Adrenergic innervation of cat adrenal medulla. Anat Rec 181: 689–704, 1975.
34. Morgan M, Pack RJ, Howe A: Structure of cells and nerve endings in abdominal vagal paraganglia of the rat. Cell Tiss Res 169: 467–484, 1976.
35. Matthews MR, Raisman G: The ultrastructure and somatic efferent synapses of small granule-containing cells in the superior cervical ganglion. J Anat 105: 255–282, 1969.
36. Mascorro JA, Yates RD: Paraneurons and paraganglia: Histological and ultrastructural comparisons between intraganglionic paraneurons and extraadrenal paraganglion cells. In: Histochemistry and cell biology of autonomic neurons, SIF cells, and paraneurons, Eränkö O, Soinila S, Pälvärenta H (eds), Raven Press, New York, 1980, pp 201–213.

37. Lu KS, Lever JD, Santer RM, Presley R: Small granulated cell types in rat superior cervical and coeliac-mesenteric ganglia. Cell Tiss Res 172: 331–343, 1976.
38. Chiba T, Willams TH: Histofluorescence characteristics and quantification of small intensely fluorescent (SIF) cells in sympathetic ganglia of several species. Cell Tiss Res 162: 331–341, 1975.

Authors' addresses:
J.A. Mascorro and R.D. Yates
Department of Anatomy
Tulane University School of Medicine
New Orleans, LA 70112, USA

A. Hervonen
Department of Biomedical Sciences
University of Tampere
Tampere 10, Finland

CHAPTER 17

Fine structure of the adrenal cortex

GASTONE G. NUSSDORFER, VIRGILIO MENEGHELLI and GIUSEPPINA MAZZOCCHI

1. Introduction

The aim of this chapter is to provide a brief survey of the current knowledge about the cytophysiology of adrenocortical cells in vivo, with a particular emphasis to the few investigations concerning the human gland. Comprehensive references can be found in several specialized review articles (1–7). The ultrastructure of adrenocortical cells cultured in vitro is discussed in the Chapter 18.

2. Ultrastructure of the normal adrenal cortex

Electron microscopic investigations indicate that mammalian adrenocortical cells possess three outstanding ultrastructural features: (1) mitochondria with tubular and vesicular cristae, (2) abundance of smooth endoplasmic reticulum membranes (SER), and (3) a variable amount of lipid droplets. It will be described firstly the ultrastructure of a typical adrenocortical cell and then the more significant differences among the parenchymal cells of the various adrenocortical zones (ultrastructural zonation).

The tridimensional architecture of the adrenal cortex and the interrelationships between adrenocortical cells and pericapillary spaces are discussed in the Chapter 19.

2.1. Adrenocortical cell organelles

Adrenocortical cells show a rather large round or oval nucleus, with 1 or 2 obvious nucleoli. According to the species examined, mitochondria are round, oval or elongated, but invariably display cristae of the tubular or vesicular type, of about 600–700 Å in diameter (Figs. 1, 2 and 3). Frequently, lipid-like inclusions can be observed in the mitochondrial matrix (Fig. 3). Earlier, vesicular cristae were assumed to originate from the inner mitochondrial membrane and to float free in the matrix of the organelles. However, more recently Allmann et al. (8) suggested that the so-called vesicular cristae are actually sections of tortuous tubular invaginations of the inner mitochondrial membranes, which are alternatively 'ballooned out' and 'squeezed-down' (Fig. 4). This contention has been confirmed by an accurate analysis of high-magnification electron micrographs (Fig. 4) (3, 5).

The SER is very abundant in all the species so far studied and is in the form of a network of anastomosing branching tubules, which are in strict topographic relation to both mitochondria and lipid droplets (Figs. 2, 5 and 6). In some species (e.g., ox and guinea pig), SER is exceedingly well developed, and in the guinea pig it is arranged both in patches of tightly packed membrane (frequently arranged as fenestrated cisternae) and in areas of loosely packed tubules (9). Fenestrated cisternae are occasionally found also in the rat (Fig. 6). Rough endoplasmic reticulum profiles (RER), free ribosomes and polysomes are scattered among the smooth tubules. However, in the human adrenal cortical cells true stacks of RER cisternae can be observed (Fig. 5) (5).

Lipid droplets are invariably present but their number varies according to the species examined. They are very numerous in the rat and human (Figs. 1 and 2), scarce in the guinea pig and mouse, and nearly absent in the ox (4, 5). On the basis of stereological evidence, Frühling et al. (10) suggested that an inverse correlation exists between the volume of the lipid droplet compartment and the quantity of SER; the functional significance of this finding will be discussed in the Section 4.2. Lipid droplets have a 40 Å-thick membrane, which has so far not been resolved as a trilaminar membrane (11).

The Golgi apparatus is always present and usually in a juxtanuclear location (Figs. 2 and 5). Frequently, it is very well developed and consists of many stacks of cisternae and of numerous vesicles, some of which, having a 'coated' appearance, seem to arise from the dilated endings of the golgian saccules.

Motta, PM (ed): Ultrastructure of endocrine cells and tissues. ISBN-13: 978-1-4613-3863-5

Numerous and uniformly scattered in the cytoplasm are pleomorphic dense bodies, displaying acid-phosphatase and arylsulfatase-β activity, which have been interpreted as primary lysosomes (Fig. 7). Furthermore, it was cytochemically demonstrated the presence, especially in close relation to SER tubules, of small peroxidase-positive dense bodies (0.2 μm in diameter), which are assumed to be peroxisomes (Fig. 8) (12). A little amount of β-glycogen particles are sparse in the cytoplasm (Fig. 2).

All adrenocortical cells possess typical microtubules and microfilaments (5). In the rat, Gabbiani et al. (13) reported the presence of a peripheral network of 40–80 Å-thick microfilaments just beneath the cell plasma membrane, but occasionally penetrating more deeply into the cytoplasm to contact the lipid droplets and the mitochondria. Immunofluorescence technique suggests that this peripheral network is composed of actin microfilaments (13).

Microvilli and coated pits (caveolae) are always present at the plasma membrane (2, 4, 5). Adjacent cortical cells display numerous cell-to-cell attachments, including gap junctions (Fig. 9). Friend and Gilula (14) also described a new type of junction, which they called 'septate-like zonulae adhaerentes' and regarded as typical of steroid producing cells (Fig. 10). Here the adjacent plasma membranes are separated by a 210 Å extracellular space penetrated

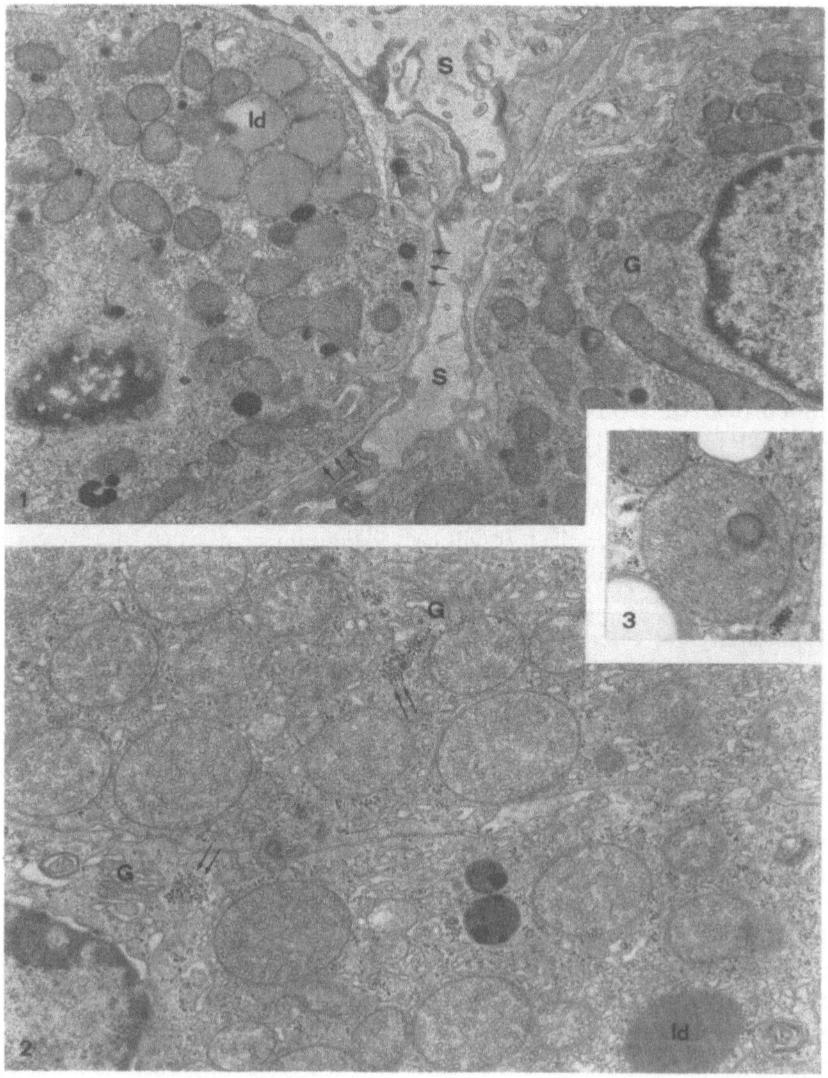

Figs. 1–3. (1) Zona glomerulosa cells of the rat adrenal cortex. Id: lipid droplet; G: Golgi apparatus; S: sinusoid lumen. The arrows indicate the basement membrane (× 8,400). (2) Zona fasciculata cells of the rat adrenal cortex. Id: lipid droplet; G: Golgi apparatus. The arrows indicate small clumps of β-glycogen particles. (× 13,650). (3) Lipid-like inclusion in the mitochondrial matrix (× 14,000).

by 100-to-150 Å particles. Further lanthanum and horseradish peroxidase permeate the extracellular space and the particles are 'encrusted' by pyroantimonate.

2.2. Ultrastructural zonation

Zona glomerulosa cells show a rather homogeneous morphology among the various mammalian species (7). Mitochondria are ovoid or more frequently elongated and display in addition to the typical fingerlike tubular cristae, also conventional laminar cristae (Fig. 1). SER is less developed that in the inner zones' cells, whereas sparse RER profiles seem to be more abundant. According to Black et al. (9), in the guinea pig freeze-fracture demonstrates the presence of only small gap junctions (0.005–0.01 μm^2) between adjacent parenchymal cells.

Zona fasciculata cells are larger than those of the other zones of the gland. In the rat, mitochondria are round or ovoid and invariably contain vesicular cristae (Fig. 2); in the human, mitochondria with short tubular cristae are also found (Fig. 5); in the mouse, rabbit and the Mongolian gerbil, tubular cristae sometimes display a tortuous convolute arrangement (Fig. 16) (4, 5). The SER is more abun-

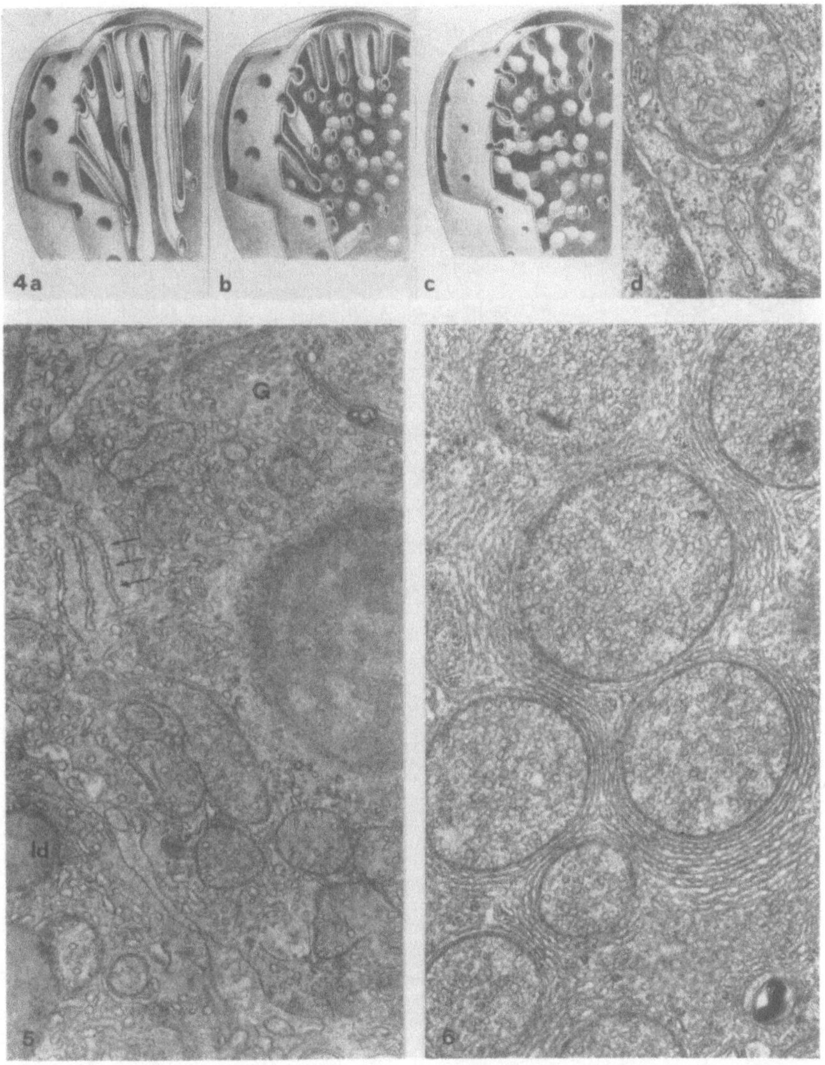

Figs. 4–6. (4) Stereogram illustrating the cristal arrangement in the adrenocortical mitochondria. a: tubular cristae; b: vesicular cristae according to the classic view; c: vesicular cristae according to the model proposed by Allmann et al. (8); d: electron micrograph supporting the hypothesis of Allmann et al. (× 17,500). (5) Zona fasciculata cells of the human adrenal cortex. Id: lipid droplet; G: Golgi apparatus. The arrows indicate some RER cisternae (× 15,400). (6) Zona fasciculata cells of the rat adrenal cortex, containing SER fenestrated cisternae, which surround mitochondria (× 16,100).

dant than in zona glomerulosa cells and, except in the human (Fig. 5) and guinea pig (9), RER profiles are invariably lacking. Microvilli and coated pits seem to be more numerous. Gap junctions are increased in size (0.01–0.1 μm^2) in the cells of the inner portion of the zone (9). In the human, some inner cells display small granules of lipofuscin pigment.

Zona reticularis cells, though rather smaller, do not differ conspicuously from those of the zona fasciculata. A distinctive feature is the presence in the rat of round or ovoid mitochondria, with tubulo-convolute cristae, filling all their matrix (Fig. 11)(4, 11); however, mitochondria with conventional vesicular cristae are also found. Several primary and secondary lysosomes, containing lipid-like inclusions, as well as lipofuscin pigment granules are observed especially in those cells located in the inner juxtamedullary portion of the zone (11). The accumulation of lipofuscin pigment is very striking in the human being (Fig. 12). Freeze-fracture studies reveal the presence in the guinea pig of very long gap junctions (about 2.5 μm^2), in addition to the small ones (0.01–0.1 μm^2) (9).

Zona intermedia. A sudanophobic layer between the zonae glomerulosa and fasciculata can be recognized in many species at the optical level. Electron

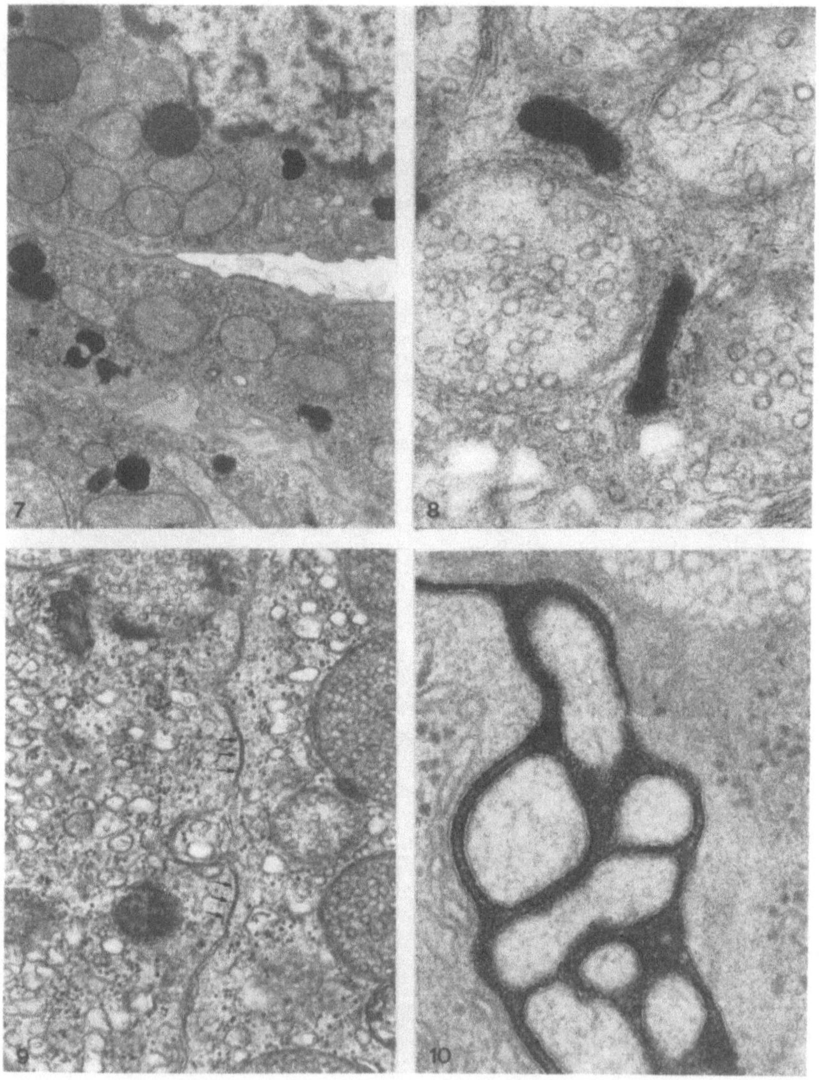

Figs. 7–10. (7) Zona fasciculata cells of the rat adrenal cortex, showing some Gomori-positive bodies (lysosomes) (× 10,500). (8) Portion of a rat zona fasciculata cell, containing some DAB-positive bodies (peroxisomes) (× 39,900). (Courtesy of M.C. Magalhães). (9) Zona fasciculata cells of the rat adrenal cortex. The arrows indicate two gap junctions (× 24,500). (10) Septate-like zonulae adhaerentes, which are fully permeated by lanthanum (× 68,600). (From Friend, Gilula, 14).

microscopic investigations have provided a description of the zona intermedia in only few species: the rat, the dog and the mouse (7). The cells have the same volume of those of the zona glomerulosa and contain pleomorphic mitochondria with tubular cristae. Except that in the mouse, SER is well developed and RER cisternae are numerous. The most striking feature is the virtual absence of lipid droplets. Nickerson (15) suggests that these are transitional cells between zona glomerulosa and zona fasciculata elements.

3. Ultrastructure of the adrenal cortex under experimental conditions

In this section the ultrastructural changes induced in the adrenal cortex by various experimental conditions will be shortly reviewed. In addition, a brief account of the current concepts on the regulation of the growth and hormonal secretion of adrenocortical cells will be provided, as a basis for an easier understanding of the experimental morphologic data.

3.1. Multifactorial regulation of adrenocortical functions

On this subject many excellent reviews have appeared, to which the readers can refer for complete discussion (16–18).

Zona glomerulosa. There are decisive proofs that the regulation of aldosterone secretion by the mammalian zona glomerulosa is chiefly mediated by: (1) the renin-angiotensin system, (2) the plasma concentration of Na^+ and K^+, and (3) the hypothalamo-hypophyseal axis (ACTH). Each of the adrenoglomerulotropic factors seems to promote aldosterone synthesis by a direct effect on the zona glomerulosa cells, as is well demonstrated by several in vitro experiments. However, interrelationships among the various adrenoglomerulotropic factors in vivo cannot be excluded at present (7).

Zonae fasciculata and reticularis. There is general agreement that the activity of the two inner adrenocortical layers is mainly controlled by the hypothalamo-hypophyseal axis (ACTH). However, it is to be mentioned that other factors have been claimed to be involved in the regulation of the adrenal cortex: ADH, serotonin, growth hormone, prolactin, prostaglandins, gonadotropins and sex hormones. Discussion of these factors is beyond the scope of this article and the readers can found appropriate references in the review of Vinson and Kenyon (17).

All the three above mentioned adrenocorticotropic factors seem to exert short-term and long-term effects on the adrenal cortex. Short-term (tropic) effects involve an immediate increase in corticosteroid hormone secretion, which starts after about 10 min and reaches a maximum after 15–30 min. Long-term (trophic) effects involve, in addition to normal growth maintenance of the gland, an increase in the adrenal weight and in steroidogenic capacity of its parenchymal cells; in other words, long-term stimulation induces a spectrum of structural changes enabling adrenocortical cells to maintain a high rate of hormonal output for longer periods (19).

3.2. Zona glomerulosa

Most of the experiments available are of chronic stimulation or inhibition and the morphologic results are rather superposable whatever was the glomerulotropic factor tested.

Long-term stimulation of zona glomerulosa was observed in renovascular hypertensive children and rat, in agiotensin II-treated rats, in sodium-depleted or potassium-loaded rat and mouse, in ACTH-administered rats as well as in mouse bearing a transplantable ACTH-secreting tumor (4, 7).

No striking qualitative morphologic changes are found, except an obvious hypertrophy of the Golgi apparatus and an increase in the number of microvilli and caveolae at the plasma membrane. Moreover, in the rat and human, chronic stimulation of the renin-angiotensin system induces the appearance of several clumps of electron dense granules (20, 21), which do not show appreciable acid-phosphatase activity and seem to arise in the hypertrophic Golgi area (see Section 4.3.). Morphometry showed that chronic stimulation provokes a time-dependent increase in the volume of the zona glomerulosa and its cells, which is mainly due to the hypertrophy of the mitochondrial compartment and SER membranes. The surface area of the mitochondrial cristae is also significantly enhanced. The volume of the lipid compartment is noticeably reduced up to the 15th day of treatment (22), and increased by the 30th day (23).

Long-term inhibition of zona glomerulosa was studied in the rat by suppressing the renin-angiotensin system with timolol maleate and the hypothalamo-hypophyseal adrenal axis with dexamethasone, and in sodium-loaded or potassium-deprived rats and mouse (4, 7).

Qualitative morphologic signs of hypofunction are the atrophy of the Golgi apparatus, and the increase in the number of lysosome-like dense bodies and β-glycogen particles. Stereology showed a significant cell atrophy, due to the decrease in the volume of the mitochondrial compartment and SER; the volume of the lipid compartment does not vary or is only slightly increased.

More recently, Nussdorfer et al. (24) observed reorganization of the tubulo-laminar cristae of zona glomerulosa mitochondria into vesicular ones after prolonged ACTH-administration to sodium-loaded rats, whose renin-angiotensin system was inhibited by timolol maleate. These authors suggest that the morphologic integrity of the inner membrane of the zona glomerulosa mitochondria requires the co-operation of all the various adrenoglomerulotropic factors.

Ultrastructural investigations on the *short-term stimulation* of the zona glomerulosa are very few: only the structural changes present 1hr after angiotensin II administration to rats, and those evident 1hr after activation of the hypothalamo-hypophyseal axis by severe stress or ACTH-treatment in the monkey and rabbit were described (7). Hypertrophy of the Golgi apparatus and a striking stereologically demonstrable lipid depletion are the only changes observed. The morphological counterpart of the *short-term inhibition* has not been yet investigated.

3.3. Zonae fasciculata and reticularis

Stimulation was obtained by administering ACTH or

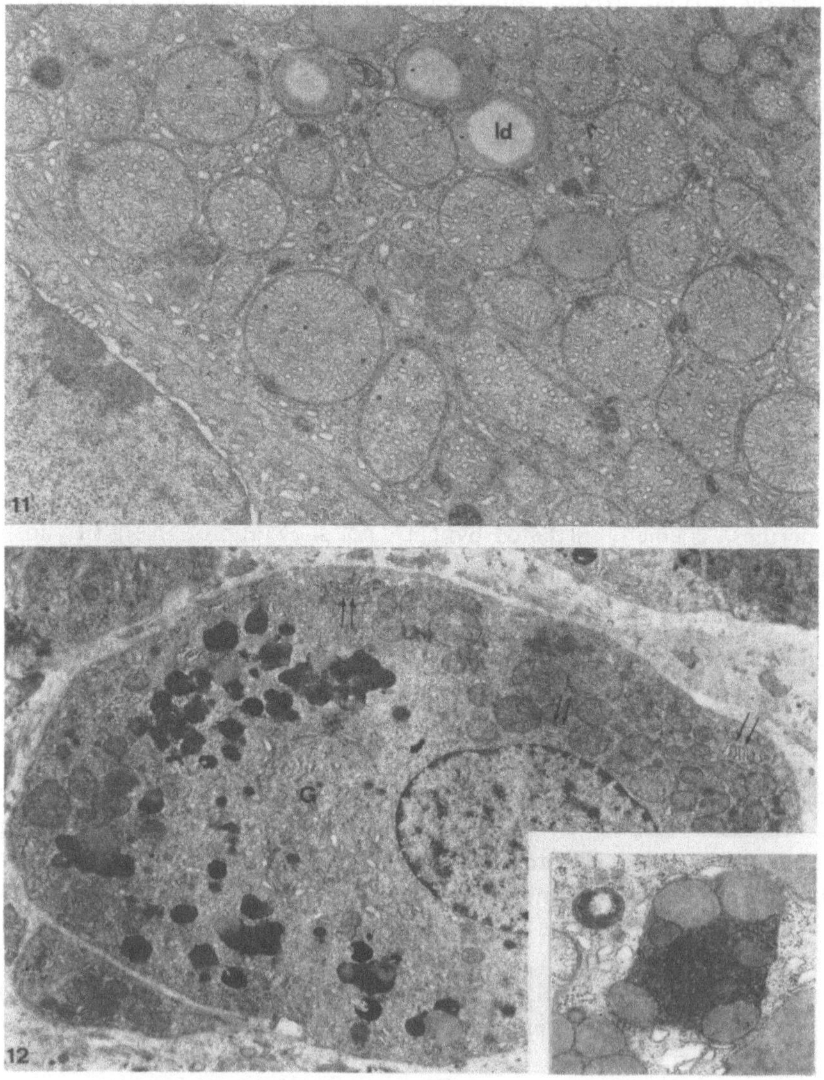

Figs. 11–12. (11) Zona reticularis cells of the rat adrenal cortex. Id: lipid droplet (× 12,600). (12) Zona reticularis cell of the human adrenal cortex. Many clumps of lipofuscin pigment are contained in the cytoplasm. The inset shows the structural heterogeneity of lipofuscin pigment bodies. G: Golgi apparatus. The arrows indicate some RER profiles (× 5,600). Inset (× 17,150).

its intracellular mediators (e.g., cyclic-AMP), or by employing various types of stressful conditions, activating the hypothalamo-hypophyseal axis; inhibition was provoked by hypophysectomy or by treating the animals with dexamethasone or other glucocorticoid hormones. In the human, the morphologic counterpart of chronic activation of the inner adrenal zones was observed in the Cushing's syndrome due to an ACTH-secreting hypophyseal tumor (25).

Long-term stimulation of the adrenal zonae fasciculata and reticularis has been studied in several mammalian species (3–5). The cellular and nuclear volumes increase significantly as a function of the duration of the stimulating treatment. Mitochondria do not display noticeable alterations, but stereology demonstates that the volume of the mitochondrial compartment undergoes to a significant enhancement, which is due to the increase both in the average volume and in the number of the organelles. Images suggesting mitochondrial division can be often observed (Fig. 13). The surface area of the mitochondrial cristae is also noticeably increased. SER membranes proliferate and frequently occupy large cytop-

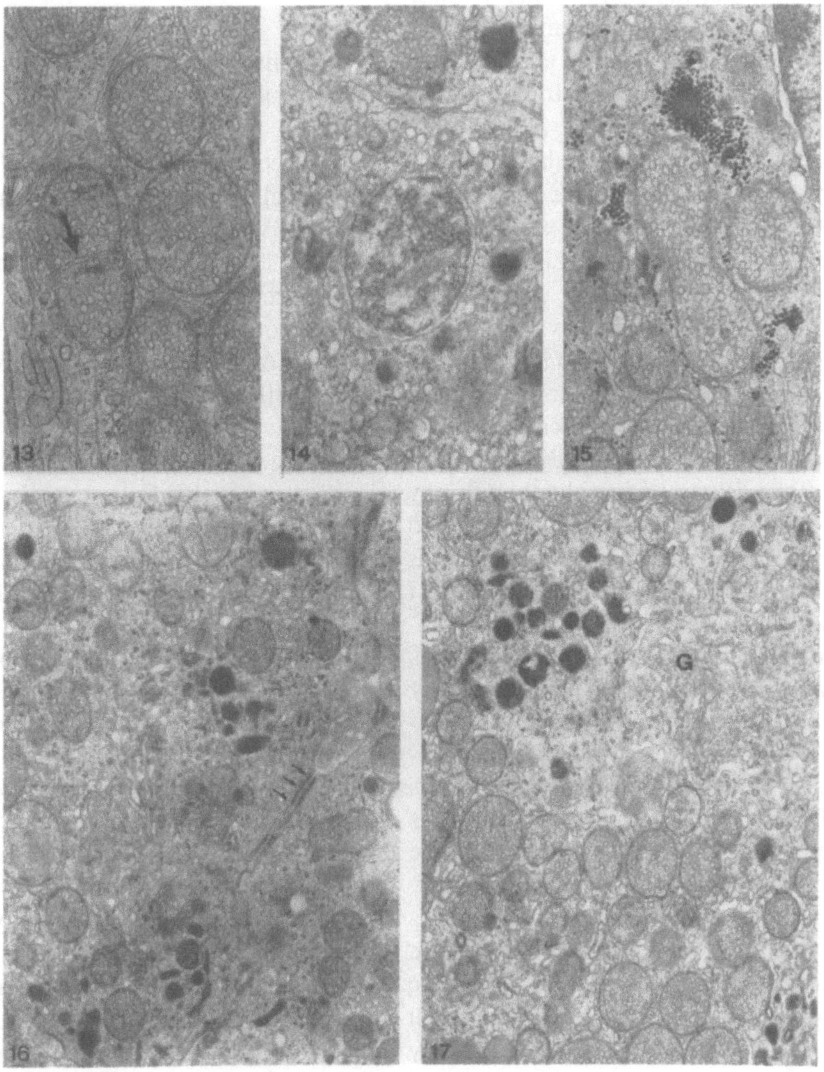

Figs. 13–17. (13) Zona fasciculata cells of a rat chronically treated with ACTH. The arrow indicates an image suggesting mitochondrial division (× 17,150). (14) A degenerate mitochondrion in a chronically suppressed adrenal cortical cell (× 17,150). (15) In chronically suppressed adrenocortical cells, large intracytoplasmic accumulation of β-glycogen particles can often be observed (× 14,700). (16) Zona fasciculata cells of a rabbit acutely treated with ACTH. Note the numerous dense bodies in the Golgi area. The two cells are attached by several desmosome-like junctions (arrows) (× 8,400). (17) Clumps of dense granules in zona fasciculata cells of a vinblastine administered rat. G: Golgi apparatus (× 9,800).

lasmic portions lacking any other organelle; stereology furnishes a clearcut demonstration of this finding. Considerable disagreement exists regarding the changes in the lipid compartment. The greatest number of investigations report a conspicuous decrease in the number and volume of the lipid droplets, but stereology discloses that this finding is only apparent, being due to the 'dilution' of the lipid droplets in the hypertrophic cytoplasm. It is now quite generally accepted that the volume of the lipid droplets increases in the zonae fasciculata and reticularis, at least after prolonged stimulation. The Golgi apparatus is commonly found to hypertrophy and possesses a high number of vesicles, many of which are coated (see Section 4.3.). Dense bodies, some of which are true lysosomes, are more numerous and the plasma membrane shows an elaborate microvillous pattern and several micropinocytotic vesicles and coated pits.

Human zonae fasciculata and reticularis cells of patients suffering Cushing's syndrome are hypertrophied and display SER hyperplasia, an apparent RER decrease and Golgi hypertrophy. Lipid droplets are variable in amount from cell to cell. Unfortunately stereological studies are not available (6).

Long-term inhibition of the hypothalamo-hypophyseal adrenal axis induces morphologic changes almost completely opposite to those described above. The cells frequently shrink, assuming irregular shapes, and their nuclei have decreased in volume and may show signs of pyknosis. Regressive changes lead to degeneration and death of some cells, and this process seems to start from the 3rd day of ACTH suppression. The volume of the mitochondrial compartment is significantly decreased and the organelles frequently assume bizarre shapes and contain mainly tubular cristae. Giant mitochondria have been described as well as degenerate organelles and autophagic vacuoles containing mitochondrial debris (Fig. 14) (5). SER and Golgi apparatus atrophy is observed. Stereology does not show noticeable changes in the volume of the lipid compartment; only a slow decrease in the late phases of suppression is reported. Subjectively, however, lipid droplets seem to be increased in number because of the cytoplasmic atrophy. β-glycogen particles accumulate in small clumps (Fig. 15). Microvilli and coated pits are only occasional features.

Atrophy of cells and SER was also found in the controlateral adrenal gland of patients bearing a hormone-secreting adrenocortical tumor (6, 25).

The morphologic counterpart of the *short-term stimulation* of the inner adrenal zones has been the object of few investigations (3–5). Some authors reported alterations in mitochondrial cristal arrangement and in SER, but quantitative stereological and autoradiographic techniques showed that the only conspicuous change is a severe lipid depletion (26, 27). Other findings are the Golgi apparatus hypertrophy and the appearance 40–60 min after ACTH administration of several dense bodies, many of which do not display acid-phosphatase reaction (see, Section 4.3.).

Short-term inhibition was studied only by Rhodin (11) in dexamethasone administered rats (16–25 hs). Lipid droplets seem to be increased in number and clumps of β-glycogen particles appear in the cytoplasm.

4. Morphological-functional correlations in adrenocortical cells

In this section the morphologic features of adrenocortical cells from normal and experimentally treated animals will be correlated with the biochemical pathways leading to steroid hormone synthesis and the subcellular topology of the enzymes involved. The hypotheses concerning the possible mechanisms of hormone release by adrenocortical cells will be also briefly discussed.

4.1. Subcellular localization of the enzymes of steroid synthesis

The bulk of evidence shows that the enzymes involved in steroid hormone synthesis are located in mitochondria and SER (28, 29). Free cholesterol molecules, taken up from the blood stream or endogenously synthesized from acetate in SER, enter the mitochondria, where side-chain cleaving enzymes transform them into pregnenolone. Pregnenolone is then transformed into progesterone by 3β-hydroxysteroid Δ^5-dehydrogenase plus Δ^{4-5} isomerase located in the SER, where 17α-hydroxylase and 21-hydroxylase convert it into 11-deoxycortisol and 11-deoxycorticosterone, respectively. The intermediate products again penetrate into the mitochondria, where 11β-hydroxylase tranforms them into cortisol and corticosterone. Corticosterone in turn is converted to aldosterone by intramitochondrial 18-hydroxylase and 18-hydroxy-dehydrogenase.

4.2. Subcellular organelles and steroidogenesis

The presence of lipid droplets can be easily explained since they contain cholesterol and cholesterol esters (30), which are the main precursors of steroid hormones. It also clearly results that intermediate product molecules must frequently switch between SER and mitochondria before being transformed into definitive hormones. Therefore, the morphological counterpart of this process may be the intimate topographic interrelationships among mitochondria, lipid droplets and SER membranes.

Mitochondria and SER. Most conspicuous changes in long-term activated or inhibited adrenocortical cells concern SER and mitochondria. Many lines of evidence show that the activity of several enzymes of steroid synthesis is enhanced or depressed in chronically stimulated or suppressed adrenocortical cells (5). Since it seems to be conceivable that, like the classic respiratory chain, the cytochrome P_{450} electron tranfer chain also requires an adequate steric arrangement for its complete activity, the hypothesis was advanced that the increase in the surface area of SER membranes and mitochondrial cristae, induced by the various adrenocorticotropic factors, provides an increased framework of basic membrane to which de novo synthesized enzymes of steroid synthesis can be inserted (5, 7).

An alternative hypothesis concerning the significance of SER proliferation in stimulated adrenocortical cells has been proposed by Black et al. (9). These investigators claim that, since SER is involved in endogenous cholesterol synthesis, it is conceivable that cholesterol may actually be incorporated into smooth membranes, which therefore would function as storage sites of steroid hormone precursors. In the guinea pig, this function would be carried out by tightly packed patches of cisternae. These last, upon stimulation, could revert to the loosely packed tubular form, which would represent the morphological arrangement of SER actively engaged in steroid synthesis. Transformation between the two forms of SER would allow the cells to be responsive to suppression and stimulation, at least in the physiologic range, without having to invoke lysosomal degradation and newly membrane synthesis at every turn. Unfortunately, these two SER forms are not present in all the mammalian species.

Lipid droplets. It is well established that in the lipid droplets are stored cholesterol and cholesterol esters (30). Frühling et al. (10) stereologically demonstrated that the number of lipid droplets is in inverse proportion to the cell capacity in endogenous cholesterol synthesis (hence in the SER quantity, see Section 2). They suggest that only cholesterol exogenously derived from the blood stream is stored in adrenocortical lipid compartment.

Nussdorfer et al. (5) pointed out that the volume of the lipid compartment in adrenocortical cells is the expression of the balance between the rates of formation and utilization of lipid droplets and therefore is the result of the following processes: (1) endogenous synthesis of cholesterol; (2) uptake of exogenous cholesterol and its esterification and storage in the lipid droplets as reserve material; (3) transformation of esterified cholesterol stored in the lipid droplets into free cholesterol; and (4) utilization of free cholesterol in steroid hormone synthesis. It seems quite well established that ACTH (and perhaps other adrenocorticotropic factors) enhances all these four processes (5).

Therefore, it is obvious that the findings concerning the behaviour of the volume of the lipid compartment in stimulated or suppressed adrenocortical cells would vary according to (1) the gland zone and the animal species investigated (in relation to their different capacity for endogenous cholesterol synthesis), and (2) the duration of the experimental procedure.

The decrease in the volume of lipid droplets at the onset of stimulating treatment can be interpreted as the morphologic expression of the increased requirement of free cholesterol for utilization in steroid synthesis (Processes 3 and 4); this would be much more evident in those species (and gland zones) in which, owing to the relative SER paucity, the cellular demand of free cholesterol cannot be satisfied by Process 1.

In the prolonged stimulating treatments, the volume of lipid droplets increases, and this may be well correlated with the SER proliferation and the consequent enhancement of Process 1, coupled with an increased uptake of cholesterol from the blood stream. Obviously, the volume of the lipid compartment would start to increase earlier in those cells (and species), which are able to adapt more rapidly their rate of endogenous cholesterol synthesis to the level required for steroid synthesis (e.g., zona fasciculata vs zona glomerulosa; guinea pig vs rat). In the suppressed adrenocortical cells the volume of lipid compartment remains in plateau and this can be easily understood by considering that in these conditions, not only the processes leading to the utilization of lipid droplets, but also those provoking their accumulation are blocked.

Golgi apparatus. The Golgi apparatus hypertrophies in hyperfunctioning adrenocortical cells and shows atrophy in the suppressed cells, so that it is currently considered to be involved in steroid synthesis and secretion. Its role, however, is still under debate. Fawcett et al. (1) suggest that the Golgi apparatus might intervene in the modulation of steroid sulfation/desulfation processes in preparation for hormone synthesis and/or secretion. More recently, autoradiographic data allowed Haddad et al. (31) to hypothesize that in adrenocortical cells this organelle is involved in the synthesis of glycoproteins, which might function as carrier molecules to transport the newly built steroid hormones to the extracellular environment (32).

Other organelles and cell inclusions. Specific investigations on the role played by lysosomes and peroxisomes in adrenocortical cell physiology are not available at present. For a discussion of this topic the readers may consult some recent review articles (4, 5).

A noticeable increase in β-glycogen particles is observed to follow both acute and chronic inhibition of adrenocortical cells. This ultrastructural change is easily explained by considering that one of the better documented mechanism of action of the ACTH is held to involve activation of glycogen phosphorylase, which, by promoting glycogenolysis, activates the pentose shunt, thus increasing the availability of NADPH, a rate-limiting co-enzyme in steroidogenesis (33).

Cell-to-cell attachments. The possible role of gap junctions and septate-like zonulae adhaerentes in adrenocortical cells deserves some discussion.

Evidence indicates that ACTH exposure increases the number and size of gap junctions in adrenocortical cells cultured *in vitro* (34). Black et al. (9) showed that these cell-to-cell attachments increase in size and number in the parenchymal cells located in the inner adrenal layers. For these investigators the exchange of small molecules (including cyclic nucleotides) is a process which seems to be a direct function of the surface area of gap junctions. On these grounds they suggest that adrenocortical cells located furthest from fresh vascular supply and therefore exposed to a lower concentration of blood-borne substances, must possess large gap junctions, which could enhance the efficiency of their response to stimulation, by serving as a pathway for the relay (and amplification?) of the hormonal signals first received at the periphery of the gland. Moreover, Black et al. (9) observed that abundant SER tubules occur near the gap junctions and advance the hypothesis that they are involved in the maintenance of the junctional permeability, by sequestering Ca^{2+}, whose intracellular concentration rises in response to the various adrenocorticotropic factors (35).

According to Friend and Gilula (14), the septate-like zonulae adhaerentes, besides functioning in adherence, may well maintain the width of the intercellular space, thus forming microchannels for the possible flow of secretory products into the blood stream. Moreover, the 100–150 Å particles may be the site of high cation concentration, presumably Ca^{2+}.

4.3. The mechanism of hormone release by adrenocortical cells

To date, ultrastructural investigations have provided little insight into the mechanism(s) of steroid hormone release. The old theories of holocrine and apocrine secretion, as well as that of endoplasmocrine secretion (11) have now only a historical interest.

Secretion by simple diffusion. The most widely accepted theory for the release of adrenocortical hormones holds that they are free to diffuse throughout the aqueous cytoplasm and the lipid phase of the plasma membrane. However, recent studies on the intracellular concentrations of many steroids vs medium concentrations demonstrate retention of the hormones against the concentration gradients at the plasma membrane. The steroid hormone molecules are located in the cytosol and the maintenance of the steroid cytosolic pool is likely to involve steroid-protein interaction (36). The possible involvement of the Golgi apparatus in the synthesis of these carrier proteins has been previously discussed (Subsection 4.2.).

Exocytotic secretion. The view that an exocytotic mechanism underlies hormone release in steroid producing cells is now gaining more and more support, although the morphological demonstration of true secretory granules and of their release has been so far very elusive. However, the presence of electron-dense granules in normal and stimulated adrenocortical cells is a rather frequent evenience (see Sections 2 and 3), but these organelles were commonly considered to be lysosomes or peroxisomes.

Gemmell et al. (37) and Mazzocchi et al. (26) showed in acutely stimulated cat and rabbit zona fasciculata cells a conspicuous increase in dense granules, which appear to arise in the Golgi area and are prevalently located near the plasma membrane (Fig. 16). These investigators exclude the possibility that these granules are lysosomes inasmuch as (1) their numerical increase is not associated with enhanced activity of certain typical lysosomal enzymes in adrenal homogenates (e.g., acid-phosphatase, β-glucuronidase and N-acetyl-β-glucosaminidase), and (2) the increase in the intracellular concentration of steroid hormones after acute stimulation is of the same order of magnitude as that of the granule volume density, as evaluated by stereology. These authors maintain that the lack of images suggesting exocytosis is due to the paucity of these putative secretory granules, which are not stored in the cytoplasm, but are released as soon as they are formed. Exocytosis is usually coupled with pinocytosis, and therefore the presence of many coated pits and vesicles associated with the plasma membrane of stimulated adrenocortical cells may well be considered an indirect support for the theory of the exocytotic mechanism of steroid hormone release.

In line with this view are the experiments of Nussdorfer et al. (38). Since microtubules are thought to be involved in the exocytosis, these workers devised to block the release of corticosterone by rat adrenocortical cells, by treating the animals with vinblastine (an antimicrotubular agent). This experimental procedure induces the appearance of several clumps of electron dense granules in both the Golgi area and the juxtasinusoidal poles of the cells (Fig. 17), coupled with a 4-5-fold rise in the intracellular concentration of corticosterone. The ACTH administration to vinblastine-treated animals provokes a further increase in the number of electron dense granules and in the intracellular hormonal concentration, without any significant rise in the plasma corticosterone level. These findings indicate that vinblastine blocks corticosterone release, without impairing its synthesis.

Finally, it seems worthy of mention a recent hypothesis advanced by Basset and Pollard (39). These investigators propose that coated vesicles, budding from the Golgi cisternae, are involved in the intracellular transport and release of corticosterone in rat adrenocortical cells.

In conclusion, we want to stress that, albeit the exocytotic theory would appear the most stimulating one, much work is still needed before the mechanism underlying hormone release by adrenocortical cells can be considered as completely elucidated.

5. Summary and conclusions

In the recent years many ultrastructural investigations, mainly employing stereological procedures, have succeeded in correlating the morphology of adrenocortical cells with the biochemical events leading to steroid hormone synthesis. The morphologic changes associated with increased or depressed activity of adrenocortical cells have been clearly recognized, so that the interrelationships between many organelles (mitochondria, SER and lipid droplets) and the cell secretory activity appear rather completely settled. However, the specific role, if any, played by the Golgi apparatus, lysosomes and peroxisomes in steroid-secreting cells is still controversial, and we think that this topic should be the task of future investigations.

One other very stimulating point of investigation is the mechanism of steroid hormone release. In our opinion the exocytotic theory is the more promising working hypothesis, for future investigations. We emphasize that (1) efforts to isolate a cellular fraction reasonably rich in dense bodies, (2) studies aimed at demonstrating the possible hormonal content of the putative secretory granules by immuno-electron microscopical techniques, as well as (3) high resolution autoradiographic investigations on the fate of 3H-precursors of steroid hormones, using specimens processed for cryoultramicrotomy to prevent steroid molecule extraction (40), would be a possible fruitful approach to resolve this problem.

References

1. Fawcett DW, Long J, Jones AL: Ultrastructure of the endocrine glands. Recent Prog Horm Res 25: 315–380, 1969.
2. Idelman S: Ultrastructure of the mammalian adrenal cortex. Int Rev Cytol 27: 181–282, 1970.
3. Malamed S: Ultrastructure of the mammalian adrenal cortex in relation to secretory function. In: Handbook of Physiology, Sect. 7: Endocrinology, Vol. 6: Adrenal gland, Blaschko H, Sayers G, Smith AD (eds), Washington DC, American Physiological Society, 1975, pp 25–39.
4. Idelman S: The structure of the mammalian adrenal cortex. In: General, comparative and clinical endocrinology of the adrenal cortex, Chester-Jones I, Henderson IW (eds), New York, Academic Press, 1978, pp 1–199.
5. Nussdorfer GG, Mazzocchi G, Meneghelli V: Cytophysiology of the adrenal zona fasciculata. Int Rev Cytol 55: 291–365, 1978.
6. Neville AM, O'Hare MJ: Aspects of structure, function and pathology. In: The adrenal gland, James VHT (ed), New York, Raven Press, 1979, pp 1–65.
7. Nussdorfer GG: Cytophysiology of the adrenal zona glomerulosa. Int Rev Cytol 64: 307–369, 1980.
8. Allmann DW, Wakabayashi T, Korman EF, Green DE: Studies of the transition of the cristal membrane from the orthodox to the aggregated configuration. I. Topology of bovine adrenal cortex mitochondria in the orthodox configuration. J Bioenergetics 1: 73–86, 1970.
9. Black VH, Robbins E, McNamara N, Huima T: A correlated thin-section and freeze-fracture analysis of guinea pig adrenocortical cells. Am J Anat 156: 453–504, 1979.
10. Frühling J, Sand G, Penasse W, Pecheux F, Claude A: Corrélation entre la morphologie et le contenu lipidique des corticosurrénales du cobaye, du rat et du boeuf. J Ultrastruc Res 44: 113–133, 1973.
11. Rhodin JAG: The ultrastructure of the adrenal cortex of the rat under normal and experimental conditions. J Ultrastruc Res 34: 23–71, 1971.
12. Magalhães MM, Magalhães MC: Microbodies (peroxisomes) in rat adrenal cortex. J Ultrastruct Res 37: 563–573, 1971.
13. Gabbiani G, Chaponnier C, Lüscher EF: Actin in the cytoplasm of adrenocortical cells. Proc Soc Exp Biol Med 149: 618–621, 1975.
14. Friend DS, Gilula NB: A distinctive cell contact in the rat adrenal cortex. J Cell Biol 53: 148–163, 1972.
15. Nickerson PA: The adrenal cortex in spontaneously hypertensive rats. A quantitative ultrastructural study. Am J Pathol 84: 545–560, 1976.
16. Müller J: Regulation of aldosterone synthesis, Berlin and New York, Springer Verlag, 1971.
17. Vinson GP, Kenyon CJ: Steroidogenesis in the zones of the mammalian adrenal cortex. In: General, comparative and clinical endocrinology of the adrenal cortex, Chester-Jones I, Henderson IW (eds), New York, Academic Press, 1978, pp 201–264.
18. Jones MT: Control of adrenocortical hormone secretion. In: The adrenal gland, James VHT (ed), New York, Raven Press, 1979, pp 93–130.

19. Kuo TH, Tchen TT: Comparison of the effects of adrenocorticotropic hormone on the steroidogenic activity and ultrastructure of the adrenal cortex. J Biol Chem 248: 6679–6683, 1973.
20. Rebuffat P, Belloni AS, Mazzocchi G, Vassanelli P, Nussdorfer GG: A stereological study of the trophic effects of the renin-angiotensin system on the rat zona glomerulosa. J Anat 129: 561–570, 1979.
21. Hashida Y, Yunis EJ: Ultrastructure of the adrenal zona glomerulosa in children with renovascular hypertension. Human Pathol 3: 301–315, 1972.
22. Mazzocchi G, Rebuffat P, Belloni AS, Robba C, Nussdorfer GG: An ultrastructural stereologic study of the effects of angiotensin II on the zona glomerulosa of rat adrenal cortex. Acta Endocrinol 95: 523–527, 1980.
23. Kasemsri S, Nickerson PA: Quantitative ultrastructural study of the rat adrenal cortex in renal encapsulation-induced hypertension. Am J Pathol 82: 143–156, 1976.
24. Nussdorfer GG, Neri G, Belloni AS, Mazzocchi G, Rebuffat P, Robba C: Effects of ACTH on the zona glomerulosa of sodium-loaded, timolol maleate treated rats: stereology and plasma hormone concentrations. Acta Endocrinol, 99: 256–262, 1982.
25. Mitschke H, Saeger W: Zur Ultrastruktur der atrophischen Nebennierenrinde bei dissoziierter, sekundärer Nebennierenrindeninsuffizienz. Virchows Arch A 361: 217–228, 1973.
26. Mazzocchi G, Belloni AS, Rebuffat P, Robba C, Neri G, Nussdorfer GG: Fine structure of the rabbit adrenal cortex and the effects of short-term ACTH administration. Cell Tissue Res 201: 165–179, 1979.
27. Sharawy M, Dirksen T, Chaffin J: Increase in free cholesterol content of the adrenal cortex after stress: radioautographic and biochemical study. Am J Anat 156: 567–576, 1979.
28. Tamaoki BI: Steroidogenesis and cell structure. Biochemical pursuit of sites of steroid biosynthesis. J Steroid Biochem 4: 89–118, 1973.
29. Sandor T, Fazekas AG, Robinson BH: The biosynthesis of corticosteroids throughout the vertebrates. In: General, comparative and clinical endocrinology of the adrenal cortex, Chester-Jones I, Henderson IW (eds), New York, Academic Press, 1976, pp 25–142.
30. Sand G, Frühling J, Penasse W, Claude A: Distribution du cholestérole dans la corticosurrénale du rat: analyse morphologique et chimique des fractions subcellulaires, isolées par centrifugation différentielle. J Microscopie 15: 41–66, 1972.
31. Haddad A, Brasileiro ILG, Pelletier G: Incorporation of L(3H)-fucose into glycoproteins of the adrenal gland of mice. Light microscope radioautographic study on semi-thin sections. Cell Tissue Res 202: 325–335, 1979.
32. Wagner RK: Extracellular and intracellular steroid binding proteins. Properties, discrimination, assay and clinical application. Acta Endocrinol 88 (Suppl 218): 7–73, 1978.
33. Haynes RC Jr: Theories on the mode of action of ACTH in stimulating secretory activity of the adrenal cortex. In: Handbook of Physiology, Sect. 7: Endocrinology, Vol. 6: Adrenal gland, Blaschko H, Sayers G, Smith AD (eds), Washington DC, American Physiological Society, 1975, pp 69–76.
34. Mattson P, Kowal J: The ultrastructure of functional mouse adrenal cortical tumor cells in vitro. Differentiation 11: 75–88, 1978.
35. Rubin RP: Calcium and the secretory process, New York, Plenum Press, 1974.
36. Goddard C, Vinson GP, Whitehouse BJ, Silbley CP: Subcellular distribution of steroids in the rat adrenal cortex after incubation in vitro. J Steroid Biochem 13: 1221–1229, 1980.
37. Gemmell RT, Laychock S, Rubin RP: Ultrastructural and biochemical evidence for a steroid-containing secretory organelle in the perfused cat adrenal gland. J Cell Biol 72: 209–215, 1977.
38. Nussdorfer GG, Mazzocchi G, Neri G, Robba C: Investigations into the mechanism of hormone release by rat adrenocortical cells. Cell Tissue Res 189: 403–407, 1978.
39. Bassett JR, Pollard IR: The involvement of coated vesicles in the secretion of corticosterone by the zona fasciculata of the rat adrenal cortex. Tissue & Cell 12: 101–115, 1980.
40. Magalhães MM: Ultrastructure of the adrenal cortex in frozen thin sections. J Cell Sci 27: 303–311, 1977.

Authors' address:
Department of Anatomy,
Laboratory of Electron Microscopy,
University of Padua,
Padua, Italy

CHAPTER 18

Ultrastructural features of adrenocortical and hypophyseal cells in culture

GYÖRGY RAPPAY

1. Introduction

Endocrine cells in primary cultures retain their capacity to produce hormones for a considerable period of time and this enables solutions of many problems involved in the control of biosynthesis, storage and release of hormones. In spite of the extended usage of cultured endocrine cells as test objects for unrolling complicated regulatory processes, the investigation of the structural features of cells grown in culture is an almost neglected topic of the endocrine cell cytology, although an excellent review on the ultrastructure of pituitary cells was published previously (1).

The structure-function relationship is unequivocally evident in peptide hormone secreting cells in which stored hormones are packed in secretory granules of various size and number. So the presence of these structures in culture may indicate the maintenance of the more or less intact specific function of a cell. In steroid hormone secreting cells, it is not easy to evidence morphologically the undisturbed specific function in culture of a given cell. In that case it is necessary to trace the hormone synthetic pathway and to correlate it to structures which are rich in processing enzymes of the steroids concerned such as for example the presence of the smooth-surfaced endoplasmic reticulum membranes in the steroid secreting cells.

Morphological changes in target cells, together with tracing hormones in the culture system, can bring evidences that parenchymal cells may properly pursue differentiated function in culture. The recent recognition of the receptor-mediated endocytosis opens a new look into the understanding of the delayed effects of hormone-receptor interactions. In addition to explaining the mechanism of lysosomal inactivation of hormones bound to their receptors and subsequent rearrangement of membrane proteins it may be helpful in unrolling events leading to morphological changes of organelles within target cells.

2. Recent development in methodology

2.1. Culture technique

Contemporary methodology involves both progress and backwardness. This ambiguity reflects the ingenuity of the pioneers and the sweep of the present-day techniques into the tissue culture laboratory. Among the most promising new techniques the significance of two procedures must first be stressed. The dispersion by harmless enzymatic mixtures of cells derived from relatively hard tissues may provide a better yield of living parenchymal cells with intact membranes capable of adhering to inert surfaces and of responding to stimuli for differentiated function. The harmless dispersing procedure seems to be particularly important in case of culturing single cells derived from endocrine tissues since they have sensitive membrane receptors. On the other hand, the extended usage of defined media is desirable indicating a new trend in the development of nutrition of cells in culture. Dr. Gordon Sato and his group demonstrated the benefits of combination of hormones, growth factors and other essential nutrients instead of serum supplements for the maintenance of many cell types in culture (2). The diversity of cell types for which the hormonal and other metabolic requirements in culture have been defined is very broad. The rapid progress made in this field to date means that the elimination or substantial reduction of serum requirements is a realizable objective for most of the mammalian cell types. The necessity of using defined media for the maintenance of endocrine cells in culture with expression of differentiated functions must be emphasized because of the complicated relationships of feed-back mechanisms existing for most of endocrine cells even in cultures. The use of conventional serum supplements favours the undesirable growth of fibroblasts in the primary cell cultures of endocrine origin. In contrast, serum-

Motta, PM (ed): Ultrastructure of endocrine cells and tissues. ISBN-13: 978-1-4613-3863-5

free media may support the growth of endocrine cells with differentiated functions and may suppress fibroblast proliferation.

Most of the endocrine organs, especially the pituitary gland, are heterogeneous tissues composed of many cell types. The investigation of the mechanism of action of any substance known to influence differentiated function requires relatively pure or at least enriched populations of target cells. A good yield of enrichment can be now achieved by the unit gravity sedimentation procedure which is based upon the sedimentation according to the cell size and at a lesser extent to the density of cells concerned. Successful attempts were made to separate and subsequently to maintain in culture somatotrophs, mammotrophs, gonadotrophs and thyrotrophs. Though time consuming the procedure is simple and offers well reproducible experimental results. Large scale cultivation of mammalian cells in monolayers on microcarrier substrate has been recently proposed. The procedure seems to be suitable for preparing long-term suspension cultures from calf anterior pituitary cells (3). It is regrettable that neither procedures have found more extended application in culturing endocrine tissues.

2.2. Microscopy and related methods

Descriptive cytology utilizing conventional light and electron microscopic techniques offered a number of information about endocrine secretory processes. In the early seventies immunocytochemistry constituted a considerable progress towards functional identification of committed cells, especially in the hypophysis of many species including man (4). A number of methods have been developed partly at the electron microscopic level to meet with specific and optimal conditions for detecting antigenic sites intracellularly. The most suitable techniques seem to be the peroxidase-antiperoxidase (PAP) method (5), the radioimmunocytochemistry by the use of radioiodinated antigen-antibody complexes (6) (mainly adopted to cultured pituitary cells, 7), the colloidal gold procedures utilizing IgG molecules and protein A-coated gold granules (8). Another important morphological probe surely is the application of radioiodinated or ferritin-labeled ligands (polypeptide hormones, growth factors and many naturally occuring substances). These in fact may interact specifically with cell surface receptors and the fate of ligand-receptor complexes can be quantitatively followed by electron microscope autoradiography (9). The majority of the techniques mentioned above are not completely exploited on cultured endocrine tissue as well as other quantitative techniques like the X-ray microanalysis.

3. Adrenocortical cells in culture

3.1. Cultures derived from embryonic materials

The pioneering studies on human fetal adrenals in culture dates back to the early sixties when Stark et al. (10) reported on steroidogenesis by adrenocortical cells grown in tissue culture. In this early study, however, the fine structure of the hormone producing cultured cells was not described. The goal to correlate morphological appearance and steroidogenic activity of human fetal adrenocortical cells has been achieved first by Johannisson (11). On a number of fetuses of different ages it was in fact revealed that the steroidogenic activity of the embryonic adrenal cells was associated with intracellular structures such as smooth-surfaced endoplasmic reticulum (SER), tubulo-vesicular mitochondria, Golgi apparatus and lipid droplets. After a prolonged intraamniotic treatment of the fetuses with ACTH these structures may proliferate except lipid droplets which diminish in number and size.

During the fetal development Johannisson (11) distinguished two zones in the adrenal cortex, viz. the outer (permanent) and the inner (fetal) zones. In culture, there is a sharp divergence in survival and fine structure of the cells belonging to these zones. While the cells of the fetal zone cannot be maintained for longer periods of time, those of the outer zone can even proliferate and produce basal levels of steroids for longer periods of culture (12). The characteristic morphological appearance of the outer zone cells differs markedly from the intact adrenocortical cells after a week in culture and in the absence of ACTH. The epithelium-like cells possess but a few microvilli, a moderate number of mitochondria which are rather tubular than vesicular. The cells contain a fair amount of rough-surfaced endoplasmic reticulum (RER) and lipid droplets (12, 13). Following ACTH administration, the cultures produce considerable amounts of steroids over the basal levels with a peak on the third or fourth day of the ACTH treatment. In spite of the marked ACTH-induced steroidogenesis, the cells retain the above morphological features, but are rich in microvilli and SER. ACTH cannot be the sole factor responsible for the structural differentiation of the embryonic adrenal cells in culture (12). In fact it was demonstrated that the survival and differentiation of the fetal zone cells with increased steroidogenic capacity can be achieved when cells are cultured in the presence of human chorionic gonadotropin (HCG) for a long while (A. Gyévai, personal communication). In these cultures the epithelium-like cells exhibit a large amount of SER, tubulo-vesicular mitochondria with electron dense matrix and less lipid droplets. The cells are inducible with ACTH to produce steroids well over the basal level (Figs. 5–7).

Regarding these experimental facts one can assume that the metabolic processes needed for specific cellular activity run their courses irrespectively of whether or not the fine structure corresponds to the features characterizing the non-cultivated cells.

Parallel to human studies there was a progress in the investigation of the fetal adrenal cells of experimental animals in culture. In the rat (14), the cortical differentiation is already apparent in the 18-day-old rat embryos, the cells of the zona fasciculata are, however, predominant even in the newborn animals. Some of the steroidogenic organelles differ in the two outer zones of the gland. The tubular aspect of the internal mitochondrial structure is highly characteristic of the cells of the zona glomerulosa which remains

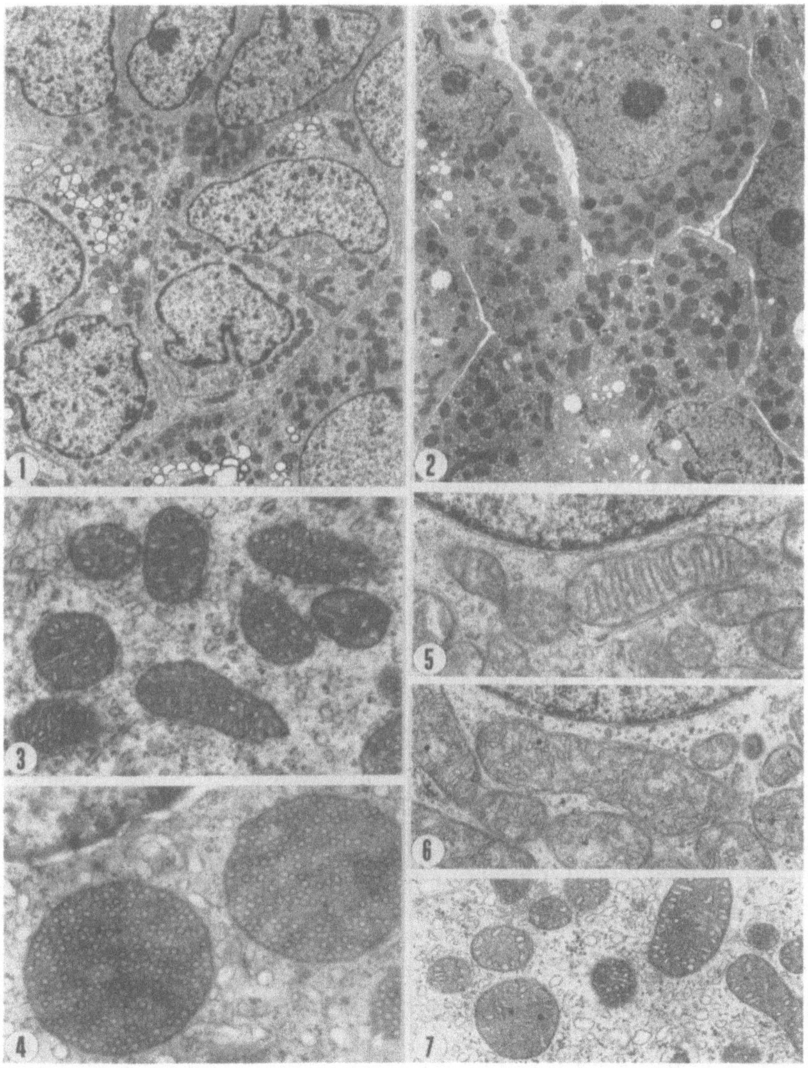

Figs. 1–7. (1–4) Cortical cells and mitochondrial profiles in tissue culture of fetal rat adrenals cultivated for 23 days in the absence of ACTH: (1) (× 2,450) and (3) (× 15,680), and after treatment with ACTH for 6 days: (2) (× 2,450) and (4) (× 15,680). Note the increase of the number of mitochondria and transformation of their inner membranes into 60 nm vesicles after ACTH treatment. From Kahri et al. (14). (5–7) Mitochondrial profiles in 11-day-old cultures of a human fetal adrenal. Cells of the permanent zone contain mainly mitochondria with cristae: (5) (× 7,000). These turn to tubulovesicular feature after treatment with ACTH for 48 hours: (6) (× 7,350). Fetal zone cells may survive in the presence of HCG in the medium: their mitochondria have tubulovesicular cristae: (7) (× 10,220). Courtesy of A. Gyévai.

unchanged in culture with or without ACTH in the culture medium. In contrast, the originally vesicular or tubulo-vesicular internal structure of the mitochondria of the cells from the zona fasciculata seen in the intact gland turns to tubular in the absence of the ACTH in the culture medium. This can be regarded as a sign of 'dedifferentiation' in culture of the mitochondria of this cell type. Following a prolonged ACTH treatment, the mitochondrial internal structure 'redifferentiates', i.e. the mitochondria become firmly vesicular (Figs. 1–4). A similar alteration of the endoplasmic reticulum can be observed. Irrespective of the zonal origin, the cytoplasm of the intact cells is rich in SER with occasional RER. In culture without ACTH in the culture medium, however, the cytoplasm is abundant in tubular SER or RER with occasional vesicular SER. After repeated ACTH administration the cytoplasm of the hormone producing cells contains large amounts of vesicular SER. The fate and the role of lipid droplets can be less well correlated to steroidogenic activity since culture conditions may influence their presence or absence within the cells (14).

The mitochondria show the most striking structural and numerical changes in culture upon the action of ACTH. Based on detailed morphometry, Salmenperä (15) suggested the hypertrophy of the mitochondrial compartment within the cultured cells. This is mainly due to mitochondrial proliferation and an increase in the surface area of the inner structures of the individual mitochondria. The change in the intramitochondrial conformation of membranes from tubular to vesicular (60 nm in diameter) is a constant feature after ACTH administration. Since the cholesterol side chain cleavage is thought to be connected with the mitochondria in the adrenal cortex it is conceivable that a stimulation of the steroid synthesis by ACTH is interrelated with the hypertrophy of the mitochondrial compartment. Salmenperä (15) produced also quantitative evidences on the prevention of mitochondrial proliferation when exogeneous corticosterone was added to cultures prior to ACTH stimulation.

Interestingly enough, the embryonic cat adrenals produce a considerable amount of steroids after ACTH stimulation in short-term incubation experiments during the very early stage of development. In spite of their steroidogenic capability the cells are undifferentiated in respect to the cytoplasmic structures thought to be involved in steroid synthetic pathways (12).

3.2. Cultures derived from adults

The fine structure of normal adult human adrenocortical cells in primary culture has been sparsely investigated. Important studies on decapsulated normal adrenal tissue showed a fair correlation upon the action of ACTH between the structure and the function of cells derived mainly from the zona fasciculata (16). In a detailed stereologic study it was demonstrated a marked increase in the mitochondrial internal membrane surface, and SER with virtually no change of RER after a short (2 days) and a long (8 days) treatment of the cultures with ACTH. Exogenous cyclic AMP caused but a slight hypertrophy of the mitochondria and of the SER whereas neither the ultrastructure nor the steroidogenic activity were influenced by cyclic GMP, the second naturally occuring 3', 5'-cyclic nucleotide, which has been reported by others (17) to stimulate the *in vitro* mammalian adrenal steroidogenic activity. In a superfusion system Pearlmutter et al. (18) demonstrated that the amount of steroid secreted in response to ACTH depends upon the total amount of hormone and not upon its external concentration. They also suggested that ACTH binds to membrane receptors. On the other hand, the hormone producing action of exogenous cyclic AMP depends on its concentration and not upon its total dose. These facts taken together with the rapid degradation of the cyclic nucleotide may indicate that trophic effects of ACTH might be interconnected with its binding to specific receptors and a presumably subsequent internalization. Besides these, possibly other factors may interplay in the growth of parenchymal cells in culture.

Tazaki et al. (19) using an organ culture system investigated normal human adrenals, adenomas and hyperplasias (Cushing's syndrome, primary aldosteronism). Adenoma cells from both Cushing's syndrome and primary aldosteronism contained a fair number of mitochondria. Their internal structure was minimally affected by ACTH when treated for a short time. In hyperplastic aldosteronism, the mitochondria were smaller and round, or even shrunken, after ACTH administration. In contrast, ACTH caused an increase in size of the mitochondria in the hyperplastic form of the Cushing's syndrome. In untreated longer-term (7 to 14 days old) cultures of the Cushing's adenoma cells mitochondrial shrinkage was observed which could be prevented when small amount of ACTH was added to the culture medium. Higher doses of ACTH in the medium caused a transformation of mitochondria for a rather rodlike shape. SER was predominant in the cytoplasms at lower doses of ACTH whereas RER became developped after higher doses of the tropic hormone. Szabó et al. (20) studied the fine structural features of a hormone-secreting adrenocortical adenocarcinoma, and the hydrocortisone production by tumorous as well as tumor-free adrenocortical cells in cultures from a patient suffering of Cushing's syndrome. They found at least two cell types in the tumorous tissue before cultivation. In one cell type, hypertrophied SER was observed in form of cisternae. The cytoplasm was rich in abnormal mitochondria and lipid droplets. The other cell type was mainly characterized by numerous cytoplasmic electron-dense granules of different sizes. After 17 days in culture, the latter cell types were predominant. It is important to note that tumorous and non tumorous cells differed markedly in hydrocortisone production. After 6 days in cultures, adenocarcinoma cells secreted about 20 times more hydrocortisone into the medium than their non tumorous counterparts. The tumor cells failed, however, to respond to ACTH and their hormone production diminished with time. In contrast, non tumorous cells could be stimulated by ACTH for a long period of time. These results are somewhat conflicting with data on non-cultured cells. Therefore more studies on tumorous cells in culture are requested to evidence that abnormal adrenocortical function may also be influenced by ACTH.

We learned from the studies of O'Hare and Neville (21) that ACTH may induce a profound and reversible retraction of adult rat adrenocortical cells in culture. This was confirmed by Ramachandran and Suyama (22) on the same cell types. They stated that the retraction of the cells is a slow process with a climax on the 4th-5th day of the ACTH administration and accompanied with the peak of the steroidogenic response given to the tropic hormone. The activation of the microfilamentous-microtubular apparatus of the parenchymal cells may occur via the adenylate cyclase-cyclic AMP system after stimulation with ACTH. The trophic action of the ACTH is evident also on the adult rat fasciculata cells in culture. Thus in cells from the zona fasciculata, mitochondrial internal structure turns to vesicular and SER hypertrophies when cultures are exposed to ACTH.

It is now common knowledge that ACTH exerts an acute influence on hormone secretion and a delayed trophic effect on the steroidogenic capacity of normal (both embryonic and adult) adrenocortical cells in culture. Its effect(s) on cell proliferation is a rather controversial issue. Armato et al. (16) observed an increase in number of cells entering the cell cycle upon the action of ACTH in cultures derived from normal adult human adrenocortex. Their findings could not be confirmed on cultures derived from other species. Ramachandran and Suyama (22) found an inhibition of replication of normal rat adrenocortical cells in culture by ACTH confirming earlier studies on a mouse adrenal tumor cell line. In studies made on a proliferating, normally functional bovine adrenocortical cell culture system it was proposed a new model for adrenocortical cell regulation (23). The model implies that ACTH acutely stimulates steroidogenesis and induces catalytic enzymes for the steroid biosynthesis. Thus ACTH evokes an increase of the steroidogenic capacity of the individual parenchymal cells in culture. At the same time ACTH interferes with the initiation of DNA synthesis arresting the cells in the G_1 phase. An accomplishment of the cellular hypertrophy requires the interplay of ACTH with one of the recently identified growth factors, viz. fibroblast growth factor and angiotensin, and/or with one of their potentiating agents (insulin, multiplication-stimulating activity, somatomedin C). Parenchymal cells may enter the mitotic cycle only if ACTH desensitization occurs. Proliferation of cells in the presence of ACTH is thus a later and limited response. From studies in cells with reduced ACTH-stimulated cyclic AMP formation it appears that higher cellular concentration of cyclic AMP are required to stimulate steroidogenesis. Full steroidogenesis may thus be maintained in desensitized cells.

4. Hypophyseal cells in culture

The pituitary gland is a complex organ composed of lobes differing in cell composition and having a double origin. In most of the vertebrates three lobes can be distinguished. The anterior and the intermediate lobes, viz. the adenohypophysis, originate from a single placode of the cranial ectoderm growing in from the surface of the embryo across the roof of the mouth. The neural lobe is a part of the developing infundibulum: a downgrowth from the floor of the diencephalon posterior to the optic chiasma. These entities of the finished gland come into intimate contact. In man, the intermediate lobe is rudimentary. The hypophysis controlled by hypothalamic hypophyseotropic hormones secretes a number of potent regulatory hormonal substances which are peptides and proteins in nature, partly conjugated with carbohydrates. According to present knowledge most of the hormones are elaborated in various cell types and stored in membrane coated secretory granules within the cytoplasm.

When culturing pituitary tissue or cells it is essential to clearly define which lobe has been explanted.

4.1. Anterior pituitary cells

The mammalian anterior pituitary contains at least six different cell types which synthetize, store and release ACTH, thyrotropic (TSH), somatotropic (GH), luteinizing (LH), follicle-stimulating (FSH), and lactotropic (PRL) hormones. It is generally believed that pituitary hormones like other secretory peptides and proteins are synthetized as pre-pro-hormones which undergo mostly enzyme-mediated maturation during travelling from the endoplasmic reticulum through the Golgi apparatus up to the secretory granules. While the maturation of most of the pre-prohormones seems to give rise to a single biologically active molecule, the formation of the biologically potent ACTH is more complicated. Its parent molecule is a large polypeptide – the pro-opiomelanocortin – whose enzymatic cleveage yields a number of other biologically significant families of peptides: melanotropins (MSHs), lipotropins (LPHs) and endorphins (ENDs). Hormone storage in secretory granules is a common feature for anterior pituitary cells. The number, size and cytoplasmic distribution of the secretory granules, however, differ markedly in the various cell types. Stored and perhaps newly synthetized pituitary hormones are released from the cells upon the action of hypothalamic hypophyseotropic hormones. Exocytosis is the sole process through which stored hormones are released from the hormone producing cells. The release process of unpacked hormones is presently unknown. The main ultrastructural features characterizing various parenchymal cells in the rat anterior pituitary are summarized in Table 1.

4.1.1. Cultures from embryonic materials

The human fetal hypophysal cultures provide conditions for hormone secretion, because ACTH (24), GH and in some cultures LH and TSH (25) could be demonstrated in the media of explants cultured on nutrient substrate. Gailani et al. (25) stated that in the majority of their cultures the rate of GH production is similar to that of the starting material for 1 to 3 weeks, after which the hormone production decreases; in some of the cultures originating from donor fetuses older than 16 weeks, hormone production is maintained at a reletively high level as long as 2 to 3 months. In monolayer cell cultures derived from fetuses of the second and third trimester of gestation, the fine structure of cells containing secretory granules was electronmicroscopically (Fig. 8) monitored (26). It has been established that during a 75 days period the cultures contained at least four granulated cell types. One of the cell types had to be somatotrophs since GH released into the medium could be demonstrated in a number of cultures up to the end of the cultivation period. It has also been established that the proportion of cells containing a

Table 1. Some cytological features *in vivo* and in culture of rat anterior pituitary parenchymal cells.

Cell type	Cell shape	Secretory granules			Identification (ref. no.) by light and electron microscopy and/or immunocytochemistry	
		Size	Shape	Distribution	*in vivo*	in culture
Corticotrophs	Stellate	150–200 nm	Round	Lining cell membrane	(4, 7)	(1, 7, 26, 34, 38)
Mammotrophs	Polygonal, ovoid or cap-shaped	600–900 nm	Irregular	Scattered throughout the cytoplasm	(4)	(1, 34, 39, 41, 42)
Thyrotrophs	Elongated, ovoid or angular	100–150 nm	Round	Throughout the cytoplasm, but tendency to group at the cell periphery	(4)	(34)
Gonadotrophs	I ovoid	about 200 nm	Round or irregular			
	II angular or stellate	200–220 nm	Round	Throughout the cytoplasm	(4)	(1, 34, 47, 39)
	III stellate	200–220 nm	Round	cell periphery		
Somatotrophs	Ovoid or pyramidal	300–350 nm	Round	Abundant throughout the cytoplasm	(4)	(1, 34, 41)

granule population with larger diameters increased with time. This fact might have several explanations. 1) The granules are spontaneously released during cultivation, and for lack of hypothalamic releasing hormones the rate of hormone production decreases; the cells with smaller granules will be gradually degranulated. 2) The granule population is changing by the aggregation of mature granules. 3) The culture conditions are unfavourable for the growth of cells containing granules with a smaller diameter. It seems reasonable to believe that morphological observations based solely on conventional electron microscopy are insufficient to draw definite conclusions as to whether hormones are actually contained in these cell types. Nevertheless, as stated earlier, a number of hormones have been demonstrated in the medium of the human embryonic hypophyseal cell cultures.

During fetal life, the various hormone producing cells appear successively in the hypophyseal primordia (27). The first appearance of committed cells

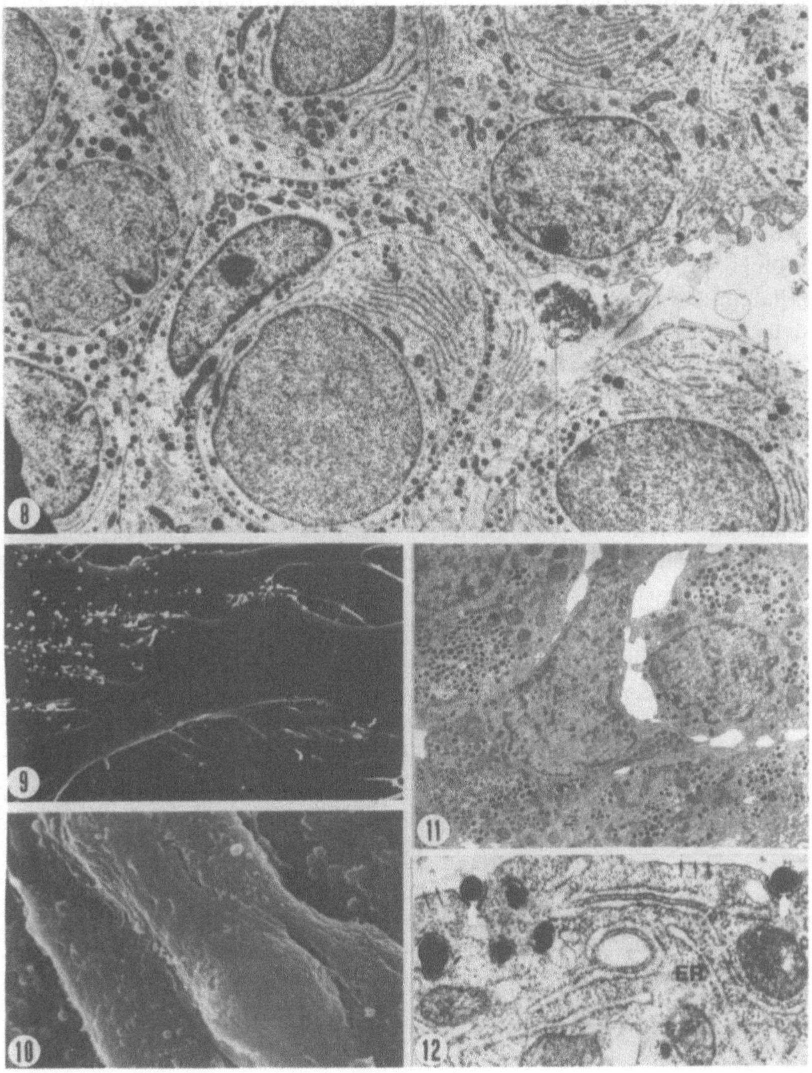

Figs. 8–12. (8) Electron microscopic montage (× 2,870) from a 59-day-old human fetal pituitary cell culture. Each cell contains many cytoplasmic secretory granules of various sizes. Most cells possess characteristically ordered RER profiles. (9–10) SEM from an 8-day-old rat anterior pituitary cell culture. (9) Fibroblasts with smooth surface and prominent cytoplasmic protrusions (× 1,645). In (10) (× 2,800) elongated epithelial cells are seen. Their surface is rich in microvilli and blebs. (11) 7-day-old rat anterior pituitary monolayer culture (× 3,920). Parenchymal cells are heavily packed with secretory granules randomly distributed in the cytoplasm. Courtesy of I. Fazekas. (12) Portion of a cell (× 7,140) from an 11-day-old rat anterior pituitary cell culture. Note cytoplasmic membrane coat (black arrows), extruded dense granules and thickened membranes at sites of extrusion (white arrows). The endoplasmic reticulum (ER) is well developed. From Rappay et al. (34).

containing cytoplasmic secretory granules filled with immunoreactive hormones occurs well before the formation of a functioning neurovascular link between the hypothalamus and the pituitary gland. This fact raised the question what is the role of the hypothalamus in the expression of differentiated functions by the hypophyseal cells during early development. To answer the question, culture techniques have been found to be indispensable.

By several authors organ cultures were prepared from hypophyseal primordia of the embryonic rat of different ages. Watanabe and Daikoku (28) explanted primordia taken from fetuses on day 12.5 and 14.5 of gestation and cultured them for 9 and 6 days, respectively. At the end of these culture periods they identified three types of anterior pituitary cells containing secretory granules with maximal diameters of 150, 200 and 350 nm, respectively. This finding indicates that uncommitted cells of the primordia possess the ability of self-differentiation, i.e. they differentiate morphologically without being controlled by the hypothalamus. They used, however, chemically defined medium supplemented with calf serum which might contain substances of hypothalamic origin. Watanabe and Daikoku (28) and Nemeskéry et al. (29) and very recently Begeot et al. (30) demonstrated that functional differentiation of hypophyseal primordia may also occur in organ culture. They showed immunoreactive cells characteristic of the adult type anterior pituitary. The functional differentiation of fetal pituitary cells in culture is further supported by the fact (31) that corticotrophs are able to respond to hypothalamic corticotropin-releasing factor (CRF). Nemeskéry et al. (29) used a chemically defined medium alone as nutrient excluding any hypothalamic substance from the culture system. The first committed cells are undoubtedly corticotrophs either in the intact or in the cultured primordia, although the significance of their earliest appearance is unknown. The successive appearance of committed cells in cultured primordia is listed in the Table 2.

4.1.2. Cultures from adult normal and adenomatous materials

Anterior pituitary cell functions are influenced by hypothalamic hypophyseotropic substances, neurotransmitters and peripheral hormones. Since the pituitary gland is closely coupled to the brain via the portal vessel system, the rate of secretion of hypophyseal hormones shows a dynamism which can be modulated by various experimental conditions. It is often difficult to determine if responses seen *in vivo* are due to direct effects on the pituitary or are mediated through extrapituitary mechanisms. Assays in culture of the hypophyseotropic substances offer isolation from such indirect effects. It has been the aim of biologists in this field to develop a culture system that would be technically simple, sensitive, reliable, accurate, reproducible and valid (32). Functional studies verify that every hormone producing cell of the adult anterior pituitary may survive and can be maintained for a shorter or a longer period of time in culture in spite of metabolic changes during cultivation. It has, namely, been established (33) that enzyme activities representing major metabolic pathways (citrate cycle, pentose cycle, and glycolysis) show changes suggesting that cell metabolism shifts to anaerobic glycolysis during a longer period of cultivation. It seems likely, however, that activities of genes responsible for hormone production are expressed even if the cells are forced to change their energy metabolism.

In the early seventies the culture technique of adult pituitary cells developed in many laboratories and it has been proved that cells are functionally insensitive

Table 2. The appearance of committed cells in intact and cultured rat hypophyseal primordia.

Cell type	Without cultivation Gestational day[a]	After cultivation for 6 to 9 days Age of donor fetuses (days)							
		12[b]	12	13	14	15[c]	13	14	15[d]
Corticotrophs	16	+	–	–	–	–	+	+	+
Mammotrophs	16–17	+	–	–	–	–	○	+	+
Thyrotrophs	17	+	○	○	○	+	–	+	+
Gonadotrophs	18	+	○	○	+	+	○	+	+
Somatotrophs	19	+	–	–	–	–	–	○	○

[a], [b], [c], [d] are references 27, 28, 30, 29 respectively
+ = immunoreactive cells are present
○ = immunoreactive cells are absent
– = not detected

to proteolytic dispersion of the gland and dispersed cells can firmly adhere to inert surfaces and form a monolayer during a relatively short period of time. The enzymatically dispersed single cells of the adult rat anterior pituitary undergo a rapid random aggregation within 24 hours after explantation. The aggregated cells spread out slowly gradually forming heaps of epithel-like cells. Beneath and among them, often at the rims of the cell aggregates, spindle-shaped fibroblast-like cells appear which grow out and intermingle slowly with their neighbours. The cells so arranged exhibit a nesting configuration which remains characteristic of the cultures near to or even at confluency. The spindle-shaped cells represent a mixed cell population consisting of randomly distributed, elongated, parenchymal cells and true fibroblasts. Elongated cells have been identified immunocytochemically as thyrotrophs, corticotrophs and prolactin cells. Further they can be distinguished from true fibroblasts by scanning electron microscopy (SEM). In SEM fibroblasts have a smooth surface and show a variable number of microvilli. Low, longitudinal wrinkles also occur on the surface of certain cells following the course of intracellular fibers. The free edge of the fibroblasts, not contacting other cells, is ruffled. In contrast, parenchymatous cells of elongated shape show multiple blebs and rugae on their surfaces which sometimes cover cytoplasmic protrusions containing one or more secretory granules (Fazekas, I., Bácsy, E., Rappay, Gy. unpublished observation, Figs. 9, 10).

Conventional transmission electron microscopy even with low power magnification reveals the presence of cells containing secretory granules varying in size and density. Epithel-like and elongated cells may equally be granulated confirming the parenchymatous origin of both cell types.

By detailed morphometry, granulated cells in monolayers derived from adult rat anterior pituitary cell suspension could be divided into four groups (34). In the first group, the granules measured 150 to 200 nm in diameter; they were mainly localized at the periphery of the cell, mostly near to the cell membrane. Cells belonging to the second and third groups frequently contained granules of 200 to 250 nm and 300 to 400 nm, and these were randomly scattered. Cells containing granules larger than 600 nm were rare. The fine structure of the majority of the granular cells shows certain pecularities: in most cells there is an external coat on the membrane surface. In the cytoplasm many free ribosomes and polysomes are visible. There is much endoplasmic reticulum, and most of it is rough-surfaced. Many mitochondria are present in the cytoplasm; even those within the same cell differ in density and in shape and all are provided with cristae. The Golgi apparatus is well developed with immature secretory granules between the lamellae and vesicles. A fair number of microtubules can be seen in the cytoplasm of cells rich in secretory granules, primarily in the more bulky cell protrusions. Lysosomes are present in varying size and number; in a number of cells, crinophagic vacuoles are seen; multivesicular bodies are rare. Most of the nuclei are irregular in shape containing but little heterochromatin. Discharge of secretory granules is often seen and at the site of exocytosis the cell membrane is thickened. This electron microscopic analysis favours the view that granule formation, hormone storage and release observed *in vivo* proceed in a closely similar fashion under conditions in culture (Figs. 11, 12).

This may be valid also for human adenomatous material in primary cultures and after subsequent passages. Landolt and Rothenbüler (35) pointed out, that a prolongation of survival of adenoma cells in culture may be an appropriate tool to unroll the involvement of functional factors operating in hormone synthesis and release. Gazsó and Pásztor (36) called the attention to the unnecessity of a fibroblastic 'feeder-layer' for the prolonged survival of adenoma cells. They are of the opinion that the presence of fibroblast-like cells in pure adenoma cell cultures is a morphological sign of the ageing of the specific cells.

Facts revealed by conventional electron microscopy either on experimental animals or human adenomas would be functionally validated by electron immunocytochemistry. However, this technique has been applied to cultured pituitary cells rather sparsely. The first attempt to analyse immunoreactivity at the ultrastructural level was made by Tixier-Vidal and her group (1) on rat anterior pituitary cultures. They used antisera against ovine LH, LHα and LHβ subunits. It has been established that depending on age of the cultures immunoreactive cells are present in the monolayer. The luteotropic cells have a single class of immunoactive secretory granules and the cytoplasms are also immunopositive sometimes in well defined areas but never in the ER cisternae. The number of immunopositive cells show a certain correlation with the immunoreactive LH released into the medium. In a subsequent study Tougard et al. (37) demonstrated that using antibodies against ovine FSH, LH and its two subunits two cell types can be distinguished in the early development of the monolayer derived from dispersed rat pituitary cells. The two types of gonadotropic cells display the same fine-structural features as *in vivo*. After 5 to 8 weeks in culture there is only one type of gonadotropic cells in the monolayers with one class of immunoreactive secretory granules, a moderately developed Golgi apparatus and large dense bodies. At the same time there is a diminished FSH and LH release into the medium. Thus it was concluded that the decrease in secretory activity may preferentially due to the cessation of the secretory activity of the gonadotropic cells rather than their reduction in number. The ultrastructural features and the corresponding functional activity of corticotrophs have been studied by Bácsy et al. (38) in rat anterior pituitary monolayer cultures stained with an antiserum raised against a 1–28 ACTH analogue. Corticotrophs of the intact anterior pituitary gland have been identified as stellate or elongated cells of irregular shape with cytoplasmic processes encircling other cell types and extending to sinusoids. In monolayer cultures, corticotrophs retain their original gross structural features. By EM, the secretory granules which form a row parallel to the cell membrane show a well-defined immunostaining (Fig. 13). The immunopositive granules display an inhomogeneous inner structure and are sometimes interconnected with tubular or saccular structures of unknown origin. As in the case of gonadotropic cells, certain areas of the cytoplasm of the corticotrophs are also immunoreactive but ER cisternae are never stained. In interpreting the extragranular immunostaining the authors consider the possibility of dislocation of ACTH owing to its solubility in water. However, it cannot be excluded that ACTH is detected during its formation and transport to the secretory granules. Mammotrophs in rat hypophyseal primary cultures as well as those of the SDI line (a cell line isolated from a primary culture of normal Sprague-Dawlay rat anterior pituitary cells) were also studied after immunostaining (39). Immunoreactive secretory granules and a particulate staining within the Golgi region were found in both cultures. A reaction deposit was sometimes seen on the inner surface of the RER membranes indicating that nascent PRL might be immunoreactive.

Substances known to influence hormone production by pituitary cells may alter the morphological appearance of cells in culture. One of the first evidences that thyrotropin releasing hormone (TRH) may change cell morphology, was described by Tashjian and Hoyt (40). In SEM preparations they found that GH_3 cells treated with TRH are flattened firmly against the inert substratum through extended and branching cytoplasmic processes. The surfaces of these cells are smoother than those of untreated cultures containing cells with rugae, prominent buds or club-like appendages. Other substances, like luteinizing-hormone releasing hormone (LH-RH), prostaglandins, oestrogens, bromoergocryptine may also affect various cytoplasmic structures as summarized in a recent review (1). In addition bivalent cations, e.g. barium ions may cause a remarkable change in hormone production and a profound retraction of pituitary cells when cultures are incubated in a medium devoid of calcium ions (41). For the explanation of the profound retraction of cells in the presence of Ba^{2+} one can assume that barium ions may occupy cellular calcium binding sites and activate contractile cytoplasmic structures via the adenylate cyclase-cyclic AMP system. Spreading and retraction are phenomena to be elucidated in future work on cultured pituitary cells.

The idea that peptide hormones act exclusively at the surface of their target cells has been accepted for some time. Recently it has been questioned particularly by the experiments on low-density lipoproteins, insulin and epidermal growth factor. After binding to specific receptor sites of the cell membranes, these molecules may be internalized. Through subsequent lysosomal breakage the internalization may serve as a tool for depriving occupied receptor sites and in addition for inactivation of biologically active signal molecules or for deliverance of breakdown products to further biological effects. Although the techniques for visualizing peptide hormones at the light and electron microscope levels are well developed the internalization of hormones has been followed but sparsely in pituitary cell cultures. Tixier-Vidal et al. (42) performed experiments on SDI cells in the presence and absence of TRH in the culture medium. They used horseradish peroxidase as an endocytotic marker and/or concanavalin A as a marker of the surface glycoproteins. From ultrastructural and cytochemical studies they came to the conclusion that TRH may exert a dual influence on the plasma membrane of SDI cells. TRH, while induces elevated PRL release, facilitates the lateral mobility of the plasma membrane components and stimulates the endocytosis followed with an accumulation of dense bodies in the Golgi region. These effects are directly interconnected with the binding of TRH to its receptors and with a subsequent internalization of plasma membrane fragments which are attacked by lysosomes as demonstrated by the detection of the acid phosphatase activity. This finding supports the view that TRH may control the balance of its own binding sites in SDI cells. LH-RH bound to its specific receptors internalizes also rapidly as was claimed by Hazum et al. (43) employing rhodamine-labeled peptide and image-intensified fluorescence microscopy. They observed also a cluster formation prior to internalization which could be prevented by the omission of Ca^{2+} from the medium. On the other hand, Conn et al. (44) clearly showed that stimulation of pituitary LH release does not require internalization of LH-RH. They firmly coupled an iodinated LH-RH-analogue to an agarose matrix which retains its full biological activity without entering the cytoplasm since the agarose beads have an average size in an order of an entire cell. How does the down-regulation of receptor sites proceed in that case, it remains unclear.

It has been long known that excess amounts of hormones not subjected to discharge might be intracellularly eliminated through a process named crinophagy. The role of the lysosomes in this process is indispensable. The occurrence of crinophagy, however, has not fully been cleared up under culture conditions of the anterior pituitary cells. There is an overall increase of acid phosphatase activity under normal culture conditions up to the 24th day of cultivation (33). The distribution of enzyme activity in various cell types has not been studied in detail. However, in the case of PRL producing cells there are some signs indicating the lysosomal participation in the elimination of surplus amount of PRL. Accumulating evidences favour the view that dopamine is one of the important inhibitory substances which reduces PRL production. Dopamine agonists, e.g. natural and synthetic ergot alkaloids, act on a similar way *in vivo* and in culture. Bromoergocryptine inhibits PRL release in pituitary cell cultures. The subsequent accumulation of the hormone inside the cytoplasm results in increased proteolysis by lysosomal enzymes. The finding that chloroquine, a presumed stabilizer of the lysosomal membrane, prevents at least partially the effect of bromoergocryptine supports such an interpretation (45). The participation of extralysosomal proteinase(s) or peptidase(s) active at a pH over 7.2 cannot be excluded, however, in the breakdown of PRL under such conditions.

4.2. Cultures from intermediate and neural lobes

The intermediate lobe of the hypophysis is the site of synthesis and conversion of a peptide hormone family. By electron microscopy the intermediate lobe is constituted by melanotropic cells characterized by a fair number of pleomorphic secretory granules randomly distributed in the cytoplasm. Signs of secretory granule formation can be observed in the Golgi region. The cytoplasm is rich in SER and RER and in mitochondria. Some parenchymal cells also occur in certain regions of the lobe in which most of the secretory granules are aligned along the cell membrane. These cells are similar to corticotrophs in the anterior lobe. Marginal cells bordering the surface of the pituitary cleft as well as follicular cells represent the non-parenchymal classes of cells whose function is unclear. A number of marginal cells are ciliated, others possess surface structures suggesting the possibility of an active transport (46).

The neural lobe is composed of pituicytes of glial origin and of neural fibers whose perikarya lie within the hypothalamus. They enter the lobe via the stalk and are packed heavily with secretory materials to be released in the neurohypophysis.

The rapid progress in culturing anterior pituitary cells was not followed by a similar, widely popular development of techniques extended over the intermediate lobe. This is unfortunate since the intermediate lobe cells may offer an excellent model for comparison of the processing of pro-opiomelanocortin in two related cell types, i.e. anterior-lobe corticotrophs and intermediate-lobe melanotropic cells. There is a strong indication that the regulation of processing and secretion of ACTH/LPH molecules differs markedly in the two lobes in culture (47).

One of the first attempts to maintain rat neurointermediate lobes in organ culture was made by Scott et al. (48) who failed to describe structural features, but evidenced an elevated ACTH release by cultured cells into the medium. Chatterjee (49, 50) explanted isolated intermediate lobes derived from fetal, neonatal, young and adult rabbits. The organs have been maintained over six weeks in culture. He observed a fairly good survival of both parenchymal and interstitial cells especially in cultures taken from fetal and perinatal animals. He did not attempt, however, to detect any hormones either in the culture medium or in the cells which contained many secretory granules.

Neurointermediate lobes taken from pituitaries of the frog (Rana pipiens) were explanted as organ cultures by Semoff et al. (51). Transmission electron microscopy showed a high degree of preservation of the parenchymal cells of the pars intermedia for a long period of time. A rapid degeneration of neuronal elements originally present in the explants was observed which indicates the early cessation of neuronal influences thought to be necessary for the control of the secretion of MSH. They could, indeed, demonstrate a continuous release of bioactive MSH into the medium, especially in higher amounts when serum-free medium was used. They suggest that MSH, at least partly, will be enzymatically degraded when the nutrient medium is supplemented with fetal calf serum. The presence of cytoplasmic lipid accumulation and of large intercellular cavities in the organoid may indicate a suboptimal nutrition, one of the drawbacks of the organ culture procedure.

Two types of culture technique suitable for long-term maintainance of adult rat pituitary intermediate lobe cells were developed by Fazekas et al. (52, 53, 54) to overcome drawbacks of the organ culture procedure and to fulfil criteria established for using anterior pituitary cell systems. In both types of cultures intermediate lobe cells are able to secrete considerable amounts of immunoreactive and bioactive ACTH and α-MSH. In the first technique, intermediate lobe cell suspensions were explanted after a gentle trypsin dispersion and cultured up to 5 weeks. During the development of a quasi-monolayer

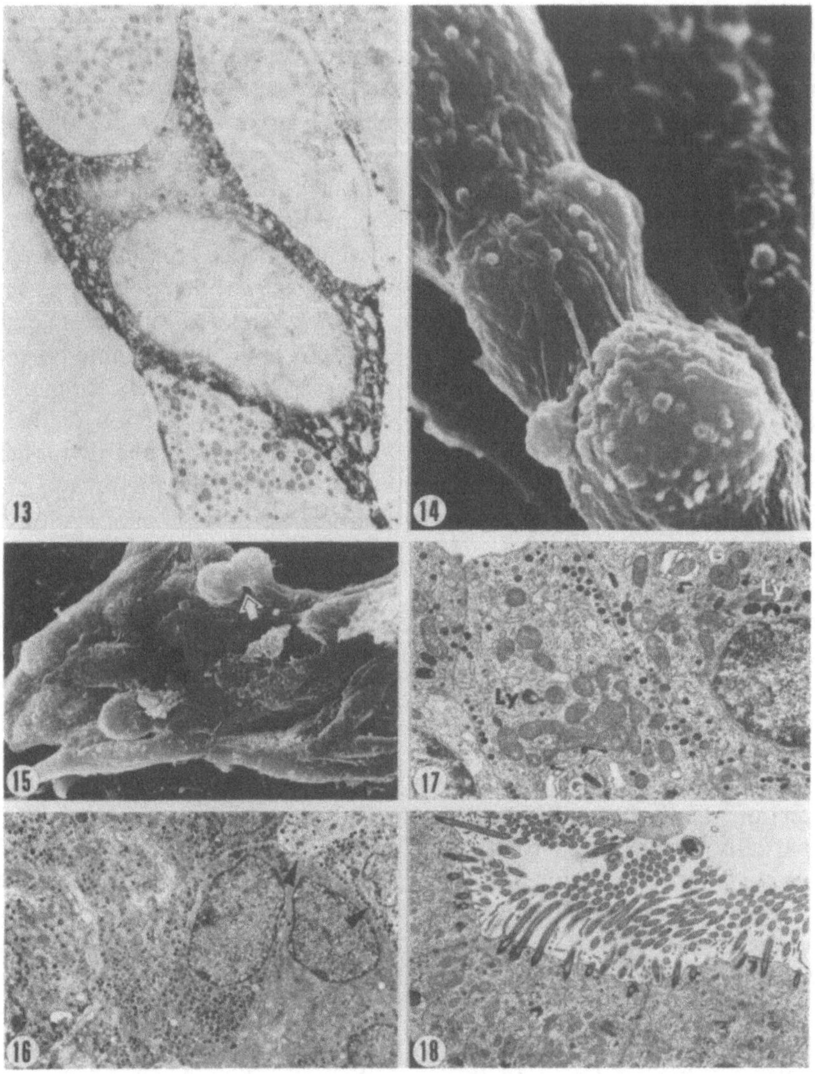

Figs. 13–18. (13) ACTH immunostaining in a 2-week-old rat anterior pituitary cell culture (× 3,500). The stellate corticotroph has an ovoid nucleus and contains many immunopositive secretory granules lining the cell membrane. Courtesy of E. Bácsy. (14–15) High power (14) (× 5,460) and low power (15) (× 770) SEM from an 8-day-old rat pituitary intermediate lobe cell culture. In (14) an epithelial cell cord is shown consisting of tightly packed cells rich in surface structures. (15) Round and elongated epithelial cells from an islet. Cell surfaces form numerous microvilli and blebs. Connections between adjacent cells are prominent. Arrow indicates a dividing epithelial cell. From Fazekas et al. (53). (16) 8-day-old culture of trypsin-dispersed rat pituitary intermediate lobes (× 2,100). Most of the cells contain many pleomorphic secretory granules randomly distributed in the cytoplasm. In others (arrows) secretory granules are lining the cell membrane. From Fazekas et al. (52). (17–18) Electron micrographs taken from a 6-week-old (17) (× 4,620) and a 5-week-old (18) (× 3,150) suspension culture from mechanically isolated rat pituitary intermediate lobes. In (17) a number of pleomorphic secretory granules are situated at the cell periphery. Note secondary lysosomes (Ly) and prominent Golgi regions (G). In (18) ciliate cells line the concave part a spherule surface. From Fazekas et al. (54).

two main cell types have been observed. Solitary or aggregated, ovoid or polygonal cells represent one type of cells. They probably originate from the parenchyma as revealed by SEM (Figs. 14, 15). The surface of these epithel-like cells is uneven with many protrusions. Solitary or multiple blebs and rugae are formed on the cell surface. Less microvilli are found on the surface of superimposed cells than on the surface of cells forming epithelial cords or bundles. Long protrusions also occur on the surface of some epithelial cells: owing to their size, they might be solitary cilia as recently described by Correr and Motta (46) in cells bordering the hypophyseal cleft. The overwhelming majority of epithel-like cells exhibit, depending on the age of the cultures, more or less pleomorphic secretory granules in the cytoplasm as judged by conventional TEM (Fig. 16). The other type of cells are true fibroblasts.

In the other technique, mechanically isolated intermediate lobe tissue fragments were kept in life in suspension cultures. From the tissue fragments under steady shaking tissue spherules measuring 0.1 to 1.0 mm in diameter are formed and found at each time in the cultures up to 6 weeks. The number of spherules slightly decrease with cultivation time. At the end of the first week in culture the spherules consisting of tightly-packed cells are encapsulated. Ultrastructurally, two types of cells containing secretory granules can be distinguished: typical melanotropic cells and a few cells reminding the corticotrophs of the anterior lobe. Fibroblasts never grow and this is the main advantage of such a type of culture. The eventual perishing of an insignificant number of cells within the larger spherules can be avoided when the nutrient medium is supplemented with horse serum (15% vv) instead of fetal calf serum. Parenchymal cells in 6-week-old cultures are duly preserved with signs of granule formation in the Golgi region. Secretory granules are, however, less in number and localized mainly at the cell periphery (Fig. 17). At the same time lysosomes occur more frequently. Sometimes, ciliated cells similar to those described by Correr and Motta (46) line a concave part of the spherule surface (Fig. 18). Both types of these cultures represent an excellent model in studying hormone processing and its regulation within the rat intermediate lobe cells.

The neural lobe alone or explanted in association with the intermediate lobe cannot be maintained in culture. When taken from adults, its cells of glial origin undergo modifications towards fibroblasts (51). In pituitary primordia, however, taken from fetuses on day 12.5 of gestation pituicytes containing lipide-like cytoplasmic bodies may survive and extend irregular glial processes but neurosecretory axons are by no means present in organ culture (28).

5. Concluding remarks

During the last decade it turned out that fetal and adult, healthy and tumoral endocrine cells in culture retain their main structural features preexisting *in vivo*. Cultured endocrine cells are able to spontaneously produce considerable amounts of hormones. They respond to stimuli specifically influencing hormone production although their metabolism changes during cultivation. Initial but promising investigations point to a new era in culture methodology by the spreading of chemically defined media. Endocrine cells in culture in combination with cytological and cytochemical techniques offer a suitable tool for finite exploration of a number of events pertinent to cell biology.

Acknowledgements

I thank Drs. E. Bácsy, A. Gyévai, G.B. Makara, E. Stark, D. Szabó for commenting on all or part of the manuscript and Mr. I. Csapó for photography.

References

1. Tixier-Vidal A: Ultrastructure of anterior pituitary cells in culture. In: The anterior pituitary, Tixier-Vidal A, Farquhar MG (eds). New York, Academic Press, 1975, pp 181–229.
2. Bottenstein J, Hayashi I, Hutchings S, Masui H, Mather J, McClure DB, Ohasa S, Rizzino A, Sato G, Serrero G, Wolfe R, Wu R: The growth of cells in serum-free hormone supplemented media. In: Methods in enzymology, Jakoby WB, Pastan IH'(eds). New York, Academic Press, 1979, vol. 58, pp 94–109.
3. Horng C-B, McLimans W: Primary suspension culture of calf anterior pituitary cells on a microcarrier surface. Biotechnol Bioeng 17: 713–731, 1975.
4. Nakane PK: Identificatior. of anterior pituitary cells by immunoelectron microscopy. In: The anterior pituitary, Tixier-Vidal A, Farquhar MG (eds). New York, Academic Press, 1975, pp 45–61.
5. Sternberger LA, Hardy PH, Cuculis Jr JJ, Meyer HG: The unlabeled antibody enzyme method of immunohistochemistry. Preparation and properties of soluble antigen-antibody complex (horseradish peroxidase-antihorseradish peroxidase) and its use in identification of spirochetes. J Histochem Cytochem 18: 315–333, 1970.
6. Larsson L-I, Schwartz TW: Radioimmunocytochemistry – A novel immunocytochemical principle. J Histochem Cytochem 25: 1140–1146, 1977.
7. Rappay Gy, Kárteszi M, Makara GB: ACTH radioimmunocytochemistry (RICH) on rat anterior pituitary cells. Histochemistry 59: 207–213, 1979.
8. Roth J, Ravazzola M, Bendayan M, Orci L: Application of the protein A-gold technique for electron microscopic demonstration of polypep-

tide hormones. Endocrinology 108: 247–253, 1981.

9. Gorden Ph, Carpenter J-L, Freychet P, Orci L: Morphologic probes of polypeptide hormone receptor interactions. J Histochem Cytochem 28: 811–817, 1980.
10. Stark E, Gyévai A, Szalay K, Ács Zs: Hypophyseal-adrenal activity in combined human foetal tissue cultures. Canad J Physiol Pharmacol 43: 1–7, 1965.
11. Johannisson E: The foetal adrenal cortex in the human. Its ultrastructure at different stages of development and in different functional states. Acta endocr (Copenh) 58: Suppl 130: 1–107, 1968.
12. Gyévai A, Bukulya B, Mihály K, Szalay K, Stark E: Fine structure and hormonal activity of intact and cultured embryonic adrenal cells of different species. Symp Biol Hung 14: 73–88, 1972.
13. Voutilainen R, Kahri AI: Functional and ultrastructural changes during ACTH-induced early differentiation of cortical cells of human fetal adrenals in primary cultures. J Ultrastr Res 69: 98–108, 1979.
14. Kahri AI, Lyytikäinen A, Pesonen S, Saure A: Comparison of the effects of ACTH on the number of mitochondrial profiles in cortical cells and on the conversion of progesteron-4-^{14}C into corticosterone and 18-OH-DOC in tissue cultures of foetal rat adrenals. In: Ultrastructural features of cells and tissues in culture, Törö I, Rappay Gy (eds). Budapest, Akadémiai Kiadó (Symposia Biologica Hungarica, vol. 14), 1972, pp 59–72.
15. Salmenperä M: Comparison of the ultrastructural and steroidogenic properties of mitochondria of fetal rat adrenals in tissue culture. A morphometric and a gas chromatographic analysis. J Ultrastr Res 56: 277–289, 1976.
16. Armato U, Nussdorfer GG, Neri G, Draghi E, Andreis PG, Mazzocchi G, Mantero F: Effects of ACTH and 3', 5'-cyclic purine nucleotides on the morphology and metabolism of normal adult human adrenocortical cells in primary tissue culture. Cell Tiss Res 190: 187–205, 1978.
17. Kitabchi AE, Sharma RK: Corticosteroidogenesis in isolated adrenal cells of rats. I. Effect of corticotropins and 3', 5'-cyclic nucleotides on corticosterone production. Endocrinology 88: 1109–1116, 1971.
18. Pearlmutter AF, Franco-Saenz R, Rapino E, Saffran M: Human adrenal tissue *in vitro*: steroidogenic properties. J Clin Endocrinol Metab 39: 150–153, 1974.
19. Tazaki H, Murai M, Baba S: Human adrenal tumors and hyperplasias *in vitro*. A bridge between morphology and function. Invest Urol 11: 288–294, 1974.
20. Szabó D, Gyévai A, Gláz E, Stark E, Péteri M, Alánt O: Changes in the fine structure and function of a hormone-secreting adrenocortical tumour investigated in tissue culture. Virchows Arch A Path Anat Histol 367: 273–280, 1975.
21. O'Hare MJ, Neville AM: Morphological responses to corticotrophin and cyclic AMP by adult rat adrenocortical cells in monolayer culture. J Endocr 56: 529–536, 1973.
22. Ramachandran J, Suyama AT: Inhibition of replication of normal adrenocortical cells in culture by adrenocorticotropin. Proc Nat Acad Sci USA 72: 113–117, 1975.
23. Gill GN, Hornsby PJ, Simonian MH: Regulation of growth and differentiated function of cultured bovine adrenocortical cell. In: Hormones and cell culture, Sato GH, Ross R (eds), Cold Spring Harbor conferences on cell proliferation, vol. 6, Cold Spring Harbor Laboratory, 1979, pp 701–715.
24. Stark E, Gyévai A, Szalay K, Pósalaky Z: Secretion of adrenocorticotrophic hormone by hypophysial cells grown in monolayer culture. J Endocr 31: 291–292, 1965.
25. Gailani SD, Nussbaum A, McDougall JW, McLimans WF: Studies on hormone production by human fetal pituitary cell cultures. Proc Soc Exp Biol Med 134: 27–32, 1970.
26. Rappay Gy, Fazekas I, Bukulya B, Gyévai A, Stark E: Fine structural evidence for hormone production by human foetal hypophyseal cell cultures. Acta Anat 89: 572–576, 1974.
27. Sétáló G, Nakane PK: Functional differentiation of the fetal anterior pituitary cells in the rat. Endocr exp 10: 155–166, 1976.
28. Watanabe YG, Daikoku S: Immunohistochemical study on adenohypophysial primordia in organ culture. Cell Tiss Res 166: 407–412, 1976.
29. Nemeskéry Á, Németh A, Sétáló Gy, Vigh S, Halász B: Cell differentiation of the fetal rat anterior pituitary *in vitro*. Cell Tiss Res 170: 263–273, 1976.
30. Begeot M, Dupouy JP, Dubois MP, Dubois PM: Immunocytological determination of gonadotropic and thyrotropic cells in fetal rat anterior pituitary during normal development and under experimental conditions. Neuroendocrinology 32: 285–294, 1981.
31. Gyévai A, Stark E, Bukulya B, Ács Zs: Basal and stimulated ACTH secretion by human pituitaries as a function of gestational age and of time *in vitro*. In: Endocrinology, Neuroendocrinology, Neuropeptides, Part II, Advances in physiological sciences, vol. 14, Stark E, Makara GB, Halász B, Rappay Gy (eds). Pergamon Press and Akadémiai Kiadó, Budapest, 1981, pp 95–99.
32. Vale W, Rivier C, Brown M, Chan L, Ling N, Rivier J: Applications of adenohypophyseal cell cultures to neuroendocrine studies. In: Hypothalamus and endocrine function, Current topics in molecular endocrinology, vol. 3, Labrie F (ed), New York, Plenum Press, 1976, pp 397–429.
33. Rappay Gy, Nagy I, Makara GB, Bácsy E, Fazekas I, Kárteszi M, Kurcz M: Major metabolic pathways and hormone production in unstimulated monolayer cultures of the rat anterior pituitary. *In Vitro* 15: 751–757, 1979.
34. Rappay G, Gyévai A, Kondics L, Stark E: Growth and fine structure of monolayers derived from adult rat adenohypophyseal cell suspensions. *In Vitro* 8: 301–306, 1973.
35. Landolt AM, Rothenbüler V: The size of growth hormone granules in pituitary adenomas producing acromegaly. Acta endocr (Copenh) 84: 461–469, 1977.
36. Gazsó L, Pásztor E: Growth characteristics of human pituitary adenomas in tissue and cell cultures. Acta Neuropathol (Berl) 49: 225–230, 1980.
37. Tougard C, Tixier-Vidal A, Kerdelhue B, Jutisz M: Étude immunocytochimique de l'évolution des cellules gonadotropes dans des cultures primaires de cellules antéhypophysaires de rat. Aspects quantitatifs et ultrastructuraux. Biol Cellulaire 28: 251–260, 1977.
38. Bácsy E, Tougard C, Tixier-Vidal A, Marton J, Stark E: Corticotroph cells in primary cultures of rat adenohypophysis: A light and electron microscopic immunocytochemical study. Histochemistry 50: 161–174, 1976.
39. Tixier-Vidal A, Tougard C, Picart R: Subcellular localization of some protein and glycoprotein hormones of the hypothalamo-hypophyseal axis as revealed by the peroxidase-labeled antibody method. In: Immunoenzymatic techniques, Feldmann G, Druet P, Bignon J, Avrameas S (eds). Amsterdam, North-Holland Publishing Company, 1976, pp 307–321.
40. Tashjian Jr AH, Hoyt Jr RF: Transient controls of organ-specific functions in pituitary cells in culture. In: Molecular genetics and developmental biology, Sussman M (ed). New Yersey, Prentice Hall, Inc. Englewood Cliffs, 1972, pp 353–387.
41. Rappay Gy, Komolov IS, Fazekas I, Bácsy E, Gudoshnikov VI, Fedotov VP: Ba^{2+} affects growth hormone and prolactin secretion as well as cell morphology in rat anterior pituitary cultures. Acta biol Acad Sci Hung 32: 137–146, 1981.
42. Tixier-Vidal A, Brunet N, Gourdji D: Plasma-membrane modifications related to the action of TRH on rat prolactin cell lines. In: Hormones and cell culture, Sato GH, Ross R (eds). Cold Spring Harbor conferences on cell proliferation, vol. 6, Cold Spring Harbor Laboratory, 1979, pp 807–825.
43. Hazum E, Cuatrecasas P, Marian J, Conn PM: Receptor-mediated internalization of fluorescent gonadotropin-releasing hormone by pituitary gonadotropes. Proc Natl Acad Sci USA 77: 6692–6695, 1980.
44. Conn PM, Smith RG, Rogers DC: Stimulation of pituitary gonadotropin-releasing hormone. J Biol Chem 256: 1098–1100, 1981.
45. Maurer RA: Bromoergocryptine-induced prolactin degradation in cultured pituitary cells. Biochemistry 19: 3573–3578, 1980.
46. Correr S, Motta PM: The rat pituitary cleft: A correlated study by scanning and transmission electron microscopy. Cell Tiss Res 215: 515–529, 1981.
47. Rosa PA, Policastro P, Herbert E: A cellular basis for the differences in regulation of synthesis and secretion of ACTH/endorphin peptides in anterior and intermediate lobes of the pituitary. J exp Biol 89: 215–237, 1980.
48. Scott AP, Lowry PJ, Ratcliffe JG, Rees LH, Landon J: Corticotrophin-like peptides in the rat pituitary. J Endocr 61: 355–367, 1974.
49. Chatterjee P: Histological and ultrastructural studies of the rabbit pars intermedia in organ culture. I. Adult and young adult tissue. Cell Tiss Res 169: 485–500, 1976.
50. Chatterjee P: Histological and ultrastructural studies of the rabbit pars intermedia in organ culture. II. Developing tissue. Cell Tiss Res 167: 387–405, 1976.

51. Semoff S, Fuller BB, Hadley ME: Secretion of melanophore-stimulating hormone /MSH/ in long-term cultures of the pituitary neurointermediate lobes. Cell Tiss Res 194: 55–69, 1978.
52. Fazekas I, Bácsy E, Rappay Gy: Monolayer cultures of rat pituitary intermediate lobe: Growth and fine structure of epithelial cells. Acta biol Acad Sci Hung 29: 273–283, 1978.
53. Fazekas I, Bácsy E, Rappay Gy: Identification of epithelial cells and fibroblasts in hypophysis intermediate lobe cultures by scanning electron microscopy. Acta biol Acad Sci Hung 29: 407–416, 1978.
54. Fazekas I, Cyévai A, Bácsy E, Rappay Gy: Rat intermediate lobe cell groups in suspension culture: Morphological and functional characteristics. Acta biol Acad Sci Hung 31: 69–80, 1980.

Author's address:
Institute of Experimental Medicine
Hungarian Academy of Sciences
Szigony u. 43
H-1083 Budapest, Hungary

CHAPTER 19

Scanning Electron Microscopy of adrenal gland in mammals

PIETRO M. MOTTA

1. Introduction

The introduction of the Scanning Electron Microscope (SEM) to the morphological sciences has offered the anatomist the opportunity to study the surface architecture of various cells and tissues (1–3). Further, a number of techniques for dissection and separation of tissues and organ components which mainly use freeze-fracturing and/or chemical dissociation followed by critical point drying have given us a great deal of information on the three-dimensional organization of different parenchymatous organs (4–6).

The adrenal gland in mammals and humans has been examined extensively by Transmission Electron Microscopy (TEM) under a number of experimental and physiopathological conditions.

Chapters 17 and 15 in this volume are two excellent reviews of current knowledge of the adrenal cortex and the adrenal medulla, respectively. Despite numerous studies by SEM on different parenchymatous organs, the three-dimensional architecture of the endocrine glands in general and that of the adrenal gland in particular have received little attention (7–10). The aim of this chapter is to review the fine three-dimensional microanatomy of the adrenal tissues in selected mammals by SEM with the hope that correlative information of this gland might provide more exact morphophysiological parameters for understanding physiopathological conditions.

Because TEM is used mainly to study intracellular aspects and SEM is particularly useful to elucidate surface cell features, in this chapter special attention will be paid to the surface of the endocrine cells and to their associated capillaries and tissue spaces. Detailed descriptions of techniques used for preparing adrenal glands, as well as other similar parenchymatous organs, can be obtained by consulting recent articles (1–6).

2. The adrenal gland

The adrenal glands in mammals are paired organs commonly located superior to each kidney and well protected by variable amounts of adipose tissue.

The single organ basically consists of two concentric layers of tissue, having different origins, structures and functions. The outer portion – *the adrenal cortex* – is composed largely of epithelioid cells arranged in cord-like structures which store and secrete different types of steroids. It is derived from the coelomic intermediate mesoderm. The inner central part – *the adrenal medulla* – is essentially a mass of connective tissue in which are contained glandular cells. These cells can be regarded as modified ganglionic sympathetic cells. In fact these – also called chromaffin cells – as true ganglionic elements, arise from the neural crest and release into the blood the neurotransmitter norepinephrine plus its derivative, epinephrine (see also Chapters 15 and 16).

A collagenous connective tissue forms a capsule to the organ and penetrates the glandular tissues, giving support to various nets of vessels which serve as conductors of nutritive and secretory products to and from the glandular cells.

3. The adrenal cortex

SEM reveals that the cortex of the species studied is made up of interconnected groups, cords and laminae of polyhedral cells. (Fig. 1). In the zona glomerulosa the secretory cells are compacted to form clusters of medium-sized elements whose surfaces, fronting the intercellular and pericapillary spaces, possess a number of short microvilli and small pits. The cells of the zona fasciculata are large and arranged in parallel cords or laminae often interconnected and closely paralleled by longitudinally running capillaries. (Fig. 2).

Motta, PM (ed): Ultrastructure of endocrine cells and tissues. ISBN-13: 978-1-4613-3863-5

In the zona reticularis the cords of cells form complicated anastomosing networks closely associated with tortuous capillaries (Figs. 3 and 4). Such features also evident in the inner areas of the zona fasciculata are strikingly similar to those displayed by hepatic tissues (11, 12). Actually, all the cells of the adrenal cortex possess polyhedral shapes. Most of their facets, covered with microvilli and possessing pits and other small invaginations, are directly or indirectly exposed to the capillary wall (Figs. 2–4). Again, as in the liver, the cells give rise to a complex labyrinthine system of spaces (lacunae) of various width in which the satellite capillaries (sinusoids) are suspended in a delicate texture of fine reticular fibers. Such intercellular and pericapillary spaces are wider and more frequent in areas of the gland such as the inner fasciculata and the reticularis where in addition the capillary networks form extensive plexuses (see also Chapter 27 in this volume). As a consequence of this close association between vessels and cells almost every single cellular facet of the polyhedral secretory cells, with the exceptions of those areas coupled by mechanical or gap junctions, must be exposed to the blood flow and its filtrate (9). In the pericapillary compartment of such a lacunar labyrinth the cortical cells project microvilli and the lacunae may contain a variable amount of fluid-like material, a discontinuous basal lamina surrounding

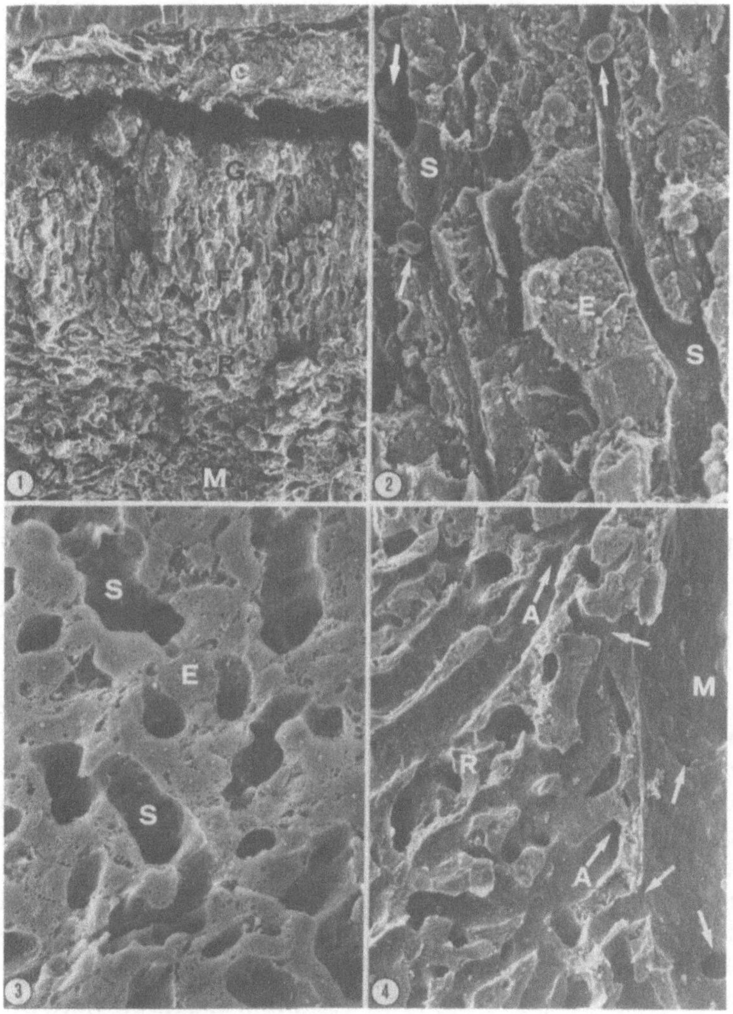

Figs. 1–4. (1) Survey of cat adrenal gland. C: capsule; G: zona glomerulosa; F: zona fasciculata; R: zona reticularis; M: medulla (× 186). (2) Rat adrenal cortex: zona fasciculata. Partially fractured endocrine cells (E) and sinusoids (S). Erythrocytes in the lumen of some sinusoids (arrows) (× 1,190). (3) Pig adrenal cortex: zona reticularis. Labyrinthic arrangement of cortical endocrine cells (E) and sinusoids (S) (× 840). (4) Pig adrenal cortex. The figure illustrates the zona reticularis (R) and portion of medullary region (M). Arrows indicate capillaries from the reticularis opening in large veins of the medulla. Larger vessels (arterioles) cross the cortex and open into a medullary vein directly (A plus arrows) (× 300). (With permission of ref. 9).

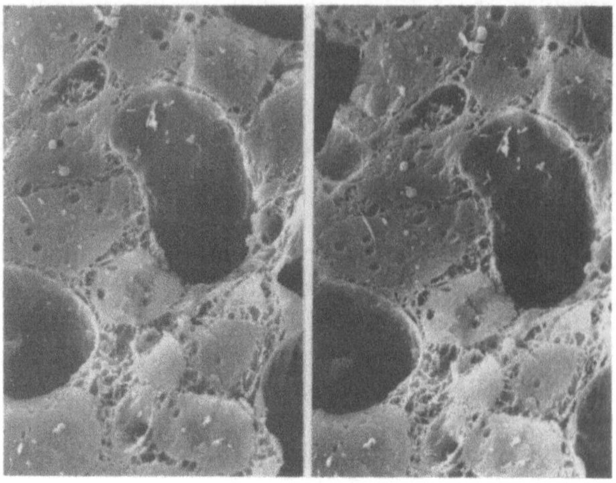

Fig. A. This stereo pair illustrates the lacunar extensions and labyrinthine arrangement of cortical cells and sinusoids in the zona reticularis of a pig adrenal (× 1,620)*.

the capillary, and a delicate fibrillar texture in which occasional cells (pericytes, macrophages) can be found (stereo-pair A). The intercellular compartment of the labyrinth is bordered only by the microvillous surface of the glandular cells and mainly is filled with fluid. SEM observations confirm the existence of this lacunar system of spaces already noted in early TEM studies (13–16) and in addition make possible a determination of the submicroscopic topographical relationship between cells and capillaries in a three-dimensional fashion (stereo-pair A). Obviously, this labyrinthine system of intercellular and pericapillary spaces greatly enhances the adsorbing and secretory surfaces of the cells and creates a current of fluids from and to the vessels into which the hormonal product must be finally released and transported.

In the adrenal gland, as well as in other endocrine tissues, this system of spaces serves also for transporting nutrients to the cells. It might be the site where both cellular and blood filtrates are continuously and temporarily accumulated before they enter the lymphatic vessels. In this regard, it is important that these fluid-filled spaces are particularly well developed and expanded to form a wide labyrinthine net in those organs (for example liver, hypophysis, adrenal gland) in which typical lymphatic capillaries have not been reported to occur among the secretory cells (16–18). Particularly, the parallelism with the liver is worthnoting at this point, in that both hepatic and adrenal cortical tissues have a similar microanatomical arrangement (cords and/or laminae of cells and wide intercellular spaces) and, curiously, both possess lymphatic vessels located only at the periphery of their respective glandular units (in the portal spaces of Mall in the liver and in the capsule in the adrenal gland) (4, 9, 12). As for the subendothelial spaces (of Disse) in the liver, in the adrenal such spaces, with their contents, might represent a sort of pericellular and pericapillary lacuna in which the lymph components are accumulated before they are conveyed in the lymphatic vessels proper of the gland.

At this point it might be pertinent to pose the question: do other endocrine glands have similar three-dimensional features regarding their capillaries and associated tissue spaces? The few available observations on other endocrine tissues such as Leydig cells of some mammals (19, 20), the corpus luteum (21), the epiphysis (22) and the hypophysis (10) showed an intriguing three-dimensional arrangement. The similarity with the adrenal cortex is most striking in the corpus luteum and the distal part of the hypophysis, especially as the latter lacks typical lymphatic capillaries.

3.1. Secretory cells of the adrenal cortex

The secretory pattern of adrenal cortical cells, like that of other steroidogenic cells in the ovary and testis (see also Chapters 20 and 21), conforms to similar structural bases, such as the occurrence of large mitochondria with villiform or tubular cristae and very abundant membranes of smooth endoplasmic reticulum with closely associated lipid droplets (15, 25–27). But despite a general consensus on the steroidogenic organization of these cells, evidence is lacking about the mechanism by which hormones accumulate within the cytoplasm and are released from the cell. It generally is accepted by most investigators that steroidogenic cells release their products at a molecular level through a process of diffusion across the plasma membrane (15, 26, 28–31).

From images provided by some early TEM studies on the adrenal cortex, it was proposed that steroids could be discharged from cells into the capillary lumen by means of cytoplasmic protrusions (a sort of apocrine mechanism) (13–15, 32, 33) or lipid droplets with associated AER membranes through a process termed 'endoplasmocrine secretion' (25). Belt et al. (34) in an early study on pelican adrenal cortical cells

* Stereo views as reproduced in this chapter can be observed with the aid of small stereo glasses (Abrams Instrument Corp, Lansing, Mich.) or alternatively as follows: 1) focus with both eyes on an object at ~30 feet; 2) without changing focus, interpose the stereo pair in the line of sight. In this way three images will be seen, and the middle one will be stereoscopic.

observed: a) small granules bounded by a smooth membrane within the cytoplasm, b) occasional small invaginations of the plasma membrane and c) granules similar in size and electron density but lacking a membrane in the subendothelial and intercellular spaces. On the basis of these findings the authors suggested originally that the dense bodies were hormones synthesized within the smooth membranes of the ER, and carried to the cell surface and secreted by a process of reverse pinocytosis. Similar bodies or granules within the cavities of the AER and coated invaginations on the plasmalemma also were reported in other studies in the opossum adrenal cortex (35). More recently an abundance of electrondense granules (0.2–0.4 μm in diameter) was described in close proximity to the Golgi apparatus and to the plasma membrane in the adrenal cortex of some mammals stimulated with ACTH (36, 37). It was concluded in these studies that the release of steroids may be linked to the appearance and release of these granules. This process of exocytosis of steroid hormones has been reported in other studies on luteal cells of different species and humans (for a full discussion of this topic consult Chapter 21). Furthermore, it was hypothesized that in adrenal cortical cells, as well as in other steroid secreting cells (Leydig cells and luteal cells), the Golgi apparatus might assemble glycoproteins which are used as carrier molecules to transport steroids to the surface of the cell (38, 39). However, before accepting the new

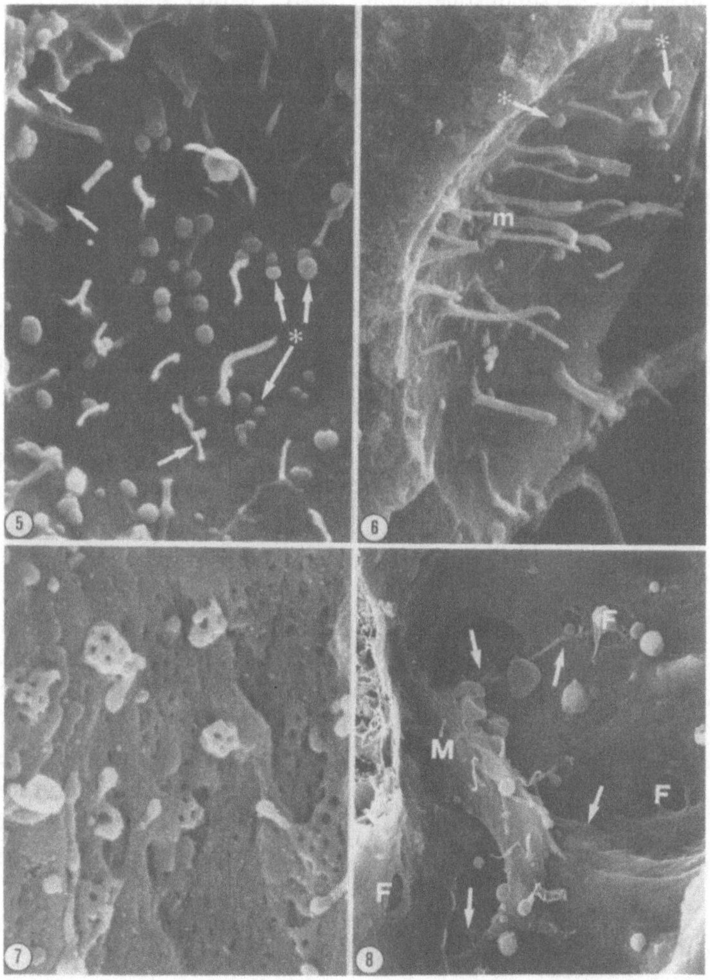

Figs. 5–8. (5–6) Surfaces of endocrine cells of the zona fasciculata of cat adrenal. The rather concave free facets show scattered microvilli (m), small invaginations (pits) (arrows) and small spheroidal droplets. Some of these latter, closely related to surface invaginations (arrows plus asterisks), might represent 'steroidogenic secretory granules' (× 15,120; × 17,160). (7) Fenestrated endothelia wall of a sinusoid in the zona reticularis of rat adrenal cortex (× 21,000). (8) This figure shows a macrophage (M) within the lumen of a large sinusoid in the reticularis of pig adrenal. Note the close relationship between macrophage extensions and large endothelial fenestrations (arrows). Other large gaps (F) are patent in adjacent areas of the endothelial wall (× 7,500). (With permission of ref. 9).

notion that steroid hormones in these cells can be contained in secretory granules and may be extruded by a process of exocytosis, it is necessary to completely exclude the possibility that these intracellularly located granules represent microperoxisomes frequently described in these cells, and verify that the released 'secretory granules' indeed contain the hormone.

Present SEM observations do not add much to the solution of this problem. They do show however that the plasmolemma of many cortical cells is provided with a number of small invaginations (pits) (Figs. 5 and 6) which appear very similar to those described by SEM on the surface of hepatocytes (4, 11, 12, 40, 41). These invaginations hypothetically are interpreted as sites of secretion of lipoproteins leaving the cell (11, 40, 41). Although as a rule these small surface pits are empty, in a few cases small spheroidal bodies were observed in relation and/or close to their surfaces. (Figs. 5 and 6). Similar droplets also were observed frequently not only on the cell surface but mixed with the fibrillar material in the pericapillary spaces and even in the capillary lumen. (Figs. 5 and 6). Although the pits and small associated granules of our SEM results seem to corroborate the data originally described by Belt et al. (34) and reported in a number of more recent experimental studies (36–39) about a mechanism of exocytosis in steroid-secreting cells, these observations must be interpreted with caution before relating them to an actual process of hormone release. Nevertheless, SEM is a useful technique, when complemented by other methods such as freeze-etching, to study particular aspects of cell secretion, in that gives a topographical view of significant events occurring on the cell surface.

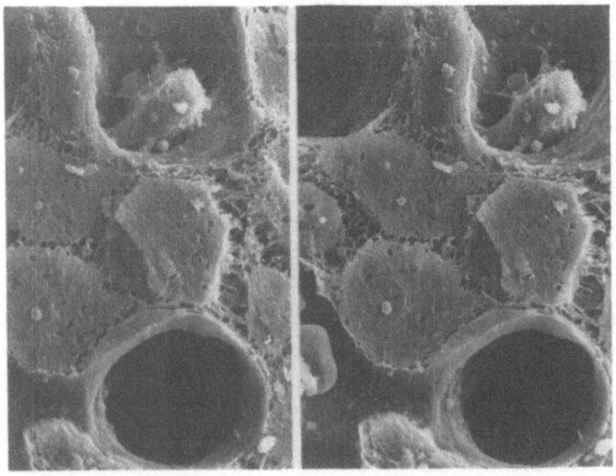

Fig. B. Stereo view of sinusoids and cortical cells of a pig zona reticularis. A macrophage within the sinusoidal lumen is evident in the upper part of the figure. The lacunar spaces among endocrine cells and sinusoids are wide and filled with microvilli (× 2,400).

4. The vessels

Rich vascular plexuses of different caliber penetrate the capsule of the gland and radiate from the cortex to the medulla where they open into a series of vessels which coalesce to form a large central vein (Fig. 4). Some of the arteries may reach the medulla directly, where they in turn branch to form a rich capillary plexus around the cells. Understanding the special and complex three-dimensional distribution of the blood vessels in the adrenal gland is essential to an understanding of the labyrinthine architecture of the gland and its characteristic zonation (Figs. 2–4) (see also Chapter 27).

4.1. Capillaries

Due to their variable caliber and irregular course, the capillaries of the adrenal cortex, as in other endocrine glands, are typical sinusoids (stereo-pair B). The endothelial wall, with the exception of the areas containing the nucleus, is extremely thin and has a variable number of fenestrations arranged in groups and intervening in zones between smooth and thickened cytoplasmic ridges (Fig. 7). SEM and correlated TEM observations have shown that adjacent endothelial cell margins might overlap and are provided with intercellular junctions (9). A few microvilli and long, isolated cilia also are encountered commonly on the luminal surface of these adrenal sinusoids. The most typical arrangement of the fenestrations (pores) in adrenal sinusoids (both cortex and medulla) is in clusters of small pores about 50–100 nm in diameter (9). Fenestrations larger than these are encountered in other areas, mainly in the inner fasciculata, in the reticularis and, to some extent, in the medulla (9). Large fenestrations have been reported in the TEM literature (13–15, 33, 35, 42) and SEM easily confirmed their presence by direct visualization of their occurrence, size, shape and distribution.

Although it is not possible to exclude that *in vivo* small pores might coalesce into large fenestrations and that their density and size might be affected by a number of physiopathological conditions, it is necessary to keep in mind that artefacts due to perfusion techniques might also produce such an appearance. Nevertheless, it is worth mentioning that a dual population of pores as shown by SEM has been reported in recent studies in which complementary methods such as freeze-etching were used, which allow direct observation of a larger specimen area than simple ultrathin sections for TEM (43). These 'en face' topographical views of the sinusoidal wall of the mouse adrenal gland show that the average diameter of pores is about 50–400 nm and the population density of pores in crowded areas is about $34.8 \pm 2.8 / \mu m^2$. Furthermore the small pores (those with a diameter up to 100 nm) comprise 80.6% while the larger ones (up to 400 nm) are 19.4% of all pores in the sinusoids of the adrenal cortex. The findings by freeze-etching obtained in these studies were not affected by different fixation methods, either perfusion or immersion (43).

The occurrence of a dual population of pores in the adrenal sinusoids is rather similar to that encountered in liver sinusoids where small and large fenestrations (sometimes very large) have been observed by SEM (4, 11, 41, 44–46) and freeze-etching (43, 47). Again, it is reasonable to propose that the presence of such large fenestrations in the liver and in the adrenal gland might be dynamically dependent upon the special permeability conditions occurring in these sinusoids in relation to various physiological situations (for example, stress to the adrenal and selective barrier to nutriens and extraneous substances from the portal circulation in the liver). Compared to the endothelia of other endocrine organs, with regard to their fenestrations, the endothelium of the adrenal gland serves as an intermediary between a common endocrine organ, such as the thyroid gland and a special one, such as the

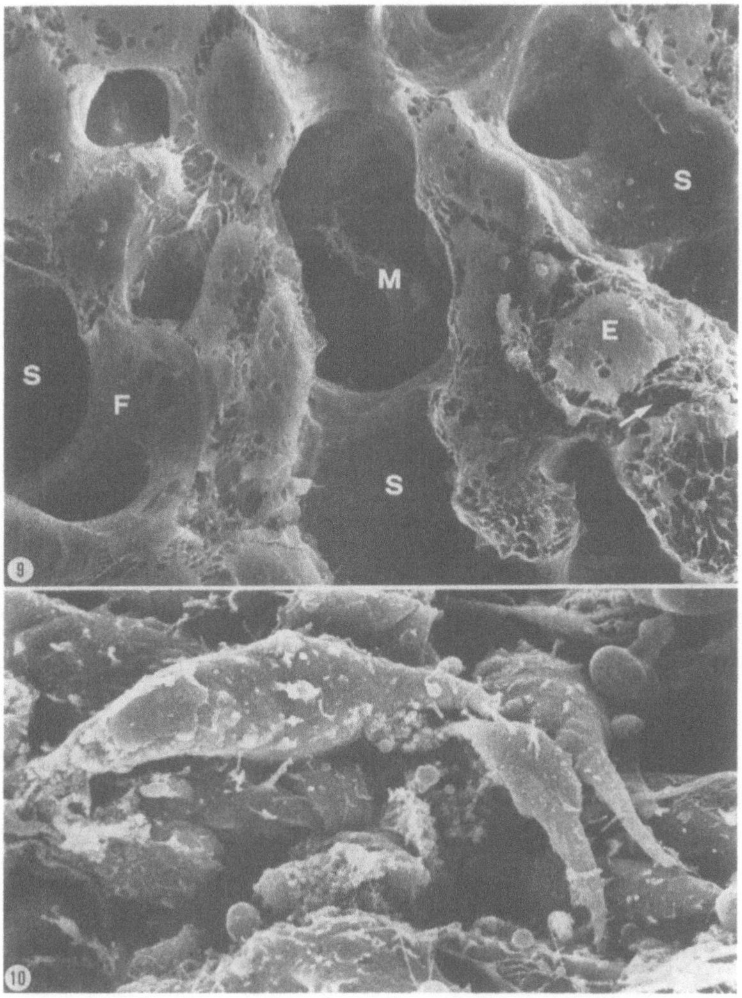

Figs. 9–10. (9) Laminar arrangement of the zona reticularis of pig adrenal. Numerous microvilli of endocrine cells (E) occupy the wide intercellular and pericapillary spaces (arrows) forming a complicated lacunar network. The sinusoids (S) may show a fenestrated wall (F) and their lumen can be occupied by occasional macrophages (M) (× 1,620). (10) Dissociated adrenal medulla. The large spindle-shaped cells at the center likely represent secretory chromaffin cells (× 2,920).

liver (43). The sinusoids of the adrenal gland are unique in nature and function as compared with other endocrine glands (9, 43).

A further aspect which makes the adrenal sinusoids somewhat comparable to those of the liver is the lumen which in some areas is occupied by large and irregularly shaped cells corresponding to macrophages (Fig. 8) (44). In some instances these cells are so closely associated with large endothelial fenestrations (Fig. 8) that it should be considered the possibility that they, as migrating elements, are implicated in producing, even temporarily and dynamically, the large openings and occasional discontinuities encountered along the endothelial wall (Figs. 8, 9 and stereo-pair B). Probably these macrophages, readily observed by SEM as intraluminally located in the sinusoids, are the same cells reported in the literature as occurring mainly in the perisinusoidal spaces (13–15) or closely associated to the endothelial wall. Like other macrophages in the body they might arise from activated monocytes in the circulating blood and then migrate out of the lumen, depending upon the local physiopathological demands on of the gland. In liver sinusoids, such macrophages clearly form a special and relatively stable population of cells (Kupffer cells) (48, 49). In the adrenal gland macrophages might help to provide the endothelial wall and the subendothelial compartment with a phagocytic function. This action might be used not only to protect the tissues (defense system) but to remove extraneous material accumulated as a result of the function of the gland. As

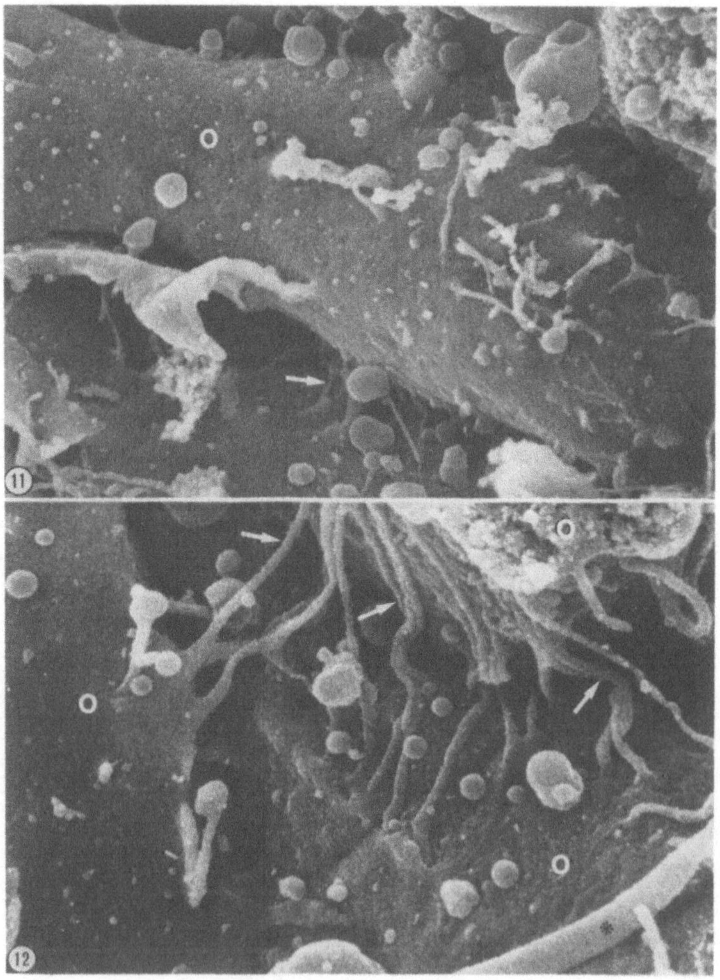

Figs. 11–12. Cat adrenal medulla. Large ovoid cells (O), likely secretory in nature, are illustrated. Their smooth surface form a number of microvilli and thiny extensions in close contact with adjacent cells (arrows). Thiny projections also arise from a smooth fiber probably corresponding to an unmielinated nerve fiber (asterisk) (× 16,920; × 21,600).

recently suggested in studies on similar steroidogenic gland (corpus luteum) macrophages, besides their role in trivial phagocytotic activity (21, 30, 50, 51), they might also play important part in modulating the gland function and maintaining hormone (progesterone) secretion (52). The intriguing possibility that macrophages in the adrenal cortex may play a similar function cannot be excluded, considering their strategical position within the glandular tissues and vessels. Finally, present SEM results clearly indicate that macrophagic cells are associated so closely with the endothelial wall (stereo-pair B), which they might cross, that it is understandable why in early studies on the adrenal gland the phagocytic activity was attributed erroneously to endothelial cells (the reticulo-endothelial system). Three-dimensional results show that in these as well as in other organs (liver, hypophysis), macrophages are present always in intimitate association with the sinusoidal wall and must be considered solely responsible for the phagocytic activity of the vessel (9, 10, 49).

5. Adrenal medulla

Because of the complex texture of the adrenal medulla in which connective tissue elements, nerve fibers, Schwann cells and occasional ganglionic cells are intermixed with the glandular cells proper (chromaffin cells) the latter cannot be distinguished as readily by SEM as those present in the cortical zones where the intervening tissues are moderate. In areas where dissociation of connective tissue occurs, large ovoid cells having a relatively smooth surface and few microvilli have been observed to correspond to chromaffin cells (Figs. 10 and 11). Of course no distinction, can be made between norepinephrine – and epinephrine – secreting cells by surface characteristics because this distinction mainly depends upon the type of granules contained in the cytoplasm (see also Chapter 15). However SEM has revealed the surface of these cells to be covered by a number of thin and long microprojections, some of which come in close contact with the plasma membrane of adjacent cells (Figs. 11 and 12). In some instances thin projections arise from elongated fibers, most likely corresponding to unmyelinated nerve fibers (Fig. 12).

In other instances, cellular adhesion occurs between glandular cells and long cellular projections or laminar processes arising from adjacent glandular and/or somewhat flattened cells (Figs. 10–12). The latter might represent Schwann cells which, as supporting elements, always are associated intimately with chromaffin cells (53).

Finally, although numerous small spherical droplets, which might be interpreted as secretory granules released by the cells, have been encountered on the surface of these cells and in the connective tissue spaces, it is difficult to exclude solely on the basis of SEM data that they are not artefactual (Figs. 10–12). Despite these limits, SEM analysis of this area further stresses the validity of such a complementary technique in obtaining topographical information such as cellular contacts and synapses and their spatial distribution on the cell surface. This information is impossible to acquire by other methods, based only on freeze-fracture and sectioned material. More observations by these and other methods are needed before a three-dimensional spatial reconstruction that incorporated from physiological and biochemical data of the gland can be obtained.

References

1. Motta PM, Andrews PM, Porter KR: Microanatomy of cell and tissue surfaces, Philadelphia, Lea & Febiger, 1977.
2. Kessel RG, Kardon RH: Tissues and organs: a text-atlas of scanning electron microscopy, San Francisco, WH Freeman and Company, 1979.
3. Fujita T, Tanaka K, Tokunaga J: SEM atlas of cells and tissues, Tokyo/New York, Igaku-Shoin, 1981.
4. Motta PM, Muto M, Fujita T: The liver. An atlas of scanning electron microscopy, Tokyo/New York, Igaku-Shoin, 1978.
5. Tanaka K, Fujita T (eds): Scanning electron microscopy in cell biology and medicine, Int Congr Series 545, Amsterdam/Oxford Princeton, Excerpta Medica, 1981.
6. Allen DJ, Motta PM, DiDio LJA (eds): Three dimensional microanatomy of cells and tissue surfaces, New York/Amsterdam/Oxford, Elsevier/North Holland, 1981.
7. Krstić R: Scanning electron microscopic aspects of some endocrine glands. Biomed Res 2, Suppl: 101–108, 1981.
8. Fujita T, Kobayashi S, Serizawa Y: Intercellular canalicule system in pancreatic islet. Biomed Res 2, Suppl: 115–118, 1981.
9. Motta PM, Muto M, Fujita T: Three dimensional organization of mammalian adrenal cortex. A scanning electron microscopic study. Cell tissue Res 196: 23–38, 1979.
10. Correr S, Motta PM: The rat pituitary cleft: a correlated study by scanning and transmission electron microscopy. Cell Tissue Res 215: 515–529, 1981.
11. Motta PM, Porter KR: Structure of rat liver sinusoids and associated tissue spaces as revealed by scanning electron microscopy. Cell Tissue Res 148: 111–125, 1974.
12. Motta PM: The three-dimensional fine structure of the liver as revealed by scanning electron microscopy. Int Rev Cytol (Suppl 6): 347–399, 1977.
13. Zelander T: Endocrine organs: the adrenal gland. In: Electron microscopic anatomy. Kurtz SM (ed), New York, Acad Press, 1964, pp 199–220.
14. Luse S: Fine structure of adrenal cortex. In: The adrenal cortex. Eisensten AB (ed), Boston, Little Brown and Co, 1966, pp 1–37.
15. Idelman S: Ultrastructure of the mammalian adrenal cortex. Int Rev Cytol 27: 181–273, 1970.
16. Brauer RW: Liver circulation and function. Physiol Rev 43: 115–213, 1963.
17. Ottaviani G: Sistema Linfatico. In: Enciclopedia medica italiana, vol 5.

Firenze, USES, pp 1877–1881, 1978.

18. Harrison RG: The adrenal circulation. Oxford, Blackwell, 1960.
19. Motta PM, Calvieri S, Palermo D: On the occurrence of spaces similar to intercellular canaliculi in the Leydig cells of mice. Experientia 29: 1120–1125, 1973.
20. Clark RV: Three-dimensional organization of testicular interstitial tissue and lymphatic space in the rat. Anat Rec 184: 203–226, 1976.
21. Van Blerkom J, Motta PM: A scanning electron microscopic study of the luteo-follicular complex. III. Formation of the corpus luteum and repair of the ovulated follicle. Cell Tissue Res 189: 131–154, 1978.
22. Krstić R: Scanning electron microscopic study of the freeze-fractured pineal body of the rat. Cell Tissue Res 201: 129–135, 1979.
23. Vila-Porcile E: Le reaseau des cellules folliculo-stellaires de l'adenohypophyse du rat (pars distalis). Z Zellforsch 129: 328–369, 1972.
24. Aguado LI, Schoebitz K, Rodriguez EM: Intercellular channels in the pars tuberalis of the rat hypophysis and their relationship to the subarachnoid space. Cell Tissue Res 218: 345–354, 1981.
25. Rhodin, JA: The ultrastructure of the adrenal cortex of the rat under normal and experimental conditions. J Ultrastruct Res 34: 23–71, 1971.
26. Nussdorfer GG, Mazzocchi G, Meneghelli V: Cytophysiology of the adrenal zona fasciculata. Int Rev Cytol 55: 291–365, 1978.
27. Nussdorfer GG: Cytophysiology of the adrenal zona glomerulosa. Int Rev Cytol 64: 307–369, 1980.
28. Christensen AK, Gillim SW: The correlation of fine structure and function in steroid secreting cells, with emphasis on those of the gonads. In: The gonads. McKerns KW (ed), New York, Appleton-Century-Crofts, 1969, pp 415–448.
29. Fawcett DW, Long JA, Jones A: The ultrastructure of endocrine glands. Rec Prog Horm Res 25: 315–368, 1969.
30. Motta PM: Electron microscopy study on the human lutein cell with special reference to its secretory activity. Z Zellforsch 98: 233–248, 1979.
31. Crisp TM, Dessouky DA: Fine structure of the primate corpus luteum. In: Biology of the ovary. Motta PM, Hafez ESE (eds), The Hague/Boston/London, Martinus Nijhoff Medical Division, 1980, pp 150–161.
32. Bloodworth JMB, Powers KL: The ultrastructure of the normal dog adrenal. J Anat 102: 457–476, 1968.
33. Kurosumi K, Fujita H: Functional Morphology of endocrine glands. An atlas of electron micrographs. Tokyo/New York, Igaku-Shoin, 1975.
34. Belt WD, Sheridan MN, Knouff RA, Hartman FA: Fine structural study of a possible mechanism of secretion by the interrenal cells of the brown pelican. Z Zellforsch 68: 864–873, 1965.
35. Long JA, Jones AL: The fine structure of the zona glomerulosa and the zona fasciculata of the adrenal cortex of the opossum. Amer J Anat 120: 463–488, 1967.
36. Gemmel RT, Laychock SG, Rubin RP: Ultrastructural and biochemical evidence for a steroid-containing secretory organelle in the perfused cat adrenal gland. J Cell Biol 72: 209–215, 1977.
37. Mazzocchi G, Belloni AS, Rebuffat P, Robba C, Neri G, Nussdorfer GC: Fine structure of the rabbit adrenal cortex and the effects of short-term ACTH administration. Cell Tissue Res 201: 165–179, 1979.
38. Haddad A, Brasileiro ILG, Pelletier G: Incorporation of L(3H)-fucose into glycoproteins of the adrenal gland of mice. Light microscope radioautographic study on semi-thin sections. Cell Tissue Res 202: 325–335, 1979.
39. Wagner RK: Extracellular and intracellular steroid binding proteins. Properties, discrimination, assay and clinical application. Acta Endocrinol 88 (Suppl 218): 7–73, 1978.
40. Motta PM, Fumagalli G: Structure of rat bile canaliculi as revealed by scanning electron microscopy. Anat Rec 182: 499–514, 1975.
41. Grisham JW, Nopanitaya W, Compagno J, Nagel AEH: Scanning electron microscopy of normal rat liver. The surface structure of its cells and tissues components. Amer J Anat 144: 295–322, 1975.
42. Sheridan MN, Belt WD: Fine structure of the guinea pig adrenal cortex. Anat Rec 149: 73–98, 1964.
43. Ishimura K, Okamato H, Fujita H: Freeze-etching images of capillary endothelial pores in the liver, thyroid and adrenal of the mouse. Arch Histol Jpn 41: 187–193, 1978.
44. Motta PM: A scanning electron microscopic study of the rat liver sinusoid. Endothelial and Kupffer cells. Cell Tissue Res 164: 371–385, 1975.
45. Itoshima T, Kobayashi T, Shimada Y, Murakami T: Fenestrated endothelium of the liver sinusoids of the guinea pig as revealed by scanning electron microscopy. Arch Histol Jpn 37: 15–24, 1974.
46. Makabe S, Motta PM: Foetal and adult liver sinusoids and Kupffer cells as revealed by scanning electron microscopy. In: The reticulo endothelial system and the pathogenesis of liver disease. Liehr H, Grun M (eds), Amsterdam, Elsevier/North-Holland, 1980, pp 11–16.
47. Montesano R, Nicolescu P: Fenestrations in the endothelium of rat liver sinusoids revisited by freeze-fracture. Anat Rec 190: 861–870, 1978.
48. Motta PM: Kupffer cells as revealed by scanning electron microscopy. In: Kupffer cells and other liver sinusoidal cells. Wisse E, Knook DL (eds), Amsterdam, Elsevier/North-Holland, 1977 pp 93–102.
49. Motta PM: Scanning electron microscopy of the liver. In: Progress in liver diseases, Vol. 7. Popper H, Schaffner F (eds), new York, Grune Stratton, 1982, pp 1–16.
50. Paavola LG: The corpus luteum of guinea pig. IV. Fine structure of macrophages during pregnancy and postpartum luteolysis, and the phagocytosis of luteal cells. Amer J Anat 154: 337–364, 1979.
51. Van Blerkom J, Motta PM: The cellular basis of mammalian reproduction, Munich/Baltimore, Urban & Schwarzenberg, 1979.
52. Kirsch TM, Friedman AC, Vogel RL, Flickinger GL: Macrophages in corpora lutea of mice: characterization and effects on steroid secretion. Biol Repr 25: 629–638, 1981.
53. Kent C, Coupland RE: On the uptake of exogenous catecholamines by adrenal chromaffin cells and nerve endings. Cell Tissue Res 221: 371–383, 1981.

Author's address:
Department of Anatomy
Faculty of Medicinc
University of Rome
Viale Regina Elena 289
00161 Roma, Italy

CHAPTER 20

Ultrastructure and stereological analysis of Leydig cells

HIROSHI MORI

1. Introduction

Shortly after the demonstration of the androgenic function of the testis by Berthold, Leydig reported in 1850 the presence of testicular interstitial cells in some species (1). These cells now bear his name. In the 1930s, progress in the study of pituitary hormones substantially established the Leydig cells to be the main source of testosterone in the male. Since the 1960s, the use of electron microscopy has revealed the morphological details of this cell type. The relation of the ultrastructure to pertinent biochemical and physiological information has opened a new dimension of studies of the male reproductive system.

In this chapter, the ultrastructure of Leydig cells is described in humans as a principal example. Several ultrastructural features of the rat (2), mouse (3), opposum (4), cat (5) and boar (6) are also cited. Another aim of this chapter is a stereological analysis of Leydig cells at electron microscope level, by which the readers would be able to understand the ultrastructure on a common basis. It may allow functional correlations with physiological and biochemical information to be more appropriate. Finally a pathologic condition is briefly reviewed. Several general reviews are available on light microscopy (7) and electron microscopy (5, 8). Reviews on the biochemistry of androgen production and its regulation (9) may be useful for functional implication.

2. Light microscopy

Leydig cells are generally mononuclear, polygonal cells with a diameter of 15–20 μm. The nucleus is vesicular, round to oval in shape, and is located eccentrically with one or two nucleoli near the nuclear rim. The chromatin is distributed predominantly toward the periphery of the nucleus, giving the nuclear membrane an appearance of moderate thickness. The cytoplasm is abundant and shows eosinophilia of various degree. Leydig cells from aged individuals or patients with a consumptive disease contain numerous brown pigments, which are lipofuscin granules. They are stained pink to reddish purple with periodic acid-Shiff reaction. Only in the human Leydig cells, rod- or needle-shaped crystals, which are called Reinke's crystals (10), occur in the cytoplasm and sometimes in the nucleus. The crystals, weakly eosinophilic with H.E., are stained strongly with Heidenhein's iron-hematoxylin, Weigert's fibrin stain, carmine and basic fuchsin.

Leydig cells occur singly or, more often, in clusters of various sizes in the interstitial tissue. They are arranged as cap-like clusters not infrequently around the seminiferous tubules. However, they tend to be located mainly around the blood vessels. The relationship of the Leydig cells with the blood and lymphatic vessels differs from species to species (11). In most of the endocrine organs, there is an extensive network of capillaries (sinusoids). The capillaries of the testicular interstitial tissue consist of intertubular and peritubular capillaries and show an intimate relationship with the seminiferous tubules rather than the Leydig cells. The testicular interstitial tissue is poorly differentiated as an endocrine organ because of poor organization and lack of intimate association with the blood vessels.

3. Electron microscopy

The ultrastructural characteristics of Leydig cells are essentially similar to those of other steroid-secreting cells, i.e., adrenocortical cells (12), luteal cells (13) and the ovarian interstitial gland cells in some species (14). Leydig cells are characterized by a well-developed smooth endoplasmic reticulum (SER), numerous mitochondria containing tubular or lamellar cristae and lipid droplets in varying amounts. Several important and prominent organelles will be described below.

Motta, PM (ed): Ultrastructure of endocrine cells and tissues. ISBN-13: 978-1-4613-3863-5

3.1. *Plasma membrane*

The outer surface of the plasma membrane is thought to be covered with a thin layer of mucopolysaccharide (glycocalyx), which, however, can not be shown by conventional electron micrography. Between the plasma membranes of the adjoining Leydig cells, there is a constant spacing of approximately 20 nm. Coated vesicles or pits are occasionally seen in association with the plasma membranes. Attachment devices between the Leydig cells consist mainly of interdigitations and gap junctions (Figs. 1, 4). The gap junction, is a close apposition of plasma membranes with a 2-nm spacing, and may serve a region of low resistance for the passage of ions between the Leydig cells in a cluster. The gap junction occurs mainly between the finger-like cytoplasmic processes of adjoining cells. Desmosomes are rare, and poorly developed. Leydig cells are generally not invested with a basal lamina. However, small patches of the basal lamina may be seen over the free surface of the Leydig cells facing the interstitial space (Fig. 2). Leydig cells may anchor in the tissue to some extent, even though they appear to float in the interstitial fluid when they occur singly. Receptors for luteinizing hormone (LH) seem to occur on the plasma membrane of the cytoplasmic processes (K. Nozu, personal communication).

3.2. *Smooth endoplasmic reticulum*

Smooth ER is the most prominent organelle in the Leydig cells. Although it is found throughout the cytoplasm, it appears to be distributed mainly in the peripheral regions because of the relative scarcity of other organelles in this area. A random network of

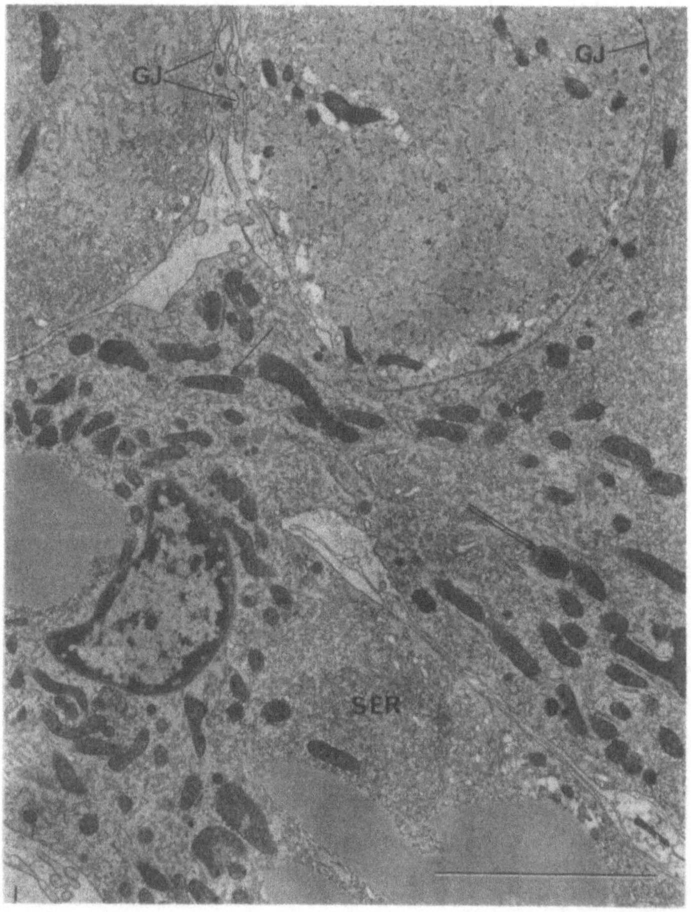

Fig. 1. A low power electron micrograph of Leydig cells from a 37-year-old human male. Smooth endoplasmic reticulum (SER) is well developed and is seen as randomly distributed tubules. Mitochondria have tubular and lamellar cristae. They contain small dense bodies (arrow) and lipid droplets (double arrow). A cell in the bottom has crystals of Reinke, whereas two cells at the top have filamentous structures, probably precursors of the crystals. GJ: gap junctions. Fixed with a mixture of 2% osmium tetroxide and 1.5% potassium ferrocyanide. Bar: 5 μm (× 6,000).

interconnected tubules with a diameter of 40–80 nm is the usual form of SER, though the tubules appear to occur independently in section (Figs. 1, 2, 3). When the tubules are cut perpendicular to their axis, they appear to be vesicles. However, when the diameter of vesicles is more than 100 nm or varies extensively, those vesicles are thought to be artifacts owing to poor fixation. The SER of steroidogenic cells is an organelle which is difficult to preserve for electron microscopy. Fixation by immersion in osmium tetroxide, the usual technique, tends to produce vesicular SER. It is preferable to perfuse with glutaraldehyde buffered to appropriate osmolarity and pH.

Regional specialization of SER has been described in some species. Although in the central cytoplasm, SER occurs as randomly distributed, interconnected tubules, it may be organized in closely packed, fenestrated cisternae in the peripheral regions. Both arrangements appear to be interconvertible. Another form of SER is a system of concentric membranous whorls. The whorls consist of flattened, interconnecting cisternae, which may surround lipid droplets or mitochondria. This system was commonly found in the Leydig cells of mouse, opposum and boar but not in the cells of human and rat.

Two specialized structures originating in SER have been found in mouse (3) and rat (15), although their function is not clear. The one in the mouse (BALB/c strain) is a double-walled tubule, part of which is enfolded by the other part which is continuous with the former like an intussusception. The other, in atrophic testes of Wistar strain rat, displays a crystalloid, cylindrical appearance, the wall of which consists of twisted tubules of SER with empty lumen.

3.3. Golgi complex

The Golgi complex consists of several stacks of closely packed, flattened and slightly curved cisternae with numerous small vesicles (Fig. 4). Coated vesicles, lysosomes and multivesicular bodies are seen intermingled with the Golgi vesicles. In mouse and cat, the Golgi complex is usually juxtanuclear. In the human, however, it is not large but Golgi components are numerous and scattered so that they can be seen almost anywhere in the cytoplasm. It is known that the Golgi complex is engaged in the concentration of secretory granules in the protein-producing cells, synthesis of polysaccharide component in the carbohydrate-elaborating cells, and transient accumulation of absorbed lipids in the intestinal mucosal cells. However, it is not clear what role the Golgi complex plays in the production and secretion of steroids. Enlargement of expansion of the Golgi area of adrenocortical cells or corpus luteum cells after stimulation with pituitary hormones suggests the involvement of the Golgi complex in steroid production and secretion (8). However, no secretory product has been demonstrated in the Golgi elements.

In this respect, it must be clarified where the testosterone produced is stored in the cell, how the testosterone is transported to the cell surface and how it is released into the extracellular space. Virtually nothing is known about these problems. The main technical difficulties are in the lipid-soluble nature of steroids and their precursors, and that they do not form visible products after fixation for electron microscopy. Lipid-solubility of steroids causes an extraction from and a diffusion within the cell during fixation and dehydration. Visualization of steroids with isotopes and use of non-aqueous fixation and embedding will help to resolve these problems.

3.4. Mitochondria

Mitochondria of human Leydig cells are generally small and numerous. They are usually elongate with a diameter of 0.3–0.4 μm and a length of 1–2 μm, but are sometimes bent or branched. They are scattered throughout the cytoplasm with a slight concentration at the center of the cell. Their internal structure shows species differences. In humans, the inner mitochondrial membrane exhibits lamellar and tubular cristae (Figs. 2, 3), whereas the mitochondria of mouse Leydig cells are packed with numerous tubular cristae of uniform size, as seen in the mitochondria of the zona fasciculata and zona reticularis of the adrenal cortex in most species.

The mitochondrial matrix is moderately electron dense in most species. The mitochondria have round, dense granules with a diameter of approximately 40 nm (Figs. 1, 2), which vary in number in different species. Occasionally mitochondria also contain lipid droplets of less electron density; these are somewhat larger than the dense granules (Fig. 1). The function of these intramitochondrial inclusions is not clear, but it is thought that they may be involved in testosterone production.

3.5. Lipid droplets

Lipid droplets are commonly present in Leydig cells, though not so numerous as in the adrenocortical cells. However, they vary much in number and size from species to species and from one cell to another even within a testis. Lipid droplets in mouse Leydig cells are numerous and of almost uniform size with a diameter of 1 μm, those in guinea pig are large in size (2 μm in diameter) but small in number, and rat Leydig cells have very few lipid droplets. In the human, they are relatively scarce. Lipid droplets appear usually as round, smooth-surfaced, homogeneous and less electron dense spheres. However, some lipid droplets show high electron density, while others are translucent at the center and opaque at the periphery. This variation seems to result from differences in the constituents of lipids (amount of unsaturated fatty acids), or differences in fixatives and dehydrating agents used.

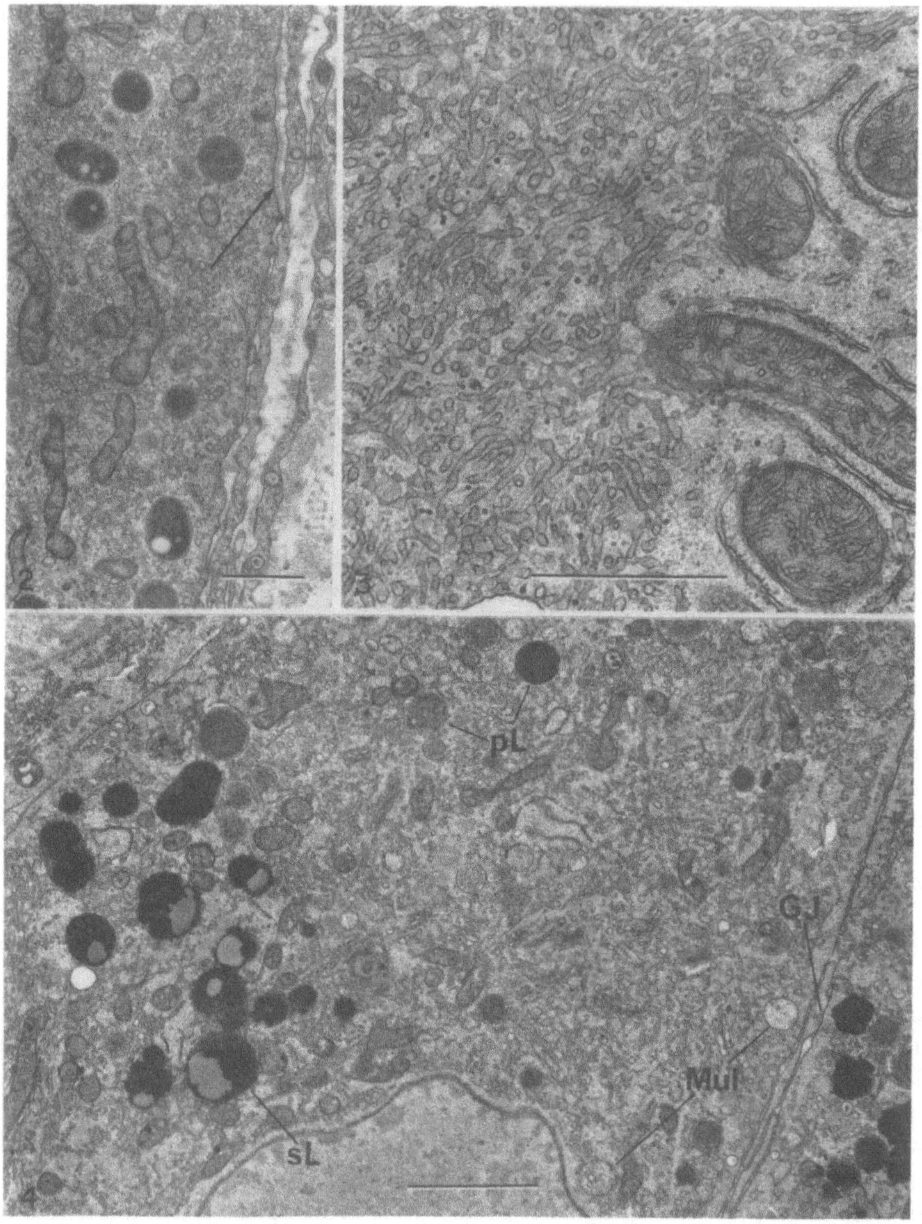

Figs. 2–4. (2) The periphery of a Leydig cell from a 27-year-old male. A part of the plasma membrane is invested with a small patch of the basal lamina (arrow). Bar: 1 μm (× 10,800). (3) Smooth ER occurs as randomly distributed, interconnected tubules with a diameter of 40–50 mm. In this cell, mitochondria have many tubular cristae and no intramitochondrial granules. Bar: 1 μm (× 25,600). (4) The Golgi complexes are widely distributed throughout the cytoplasm. In and around the Golgi area, numerous primary lysosomes (pL), secondary lysosomes (sL) and multivesicular bodies (Mul) are seen. Some of the secondary lysosomes appear as lipofuscin granules. GJ: gap junction. From a 74-year-old male, fixed with the same fixative as in Figure 1. Bar: 2 μm (× 8,560).

The lipid droplets are often closely associated with tubules or cisternae of SER, and may be enveloped by flattened cisternae of concentric membranous whorls of SER in Leydig cells of species such as mouse and cat. This relationship may facilitate the transfer of materials between two organelles. A limiting membrane of droplets is not usually present. However, some droplets appear to be wrapped by a membrane-like structure 4–5 nm in thickness, thinner than the unit membrane of other organelles. It is possible that this membrane-like structure is a half-membrane consisting of a protein layer on the cytoplasmic face and a phospholipid layer on the inner face; it is two-thirds of the unit membrane in thick-

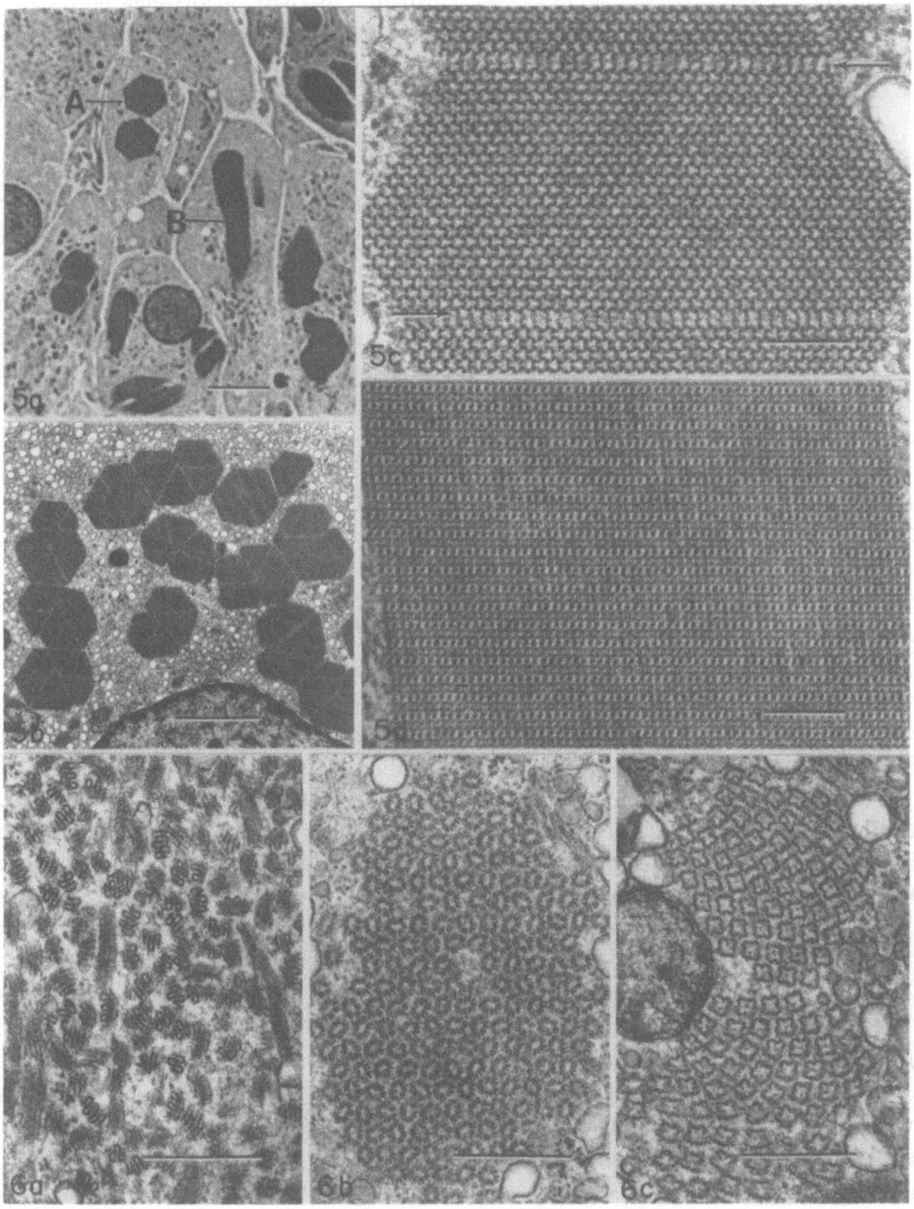

Figs. 5–6. (5) Crystals of Reinke. (a) Crystals are clearly visible on light microscope section embedded in Epon and stained with Heidenhein's iron-hematoxylin. A is a cross section of crystals and B is a longitudinal section. Bar: 10 μm (× 800); (b) A cross section of numerous small crystals, corresponding to A of Figure 5a. Bar: 2 μm (× 5,440); (c) A higher power magnification of Figure 5b, showing a honeycomb pattern. Arrows indicate a dislocation. Bar: 0.2 μm (× 53,440); (d) A longitudinal section of a Reinke's crystal, corresponding to B of Figure 5a. Bar: 0.2 μm (× 53,440). (6) (a, b, c) Various forms of filamentous structures, presumably precursors of the crystals. These filamentous structures occur more frequently than the crystals and usually do not co-exist with the crystals. Bar: 0.5 μm (× 3,120), respectively.

ness. It is not known how lipid droplets arise. Whether they accumulate freely in the cytoplasmic matrix, or originate in the cavity of SER and then dissolve the membrane of SER remains to be clarified. The half-membrane on the surface of the lipid droplets might favour the latter view.

In the Leydig cells of aged humans, lipid droplets in their original form are fairly few. More often they occur as a part of lipofuscin granules described below.

3.6. Lysosomes

There are a considerable number of lysosomes in human Leydig cells (Figs. 2, 4). They consist of primary and secondary ones. The primary lysosomes are membrane-bounded granules which are less than 0.8 μm in diameter and contain a homogeneous material, though varying in electron density. They are frequently found in and near the Golgi area. When the primary lysosomes fuse with expendable or noxious materials taken up by the cell or produced within the cell, and digest them by their hydrolytic enzymes, secondary lysosomes are formed. The secondary lysosomes vary in size, shape and electron density and are up to 4 μm in diameter. When most of the material in the secondary lysosomes has been digested, a digestive vacuole is formed. If most of the material remains undigested, it is called a residual body. A special form of residual body is an accumulation of pigment called lipofuscin or ceroid granule (Fig. 4). The lipofuscin granules consist of varying amounts of lipid droplets of low density and granular, heterogeneous material of high density. They are bounded by a single membrane, and tend to fuse wich each other. Lipofuscin granules are thought to form when lipid in the Leydig cells goes unutilized for a long period and is subject to autoxidation to peroxides. The peroxided lipid thus formed is noxious to the living cell and consequently comes into contact with the primary lysosomes, resulting in lipofuscin granules (16). This will account for the abundance of lipofuscin granules in the Leydig cells of aged humans, as shown in Figure 4. This seems common to myocardial cells, ganglion cells and adrenocortical cells (particularly the zona reticularis cells), since these cells are rich in lipids and scarcely undergo regeneration.

3.7. Crystals

Crystals of Reinke are a unique inclusion characteristic to human Leydig cells (Figs. 1, 5). They are hexagonal prisms which appear differently according to the plane of sectioning (Figs. 5a, 5b). It has been shown by optical diffraction that the crystals consist of a meshwork of 50-Å-thick filaments (17). A cross section of prisms shows a hexagonal honeycomb pattern of 200 Åon each side (Fig. 5c). A longitudinal section shows repeated sets of two parallel straight lines and a row of dense dots, all 150 Å apart (Fig. 5d). Reinke's crystals occur mostly in the cytoplasm and sometimes in the nucleus. The frequency of their occurrence varies in each individual and from cell to cell. Filamentous structures, which may appear as rice-form bodies by light microscopy (17) and are probably precursors of the crystal, are more frequent than the crystals in the cytoplasm. They can be seen in various forms in the sections (Figs. 6a-6c). Usually the filamentous structures and the crystals do not co-exist within the cell.

Reinke suggested that the crystals were involved in spermatogenesis and libido (10). However, it does not seem probable that they are essential for testosterone production in human Leydig cells, because they are not present in other species and there is no evidence of correlation with fertility. It has been recently reported that the crystals may be a sort of degenerative product, since their occurrence does not relate to the serum testosterone level but parallels the age of individuals (18).

4. Stereological analysis

Hitherto the fine structural characteristics of Leydig cells has been described mainly in humans. It can easily be supposed that there may be a quantitative difference in Leydig cell ultrastructure according to differences in hormonal circumstance or between species, even though there would be no qualitative difference. In general as well as in the last section of the present chapter, ultrastructural characteristics have been described quantitatively as 'large or small in number of volume' or 'well or poorly developed'. However, these seem to be subjective descriptions which may not mean the same thing to all the readers. The quantification of the ultrastructure will serve for a better understanding of Leydig cell biology. It may allow functional correlation with pertinent physiological and biochemical data available at present as well as in the future.

4.1. Stereology at light microscope level

In earlier studies, simple quantitative analysis of Leydig cells was performed only to estimate the number of Leydig cells per seminiferous tubule or

Sertoli cell (19). In the last decade, improved quantitative techniques in microscopy have been widely applied to various tissues and organs. These methods, collectively called 'morphometry' or 'stereology', allow the number, volume and surface area of structures in three dimensions to be assessed from two-dimensional, conventionally prepared light and electron microscope sections. Principles and practical applications have been detailed elsewhere (20).

Using a point-count method, which is the most popular in morphometry, the volume and/or the number of Leydig cells within a unit volume of testis tissue or per a whole testis have been estimated at light microscope level (19). In the human, volume occupied by Leydig cells in the testis tissue is reported to be 12% (21) and a total volume per testis of normal young adults to be 0.8 ml (22). The relative volume of Leydig cells in the testis tissue of dog, hamster, rat, rabbit and guinea pig is reported to be 3.2, 2.8, 2.7, 2.2 and 1.9%, respectively (23). An extremely large value has been reported in the boar in which Leydig cells occupy 37% of the testis tissue (8).

The development of Leydig cells in human testes of embryos and fetuses was stereologically studied by Holstein et al. (24). They found that the volume percentage of Leydig cells in the testis increased rapidly from the 7th gestational week and reached a maximum (41.7%) between the 12th and 14th weeks of gestation.

Changes of Leydig cells under stimulation by gonadotropin (LH or exogeneous hCG) were morphometrically studied. In a study of seasonally-breeding rock hyraxes, Neaves (25) observed that the total volume of Leydig cells per testis increased by a factor of 2.6 during the breeding season, whereas the number of Leydig cells per testis scarcely changed. On the contrary, Christensen and Peacock (19) showed that a striking increase in volume of Leydig cell clusters in testis tissue treated with excess hCG resulted mainly from an increase in number of Leydig cells. They found that Leydig cells of rat testis increased in volume percentage by a factor of 4.7 and increased in number per cm^3 of testis to 3 times the control value by 5 weeks of hCG treatment, whereas the average volume of a single Leydig cell enlarged only 1.6 times.

Alterations of Leydig cells in cryptorchism was analyzed by Bergh and Damber (26) in neonatally unilateral cryptorchid rats. They observed that the volume (and the number) of Leydig cells per unit volume of testis increased in the abdominal testis at 100 days of age, as compared with the scrotal testis. This would appear proliferation of Leydig cells in cryptorchism. However, the total number of Leydig cells remained unchanged, and the volume of a single Leydig cell decreased in the abdominal testis. Consequently, the total volume of Leydig cell was about one half of that of the scrotal testis. On the other hand, Kerr et al. (27) reported hypertrophy of Leydig cells 4 weeks after inducing cryptorchism in adult rats. Morphometric analysis at electron microscope level showed that the hypertrophy of the Leydig cells was accompanied with an increase in volume of organelles, particularly the mitochondria, Golgi

Table 1. Stereological values for human Leydig cells.*

Component	Parameter	Per cm^3 testis			Per Leydig cell		
		Mean	SE	Unit	Mean	SE	Unit
Leydig cells	Number	7.73×10^6	0.97×10^6	1/cm^3	1		
	Volume	31.5	4.86	mm^3/cm^3	4,080	629	μm^3
	Surface	188	8.68	cm^2/cm^3	2,430	112	μm^2
Nuclei	Volume	2.88	0.77	mm^3/cm^3	373	100	μm^3
Cytoplasm	Volume	28.6	4.37	mm^3/cm^3	3,710	565	μm^3
Smooth ER	Volume	4.24	0.23	mm^3/cm^3	548	30	μm^3
	Surface	2,474	150	cm^2/cm^3	32,000	1,940	μm^2
Rough ER	Volume	0.122	0.012	mm^3/cm^3	16	1.6	μm^3
	Surface	59.7	8.18	cm^2/cm^3	772	106	μm^2
Mitochondria	Volume	1.82	0.077	mm^3/cm^3	236	10	μm^3
Inner membrane	Surface	433	19.7	cm^2/cm^3	5,600	255	μm^2
Outer membrane	Surface	201	12.1	cm^2/cm^3	2,600	157	μm^2
Lipid droplets	Volume	0.128	0.064	mm^3/cm^3	17	8	μm^3
Lysosomes	Volume	1.51	0.17	mm^3/cm^3	195	23	μm^3
Reinke's crystals	Volume	0.593	0.152	mm^3/cm^3	77	20	μm^3

* Evaluated from 6 individuals 74 years old on average (65–87 y.o.), whose testes are 11.1 ml in volume on average
Values are corrected for fixation, sectioning and section thickness effects.

complex and SER in Leydig cells.

However, there are very few papers on morphometric analysis at electron microscope level (28), except our reports (29–32) and the one cited above (27). In the following section, the stereological values of human Leydig cells will be shown, and the values for several important organelles will be compared between species. Finally the differences in stereological data between Leydig cells and adrenocortical cells will be discussed in relation to biochemical information.

4.2. Human Leydig cells

Stereological analysis was performed on perfusion-fixed testes from individuals 74 years old on average (65–87 y.o.). Since the following values were corrected for systematic errors such as section thickness effect (which usually causes overestimate), they may give an impression of considerably smaller values than the survey view of electron micrographs (Table 1).

The decapsulated testis tissue of the aged in-

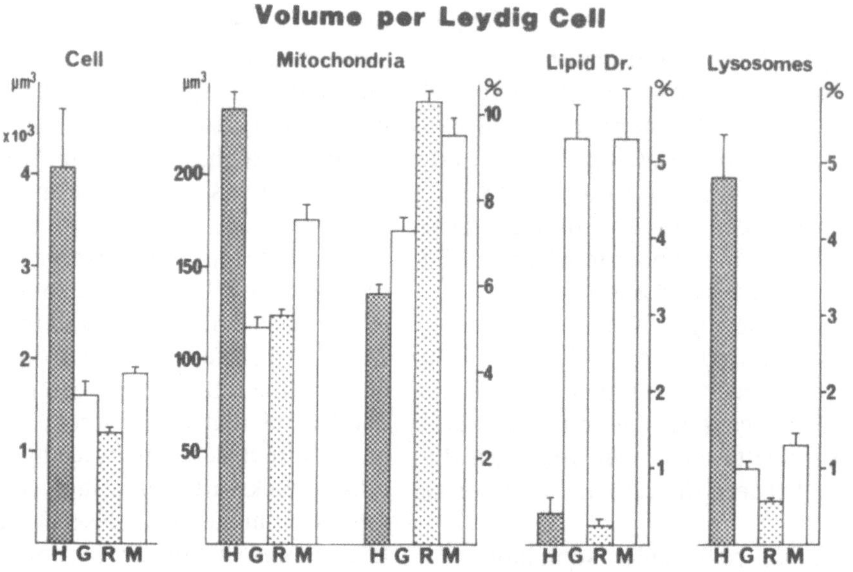

7

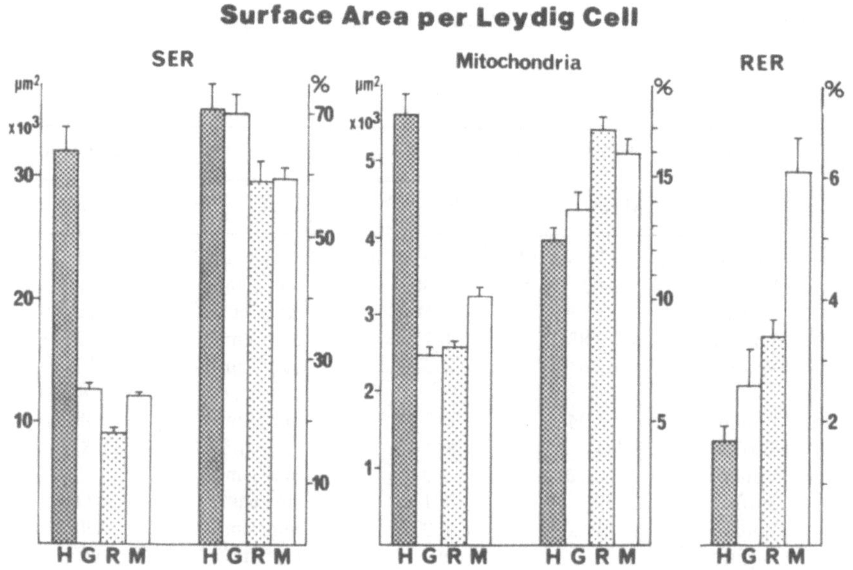

8

Figs. 7–8. Graphs of between-species comparison of Leydig cell stereology. H: human; G: guinea pig; R: rat; M: mouse. T bar shows a standard error of mean. In (7) the scales on the left indicate absolute volume and those on the right volume percentage in the cell. In (8) the scales on the left indicate absolute surface area and those on the right percentage of surface area of each organelle in the total membrane surface area of the cell.

dividuals consisted of 61% seminiferous tubules and 39% interstitial tissue. Leydig cells occupied 8% of the interstitial tissue volume or 3.1% of the whole testis tissue. There were 7.7 million Leydig cells per cm^3 of testis tissue. An average Leydig cell has a volume of 4,080 μm^3 and a surface area of 2,430 μm^2. The most frequent organelle was the SER, which occupied 13.4% of the cell volume or 14.8% of the cytoplasmic volume. These values are measurements of the SER itself (membranes plus luminal contents) and do not refer merely to the general areas of the cytoplasm in which the SER predominates. If they referred to the general areas, they would be up to 60–70% of the cell volume. The surface area of SER was 32,000 μm^2 per cell, i.e., 13 times the surface area of the plasma membrane and constituted 71% of the total membrane surface area of the Leydig cell. Mitochondria occupied 5.8% of the cell volume. The surface area of the mitochondrial inner membrane (including cristae) was 5,600 μm^2/cell, 2.3 times the plasma membrane and constituted 12% of the total membrane surface area of the cell. Lysosomes occupied 4.8% of the cell volume, two-thirds of which were lipofuscin granules. Lipid droplets occupied only 0.4% of the cell volume. This is probably due to the fact that most of the lipid droplets had changed into lipofuscin granules because of the old age of the individuals. Reinke's crystals constituted 1.9% of the cell volume.

Kaler and Neaves (33) reported that, with aging, human Leydig cells decreased in number and in total volume per individual, whereas the volume of an average Leydig cell remained unchanged. Their results showed that there were about 700 million Leydig cells in a pair of testes per individual at 20 years of age. Almost the same value was estimated from our data on aged humans by using the regression line of these authors and supposing a testis volume of young humans to be 20 ml. Leydig cells of young humans at the age of 20 would occupy about 7% of the testis volume. The SER of Leydig cells would have a volume of 9.6 mm^3 and a surface area of 5,600 cm^2 per cm^3 of testis tissue. Mitochondria would have a volume of 4.1 mm^3 and an inner membrane surface area of 980 cm^2 per cm^3 tissue.

4.3. Between-species comparison

A between-species comparison of stereological values for several organelles of Leydig cells is shown in Figures 7 and 8. As mentioned above, testes of humans were obtained from aged individuals, whereas those of guinea pigs, rats and mice came from normal young adults. On the other hand, the stereological analysis was performed by an almost identical method in all species examined, i.e., the tissues were perfused with the same fixatives and the primary data were corrected for systematic errors in the same way (29, 30).

A single average Leydig cell of the human had a volume 2–3 times greater than that of the other three species. Smooth ER of the human had a volume of 550 μm^3/cell, 3–4 times greater than that of the other species. Smooth ER per Leydig cell of the guinea pig had a volume approximately 1.5 times greater than that of the rat and the mouse. However, there was no significant difference among the three species in volume percentage of SER (11.8–13.4%) except the mouse (7.3%). The surface area of SER in the human was 32,000 μm^2/cell, around 3 times greater than in the other species, with the smallest value in the rat. The surface area of SER in the total membrane surface area of the cell was 60–70% in all species examined. On the contrary, volume occupied by rough ER (RER) in the cells of rat and mouse (respectively 1%) was greater than that of the human (0.4%) and the guinea pig (0.6%). In the mouse, surface area of RER constituted 6% of the total membrane surface area of the cell. This was more than 3 times greater than that of the human and around 2 times greater than in the guinea pig and rat.

Volume of mitochondria per cell of the human was largest (236 μm^3), approximately 2 times that of the guinea pig (117 μm^3) and the rat (124 μm^3), whereas the volume percentage of the mitochondria was smallest in the human (5.8%) and largest in the rat (10.3%). Surface area of mitochondrial inner membrane also showed a similar trend of differences. Lipid droplets occupied approximately 5% of the cell volume in the guinea pig and the mouse. However, the number of droplets in a cell differed considerably between these two species (mouse 123, guinea pig 13) because of the difference in their size. Volume occupied by lysosomes was approximately 5% of the cell volume in the human, whereas that in other species was around 1%. This seems to result from the advanced age of the human individuals.

Testosterone secretion rate for human is reported to be 7 mg/day/individual (9) and that for rat is 6.7 ng/g tissue/min (34). This means that the testosterone secretion rate of human testis is about 120 ng/g tissue/min (supposing a testis volume to be 20 ml), and is 18 times higher than that of rat testis. What does this difference result from, the difference in number and/or volume of Leydig cells within a unit volume of testis tissue, or the difference in amount of Leydig cell organelles necessary for steroidogenesis? Recently, Ewing et al. (23, 28) estimated the relative volume and the surface area of SER, RER and mitochondria in several species, and related those stereological values to testosterone secretion rate. Their findings indicated that testosterone secretion rate/g Leydig cells paralleled the volume or surface area of SER/cm^3 Leydig cell cytoplasm. They concluded that the between-species differences in testosterone secretion rate by testes stimulated with hCG could be accounted for by the between-species differences in amount of SER. Our stereological data seem to suggest that the difference between species in testosterone secretion rate results from not only the differences in the amount of Leydig cells and the amount of their organelles but also from the differences in steroidogenic activity of organelles.

4.4. Comparison with adrenocortical cells

A stereological analysis of the adrenocortical cells at electron microscope level has been reported on rat (35) and guinea pig (36). The values of those reports, however, were not corrected for systematic errors. In the following, our stereological values of adrenocortical cells in rats and guinea pigs will be compared with those of Leydig cells in relation to the activity of cholesterologenesis, all the values being corrected for systematic errors in the same way.

As shown in Figure 9, amount of SER and mitochondria differs considerably between the Leydig cells and the adrenocortical cells, although these organelles are important for steroid synthesis. In rat, the volume of SER per cm^3 Leydig cells is 2 times greater than that of the zona fasciculata and the zona reticularis cells per cm^3 zonal cells of the adrenals, whereas the volume of mitochondria of Leydig cells is smaller than that of zona fasciculata and zona reticularis cells by a factor of about 3 and 2, respectively. Differences in surface area of both organelles are almost the same. Volume of lipid droplets in the adrenals of rats are remarkably larger than in Leydig cells. These ultrastructural differences between Leydig cells and the adrenocortical cells are of much interest, when considering the involvement of SER in the production of cholesterol as well as steroid hormones, and of the mitochondrial inner membranes in the side-chain cleavage of the cholesterol. This is consistent with biochemical information that the rat adrenal takes virtually all (>92%) of its cholesterol from the plasma, whereas the rat testicular tissue synthesizes 60% of its cholesterol by itself (37). Taking into account a partition for spermatogenesis, rat Leydig cells seem to synthesize most of the cholesterol for androgen production by themselves.

In the guinea pig, the volume of SER of Leydig cells is 2.3 times larger than that of the zona fasciculata cells and slightly smaller than that of the zona reticularis cells. In contrast, the volume as well as the inner membrane surface area of mitochondria of Leydig cells is approximately half that of the zona fasciculata cells and somewhat smaller than that of the zona reticularis cells. The volume of lipid droplets in Leydig cells is about one-sixth that of the zona fasciculata cells but is twice that of the zona reticularis cells. Namely, the ultrastructure of the zona reticularis cells is stereologically close to that of the Leydig cells in the guinea pig, in contrast to the rat in

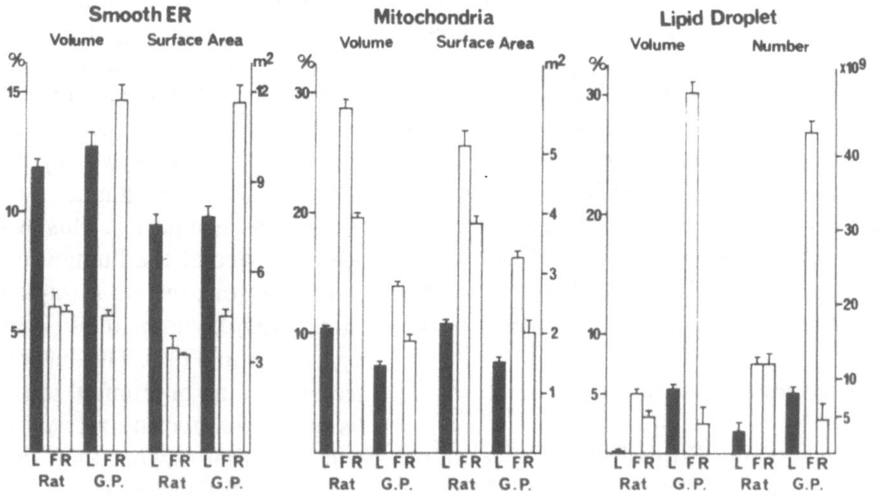

9

Fig. 9. A graph showing a comparison of stereological values between Leydig cells and the adrenocortical cells in rat and guinea pig. Each value is shown per cm^3 Leydig cells or per cm^3 zonal cells of the adrenal cortex. L: Leydig cells, F: zona fasciculata cells, R: zona reticularis cells.

which the ultrastructure of the zona reticularis is similar to that of the zona fasciculata. The rate of *de novo* synthesis of cholesterol in guinea pigs is 40% for the adrenal and 87% for the testis (38). The great difference between rats and guinea pigs in *de novo* synthesis rate of cholesterol in the adrenals seems to be accounted for by the difference between the two species in the ultrastructure of the zona reticularis.

What is the significance of these differences in ultrastructure and cholesterologenesis between the Leydig cells and the adrenocortical cells? It seems probable that the ultrastructural differences reflect the differences in importance for maintenance of the individuals and in reserve capacity of response to stress between the two types of cells. Although the testosterone produced by the Leydig cells is necessary for the maintenance of species, it is less important for the maintenance of the individual. It seems enough for the Leydig cells to produce testosterone from acetate little by little as in a home industry. On the other hand, glucocorticoids produced by the adrenocortical cells are very important for the maintenance of the individual. The adrenocortical cells are required to have a larger reserve capacity of glucocorticoid production, and in turn, it seems efficient for the adrenocortical cells to take the cholesterol as an intermediate substrate from the plasma, store it in the lipid droplets and metabolize it to glucocorticoids. Smooth ER of the adrenocortical cells, on which enzymes for cholesterologenesis are located, does not seem to be required to develop so much. Consequently, mitochondria and lipid droplets would be able to occupy larger volume in the adrenocortical cells, as compared with Leydig cells.

5. Cells identical to Leydig cells and ectopic localization

Cells which are considered to be identical to human Leydig cells are known to occur, though rarely, at sites other than the testicular interstitial tissue. In the ovaries, they are called ovarian hilus cells and are usually found as a hilus cell tumor or hyperplasia which causes virilism (39). In a female patient with ganglioneuroma originating in the adrenal medulla, Leydig-like cells with Reinke's crystals were found in a non-tumorous part of the adrenal medulla (40). Magalhães (41) observed cells with clear cytoplasm containing Reinke's crystals and numerous microfilaments in the periendothelial spaces of the adrenal cortex of human males and suggested a possibility that what appeared to be identical to Leydig cell was a normal constituent of the adrenal cortex. Leydig cells were observed in and around the myelinated bundles along the spermatic cord or intraabdominal testicular vessels (42). Recently, Leydig cells within the lamina propria of the seminiferous tubules or even in direct contact with Sertoli cells have been reported in infertile humans (43, 44, and Fig. 10). Identification of these ectopically-situated cells with Leydig cells requires certain discrimination from the adrenocortical cells. The reason is that remnants of the adrenal cortex are present not infrequently in and

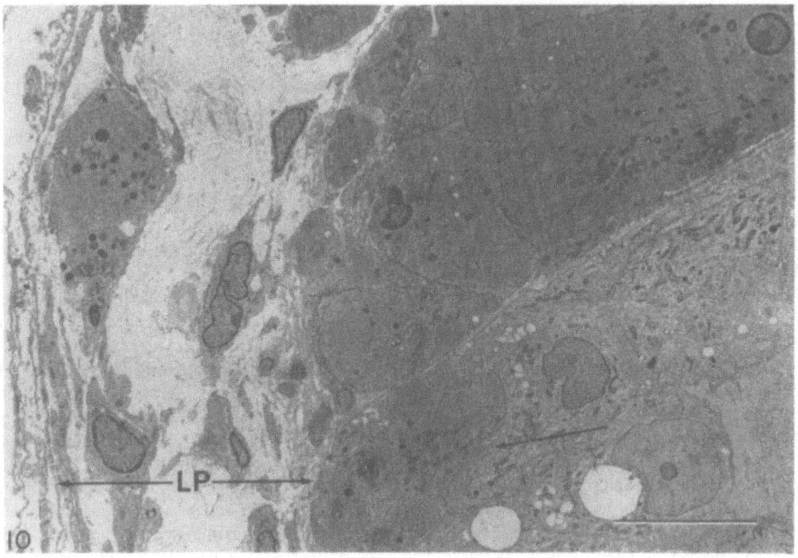

Fig. 10. Leydig cells in the lamina propria (LP) of the seminiferous tubules, seen in a 27-year-old male with azoospermia. Several Leydig cells in the bottom (arrow) lie beneath the basal lamina and in direct contact with Sertoli cells. Bar: 10 μm (× 1,890).

around the gonads (45), and that the ultrastructure of both cell types show a close resemblance. In the absence of Reinke's crystals, the following two characteristics may serve for the distinction between the two cell types: morphologically the presence of mitochondria with vesicular cristae in the cell and fenestrated capillaries in the tissue, and biochemically the activity of 11β-hydroxylase. Leydig cells are devoid of these characteristics. However, these are not always effective, since tumorigenesis may alter the property of the cells.

Acknowledgement

The author appreciates the permission of reproducing illustrations by Springer-Verlag KG (Heidelberg), New Sciences (Tokyo), Wavery Press Inc (Baltimore) and Asakura Shoten Co. (Tokyo). Figure 5a is reproduced from *Virchows Archiv* A 380: 1–9, 1978; Figures 5a, 5c, 5d and 10 are from *The Cell* 12: 92–100, 1980 (in Japanese); Table 1 is from *J Clin Endocrinol Metab* 55: 634–641, 1982; Figures 5a, 5b, 5c, 5d, 6a, 6b, 6c and 10 are from *Histology of Humans*, in press (in Japanese).

References

1. Leydig F: Zur Anatomie der männlichen Geschlechtsorgane und Analdrüsen der Säugethiere. Z Wiss Zool 2: 1–57, 1850.
2. Leeson CR: Observations on the fine structure of rat interstitial tissue. Acta Anat 52: 34–48, 1963.
3. Christensen AK, Fawcett DW: The fine structure of testicular interstitial cells in mice. Am J Anat 118: 551–572, 1966.
4. Christensen AK, Fawcett DW: The normal fine structure of opossum testicular interstitial cells. J Biophys Biochem Cytol 9: 653–670, 1961.
5. Burgos MH, Vitale-Calpe R, Aoki A: Fine structure of the testis and its functional significance. In: The testis. Johnson AD, Gomes WR, Vandemark NL (eds), New York, Academic Press, 1970, vol 1, pp 551–649.
6. Belt WD, Cavazos LF: Fine structure of the interstitial cells of Leydig in the boar. Anat Rec 158: 333–349, 1967.
7. Hooker CW: The intertubular tissue of the testis. In: The testis Johnson AD, Gomes WR, Vandemark NL (eds), New York, Academic Press, 1970, vol. 1, pp 483–550.
8. Christensen AK: Leydig cells. In: Handbook of Physiology. Hamilton DW, Greep RO (eds), Am Physiol Soc, Washington DC, 1975, sec 7, vol V pp 57–94.
9. Hall PF: Endocrinology in the testis. In: The testis. Johnson AD, Gomes WR, Vandemark NL (eds), New York, Academic Press, 1970, vol 2, pp 1–71.
10. Reinke Fr: Beiträge zur Histologie des Menschen. I. Ueber Krystalloidbildungen in den interstitiellen Zellen des menschlichen Hodens. Arch Mikrosk Anat Entwicklungsgesch 47: 34–44, 1896.
11. Fawcett DW, Neaves WB, Flores MN: Comparative observations on intertubular lymphatics and the organization of the interstitial tissue of the mammalian testis. Biol Reprod 9: 500–532, 1973.
12. Idelman S: Ultrastructure of the mammalian adrenal cortex. Int Rev Cytol 27: 181–281, 1970.
13. Crisp TM, Dessouky DA, Denys FR: The fine structure of the human corpus luteum of early pregnancy and during the progestational phase of the menstrual cycle. Am J Anat 127: 37–70, 1970.
14. Mori H: The fine structure of interstitial gland cells in rabbit ovaries. Med J Osaka Univ 20: 215–233, 1970.
15. Murakami M, Kitahara Y: Cylindrical bodies derived from endoplasmic reticulum in Leydig's cell of the rat testis. J Electron Microsc 20: 318–323, 1971.
16. Frank AL, Christensen AK: Localization of acid phosphatase in lipofuscin granules and possible autophagic vacuoles in interstitial cells of the guinea pig testis. J Cell Biol 36: 1–13, 1968.
17. Nagano T, Ohtsuki I: Reinvestigation of the fine structure of Reinke's crystal in the human testicular interstitial cell. J Cell Biol 51: 148–161, 1971.
18. Mori H, Fukunishi R, Fujii M, Hataji K, Shiraishi T, Matsumoto K: Stereological analysis of Reinke's crystals in human Leydig cells. Virchows Arch A 380: 1–9, 1978.
19. Christensen AK, Peacock KC: Increase in Leydig cell number in testes of adult rats treated chronically with an excess of human chorionic gonadotropin. Biol Reprod 22: 383–391, 1980.
20. Weibel ER, Bolender RP: Stereological techniques for electron microscopic morphometry. In: Principles and techniques of electron microscopy. Hayat MA (ed), New York, Van Nostrand Reinhold Co, 1973, vol 3, pp 237–296.
21. Dykes JRW: Histometric assessment of human testicular biopsies. J Path 97: 429–440, 1969.
22. Ahmad KN, Lennox B, Mack WS: Estimation of the volume of Leydig cells in man. Lancet 2: 461–464, 1969.
23. Ewing LL, Zirkin BR, Cochran RC, Kromann N, Peters C, Ruis-Bravo N: Testosterone secretion by rat, rabbit, guinea pig, dog, and hamster testes perfused *in vitro*: Correlation with Leydig cell mass. Endocrinology 105: 1135–1142, 1979.
24. Holstein AF, Wartenberg H, Vossmeyer J: Zur Cytologie der pränatalen Gonadenentwicklung beim Menschen. III. Die Entwicklung der Leydigzellen im Hoden von Embryonen und Feten. Z Anat Entwickl-Gesch 135: 43–66, 1971.
25. Neaves WB: Changes in testicular Leydig cells and in plasma testosterone levels among seasonally breeding rock hyrax. Biol Reprod 8: 451–466, 1973.
26. Bergh A, Damber J-E: Morphometric and functional investigation on the Leydig cells in experimental unilateral cryptorchism in the rat. Int J Androl 1: 549–562, 1978.
27. Kerr JB, Rich KA, de Kretser DM: Alterations of the fine structure and androgen secretion of the interstitial cells in the experimentally cryptorchid rat testis. Biol Reprod 20: 409–422, 1979.
28. Zirkin BR, Ewing LL, Kromann N, Cochran RC: Testosterone secretion by rat, rabbit, guinea pig, dog, and hamster testes perfused in vitro: Correlation with Leydig cell ultrastructure. Endocrinology 107: 1867–1874, 1980.
29. Mori H, Christensen AK: Morphometric analysis of Leydig cells in the normal rat testis. J Cell Biol 84: 340–354, 1980.
30. Mori H, Shimizu D, Takeda A, Takioka Y, Fukunishi R: Stereological analysis of Leydig cells in normal guinea pig testis. J Electron Microsc 29: 8–21, 1980.
31. Mori H, Kadota A, Fukunishi R, Kukita H, Takeuchi N, Matsumoto K: Effects of a cholesterol-rich-diet and a hypolipidemic drug (Clofibrate, CPIB) on Leydig cells in rats. Stereological and biochemical analysis. Andrologia 12: 281–291, 1980.
32. Mori H, Hiromoto N, Nakahara M, Shiraishi T: Stereological analysis of Leydig cell ultrastructure in aged humans. J Clin Endocrinol Metab 55: 634–641, 1982.
33. Kaler LW, Neaves WB: Attrition of the human Leydig cell population with advancing age. Anat Rec 192: 513–518, 1978.
34. Free MJ, Tillson SA: Secretion rate of testicular steroids in the conscious and halothane-anesthetized rat. Endocrinology 93: 874–879, 1973.
35. Rohr HP, Bartsch G, Eichenberger P, Rasser Y, Kaiser Ch, Keller M: Ultrastructural morphometric analysis of the unstimulated adrenal cortex of rats. J Ultrastruct Res 54: 11–21, 1976.
36. Black VH, Russo JJ: Stereological analysis of the guinea pig adrenal: Effects of dexamethasone and ACTH treatment with emphasis on the inner cortex. Am J Anat 159: 85–120, 1980.
37. Morris MD, Chaikoff IL: The origin of cholesterol in liver, small intestine, adrenal gland, and testis of the rat: Dietary versus endogenous contributions. J Biol Chem 234: 1095–1097, 1959.
38. Werbin H, Chaikoff IL: Utilization of adrenal gland cholesterol for

synthesis of cortisol by the intact normal and the ACTH-treated guinea pig. Arch Biochem Biophys 93: 476–482, 1961.
39. Sternberg WH: The morphology, androgenic function, hyperplasia, and tumors of the human ovarian hilus cells. Am J Pathol 25: 493–521, 1949.
40. Scully RE, Cohen RB: Ganglioneuroma of adrenal medulla containing cells morphologically identical to hilus cells (extraparenchymal Leydig cells). Cancer 14: 421–425, 1961.
41. Magalhães MC: A new crystal-containing cell in human adrenal cortex. J Cell Biol 55: 126–133, 1972.
42. Peters KH: Die Ultrastruktur heterotoper Leydigzellen beim Menschen. Andrologia 9: 337–348, 1977.
43. Mori H, Shiraishi T, Matsumoto K: Ectopic Leydig cells in seminiferous tubules of an infertile human male with a chromosomal aberration. Andrologia 10: 434–443, 1978.
44. Schulze C, Holstein A-F: Leydig cells within the lamina propria of seminiferous tubules in four patients with azoospermia. Andrologia 10: 444–452, 1978.
45. Mori H, Matsumoto K: Constant occurrence of adrenocortical tissue in the juvenile rabbit ovary. Amer J Anat 141: 73–90, 1974.

Authors' address:
Department of Pathology
Osaka University School of Medicine
Kitaku, Osaka 530
Japan

CHAPTER 21

Fine structure of the luteal tissue

BELA J. GULYAS

1. Introduction

The corpus luteum is unique among endocrine organs, because it has a transient existence, whether it be short lived – as in case of the extrous and menstrual cycles, or long lived – as in case of pregnancy and pseudopregnancy. Furthermore, the gradual demise of one population of organ(s) heralds the potential for the development of a new set of similar organ(s). Yet, through its secretions the corpus luteum exerts a major influence on the reproductive function of the individual during its lifespan.

1.1. Historical background

The first to recognize and describe the ovarian follicles and possibly the corpus luteum (although he did not name it as such), was A. Vesalius in his 1555 contribution to 'Fabrica' (1). There he described a 'rather large pea full of yellow fluid, colouring the adjacent tissues'. Half a century later, Fabricius in his book 'De Formato Foetu' (1604), in the legends of one of his illustrations, referred to the corpus luteum as numerous conjoined little glands (1). It was a young Dutchman, Regnier de Graaf who realized in the rabbit that the number of 'glandular bodies' formed on the ovaries after coitus was indicative of the number of embryos present (1). His drawing of the sheep ovary (Text Fig. I) depicted various stages of corpus luteum formation. The term 'corpus luteum' was coined by Malpighi in 1697, when he was working with cow ovaries, in which the corpus luteum is truly a *yellow body*.

Historically, the names of von Baer and Bischoff (1) have been credited as the original proponents of the chief doctrines concerning the formation of the corpus luteum. These papers concerned mostly with the controversy about the thecal or granulosal origin of the organ. Although both were incorrect in part, their efforts laid the foundation, a decade before the enunciation of Schleiden and Schwann's cell theory, for a vast amount of work that has been forthcoming on the corpus luteum. Not until the first publications of Sobotta in 1895 and 1896 on the mouse corpus luteum had anyone offered a view and supporting evidence favoring granulosa cells as the origin of the lutein cells; this discovery marked the onset of modern work on the corpus luteum.

A number of experiments performed toward the

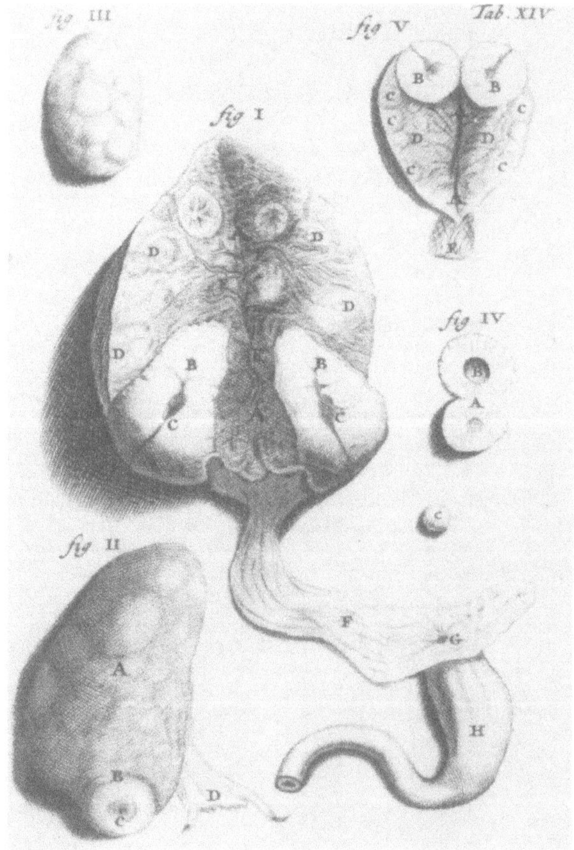

Fig. 1. de Graaf's drawing of the sheep ovary, showing various forms of the corpus luteum. ('De Mulierum Organis Generationi Inservientibus', 1672) (1).

Motta, PM (ed): Ultrastructure of endocrine cells and tissues. ISBN-13: 978-1-4613-3863-5

end of the 19th century provided the first indication that the corpus luteum is one of the endocrine glands of internal secretions which releases its products directly into the circulation (1). This perspective was soon confirmed after the turn of the century. The accomplishments during the latter part of the last century are well summarized by Corner (2), who himself contributed substantially to our current understanding of formation, cytology and regression of pig and monkey corpus luteum, demonstrating beyond doubt its granulosa cell origin (3). The isolation of progesterone from rabbit corpus luteum was the first major step toward a concerted effort to correlate structure and function in this endocrine organ. Others advanced this field by taking advantage of newly developed histochemical techniques and demonstrated the presence of enzymes essential in intermediate steps of steroidogenesis (3).

1.2. Occurrence of corpus luteum

Corpora lutea are apparently ubiquitous organs of vertebrate ovaries and have been described in nearly all classes of vertebrates (4). Originally, corpora lutea of vertebrates have been classified either as corpus luteum of atresia, formed by atretic follicles, and corpus luteum of ovulation formed from the follicle after ovulation (4). It is the latter corpus luteum that this chapter shall be concerned with, examining its fine structural features in mammals as they relate to their functional significance.

2. Formation of corpus luteum

2.1. Structure of the preovulatory follicle

The corpora lutea of the estrous or menstrual cycle develop from the recently ovulated Graafian follicles. During the last half of the preovulatory phase, the diameter of the largest follicle(s), presumably destined to ovulate, increases at a rapid rate. The follicular hypertrophy, ovulation, and the morphologic and functional transformation of the follicle(s) into a corpus luteum occurs apparently under pituitary gonadotropin influence, especially luteinizing hormone. In reality, the preovulatory follicle (Fig. 1) performs a dual function; it is the site of oocyte maturation, and after rupture of the mature follicle the remaining cells of the follicle transform into a corpus luteum that produces steroids – predominantly progesterone, which is known to be essential, at least, for the establishment of pregnancy after fertilization of the ovum.

The preovulatory large antral follicle is constituted by several cell types that differ from each other with respect to origin, topographical location and their secretion before and/or after ovulation. The attenuated cells outside of the follicular basement membrane comprise the theca folliculi, which can be discerned into a theca interna and a theca externa. There is rarely a distinct boundary between the theca interna and externa. The cells within the follicular basement membrane are the follicular or granulosa cells which comprise several layers of cells. In the area of the oocyte the granulosa cells form a mound which is discharged with the oocyte at ovulation; frequently, these are called mural granulosa cells.

2.2. Granulosa cells of the large antral follicle

The granulosa cells of the preovulatory and ovulatory follicles are polyhedral in shape (Fig. 1) and measure approximately 10 μm in diameter (5, 6). Typically, the nucleus of the granulosa cells depicts dense heterochromatin and at least one prominent nucleolus. Granulosa cells contain sparce to moderate quantities of granular endoplasmic reticulum (GER) that are dispersed throughout the cells, often associated with mitochondria. Agranular endoplasmic reticulum (AER) is hardly detectable in these cells (5, 6, 7, 8). Mitochondria are small, round or rod-shaped, for the most part, the cristae are lamelliform (5, 6). The mitochondrial matrix of monkey granulosa cells contains one-to-several small dense granules. The Golgi complex exhibits a degree of polarity, being oriented toward the oocyte or the antrum (6), sometimes, it assumes a vascular pole position (8). Coated vesicles (600–800 Å) may become abundant in the vicinity of the Golgi.

Lipid inclusions, free ribosomes and glycogen particles are present to a varying extent in different species. In the rabbit parallel bundles of microfilaments (40–70 Å) are located mainly in the outer zone of the granulosa cells which extend into the entire length of cytoplasmic projections (6). In some species, such as the pig, granulosa cells enlarge and begin to demonstrate characteristics of luteinization: i.e. changes in the mitochondria, AER, and lipid accumulation occur shortly before ovulation (9).

2.3. Vascularization prior to and upon ovulation

Whereas the process of vascularization plays an important role in the formation of this endocrine organ, the actual process is still incompletely understood. Because the blood capillaries do not penetrate beyond the thick basement membrane, the granulosa cells of mature follicles receive their nutrients by osmosis. The capillary endothelium at the follicular apex just prior to ovulation in the rat and ewe have been described as having gaps, permeable to such tracer molecules as india ink or ferritin (10). Small fenestrations are present in the endothelium of capillaries in the rabbit ovary the number of which can be increased by gonadotrophin injection (11). Furthermore, in the rat, intercellular clefts (200 Å) are present between adjacent endothelial cells which are permeable to peroxidase (12). More recent observations indicate that near ovulation the capillaries at the basement membrane expand to a point that they take on the appearance of sinusoids (13).

The first sign indicative of vascularization of the membrane granulosa is the initial breaking down of

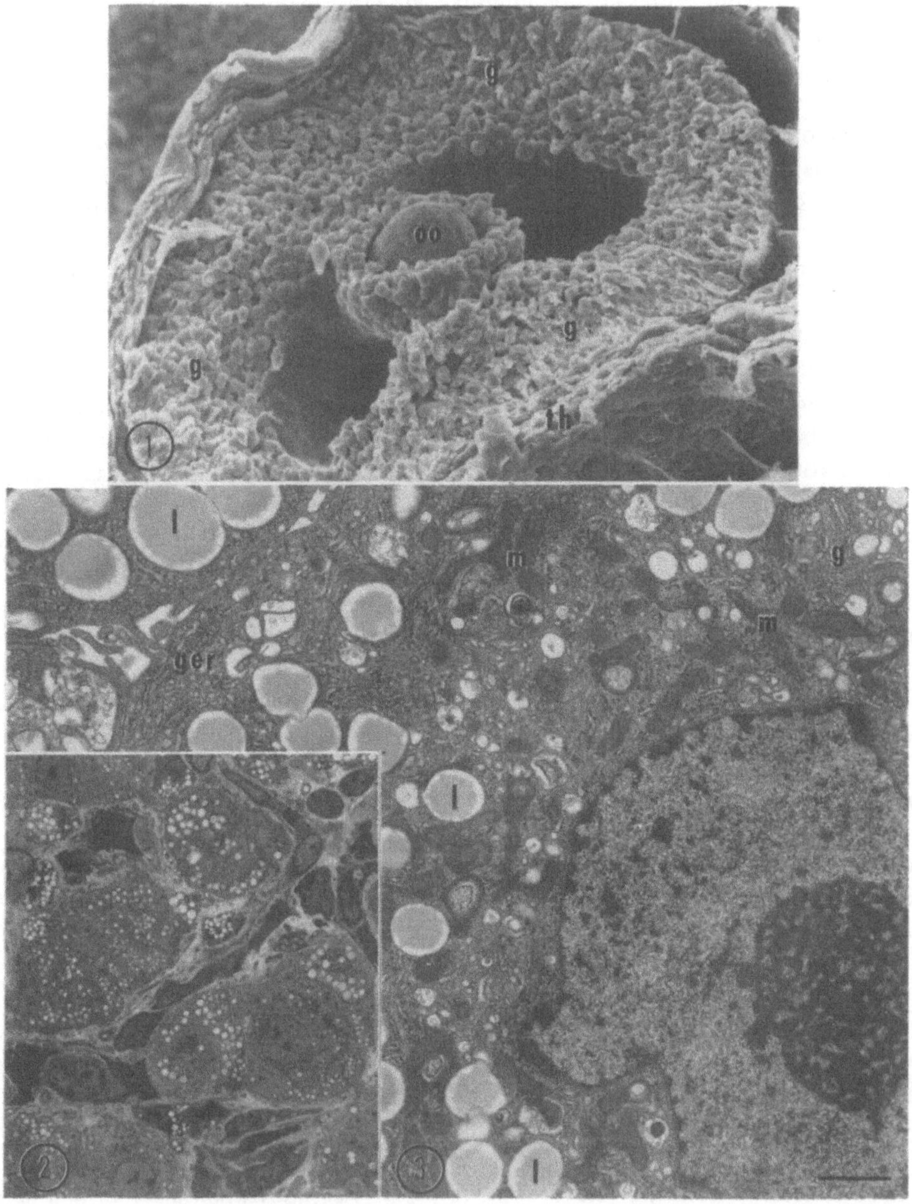

Figs. 1–3. (1) SEM of a developing follicle demonstrating oocyte (oo), granulosa cells (g) and theca cells (th). The surface of the oocyte is covered by a thick, amorphous material corresponding to the zona pellucida. (Kindly contributed by Dr. P. Bagavandos, The University of Michigan). (2) One μm Epon section of monkey luteal cells of intact corpus luteum from day 18 of the menstrual cycle (midluteal phase). (3) Granulosa lutein cell of monkey corpus luteum from day 18 of the menstrual cycle. Note the abundance of all cellular organelles: mitochondria (m), granular endoplasmic reticulum (ger), Golgi complex (g), lipid (l) and other dispersed structures.

the basement membrane and the disappearance of the sharp division line between granulosa and theca interna cells (3). In the pig, the connective tissue from the theca externa penetrates into the granulosa cell mass forming intercalating septae. This gives rise to rapid sprouting of blood capillaries that spread to form an elaborate channel system. On the other hand, in the dog no thecal or connective tissue cells are observed to penetrate the basal lamina (8). Upon development of the corpus luteum the granulosa lutein cells are supplied by well developed sinusoidal capillary plexuses.

3. Luteinization and early pregnancy

Originally, luteinization was meant to describe the structural transformation of follicular cells following ovulation. However, in its more modern usage, it has also become a synonym for renewed (transformed) steroidogenesis in addition to structural transformation. Considerable evidence indicates that luteinization can occur without ovulation, but this topic is beyond the scope of discussion of this chapter. At the very onset of corpus luteum formation in most species examined, two kinds of lutein cells can be distinguished. Peripherally are the theca interna cells giving rise to the theca lutein cells. These may be discernable only during early stages of luteinization. In the guinea pig and rat, for example, the theca cells become indistinguishable after luteinization (7, 14). Centrally are the granulosa lutein cells which develop from the granulosa cells and encompass the overwhelming bulk of the corpus luteum.

3.1. Cellular transformation

Following rupture of the follicle, the granulosa lutein cells tend to separate into small groups. Hyperplasia of the granulosa cells ceases near the time, or shortly after ovulation. The overall growth in the mass of the differentiating corpus luteum occurs as a result of 1) hypertrophy of the lutein cells, and 2) invasion and expansion of the vasculature. During luteinization luteal cells increase several folds in size, and accompanying the cellular hypertrophy is an increase in cytoplasmic-nuclear ratio.

The nuclear matrix gradually transforms from a dense heterochromatin of the granulosa cells to a lighter staining homogeneous euchromatin pattern in the luteinized cells (Fig. 2). One to three nucleoli may form in the nuclei. The AER demonstrates extensive development in some species, whereas only a moderate one in others. The development of the GER is much less extensive usually limited to several small stacks located peripherally (Fig. 3). The Golgi complex also demonstrates elaborate proliferation during granulosa cell transformation which then desperses into isolated regions of the cytoplasm. The changes

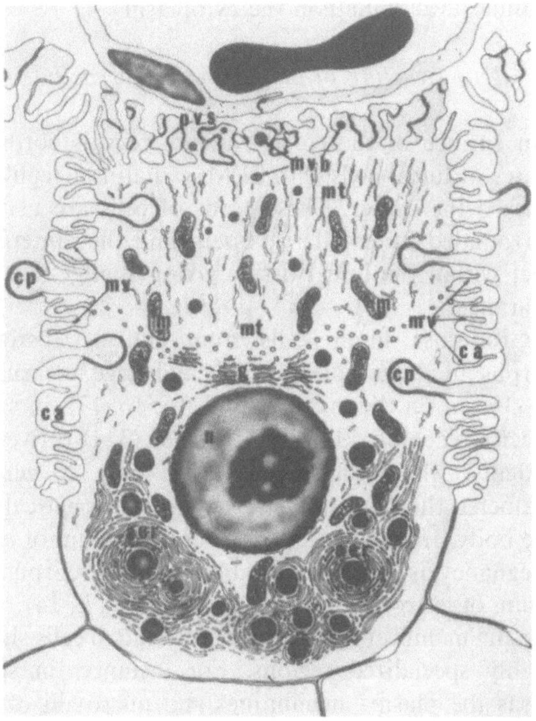

Fig. 11. Schematic diagram of the basic pattern of luteal cell organization in the bitch. At the bottom of the cell, or the most avascular pole, whorled and stacked cisternae of agranular endoplasmic reticululm (aer) develop. The nucleus (n) is also generally displaced in this end of the cell. The Golgi apparatus (g) is centered over the pole of the nucleus that faces the pericapillary space (prs) and is surrounded by abundant small coated vesicles. Numerous mitochondria (m) and closely associated anastomosing tubules of AER, in turn, surround the Golgi apparatus. They are oriented principally with their long axis towards the capillary space in concert with ubiquitous microtubules (mt) that course through this half of the cell. Microfilaments course directly into the highly infolded basal surface. These folds are joined to each other by small gap junctions and possess numerous multivesicular bodies (mvb). The multivesicular bodies in turn are surrounded by and attached to abundant small coated vesicles. The lateral surface of the cell possesses numerous long and tenuous microvilli (mv) that interdigitate with similar structures from adjacent cells, and thus form narrow tortuous intercellular channels or canaliculi (ca) which extend down to the level of the stacked and whorled AER. The intercellular channels are interrupted at frequent intervals by cytoplasmic processess (cp) that project from one cell well into the cytoplasm of the next. Wherever these processes touch the cytoplasm of adjacent cells they are surrounded by continuous pentalaminar gap junctions. Adjacent cells are joined at their avascular extremities by long continuous uninterrupted pentalaminar gap junctions. From Abel et al. (21).

occurring in the mitochondria are not uniform in all species examined. In general, the small rod-shaped mitochondria with lamelliform cristae enlarge and often become pleomorphic. The cristae of the luteal cell mitochondria vary from a mixed lamelliform-villiform to predominantly villiform or tubulovesicular. In addition to these changes, various quantities of lipid inclusions, lysosomes and membrane-limited granules accumulate in the cytoplasm.

3.2. Fine structure of the functional corpus luteum

Cut surface of the corpus luteum reveals cords or plates of luteal cells with SEM, radiating irregularly toward the center. These cords of cells are usually surrounded by capillary vessels, and they interconnect within the luteal tissue, giving sponge-like appearance.

Granulosa lutein cells from actively secreting corpus luteum are pleomorphic and large, measuring 20–35 μm on the average (5, 7, 15, 16, 17), in some species reaching 40 μm in diameter (3, 18). Even in rodents, where luteal cells reach only 20 μm in diameter, they constitute the largest endocrine cells in the body. In most species the corpus luteum of early pregnancy is indistinguishable from the corpus luteum of estrous or menstrual cycles (7, 14). The plasma membrane of the granulosa lutein cells shows highly specialized regions. For instance, in some areas the plasma membranes and microvilli of adjacent cells are highly interdigitated, occasionally interrupted by tight or intermediate junctions. In other areas the plasma membranes of adjacent cells form short segments of intercellular channels or canaliculi (3, 5, 15, 19, 20, 21). In these areas, microvilli project into the canaliculi and interdigitate; occasionally they are held together by macular intermediate junctions (Text Fig. II).

Lateral projections of the plasma membrane with dilated distal ends extend deeply into the adjacent cell. The dilated distal portions communicate with corresponding complementary infoldings of the other cells (Text Fig. II). These projections are best examined in half micron thick sections in the high voltage electron microscope (21). In thin sections, these structures appear as 'spherical inclusion', 'annular nexus', or 'circumscribed tight junction' (3, 8, 19, 20). The distended portions of the projections may detach from the plasma membrane and become internalized by the cell.

Luteal cells of most species are described as depicting a limited or variable degree of polarity. TEM observations of human corpus luteum (19), and high voltage electron microscopic examination of dog luteal cells revealed a much higher degree of polarity of organelles (8, 21). The lateral margins and the avascular pole of these cells are occupied by stacks and whorls of AER (Text Fig. II). Anastomosing tubular AER and mitochondria are the predominant structures of the central medial areas, whereas the Golgi apparatus is large and prominent and occupies a paranuclear position, facing the vascular pole of the cell (Text Fig. II). The microfilaments are arranged in a longitudinal fashion directed toward the pericapillary, thus spatially comparmentalizing the cell. The basal or vascular pole surface depicts great numbers of folds; these are interconnected and held together by small gap junctions. The lateral surfaces form numerous microvilli which intertwine with microvilli of adjacent cells. Together, they form intercellular channels or canaliculi (Text Fig. II).

3.2.1. Nucleus

The usually eccentric nucleus of the granulosa lutein cells is large, round or somewhat elongated in shape (Figs. 2, 3, and Text Fig. 2). The nuclear envelope has a relatively smooth contour which is interrupted at irregular intervals by nuclear pores (Figs. 3, 4, 5). The nucleus generally exhibits a relatively homogeneous matrix of low electron density. The heterochromatin lacks any characteristic pattern, instead an irregularly shaped layer of thin heterochromatin, which is interrupted only by nuclear pores, is distributed around the periphery of the nucleus. In the human (17), monkey (15) and guinea pig (14) large (500–600 Å) perichromatin granules with electron lucent halo around them are located in the peripheral heterochromatin (Figs. 3 and 4). The function of these granules in luteal cells has not been determined.

3.2.2. Mitochondria

Although mitochondria undergo morphogenesis during luteinization, a considerable variation in size and shape is frequent, not only within the same cell, but also among luteal cells of different species. Aside from the more common spherical or elongated mitochondria, some tortuous branching or cupshaped configurations are also commonly encountered (Figs. 3, 4, 5, 6, 7, 8). Thus polymorphism is a rule, rather than an exception among mitochondria of luteal cells (3). Polymorphism is also characteristic of the cristae among mitochondria of luteal cells from different species. Whereas in the guinea pig, armadillo and bat mitochondrial cristae are largely lamellar, in the rabbit (22) and rat (Fig. 5) they are mixed lamellar and tubular. In the rat and hamster the trend is toward lamelliform cristae (7) with increasing gestational age. In human (Fig. 6) and cow the mit-

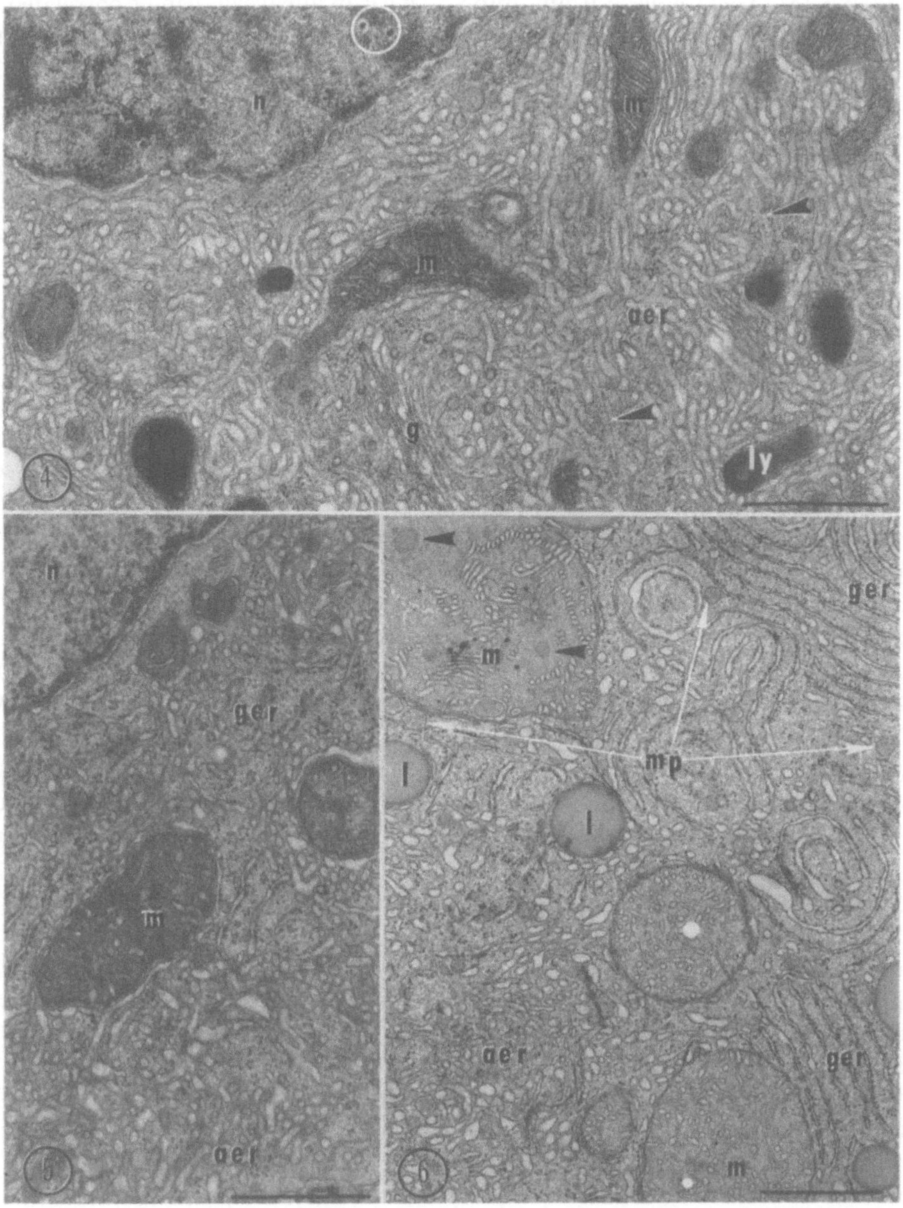

Figs 4–6. (4) Guinea pig luteal cell, 26 days of pregnancy. At this stage, luteal cells characteristically contain abundant profiles of the agranular endoplasmic reticulum (aer), numerous rod-shaped mitochondria (m), a well-developed Golgi complex (g), occasional lysosome-like bodies (ly), many free ribosomes and microtubules (arrowheads). The agranular endoplasmic reticulum is commonly in the form of densely-packed, interweaving tubules that occupy much of the available cytoplasms. The large, spherical nucleus (n) usually displays a number of perichromatin granules (circle). From Paavola (14). (5) Granulosa lutein cell from rat at 12 days of gestation. Stacked cisternae of granular endoplasmic reticulum (ger) and tubular agranular endoplasmic reticulum (aer) are abundant. Mitochondria (m) have mixed lamellar and tubular cristae. Nucleus (n). From Long (7). (6) Part of a granulosa lutein cell from human at nine weeks of gestation. The dell possesses abundant granular and agranular endoplasmic reticulum. The granular form (ger) appears as whorls or parallel stacks of ribosome-studded membranes. The agranular variety (aer) of this organelle consists of vesicles and a reticulum of anastomosing tubules. Large mitochondria (m) possess tubular cristae and a few inclusions (arrow heads). Lipid droplets (l) are seen throughout the cytoplasm. Microperoxisomes (mp) of moderate electron density are usually associated either with the granular or agranular endoplasmic reticulum. From Crisp et al. (17).

ochondrial cristae of progestational luteal cells are tubular or villiform (3, 17), whereas in the dog they are tubulo-vesicular (8). There is little if any information available to explain the polymorphism of mitochondrial cristae of luteal cells among different species, nor is there a good way to correlate morphological polymorphism to levels of steroidogenic activity. The mitochondrial matrix of most species is described as homogeneous and moderately electron dense. In the monkey (15), human (17, 19), pig (9) and mare (23) luteal cells the mitochondrial matrix encloses irregularly shaped electron-dense (or osmiophilic) inclusions often taking up a large portion of the matrix. It has been postulated that these mitochondrial inclusions may contain precursor substrates of steroid hormones (24).

3.2.3. Endoplasmic reticulum

In many species the most conspicuous organelle in luteal cells appearing concurrent with maximal progesterone secretion is the AER (Text Fig. 2 and Figs. 4, 6, 7). This also appears to be the most consistent feature of luteal cells and has been implicated in steroid hormone production (25). However, the abundance of AER varies not only from cell to cell within the same corpus luteum (Figs. 3, 4, 5, 6), but also it demonstrates characteristic variation among species (3, 8, 14, 15, 17, 19, 26). Also, depending on the methods of preservation employed, the AER may depict various configurations. Under what appears to be the most improved fixation methods, the AER has been identified in the forms of anastomosing or interconnecting network of tubules (5, 14, 23), systems of concentric membranous whorls (18), and tubular fenestrated sheets. The tubular sheets and fenestrated cisternae (15, 18) are variably in ordered stacks of lamellae. Ubiquity of vesicular AER in luteal cells from different species suggests that vesicular forms of AER are not always fixation artefacts, but represent AER at different stages of functional development. For example, in the 2-day rat luteal cells the AER is tubular and vesicular, whereas at 12 days the AER assumes long anastomosing tubular profiles; this profile is contemporaneous with peak serum progesterone levels. Areas of high concentration of AER are often, but not exclusively, located peripherally in the cells. In these instances the areas may be void of other cellular organelles. Unusual arrangements of the AER into folded membrane arrangements have been noticed in human and monkey luteal cells (17, 19, 27). To date, the different, often bizarre arrangements of the AER and their varying configurations cannot be associated with well-defined functions of the luteal cells other than their general association with steroidogenesis. It may be said that an increase in the amount of the AER in luteal cells can be correlated with an increase in steroid hormone secretion. At the time of maximal progesterone levels the cells are packed with AER, which became disorganized when steroidogenesis declined; however, in some species extensive amounts of AER may remain intact during the period of decreasing steroidogenesis (3).

The enzyme 3β-hydroxysteroid dehydrogenase (HSD) has been localized by EM histochemistry in AER that are in contact with lipid droplets or mitochondria (28) of human luteal cells. In contrast, the enzyme activity is widely distributed in the rat (29). In ovine corpus luteum HSD activity is localized in the microsomal fraction (30). These observations support biochemical data that these organelles are the site of steroidogenesis and that HSD is closely related to luteinization.

In general, granulosa lutein cells possess moderate to abundant quantities of GER (3, 9, 14, 15, 17, 26). Short segments of GER are often associated with other organelles, such as mitochondria or occasionally lipid droplets of the developing corpus luteum. Small stacks of the organelle may also be dispersed through the cytoplasm (Figs. 3, 5, 6). Large parallel arrays of GER characteristically occupy a portion of the peripheral cytoplasm; not uncommonly, these are furthest from the nucleus (Figs. 3, 6). It is possible that the abundance of GER in luteal cells during maximal progesterone secretion in pregnancy may indicate increased synthesis of steroidogenic enzymes in response to gonadotropic hormones (27). Aside from production of these enzymes, large quantities of GER are commonly associated with protein synthesis for export, a process further described below.

3.2.4. Golgi complex

The Golgi complex is distributed throughout the cytoplasm (Figs. 3, 4, 7 and Text Fig. II). In monkey granulosa lutein cells it is more concentrated in the central or paranuclear region (15), whereas in the dog, it occupies a vascular pole position (8). The organelle consists of five to seven slightly curved saccules, the terminal ends of which are slightly distended. Characteristically, various sizes of smooth and coated vesicles and granules are associated with the periphery of the flattened saccules (3, 8).

The Golgi complex is greatly enlarged in the functional corpus luteum. In addition, a series of Golgi associated lamellar endoplasmic reticulum (GERL) has been described in rat (7) and guinea pig (14) luteal cells, which are located at the inner face of the Golgi complex (Fig. 7). At the time of maximal steroidogenesis in guinea pig luteal cells the Golgi complex displays acid phosphatase (ACPase) activity in its cisternae (31), although the GERL at this stage shows low levels or no enzyme activity. The inner element and one or more of the remaining elements of the Golgi complex manifest thiamine pyrophosphatase (TPPase) activity, although the GERL is negative for this enzyme.

Coated vesicles that range between 600 Å and 1000 Å in size (3, 8, 15, 21) are associated with the periphery of the flattened saccules of Golgi complexes (Text Fig. II and Figs. 4, 7). They are believed to have an active role in endocytosis or plasma membrane generation and resorbtion (8, 21). The exact role of these cytoplasmic structures in luteal cells has not been studied extensively, although preliminary results suggest that they might be involved in the endocytosis of gonadotropin receptors, at least in some species.

3.2.5 Lipid inclusions

Lipid inclusions vary in size and in electron density (Figs. 2, 3, 6 and Text Fig. II). They are dispersed through the cell, although clusters of lipid droplets are not unusual (3, 7). Frequently, AER adhere closely to portions of the lipid surface, giving the appearance of a membrane limited organelle (22, 23, 26). The lipid content of granulosa lutein cells increases gradually during luteinization. Nevertheless, the lipid content of luteal cells varies appreciably with different species and the physiological state of the cells. For example, in the rat (7) numerous lipid droplets accumulate at the onset of gestation; later they decline considerably as the corpus luteum reaches its nadir of progesterone synthesis.

3.2.6. Microfilaments and microtubules

Cytoplasmic microfilaments (Text Fig. II and Fig. 8) are ubiquitous in luteal cells of most species examined (15, 17, 19, 23, 32). In some species, particularly during granulosa cell transformation, bundles of microfilaments are associated closely with rudimentary desmosomes and surface areas of cells extending into microvillous projections of the plasma membrane. In other species, in addition to surface microfilaments, there is an abundance of microfilaments amongst other cellular organelles. In the human (19, 32), and particularly in the dog (21), the bundles of microfilaments tend to compartmentalize the lutein cells into central and peripheral regions. Intracellular lacunae are usually circumscribed by bundles of microfilaments (15). Because granulosa lutein cells are considerably larger than most other cells, it has been suggested that the microfilament system along with microtubules are essential components of the cellular skeletal system that enable to maintain cell shape and size (15). Microtubules, on the other hand, have been reported only in a number of species (Text Fig. II and Figs. 4, 8). In these instances they were not numerous and were not observed in bundles.

3.2.7. Secretory granules

More recent fine structural, endocrinological and histochemical evidence suggests that secretion by luteal cells occurs not only through simple diffusion of steroids, but secretory products may also be released by apocrine or eccrine mechanisms. Granulosa lutein cells have a moderate quantity of GER and well developed Golgi complex, the cellular organelles essential for production and packaging substances for secretion. In the rat, a class of membrane limited granules (270 nm) appears in luteal cells during the latter portion of pregnancy (8). They are transitory because they disappear after parturition. Indirect antibody techniques suggested that the content of these granules is relaxin, a polypeptide hormone. Although the presence of relaxin granules in luteal cells has been suggested for porcine, human and monkey (3), none of these studies presented solid evidence that these granules are not lysosomes. Before a definitive statement can be made concerning the production and localization of relaxin in luteal cells of these species, considerable work still needs to be done.

Whereas correlation of steroidogenesis and cell ultrastructure has been reported for the corpus luteum in a number of species, less is understood about the actual mechanism of steroid release by the luteal cells. Secretion of steroids is assumed to occur through diffusion to the extracellular space, then into the circulation (3) where binding proteins transport the steroids. More recently, several papers indicated formation and release of secretory granules by luteal cells of several species via an exocytotic mechanism. The chronologic appearance of the granules closely reflected the pattern of maximal progesterone synthesis in ovine (16, 33, 34, 35), bovine (36), and goat (34) luteal cells. For example, in the sheep corpus luteum, the densely staining granules (0.2 μm) are first seen on days 2 and 3 of the estrous cycle (fig. 9). Secretion of granules is maximal at days 10 and 11, coinciding with maximum progesterone secretion. During the secretory process the limiting membranes of several granules may merge along the plasma membrane or several deeply located granules may fuse (37), forming a channel to the plasma membrane (Fig. 10). In this manner the content of several granules are released simultaneously. From these observations it was suggested that at least in some species a portion of progesterone secretion may occur through an active secretion. Similar densely staining granules were observed in guinea pig and to a lesser extent rabbit luteal cells (38), although they could not be correlated with progesterone secretion. Thorough examination of guinea pig luteal cells failed to reveal secretory granules similar to those in the sheep and cow (14, 39).

In several species the dense secretory granules are clearly distinguishable from lysosomes and microperoxisomes by histochemical means (35, 36, 38, 40).

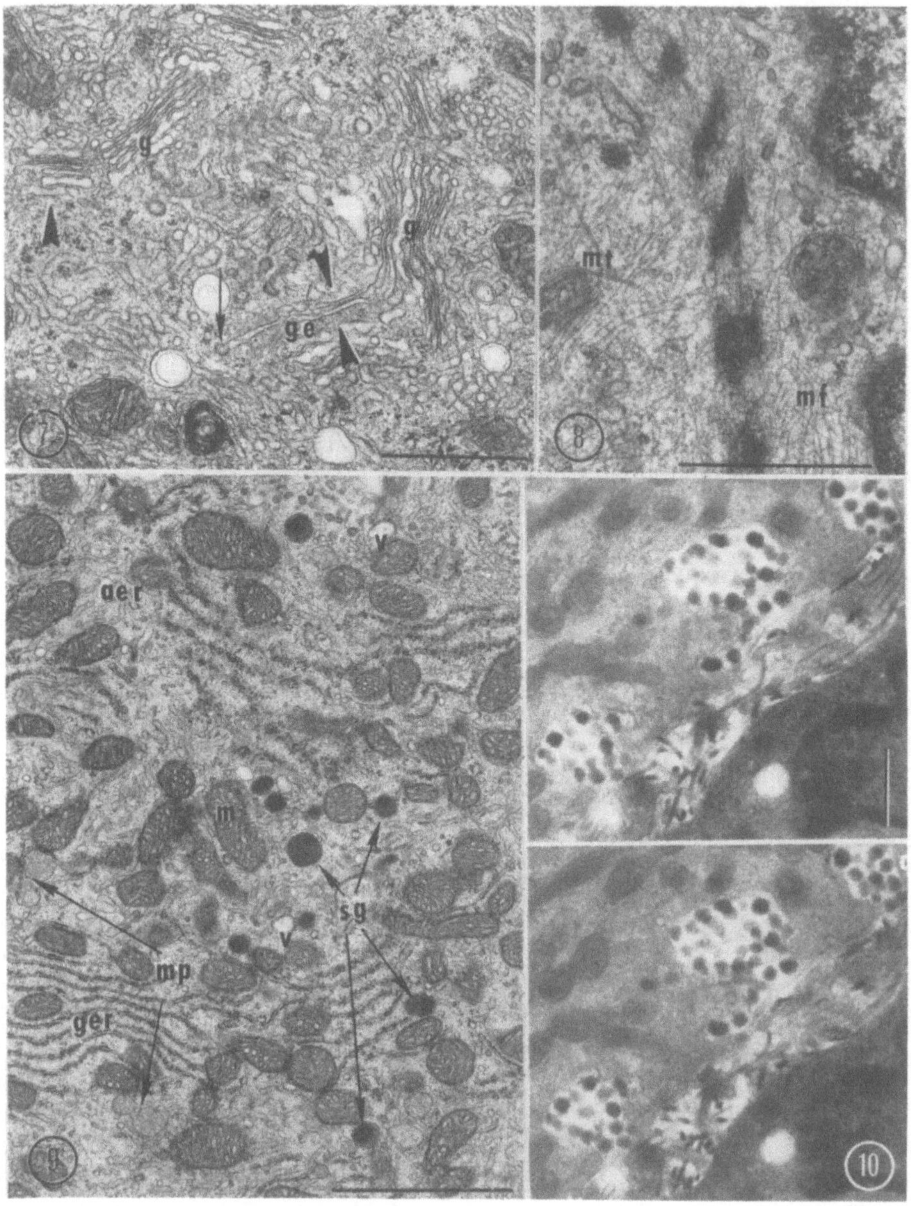

Figs. 7–10. (7) GERL (ge) arrangement in guinea pig corpus luteum. Although GERL is frequently separated from the Golgi stack (g) by a small region of cytoplasm (double long arrowheads), in other areas it may lie closely apposed to the inner element of the Golgi complex (single arrowhead). GERL may have coated vesicles (arrow) attached to its ends. (original kindly contributed by Dr. L.G. Paavola). (8) Luteal cells from monkey. Note the abundance of fine filaments (mf), dense areas within filament bundle, and some microtubules (mt). From Gulyas (15). (9) Sheep luteal cell from day 8 of cycle illustrating secretory granules (sg) and microperoxisomes (mp). Also present are agranular (aer) and granular (ger) endoplasmic reticulum, mitochondria (m) and some empty appearing vacuoles (v). From Paavola and Christensen (35). High voltage stereoelectron micrographs of sheep luteal cells demonstrating canaliculi that contain the contents of a dozen or more secretory granules at intervals along the cell surface. From Sawyer et al. (33).

It appears that dense secretory granules are formed in the Golgi region, then transported to the cell surface where they are released. Colchicine reduces progesterone secretion (33, 41) and causes accumulation of pleomorphic saccules that contain electron dense material. Colchicine treatment inhibits formation of granules at the Golgi level, although it has little effect on the release on preformed secretory granules from ovine luteal cells (33). Thus, colchicine blocks transport of newly synthesized granules and induces accumulation of protein at the level of Golgi apparatus. Furthermore, additional information was gained from cell fractionation studies which showed a correlation between the presence of densely staining granules and progesterone content of cell fractions obtained by differential centrifugation of ovine (16, 33) and bovine (36) luteal cells. Progesterone is present in greatest amounts in the 10,000 g fraction which contains the densely staining granules. In the bovine corpus luteum at least 30% of the total progesterone is particulate (36). It is likely that a protein is present within the granules, hence their electron dense appearance in the electron microscope after proper staining. The proposed scheme of progesterone release in those luteal cells that possess dense granules distinguishable from lysosomes and peroxisomes is as follows (40): A carrier protein is packaged into secretory granules in the Golgi apparatus. The secretory granules accumulate progesterone as they traverse to the cell surface; and they release their content into the extracellular space via exocytosis. However, that such a mechanism of progesterone secretion is indeed a significant one in certain luteal cells needs further clarification. Additional data are needed to determine the exact content of the secretory vesicles, and on the exact mode of triggering of granule release, before the molecular aspects of steroid hormone secretion can be fully understood in luteal cells of these species.

3.2.8. Lysosomes, peroxisomes and multivesicular bodies

The membrane bound granules, other than secretory granules, described in luteal cells are either lysosomes or microperoxisomes (3, 42, 43). Membrane bound dense bodies resembling lysosomes (Figs. 4, 11) have been identified in luteal cells of bovine (36), monkey (15), guinea pig (31). However, definite histochemical or biochemical identification is available only for a couple of species (31, 36, 42, 43). During the period of peak steroidogenesis lysosomes occur in modest numbers and with varied morphology in all species examined. Lysosomes of luteal cells have a homogeneous content, some displaying myelin figures. In the guinea pig corpus luteum, lysosomes react positively for acid phosphatase and aryl sulfatase with varying quantities (31). Multivesicular bodies give limited or scanty reaction product for either enzyme. Lysosomes are particularly abundant in degenerating cells during luteolysis, discussed in more detail later.

Microperoxisomes (Figs. 6, 9, 11) are small cellular organelles that have a single limiting membrane, and contain a finely particulate, moderately electrondense matrix without a nucleoid. Microperoxisomes measure 0.1–0.2 μm and can be distinguished from other organelles, because they give a positive reaction with 3, 3′-diaminobenzidine (DAB) following aldehyde fixation of the tissue. Microperoxisomes have been clearly identified in granulosa lutein cells from rhesus monkey (42), goat (38), ewe (35), cow (36, 40) and guinea pig (14, 39). Often microperoxisomes are attached to AER by means of a short, smooth surfaced tubular extension. In most instances direct communication can be demonstrated between these two organelles (44), which is depicted diagrammatically as Text Figure III. In the monkey luteal cells, microperoxisomes are often contiguous with lipid droplets and in some instances the DAB staining substance is detectable on the surface of the lipid droplets (42). The functional significance of microperoxisomes in luteal cells is not at all clear. Microperoxisomes have been

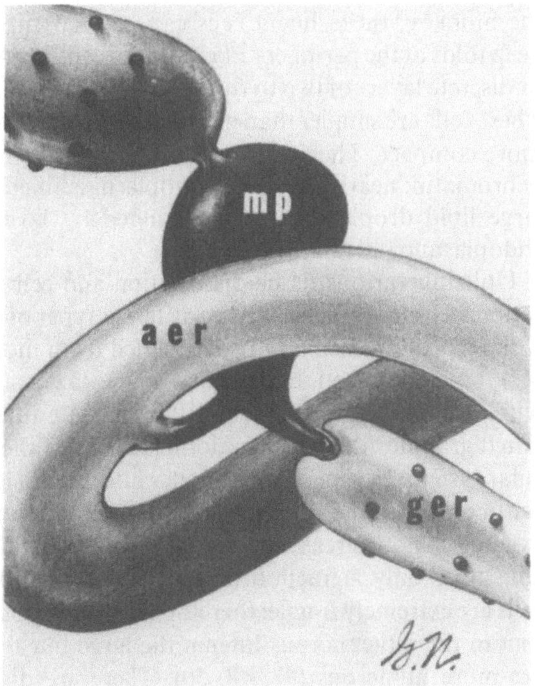

Fig. III. Schematic presentation of communication between microperoxisome (mp), agranular endoplasmic reticulum (aer) and nongranular portion of granular endoplasmic reticulum (ger). (This represents a composite of serial sections prepared by L.C. Yuan).

implicated to function in lipid metabolism, however, this is strictly conjectural.

Multivesicular bodies (MVB) are usually spherical structures measuring 0.4 μm in diameter and contain one to numerous small (40 nm) vesicles. MVB have moderately electron dense matrix, and in general they are unreactive for catalase and ACPase, although the matrix or vesicles occasionally display scant positive ACPase reaction. MVB have been described in human, monkey and dog luteal cells (19, 20, 21). In the human and dog they are heavily concentrated in the cytoplasmic folds that fill the pericapillary space and in the monkey they are also found in the intercellular space. There is no known function of the MVB in luteal cells at this time.

3.3. Cell types and size

As mentioned earlier, once the corpus luteum has developed to its fullest extent, it is hard to distinguish between theca and granulosa lutein cells in many species. For example, in the rat the theca interna cells are incorporated into the body of the corpus luteum. With the electron microscope, they do not appear different from granulosa lutein cells (7). On the other hand, in sow, human and dog, theca luteal cells are distinguishable because they are smaller, have fewer lipid droplets (17, 21, 32) and lack large mitochondria. Also, in the monkey, theca luteal cells can be recognized in deep folds at the periphery of granulosa cells in groups or discrete layers of two to three cell thickness (45, 46). These cells are smaller than granulosa cells and appear more compact. Their nucleus shows extensive heterochromatin; nearly the entire cytoplasm is filled with large lipid droplets, often surrounded by layers of endoplasmic reticulum.

Utilizing various tissue dissociation and cell separation techniques, at least three different types of cells, with respect to size, have been isolated from monkey (47), bovine (48, 49), rat (50) and sow (51) corpus luteum. The smallest of these ($< 10\ \mu$m) are undifferentiated granulosa cells and endothelial cells of the vascular system. In general, the smaller luteal cells (15–25 μm) constitute the greatest portion (70–90%) of the isolated cells; whereas the large luteal cells (25–50 μm) constitute only a fraction (2–30%). The large luteal cells are extremely fragile; therefore, it is quite possible that in the intact corpus luteum the large luteal cells are more numerous (47, 48, 50). There are distinct ultrastructural differences between large and small luteal cells in each species. Moreover, fine structural characteristics of small or large luteal cells do not correspond from species to species. For example, whereas small luteal cells of the cow (49) and monkey (44) are described as having fewer steroidogenic characteristics, in the rat, small luteal cells are described as having the more typical ultrastructural morphology, characteristic of steroidogenesis (50). In contrast, large luteal cells of monkey and cow have: (a) nuclei with dispersed heterochromatin, (b) extensive AER, and (c) mitochondria mostly with villiform cristae.

Less is understood concerning the origin (theca or granulosa) of the two different sizes of luteal cells. At best, all reports are inconclusive as to the origin of the two cell types. According to their appearance and localization in the intact corpus luteum, it has been suggested that in the sow (51) the larger cells are of granulosa, and the smaller cells are of theca origin, thus seemingly supporting the notion of two cell theory in the corpus luteum. However, without definitive markers for identifying the two cell types, they cannot be traced back to their origin with any certainty.

The information on secretory function of large and smaller luteal cells is equally inconclusive. In all experiments performed to date, this evaluation is greatly hampered by the inability to separate clumps of small cells from large luteal cells during separation procedures. In the cow and sow, both large and smaller luteal cells make progesterone (49, 51); however, the two luteal cell types show dramatically different responses to a luteinizing hormone (LH), with the smaller cells responding to a greater degree than the large ones. This suggests that the smaller luteal cells might play a major role in steroidogenesis. Clearly, until cell dissociation and separation techniques can be improved to get better separation of different size cells without fragmenting the cells, the functional diversity (if any) of these cells cannot be established. It is conceivable that the large cells are derivatives of the smaller luteal cells (50) manifesting similar functions.

3.4. In vitro studies on luteinization

With the advent of tissue culture techniques applicable to granulosa cells, it has been possible to study some of the regulatory mechanisms of luteinization *in vitro*, in the absence of the complex *in vivo* milieu. It was reasoned that this approach should enable one to study and define the hormonal conditions required for luteinization of granulosa cells and correlate fine structure with progestin secretion during hormone mediated luteinization.

Granulosa cells harvested from large preovulatory follicles of the rhesus monkey at days 12–14 of the menstrual cycle accumulate lipid droplets and osmiophilic granules in culture. These cells luteinize spontaneously in culture and secrete high levels of progesterone. Similarly, luteinization *in vitro* appears to be a spontaneous phenomenon for granulosa cells recovered from preovulatory follicles from human, horse, pig, rabbit, rat and mouse (52). The reason for spontaneous luteinization of granulosa cells *in vitro*, or for that matter *in vivo*, may well be due to high levels of estrogen that forestall luteinization. That the follicular fluid has a profound influence on the lutenization process has been suggested.

The morphological aspects of *in vitro* luteinization closely resemble those occurring *in vivo*. Namely: 1) development of the characteristic cell surface microvillous projections; 2) transformation of the nuclear heterochromatin to a lighter staining euchromatin pattern; 3) accumulation of lipid inclusions; 4) transformation of the small mitochrondria into large often pleomorphic mitochondria with mixed lamelliform-villiform cristae; 5) a gradual increase in the anastomosing AER content concomitant with a decrease in free ribosomes; and 6) hyperplasia and hypertrophy of the Golgi complex (5, 53). Luteinization *in vitro* is, in general, closely correlated with elevated amounts of progesterone secretion.

In vitro culture requirements are dissimilar for granulosa cells from different species; furthermore, the hormonal requirements for continued maintenance of luteal cells *in vitro* have not been well elucidated. Maintenance of rat granulosa cells, for example, are complicated by the apparent fluctuations in the *in vivo* levels of required gonadotropins, as well as possible synergism between several gonadotropins and steroids (54). *In vitro* studies corroborate *in vivo* studies demonstrating that in the rat prolactin, rather than luteinizing hormone or follicle stimulating hormone, is the primary luteotropic stimulus for early maintenance of luteal function (5, 54, 55). Mitochondria and extensive AER characteristic of luteinized cells are present in cells only after prolactin stimulation. Rat granulosa cells cultured with prolactin alone not only showed enhanced levels of progesterone secretion, but SEM observations illustrated that they have an increased number of microvilli, as compared to control cells. Under different culture conditions, and in the presence of FSH, flattened rat granulosa cells assume nearly spherical shape; further, bundles of microfilaments characteristic at the cortical and basal regions of the flattened cells disappear after FSH stimulation. The time course of *in vitro* luteizination varies in different species. In monkey it takes, 4 to 6 days for complete morphological luteinization, whereas *in vitro* progestin secretion in the rat reaches maximal levels within 2–4 days. It has been suggested that at least in some species initial fine structural changes occur in the granulosa cells shortly prior to ovulation (9, 26); however, similar changes have not been reported to occur in granulosa cells of most species examined.

4. Corpus luteum of mid and late pregnancy

4.1. Endocrinology

In such species as the rat, the corpus luteum is an essential steroidogenic organ throughout pregnancy; its removal at any point prior to parturition leads to termination of gestation. In other species, such as the guinea pig, monkey and human, the corpus luteum can be removed after the initial critical period without having an adverse effect on gestation (56). The requisite or essential life span of the corpus luteum of pregnancy in the monkey and human has been the subject of considerable debate. In the pregnant monkey the corpus luteum reaches the zenith of its progestational activity around 14th day of gestation; then, the placenta assumes the dominant role of progesterone secretion. After the third week of gestation, the corpus luteum regresses in size and its vascularity decreases. For that matter, in the monkey the corpus luteum becomes dispensable during the third week of gestation (57). Both in monkey and human the corpus luteum is unnecessary during the mid and late gestation. Nevertheless, by marking the corpus luteum of early pregnancy with india ink, it was possible to demonstrate that it is maintained throughout gestation in monkeys. During midpregnancy the corpus luteum regresses in size, its vascularity decreases considerably and morphologically, it appears quiescent. The ovarian vein progesterone level of the monkey ovary bearing the corpus luteum drops to 19 ng/ml at day 40 of gestation, as compared to 63 ng/ml and 54 ng/ml at 13 and 19 days of age of the corpus luteum (46), respectively. The corpus luteum remains in a refractory state until about 150–155 days of gestation, at which time it becomes rejuvenated structurally and functionally, and is maintained into the post partum interval in suckled monkeys. The factor responsible for the recrudescence of corpus luteum activity during later pregnancy remains undefined. The morphological changes of the luteal cells correlate well with the elevated ovarian vein level progesterone (149–2,300 ng/ml) (58). These endocrine observations are complemented by histochemical observations which show an increase in 3β-hydroxysteroid dehydrogenase activity in the corpus luteum prior to parturition (59) and an elevation of 20α-dihydroprogesterone in ovarian vein of the ovary bearing the corpus luteum (60). Similarly, in the human the corpus luteum of pregnancy at term secretes progesterone (61).

Yet, just as in the monkey, the purpose of maintaining the corpus luteum for such an extended period of time remains enigmatic.

4.2. Ultrastructure

In the monkey the rejuvenated corpus luteum at 150 days of gestation can occupy 1/3 to 1/2 of ovarian volume. It is extensively vascularized and it may be confined within the ovary or it may be everted (protruding) above the ovarian surface, measuring 3–5 mm in length and 2–3 mm in width. The granulosa lutein cells are aligned in rows with sinusoids running between them.

The luteal cells are characteristically large (25–30 μm) and pleomorphic (15, 20, 58). At the light microscopic level the borders of these cells are not well defined because neighboring cells form extensive canaliculi. A clear area, or lacuna measuring 5–15 μm in diameter is observed in many cells.

In general, in the monkey the fine structure of luteal cells at late pregnancy, or at term, resembles the granulosa lutein cells of early pregnancy, with a few minor exceptions. The plasma membranes of adjacent cells show a more highly specialized, elaborate canaliculi, and microvilli that interdigitate with each other (58) and project into the canaliculi. In those regions of apposing plasma membranes where intercellular channels are absent the parallel plasma membranes are interrupted by intermediate junctions. Canaliculi of neighboring cells merge, then enter into the perivascular space of the fenestrated capillaries (15, 20).

Intracellular lacunae of monkey luteal cells are more numerous; they appear slightly larger (up to 12 μm in diameter) than in the earlier stage corpus luteum. The cytoplasmic margin of the lacunae are circumvented by bundles of microfilaments (50 A). The lumen of the lacunae are lined by microvilli; some of which are 100 μm wide and 2.2. μm long (15).

The appearance of the nucleoplasm, the distribution of heterochromatin and the general features of the nucleus are hardly distinguishable from luteal cells of early pregnancy in those species in which the corpus luteum is retained (7, 15, 20, 27).

Monkey and human granulosa lutein cells of late pregnancy have an abundance of pleomorphic mitochondria, predominantly with villiform cristae (15, 20, 27). The number and size of electron-dense mitochondrial inclusions are greater than in earlier stages of corpus luteum.

Both the AER and the Golgi complex are elaborate in monkey luteal cells of late pregnancy (58), often depicting unique arrangements of the Golgi complex and an array of tubular cisternae (15). This confirms earlier observations that in steroidogenic cells these organelles have direct communications. Smooth and coated vesicles are associated with the periphery of the flattened saccules of the Golgi complex. The AER of these luteal cells is comprised of both ansstomosing, fenestrated tubular elements and some vesicular forms. GER usually occupies the peripheral cytoplasm, furthest away from the nucleus (15).

The corpus luteum of pregnancy in the monkey is maintained even after maternal or fetal hypophysectomy, and after fetectomy (20), suggesting that the luteotropic substance that mediates the rejuvenation process is of placental origin. Granulosa lutein cells from these animals have fine structural characteristic of functional steroidogenic cells (Figs. 12, 13), much like those from intact pregnant monkeys at term of pregnancy (20). Morphological indications of the onset of luteolysis are most noticable in intact and hypophysectomized animals and least noticed in fetectomized monkeys in which the placenta was maintained to term. Luteal cells from hypophysectomized rats (20), on the other hand, show less developed AER, mitochondria with irregular outlines and relatively lucid matrix, as opposed to corpus luteum of normal pregnancy or animals that were treated with mammotropic hormone following hypophysectomy.

5. Luteolysis

In the nonfertile cycle the corpus luteum becomes a transient organ, as demonstrated by the decline in the levels of circulating progesterone and a morphological regression of the corpus luteum. The latter is a pivotal event in the regulation of cyclic ovarian function, because in primates only after the demise of the corpus luteum does timely new follicle development resume.

Luteolysis is the least studied and among the least understood segments of the life span of the corpus luteum. Cells of the infolding corpus luteum are generally hard to preserve for ultrastructural studies making it difficult to distinguish different cell types, thus leading to some confusing interpretations. Luteolysis at the cellular level is thought to involve either self destruction or digestion by macrophages. However, there is considerable disagreement as to which one of these processes is predominant or whether this phenomenon shows species specificity. Indications are that luteolysis of the ovarian cycle may differ from luteolysis after parturition.

It is a general, but not universal observation that morphological luteolysis is preceeded by a drop in circulating progesterone level before any significant morphological change occurs. There are several hypothesis concerning the lifespan of the corpus luteum. Originally, functional luteolysis in women and rhesus monkeys was believed to be a consequence of elevated circulating levels of estrogen (62), whereas in the sheep the decline in progesterone secretion that signals the end of estrus is associated with the first bursts of prostaglandin $F_2\alpha$(PGF) secretion in the uterus (63). Furthermore, pharmacologic doses of

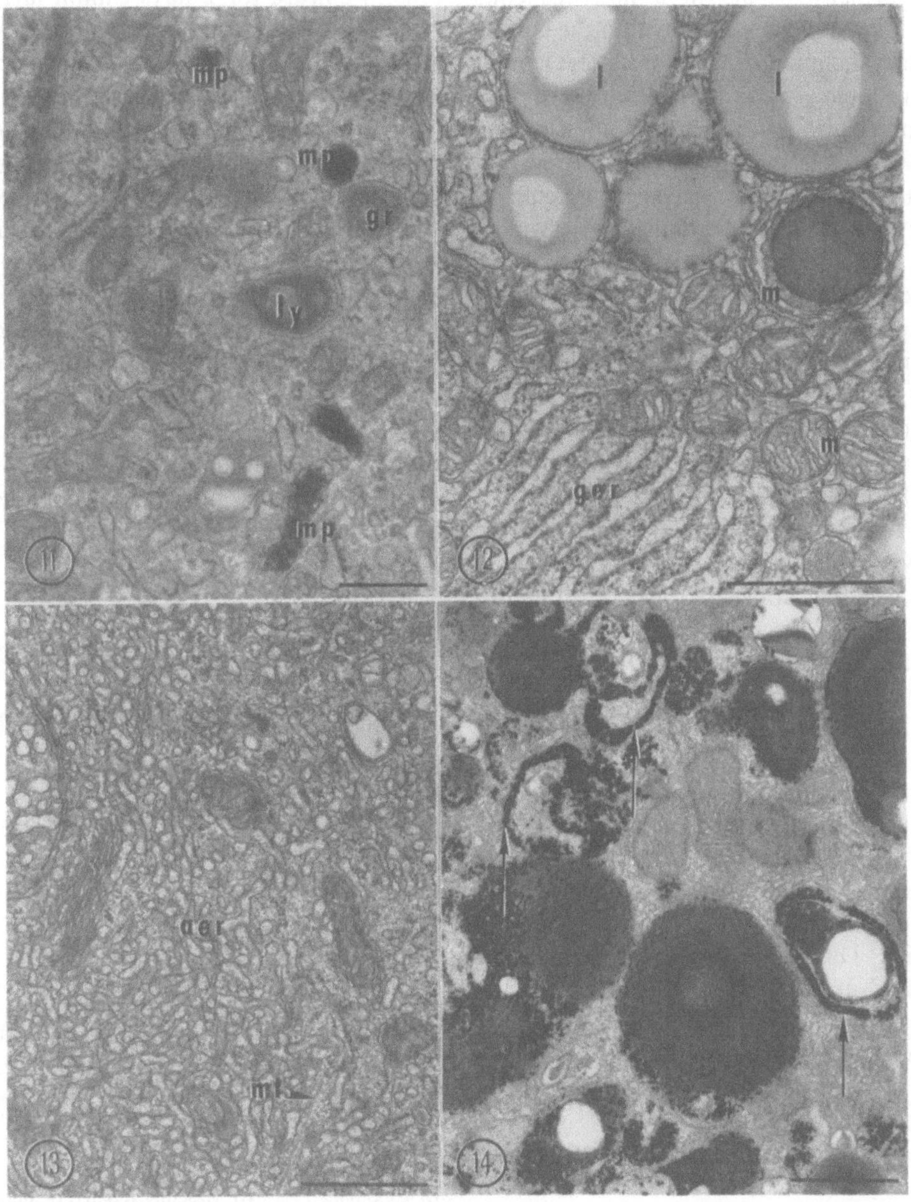

Figs. 11–14. (11) Monkey granulosa lutein cell fixed in 3% glutaraldehyde, incubated in diaminobenzidine. The small granules with the loosely packed particulate matrix are microperoxisomes (mp). The granules with homogeneous matrix and electron-lucent rim between the matrix and the limiting membrane (gr), lysosomes (ly) and Golgi complex do not stain with DAB. From Gulyas, Yuan (42). (12) Monkey corpus liteum at term from mother that was fetectomized on day 110 of gestation. The cytoplasm is rich in well preserved stacks of granular endoplasmic reticulum (ger), mitochondria (m), and heterogeneous lipid (l) inclusions. Note the large osmiophilic inclusion in a mitochondrion. From Gulyas et al. (20). (13) Fetectomized monkey. Peripheral regions of adjacent cells are abundant in tubular and vesicular agranular endoplasmic reticulum (aer). Microtubule (mt). From Gulyas et al. (20). (14) Autophagic vacuoles in regressing guinea pig luteal cells. The marked accumulation of autophagic vacuoles that takes place in luteal cells during involution is clearly illustrated in this large field from an ACPase preparation. Note that in these cells the predominant type of autophagic vacuole is enclosed by double walls (arrows). From Paavola (43).

injected PGF have luteolytic effect in human, monkey, sheep and rabbit (64, 65, 66, 67), leading to premature demise of the corpus luteum. It has been suggested that luteolysis is initiated by decreased blood flow, possibly induced by PGF, to the corpus luteum just when progesterone levels begin to decline (67).

Luteolysis normally occurs in a heterogeneous manner rather than in isolated loci within the corpus luteum. In general, lipid droplets are prevalent in luteal cells; thus, luteolysis is not due to exhaustion of progesterone precursors. At the morphological level luteolysis may be characterized as follows: 1) decrease in cell size; 2) invasion of the corpus luteum by collagen; 3) crenated or folded nuclei; 4) swollen mitochondria with greater electron density; 5) degeneration and vesiculation of the AER and Golgi complex, except in guinea pig; 6) gradual accumulation of lipid inclusions; 7) increase in the number of heterogeneous lysosomes; 8) accumulation of authophagic vacuoles; 9) appearance of myelin figures; and 10) general cellular disorganization (14, 23, 64, 65, 66, 67, 68). Accumulation of lipid suggests that the cells are retaining lipidogenesis, but not the ability to convert lipid to steroid hormones.

Autophagic vacuoles appear at the sites of cytoplasmic involution. The vacuoles contain cellular organelles which depict disorganization. At first, ACPase activity is confined only to small areas of the luteal tissue. It is believed that lysosomes become fragile, first fusing with autophagocytic vesicles, then they release their enzymes that spread to larger areas. Autography plays an important role in the regression of guinea pig (Fig. 14) luteal cells (14, 43). In this species, autophagic vacuoles seem to originate from GERL as a cup-shaped structure and they contain ACPase activity. This entire process is diagramaticallu summarized in Text Fig. IV. Therefore, in the guinea pig the GERL plays a major role in the early stages of luteolysis. In the aging corpus luteum, it shows intense lytic enzyme activity and functions in the packaging of acid hydrolases.

Macrophages have been mentioned briefly in the degenerating corpus luteum of human, hamster and sheep (19, 65), although no major function was assigned to them in the initial process of luteolysis. In the guinea pig corpus luteum macrophages become abundant only at the advanced stages of regression (43, 68). Macrophages phagocytize small pieces of luteal cells or entire cells, thus reducing corpus luteum volume. The emerging pattern of luteolysis appears as follows. At the onset of luteolysis autophagic vesicles arise which set in the initial steps of luteolysis. At a later time both autophagy and heterophagy become involved in reducing the size of the corpus luteum.

Because the cellular processes that participate either in the maintenance or the demise of the corpus luteum are incompletely understood, luteolysis was studied utilizing dissociated monkey luteal cells maintained in culture (47, 69, 70). Monkey luteal cells in culture for 2 days exhibit large numbers of small spherical structures. The content of the cell surface protrusions consist of lipid inclusions, AER and aggregates of small vesicles (70). Cells in culture for longer than 2 days depict diminished cell surface blebs and drastic reduction of cytoplasmic lipid and AER content. These observations are paralleled with declining patterns of progesterone secretion regardless of what nutrients the culture medium was supplemented with (47, 69, 70). Thus, in view of the TEM and SEM observation, it is evident that luteal cells in culture lose their steroidogenic ability because once they are introduced into the culture system they dispose of their lipid and AER, both of which are essential for steroidogenesis. It is quite evident that before luteolysis can be studied *in vitro*, a more suitable environment must be provided for them in culture.

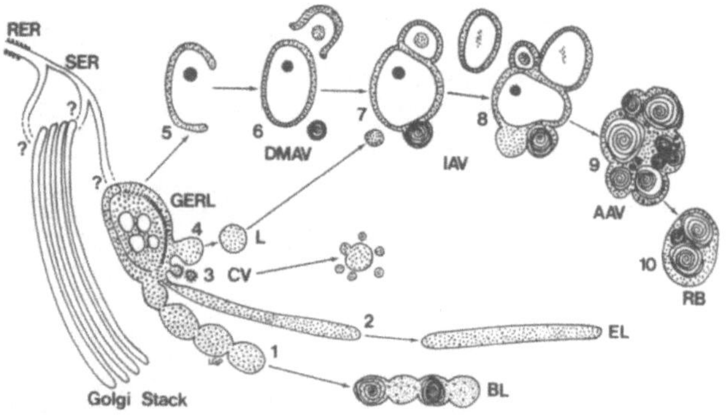

Fig. IV. This diagram summarizes the ways in which GERL appears to participate in the formation of lytic bodies during postpartum luteolysis in guinea pigs. Although the pathway that lytic enzymes follow to reach GERL remains uncertain (?, rer-rough er; ser-smooth er), GERL is responsible for packaging acid hydrolases into lytic bodies in these cells. Lysosomes are thought to arise from GERL by the growth and detachmant of beaded (1, BL), elongated (2, EL), vesicular (coated or smooth) (3, CV), and rounded vacuolar (4, 1) structures. The small vesicles may possibly fuse, forming larger lysosomes. As cup-shaped GERL cisternae (5) wall off areas of cytoplasm by becoming completely closed spheres, they produce double-membrane type autophagic vacuoles (6, DMAV). These double-membrane type autophagic vacuoles may then enlarge, possibly developing into autophagic vacuoles of the intricate variety (7, 8, IAV). The inner limiting membrane of double-walled autophagic vacuoles apparently breaks down and digestion ensues, converting the vacuoles into an aggregate inclusion (a lysosome) (9, AAV). Further digestion may take place, ultimately producing a residual body (10, RB), filled with myelin figures. From Paavola (43).

6. Concluding remarks

The objective of this chapter was not to oversimplify our concept of mammalian corpus luteum. Instead, an effort was made to point out ultrastructural similarities, as well as diversities of corpus luteum from different species. It can no longer be said that the corpus luteum is the most misunderstood structure there is, yet there are still numerous questions that need further probing. It was demonstrated that the mammalian corpus luteum exhibits striking changes in its structure, that for the most part can be correlated with functional changes throughout its lifespan. The onset of initial luteinization *in vivo* deserves further biochemical and ultrastructural studies. The process of vascularization following ovulation is still perhaps the least understood and certainly the least studied area.

Polymorphism has been demonstrated by a number of organelles including AER and mitochondria, the essential organelles of steroidogenesis. It is unclear why the variation in the structure of these organelles. It is uncertain whether different configurations manifest production of different steroids or they represent different stages in the evolution of steroidogenic organelles. Present observations seem to indicate polymorphism even in the mode of steroid release from the cells: simple diffusion vs release by active secretion. The significance of packaging at least some of the progesterone into secretory granules in certain species is puzzling. The exact process of packaging of progesterone in luteal cells is also yet to be demonstrated. The significance of such structures as the intracellular lacunae, mostly in primate luteal cells, and the function of microperoxisomes and multivesicular bodies is yet to be elucidated, and deserve further attention.

It is unclear whether smaller luteal cells give rise to large luteal cells indiscriminately, or only certain cells are destined to become large cells. If both sizes of luteal cells produce progesterone, than the significance for having different sizes is yet to be revealed. The exact origin of these cells is an equally important question that must be examined with great care.

Acknowledgements

The author is deeply indebted to Ms. Linda Baldwin for het expert clerical assistance in the preparation of this manuscript. Also, the author is grateful to Dr. J. Abel, Dr. P. Bagavandos, Dr. T. Crisp, Dr. J. Long, Dr. L. Paavola, Dr. H. Sawyer, Dr. R. Short, and Mrs. L. Yuan for providing micrographs of either unpublished or previously published observations and granting permission of their use.

References

1. Short RV: The discovery of the ovaries. In: The Ovary. Zucherman L, Weir BJ (eds) 1, New York, Academic Press, 1977, pp 1–39.
2. Corner GW, Hartman CG, Bartelmez GW: Development, organization and breakdown of the corpus luteum in the rhesus monkey. Cont Embry Carneg Inst 31: 117–146, 1945.
3. Enders AC: Cytology of the corpus luteum. Biol Reprod 8: 158–182, 1973.
4. Browning HC: The evolutionary history of the corpus luteum. Biol Reprod 8: 128–157, 1973.
5. Crisp TM, Denys FR: The fine structure of rat granulosa cell cultures correlated with progestin secretion. In: Electron Microscopic Concepts of Secretion; Ultrastructure of Endocrine and Reproductive Organs. Hess M (ed), John Wiley and Sons, New York, 1975, pp 1–33.
6. Motta P, van Blerkom J: Structure and ultrastructure of the Graafian follicles. In: Human Ovulation. Hafez ESE (ed), The Netherlands, North Holland/Elsevier, 1979, pp 17–38.
7. Long JA: Corpus luteum of pregnancy in the rat- Ultrastructural and cytochemical observations. Biol Reprod 8: 87–99, 1973.
8. Abel JH, Jr., Verhage HG, McClellan MC, Niswender GN: Ultrastructural analysis of the granulosa-luteal cell transition in the ovary of the dog. Cell Tissue Res 160: 155–176, 1975.
9. Bjersing L: On the ultrastructure of granulosa lutein cells in porcine corpus luteum. Z Zellforsch 82: 187–211, 1967.
10. Morris B, Sass MB: The formation of lymph in the ovary. Proc Roy Soc London Ser B, 164: 577–591, 1966.
11. Bjersing L, Cajander S: Ovulation and tha mechanism of follicle rupture. IV. Ultrastructure of the theca interna and the inner muscular network surrounding rabbit Graafian follicles prior to induced ovulation. Cell Tissue Res 153: 31–44, 1974.
12. Anderson WA: Permeability of ovarian blood vessels and follicles of juvenile rats. Microvasc Res 4: 348–373, 1972.
13. Ellinwood WE, Nett TM, Niswender GD: Ovarian vasculature: structure and function, In: The Vertebrate Ovary. Jone RE (ed), New York, Plenum Press, 1978, pp 583–614.
14. Paavola, LG: The corpus luteum of the guinea pig. Fine structure at the time of maximal progesterone secretion and during regression. Amer J Anat 150: 565–604, 1977.
15. Gulyas BJ: The corpus luteum of the rhesus monkey (Macaca mulatta) during late pregnancy. An electron microscopic study. Amer J Anat 139: 95–122, 1974.
16. Gemmell RT, Stacy BD, Thorburn GD: Ultrastructural study of secretory granules in the corpus luteum of the sheep during the estrous cycle. Biol Reprod 11: 447–462, 1974.
17. Crisp TM, Dessouky DA, Denys FR: The fine structure of the human corpus luteum of early pregnancy and during the progestational phase of the menstrual cycle. Amer J Anat 127: 37–70, 1970.
18. Sinha AA: Comparative ultrastructure of the corpus luteum of implantation and pregnancy in carnivora. In: Electron Microscopic Concepts of Secretion; Ultrastructure of Endocrine and Reproductive Organs. Hess, My (ed); John Wiley and Sons, New York, 1975, pp 53–69.
19. Adams EC, Hertig AT: Studies on the human corpus luteum. I. Observations on the ultrastructure of development and regression of the luteal cells during the menstrual cycle. J Cell Biol 41: 696–715, 1969.
20. Gulyas BJ, Yuan L, Tullner WW, Hodgen GD: The fine structure of corpus luteum from intact, hypophysectomized and fetectomized pregnant monkeys (Macaca mulatta) at term. Biol Reprod 14: 613–626, 1976.
21. Abel JH Jr., McClellan MC, Verhage HG, Niswender GH: Subcellular compartmentalization of the luteal cell in the ovary of the dog. Cell Tiss Res 158: 461–480, 1975.
22. Blanchette EJ: Ovarian steroid cells. II. The lutein cell. J Cell Biol 31: 517–542, 1966.
23. Levine H, Wight T, Squires E: Ultrastructure of the corpus luteum of

the cycling mare. Biol Reprod 20: 492–504, 1979.

24. Christensen AK: The fine structure of testicular interstitial cells in guinea pigs. J Cell Biol 26: 911–935, 1965.
25. Christensen AK, Gillim SW: The correlation of fine structure and function in steroid secreting cells with emphasis on those of the gonads. In: The Gonads. McKerns KW (ed), New York, Appleton, 1969, pp 415–488.
26. Blanchette EJ: Ovarian steroid cells I. Differentiation of lutein cells from the granulosa follicle cell during the preovulatory stage and under the influence of exagenous gonadotrophins. J Cell Biol 31: 501–516, 1966.
27. Crisp TM, Dessouky DA, Denys FR: The fine structure of the human corpus luteum of term pregnancy. Amer J Obstet Gynecol 115: 901–911, 1973.
28. Laffargue P, Chamlian AC, Adechy-Benkoel L: Localisation probable en microscopie electronique de la 3β-hydroxysteroide deshydrogenase, de la glucose-6-phosphate deshydrogenase et de la HADH diaphorase dans le corps jaune ovarien de la femme. J Microsc 13: 235–246, 1972.
29. Bara G, Anderson WA: Fine structural localization of 3β-hydroxysteroid dehydrogenase in rat corpus luteum. Histochem J 5: 437–449, 1973.
30. Caffrey JL, Nett TM, Abel JH, Jr, Niswender GD: Activity of 3β-hydroxy-Δ^5-steroid dehydrogenase/Δ^5-Δ^4-isomerase in the ovine corpus luteum. Biol Reprod 20: 279–287, 1979.
31. Paavola LG: The corpus luteum of the guinea pig. II. Cytochemical studies on the Golgi complex, GERL, and lysosomes in luteal cells during maximal progesterone secretion. J Cell Biol 79: 45–58, 1978.
32. Adams EC, Hertig AT: Studies on the human corpus luteum. II. Observation on the ultrastructure of luteal cells during pregnancy. J Cell Biol 41: 716–735, 1969.
33. Sawyer HR, Abel, JH JR., McClellan MC, Schmitz M, Niswender GD: Secretory granules and progesterone secretion by bovine corpora lutea in vitro. Endocrin 104: 476–486, 1979.
34. Gemmell RT, Stacy BD, Nancarrow CD: Secretion of granules by the luteal cells of the sheep and the goat during the estrous cycle and pregnancy. Anat Rec 189: 161–168, 1977.
35. Paavola LG, Christensen AK: Characterization of granule types in luteal cells of the sheep at the time of maximum progesterone section. Biol Reprod 25: 203–215, 1981.
36. Quirk SJ, Willcox DL, Parry DM, Thorburn GD: Subcellular localization of progesterone in the bovine corpus luteum: A biochemical, morphological and cytochemical investigation. Biol Reprod 20: 1133–1145, 1979.
37. Gemmell RT, Stacy BD: Granule secretion by the luteal cell of the sheep: The fate of the granule membrane. Cell Tiss Res 197: 413–419, 1979.
38. Gemmell R, Stacy BD: Ultrastructural study of granules in the corpora lutea of several mammalian species. Amer J Anat 155: 1–14, 1979.
39. Paavola LG, Boyd CO: Cytoplasmic granules in luteal cells of pregnant and non-pregnant guinea pigs. A cytochemical study. Anat Rec 201: 127–140, 1981.
40. Barry DM, Willcox DL, Thorburn GD: Ultrastructural and cytochemical study of the bovine corpus luteum. J Reprod Fertil 60: 349–357, 1980.
41. Gemmell RT, Stacy BD: Effects of colchicine of the ovine corpus luteum: Role of microtubules in the secretion of progesterone. J Reprod Fertil 49: 115–117, 1979.
42. Gulyas BJ, Yuan LC: Microperoxisomes in the late pregnancy corpus luteum of rhesus monkeys (Macaca mulatta). J Histochem Cytochem 23: 359–368, 1975.
43. Paavola LG: The corpus luteum of the guinea pig. III. Cytochemical studies on the Golgi complex and GERL during normal postpartum regression of luteal cells, emphasizing the origin of lysosomes and autophagic vacuoles. J Cell Biol 79: 59–73, 1978.
44. Gulyas BJ, Yuan LC: Association of microperoxisomes with the endoplasmic reticulum in the granulosa lutein cells of the rhesus monkey (Macaca mulatta). Cell Tiss Res 179: 357–366, 1977.
45. Koering MJ: Comparative morphology of the primate ovary. Contributions to Primatology 3: 38–81, 1974.
46. Koering MJ, Wolf RC, Meyer RK: Morphological changes in the corpus luteum correlated with progestin levels in the rhesus monkey during early pregnancy. Biol Reprod 9: 254–271, 1973.
47. Gulyas BJ, Stouffer RL, Hodgen GD: Progesterone synthesis and fine structure of dissociated monkey (Macaca mulatta) luteal cells maintained in culture. Biol Reprod 20: 779–792.
48. Ursely J, Leymarie P: A comparison of the LH control of progesterone synthesis in small and large cells from pregnant cow corpus luteum. Adv Exp Med Biol 112: 545–548, 1979.
49. Koos RD, Hansel W: The large and small cells of the bovine corpus luteum: Ultrastructural and functional differences. In: Dynamics of Ovarian Function. Schwartz NB, Hunzicker-Dunn M (eds), New York, 1981, pp 197–203.
50. Wilkinson RF, Anderson E, Aalberg J: Cytological observations of dissociated rat corpus luteum. J Ultrastruct Res 57: 168–184, 1976.
51. Lemon M, Loir M: Steroid release in vitro by two luteal cell types in the corpus luteum of the pregnant sow. J Endocrin 72: 351–359, 1977.
52. Channing CP: Intraovarian inhibitors of follicular function. In: Ovarian Follicular Development and Function. Midgley, AR Jr., Sadler WA (eds), New York, Raven Press, 1979, pp 59–64.
53. Channing CP, Crisp TM: Comparative aspects of luteinization of granulosa cell cultures at the biochemical and ultrastructural levels. Gen Comp Endocrin Suppl 3: 617–625, 1972.
54. Crisp TM: Hormone requirements for early maintenance of rat granulosa cell cultures. Endocrin 101: 1286–1297, 1977.
55. Centola GM: Hormone requirements for long-term maintenance of rat granulosa cell cultures. In: Ovarian Follicular and Corpus Luteum Function. Channing CP, Marsh JM, Sadler WA (eds), New York, Plenum Press, 1979, pp 225–233.
56. Hodgen GD: Patterns of secretion and antigenic similarities among primate chorionic gonadotropins: Significance in fertility research. In: Chorionic Gonadotopin. Segal SJ (ed), New York, Plenum Press, 1980, pp 53–63.
57. Goodman AL, Hodgen GD: Corpus luteum-conceptus-follicle relationships during the fertile cycle in rhesus monkeys: Pregnancy maintenance despite early luteal removal. J Clin Endocrin Metab 49: 469–471, 1979.
58. Koering MJ, Wolf RC, Meyer RK: Morphological and functional evidence for corpus luteum activity during late pregnancy in the rhesus monkey. Endocrin 93: 686–693, 1973.
59. Sholl SA, Wolf RC, Colas AE: Δ^5-3β-hydroxysteroid dehydrogenase/Δ^5-Δ^4-isomerase activity in the corpus luteum of the pregnant rhesus monkey. Endocrin 94: 908–910, 1974.
60. Sholl SA, Wolf RC: Quantification of 20$^{\alpha}$- and 20β-dihydroprogesterone in plasma of the pregnant rhesus monkey. Steroids 23: 269–287, 1974.
61. LeMaire WJ, Conley PW, Moffett A, Cleveland WW: Plasma progesterone secretion by the corpus luteum of term pregnancy. Amer J Obstet Gynecol 108: 132–134, 1970.
62. Knobil E: On the regulation of the primate corpus luteum. Biol Reprod 8: 246–258, 1973.
63. Thorburn GD, Cox RI, Curri WB, Restole BJ, Schneider W: Prostaglandin F_2 concentration in the utero-ovarian venous plasma of the ewe during the oestrous cycle. J Endocrin 53: 325–326, 1972.
64. Kirton KT, Koering MJ: Prostaglandin F_2 and primate corpus luteum: A correlation of structure and function. Fertil Steril 24: 926–934, 1973.
65. Stacy BD, Gemmell RT, Thorburn GD: Morphology of the corpus luteum in the sheep during regression induced by prostaglandin F_2. Biol Reprod 14: 280–291, 1976.
66. Koering MJ, Kirton KT: The effects of prostaglandin F_2 in the structure and function of the rabbit ovary. Biol Reprod 9: 226–245, 1973.
67. Koering MJ: Luteolysis in normal and prostaglandin F_2-treated pseudopregnant rabbits. J Reprod Fertil 40: 259–267, 1974.
68. Paavola LG: The corpus luteum of the guinea pig. IV Fine structure of macrophages during pregnancy and postpartum luteolysis, and the phagocytosis of luteal cells. Amer J Anat 154: 337–364, 1979.
69. Gulyas BJ, Yuan LC, Hodgen GD: Synthesis of progesterone and estradiol by monkey luteal cells in culture: Effects of insulin, thyroxine, cortisol, and cholesterol with and without hCG. Biol Reprod 23: 21–28, 1980.
70. Gulyas BJ, Hodgen GD: *In vitrò* studies on luteolysis in the rhesus monkey. In: Dynamics of Ovarian Function. Schwartz NB, Hunzicker-Dunn M, (eds), Raven Press, New York, 1981, pp 191–196.

Author's address:
Pregnancy Research Branch
National Institute of Child Heath and Human Development
National Institutes of Health
Bethesda, MD 20205, U.S.A.

CHAPTER 22

Localization of gonadotropin receptors in the gonads

ABRAHAM AMSTERDAM and HANS R. LINDNER

1. Introduction

Gonadotrophic hormones play an important role in the regulation of ovarian and testicular function. In the ovary, this control is exerted in a sequential manner: follicular stimulating hormone (FSH) is primarily responsible for regulation of follicular development, while luteinizing hormone (LH) controls the later processes of ovulation and luteinization, including the formation of progesterone (1, 2).

In the testis, LH regulates the production of testosterone, while FSH regulates spermatogenesis (2–4).

The regulatory action of testicular and ovarian function by gonadotropic hormones is initiated by binding to specific, high-affinity receptors that are located in the cell membrane of the respective target cells, and subsequent stimulation of these cells is mediated by adenylate cyclase. Receptor sites for LH and FSH in testis and ovary have been identified and characterized by *in vivo* and *in vitro* binding studies with various labelled forms of the hormone, most commonly the radioactive derivatives formed by radioiodination or tritiation of highly purified gonadotropin preparations (1–4). The precise localization of the receptors to LH and FSH in the various cell types in the gonads was achieved mainly by using labelled hormones that bound to the receptor, thus permitting its identification at the level of the light and/or the electron microscope.

2. Method of choice

A common feature of receptors to gonadotropins is their extremely high affinity for the respective ligand, the association constant being about $10^{10}M^{-1}$. On the other hand the rate of dissociation is very low. For example, the half-life of the hormone-receptor complex for hCG is about 24 hours at 24° C, and the dissociation reaction proceeds even more slowly at 4° C. The gonadotropins are large molecules (glycoproteins in the range of about 30,000–40,000 daltons) which can be tagged with specific markers without losing their affinity to the receptor, and thus visualized at the level of the light or the electron microscope. For localization of the gonadal receptors for LH, labelled human chorionic gonadotropin (hCG) has been more frequently employed than labelled LH. The placental gonadotropin is more readily available in highly purified form and exhibits higher stability after being labelled with radioactive iodine (3). In addition, LH is readily stripped from its receptor by homologous antibody, whereas hCG is not (5), so that antibodies to hCG are more suitable for localization of the receptor by immunocytochemical methods. The biological properties of hCG appear to be very similar to those of LH, and the two glycoprotein hormones bind to the same receptor sites in the testis and ovary. Therefore, receptor sites identified by binding studies with labelled hCG will be referred to as LH or LH/hCG receptors. As for FSH receptors, human FSH was more frequently used for localization studies since it could be achieved in highly purified form. One of the difficulties in the localization of LH and FSH in target cells in the gonads is the relatively low concentration of the receptors, which normally does not exceed 20,000 receptors per cell, as calculated from binding experiments (1–4).

3. Localization of receptors to gonadotropins in the ovary

The existence of ovarian LH/hCG receptor sites was evident from numerous *in vitro* data demonstrating that specific hormone binding sites are present in slices, cell suspensions, and homogenates of the ovary. By such methods, LH receptors have been demonstrated in the rat, mouse, pig, cow, rhesus, and human ovary (1–3).

Motta, PM (ed): Ultrastructure of endocrine cells and tissues. ISBN-13: 978-1-4613-3863-5

The initial morphological evidence for the localization of LH receptors in the different compartments of the ovary were obtained by isotopic tracer methods by *in vivo* localization of radioiodinated LH or hCG or by topical application of the labelled hormone on ovarian slices (6–8). It was shown that LH receptors of the rat ovary are present in the corpus luteum and in the interstitial tissue and theca cell of the developing follicle, and that LH receptors also begin to appear in granulosa cells at the time of antrum formation (7–8).

The site of the LH binding to its receptor in the rat preovulatory follicles was studied in more detail using ^{125}I-hCG, and autoradiography. Subsequent to *in vivo* administration of the hormone to proestrous rats, labelled molecules could be demon-

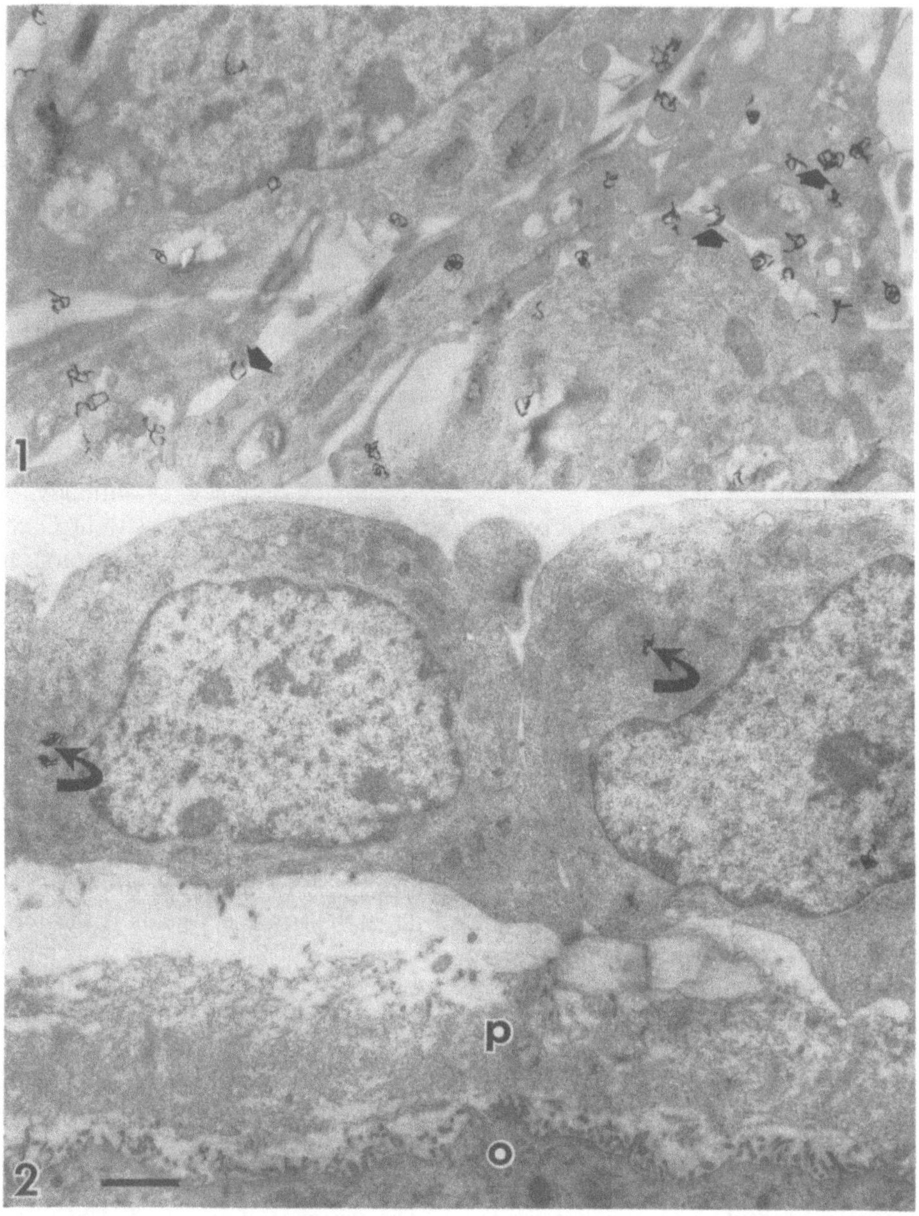

Figs. 1–2. High resolution autoradiography of membrana granulosa cells (1) and cumulus oophorus cells (2) obtained from preovulatory follicle after *in vitro* incubation with ^{125}I-hCG for 1 h at 37° C. Note the high density of silver grains over the cell membrane of membrana granulosa cells (arrows). In contrast, only few silver grains can be detected over corona radiata cells, some of which probably represent internalized hormone molecules (curved arrows in 2). The zona pellucida (p) and the oocyte (o) are devoid of label. Bar represents 0.5 μm for both figures (By permission from reference 10).

strated only in the periphery of the follicle, in the theca and mural granulosa cells (2, 8–11). No significant labelling was found over the oocyte and the cumulus granulosa cells in contact with the ovum (corona radiata). Following *in vitro* incubation of isolated compartments of the follicles with the labelled hormone, it was found (10, 12) that the number of LH receptor sites in the outer granulosa cell layers (membrana granulosa) was 7 to 10 times greater than in the inner layers (cumulus oophorus); the oocyte and zona pellucida still showed no specific labelling (Figs. 1, 2). It is, therefore, conceivable that not only those cells that bind LH or hCG respond to the hormone, but that the signal generated by the primary cell-hormone interaction is translated into a chemical message or electric impulse that can be

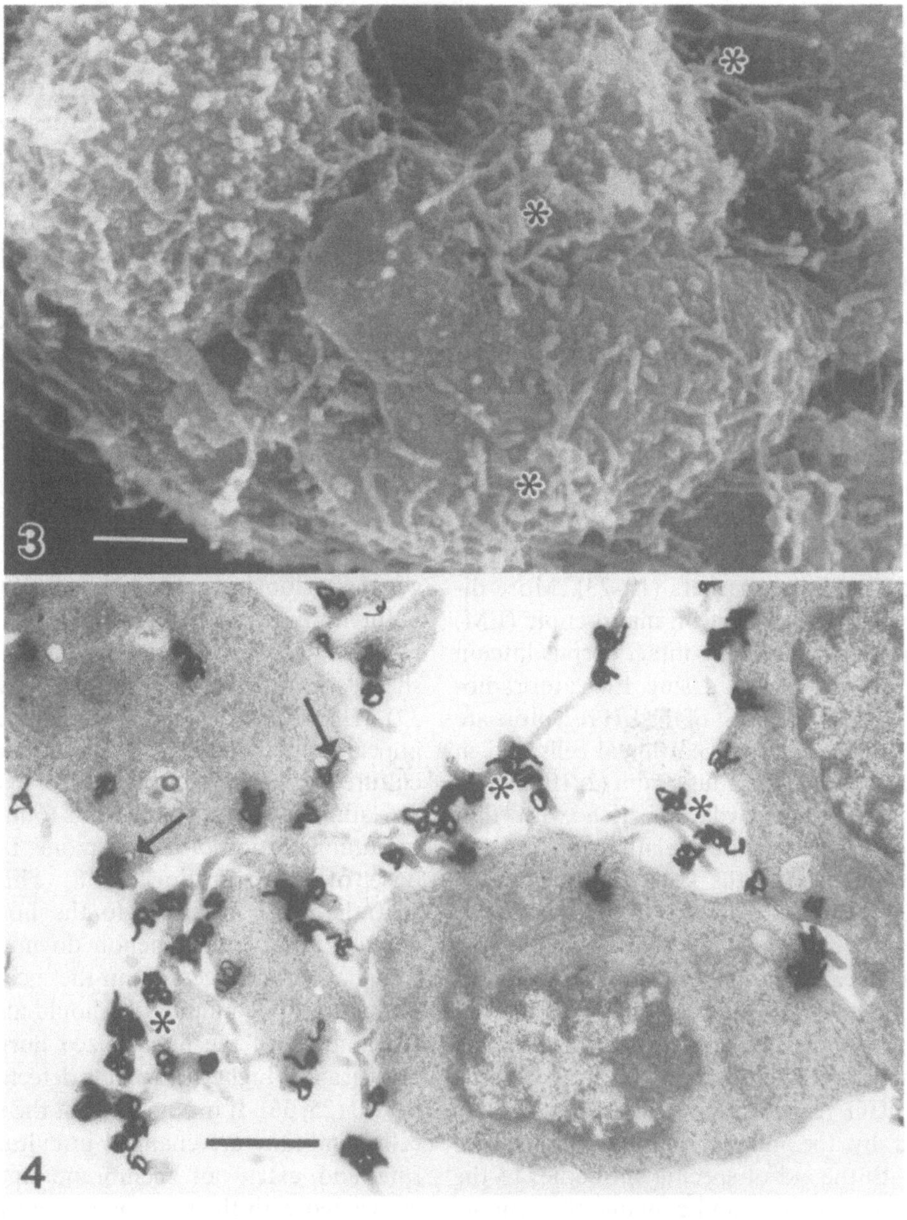

Figs. 3–4. Localization of LH/hCG receptors in granulosa cells obtained from hypophysectomized diethylsibestrol treated rats and cultured in the presence of FSH for 48 h, in the absence of serum. (3) Scanning electron microscopic view of the cultured cells which are highly aggregated and covered with numerous microvilli (asterisks) (From Amsterdam, A., Phillips, D.M., Knecht, M. & Catt K.J., in preparation). (4) Electron microscope autoradiogram of the same culture after 90 min incubation with ^{125}I-hCG at 37° C. Most of the label is confined to the microvilli of the cells. Few of grains seem to be associated with coated area of the cell membranes and with coated pits (arrows). Bar, 0.5 μm for both figures. (By permission from ref. 2).

propagated towards the interior of the follicle and oocyte. Gap junctions have been characterized between follicular cells (13, 14) and similar junctions have also been identified between follicular cells and the oocyte (14, 15). Gap junctions elements may permit intercellular transfer of ions and small molecules such as cyclic AMP (16). Therefore, the network of follicular gap junctions may provide a structural basis for a flow of electrical or chemical information from follicular cells to the oocyte. This transfer of information may be modified by LH, permitting the hormone to control ovum maturation, in spite of the lack of LH receptors on the oocyte. It was recently found that in cultured granulosa cells obtained from hypophysectomized immature rats, formation of LH receptors coincides with the appearance of gap junctions between the cultured cells following FSH stimulation (17). A causal relationship between these two developmental events has not been established, but both appear to depend on the action of FSH, as does the formation of follicular antrum (1, 18). *In vivo*, follicular gap junctions are largely confined to the more mature follicles, viz. the antral follicles (13, 14).

Evidence for the location of LH receptors in the plasma membrane of luteinized cells of the corpus luteum and in the follicular cells was provided by density gradient centrifugation followed by electron microscopy and enzyme markers (19–23). More direct evidence came from electron microscopic (EM) autoradiography (24–27) using intact corpus luteum or cultured granulosa cells. Using EM autoradiography, it was found that most of the LH receptors are concentrated on the microvilli of luteal cells and in granulosa cells cultured without serum (2, 10, 28, 29) (Figs. 3, 4). Since microvilli are rich with thin filaments (actin), a possible interrelationship between cytoskeletal elements and the LH receptor must be considered (30, 31).

Important steps towards the understanding of gonadotropin receptor function are the study of receptor localization and its possible redistribution after hormone binding and attempts to correlate the changes with the cellular response to the hormone. Mobility of hCG receptors on granulosa cells was demonstrated by the indirect immunofluorescence technique. With the aid of specific antibodies to the bound hormone, patching and capping of hormone receptor complexes was demonstrated concomitant with the loss of cellular response to the hormone (27, 32). Radiolabelled hormone-receptor complexes were shown by high resolution autoradiography to the internalized (Figs. 5–12), accounting for the loss of surface receptors or 'down-regulation' observed after prolonged hormonal stimulation (2, 25). However, desensitization of the adenylate cyclase system preceded extensive receptor internalization (25). This uncoupling of receptor and cyclase may in part be due to the immobilization of receptor molecules by aggregation (25–27, 32). However, a biochemical uncoupling mechanism has also been identified in desensitized ovarian cells (33, 34). Internalization of clustered receptors seems to be initiated by cytoskeletal elements, since microfilaments were found (25) to be associated with invaginations of the plasma membrane carrying aggregated receptor molecules (Fig. 6). Much of the internalized hormone-receptor complex was incorporated into lysosomes (Figs. 7, 8, 9), and evidence of hormone degradation was obtained (25). Limited labelling over the Golgi area suggests the possibility of receptor recycling (Fig. 10). Significant labelling was also found over the nuclei 8 h after introduction of the labelled hormone (25) (Figs. 11, 12), but it is not clear whether this represented intact hormone, a peptide fragment, or recycled radio-tyrosine. The penetration of protein hormones into the interior of the cell opens up the possibility that these hormones, believed to interact exclusively with cell surface receptors, may have additional sites of action unrelated to adenylate cyclase. Significant binding of hCG was also observed in nuclear fractions obtained from homogenates of bovine corpora lutea (35).

Internalization of receptor-bound hCG was also revealed by EM autoradiography in the luteinized sheep and rat ovary following *in vivo* labelling (36, 37). However, the rate of internalization *in vivo* appears significantly lower than that observed in cultured cells. Significant loss of receptors after *in vivo* administration of hCG was found only after 24 h of administration of the hormone, i.e. long after the onset of desensitization (2, 38). This would suggest that the acute response to the hormone, and the desensitization phenomenon do not depend on internalization of receptors in these cells following the initial hormone binding. It should also be noted that the amount of the internalized hormone *in vitro* is significantly higher than that detected in luteal cells *in vivo* (25, 36). It is possible that the properties of the cell membrane are changed in culture, and that the rate and extent of membrane uptake in regions associated with the hormone-receptor complexes are accentuated. During the process of internalization, labelled hormone was only occasionally found to be associated with coated pits or coated vesicles (Fig. 4), so that it is still unclear whether these structures may play a major role in the uptake of the hormone-receptor complexes into the cells.

The demonstration of receptor sites for FSH in the ovary has been performed by binding studies with tritiated or, more commonly, radioiodinated human FSH (1–3). The ovarian receptors for FSH are situated upon granulosa cells (7, 39), an appropriate location to mediate the characteristic action of FSH on follicular growth and maturation. In the rat, FSH receptors have been demonstrated in granulosa cells of small follicles, prior to the development of LH receptors, and treatment with FSH has been followed by the induction of LH receptors both *in vivo* (40) and in cultures (17, 41–43), chiefly on microvilli of aggregated cells (Fig. 4) which show numerous gap junctions between them (2, 10, 17). Attempts have been made to localize the LH-receptor in the ovary by the immunoperoxidase technique using antibodies to hCG (44). From these studies, there is suggestive evidence for an intracellular location of the hCG receptors in the cytoplasm and nucleus of luteal cells. Using ovine LH coupled to ferritin, the location of LH/hCG binding sites on the cell membrane of luteal cells was confirmed (45).

4. Localization of gonadotropin receptors in the testis

There is some, but not complete, analogy between the

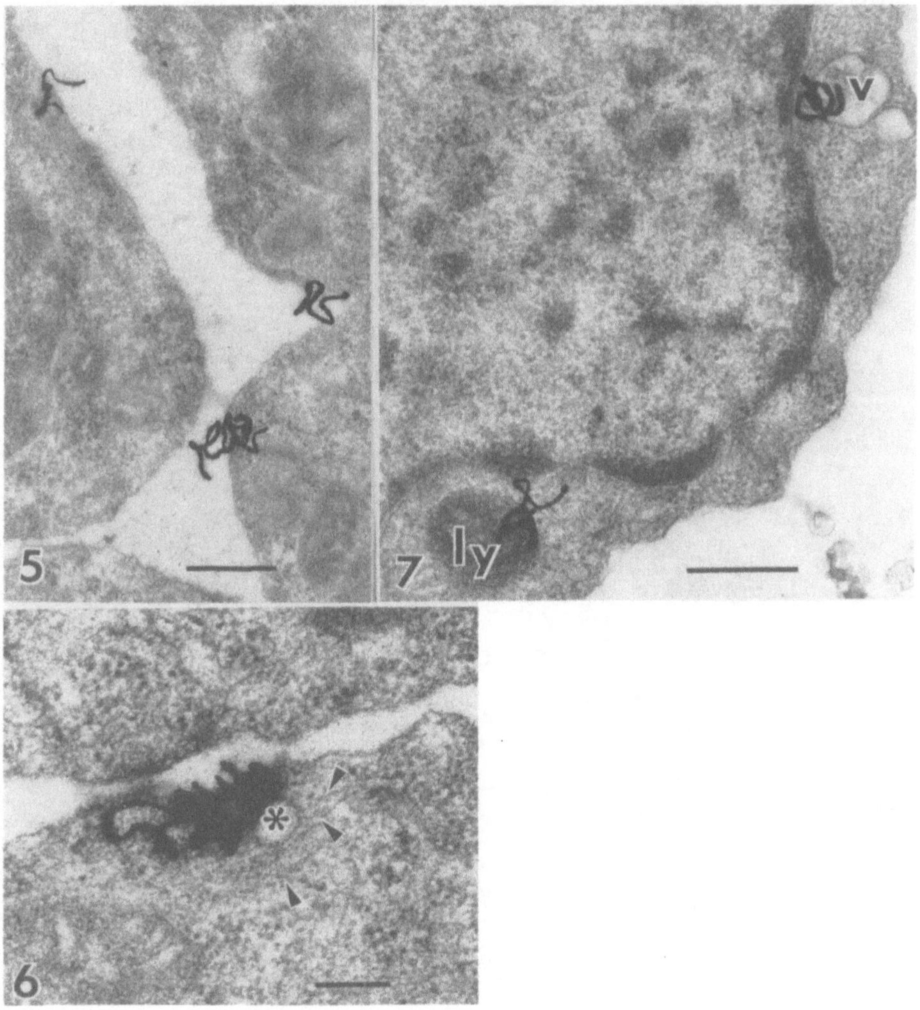

Figs. 5–7. Electron microscopic autoradiograms of cultured granulosa cells after 3 h incubation with 2 IU/ml of ^{125}I-hCG. (5) Individual silver grains as well as clustered grains are localized over cell membrane. Bar, 0.5 μm. (6) Cluster of silver grains is evident on plasmalemma of granulosa cell. Radioactivity is associated with invagination of membrane (asterisk). Filamentous material appears to be associated with infolding of membrane (arrowheads), and this area is devoid of ribosomes; bar, 0.2 μm. (7) Part of granulosa cell. Labeling is confined to intracellular smooth vesicle (v) and lysosome-like structure (ly) adjacent to plasma membrane; bar, 0.5 μm (By permission from ref. 25).

location of the receptors of LH and FSH in the ovary and the testis. In contrast to the situation in the ovary, testicular LH/hCG receptors are confined to a single cell type, the Leydig cells. The structure of the rat testis permits extensive dissection of the tubules and interstitial elements, and simply teasing apart the seminiferous tubules causes the release of large numbers of Leydig cells which are suitable for binding experiments (3, 4). When administration of the labelled hormone was followed by light microscopic autoradiographic location of the radioactive tracer, the label was confined to Leydig cells as expected

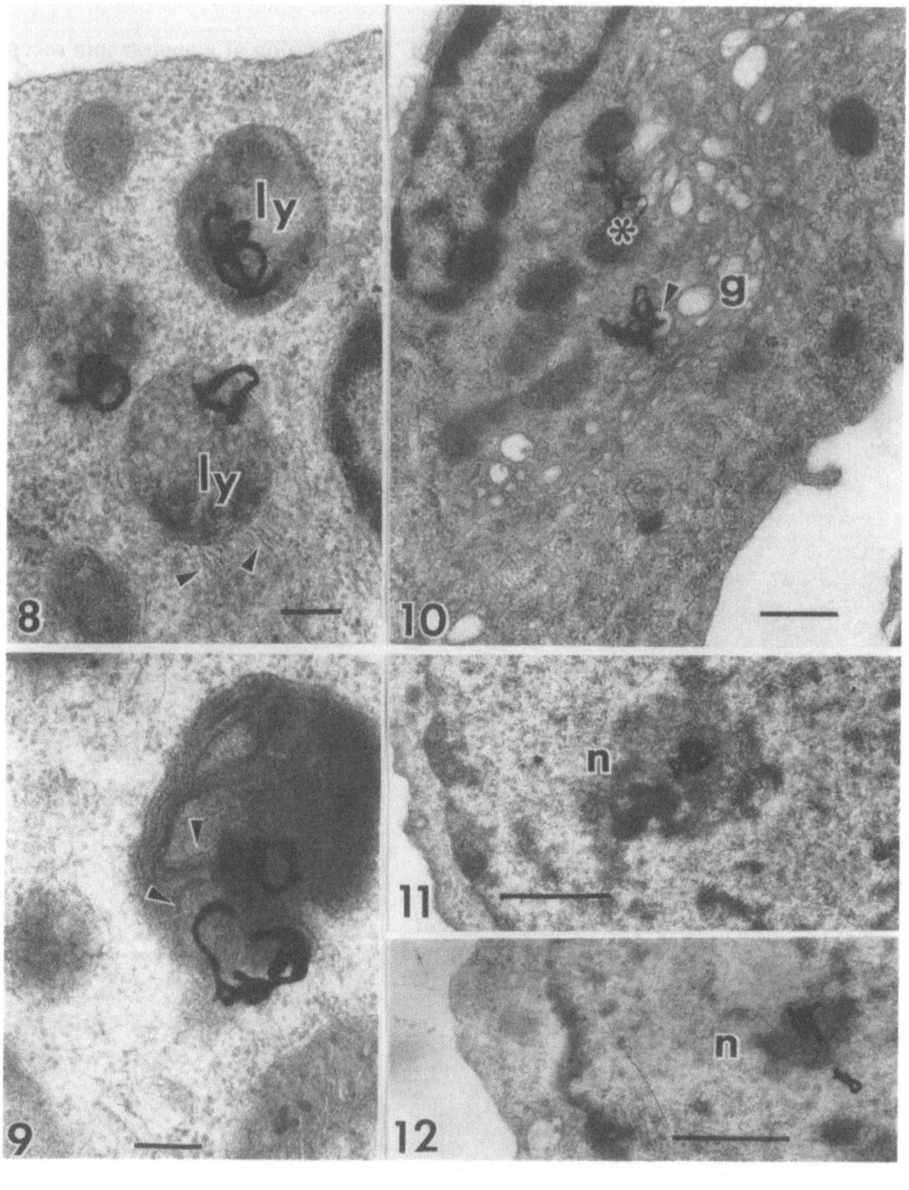

Figs. 8–12. Electron microscopic autoradiograms of cultured granuloca cells after 8 h of incubation with 2 IU/ml of ^{125}I-hCG followed by 5 h of incubation with 20 IU/ml of unlabeled hormone. (8) Part of granulosa cell. Silver grains are localized over lysosome-like structures (ly). Note that 10 nm filaments appear to be associated with lower lysosome (arrow heads); bar, 0.2 μm. (9) Note that membrane fragments (arrow heads) are visible within lysosome; bar, 0.2 μm. (10): part of granulosa cell showing Golgi complex (g). Silver grains are visible on smooth membrane vesicle (arrow head), which seems to be part of Golgi complex. Another cluster of silver grains is seen in close proximity to electron dense granules (asterisk); bar 0.5 μm. (11) Silver grains over heterochromatin area of granulosa cell nucleus (n); bar, 1 μm. (12) Part of another granulosa cell showing labeling in a similar position to 11; and also in less electron dense substance within the nuclear area: bar, 1 μm (By permission from ref. 25).

(46–48). Morphological analysis of gonadotropin uptake was also performed by immunohistochemical and ferritin-labelling approaches, which gave evidence for the *in vivo* uptake of LH and FSH in Leydig and Sertoli cells, respectively (49, 50). It is of interest to note that the development of LH receptors in the fetal testis occurs at an extremely early stage of embryonic life. These specific binding sites appear in the fetal rabbit testis on about day 18 of development concomitant with the morphological and biochemical maturation of Leydig cells (51). LH receptors have been visualized in individual Leydig cells by

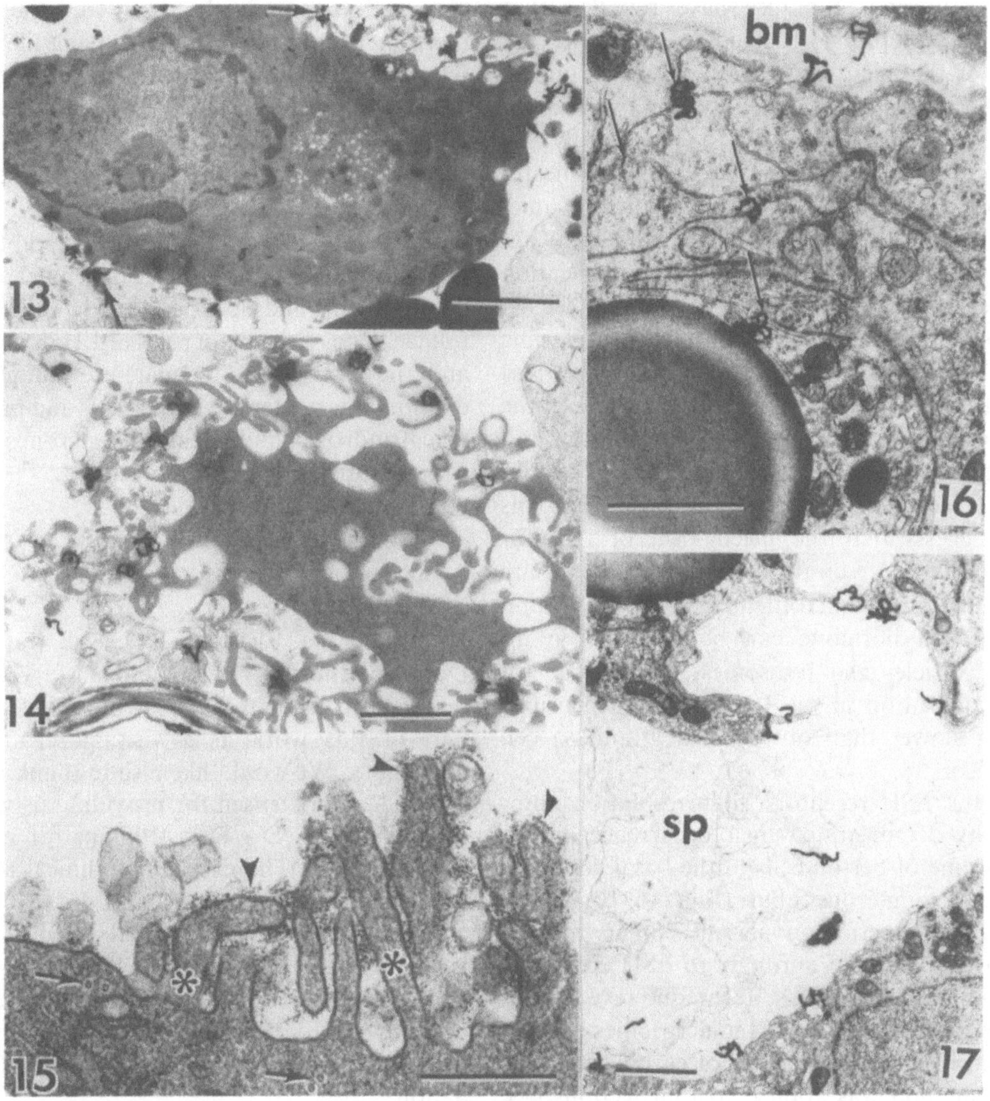

Figs. 13–17. (13) Localization of ^{125}I-hCG binding sites over a Leydig cell (by high resolution autoradiography) after 3 h of incubation with the labeled hormone at 37° C. Most of the label (arrows) is found at the circumference of cell, associated with microvilli. Bar represents 1 μm. (14) Tangential out of Leydig cell preincubated with ^{125}I-hCG, at higher magnification. The localization of silver grains, representing the hormone binding sites, is on the microvillous processes of the cell. Bar, 0.5 μm. (15) Aggregation of hCG receptors over the microvilli of a Leydig cell following incubation with the hormone (90 min at 37° C) and with specific antibodies to the hormone (30 min at 37° C) visualized by the indirect immunoferritin technique. The microvilli are heavily covered with aggregated ferritin particles (arrow heads) resembling aggregated receptor molecules, and are loaded with microfilaments (asterisks). Bundles of microtubules are seen in cross-section beneath the cell membrane (arrows). Bar, 0.5 μm. (Figs. 13–15 by permission from ref. 2). (16–17) Illustration of two instances where a Sertoli plasma membrane is labeled specifically with ^{125}I-hFSH after injection of the hormone to rats. (courtesy Drs. J. Orth and A.K. Christensen; for experimental details see ref. 57). (16) Interdigitating processes of Sertoli cells at the base of the seminiferous eptithelium, near the basement membrane (bm). All the membranes seen in the field belong to Sertoli cells. Several grains (arrows) are seen over these Sertoli plasma membranes. (17) Sertoli plasma membranes bounding the intercellular space (sp) show several associated grains. Bar, 1 μm for both (16) and (17) (By permission from ref. 57).

immunofluorescence (52) and more recently by autoradiography with ferritin-labelled hCG, and with the indirect immuno-ferritin technique (2, 53). Each of these methods has demonstrated a predominantly surface location of the hormone-receptor complexes, consistent with the biochemical characteristics of the sites analyzed in subcellular interstitial tissue fractions and in isolated Leydig cells (2–4, 54).

High resolution techniques have revealed that most of the receptors occupied by hormone are located over the microvillous area at the circumference of these cells (Figs. 13, 14). Moreover, application of antibodies to hCG with the immuno-ferritin technique results in massive aggregation of the receptor-bound hormone, with accumulation of microfilaments and microtubules beneath such aggregates (Fig. 15). This would suggest an association of cytoskeletal elements with the receptor molecule. Indeed, several studies have implicated cytoplasmic contractile elements in receptor mobility (55). It was recently found that LH can acutely increase the number of its own receptors in the Leydig cell and that the up-regulation observed 1 h after LH administration is blocked by cytochalasin B (2, 56). This finding suggests that the mechanism of receptor insertion or exposure on the cell surface may require the participation of microfilaments. Internalization of the labelled hormone can be observed in intracellular vesicles and lysosomes, but the rate of such internalization in rat Leydig cell suspensions was much slower than observed in granulosa cell cultures (53).

As for the FSH receptors, high resolution autoradiography demonstrated their occurrence on the cell membrane of Sertoli cells in the basal compartment of the seminiferous tubule (Figs. 16, 17) and on the surface of spermatogonia at similar concentration (57, 58). Use of antibody to FSH combined with the immuno-peroxidase technique revealed intracellular localization of FSH in acrosomes of spermatids and intra-nuclear bodies of early spermatids when the method was applied to thin sections (59, 60). It is not clear whether such a localization exhibits intracellular receptor to FSH or non specific accumulation, or biosynthesis of the hormone-like molecules within these cells (59, 60),.

5. Concluding remarks

There are similarities in the localization of receptors to gonadotropins in the ovary and the testis. FSH receptors were found in cells nursing the gametes: in the ovary, the granulosa cells and in the testis, Sertoli cells. Differences, however, were found with respect to the LH receptor. While in the testis this receptor was confined to a single type of cells, the interstitial Leydig cell, in the ovary it was found in interstitial, thecal, granulosa and luteal cells, though its appearance on granulosa cells depended on stage of development and location of the cell within the follicle. One of the main roles of LH in both the testis and the ovary is to regulate steroidogenesis; production of testosterone in the testis and production of thecal androgen and luteal progesterone in the ovary, – by occupying specific receptors.

Acknowledgements

We would like to thank Drs. K.J. Catt and M.L. Dufau of the N.I.H., Bethesda, Maryland, for the opportunity given to Dr. A. Amsterdam to carry out part of this work as a visiting scientist in their laboratory. We would like also to thank Drs. J. Orth and A.K. Christensen for providing us with Figures 16 and 17 and Dr. D.M. Phillips for providing us with Figure 3. The excellent technical assistance of Mrs. A. Azrad, Mrs. R. Levin, Mrs. M. Kopelowitz and Mr. S. Gordon is also appreciated. H.R. Lindner is the Adlai E. Stevenson III Professor of Endocrinology and Reproductive Biology at the Weizmann Institute of Science. The work was supported by the Ford Foundation and the Rockefeller Foundation.

References

1. Richards JS: Maturation of ovarian follicle: Actions and interactions of pituitary and ovarian hormones of follicular cell differentiation. Phys Rev 60: 51–89, 1980.
2. Amsterdam A, Naor Z, Knecht M, Dufau ML, Catt KJ: Hormone action and receptor redistribution in endocrine target cells: Gonadotropins and gonadotropin-releasing hormone. In: Receptor-mediated binding and internalization of toxins and hormones. Middlebrook JL, Kohn LD (eds) Academic Press, 1981, pp 283–310.
3. Dufau ML, Catt KJ: Gonadotropin receptors and regulation of steroidogenesis in the testis and ovary. Vitam Hormone 36, Academic Press 1978, pp 461–592.
4. Catt KJ, Harwood JP, Clayton RN, Davies FD, Chan V, Katikineni M, Nozu K, Dufau ML: Regulation of peptide hormone receptors and gonadal steroidogenesis: Recent Progress in Hormone Research, 36, Academic Press, 1980, pp 557–622.
5. Koch Y, Zor U, Chobsieng P, Lamprecht S, Pomerantz S, Lindner HR: Binding of LH and HCG to ovarian cells and activation of adenylate cyclase. J Endocr 61: 179–191, 1974.
6. Lunenfeld B, Eshkol A: Immunology of human chorionic gonadotropin (HCG). Vitam Hormone 25: 137–190, 1967.
7. Midgley Jr AR: Autoradiographic analysis of gonadotropin binding to

rat ovarian tissue sections. In: Receptor for reproductive hormones. O'Malley BW, Means AR (eds): Plenum Press, New York, 1973, pp 365–378.

8. Amsterdam A, Koch Y, Lieberman ME, Lindner HR: Distribution of binding sites for human chorionic gonadotropin in the preovulatory follicle of the rat. J Cell Biol 67: 894–900, 1975.
9. Lindner HR, Amsterdam A, Salomon Y, Tsafriri A, Nimrod A, Lamprecht SA, Zor U, Koch Y: Intraovarian factors in ovulation: determinants of follicular response to gonadotrophins. J Reprod Fert 51: 215–235, 1977.
10. Amsterdam A, Knecht M, Catt KJ, Lindner HR: Regulation of ovarian function by gonadotrophic hormones. In: Hormone cell interactions in reproductive tissues. Proc. 2nd Innsbruck Winter Conference. Wittliff J, Daxenbichler D (eds) Masson Publ. USA in press 1983.
11. Amsterdam A: Function, turnover and distribution analyses of plasma membrane – Associated receptors. Fresenius Z Anal Chem 311: 338–339, 1982.
12. Amsterdam A, Tsafriri A: In vitro binding of ^{125}I-human chorionic gonadotrophin (hCG) in the preovulatory follicle: Absence of receptor sites on oocyte. J Cell Biol 83: 255a, 1979.
13. Albertini DF, Anderson E: The appearance and structure of intercellular connection during the ontogeny of the rabbit ovarian follicle with particular reference to gap junctions. J Cell Biol 63: 234–250, 1974.
14. Amsterdam A, Josephs R, Lieberman ME, Lindner HR: Organization of intramembrane particles in freeze-cleaved gap junctions of rat Graafian follicles: optical-diffraction analysis. J Cell Sci 21: 93–105, 1976.
15. Gilula NB, Epstein ML, Beers W: Cell-to-cell communication and ovulation: a study of the cumulus-oocyte complex. J Cell Biol 78: 58–75, 1979.
16. Sheridan JD: Dye movement and low resistance junctions between reaggregated embryonic cells. Devl Biol 26: 627–636, 1971.
17. Amsterdam A, Knecht M, Catt KJ: Hormonal regulation of cytodifferentiation and intercellular communication in cultured granulosa cells. Proc Natl Acad Sci USA 78: 3000–3004, 1981.
18. Goldenberg RL, Vaitukaitis JL, Ross GT: Estrogen and follicular stimulating hormone interaction on follicular growth in rats. Endocrinology 90: 1492–1498, 1980.
19. Rajaniemi H, Vanha-Perttula T: Specific receptors for LH in the ovary: Evidence by autoradiography and tissue fractionation. Endocrinology 90: 1–9, 1972.
20. Rajaniemi H, Vanha-Perttula T: Attachment to the luteal plasma membranes: an early event in the action of luteinizing hormone. J Endocrinol 57: 199–206, 1973.
21. Gospodarowicz D: Properties of the luteinizing hormone receptor of isolated bovine corpus luteum plasma membranes. J Biol Chem 248: 5042–5049, 1973.
22. Rajaniemi HJ, Hirshfeld AN, Rees Midgley Jr AR: Gonadotropin receptors in rat ovarian tissue. I. Localization of LH binding sites by fractionation of subcellular organelles. Endocrinology 95: 579–588, 1974.
23. Minz Y, Amir Y, Amsterdam A, Lindner HR, Salomon Y: Properties of LH-sensitive adenylate cyclase in purified plasma membranes from rat ovary. Mol Cell Endocrinol 11: 256–283, 1978.
24. Han SS, Rajaniemi HJ, Cho MI, Hirshfeld An, Midgley Jr AR: Gonadotropin receptors in rat ovarian tissue. II. Subcellular localization on LH binding sites by electron microscopic radioautography. Endocrinology 95: 589–598, 1974.
25. Amsterdam A, Nimrod A, Lamprecht SA, Burstein Y, Lindner HR: Internalization and degradation of receptor-bound hCG in granulosa cell cultures. Am J Physiol 5(2): E129–E138, 1979.
26. Amsterdam A, Kohen F, Nimrod A, Lindner HR: Lateral mobility and internalization of hormone receptors to human chorionic gonadotropin in cultured rat granulosa cells. Adv Exp Med Biol 112: 69–75, 1979.
27. Amsterdam A, Nimrod A, Kohen F, Lindner HR: Redistribution of receptors for human chorionic gonadotropin in cultured rat granulosa cells in relation to the cellular response of the hormone. In: Molecular mechanisms of biological recognition. Balaban M (ed) Elsevier/North Holland, Biomedical Press, Amsterdam, 1979, pp 419–428.
28. Anderson W, Kang Y, Perotti ME, Bramley TA, Ryan RJ: Interactions of gonadotropins with corpus luteum membranes. III Electron microscopic localization of [^{125}I]-hCG binding to sensitive and desensitized ovaries seven days after PMSG-hCG. Biol Reprod 20: 362–376, 1979.
29. Markkanen SO, Rajaniemi HJ: Role of internalization and degradation in the removal of receptor-bound human chorionic gonadotropin in rat luteal cells in vivo. Endocrinology 107: 1153–1161, 1979.
30. Zor U, Strulovici B, Lindner HR: Implication of microtubules and microfilaments in the response to the ovarian adenylate cyclase-cyclic AMP system to gonadotropins and prostaglandin E_1. Biochem Biophys Res Commun 80: 983–992, 1978.
31. Zor U, Strulovici B, Lamprecht SA, Amsterdam A, Oplatka A, Lindner HR: Effect of modulators of cytoskeletal function on desensitization and recovery of PGE_1-responsive ovarian adenylate cyclase. Prostaglandins 18: 869–882, 1979.
32. Amsterdam A, Berkowitz A, Nimrod A, Kohen F: Aggregation of luteinizing hormone receptors in granulosa cells: a possible mechanism of desensitization to the hormone. Proc Natl Acad Sci USA 77: 3440–3444, 1980.
33. Salomon Y, Ezra E, Nimrod A, Amir-Zaltsman Y, Lindner HR: The interaction of the LH/HCG receptor with adenylate cyclase in the rat ovary. In: Chorionic Gonadotropin. Segal SJ (ed) Plenum Publishing Corporation, NY 1980, pp 345–369.
34. Salomon Y, Ezra E, Amir-Zaltsman Y: The role of GTP in luteoropin-induced desensitization of the GTP regulatory cycle and adenylate cyclase in the rat ovary. Adv Cycl Nucl Res, Vol 14, Dumont JE, Greengard P, Robison GA (eds) Raven Press NY 1981 pp 101–109.
35. Rao ChV, Mitra S: Gonadotropin and prostaglandins binding sites in nuclei of bovine corpora lutea. Biochem Biophys Acta 584: 454–466, 1979.
36. Chen TT, Abel Jr JA, McCellan MI, Sawyer HR, Diekman MA, Niswender GD: Localization of gonadotropic hormones in lysosomes of ovine luteal cells. Cytobiologie 14: 412–420, 1977.
37. Conn PM, Conti M, Harwood JP, Dufau ML, Catt KJ: Internalization of gonadotrophin-receptor complex in ovarian luteal cells. Nature 274: 598–600, 1978.
38. Conti M, Harwood JP, Dufau ML, Catt KJ: Regulation of luteinizing hormone receptors and adenylate cyclase activity by gonadotropin in the rat ovary. Mol Pharmacol 13: 1024–1032, 1977.
39. Midgley Jr AR: Autoradiographic analysis of gonadotropin binding to rat ovarian tissue sections. Adv Exp Med Biol 36: 365–378, 1973.
40. Zeleznik AJ, Midgley Jr AR, Reichert Jr LE: Granulosa cell maturation in the rat: Increased binding of human chorionic gonadotropin following treatment with follicle-stimulating hormone in vivo. Endocrinology 95: 818–825, 1974.
41. Erickson GF, Wang C, Hseuch AJW: FSH induction of functional LH receptors in granulosa cells cultured in a chemically defined medium. Nature 279: 336–338, 1979.
42. Nimrod A, Lindner HR: Heparin facilitates the induction of LH receptors by FSH in granulosa cells cultured in serum-enriched medium. FEBS Lett 119: 155–157, 1980.
43. Knecht M, Amsterdam A, Catt KJ: The regulatory role of cyclic AMP in hormone-induced granulosa cell differentiation. J Biol Chem 256: 10628–10633, 1981.
44. Petrusz P: Gonadotropin – target cell interactions: a model based on morphological localization. In: Structure and function of the gonadotropins. McKerns KW (ed), Plenum Publishing Corporation 1978, pp 577–589.
45. Luborsky JL, Behrman HR: Localization of LH receptors on luteal cells with a ferritin-LH conjugate. Molec Cell Endocrin 15: 61–78, 1979.
46. Espeland DH, Naftolin F, Paulsen CA: Metabolism of labelled ^{125}I-HCG ovary. In: Gonadotropins 1968. Rosenberg E (ed) Gerom-x Inc, Los Altos, California, 1968, pp 177–184.
47. DeKrester DM, Catt KJ, Burger HG, Smith GC: Radioautographic studies on the localization of ^{125}I-labelled human luteinizing and growth hormone in immature male rats. J Endocrinol 43: 105–111, 1969.
48. DeKrester DM, Catt KJ, Paulsen CA: Studies on the in vitro testicular binding of iodinated luteinizing hormone in rats. Endocrinology 88: 332–337, 1971.
49. Castro AE, Alonso A, Mancini RE: Localization of follicle-stimulating and luteinizing hormones in the rat testis using immunohistological tests. J Endocrinol 52: 129–136, 1972.
50. Mancini RE, Castro A, Seigeur AC: Histologic localization of follicular-stimulating and luteinizing hormones in the rat testis. J Histochem Cytochem 15: 516–525, 1967.
51. Catt KJ, Dufau ML, Neaves WB, Walsh PC, Wilson JD: LH-hCG receptors and testosterone content during differentiation of the testis in the rabbit embryo. Endocrinology 97: 1157–1165, 1975.

52. Hseuch AJW, Dufau ML, Katz SI, Catt KJ: Immunofluorescence labelling of gonadotropin receptors in enzyme-dispersed interstitial cells. Nature 261 710–711, 1976.
53. Amsterdam A, Dufau ML, Catt KJ: Regional distribution of receptors to luteinizing hormone in dispersed testicular Leydig cells. J Cell Biol 87: 160a, 1980.
54. Dufau ML, Horner KA, Hayashi K, Tsuruhara T, Conn PM, Catt KJ: Actions of choleragen and gonadotropin in isolated Leydig cells. J Biol Chem 253: 3721–3729, 1978.
55. Silverstein SC, Steinman RM, Conn ZA: Endocytosis. Ann Rev Biochem 46: 669–722, 1977.
56. Huhtaniemi IT, Katikineni M, Chan V, Catt KJ: Gonadotropin-induced positive regulation of testicular luteinizing hormone receptors. Endocrinology 108: 58–65.
57. Orth J, Christensen AK: Localisation of ^{125}I labeled FSH in the testes of hypophysectomized rats by autoradiography at the light and electron microscope levels. Endocrinology 101: 262–278, 1977.
58. Orth J, Christensen AK: Autoradiographic localization of specifically bound ^{125}I-labeled follicle-stimulating hormone on spermatogonia of the rat testis. Endocrinology 103: 1944–1951, 1978.
59. Huston JC, Gardner PJ, Moriarty GC: Immunocytochemical location of a follicular stimulating hormone like molecule in the testis. J Histochem Cytochem 25: 1119–1126.
60. Childs GV: Immunocytochemical demonstration of endogenous gonadotropin binding sites in the fetal rat testis. In: Structure and Function of the gonadotrophins. McKerns KW (ed) Plenum Press New York, 1978, pp 553–575.

Authors' address:
Department of Hormone Research
The Weizmann Institute of Science
Rehovot 76100, Israel

CHAPTER 23

Fine structure of the thyroid follicle

HISAO FUJITA

1. Introduction

Throughout the vertebrates, the thyroid gland is a morphologically simple organ consisting of numerous ball-like structures called follicles, and of interfollicular connective tissues with blood capillaries. Each follicle is composed of numerous follicle epithelial cells arranged as a simple epithelium, a follicle lumen surrounded by the epithelial cells, and a few parafollicular cells located singly or in groups in the outside of the follicle epithelium. The follicle lumen stores a high molecular glycoprotein called thyroglobulin, which is secreted from the follicle epithelial cell. Each follicle epithelial cell shows an exocrine type polarity, though it is an endocrine cell. The apical pole facing the follicle lumen and that of the basal pole facing the interfollicular connective tissue are different in structure and function from each other, and the zonula occludens is well developed at the apical end of the lateral surface (1, 2), though the cells of other most endocrine organs lack this structure. The rough endoplasmic reticulum (RER) is well developed in the basal and lateral cytoplasm, the Golgi apparatus is generally located in the supranuclear region, and the secretory granules are localized in the subapical cytoplasm. Microvilli are usually present at the apical surface of the cell (Figs. 1–4).

The follicle epithelial cell secretes thyroglobulin from the apical part of the cell into the follicle lumen, and thyroxine (T_4) and triiodothyronine (T_3) from the basal part of the cell into the interfollicular connective tissue, while the parafollicular cell secretes thyrocalcitonin into the connective tissue.

The present review deals with the functional morphology of the follicle epithelial cell.

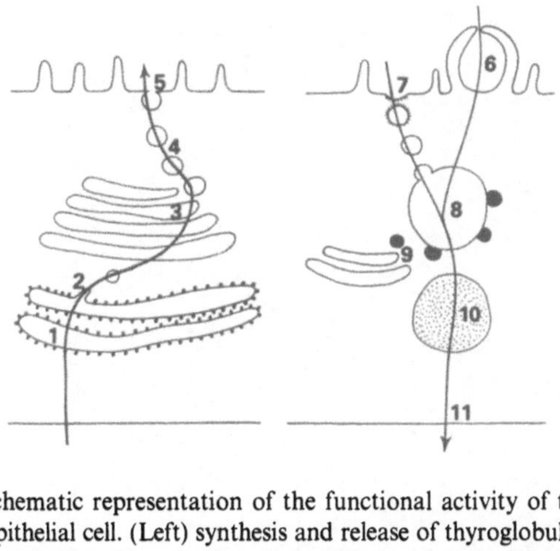

1

Fig. 1. Schematic representation of the functional activity of the thyroid epithelial cell. (Left) synthesis and release of thyroglobulin. 1. rough endoplasmic reticulum; 2: budding to be transporting vesicle; 3: Golgi apparatus; 4: secretory granule containing thyroglobulin. (Right) reabsorption of colloid and release of T_4 and T_3. 6: phagocytosis; 7: micropinocytosis; 8: large colloid droplet; 9: primary lysosome; 10: phagolysosome consisting of primary lysosome and colloid; 11: release of T_4 and T_3.

2. Functional morphology of the thyroid follicle

2.1. Characteristics of thyroid hormones, T_4 and T_3

Though T_4 as well as T_3 have hormonal activities, T_4 is believed to be a main hormone secreted from the follicle epithelial cell. The hormones are not peptides but characteristic amino acid derivatives, derived from two molecules of iodotyrosine linked by ether bond (–O–). Though all the peptide hormones in most endocrine organs are synthesized and stored in situ in the form of secretory granules which are produced by the RER and Golgi apparatus, formation of the thyroid hormones, T_4 and T_3 requires more complicated steps. The hormones are not synthesized directly from their constituents, tyrosine and iodine in the cell. The secretory processes of T_4 and T_3 are as follows (3); 1) synthesis of a large molecular glycoprotein named thyroglobulin in the follicle epithelial cell and release of it into the follicle lumen; 2) iodination of the tyrosyl residues of thyroglobulin; 3) coupling of the iodotyrosyl residues of thyroglobulin; 4) reabsorption of thyroglobulin from the follicle

Motta, PM (ed): Ultrastructure of endocrine cells and tissues. ISBN-13: 978-1-4613-3863-5

lumen into the follicle epithelial cell; 5) hydrolysis of thyroglobulin reabsorbed for liberation of T_4 and T_3, and 6) release of T_4 and T_3 from the follicle epithelial cell into the interfollicular connective tissue space.

2.2. Synthesis and release of thyroglobulin

The thyroid follicular epithelial cell does not synthesize T_4 and T_3 directly, but first produces thyroglobulin, highmolecular-weight glycoprotein. Though there are several kinds of subunits of thyroglobulin, 3–6 S, 12 S, 19 S, 27 S and 33 S, the 19 S protein, which is 660,000 in molecular weight and occupies about 90% of thyroglobulin in this organ, is thyroglobulin in a common sence.

Like all the exportable proteins in the peptide or protein secretory cells, the proteinous part of thyroglobulin is synthesized on the attached ribosomes of the RER and stored in its cisternae. The RER is very well developed occupying the over half region of the

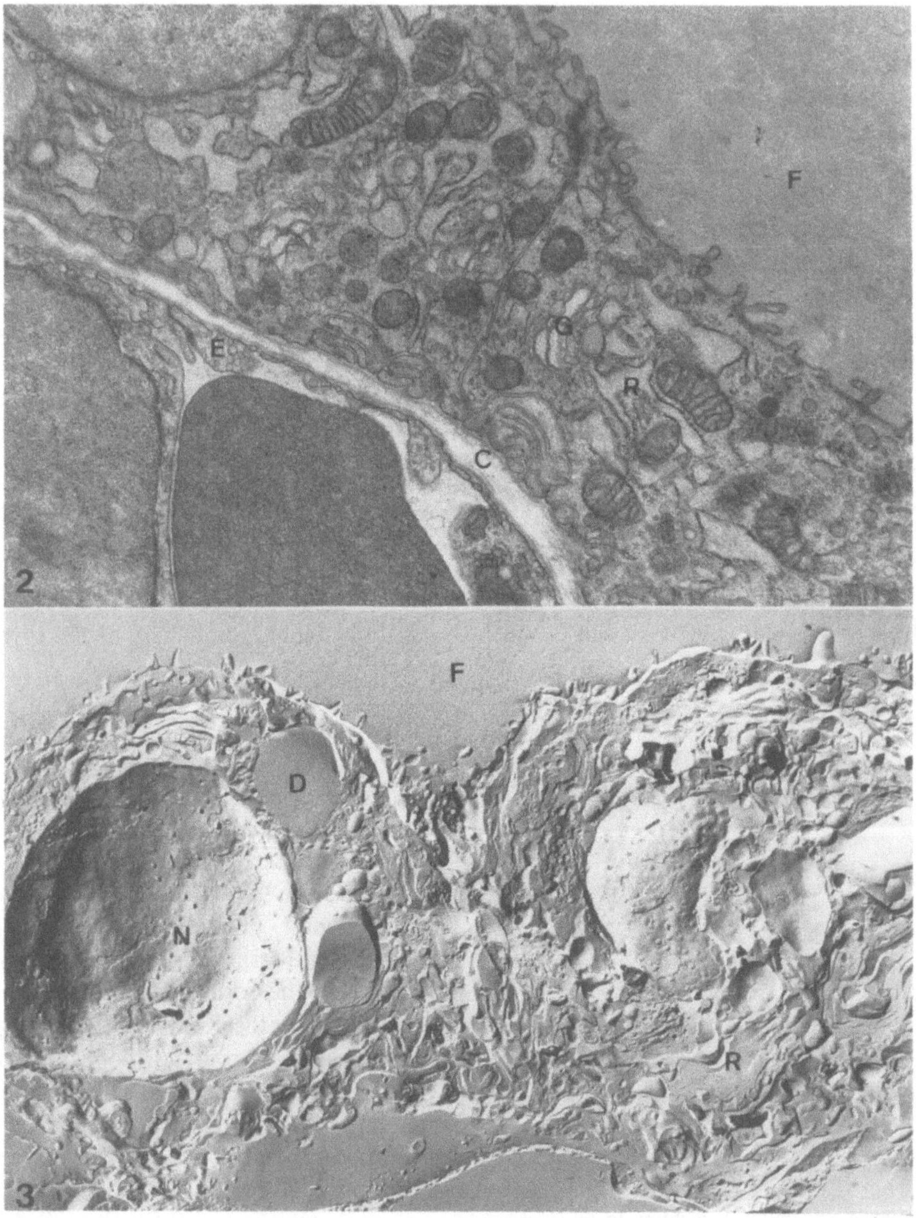

Figs. 2–3. (2) An electron micrograph of a part of mouse thyroid. Notice well developed rough endoplasmic reticulum (R) and Golgi apparatus (G) in the follicle epithelial cell. F: follicle lumen; C: connective tissue space; E: endothelial cell (× 11,200). (3) A freeze-replica image of a part of mouse thyroid. N: nucleus; F: follicle lumen; D: reabsorbed colloid droplet; R: rough endoplasmic reticulum (× 7,200).

cell and the cisternae are almost always dilated in the follicle epithelial cell. This means that synthesis of the exportable protein, thyroglobulin is very active in this type of cells.

Thyroglobulin is bound to carbohydrate elements to be a glycoprotein. It is now well known that mannose and N-acetylglucosamine are incorporated into thyroglobulin in the RER (4, 5). Both mannosyl transferase and N-acetylglucosaminyl transferase have biochemically been demonstrated in rough microsomes of sheep thyroid (5). Thyroglobulin stored in the cisternae of RER is transferred to the Golgi apparatus, which is usually localized at the supranuclear region. The membrane of the endoplasmic reticulum facing the Golgi apparatus is smooth-surfaced. Like in other protein secretory cells, the exportable protein is transported to the Golgi apparatus by mechanism of budding from the RER. The sugar components of thyroglobulin, such as fucose, galactose and sialic acid are known to be bound to thyroglobulin in the Golgi apparatus (4, 6). The galactosyl transferase activity has been demonstrated in the Golgi rich fraction of sheep thyroid cells (7).

The Golgi apparatus is considered to have an internal polarity. The thiamine pyrophosphatase (TPPase) activity and acid phosphatase (AcPase) activity are localized in the cisternae of 1–3 stacks and vesicles on the trans-side of the Golgi apparatus (8). In addition, rigid lamellae, coated vesicles, multivesicular bodies and lysosomes are also positive for AcPase activity. To the silver methenamine method for the localization of carbohydrates and glycoproteins, the cisternae of the RER and the outer (cis-side) saccules of the Golgi apparatus show a weak but positive reaction, while the intermediate to inner (trans-side) saccules are intensely positive (9). The secretory granule containing thyroglobulin and the luminal colloid are also strongly positive (Figs. 5, 6). These findings coincide with the fact that mannose and N-acetylglucosamine are bound to protein in the RER, and galactose, fucose and sialic acid in the Golgi apparatus. These three might be incorporated into thyroglobulin in the inner-saccules of the Golgi apparatus (9).

Like in other protein secretory cells, the secretory granules 150–200 nm in diameter, containing thyroglobulin are derived from the Golgi apparatus, especially from its inner saccules. They are transported to the apical surface of the cell and released into the follicle lumen by exocytosis.

By the electron microscopic autoradiography of ^{3}H-leucine, the intracellular route for the synthesis and release of thyroglobulin has been clarified (3, 10, 11). The labelled amino acid first appears in the cisternae of RER within 15 min after the injection and in the Golgi apparatus and secretory granules after 30–45 min and in the follicle lumen after 45 min (3, 10).

2.3. Iodination of thyroglobulin

Inorganic iodide is taken up very quickly into the thyroid gland from the blood capillary. The mechanism is not clear but the iodide ion is believed to be taken up into the follicle epithelial cell by active transport (12). Quabain-sensitive ATPase is known to be implicated in the taking up of iodide ion into the cell. The ATPase activity has been demonstrated in the basal as well as the apical plasma membrane in the follicle epithelial cell of the guinea pig (13). The ATP activity in the basal plasma membrane of the follicle epithelial cell is necessary for iodide trapping, but the mechanism by which iodide is specifically accumulated in the thyroid gland is obscure.

The site of the iodination of tyrosyl residue of thyroglobulin is an important subject. The autoradiography of radioactive iodine is one of the useful methods to solve this problem. Inorganic iodide is washed away by autoradiographic procedures such as washing, fixation and dehydration of the tissue, and only iodide which is bound to protein is detected by this method. In various vertebrates it was showed that silver grains appear almost entirely in the follicle lumen within a few min or a few hr after the injection of ^{125}I (3, 14, 15, 16, 17, 18). In the mouse thyroid cultured in the medium containing ^{125}I, silver grains appear in the follicle lumen within 30 sec and increase rapidly in number with time (Fig. 7). It is considered that the iodination of tyrosyl residue in thyroglobulin takes place almost entirely in the follicle lumen, especially at the periluminal region.

In our study, light microscopic autoradiography of freeze-dried sections shows that silver grains for inorganic iodide are accumulated chiefly over the follicle lumen after the injection of ^{131}I into mercaptoimidazole-treated rats, and the silver grains disappear in the fixed, dehydrated, and paraffin-embedded sections obtained from the same animal (19). Mercaptoimidazole inhibits the iodination of tyrosyl residue of thyroglobulin, and the fact that the silver grains disappeared in the fixed, dehydrated, and paraffin-embedded sections means that there was no radioactive iodide which was bound to tyrosyl residue of thyroglobulin in the thyroid of this animal. Thus, it is concluded that inorganic iodide taken up is mostly transported to the follicle lumen passing through the epithelial cell.

The enzyme for iodination of tyrosyl residue of thyroglobulin is peroxidase. Iodide ion taken up into the follicle is oxidized to become I_2, active form of iodine. In this reaction H_2O_2 acts as the electron acceptor. Reactions are summarized as follows:

$NADH$ (or $NADPH$) $+ H^+ + O_2 \xrightarrow{peroxidase} NAD^+$ (or $NADP^+$) $+ H_2O_2$

$H_2O_2 + I^- \xrightarrow{peroxidase}$ Oxidized iodide $+ H_2O_2$

Oxidized iodide + tyrosine $\xrightarrow{peroxidase}$ Iodotyrosine

The localization of endogeneous peroxidase has been demonstrated cytochemically using 3,3′-diaminobenzidine tetrahydrochloride (DAB) method. Reaction products for peroxidase are localized in the perinuclear cisternae, RER cisternae, Golgi saccules, subapical vesicles and around the external surface of the luminal colloid (3, 20, 21, 22, 23). However, Tice

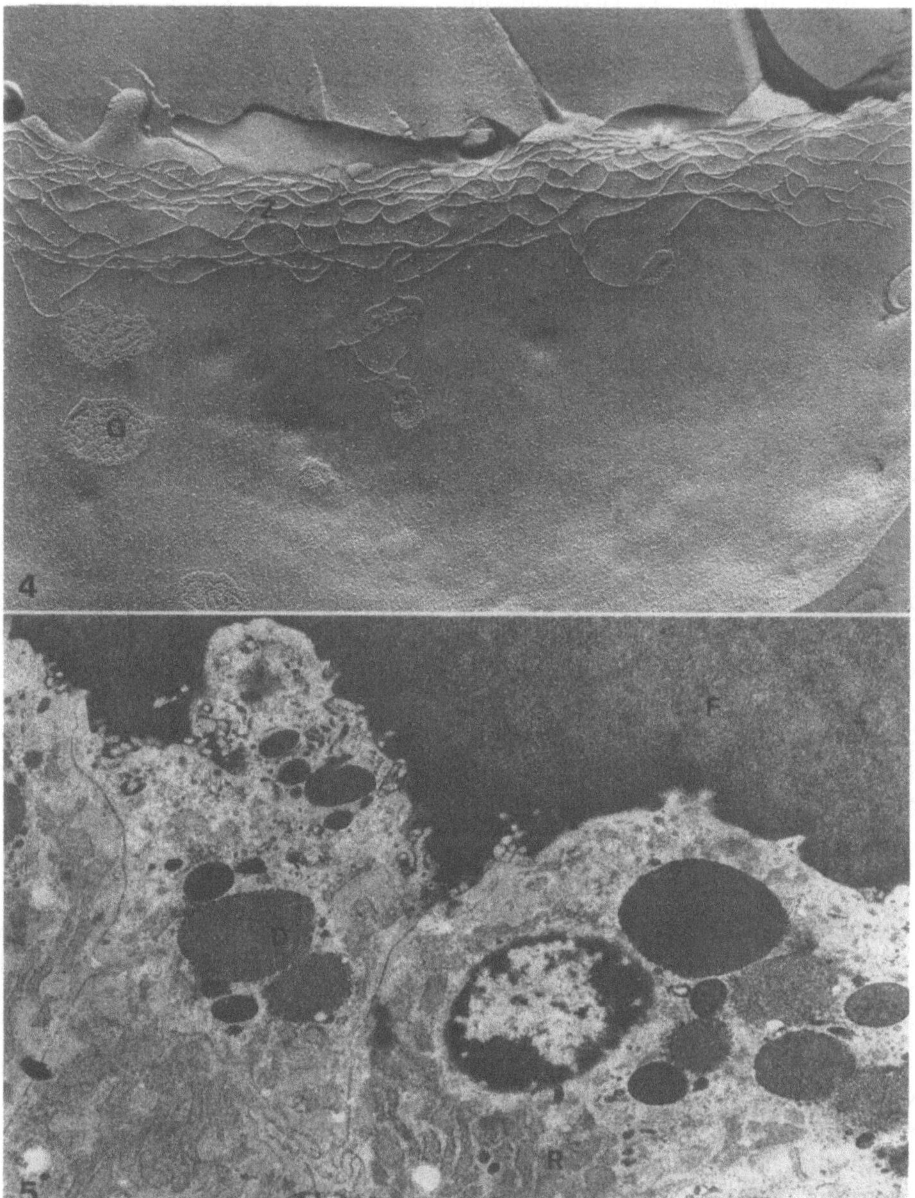

Figs. 4–5. (4) A freeze-replica image of zonula occludens (Z) and gap junctions (G) between follicle epithelial cells in a 19 day-old chick embryo. Zonula occludens consisting of anastomosing strands is located at the apical end of the lateral surface of the cell. Reprinted with permission from Ishimura, K. and Fujita, H. (52). (× 28,800). (5) Localization of complex carbohydrates in the follicle epithelial cell of a mouse thyroid. Silver methenamine method. Cisternae of rough endoplasmic reticulum (R) is weakly positive and luminal colloid (F) and reabsorbed colloid droplet (D) are strongly positive. From Tamura, S. and Fujita, H.) (× 11,200).

and Wollman (23), who tried peroxidation of thyroid tissue using hydrogen peroxide, glucose-glucose oxidase, or potassium ferricyanide before immersion in DAB solution, emphasized that the peroxidase activity is localized on the apical plasma membrane instead of periluminal colloid. Recently, cytochemical localization of hydrogen peroxide (H_2O_2) has been studied by Björkman et al., using NADPH oxidase in isolated rat thyroid follicles (24). They, who stated that H_2O_2 is generated on the apical surface of the follicle cell by NADPH oxidase residing in the apical plasma membrane, have considered that iodination of thyroglobulin takes place on the apical surface of the follicle cells.

Two molecules of tyrosyl residue in thyroglobulin are coupled with each other to be a thyronine residue.

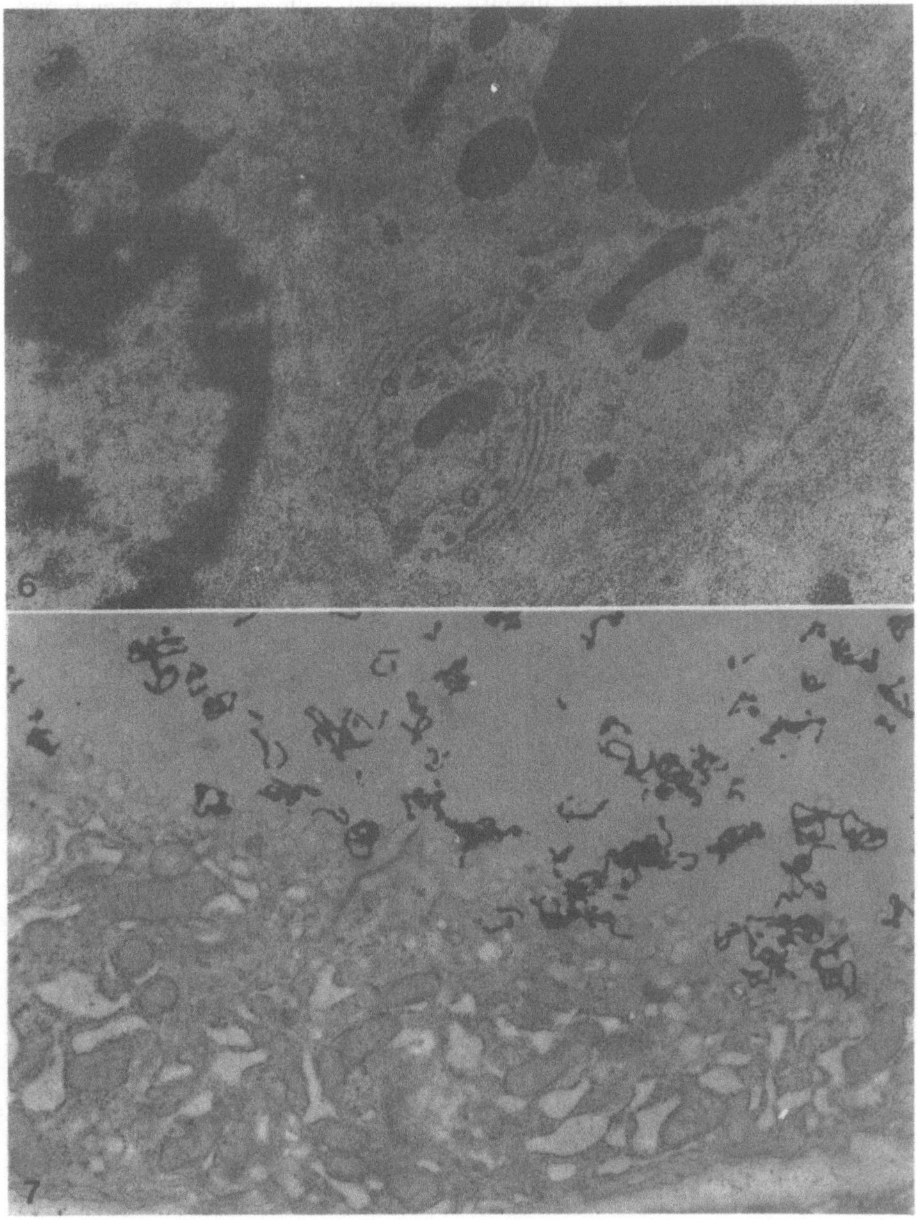

Figs. 6–7. (6) Carbohydrate moiety in the Golgi apparatus (G) and reabsorbed colloid droplet of a mouse thyroid follicle epithelial cell stained with silver methenamine method. The outer saccules (cis-side) of the Golgi apparatus are weakly positive, and the inner saccules (trans-side) are strongly positive. (Reprinted with permission from Tamura, S. and Fujita, H. (9) (× 33,600). (7) An autoradiography of the cultured mouse thyroid 30 seconds after immersion in a medium containing 600 μCi/ml of $Na^{125}I$ (× 9,600).

This coupling is considered to take place after the iodination of tyrosine by the data obtained from the autoradiography of ^{131}I in the paper chromatogram of the mouse thyroid. Since the iodination of tyrosyl residue occurs almost entirely in the follicle lumen, it is reasonable to presume that the coupling of iodotyrosine residues in a thyroglobulin molecule occurs in the follicle lumen.

2.4. Reabsorption of colloid

Thyroglobulin in the follicle lumen is taken up into the follicle epithelial cell by micropinocytosis as well as phagocytosis (25). Thyroid stimulating hormone (TSH) secreted from the anterior pituitary is an important agent to stimulate the reabsorption of the luminal colloid into the cell. After injection of TSH, features of micropinocytosis and phagocytosis are often observed in the follicle epithelium of many kinds of vertebrates. Until the description of micropinocytosis by Seljelid (25, 26), an appearance of cytoplasmic protrusions like pseudopods containing large colloid droplets has been observed by numerous investigators. The features of engulfment of the luminal colloid by the pseudopod correspond to phagocytosis. This mechanism seems to play a role in

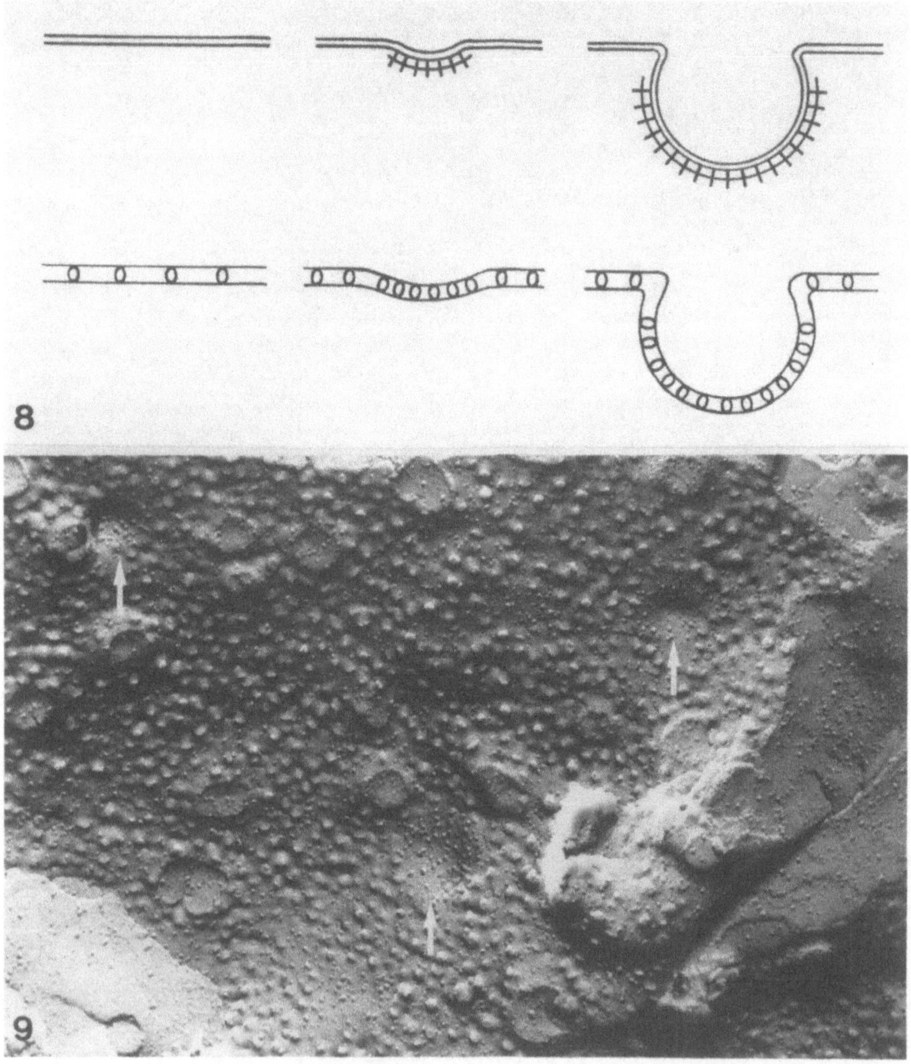

Figs. 8–9. (8) Schematic representation of the micropinocytosis of luminal colloid. Notice the coated pits (upper) and the aggregation of membrane particles (lower). (9) A freeze-replica image of the PF of apical plasma membrane of filipin-treated mouse thyroid follicle cell 45 min after injection of TSH. Numerous filipincholesterol complexes are distributed on the fractured membrane. The complexes are absent on the shallow depressions or pits where membrane particles are aggregated (arrow). Reprinted with permission from: Fujita, H., et al. (38) (× 60,000).

a hyperfunctional state of the gland induced by some stimulations such as the injection of a large dose of TSH. Similar findings have been reported after the treatment with dbc-AMP (27, 28). TSH is now known to stimulate adenylate cyclase in the plasma membrane of the thyroid epithelial cell to produce cAMP (29). In our data the reabsorption of colloid seems not to be affected by calcium influx induced by the treatment with A23187, and even the thyroid tissue cultured in the Ca^{++}-free medium is well reacted to TSH (30). Microtubules and microfilaments are believed to play a role in the phagocytosis of the luminal colloid. Colchicine, vinblastin and hexylen glycol inhibit the colloid phagocytosis after treatment with TSH (31, 32, 33).

By Zeligs and Wollman, it has been reported that the phagocytotic activity of the follicle epithelial cell is not always specific to the luminal colloid, and the erythrocytes accumulated in the follicle lumen in the thiouracil treated mouse are always phagocytosed by pseudopods into the follicle epithelial cell (34). There are various grades in iodination and coupling of tyrosyl residues in luminal thyroglobulin. Though it is not clear whether or not thyroglobulin is phagocytosed selectively according to the grade of maturation or randomly. The occurrence of the phagocytosis of erythrocytes suggests that by this mechanism the apical plasma membrane takes up thyroglobulin randomly into the cell.

In normal conditions, the luminal colloid seems to be reabsorbed mainly by micropinocytosis into the cell (Fig. 8). Coated pits and vesicles appear in the apical part of the cell after the injection of TSH. In our study, at the small depression of the apical plasma membrane which suggests the initial site for the micropinocytosis of the luminal colloid, membrane particles on the protoplasmic face (PF) form aggregates in freeze-replica images (35). A similar phenomenon has been reported by Orci et al. in the fibroblast, showing the pinocytic activity for very low density lipoprotein (VLDL), and the aggregated particles are considered to be receptor for VLDL (36). The functional significance of particle-aggregates on the apical surface of the thyroid cell at the initial site for micropinocytosis of the luminal colloid is not clear. It is speculated that the aggregates of the membrane particles might have a receptor activity to recognize the maturation grade of thyroglobulin in iodination and coupling. By double labelling experiments with ^{125}I (old label) and ^{131}I (new label), and measurement of specific radioactivity of $^{131}I/^{125}I$, Miquelis and Simon demonstrated biochemically that old iodinated thyroglobulin (labelled with ^{125}I) as well as newly iodinated thyroglobulin (labelled with ^{131}I) are randomly reabsorbed into the cell, and the newly iodinated one is not preferentially taken up, while newly iodinated thyroglobulin is hydrolysed preferentially (37). Recently a polyene antibiotic, filipin has been used for the localization of cholesterol in the membrane. This agent reacts specifically with membrane cholesterol and form filipin-sterol complexes which are easily recognized in freeze-fracture images as numerous protuberances or pits (Fig. 9). Using freeze-fracture images of the filipin-treated thyroid of TSH-treated mice, it becomes clear that only the depressions showing aggregates of membrane particles lack the cholesterol contents though the other regions of the apical plasma membrane are very rich in cholesterol (38). Nakajima and Bridgman reported a similar finding in the membranes of active zone and acetylcholine receptor aggregates in the neuromuscular junction (39). Micropinocytosis might also depend on the presence of microfilaments (30). In our study, actin filaments become visible bound to the coated pits as well as vesicles in the heavy meromyosin treated tissues (30). Nève et al. reported that cytochalasin B which is the inhibitor of contractile function of microfilaments inhibit the phagocytosis of colloid (32).

2.5. Liberation and release of thyroxine

The reabsorbed colloid droplets are fused with primary lysosomes thus becoming phagolysosomes. Sometimes a large colloid droplet is enwrapped by the elongated lysosome, as reported in the histiocyte (30). The acid phosphatase (AcPase) activity is positive in primary as well as phagolysosomes, multivesicular bodies, rigid lamellae and the cisternae of 1–2 saccules of trans-side of the Golgi apparatus. The large colloid droplets change from low to high in electron density by the fusion of lysosomes derived from the trans-side of the Golgi apparatus and from negative to positive in AcPase reaction. The hydrolysis of thyroglobulin takes place in the phagolysosomes, and T_4, T_3 and various kinds of aminoacids are liberated (Figs 5, 6).

T_4 and T_3, which are amino acid derivatives and very small in molecular weight, are not visible by electron microscopy of ultrathin sections. It is speculated that the hormones are liberated into the cytoplasmic matrix from the phagolysosomes and transported to the basal part of the cell passing through the cytoplasmic matrix (3). Granule-like structures suggesting to contain the secretory substance (hormones) are difficult to find in the basal cytoplasm.

The release mechanism of the hormones from the follicle epithelial cell into the interfollicular connective tissue is also unknown. If T_4 and T_3 are transported through the cytoplasmic matrix in free form and released passing through the plasma membrane without making any visible structural change, the releasing mechanism could be classified into the diacrine type of Kurosumi (40). Inoue et al. (41) and Herzog and Miller (42) succeeded to make inside-out follicles in the porcine cultured thyroid. The microvilli and zonula occludens are localized at the outer side of the follicle and thyroglobulin and hormones are secreted into the cultured medium in this system. Using this follicle it became clear that the amount of hormones liberated into the cultured medium depends on the concentration of thyroglobulin in the cultured medium and on the stimulation of endocytosis by TSH (42).

As to the membrane retrieval of the reabsorbed colloid droplet, Herzog and Miller (42) clarified that cationic ferritin, binding to the luminal plasma membrane and taken up by micropinocytosis as coated

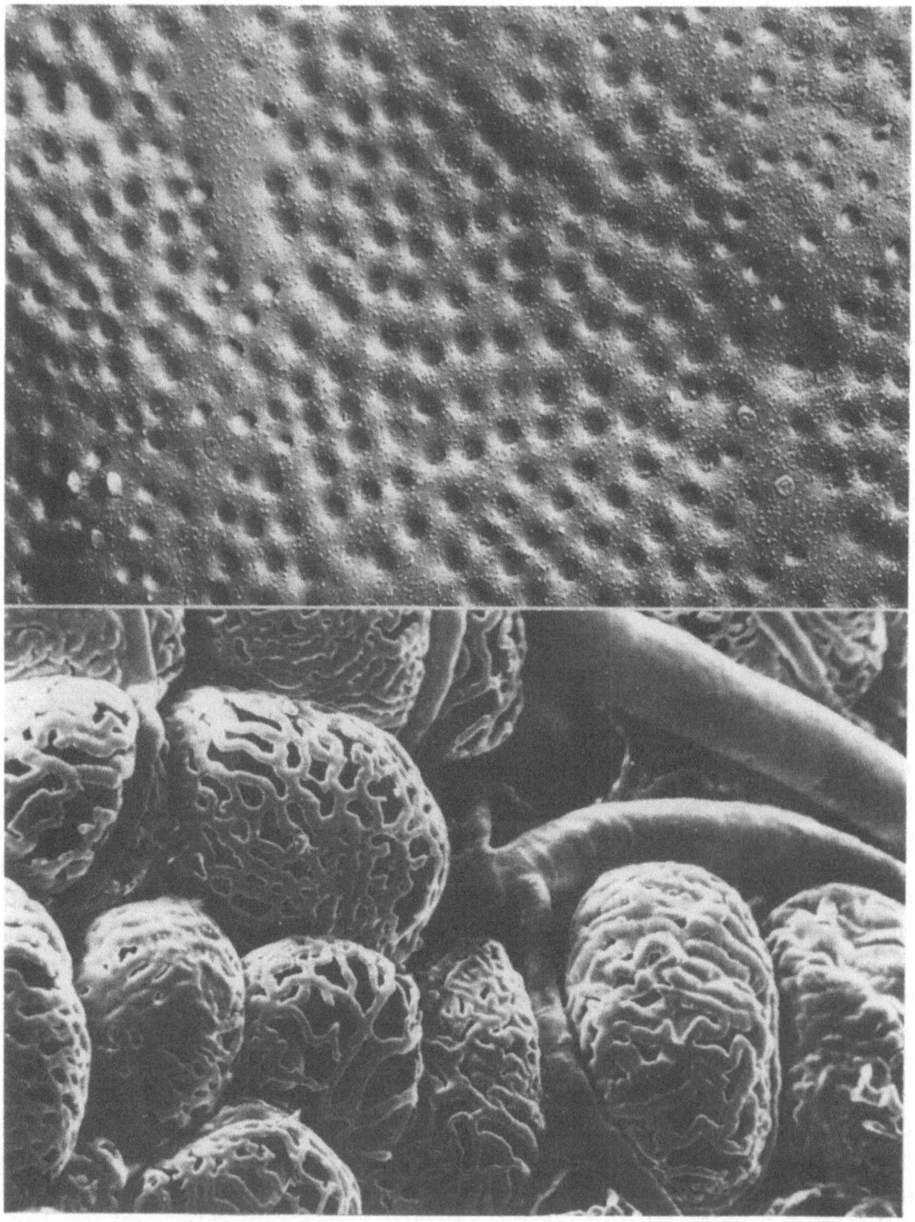

Figs. 10–11. (10) A freeze-fracture image of the capillary endothelial fenestrations of a mouse thyroid. (From: Ishimura, K. and Fujita H.) (× 33,600). (11) Blood vessels of monkey thyroid. Each follicle is encapsulated with a capillary network. (Reprinted with permission from: Fujita, H. and Murakami, T. (44) (× 264).

vesicles appears in the lysosomes after 5 min and Golgi cisternae after 30 min, and uncharged dextran which does not bind to the plasma membrane is taken up in small amounts and appears only in the Golgi stacks, while anionic ferritin which does not bind to the cell membrane reaches only the lysosomes (4). From these findings, they considered that the limiting membrane of the reabsorbed colloid droplet may be reused in the Golgi apparatus for the formation of the secretory granules (4).

2.6. Interfollicular connective tissue and blood capillary

Among the follicles is an interfollicular connective tissue containing well ramified blood capillaries coming from the thyroid arteries.

There are two basal laminae in the connective tissue; one belongs to the capillary endothelium and the other to the follicle epithelium. The interfollicular connective tissue, which is sometimes called pericapillary space or connective tissue space, is loose and consists of tissue fluid, connective tissue cells, connective tissue fibrils, and nerves. The hormones are released from the follicle epithelial cell into the connective tissue and then enter the blood capillary through the endothelium.

The distribution of the blood capillary is easily understood three dimensionally by the scanning electron microscope using corrosion casts prepared with methylmethacrylate (Fig. 11). The blood capillaries make a complex network encapsulating each follicle like a basket (44). In most follicles capillaries for one follicle are well ramified and anastomosed with one another (44). In the monkey, the capillary network of each basket is completely independent of that of the adjacent one, making a round or oval complete structure, while in the dog and rat, the network of one follicle is often common with that of adjacent follicles (44).

As in other endocrine organs, the capillary endothelium of the thyroid gland possesses numerous fenestrations throughout the vertebrates except for cyclostomes. The population density of the endothelial fenestrations can easily be calculated using the freeze-fracture images (Fig. 10). In our data the number of the fenestrations and their population density in the thyroid gland are increased in long term TSH-treated mice which are in a hyperactive state, and decreased in long term Thyradin-treated animals which are in lower active state (45). In the most primitive vertebrate, lamprey and hagfish whose secretory activity is very low, no fenestrations were found in the thyroid capillary endothelium (46, 47). It is considered that the number of endothelial fenestrations changes according to the functional activity of the gland. The hormones released into the connective tissue space might enter the capillary lumen through the endothelial fenestrations.

3. Morphological characteristics of the follicle and polarity of secretory cells

In most endocrine organs with the exception of thyroid and enteroendocrine cells, there are no zonulae occludentes, though gap junctions are usually and maculae occludentes are sometimes present. On the other hand, the thyroid epithelial cell has an exocrine cell-like polarity and very tight zonulae occludentes are well developed (1, 2). By this structure, the luminal colloid is strongly sealed out from the intercellular space and the interfollicular connective tissue (Fig. 4).

As to the formation of the zonulae occludentes, it has been reported that the terminal bar is already present between the cells when the primitive lumen appears (48, 49, 50, 51). Our studies showed that the primitive follicle lumen appears in the region of a macula occludens, which has already been formed between two adjacent cells of the cell cords in an 8 day-old chick embryo, and the macula occludens becomes the zonula occludens when the primitive follicle lumen first forms (52). Thyroglobulin stored within the follicle lumen is isolated from the mesenchymal tissue even at the first appearance of the primitive cavities. The Golgi apparatus is located near the primitive follicle lumen to which the microvilli protrude (49). Thus, it is reasonable to conclude that the exocrine cell-like polarity of the apical pole of the thyroid cell is formed at the time when the primitive follicle lumen appears. By the invasion of the mesenchyme tissue, the cell cord is separated into some typical follicle unit structures at 14 days of incubation in the chick embryo (49), when the basal pole of the cell is formed and the cell polarity is completed (Fig. 4). Gap junctions are well developed between the thyroid follicle epithelial cells, as those in other endocrine as well as exocrine organs (1) and this might serve to synchronized cell activities. These structures already occur in the thyroid of the 7 day old embryo (52). The intimate communication between these cells already exists at the time of their functional differentiation. Recently, using cultured and isolated thyroid cells which lose their polarity, the factors necessary for the formation of the follicle structure have been studied. Aggregation and follicle reconstruction from the isolated follicles are induced by TSH (53, 54) and cAMP (55, 56) in the cultured medium. Recently it has been demonstrated that collagen is necessary for the differentiation of the basal part of the cell. Chambaud et al. clarified that, when the isolated cells are cultured on the surface of a collagen gel, the basal pole of the thyroid cell is formed in the area of contact between the cell membrane and collagen, while the apical pole is in contact with the cultured medium (57). Inoue et al. (41), using the inside-out follicles formed in the rotated suspension culture in the absence of TSH, reported that the normal follicle having a usual polarity is reformed in the presence of TSH.

References

1. Fujita H, Mishima H, Otsuka N: Freeze etching images of rabbit thyroid glands. Arch histol Jap 38: 275–284, 1975.
2. Tice LW, Wollman SH, Carter RC: Changes in tight junctions of thyroid epithelium with changes in thyroid activity. J Cell Biol 66: 657–663, 1975.
3. Fujita H: Fine structure of the thyroid gland. Int Rev Cytol 40: 197–280, 1975.
4. Whur P, Herscovics A, Leblond CP: Radioautographic visualization of the incorporation of galactose-^{3}H and mannose-^{3}H by rat thyroids *in vitro* in relation to the stages of thyroglobulin synthesis. J Cell Biol 43: 289–311, 1969.
5. Bouchilloux S, Chabaud O, Michel-Béchet M, Ferrand M, Athouel-Haon AM: Differential localization in thyroid microsomal subfractions of a mannosyltransferase, N-acetylglucosaminyltransferase and a galactosyltransferase. Biochem Biophys Res Commun 40: 314–320, 1970.
6. Haddad A, Smith MD, Herscovics A, Nadler NJ, Leblond CP: Radioautographic study of *in vivo* and *in vitro* incorporation of fucose-^{3}H into thyroglobulin by rat thyroid follicular cells. J Cell Biol 49: 856–882, 1971.
7. Bouchilloux S, Ferrand J, Gregoire J, Chabaud O: Localization in smooth microsomes from sheep thyroid of both a galactsyltransferase and a N-acetylhexosaminyltransferase. Biochem Biophys Res Commun 37: 538–544, 1969.
8. Sawano F, Fujita H: Cytochemical studies on the internal polarity of the Golgi apparatus and the relationship between this organelle and GERL. Histochemistry 71: 335–348, 1981.
9. Tamura S, Fujita H: Cytochemical localization of complex carbohydrates in the thyroid gland of normal and TSH-treated mice. Histochemistry 58: 57–64, 1978.
10. Nadler NJ, Young BA, Leblond CP, Mitmaker B: Elaboration of thyroglobulin in the thyroid follicle. Endocrinology 74: 333–354, 1964.
11. Fujita H: Outline of the fine structural aspects on the synthesis and release of the thyroid hormone. Gunma Symp Endocrinol 7: 49–63, 1970.
12. Doniach I, Logothetopoulos JH: Radioautographs of inorganic iodide in the thyroid. J Endocrinol 13: 65–69, 1955.
13. Fujita H, Nanba H: Fine structural localization of ATPase in the guinea pig thyroid. Histochemistry 40: 301–304, 1974.
14. Stein O, Gross J: Metabolism of ^{125}I in the thyroid gland studies with electron microscopic autoradiography. Endocrinology 75: 787–798, 1964.
15. Ibrahim MS, Budd GC: An electron microscopic study of the site of iodine binding in the rat thyroid gland. Exp Cell Res 38: 50–56, 1965.
16. Fujita H: Studies on the iodine metabolism of the thyroid gland as revealed by electron microscopic autoradiography of ^{125}I. Virchows Arch Abt B Cellpathol 2: 265–279, 1969.
17. Fujita H: Morphological aspects on the site of iodination of thyroglobulin in the thyroid gland. Arch histol jap 34: 109–141, 1972.
18. Nadler NJ: The application of radioautography to the localization of the binding of iodine to thyroglobulin in rat thyroid follicles. Anat Rec 169: 384–385, 1971.
19. Fujita H: Application of electron microscopic autoradiography of radioactive iodine for a comparative study of the thyroid gland. In: Recent progress in electron microscopy of cells and tissues. Yamada E, Mizuhira V, Kurosumi K, Nagano T (eds). Tokyo, Igaku-shoin, 1976, pp 175–188.
20. Nakai Y, Fujita H: Fine structural localization of peroxidase in the rat thyroid. Z Zellforsch 107: 104–110, 1970.
21. Strum JM, Karnovsky MJ: Cytochemical localization of endogeneous peroxidase in thyroid follicular cells. J Cell Biol 44: 655–666, 1970.
22. Sawano F, Fujita H: Some findings on the cytochemistry of the thyroid follicle epithelial cell in rats and mice. Arch histol jap 44: 439–452, 1981.
23. Tice LW, Wollman SH: Ultrastructural localization of peroxidase activity on some membranes of the typical thyroid epithelial cell. Lab Invest 26: 63–73, 1972.
24. Björkman U, Ekholm R, Denef JF: Cytochemical localization of hydrogen peroxide in isolated thyroid follicles. J Ultrastr Res 74: 105–115, 1981.
25. Seljelid R: Endocytosis in thyroid follicle cells. II. A microinjection study of the origin of colloid droplets. J Ultrastr Res 17: 401–420, 1967.
26. Seljelid R, Reith A, Nakken KF: The early phase of endocytosis in rat thyroid follicle cells. Lab Invest 23: 595–605, 1970.
27. Pasta I, Wollman SH: Colloid droplet formation in dog thyroid *in vitro*. J Cell Biol 35: 262–266, 1967.
28. Nève P, Dumont JE: Effects *in vitro* of thyrotropin, cyclic 3′, 5′-AMP, dibutyryl cyclic 3′-5′-AMP, and prostaglandine E_1, on the ultrastructure of dog thyroid slices. Exp Cell Res 63: 285–292, 1970.
29. Gilman AG, Rall TW: Factors influencing adenosine 3′, 5′-phosphate accumulation in bovine thyroid slices. J Biol Chem 243: 5867–5871, 1968.
30. Miyagawa J, Ishimura K, Fujita H: Fine structural studies on the reabsorption of colloid and fusion of colloid droplets in thyroid glands of TSH-treated mice. Cell Tiss Res (in press).
31. William JA, Wolff J: Colchicine-binding protein and the secretion of thyroid hormone. J Cell Biol 54: 157–165, 1972.
32. Nève P, Ketelbant-Balasse P, Willems C, Dumot JE: Effect of inhibitors of microtubules and microfilaments on dog thyroid slices *in vitro*. Exp Cell Res 74: 227–244, 1972.
33. Ekholm R, Ericson LE, Joseffson J-O, Melander A: *In vivo* action of vinblastine on thyroid ultrastructure and hormone secretion. Endocrinology 94: 641–649, 1974.
34. Zeligs JD, Wollman SH: Ultrastructure of erythrophagocytosis and red cell fission by thyroid epithelial cells *in vivo*. J Ultrastr Res 59: 57–69, 1977.
35. Ishimura K, Okamoto H, Fujita H: Freeze-etching observations on the characteristic arrangement of intramembranous particles in the apical plasma membrane of the thyroid follicular cell in TSH-treated mice. Cell Tiss Res 171: 297–303, 1976.
36. Orci L, Carpentier J-L, Perrelet A, Anderson RGW, Goldstein GL, Brown MS: Occurrence of low density lipoprotein receptors within large pits on the surface of human fibroblasts as demonstrated by freeze-etching. Exp Cell Res 113: 1–13, 1978.
37. Miquelis R, Simon C: The thyroid lysosomal system: dynamic state of the organelles in relation to iodine release. Eur J Cell Biol 24: 70–73, 1981.
38. Fujita H, Ishimura K, Matsuda H: Freeze-fracture images on filipin-sterol complexes in the thyroid follicle epithelial cell of mice with special regard to absence of cholesterol at the site of micropinocytosis. Histochemistry 73: 57–63, 1981.
39. Nakajima Y, Bridgman PC: Absence of filipin-sterol complexes from the membranes of active zones and acetylcholine receptor aggregates at frog nonmuscular junctions. J Cell Biol 88: 453–458, 1981.
40. Kurosumi K: Electron microscopic analysis of the secretion mechanism. Int Rev Cytol 11: 1–124, 1961.
41. Inoue K, Horiuchi R, Kondo Y: Effect of thyrotropin on cell orientation and follicle reconstruction in rotated suspension culture of hog thyroid cells. Endocrinology 107: 1162–1168, 1980.
42. Herzog V, Miller F: Structural and functional polarity of inside-out follicles prepared from pig thyroid gland. Eur J Cell Biol 24: 74–84, 1981.

43. Herzog V, Miller F: Membrane retrieval in epithelial cells of isolated thyroid follicles. Eur J Cell Biol 19: 203–215, 1979.
44. Fujita H, Murakami T: Scanning electron microscopy on the distribution of the minute blood vessels in the thyroid gland of the dog, rat, and rhesus monkey. Arch histol jap 36: 181–188, 1974.
45. Ishimura K, Okamoto H, Fujita H: Freeze-etching studies on ultrastructural changes of endothelial cells in the thyroid of normal, TSH-treated and Thyradin-treated mice. Cell Tiss Res 175: 297–313, 1976.
46. Fujita H, Honma Y: Some observations on the ammocoetes of Lampetra japonica. Gen Comp Endocrinol 11: 111–131, 1968.
47. Fujita H, Shinkawa Y: Electron microscopic studies on the thyroid gland of the hagfish, Eptatretus burgeri. (A part of phylogenetic studies of the thyroid gland). Arch histol jap 37: 277, 1975.
48. Hilfer SR: Follicle formation in the embryonic chick thyroid. I. Early morphogenesis. J Morphol 115: 135–151, 1964.
49. Fujita H, Tanizawa Y: Electron microscopic studies on the development of the thyroid gland of chick embryo. Z Anat Entw Gesch 125: 132–151, 1966.
50. Calvert R, Pusterla A: Formation of thyroid follicular lumina in rat embryos studied with serial fine sections. Gen Comp Endocrinol 20: 584–597, 1973.
51. Tice LW, Carter RL, Cahill MC: Tracer and freeze fracture observations on developing tight junctions in fetal rat thyroid. Tissue Cell 9: 395–417, 1977.
52. Ishimura K, Fujita H: Development of cell-to-cell relationships in the thyroid, gland of the chick embryo. Cell Tiss Res 198: 15–25, 1979.
53. Fayet G, Michel-Béchet M, Lissitzky S: Thyrotropin-induced aggregation and reorganization into follicles of isolated porcine thyroid cells. 2. Ultrastructural studies. Eur J Biochem 24: 100–111, 1971.
54. Lissitzky S, Fayet G, Giraud A, Verrier B, Torresani J: Thyrotropin-induced aggregation and reorganization into follicles of isolated porcine-thyroid cells. 1. Mechanism of action of thyrotropin and metabolic properties. Eur J Biochem 24: 88–99, 1971.
55. Fayet G, Lissitzky S: Cyclic 3′, 5′-adenosine monophosphate mediated follicular reorganization of isolated thyroid cells in culture. Fed Eur Biochem Sco Letter 11: 185–188, 1970.
56. Winand RJ, Kohn LD: Thyrotropin effects on thyroid cells in culture. Effects of trypsin on the thyrotropin receptor and on thyrotropin-mediated cyclic 3′, 5′-AMP changes. J Biol Chem 250: 6534–6540, 1975.
57. Chambaud M, Galrion J, Mauchamp J: Influence of collagen gel on the orientation of epithelial cell polarity: follicle formation from isolated thyroid cells and from preformed monolayers. J Cell Biol 91: 157–166, 1981.

Author's address:
Osaka University
Medical School, Nakanoshima, Kitaku,
Osaka, 530, Japan

CHAPTER 24

Thyroid parafollicular cells: Ultrastructural and functional correlations

LARS E. ERICSON and FRANK SUNDLER

1. Introduction

The existence of a second plasma calcium regulating hormone besides the parathyroid hormone was demonstrated in 1962 (1). The thyroid origin of this hormone, calcitonin, was rapidly established. The detection of calcitonin initiated studies in many laboratories in order to characterize its structure and its biological activities. Calcitonin is a polypeptide consisting of 32 amino acid residues. The most thoroughly studied effect of calcitonin has been its ability to lower plasma calcium concentrations by suppressing bone resorption (2).

There is convincing evidence that calcitonin originates from the thyroid parafollicular, or C cells. The existence of parafollicular cells distinguishable from follicle cells was suggested more than 100 years ago. They were later described by several investigators and several names and putative functions were attributed to them. An important contribution was that of Nonidez (3) who described in the dog thyroid large epithelial 'parafollicular cells' containing argyrophilic granules and regarded by him as endocrine cells. The linking of parafollicular cells to calcitonin (4) renewed the interest for these cells and they have now been ultrastructurally characterized in a large number of species (5). With the availability of techniques for the histochemical demonstration of monoamines, it was soon established that the 'parafollicular cells' have the ability to form and store 5-hydroxytryptamine (5-HT) or dopamine either endogenously or after administration of the corresponding precursor amino acid (6). The parafollicular cells became one of the first members of the now large, and still growing, family of cells sharing the properties of producing polypeptide hormones and forming monoamines and often referred to as APUD cells, an acronym that stands for *a*mino and *p*recursor *u*ptake and *d*ecarboxylation (7).

2. Origin and development of the parafollicular cells

The calcitonin-producing parafollicular cells originate from the ultimobranchial body, in most species a derivative from the fourth branchial pouch, which in mammals merges during embryological life with the median thyroid anlage from which the majority of the follicle cells arise. The ultimobranchial body also contains other types of epithelial cells which in the adult thyroid form follicles or ducts composed of cells ultrastructurally different from follicle cells. This second kind of follicles is sometimes associated with parafollicular cells (8). Not all calcitonin producing cells follow the ultimobranchial bodies into the thyroid but some may be found also in other derivates of the endoderm of the branchial pouches as parathyroid IV and thymus IV (8). In non-mammalian vertebrates the ultimobranchial body remains as an organ separated from the thyroid throughout life and in such animals the ultimobranchial body is a rich source of calcitonin and contains cells ultrastructurally and functionally very similar to mammalian parafollicular cells (9). There is convincing evidence that in birds calcitonin cells, which like their counterparts in mammals form and store monoamines (9, 10), are derived from the neural crest (10). There is evidence that this is the case also in mammals (11). In the mouse precursors of parafollicular cells follow a multistep pathway during fetal development first migrating into the fourth branchial arch mesenchyme, known to be derived mainly from the neural crest, and then into the fourth branchial pouch endoderm from which the ultimobranchial body arises (11). An origin from neural crest was originally suggested for all APUD cells (7). However, at present there is convincing evidence for a neural crest origin for only a limited number of APUD cells, including the parafollicular cells.

Motta, PM (ed): Ultrastructure of endocrine cells and tissues. ISBN-13: 978-1-4613-3863-5

3. Intrathyroidal distribution of parafollicular cells

A large number of staining methods, including various types of silver stains, masked metachromasia and lead hematoxylin, have been used for the light microscopic demonstration of parafollicular cells (5). The two most specific and now widely used methods are the immunohistochemical demonstration of calcitonin (Fig. 1) or the demonstration of monoamines by fluorescence microscopy, often after administration of the monoamine precursors (Fig. 2). As shown in Figure 1 the parafollicular cells are concentrated in the central portions of the lobes, leaving the peripheral parts and often also the isthmus devoid of parafollicular cells. This concentration to central parts of the lobe is even more pronounced during development and probably reflects the way the parafollicular cells are brought into the gland by fusion of the ultimobranchial body with the median thyroid anlage (8).

As shown in Figures 3–5 the relation between parafollicular cells and follicle cells differs considerably in different species. The parafollicular cells may be situated predominantly in the follicle wall (Fig. 3; pig), predominantly in groups between the follicles (Fig. 4; cat) or in both these locations (Fig. 5; rat).

The relative number of parafollicular cells as compared to follicle cells varies not only in different parts of the thyroid lobe but also with species and with the age of the individual. In the 3 months old rat the parafollicular cells occupy about 4% of section area of a midlongitudinal section of a lobe (the corresponding figure for the follicle cells is about 30%) but this figure increases with age (12). In the sheep thyroid the parafollicular cells appear to be particularly abundant and occupy on average about 20% of the total volume of the follicle cells; in central portions of the lobe the relative volume of parafollicular cells is considerably higher, about 50% (13).

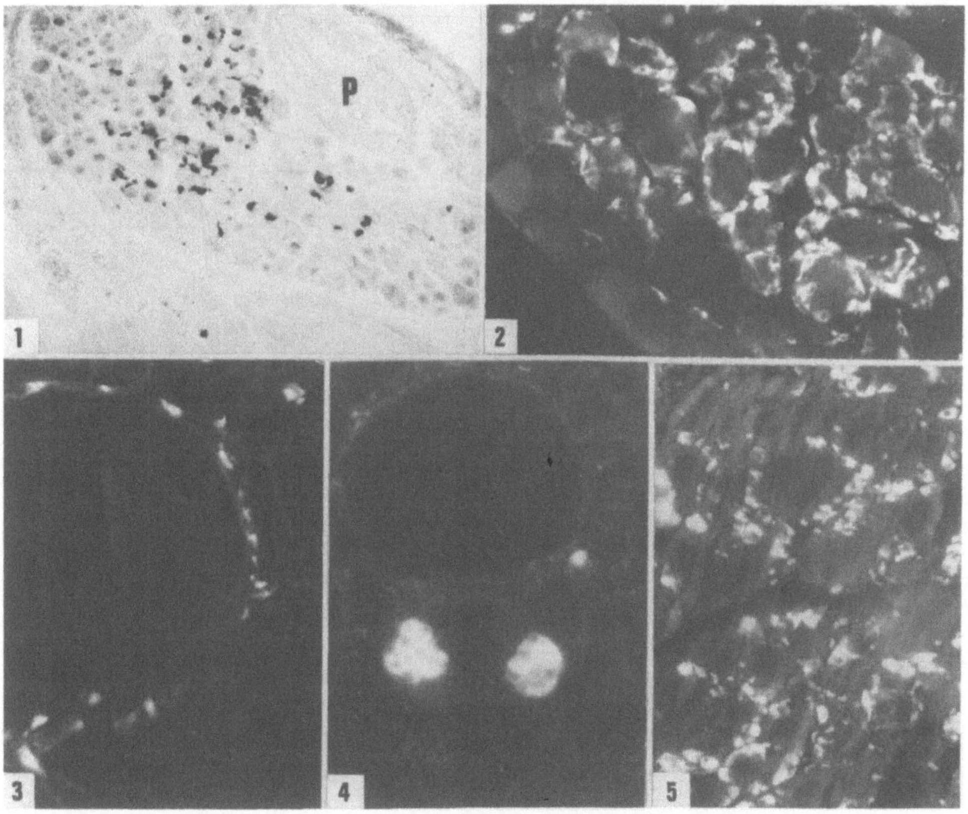

Figs. 1–5. (1) Section through a thyroid lobe of a one week old rat. Section stained for calcitonin using the immunoperoxidase (PAP) technique. The densely stained calcitonin-containing parafollicular cells are concentrated in the central portion of the lobe. The parathyroid (P) is embedded in thyroid tissue (× 85). (2) Dopamine fluorescence in thyroid parafollicular cells in a mouse injected with L-DOPA (100 mg/kg bw) 1 h before sacrifice (× 135). (3–5) Cells displaying calcitonin immunofluorescence in the thyroid of pig (3) (× 255), cat (4) (× 170) and rat (5) (× 170). In the pig (3) calcitonin cells are distributed predominantly as single cells within the follicle epithelium, whereas in the cat (4) the cells occur in clusters among the follicles. In the rat (5) calcitonin cells occur both as single cells and clusters.

Also hibernators appear to have relatively large numbers of parafollicular cells (5). In the adult human thyroid the parafollicular cells seem to be much more sparse than in most animals examined. However, their number as well as the thyroid calcitonin content is appreciably higher in the neonatal period than in the adult gland (14).

4. Ultrastructure of the parafollicular cells

During the last decades the ultrastructure of parafollicular cells has been described in a very large number of species (5). In the earliest ultrastructural studies on the thyroid these cells were to a great extent overlooked. One reason for this is that the parafollicular cells, especially their granules, are poorly preserved after osmium tetroxide fixation by immersion alone which was the only fixative used in these early studies. An adequate fixation of parafollicular cells requires fixation with glutaraldehyde, preferentially by perfusion, followed by immersion in osmium tetroxide (5, 15).

The general ultrastructure of parafollicular cells is very similar in all species examined (5). Thus, the description given here, can probably be generalized to most mammalian species, including human parafollicular cells (16).

4.1. Normal ultrastructure

The parafollicular cells most often occur in groups which often appear as outbuddings from the thyroid follicle. However, as described in section 3, in many species a significant proportion of the parafollicular cells is located in between the follicle cells in the follicle wall. In this location they never reach the follicle lumen, although the rim of follicle cell cyto-

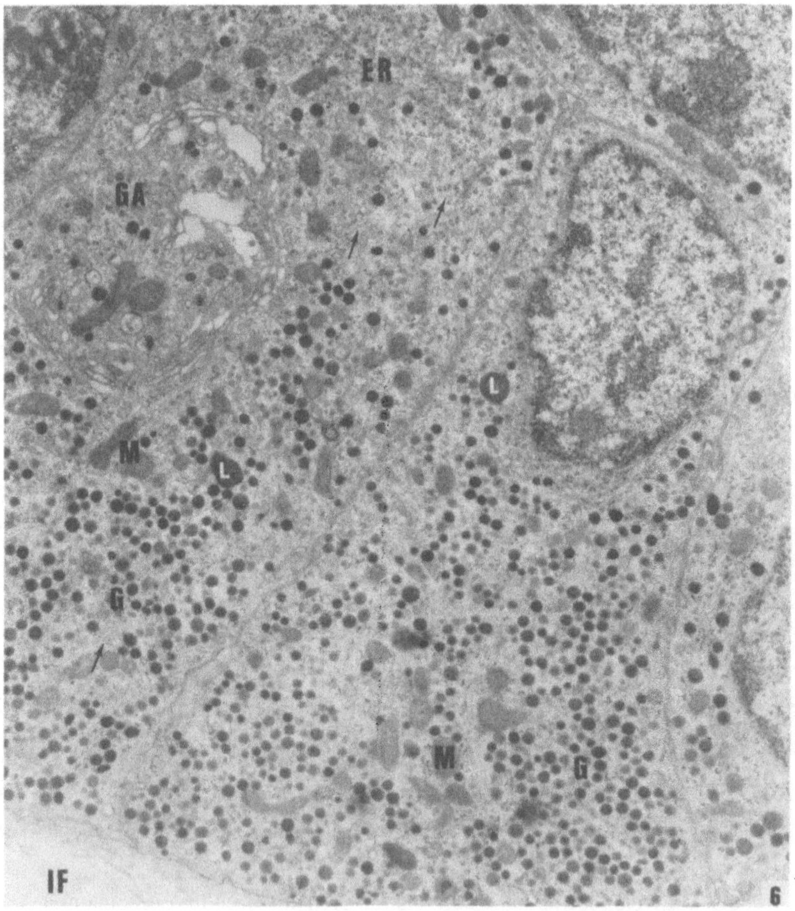

Fig. 6. Rat thyroid parafollicular cells. The cytoplasm contains numerous, dense granules (G) distributed towards the interfollicular space (IF). The part of the cytoplasm devoid of granules is occupied by a Golgi apparatus (GA) and rough endoplasmic reticulum (ER). In the Golgi apparatus granules in various stages of maturation are present. Mitochondria (M), some of which are labelled, are distributed in all parts of the cell. Lysosomes (L) and microtubules (arrows) are also present (× 9,800).

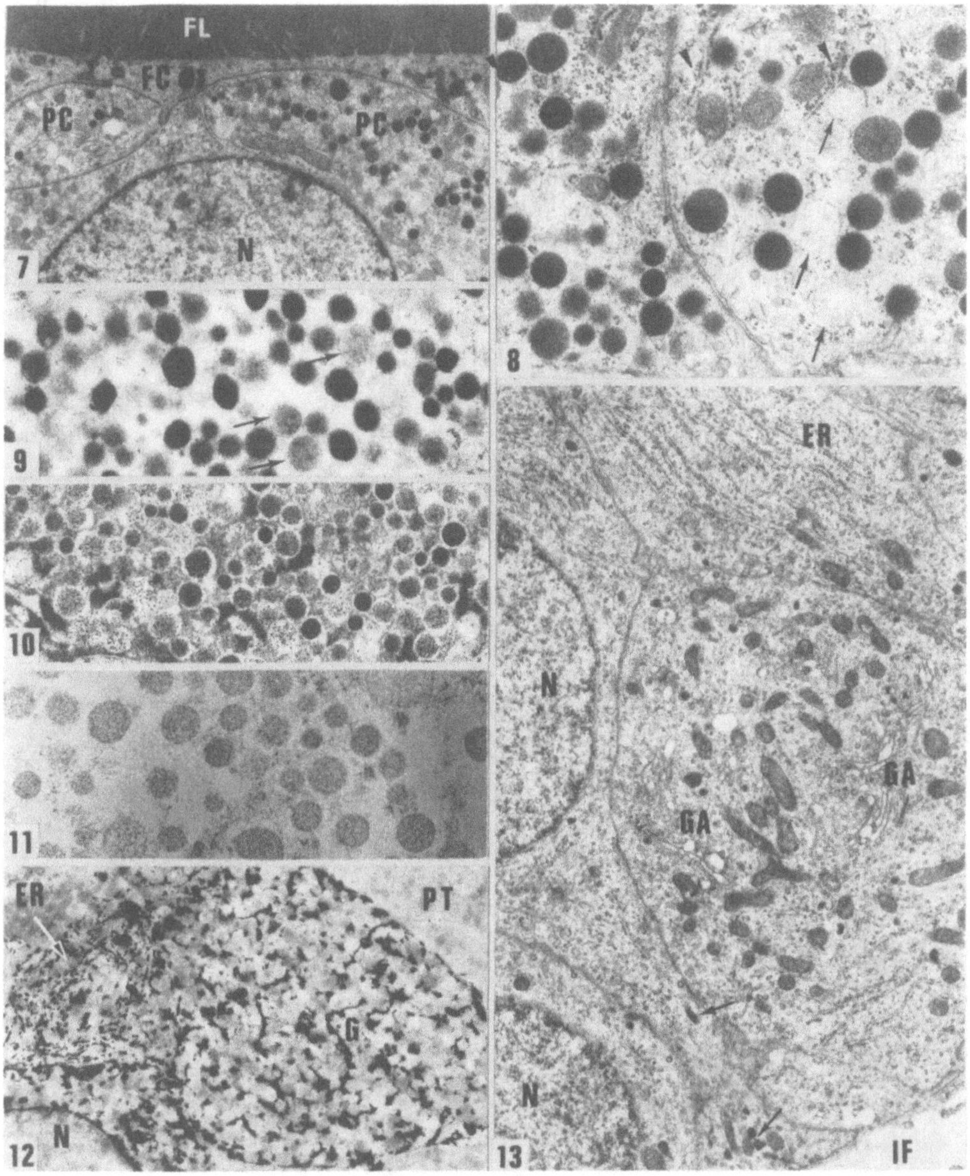

Figs. 7–13. (7) Rabbit thyroid. Only a thin rim of follicle cell (FC) cytoplasm separates the parafollicular cells (PC) from the follicle lumen (FL). N: nucleus of parafollicular cell (× 6,800). (8) Portions of two rabbit thyroid parafollicular cells. Besides the dense cytoplasmic granules, others with a content of low density are present (arrows). Free ribosomes as well as single cisternae of rough endoplasmic reticulum (arrow-heads) are located between the granules (× 20,400). (9–11) Sheep thyroid parafollicular cells. These cells contain 5-HT located in the cytoplasmic granules. (9) Conventional glutaraldehyde-osmium tetroxide fixation. Many of the granules have a very dense content while others have a content of fairly low density. (arrows) (× 17,850). (10) Incubated for the demonstration of argyrophil reaction according to Grimelius. Silver deposits are mainly located in the dense granules (× 17,000). (11) Incubated for the demonstration of argentaffin reaction. Finely granular silver deposits in the granules indicates the presence of a reducing substance, in this case 5-HT (× 19,550). (12) Internal rabbit parathyroids incubated for the demonstration of acetylcholinesterase activity according to Lewis ans Shute. Section unstained. Reaction product is present in nuclear (N) membrane, rough endoplasmic reticulum (ER) and cell membrane. The cytoplasmic granules (G) are unreactive but reaction product is located between the granules probably in cisternae of rough endoplasmic reticulum (cf. Fig. 8). The parathyroid chief cell (PT) is unreactive. An identical cytochemical reaction can also be demonstrated in cells located in the thyroid (× 5,950). (13) Thyroid parafollicular cells in a rat injected with vitamin D_2 (250,000 units/d) for 5 days. Only a few, small granules (arrow) remain in the cytoplasm. The rough endoplasmic reticulum (ER) and Golgi areas (GA) occupy a large portion of the cytoplasm. N: nuclei; IF: interfollicular space (× 11,900).

plasm separating the parafollicular cells from the lumen is often very narrow (Fig. 2). When occurring in clusters parafollicular cells have a polyhedral shape while those located single in the follicle wall are spherical or oval. In some species the parafollicular cells have been observed occasionally to form small follicles. The basement membrane of the parafollicular cells is continuous with that of the follicle cells. Desmosomes are often found between parafollicular cells but are rare, and in several species absent (15), between parafollicular cells and follicle cells. Gap junctions between parafollicular cells were not observed in a study using freeze-fracture technique (17). The nucleus is spherical or oval (Figs. 6, 7) and tends to be larger than that of the follicle cells.

The most characteristic feature of the parafollicular cell is the spherical cytoplasmic granules which in the adult mouse occupy about 40% of the cytoplasmic volume and which in general are accumulated in the part of the cytoplasm located closest to the interfollicular space (Fig. 6). The diameter of the granules varies somewhat between species but their

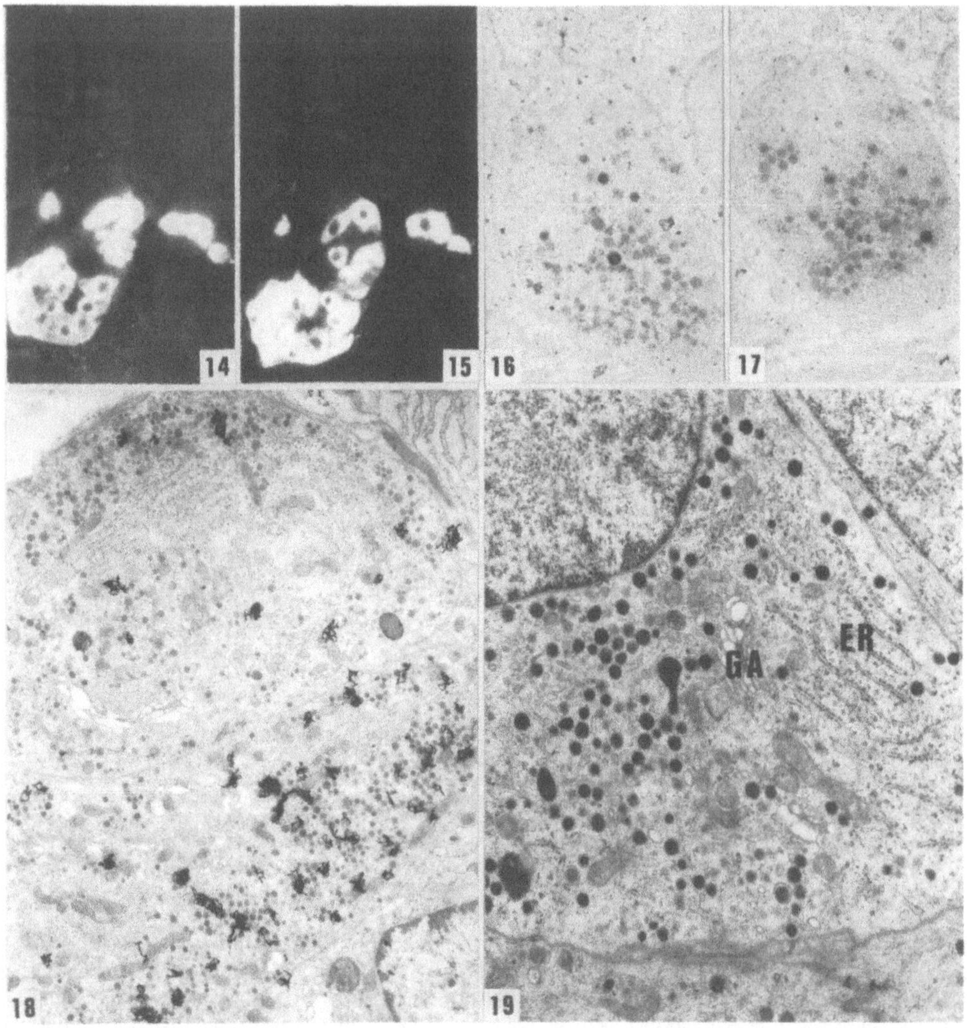

Figs. 14–19. (14) Rabbit thyroid parafollicular cells in two consecutive, 1 μm thick, Epon sections immunostained for calcitonin (14) and somatostatin (15). Most parafollicular cells contain both immunoreactive calcitonin and somatostatin (× 340). (16–17) Rabbit thyroid parafollicular cells. Consecutive sections were incubated for the immunocytochemical demonstration of calcitonin (16) and somatostatin (17) (immunoperoxidase technique) in the same parafollicular cell. Virtually all granules are stained with both antisera indicating the presence of immunoreactive calcitonin and somatostatin in the same granules (× 3,400). (18) Electron microscopic autoradiography. Parafollicular cells in a thyroid of a mouse injected with ^{3}H-5-hydroxytryptophan 4 h before fixation. The autoradiographic silver grains are mainly located over the cytoplasmic areas containing cytoplasmic granules. (19) Human medullary carcinoma. Dense, cytoplasmic franules similar to those present in normal parafollicular cells are scattered in the cytoplasm. A Golgi area (GA) and rough endoplasmic reticulum (ER) are also present. The narrow cisternae of the rough ER resemble those in non-malignant parafollicular cells (× 11,000).

average diameter is in the range between 100 nm-200 nm (Figs. 6–12). In adequately fixed tissue the majority of the granules have a dense core, in some species separated from the limiting membrane by a less dense rim (Fig. 6). Granules containing a monoamine (see section 6) tend to have in general a very dense content (Fig. 9). However, in most species, granules of a larger size and with a less dense content are also present (Figs. 7–9). If the different types of granules reflect a different composition of the granule content or are merely a fixation artifact remains to be elucidated (see further section 5). It is generally assumed that the granules release their content into the extracellular space by exocytosis although images indicating a fusion between the plasma membrane and the granule membrane are rarely encountered during normal conditions. The granules are argyrophilic (Fig. 10) and when containing a monoamine they are argentaffin (Fig. 11) or chromaffin.

The cytoplasmic regions free of granules are occupied by rough endoplasmic reticulum, numerous free ribosomes and by an often prominent Golgi area (Fig. 6). The rough endoplasmic reticulum consists of narrow, flattened cisternae (Fig. 13), similar in structure to that found in other polypeptide-producing cells and distinctly different from the dilated cisternae of the follicle cells. In addition, free ribosomes and single cisternae are present in the cytoplasm in between granules (Fig. 8). The Golgi apparatus (Fig. 6), described in detail elsewhere (15), is prominent and displays a close relation to granules in that intermediates are found between Golgi saccules, containing varying amounts of dense material, and mature granules (Fig. 6).

Lysosomes are present although in an appreciably lower number than in the follicle cells (Fig. 6). The slender mitochondria do not display any particular features (Fig. 6). The cytoplasm also contains numerous microtubules (Fig. 6) which appear to increase in number when the cells are activated. Filaments with a diameter of about 6 nm form, as in most other endocrine cells, a more or less distinct layer beneath the parts of the plasma membrane facing the interfollicular space. The parafollicular cells also often contain bundles of 10 nm filaments (15). Both microtubules and 6 nm filaments have been implicated in the secretory process in many endocrine and exocrine cells, but there is not firm evidence for such a function in parafollicular cells.

4.2. Experimental changes in the ultrastructure

Calcitonin is secreted in response to hypercalcemia and a direct proportional relationship has been found between plasma calcium concentration and secretion rate (2). Calcitonin secretion is also stimulated by a number of other agents including dibutyryl cyclic AMP, β-adrenergic agonists and certain polypeptide hormones. Particular attention has been paid to gastrin and its rôle in stimulation of calcitonin release for the prevention of postprandial hypercalcemia (2).

Acute hypercalcemia has been reported to induce a transient decrease in the number of secretory granules in parafollicular cells. Long-term hypercalcemia, induced by pharmacological doses of vitamin D, leads to a progressive depletion of granules as well as thyroid calcitonin content and after 4–5 days only scattered, small, dense granules are present (18–20; Fig. 13). Similar changes are induced in calcitonin cells located in the bird ultimobranchial body (9). At the same time the parafollicular cells display signs of increased protein synthesis such as an increase in rough endoplasmic reticulum and free ribosomes and enlargement of Golgi areas (Fig. 13), indicating an activation of protein synthesis to compensate for the increased demand of calcitonin.

Chronic hypocalcemia induced by parathyroidectomy causes a progressive rise of thyroid calcitonin content and increase in the number of parafollicular cells which appear to contain an increased number of granules (21).

In addition, ultrastructural alterations have been described in parafollicular cells following administration of various substances, including calcitonin, cortison, propylthiouracil, and TSH. However, the lack of quantitative information and/or not optional quality of tissue preservation in these studies make it difficult to evaluate the significance of these observations.

4.3. Ultrastructure of parafollicular cells in hibernating animals

The parafollicular cells are particularly abundant in hibernators. They undergo drastic ultrastructural changes during the different stages of the yearly cycle of hibernation (5). During the active phase of the cycle the parafollicular cells have an ultrastructure similar to that of ther mammals. In the prehibernation phase ultrastructural changes, such as accumulation of secretory granules and appearance of intracisternal granules in the rough endoplasmic reticulum, suggest a slow-down in function. At the beginning of hibernation the parafollicular cells undergo degranulation, probably by crinophagy. The granules reappear again just before the arousal phase and during this phase signs of exocytosis are frequent (5).

4.4. Ultrastructure of parafollicular cells during development

The ultrastructure of the parafollicular cells in developing thyroid gland has been studied in several species including man, dog, rat, and mouse (5). Typical cytoplasmic granules, as well as the ability to form and store monoamines, appear relatively early during development. For instance in the mouse this seems to happen at about the 14th day of gestation (11). Granules start to form in association with the Golgi apparatus and then become more widely distributed in the cytoplasm. In the early development parafollicular cells appear to have more rough endoplasmic reticulum than at later stages. By the end of the fetal life the parafollicular cells resemble those of adult animals. During the early postnatal period ultrastructural signs of exocytosis are frequent and the parafollicular cells contain fewer granules. This observation, together with a rapid decrease of thyroid calcitonin levels after birth, suggests that the parafollicular cells start to secrete actively during the neonatal period (5).

5. Intracellular location of polypeptide hormones

The concomitant depletion of granules in the parafollicular cells and of calcitonin from the thyroid gland caused by chronic hypercalcemia indicates that the granules represent the storage site of calcitonin. This is supported by results of subcellular fractionation studies (22). As mentioned above, the parafollicular cells constitute in most species only a small fraction of the thyroid gland. However, in central portions of the sheep thyroid the relative volume of parafollicular cells is very high, about 50% of that of follicle cells (13). Taking advantage of this circumstance, it has been possible to isolate a fraction, using a combination of continuous and discontinuous sucrose gradients, which contains more than 90% of the biologically active calcitonin present in the original homogenate (22). This fraction consists to large extent of granules identical to those found in intact parafollicular cells, indicating that the granules represent the storage site of calcitonin and that only small amounts of calcitonin are normally present in other cellular compartments (22). The location of calcitonin to cytoplasmic granules can also be demonstrated by immunocytochemistry (Fig. 16).

In addition to calcitonin, the presence of immunoreactive somatostatin has been demonstrated in parafollicular cells of several mammalian species including man (23–27). In the rabbit thyroid virtually all calcitonin cells contain immunoreactive somatostatin throughout life (Figs. 14, 15, 25) whereas in the rat somatostatin immunoreactive cells are frequent only during the first weeks postnatally when also thyroid somatostatin content is high (26). In the adult rat only a few somatostatin immunoreactive cells located centrally in the thyroid lobe remain (26). The dog thyroid, contrary to the thyroid of rats and rabbits, contains somatostatin immunoreactive cells which seem to be distinct from those storing calcitonin (27). It has been pointed out that the parafollicular cells in rat and rabbit thyroids which contain somatostatin are characterized by granules, which are larger and less dense (see Figs. 7, 8) than granules present in parafollicular cells containing calcitonin alone (25, 26). Ultrastructurally the predominating granules in somatostatin containing parafollicular cells resemble those present in somatostatin containing cells in pancreatic islets and gastrointestinal tract (25, 26). Our own observations on the rabbit thyroid using electron microscopic immunocytochemistry indicate that calcitonin and somatostatin are stored in the same granules and that all granules are stained irrespective of their morphology (Fig. 16, 17).

In conclusion, it is well established that somatostatin is present in a population of parafollicular cells of several species. At least in the rabbit somatostatin and calcitonin occur in the same granule. This suggests that the two polypeptides are released together upon stimulation of the parafollicular cells.

6. Intracellular location of monoamines

The parafollicular cells of bat, goat, sheep, horse, and callithricid primates (5, 28) as well as the corresponding cells in the ultimobranchial body of the duck (29) contain 5-HT while dopamine is present in bovine parafollicular cells (29) and in calcitonin cells in the chicken ultimobranchial body (9). However, in many other species an amine cannot be demonstrated unless the immediate precursor, L-5-hydroxytryptophan (L-5-HTP) or L-dihydroxyphenylalanine (L-DOPA), is administered (6, 7, 28). These amino acids are taken up by the parafollicular cells, decarboxylated to the corresponding amine which is then stored in the cytoplasm by a mechanism sensitive to reserpine. The thyroid contains high levels of aromatic amino acid decarboxylase, operative in the formation of monoamines from their precursors, and this enzyme is distributed within the gland in the same way as the parafollicular cells (30). The amines formed are metabolized by monoamine oxidase (5, 28, 29). The amines themselves are very poorly taken up by the parafollicular cells (5, 28, 29).

Ultrastructural studies using methods based on the reduction and precipitation of heavy metals, such as the chromaffin and argentaffin reactions (Fig. 11), indicate that the monoamines are stored in the same granules that contain calcitonin. Studies on the sheep thyroid using cell fractionation technique (22) also demonstrate that endogenous 5-HT is almost exclusively located in the granules.

The intracellular location of 5-HT or dopamine formed from the labelled 5-HTP or L-DOPA can be quantitatively evaluated by electron microscopic autoradiography (Fig. 18; 31–33). There is experimental evidence that the autoradiographic silver grains represent the labelled amine which is retained to a large extent in the tissue following fixation with glutaraldehyde (32). As seen in Figure 18 the autoradiographic grains are located almost exclusively over cytoplasmic regions containing granules and quantitative evaluation indicates that the concentration of label over granules is at least 10–12 times higher than over any other cytoplasmic constituent (31–33). The mechanism by which the monoamine is bound within the granules is poorly understood. Recently, the presence of a high-affinity 5-HT binding protein in sheep parafollicular cells has been reported (13, 34) suggesting that the amine may be bound to a granular component which is distinct from the polypeptide hormone.

The functional significance of intragranular amines in parafollicular cells is not known. One possibility is that the amine influences the storage or release of the polypeptide hormone. Such an effect of intracellular DA and 5-HT has been observed in other cells, as for instance pancreatic beta cells also storing a monoamine and a polypeptide hormone in the same granules (29). Another possibility is that the amine is released together with the polypeptide hormone to act locally on other cells (35). Thus, for instance it is well known that 5-HT and dopamine stimulate endocytosis and thyroid hormone release in follicle cells (28).

The parafollicular cells in many species display a prominent cholinesterase activity (Fig. 12). The reaction product is confined to the nuclear membrane, rough endoplasmic reticulum and the plasma membrane but is not present in the granules. The significance of this enzyme, also present in many other polypeptide-producing cells (6), is unknown.

7. Relation between parafollicular cells and follicle cells

The mere fact that parafollicular cells are located within the thyroid and often in direct contact with the follicle cells suggests that these cell types are in some way functionally related. Both cell types are also supplied by the same capillary system (8). There are several reports that alterations in the thyroid follicle cell function may influence the number and morphology of the parafollicular cells and also that TSH may directly influence their morphology. Many of these observations are not supported by any quantitative data. Further, it cannot be excluded that some of these findings are secondary to for instance changes in the calcium metabolism caused by the altered follicle cell function. Conversely, an influence of the parafollicular cells on the follicle cell is also possible. In the previous section the possible effect of 5-HT released from parafollicular cells on endocytosis in follicle cells was mentioned (35). Although calcitonin does not seem to have an effect on follicle cells it is possible that somatostatin released from parafollicular cells may influence follicle cells. In fact there are several reports that infusion of somatostatin inhibits iodothyronine release in man and laboratory animals. However, at present no unifying hypothesis on the functional relation between parafollicular cells and follicle cells can be put forward. The varying content of for instance somatostatin and monoamines in parafollicular cells suggest that such a relation is dependent on species.

8. Relation between parafollicular cells and autonomic nerves

The thyroid is supplied with adrenergic and cholinergic nerves (28). Also peptidergic nerves are present and in the mouse thyroid for instance vasointestinal polypeptide (VIP); substance P and gastrin-cholecystokinin have a neuronal localization (36; Sundler, unpublished observations). Ultrastructurally nerve terminals are found in close relation to epithelial cells, including parafollicular cells (5, 28), but there seems not to be a more abundant innervation of parafollicular cells than of follicle cells. Autonomic nerves are probably of importance in the regulation of hormone formation and secretion in follicle cells (28) and it appears quite possible that they may have a similar rôle in the regulation of calcitonin secretion. In support of this possibility are observations that gastrin, acetylcholine as well as norepinephrine, which are all present in thyroid nerves, can influence calcitonin secretion (2).

9. Medullary carcinoma

It is now well established that thyroid medullary carcinoma originates from the parafollicular cells (37). In man medullary carcinoma occurs both sporadically and as a genetic disorder with an autosomal dominant mode of inheritance. The familial form is often accompanied by phaeochromocytoma, parathyroid disorders and multiple neuroma. In family members at high risks of developing a medullary carcinoma a spectrum of parafollicular cell abnormalities ranging from hyperplasia to invasive carcinomas has been observed (16). Medullary carcinomas may also occur in certain animal species. In the rat a stage of parafollicular cell hyperplasia precedes the appearance of tumour (12). Medullary carcinoma cells have an ultrastructure which in many aspects is similar to that of normal parafollicular cells including the presence of numerous dense cytoplasmic granules and a rough endoplasmic reticulum consisting of narrow cisternae (Fig. 19). As in normal parafollicular cells the granules may be either fairly small with a dense content or larger with a less dense core. In addition, different types of cells within the same tumour have been distinguished by the ultrastructure of the cytoplasmic granules (38). Besides calcitonin medullary carcinomas often contain a number of other substances, including polypeptides like ACTH and somatostatin as well as monoamines, and it has been suggested that cells containing such substances may differ ultrastructurally from cells containing calcitonin (38). It should also be pointed out that calcitonin may occur also in other foregut endocrine tumours, such as bronchial carcinoids and pancreatic endocrine tumours (39).

10. Concluding remarks

It is well established that the thyroid parafollicular cells secrete calcitonin. This hormone, as well as several other polypeptide hormones, is also produced in medullary carcinomas which originates from the parafollicular cells. Recent studies have revealed that parafollicular cells in certain species contain somatostatin. More work is needed to clarify the cellular and intracellular location of somatostatin in different species as well as to clarify its function in the thyroid gland.

The parafollicular cells form monoamines, like 5-HT and dopamine, which are stored together with calcitonin in the specific, cytoplasmic granules. At present it is not known if the amines exert their function inside the parafollicular cell by for instance influencing the secretion of calcitonin or outside the parafollicular cells by for instance influencing the secretion in thyroid follicle cells.

Other areas in which information to a large extent is lacking include for instance the cellular pathway of calcitonin (and precalcitonin), biosynthesis, cellular mechanism of hormone secretion and the rôle of cholinesterase in parafollicular cell function. Also the functional relation between parafollicular cells and follicle cells represents a largely unexplored area of research.

An inherent problem in the study of the parafollicular cell is that these cells constitute a minority population of cells in the thyroid gland. This circumstance makes detailed studies on the parafollicular cells a difficult task. As an example, purification and biochemical characterization of individual cell components, such as the specific granules, is hampered by contaminants from other thyroid cells, mainly the follicle cells. One way to overcome this problem appears to be to isolate parafollicular cells from other cells by centrifugation after mild enzymatic digestion of thyroid tissue. The feasibility of this approach was recently demonstrated by Bernd et al. (34). Observations on isolated parafollicular cells, either freshly prepared or cultured, will probably be of great importance in future studies.

References

1. Copp DH, Cameron EG, Cheney BA, Davidson AGE, Henze KG: Evidence for calcitonin – a new hormone from the parathyroid that lowers serum calcium. Endocrinology 70(5): 638–649, 1902.
2. Munson PL: Physiology and pharmacology of thyrocalcitonin. In: Handbook of Physiology, vol 7, sect 7. Aurbach GD (ed), Washington DC, American Physiological Society, 1976 pp, 443–464.
3. Nonidez JF: The origin of the parafollicular cells, a second epithelial component of the thyroid of the dog. Am J Anat 49(3): 479–505, 1932.
4. Bussolati G, Pearse AGE: Immunofluorescent localization of calcitonin in the 'C' cells of pig and dog thyroid. J Endocr 37(1): 205–209, 1967.
5. Nunez EA, Gershon M: Cytophysiology of thyroid parafollicular cells. Int Rev Cytol 52: 1–80, 1978.
6. Larson B, Owman Ch, Sundler F: Monoaminergic mechanism in parafollicular cells of the mouse thyroid gland. Endocrinology 78(6): 1109–1114, 1966.
7. Pearse AGE: The cytochemistry and ultrastructure of polypeptide hormone producing cells of the APUD series and the embryologic, physiologic and pathologic implications of the concept. J Histochem Cytochem 17(5): 303–313, 1969.
8. Kameda Y: Electron microscopical and immunohistochemical study of parafollicular cell complex with reference to parafollicular cell as a paraneuron. Arch Histol Japon 40 (suppl): 133–145, 1977.
9. Melander A, Owman Ch, Sundler F: Concomitant depletion of dopamine and secretory granules from cells in the ultimobranchial gland of vitamin D_2-treated chicken. Histochemie 25(1): 21–31, 1971.
10. LeDouarin N, Fontaine J, LeLièvre C: New studies on the neural crest origin of the avian ultimobranchial glandular cells. Interspecific com-

binations and cytochemical characterization of C cells based on the uptake of biogenic amine precursors. Histochemistry 38(4): 297–305, 1974.
11. Fontaine J: Multistep migration of calcitonin cell precursors during ontogeny of the mouse pharynx. Gen Comp Endocrinol 37(1): 81–92, 1979.
12. DeLellis RA, Nunnemacher BA, Bitman WR, Gagel RF, Tashjian AH, Blount M, Wolfe HJ: C-cell hyperplasia and medullary thyroid carcinoma in the rat. Lab Invest 40(2): 140–154, 1979.
13. Bernd P, Gershon MD, Nunez EA, Tamir H: Localization of a highly specific neuronal protein, serotonin binding protein, in thyroid parafollicular cells. Anat Rec 193(2): 257–268, 1979.
14. Wolfe HF, DeLellis RA, Voelkel EF, Tashjian AH Jr: Distribution of calcitonin-containing cells in the normal neonatal human thyroid gland: A correlation of morphology with peptide content. J Clin Endocrinol Metab 41(6): 1076–1081, 1975.
15. Ekholm R, Ericson LE: The ultrastructure of the parafollicular cells of the thyroid gland in the rat. J Ultrastruct Res 23(5–6): 378–402, 1968.
16. DeLellis RA, Nunnemacher G, Wolfe HJ: C-cell hyperplasia. An ultrastructural analysis. Lab Invest 36(3): 237–248, 1977.
17. Thiele J: Parafollikuläre Zellen and Gefässe der Schilddrüse in Darstellungen der Gefrierätztechnic. Verh Dtsch Ges Path 61: 305–310, 1977.
18. Ericson LE: Degranulation of the parafollicular cells of the rat thyroid by vitamin D_2-induced hypercalcemia. J Ultrastruct Res 24(1): 145–149, 1968.
19. Capen CC, Young DM: Fine structural alterations of thyroid parafollicular cells of cow in response to experimental hypercalcemia induced by vitamin D. Amer J Path 57(2): 365–382, 1967.
20. Håkanson R, Melander A, Owman Ch, Sundler F: Depletion of secretory granules, calcitonin, and formaldehyde-ozone-induced fluorescence from cat thyroid C cells by vitamin D_2 treatment. Histochemie 36(1): 89–96, 1973.
21. Peng TC, Garner SC: Hypercalcitonemia associated with return of calcium concentration toward normal in chronically parathyroidectomized rats. Endocrinology 104(6): 1624–1630, 1979.
22. Atack CV, Ericson LE, Melander A: Intracellular distribution of amines and calcitonin in the sheep thyroid gland. J Ultrastruct Res 41(5–6): 484–498, 1972.
23. Van Noorden S, Polak JM, Pearse AGE: Single cellular origin of somatostatin and calcitonin in the rat thyroid gland. Histochemistry 53(3): 243–247, 1977.
24. Yamada Y, Ho S, Matsubara Y, Kobayashi S: Immunohistochemical demonstration of somatostatin-containing cells in the human, dog and rat thyroid. Tohoku J Exp Med 122(1): 87–92, 1977.
25. Buffa R, Chayvialle JA, Fontana P, Usellini L, Capella C, Solcia E: Parafollicular cells of rabbit thyroid store both calcitonin and somatostatin and resemble gut D cells ultrastructurally. Histochemistry 62(3): 281–288, 1979.
26. Alumets J, Håkanson R, Lundqvist G, Sundler F, Thorell J: Ontogeny and ultrastructure of somatostatin and calcitonin cells in the thyroid gland of the rat. Cell Tissue Res 206(2): 193–201, 1980.
27. Kusumoto Y: Calcitonin and somatostatin are localized in different cells in the canine thyroid gland. Biomed Res 1(3): 237–241, 1980.
28. Melander A, Ericson LE, Sundler F, Westgren U: Intrathyroidal amines in the regulation of thyroid activity. Rev Physiol Biochem Pharmacol 73: 39–71, 1975.
29. Sundler F, Håkanson R, Lorén I, Lundquist I: Amine storage and function in peptide hormone-producing cells. Invest Cell Pathol 3(1): 87–103, 1980.
30. Håkanson R, Owman Ch, Sundler F: Aromatic L-amine acid decarboxylase in calcitonin-producing cells. Biochem Pharmacol 20(9): 2187–2190, 1971.
31. Ericson LE: Quantitative electron microscopic autoradiography on the mouse thyroid gland after administration of ^{3}H-L-DOPA. Z Zellforsch 126(1): 82–192, 1972.
32. Ericson LE: Formation and storage of 5-hydroxytryptamine in thyroid parafollicular cells. J Ultrastruct Res 41(5–6): 467–483, 1972.
33. Nunez EA, Gershon MD: Synthesis and storage of serotonin by parafollicular (C) cells of the thyroid gland of active, prehibernating and hibernating bats. Endocrinology 90(4): 1008–1024, 1972.
34. Bernd P, Gershon MD, Nunez EA, Tamir H: Separation of dissociated thyroid follicular and parafollicular cells; association of serotonin binding protein with parafollicular cells. J Cell Biol 88(3): 499–508, 1981.
35. Gershon MD, Kanarek D, Nunez EA: Calcium induced release of 5-hydroxytryptamine from thyroid lobes in vitro and accompanying ultrastructural changes in parafollicular and follicular cells. Endocrinology 103(4): 1128–1143, 1978.
36. Ahrén B, Alumets J, Ericson M, Fahrenkrug J, Fahrenkrug L, Håkanson R, Hedner P, Lorén I, Melander A, Rerup C, Sundler F: VIP occurs in intrathyroidal nerves and stimulates thyroid hormone secretion. Nature 287(5780): 343–345, 1980.
37. Hazard JB: The C cells (parafollicular cells) of the thyroid gland and medullary thyroid carcinoma. A review. Am J Pathol 88(1): 213–250, 1977.
38. Capella C, Bordi C, Monga G, Buffa R, Fontana P, Bonfanti S, Bussolati G, Solcia E: Multiple endocrine cell types in thyroid medullary carcinoma. Virchows Arch A Path Histol 377(1): 111–128, 1978.
39. Sundler R, Alumets J, Håkanson R: Majority and minority cell populations in GEP and bronchial endocrine tumours. Scand J Gastroent 14 (suppl 53): 1979.

Authors' addresses:
Lars E. Ericson
Department of Anatomy
University of Göteborg
Box 33031
S-400 33 Göteborg, Sweden

Frank Sundler
Department of Histology
University of Lund
Biskopsgatan 5
S-223 62 Lund, Sweden

CHAPTER 25

Ultrastructural pathology of the thyroid gland

AUREL LUPULESCU

1. General outlines

1.1. Ultrastructure of the normal human thyroid gland

Morbid changes which occur in the human thyroid gland can be understood only from the standpoint of normal histology and physiology of the thyroid. Conventional transmission electron microscopy (TEM), freeze-etching and scanning electron microscopy (SEM) have revealed that the ultrastructural pattern of the normal human thyroid gland is quite similar to normal patterns already described in other mammals (mouse, rat, rabbit, guinea pigs, dogs), resembling especially the thyroid glands of guinea pigs and dogs (1–4). The functional unit is the follicle (Fig. 1). The human thyroid gland is composed of follicles separated by capillaries, lymph vessels, large areas of connective tissue and interfollicular cells (mast cells, fibroblast). Thyroid follicles are heterogenous; some are large, composed of flat cuboidal cells and filled with a homogenous dense colloid; others are small and outlined by columnar cells. No vacuoles can be seen at the periphery of the colloid (Fig. 2). Light microscopic autoradiography further delineates the heterogeneity of the thyroid gland. Radioiodine (^{131}I or ^{125}I) is incorporated differently; fewer grains are seen in the larger follicles compared with smaller, in which a more isotopic incorporation can be seen as mottled grains within the follicular colloid.

Considerable attention has been paid in the past to the types of thyroid cells. At present, these types are classified as follows: 1) follicular cells; 2) parafollicular cells, and 3) some rare thyroid cells.

1.1.1. Follicular cells

Follicular cells have microvilli protruding into the follicular lumen and occasionally pseudopodia. A large population of small apical vesicles also is present close to the apical membrane; a few large oval granules or colloid droplets (phagosomes) and lysosomes sometimes fused together (phagolysosomes or multivesicular bodies). Some of those previously described as phagolysosomes in human thyroid are more accurately lipofucsin granules. The Golgi complex consists of piled cisternae. The rough endoplasmic reticulum is well developed and shows distended cisternae containing a mottled granular material. Mitochondria have typical cristae and show regular features (Fig. 3). Nuclei have oval profiles and are outlined by a nuclear membrane perforated by a number of pores. The cytoplasmic matrix is finely granular and contains, in addition to polysomes, occasional filaments and microtubules. Lateral cell membranes are folded and often separated by intercellular spaces. The cells are joined by typical junctional complexes in the vicinity of the lumen. Follicular cells are oriented with their apical pole toward the follicular lumen and with the other pole towards the basal lamina. The basal lamina is well developed and has a fine fibrillar structure. It is continuous and sometimes, especially in adult thyroids, displays several foldings, being reduplicated over quite large areas.

Cytochemical studies have revealed that thyroid cells possess several enzymes, such as: E-600 esterase, acid and alkaline phosphatase, proteases and peroxidase-iodinases (5). Occasionally, follicular cells are provided with 1–2 typical cilia.

1.1.2. Parafollicular cells

Parafollicular cells, the second type of cell found in the human thyroid gland, resemble those found in other mammalian thyroid glands. They are of different embryologic origin, are larger than follicular cells, and do not reach the follicular colloid (Fig. 2). Their nuclei are large and sometimes vesicular. The parafollicular cells contain dense granules which are the source of calcitonin, and for this reason are also called C-cells. Parafollicular cells occur only rarely in human thyroids and do not incorporate radioiodine. Immunohistochemical and ultrastructural studies

Motta, PM (ed): Ultrastructure of endocrine cells and tissues. ISBN-13: 978-1-4613-3863-5

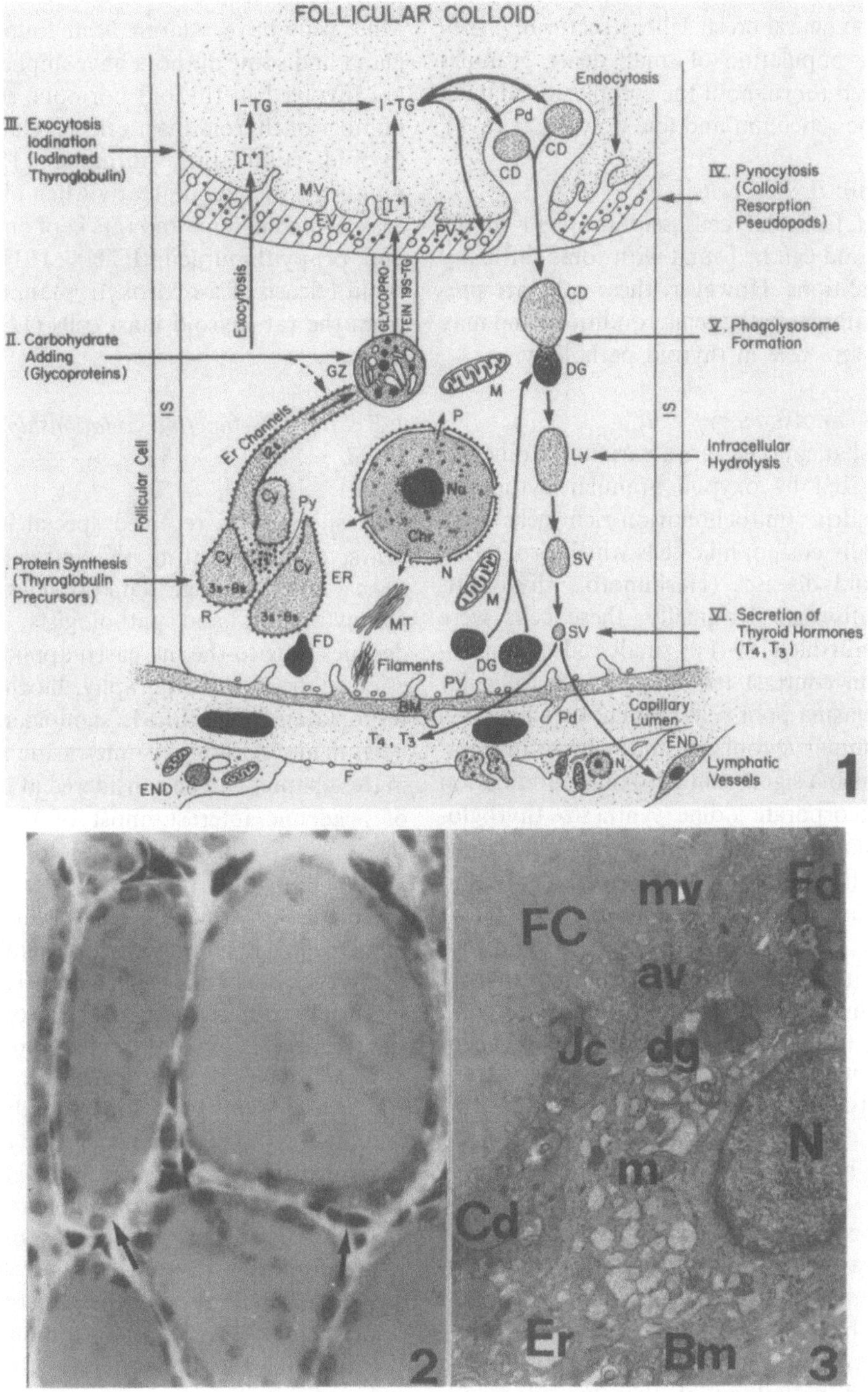

Figs. 1–3. (1) A schematic representation of the ultrastructural organization of a follicular thyroid cell and the phases of secretion of thyroid hormones. Pd: pseudopodia; CD: colloid droplets; MV: microvilli; EV: exocytotic vesicles; I-TG: iodinated thyroglobulin; GZ: Golgi zone; M: mitochondria; Ly: lysosome; DG: dense granules; MT: microtubules; R: ribosomes; ER: endoplasmic reticulum; P: pores; BM: basement membrane; Cy: cysternae; Py: polysomes; PV: pinocytotic vesicles; SV: small vesicles; FD: fat droplets; N: nucleus; F: fenestrae; END: endothelial cell; IS: intercellular spaces. (2) Normal human thyroid. Large follicles lined by cuboidal cells and filled with a dense homogeneous colloid without resorptive vacuoles; C cells are visible (arrows). 1μ thick section, Epon embedded and stained with toluidine blue (× 320). (3) Electron micrograph of normal human thyroid. Follicular cells with microvilli (mv) protruding into follicular colloid (FC). Junctional complex (JC), apical vesicles (av), colloid droplets (Cd) dense granules (dg), endoplasmic reticulum (Er), nucleus (N), mitochondria (m), fat droplets (Fd) and basement membrane (Bm) can be seen (× 6,400).

have shown that C-cells are a distinct type of cell arising from the neural crest. Ultrastructurally, they exhibit a large population of small dense granules (Fig. 5) scattered thorughout the cytoplasm, a lamellar endoplasmic reticulum and few polysomes (6–8).

1.1.3. Some rare thyroid cells

Other types of follicular cells seldom occur in the thyroid gland and can be found with some difficulty in normal conditions. However, these cells are predominant in different pathogenic conditions and may play an important role in thyroid pathology.

1.1.4. Hürthle (or Askanazy) cells

Hürthle (or Askanazy) cells are a variant of follicular cells characterized by oxyphil granular cytoplasm and mitochondria (mitochondrion-rich cells) (9). These are usually eosinophilic cells which proliferate in some thyroid diseases (Hashimoto's thyroiditis and Graves' disease). Originally, these cells were described by Hürthle (1894) as small and protoplasma-rich cells in contrast to tall principal follicular cells or protoplasma-poor cells (9). These cells have a dense and granular cytoplasm and a large nucleus. They possess also a significant peroxidase-iodination activity, can incorporate iodine, synthesize thyroglobulin and thus are involved in thyroid hormone synthesis. Hürthle assumed that they were identical to so-called parenchymatous cells described by Baber (9). These cells, also termed oncocytes, are morphologically similar to those found in the parathyroid gland (oxyphilic cells) and salivary glands. Thyroid tumors originating from them are called oncocytomas. Recent electron microscopic observations suggest that Hürthle cells arise from follicular cells.

1.1.5. Neurohormonal cells

Neurohormonal cells (Neurohormonale or Vagale Zellen) have been described in human thyroid with goiter and Graves-Basedow's disease, as well as in some experimental conditions. These cells are macrothyreocytes, which synthesize acethylcholine. They have been reported to play an important role in the physiology and pathology of the thyroid gland, especially in colloid resorption and pathogenesis of goiter and Graves-Basedow's disease (10).

1.1.6. Ciliated cells, U-cells

Ciliated cells, or U-cells (resembling those found in the ultimobranchial follicle) and atypical extrafollicular cells, have also been described in mouse, rat and bat thyroids. Their role in thyroid physiology is unknown (11).

1.1.7. Mast cells

Mast cells have seldom been found in the thyroid gland and some authors have implied that these cells are involved in thyroid hormone secretion (2). The number of thyroid mast cells is related to the plasma thyroid stimulating hormone (TSH-)level. Their number is increased greatly when plasma TSH-levels are elevated in mice and rats kept on a low iodine diet and propylthiouracil (PTU). TSH also induces a rapid release of 5-hydroxytryptamine and histamine from the rat thyroid mast cells (12).

1.2. Structure-function relationship in the thyroid gland

This subject has received special interest in recent years, and correlating the ultrastructural, physiological and pathologic data provides a great challenge for cytologists and pathologists. In the last two decades, due to the increased application of electron microscope autoradiography, biochemistry, immunofluorescence and SEM, significant advances have been made. Although some of the aspects discussed in this chapter can be considered at present to be only of academic interest, most of them have notable relevance to the pathophysiology of the thyroid gland. Basically, there are two distinct types of thyroid cells; follicular cells which are primarily involved in the synthesis of thyroglobulin and release of thyroid hormones (thyroxine, T_4 and triiodothyronine, T_3) and the parafollicular cells which synthesize the peptide hormone calcitonin in normal conditions and other peptides (β-endorphin, somatostatin and ACTH) in pathologic conditions. Therefore these two cell types can proliferate into two different types of tumors, one from follicular cells (follicular adenoma or carcinoma and papillary carcinoma) and another from C-cells (adenomas and carcinomas).

The follicular thyroid cells are bipolar cells. At their apical pole they release as small vesicles (exocytosis) molecules of newly synthesized thyroglobulin into the follicular lumen, where these are iodinated and stored as homogenous material (colloid). Then colloid droplets are reabsorbed by pseudopodia and microvilli (endocytosis) and intracellularly degraded by lysosomal enzymes and finally released into circulation as T_4 and T_3 (13). This secretory polarity of thyroid cells is preserved in differentiated thyroid tumors but is mainly lost in undifferentiated tumors. Accordingly, an intact ultrastructural organization of thyroid cells is very important for normal thyroid function. This close correlation between thyroid cytology and physiology may be almost preserved in some thyroid diseases and most differentiated tumors, whereas it is distorted or lost in undifferentiated or malignant thyroid tumors. In some cases abnormal iodoproteins also occur, which can be detected by immunological methods.

The structure and function of thyroid follicular cells are under the control of TSH or thyrotropin. This is in turn regulated by thyrotropin releasing factor (TRF) synthesized in hypothalamic centers. This hypothalamic-pituitary-thyroid axis is very important in the maintenance of the structure and function of the thyroid gland. It is impaired in most pathologic conditions and, to a great extent, the loss of thyroid function goes along with reduction in number of highly specialized organelles, such as colloid droplets, lysosomes, endoplasmic reticulum, mitochondria and phagolysosomes. The integrity of the ultrastructural pattern is also important as a reflection of the ability of thyroid cells to incorporate radioiodine and thus is a good 'marker' for radioiodine therapy.

Columnar and well differentiated cells exhibit a marked capacity for radioiodine accumulation, whereas in the poorly differentiated or undifferentiated (anaplastic) cells this capacity is lost and they are unresponsive to radioiodine therapy. Microfibrils and microtubules which are scarce in normal cells are often increased considerably in hypersecretory states or some thyroid tumors. Recently, it has been postulated that the cytoplasmic filament-microtubular system may play an important role in migration and secretion of thyroid hormones. Microtubules act via microfilaments on the apical membrane, in colloid droplet formation and migration. TSH administration increases the number of microtubules and filaments, whereas colchicine has the opposite effect, acting only on the microtubules. Cytochalasin-B affects only microfilaments, whereas chloropromazine, which also abolishes colloid droplet formation and hormone secretion, may act only on the cell membrane (14).

Because it is somewhat difficult to define the 'normal human thyroid gland' a correlation between structure and function is clearly exhibited only in experimental thyroid conditions.

Freeze-etching and SEM techniques allow greater visualization of the vast cell membranous system of the thyroid cells and their dynamics in hyper-or hypofunctional states. The formation of microvilli, pseudopodia, zonulae occludentes, lysosomes and the fenestrae or pores of capillary endothelium are markedly changed during the functional activity of the thyroid gland. These features correlate well with hormone secretion. A significant increase in pseudopodia formation and microvilli in follicular cells and an increase in number of endothelial pores following TSH administration has been demonstrated. Short term administration of propylthiouracil (PTU) increased pseudopodia formation; however, this response is different from the response to TSH. Thyroxine administration, inducing a thyroid hypofunction, produces the opposite effects, decreasing the formation of organelles as well as endothelial pores in the mouse thyroid gland (15–17).

1.3. The mechanism of synthesis and secretion of thyroid hormones in normal and pathologic conditions

The application of modern methods of research markedly changed previous conceptions of the synthesis of thyroglobulin, iodination, and intracellular transport of thyroid hormones towards the capillary lumen. The transport of colloid is mainly an intracellular transport. Other forms of transport (intercellular, etc.) are suggested but not yet confirmed. In normal conditions, when follicular cell membranes are preserved, colloid cannot escape through intercellular spaces. It is well known that some thyroid diseases are due to congenital or acquired defects in the synthesis and release of thyroid hormones.

Thyroid follicular cells are bipolar cells in which both synthesis and release of thyroid hormones have been demonstrated to take place continuously and simultaneously. The first three steps of thyroglobulin synthesis are directed towards the apical pole of the cells. The storage of thyroglobulin and its iodination in the follicular colloid, and the last 3 steps of hydrolysis and secretion of active thyroid hormones are directed towards the basal pole and the capillary lumen (Fig. 1).

First step. The major protein synthesized in thyroid cells is a glycoprotein with a molecular weight of approximately 660,000 daltons and a sedimentation constant of 19S. It was originally called thyroglobulin by Oswald (18). He also postulated an important role of thyroglobulin in the physiology of thyroid gland as well as in goiter development. This glycoprotein is a polymer made up of subunits. The 12S half molecules are held by noncovalent bonds and the 6S subunits are linked together by disulfide bonds. 19S molecules can polymerize into heavier proteins, i.e. 27S and 35S (19). With ^{3}H-leucine it has been shown that the protein moiety of the thyroglobulin molecule is synthesized by the ribosomes attached to the rough endoplasmic reticulum (RER) membranes and also in the free ribosomes (or polysomes). TSH administration increases protein synthesis. Kinetic studies revealed that the protein synthesis starts very rapidly. In approximately ten minutes, the synthesis is continous and takes place in all follicular cells at the same rate. Immediately after amino acid incorporation, thyroglobulin precursors (3S–8S) are detectable. Later, another component, 12S, is formed, which by dimerization is converted into 19S-thyroglobulin. There are some phylogenetic differences in thyroglobulin synthesis. For example, in the lamprey only the 12S moiety is present. It seems that 19S thyroglobulin is the major component only in higher vertebrates. From ribosomes and polysomes the thyroglobulin precursors are polymerized and condensed into a highly viscous material which, viewed with the electron microscope, appears as a fine textural material within the cisternae of endoplasmic reticulum. In experimental thyroid tumors as well as in some human thyroid cancers this proteinaceous material appears as a filamentous substance. The rate of thyroglobulin polymerization is different in different animal species and can be related to thyroid activity. Thus, the conversion of 12S into 19S is slow in the guinea pig thyroid gland, whereas it is more rapid in the rat thyroid gland. Also, the iodine level in the thyroid gland affects thyroglobulin polymerization. In the thyroid glands of iodine-deficient rats the formation of 19S-thyroglobulin is markedly impaired. Administration of propylthiouracil (PTU) inhibits thyroglobulin synthesis, whereas puromycin and actinomycin-D only slightly impaired the thyroglobulin formation. This evidence suggests that thyroglobulin synthesis is not RNA-dependent.

Second step. The protein moiety becomes glycoprotein by attachment of carbohydrate units in the second step of thyroglobulin synthesis. Electron microscopic autoradiographic studies using ^{3}H-leucine, ^{3}H-glucose and ^{3}H-fucose have demonstrated that the protein molecule move from the cisternae of the endoplasmic reticulum and is found after one hour in the neighborhood of the Golgi zone, where it is packed and then converted into glycoproteins (PAS-positive) by the attachment of glucidic moieties already present in this area. The thyroglobulin molecule then travels towards the apical zone where it is found within small dense apical vesicles and extruded into the follicular lumen by exocytosis.

Third step. At this stage the iodination of the glycoprotein matrix formed in the Golgi zone takes place. It has been postulated, based mostly on biochemical findings, that the thyroglobulin iodination takes place mainly intracellularly. However, EM autoradiography using ^{125}I or ^{131}I clearly demonstrated that the iodination process occurs mainly extracellularly. Thus, the microvillous apical border of the cell and the peripheral follicular colloid are the active zones of thyroglobulin iodination. Also, the presence of an enzymatic system catalyzing the peroxidation of iodide is assumed to be there. Theoretically, only the mature thyroglobulin (19S) should be present in the apical cellular zone. However, some studies have shown that incompletely formed thyroglobulin molecules (immature thyroglobulin) can be iodinated. In the thyroid glands of iodine-deficient rats iodination occurs also on 12S thyroglobulin.

Fourth step. The resorption of iodinated thyroglobulin (I-TG) from the follicular colloid involves three sequential events; the uptake of colloid (by pseudopodia or by small pinocytotic vesicles) the occurrence of colloid droplets and the hydrolysis of intracellular droplets. Transmission and scanning electron microscopy have provided further insights into this process in normal, TSH-stimulated and pathologic thyroid glands. Findings have revealed two different types of endocytotic processes. One type is micropinocytosis and results in the formation of small bristle-coated vesicles (pinocytotic vesicles) on the apical border. Another type consists of the formation of large pseudopodia which are inserted deeply into the follicular colloid and engulf it by a sort of phagocytosis thus forming colloid droplets. Micropinocytosis is predominant in normal thyroid glands whereas the pseudopodia formation is more evident in TSH-stimulated thyroid glands. Therefore the colloid droplets are endocytotic vacuoles and contain the resorbed iodinated thyroglobulin from the follicular colloid, since radioiodine (^{125}I or ^{131}I) is found in colloid droplets only after the labeling of follicular colloid.

Fifth step. Intracellular migration and simultaneous hydrolysis comprise this step. The colloid droplets, moving towards the basal zone, encounter lysosomes and, fusing with them, form large structures termed phagolysosomes. At this stage the colloid droplets acquire a positive reaction for enzymatic activity and an increased electron density. As they move towards the basal area the colloid droplets become smaller and denser. The intracellular movement lasts between 90 and 180 minutes.

Sixth step. Release of active thyroid hormones (T_3 and T_4) into capillaries is the final step, but it is not well understood. As the small colloid droplets approach the basement membrane, they are released through the pores into the capillary lumen as small vesicles (endothelial micropinocytotic vesicles); in other instances emission of pseudopodia (merocrine secretion) can be seen. EM autoradiographic studies support the idea that thyroid hormones are secreted into the capillary lumen (mostly blood capillaries). Small amounts of thyroid hormones can be found also in lymphatic capillaries (Fig. 1). The occurrence of thyroglobulin in the circulation is due mostly to cellular damage or injury. Each of these several steps could be impaired either by congenital errors or by acquired factors. At present, with the aid of modern methods, we are able to detect earlier and more accurately these congenital abnormalities (20–22).

1.4. Advantages of EM, autoradiography and SEM for the diagnosis of thyroid diseases

Electron microscopy, autoradiography, freeze-etching, immunochemistry and scanning electron microscopy have improved significantly our knowledge of the cellular physiology and pathology of the thyroid gland. Thus, an 'in depth' study of the thyroid gland is possible and enables us to make a more accurate diagnosis of its diseases. By using these methods we can detect a disease in its early stages (incipient phases) and follow a benign tumor through its transition stages (intermediate phases) toward a malignant or neoplastic one. We can detect the origin of neoplastic cells by using ultrastructural characteristics of follicular and parafollicular cells, as well as cellular differentiation. A comparison of normal, experimentally treated or diseased thyroid glands reveals pronounced and characteristic ultrastructural changes.

2. Experimental ultrastructural pathology of the thyroid gland

2.1. Effects of acute and chronic TSH administration

2.1.2. Effect of acute TSH administration

Acute administration of TSH has been investigated in different species (rat, mouse, guinea pig, dogs, chicken, teleost fish and salamander) and the results are almost identical (2). Salient findings gleaned from the ultrastructural analysis of these glands reveal an increase of microvilli and large pseudopodia in the cell apical zone. These structures can reach 1–2 μm in length. They deeply penetrate the follicular lumen and do not contain cell organelles but only large granules with a fine granular material identical to follicular colloid (colloid droplets). Light and electron microscopic autoradiography using ^{3}H-leucine and radioiodine (^{125}I or ^{131}I) reveals that these droplets do not take up ^{3}H-leucine but incorporate radioiodine. This finding strongly suggests that they are resorbed formations. Small vesicles (exocytotic vesicles) are also seen in the apical cytoplasm. Therefore, TSH administration markedly increases the synthesis and release of thyroglobulin into the follicular lumen (exocytosis) as well as the resorption of follicular colloid as iodinated thyroglobulin (endocytosis). The Golgi complex is expanded. No ultrastructural changes of mitochondria are visible after short-term TSH administration. In addition, no significant changes are seen in the parafollicular cells. Radioautography shows an intense incorporation of radioiodine in the follicular colloid, mostly in the peripheral colloid and apical zone of the cells. These dramatic apical cell changes can be better visualized in rat or dog TSH-stimulated thyroid glands by using SEM. Thus, several pseudopodia appear as large cylindroid formations protruding deeply into the follicular lumen and engulfing the colloid. The formation of pseudopodia is completely inhibited by cytochalasin-B (14). Ultrastructural, autoradiographic and cell surface changes appear immediately after TSH administration, last 10–12 hours and they return to normal after approximately 24 hours.

2.1.3. Effect of chronic TSH administration

Most ultrastructural findings after chronic TSH administration are similar to those observed after acute TSH administration but only in moderate degree; e.g., hyperplasia of microvilli, a moderate increase in colloid droplets and hypertrophy of the Golgi complex. The nuclei are moderately enlarged and exhibit several indentations. Intercellular spaces are enlarged and lateral membranes are irregular. Ultrastructural changes occur also at the basal zone and consist of enlargement of the basement membrane with several infoldings and reduplications. Autoradiography reveals a moderate accumulation of radioiodine and some scattered filaments may be seen over the cisternae of ER. Also, an increased ^{3}H-thymidine incorporation occurs in the nuclei of thyroid cells following long-term TSH administration. Viewed with the SEM apical cell surface changes with pseudopodia are seen only seldom. Radiochromatography shows a predominance of diiodothyrodine (DIT) and thyroxine (T_4) over monoiodothyrosine (MIT) and triiodothyronine (T_3.

2.2. Experimental goiter: role of calcitonin and prostaglandins

Our investigations regarding the role of hormones on goiter induction and development (goitrogenesis) revealed the effects of calcitonin and prostaglandins (E_1, E_2 and $F_{2\alpha}$) on goiter and their interference in thyroid hormone synthesis. Experimental goiter is an interesting model and several findings regarding the histogenetic mechanism of experimental goiter can be extrapolated to endemic goiter. In particular, calcitonin and prostaglandins induce pronounced and characteristic thyroid ultrastructural and autoradiographic changes, especially on the cell surface, and can exert an important effect in the goiter development of some laboratory animals (rat, rabbit).

2.2.1. Parenchymatous hyperplastic goiter induced by calcitonin

TEM, high resolution autoradiography and SEM studies reveal that prolonged administration of calcitonin in rabbits and rats induced characteristic ultrastructural changes in radioiodine metabolism and the thyroid cell surface (23–25). Ultrastructural micrographs of follicular cells show large pseudopodia and microvilli, a heterogenous population of dense granules, colloid droplets, dilated cisternae of endoplasmic reticulum, hypertrophied Golgi complex and free polysomes (Fig. 4). Ultrastructural micrographs of parafollicular cells reveal that these cells are increased in number and their cytoplasm is filled with small dense granules (Fig. 5). A decrease of ^{125}I uptake into follicular colloid and an increase of intracellular distribution over the cytoplasm of follicular cells was noted. Further, protein bound iodine (PBI), thyroxine (T_4) and triiodothyronine are lower in calcitonin-treated rabbits and rats as compared to controls. Intracellular iodination, which is insignificant in normal conditions, is increased within endoplasmic reticulum cisternae (Fig. 15). Interesting findings regarding colloid resorption in rabbit calcitonin-treated thyroids were revealed by SEM. Several large pseudopodia and small exocytotic vesicles are seen in follicular cells of rabbit thyroid glands treated with calcitonin, whereas only microvilli are present in the thyroid gland of control animals. Studies regarding the ultrastructure and intracellular kinetics of radioiodine in experimental goiter induced by an iodine-deficient diet in rats revealed also an increased intracellular iodination within endoplasmic reticulum cisternae, colloid and dense droplets. Ultrastructurally, an extensive enlargement and disorganization of endoplasmic reticulum with hypertrophic microvilli and Golgi apparatus are the predominant changes. Radiochromatograms of iodine-deficient goiters revealed the predominance of MIT over DIT and T_3 and T_4, which suggests a defective transport of thyroglobulin through the ER channels (21).

2.2.2. Goiter induced by prostaglandins

Long-term administration of PGE_1 and PGE_2 markedly stimulated the thyroid gland, inducing a hyperplastic microfollicular goiter with a high radioiodine (^{131}I) uptake, increased endocytosis, a heavy autoradiographic (^{125}I) distribution and a moderate increase of thyroid hormones (T_3, T_4), thyroxinbinding globulin (TBG) and thyrotropin (TSH) concentration in adult rats. Ultrastructural studies of the thyroid gland following PGE_1 and PGE_2 revealed a marked cellular activity with increased number of microvilli, of colloid droplets and a marked agglomeration of dense granules and polysomes; in some follicular cells the dense granules occupy the entire cytoplasm (Fig. 6). Conversely, a thyroid hypofunction with low radioiodine (^{131}I) uptake, autoradiographic (^{125}I) distribution and a moderate decrease in T_4, T_3, TGB and TSH concentrations occurred following $PGF_{2\alpha}$ administration. The ultrastructural analysis indicated a decrease of colloid droplet and dense granule populations and occurrence of degenerative mitochondria. However, this goitrogenic effect of prostaglandins does not occur in hypophysectomized rats, suggesting a direct effect of prostaglandins on TSH secretion from the anterior pituitary gland. The occurrence of a marked hyperplasia of parafollicular cells and an increase number of their characteristic granules following $PGF_{2\alpha}$ administration suggests a relationship between $PGF_{2\alpha}$ and calcitonin synthesis in rat thyroids. Hence these findings reveal that prostaglandins exert significant effects on thyroid gland and goiter formation, radioiodine metabolism and thyroid hormone synthesis and most of these effects can be mediated by TSH (26).

2.3. Ultrastructure and autoradiography of experimental thyroid tumors

Most thyroid neoplasms are classified as follicular neoplasms (adenomas, carcinomas) papillary neoplasms (adenomas, carcinomas) epitheliomas or sarcomas. In most instances thyroid neoplasms originate from follicular cells, which are more reactive than other thyroid cells to carcinogenic stimuli. Only occasionally do parafollicular cells proliferate as 'C-cell hyperplasia' or 'C-cell adenomas'. Most studies regarding thyroid neoplasms were carried out with the light microscope, a situation which greatly hampered our understanding of their histogenetic mechanism and classification. Many findings from experimental thyroid neoplasms can be extrapolated to human thyroid cancers.

Electron microscopy of thyroid neoplasms reveals mainly in follicular adenomas the occurrence of microvilli which penetrate into the follicular lumen and contain a reduced amount of colloid having a filamentous texture. The subapical layer is also reduced in size. The predominant ultrastructural lesions consist in a marked expansion of endoplasmic reticulum with dilated and confluent cisternae, which give a 'lacy' cytoplasmic pattern. The Golgi complex is slightly hypertrophic; several mitochondria and a large population of free ribosomes (polysomes) can be seen. The nuclei are atypical. The basement membrane is thick and exhibits several folds and reduplications. The ultrastructural pattern of papillary tumors reveals that the microvilli are absent and the apical membrane is barely visible. However, the

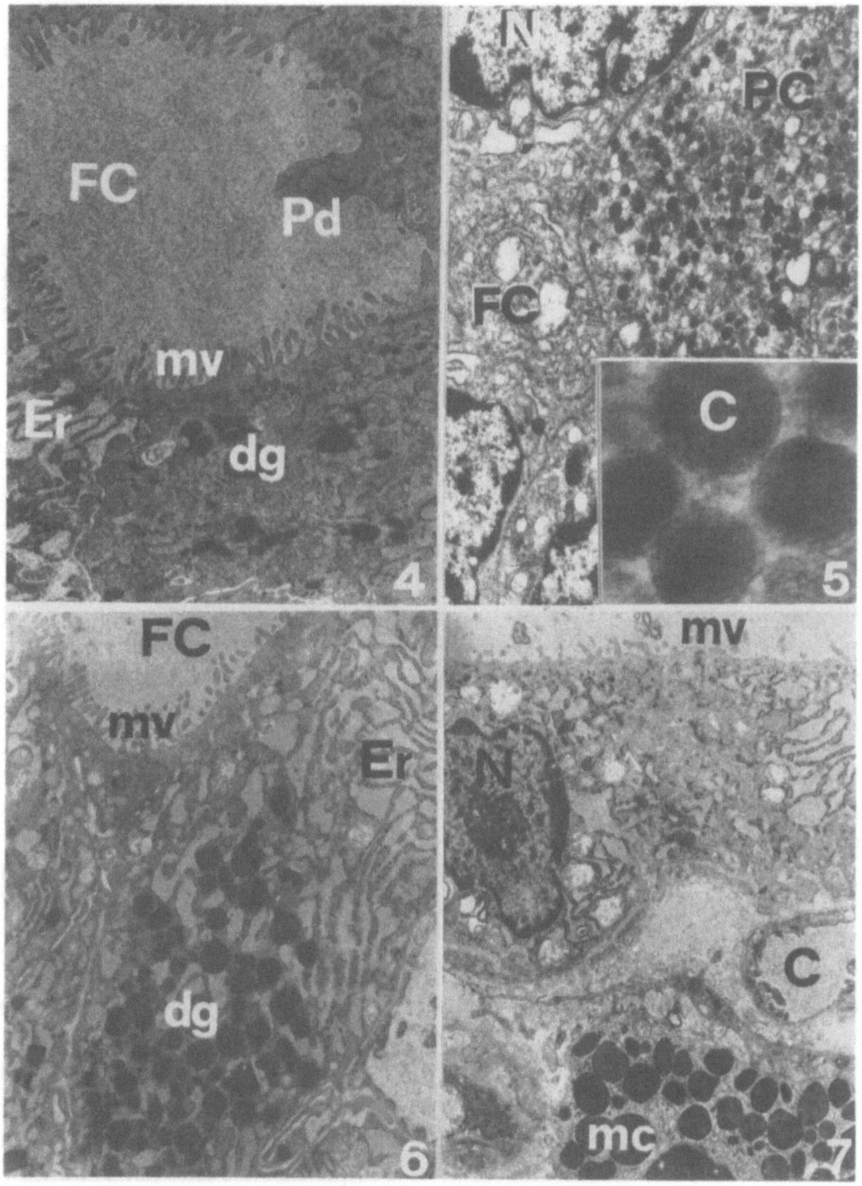

Figs. 4–7. (4) Parenchymatous goiter following calcitonin administration in rabbit. Follicular cells exhibit pseudopodia (Pd) and hypertrophied microvilli (mv) protruding into follicular colloid (FC); dense granules (dg) and dilated endoplasmic reticulum (Er) are seen (× 8,000). (5) Hyperplastic follicular cells (FC) and parafollicular cells (PC) can be seen in calcitonin induced goiter in rabbits. Parafollicular cells exhibit a large population of dense characteristic granules (C) enveloped by a single membrane (see inset); nuclei (N) (× 8,000 and inset × 64,000). (6) Parenchymatous-microfollicular goiter induced by chronic prostaglandin (PGE_2) administration in rats. Follicular cells with elongated microvilli (mv) protruding into follicular colloid (FC); a large population of dense granules (dg) occupying entire cytoplasm and markedly distended endoplasmic reticulum (Er) (× 8,000). (7) Thyroid of hypophysectomized rat. Flat cuboidal follicular cells with short microvilli (mv); nuclei (N) and scarse endoplasmic reticulum; mast cells (mc) with dense granules; capillary (C) (× 8,000).

cytoplasm exhibits the same lacy pattern with dilated ER cisternae, tumefied mitochondria and nuclear monstrosities. Autoradiography using radioiodine (^{125}I or ^{131}I) reveals also characteristic findings, e.g. in follicular neoplasm a marked radioiodine incorporation mostly over the follicular colloid. The reaction is less visible in malignant or follicular carcinomas. The autoradiographic reaction is also poor or lacking in papillary neoplasms, indicating that the follicular cells are losing their capacity to incorporate iodine.

2.4. Ultrastructural thyroid changes in various conditions

2.4.1. Ultrastructural changes following administration of antithyroid compounds (Propylthiouracil, PTU and Methylthiouracil, MTU).
Ultrastructural studies of thyroid glands from rats chronically treated with Methylthiouracil (MTU) showed development of predominant microfollicular hyperplasia, marked hyperemia, and very little colloid. The apical cell membrane exhibits hypertrophic microvilli, pseudopods and a large population of small dense granules. Large formations of conglomerate membranes can be found in the follicular lumen. Ergastoplasmic sacs are dilated and ribosomes and polysomes are increased in numbers suggesting an increased protein synthesis. The Golgi complex is hypertrophied and mitochondria are tumefied, but the cristae are preserved. Colloid droplets are decreased. Nuclei are enlarged and indented; a marked expansion of the basement membrane with lamelliform folds and invaginations can be seen. Autoradiographic studies revealed a marked reduction of radioiodine in the thyroid cells. SEM revealed that short term PTU administration (1.5–4 hrs) increased the pseudopod formation and colloid resorption. However, most pseudopods produced by PTU are located at the periphery of follicular cells. Some biochemical data also show that thyroglobulin incorporation in lysosome fractions increases after PTU treatment comparable to TSH administration in rats (27).

2.4.2. Ultrastructure of the thyroid gland after hypophysectomy
Ultrastructural changes after hypophysectomy are opposite to those found in TSH-stimulated thyroid glands. Thus, cells are smaller with hypotrophic microvilli. The lumen of ergastoplasmic sacs is empty; ribosomes and polysomes are also significantly decreased in number. The Golgi complex is hypotrophic. Colloid droplets are markedly decreased in both size and number but dense granules can be found. Mitochondria are normal in number but their size is reduced. Nuclei are slightly smaller. The basement membrane is not affected. Mast cells are normally present and their ultrastructure is not altered (Fig. 7). Autoradiography showed a marked reduction in radioiodine accumulation.

2.4.3. Ultrastructural thyroid changes following thyroxine (T_4) and triiodothyronine (T_3) treatment
Acute administration of thyroxine induced an inhibition of cell growth and a decrease in the number of cellular organelles. Microvilli are hypotrophic. Colloid droplets are decreased in number and mitochondria have blurred cristae. The Golgi complex is hypotrophic. There are also fewer ribosomes. Freeze-etching of thyroxin-treated mice revealed a capillary endothelium with fewer pores. Ultrastructural changes after a chronic T_3 administration show rare microvilli and a very thin subapical layer. The endoplasmic reticulum and Golgi apparatus are hypotrophic. Nuclei are small.

2.4.4. Ultrastructure of thyroid gland following cold exposure
Electron microscopy of rat thyroid gland exposed to cold (48 hrs) revealed a hyperactivity with microfollicles, less follicular colloid and hyperemia. The endoplasmic reticulum has dilated cisternae. The apical cell membrane exhibits elongated microvilli and small apical vesicles. The ribosome population is increased. Colloid droplets are large and the Golgi complex is moderately hypertrophied. Several mitochondria and occasional intramitochondrial granules are visible. Nuclei exhibit a dentated pattern. Autoradiography revealed an intense radioiodine accumulation.

3. Human thyroid ultrastructural pathology

The ultrastructural findings obtained from the thyroid gland of laboratory animals cannot always be extrapolated to human thyroid glands. Due to more complex environmental conditions in which man is living, the demands on and responses by the thyroid gland are more diversified, which can explain the heterogeneity of the human thyroid gland.

3.1. Ultrastructure of Graves-Basedow's disease

Electron microscopy reveals the occurrence of hypertrophic microvilli, giving the typical 'brush' aspect of the apical membrane, and occasionally large pseudopodia containing colloid droplets. A large population of small vesicles is present in the subapical zone and below this layer several dense granules, sometimes fusing with large colloid droplets, can be seen. The endoplasmic reticulum is expanded and pleomorphic. The Golgi apparatus is hypertrophied and the nuclei are enlarged with 2–3 nucleoli. An expansion of the basement membrane with several reduplications and introflexions that penetrate into the basal cytoplasm are visible. Parafollicular cells are occasionally noted and exhibit a slight hypertrophy with characteristic granules.

3.2. Ultrastructure of goiter and endemic cretinism

3.2.1. Multinodular (colloidocystic) goiters

Light microscopy of plastic sections reveals the predominance of large follicles (macrofollicles) lined by a flat cuboidal epithelium and containing a homogenous dense colloid. Electron microscopy shows large follicles with hypotrophic microvilli; the endoplasmic reticulum is expanded with disrupted membranes. The number of phagolysosomes is moderately increased and it has been assumed that they play an important role in colloid hydrolysis and transport (1); some of these are lipid and/or lipofuscin granules. The Golgi complex is hypotrophic and nuclei are poor in chromatin.

3.2.2. Ultrastructure of simple goiter

The ultrastructural pattern of simple goiter is characterized by a reduced amount of endoplasmic reticulum, an increase in the number of mitochondria and phagolysosomes. The basement membrane is clearly visible and cells with a long isolated cilium protruding into the follicular lumen can be seen. The reduction of endoplasmic reticulum and of polysomes may suggest a decrease in thyroglobulin synthesis. The increased number of mitochondria in simple goiter is similar to that reported in thyrotoxic goiter or in some types of thyroid tumors. Light and electron microscopic autoradiography reveal a moderate decrease in the radioiodine cellular distribution. These ultrastructural and autoradiographic findings might suggest that in simple goiter protein synthesis and the iodination process are slightly decreased, whereas colloid resorption and intracellular hydrolysis are enhanced. The study of soluble iodoproteins reveals an increased amount of the slow sedimenting component (4S-7S). These abnormalities in the thyroglobulin synthesis and its iodination can be related to the ultrastructural changes (28).

3.2.3. Ultrastructural changes of the thyroid gland in endemic cretinism

Light microscopy shows a disorganization of follicular structure and its replacement by cellular cords and sclerosis. Electron microscopy reveals complex ultrastructural changes, which are different in nodules and in paranodular parenchyma. Thus, irregular follicles with little colloid and hypotrophic microvilli are present in the nodules. Several dense granules containing a homogeneous dense material are present. The Golgi complex is hypotrophic. Mitochondria are swollen, nuclei are poor in chromatin and the basement membrane is enlarged but no introflexions are visible. In the paranodular parenchyma predominant macrofollicles with few hypotrophic microvilli are present. The endoplasmic reticulum is markedly reduced in size. Sometimes large areas of sclerosis and calcarous deposits are visible. Autoradiographic studies reveal that most are redundant and non-functional nodules which incorporate the radioiodine in very small amounts. The perinodular parenchyma moderately incorporates the radioiodine. Radiochromatograms reveal a defect in the coupling of iodothyrosines (MIT and DIT) and a marked decrease of iodothyronines (T_3 and T_4) with the occurrence of DIT in blood.

3.3. Ultrastructure of thyroid tumors

Cytology of normal human thyroid gland reveals different cell types: follicular cells, parafollicular (or C) cells and some rare cell types. Accordingly, thyroid cells exhibit a different neoplastic transformation and thus different types of thyroid tumors are described. The most frequently occurring tumors are follicular adenoma and papillary adenoma. Among the malignant tumors are follicular carcinoma and papillary carcinoma. Rare thyroid tumors include medullary thyroid tumors (C-cell tumors) and oxyphil cell tumors (Hürthle cell tumors). Anaplastic thyroid carcinoma, embryonal or fetal thyroid tumors, sarcoma, fibrosarcoma, reticulohistiosarcoma, plasmocytoma and squamous cell carcinoma are occasionally reported. On the thyroid tumors there are no comprehensive and systematic ultrastructural studies.

3.3.2. Follicular adenomas

Follicular adenomas are of two types: microfollicular and macrofollicular. Light microscopy of microfollicular adenoma reveals the predominance of small follicles with sparse colloid and lined by tall columnar epithelium with several infoldings. A multitude of colloid droplets is visible at the apical cell pole (Fig. 8). Macrofollicular adenoma is composed of large follicles with dense colloid and lined by low cuboidal cells. Only few colloid droplets can be seen.

3.3.3. Ultrastructure of microfollicular adenomas

The ultrastructure of microfollicular adenoma is characterized by the predominance of high columnar cells with large nuclei and nucleoli and several microvilli and pseudopodia at the apical border. A heterogeneous population of large colloid droplets (phagosomes) and lysosomes, occasionally fusing to form phagolysosomes, can be seen in the cytoplasm (Fig. 9). Enlarged and pleomorphic endoplasmic reticulum containing a fine granular material and

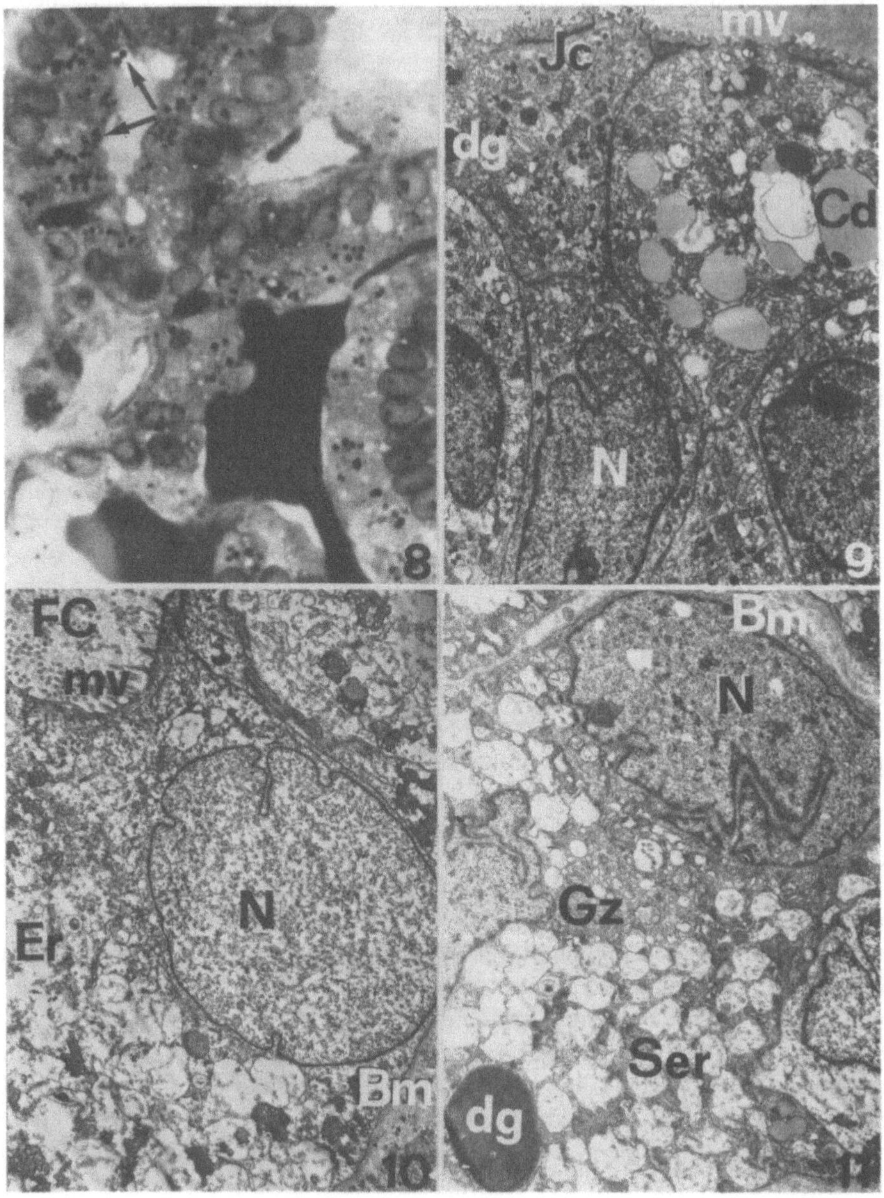

Figs. 8–11. (8) Human thyroid-microfollicular adenoma. Microfollicles composed of columnar cells which exhibit a multitude of colloid droplets (arrows), mostly at the apical pole; nuclei with nucleoli and little colloid are seen. 1 μ thick section, Epon embedded and toluidine blue stain (× 320). (9) Microfollicular adenoma. Columnar cells with numerous colloid droplets (Cd) fusing with dense granules (dg); several microvilli (mv), junctional complexes (Jc) and nuclei (N) are visible (× 6,400). (10) Follicular carcinoma. High follicular cells with numerous microvilli (mv) protruding into follicular colloid (FC). Enlarged nuclei (N) and disorganized endoplasmic reticulum (Er); intact basement membrane (Bm) is visible (× 6,400). (11) Papillary carcinoma. No follicular structures can be seen. Nuclei (N) are enlarged and dentated; Golgi complex (Gz), large dense granules (Dg), smooth endoplasmic reticulum (Ser) and several polysomes; basement membrane (Bm) is still intact (× 6,400).

several mitochondria are predominant. The Golgi complex is hypertrophic with many dense granules and dilated vacuoles. The basement membrane is intact and regular. Occasionally, cytoplasmic filaments and microtubules can be seen.

3.3.4. Ultrastructure of macrofollicular (colloid) adenomas

The ultrastructure of macrofollicular (colloid) adenomas is characterized by a predominance of small cuboidal cells, with oval nuclei and few colloid droplets. The endoplasmic reticulum is flattened, with a reduced number of ribosomes. Mitochondria are smaller and the Golgi complex is inconspicuous. Autoradiographic studies reveal an increased accumulation of radioiodine in microfollicular adenomas, mostly over the follicular colloid, whereas a decreased autoradiographic reaction is visible over the colloid of macrofollicular adenomas. By SEM, follicular cells in microfollicular adenomas appear as rounded cells with smooth surfaces and are arranged in regular pattern. Short and round projections can occasionally be seen on the cell surface and the intercellular spaces are enlarged. SEM of macrofollicular adenomas shows large cystic follicles lined by cuboidal flat cells; occasionally cilia protruding into the lumen are visible. Therefore, ultrastructural, autoradiographic and SEM studies demonstrate that microfollicular adenomas are hyperfunctional structures in which thyroglobulin synthesis, transport and intracellular hydrolysis are increased. Hypofunctional adenomas are those tumors containing macrofollicles in which thyroglobulin synthesis and hydrolysis are reduced. Cell growth and intercellular connections are preserved in both types of adenomas and are an indication of their benign nature (29).

3.3.5. Ultrastructure of thyroid carcinomas

Ultrastructural studies of thyroid carcinomas reveal marked differences between follicular carcinoma cells and papillary carcinoma cells.

3.3.6. Follicular carcinoma cells

Follicular carcinoma cells exhibit a modified or distorted ultrastructural organization. Microvilli are hypertrophied and protude into the follicular lumen. The endoplasmic reticulum is pleomorphic and possesses dilated cisternae. Colloid droplets are few but the number of lysosomes is increased. The Golgi complex is slightly hypertrophic. Mitochondria are increased in number and swollen with irregular cristae. Nuclear abnormalities are frequently seen (Fig. 10). Ultrastructural investigations into lymph node metastases originating from follicular carcinoma revealed a similar follicular pattern with colloid and occasional dense 'colloid bodies'.

3.3.7. Papillary carcinoma cells

The ultrastructural pattern of papillary carcinoma cells is more disorganized and only in a few instances follicles can be seen. Extensive ultrastructural changes are observed in the endoplasmic reticulum and mitochondria. A filamentous material is contained in several cisternae of RER and occasionally this is continuous with the smooth endoplasmic reticulum (SER). Colloid droplets are few, but large dense bodies are frequent. The mitochondria are numerous and tumefied and occasionally show focally distended cristae. The nuclei exhibit irregular and deeply indented profiles, suggesting an advanced degree of malignancy (Fig. 11). The distension of RER, the decrease in attached ribosomes and the increase in polysomes suggest a reorientation of cell organelles and possibly an increase in protein synthesis in the neoplastic cells. It is possible that the mitochondrial abnormalities indicate an impaired cellular respiratory function and ATP production. The loss of basal cell membrane infoldings also suggests a modified cell-vessel relationship in cancerous cells. The pronounced disorganization of neoplastic thyroid cells coincides with a striking decrease of 19S thyroglobulin which is replaced by a heterogeneous slowly sedimenting material (30). Autoradiography using ^{131}I reveals a marked parallelism between the disorganization of the ultrastructural pattern and the loss of capacity to incorporate radioiodine. The more differentiated thyroid tumors incorporate more actively the radioiodine, whereas the less differentiated thyroid tumors incorporate only a small amount. These are important diagnostic criteria and also valuable for the treatment of thyroid cancer. In some studies the ultrastructure correlated well with thyroglobulin synthesis and the capacity to incorporate iodine. Only the well differentiated tumors are capable of synthesizing thyroglobulin. Undifferentiated or poorly differentiated tumor cells partially lose this ability. Thus, thyroglobulin may be used as a 'marker' for potential function of thyroid carcinoma (30, 31). Other sensitive markers, such as TSH-receptors and adenylate cyclase activity have been used for the diagnosis of differentiated and undifferentiated thyroid carcinomas (32). Some authors have not found marked differences in thyroglobulin and other soluble iodoproteins in papillary carcinomas, but a lower release in follicular adenomas (33). Thus, ultrastructural studies provide more accurate criteria for the diagnosis of transitional stages between benign and malignant thyroid tumors.

3.4. *Oxyphylic (Hürthle) cell tumors*

Oxyphylic cell tumors (adenomas and carcinomas) have also been reported. These neoplastic follicular cells are characterized by an oxyphilic basal area, a cytoplasm filled with mitochondria, a central nucleus and the presence of dense bodies in the existing microfollicles. In some cells, almost the entire cytoplasm is occupied by mitochondria. It is not clear if this tremendous increase in mitochondria is due to increase in their formation or decrease in their breakdown (34). Electron microscopic, histochemical and biochemical techniques have revealed in a large series of Hürthle cell tumors (oncocytomas) characteristic cytological features, such as a mitochondrion-rich cytoplasm and an increased number of dense bodies or lysosomes. These neoplastic cells possess a high enzymatic activity and can synthesize thyroglobulin; however, this thyroglobulin is poorly iodinated. Ultrastructurally, mitochondria are of normal appearance but exhibit a higher enzymatic activity. Ultrastructural studies reveal that Hürthle cells are a variant of follicular cells and Hürthle cell tumors can be classified as follicular, papillary or undifferentiated (35). These tumors are also described as oncocytomas. Thyroglobulin is absent in some Hürthle cell carcinomas. A 'clear cell' variant of these oxyphylic tumors is characterized by the predominance of a clear-water cytoplasm, due to an increased number of mitochondria, which are markedly distended and appearing as empty vacuoles (36).

3.5. *Medullary (C-cell) thyroid tumors*

The pathology of C-cells in the human thyroid gland can be manifested as C-cell carcinomas, C-cell hyperplasia, and C-cell adenomas. Ultrastructural studies made possible an accurate diagnosis of C-cell tumors, i.e. the predominance of parafollicular cells containing characteristic granules and their capacity to synthesize large amounts of calcitonin (CT), as well as some other biogenic amines. Recent immunohistological studies using immunoperoxidase histology demonstrated, in a large series of thyroid C-cell neoplasms, a notable incidence of calcitonin, β-endorphin, ACTH and somatostatin. It is postulated that there is a functional relationship among these four peptides (37). In all cases, medullary tumors originate from parafollicular (C) cells and are now termed C-cell carcinomas. In some instances an increased extracellular amyloid formation has been observed (amyloid-forming medullary carcinomas) (38). C-cell adenomas of the human thyroid gland with increased amounts of calcitonin have also been reported (39). C-cell hyperplasia has occasionally been reported in human thyroid glands as well as in experimental conditions. C-cell hyperplasia can also develop in residual thyroid glands following resection for medullary carcinoma.

3.6. *Ultrastructure of thyroid gland in myxedema*

Electron microscopy shows an advanced cellular atrophy and disorganization. Follicular cells are atrophic and exhibit few short microvilli. The apical membrane is regular and no apical vesicles can be seen. The endoplasmic reticulum is markedly reduced with few attached ribosomes. Mitochondria are decreased in number and severely damaged. Colloid droplets are visible only in few instances, but the amount of lipid material (or lipfucsin granules) is increased. Nuclei are small and poor in chromatin. The basement membrane is thin and no introflexions can be detected. These findings indicate an atrophy and hypoactivity of thyroid cells as regards their protein synthesis, colloid resorption and thyroglobulin synthesis.

3.7. *Ultrastructure of Hashimoto's thyroiditis*

Electron microscopic studies reveal the occurrence of normal thyroid cells intermingled with cells in different stages of necrobiosis. Also, a proliferation of oxyphilic or mitochondrion-rich cells (Hürthle cells) can be seen; occasionally, squamous cell metaplasia of thyroid cells is visible. Lymphocytes and plasma cells penetrating into follicular epithelium are frequently seen. Ultrastructural studies have focused on the follicular basement membrane and its alterations. Thus, distinct ultrastructural changes of follicular basement membranes, such as the presence of electron-dense deposits, similar to immune complexes, and their close association with plasma cells strongly suggest the autoimmune origin of this disease. Occasionally, gaps in the follicular basement membrane can also be seen. Leukocytes and lymphocytes can escape through these gaps. Extensive processes of follicular cells encompassing the lymphocytes or plasma cells are frequently seen. However, no reduplications or introflexions of the basement membrane can be detected. Intercellular spaces are enlarged. Similar electron-dense deposits have also been found in the follicular basal lamina of obese-strain chickens with spontaneous hereditary autoimmune thyroiditis, which suggest that these deposits in the chicken are immune complexes (antigen-antibody complexes) and proved the autoimmune origin of Hashimito's thyroiditis (2, 40).

4. Comments and conclusions

In recent years, the use of modern methods such as conventional TEM, autoradiography, SEM, immunohistochemistry and freeze-etching microscopy for the study of normal and pathologic thyroid cytology has been tremendously increased. Material has been accumulated which has changed our views significantly regarding the thyroid cell organization, the mechanism of synthesis and secretion of thyroid hormones, and the intrinsic histogenetic mechanism of thyroid diseases.

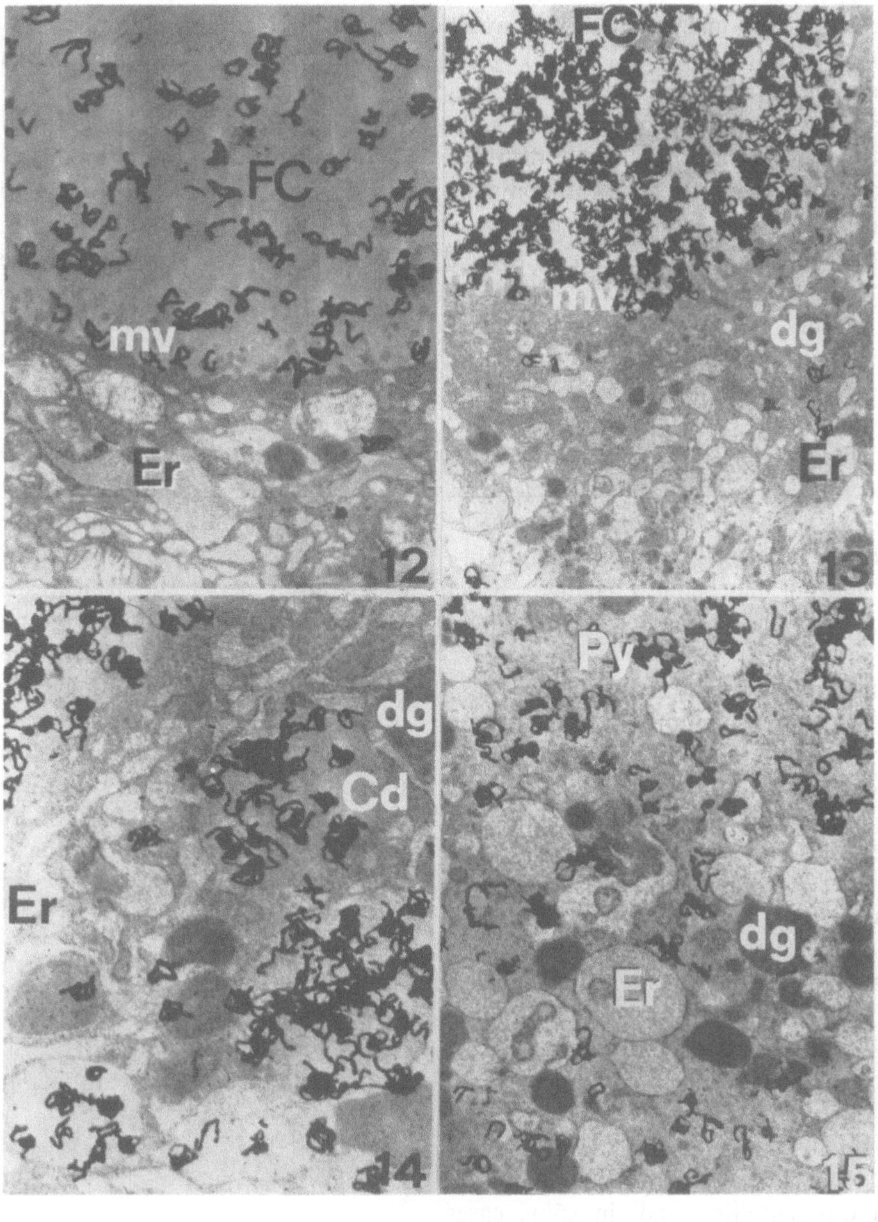

Figs. 12–15. (12) Electron microscopic autoradiogram of rat thyroid. Radioiodine (^{125}I) is mostly accumulated as filaments into the follicular colloid (FC) and over the microvilli (mv); no incorporation is seen over the cytoplasmic organelles. Endoplasmic reticulum (Er) (× 9,600). (13) EM autoradiogram of hyperplastic goiter induced by PGE_2 in rat. A marked radioiodine (^{125}I) incorporation is evident only in the follicular colloid (FC) and microvilli (mv); no developed grains are seen over dense granules (dg) but these are a few in the endoplasmic reticulum (Er) (× 8,000). (14) EM autoradiogram of hyperplastic goiter in rat following RGE_2 administration showing a specific incorporation after 6 h over colloid droplets (Cd); no grains are visible over dense granules (dg) or endoplasmic reticulum (Er) (× 11,200). (15) EM autoradiogram of a rat thyroid-parenchymatous goiter induced by calcitonin and 2 h after ^{125}I administration shows an intracellular distribution, mostly over polysomes (Py) and endoplasmic reticulum (Er); dense granules (dg) (× 10,400).

The purpose of this chapter was 1) to review the ultrastructural, autoradiographic and cell surface changes of the thyroid gland in different experimental and pathologic conditions; 2) to provide more accurate diagnostic criteria for thyroid diseases, and 3) to correlate ultrastructural findings with biochemical and immunological abnormalities. This might enable the student of ultrastructural pathology to diagnose the early stages and the gradual transitional forms between benign and malignant tumors, which cannot be detected by light microscopy only. Thyroid cells have a definite secretory polarity, in that they are bipolar cells. The cell polarity is preserved in the differentiated thyroid tumors, but is lost or distorted in the undifferentiated or poorly differentiated ones. There is a parallelism between the ultrastructural morphology and biochemical function of thyroid cells. When the ultrastructural organization is lost, abnormal or morbid biochemical findings occur. Thus, ultrastructural, autoradiographic and SEM findings are precise characteristics for the diagnosis of thyroid neoplasms, their origin as well as their functional capacity. Hence, correlations between electron microscopy, autoradiography and biochemistry will also extend our knowledge of the histogenesis of endemic goiter, thyroid tumors or myxedema. The morbid biochemical changes which occur in the thyroid gland can be understood only in the light of this ultrastructural changes. A characteristic organization and orientation of cellular organelles is required for a normal function of the thyroid gland. Incorporation of iodine and synthesis of thyroid hormones require an integrity in its ultrastructural organization. High-resolution autoradiography is a very sensitive method in order to detect early changes in the radioiodine localization, thyroglobulin synthesis, iodination and its disorders (Figs. 12–15). During neoplastic transformation, a reorientation of cellular organelles and their secretory polarity occurs. Thus, protein synthesis as well as thyroglobulin synthesis and its iodination are markedly impaired in poorly differentiated or undifferentiated thyroid carcinomas as compared to that of well differentiated tumors. Even in well differentiated thyroid tumors there are different stages of cell differentiation and, in some cases, abnormal thyroglobulin(s) or iodoproteins are frequently observed. Also, the origin of neoplastic cells derived from follicular or parafollicular cells can be determined more precisely with these techniques. Hürthle cell carcinomas (oncocytomas) and other oxyphilic cell tumors originate from follicular cells, whereas medullary carcinomas originate from parafollicular or C-cells; these C-cell tumors secrete immunoreactive calcitonin and in some cases other peptides such as ACTH, β-endorphins and somatostatin. Some medullary carcinomas of the thyroid (MCT) secrete excessive amounts of ACTH, causing Cushing's syndrome. Immunohistochemical studies demonstrate two cell lines: one containing both calcitonin and ACTH and another containing only calcitonin (41).

A few cases of primary squamous cell carcinoma of the thyroid gland have been reported recently. These tumors originate from benign squamous cell metaplasia and are associated with a marked leukocytosis and hypercalcemia (42). Recently an immunoreactive thyroglobulin or C-thyroglobulin was isolated from canine C-cells and from human thyroid medullary carcinomas. This C-thyroglobulin is predominantly a 27S glycoprotein (43). It has been also suggested that MCT are APUDomas or tumors arising from APUD (Amine Precursor Uptake Decarboxylase) cell system which includes C-cells. These neoplastic cells exhibit a pluriotential secretory capacity and can synthesize a variety of ectopic hormones and substances, such as: ACTH, MSH, endorphins, serotonin, kallikrein, prostaglandins, somatostatin, substance P, nerve-growth factor and, surprisingly, melanin (44). These findings indicate a pluripotential capacity of thyroid cells during neoplastic transformation. Therefore, pathologists always should require more sensitive and accurate methods (electron microscopy, autoradiography, SEM, immunocytochemistry and estimation of cellular receptors) for diagnosis of thyroid diseases. Also, the responsiveness of thyroid tumors to therapy (radioiodine, x-ray, chemotherapy) can be more precisely evaluated and conditions for improvement can be dramatically changed when potentially aggressive tumors are recognized in early stages.

Acknowledgements

The skillful help of Daniel Stachelski, a student in the Biological Sciences is greatly acknowledged.

References

1. Heimann P: Ultrastructure of human thyroid. Acta Endocrin 53 (Suppl 110): 1–102, 1966.
2. Lupulescu A, Petrovici A: Ultrastructure of the thyroid gland. Baltimore, The Williams & Wilkins Co, 1968.
3. Klink G, Oertel J, Winship T: Ultrastructure of normal human thyroid. Lab Invest 22: 2–22, 1970.
4. Fujita H: Fine structure of the thyroid gland. Int Rev Cytol 40: 197–280, 1975.
5. Strum J, Karnovsky M: Cytochemical localization of endogenous peroxidaze in thyroid follicular cells. J Cell Biol 44: 655–666, 1970.
6. Nonidez J: The origin of the 'parafollicular' cell: a second epithelial

component of the thyroid gland of the dog. Am J Anat 49: 479–505, 1932.
7. Teitelbaum S, Moore K, Shieber W: Parafollicular cells in the normal human thyroid. Nature 230: 334–335, 1971.
8. McMillan P, Hooker W, Deftos L: Distribution of calcitonin-containing cells in the human thyroid. Am J Anat 140: 73–79, 1974.
9. Hürthle K: Beitrage zur Kenntniss der Secretionsvorgangs in der Schilddrüse. Arch Ges Physiol 56: 1–44, 1894.
10. Sunder-Plassmann P: Basedow-studien. Berlin, Springer, 1941.
11. Nève P, Wollman S: Fine structure of a fifth type of epithelial cell in the thyroid gland of the C_3H mouse. Anat Rec 172: 37–44, 1972.
12. Melander A, Sundler F: Significance of thyroid mast cells in thyroid hormone secretion. Endocrinology 90: 802–807, 1972.
13. Lupulescu A, Andreani D, Andreoli M: Thyroglobulin: synthesis, iodination and hydrolysis. Fol Endocr 20: 385–405, 1967.
14. Wolff J, Williams J: The role of microtubules and microfilaments in thyroid secretion. Rec Progr Horm Res 29: 229–285, 1973.
15. Ketelbant-Balasse P, Rodesch F, Nève P, Pasteels J: Scanning electron microscopic observations of apical surfaces of dog thyroid cells. Exp Cell Res 79: 111–119, 1973.
16. Fujita H, Mishima H, Otsuka N: Freeze-etching images of rabbit thyroid gland. Arch Histol Jap 38: 275–284, 1975.
17. Ishimura K, Okamoto H, Fujita H: Freeze-etching studies on ultrastructural changes of endothelial cells in the thyroid of normal, TSH-treated and thyradin-treated mice. Cell Tiss Res 175: 313–317, 1976.
18. Oswald A: Die Eiweisskörper der Schilddrüse. Hoppe-Seylers Z Physiol Chem 27: 14–49, 1899.
19. Salvatore G, Vecchio G, Salvatore M, Cahnmann H, Robbins J: 27S thyroid iodoprotein isolation and properties. J Biol Chem 240: 2935–2938, 1965.
20. Haddad A, Smith M, Herscovics A, Nadler N, Leblond C: Radioautographic study of *in vivo* and *in vitro* incorporation of fucose-^{3}H into thyroglobulin by rat thyroid follicular cells. J Cell Biol 49: 856–889, 1971.
21. Lupulescu A: Experimental goiter: ultrastructure and autoradiography. Experientia 26: 76–78, 1970.
22. Fujita H: Morphological aspects on the site of thyroglobulin in the thyroid gland. Arch Histol Jap 34: 109–141, 1972.
23. Lupulescu A: Effect of calcitonin on rabbit thyroid glands. Endocrinology 90: 1046–1054, 1972.
24. Lupulescu A, Stebner F: Electron microscopic autoradiography of rabbit and rat thyroid glands following calcitonin. In: Hess M (ed) New York, Wiley, 1975, pp 357–377.
25. Lupulescu A: Colloid resorption in thyroid glands as revealed by scanning electron microscopy. Endokrinologie 72: 347–353, 1978.
26. Lupulescu A: Goiter formation following prostaglandin administration in rats. Am J Pathol 85: 21–36, 1976.
27. Kawada J, Naito S: Morphological changes on the free surface of thyroid cells at the hormone secretion by scanning electron microscopy. Endocrinol Jap 25: 217–223, 1978.
28. Andreoli M, Negri M, Lupulescu A, Monaco F, Badalamenti G: Electron microscopy, autoradiography and the soluble iodoproteins in simple goiter. Fol Endocr 20 721–735, 1967.
29. Lupulescu A, Boyd C: Follicular adenomas: an ultrastructural and scanning electron microscopic study. Arch Pathol 93: 492–502, 1972.
30. Lupulescu A, Andreani D, Monaco F, Andreoli M: Ultrastructure and soluble iodoproteins in human thyroid cancer. J Clin Endocr Metab 28: 1257–1268, 1968.
31. Valenta L, Michel-Bechet M: Ultrastructure and biochemistry of thyroid carcinoma. Cancer 40: 284–300, 1977.
32. Abe Y, Ichikawa Y, Muraki T, Ito K, Homma M: Thyrotropin (TSH) receptor and adenylate cyclase activity in human thyroid tumors: absence of high affinity receptors and loss of TSH responsiveness in undifferentiated thyroid carcinoma. J Clin Endocr Metab 52: 23–28, 1981.
33. Sinadinovic J, Mivic J, Kraincanic M: Thyroglobulin and other soluble thyroid iodoproteins in papillary carcinoma and follicular adenoma: properties and biosynthesis *in vitro*. Endokrinologie 76: 143–151, 1980.
34. Feldman P, Horvath E, Kovacs K: Ultrastructure of three Hürthle cell tumors of the thyroid. Cancer 30: 1279–1285, 1972.
35. Valenta L, Michel-Bechet M, Maloof F: Human thyroid tumors composed of mitochondria rich cells: Electron microscopic and biochemical findings. J Clin Endocr Metab 39: 719–733, 1974.
36. Dickersin G, Vickery A, Smith S: Papillary carcinoma of the thyroid, oxyphil cell type 'clear cell' variant. A light and electron microscopic study. Am J Surg Pathol 4: 501–509, 1980.
37. Deftos L, Bone H, Parthemore J: Immunohistological studies of medullary thyroid carcinoma and C-cell hyperplasia. J Clin Endocr Metab 51: 857–862, 1980.
38. Meyer J: Fine structure of two amyloid-forming medullary carcinoma of the thyroid gland. Cancer 21: 406–425, 1968.
39. Beskid M: C-cell adenoma of the human thyroid gland. Oncology 36: 19–22, 1979.
40. Kalderon A, Bogaars H, Jolly G, Diamond I: Electron-dense deposits in the follicular basal lamina of obese strain chickens with spontaneous hereditary autoimmune thyroiditis. An electron microscopic study. Lab Invest 37: 487–496, 1977.
41. Jolivet J, Beauregard H, Somma M, Band P: ACTH-secreting medullary carcinoma of the thyroid. Cancer 46: 2667–2670, 1980.
42. Saito K, Kuratomi Y, Yamamoto K, Saito T, Kuzuya T, Yoshida S, Moriyama S, Takahashi A: Primary squamous cell carcinoma of the thyroid associated with marked leukocytosis and hypercalcemia. Cancer 48: 2080–2083, 1981.
43. Kameda Y, Ykeda A: C-cell (parafollicular cell) immunoreactive thyroglobulin: Purification, identification and immunological characterization. Histochemistry 60: 155–168, 1979.
44. Marcus J, Dise C, Livolsi V: Melanin production in a medullary thyroid carcinoma. Cancer 49: 2518–2526, 1982.

Author's address:
Wayne State University
School of Medicine
Detroit, MI 48201, USA

CHAPTER 26

Fine structure of the normal human parathyroid gland as revealed by thin section and freeze-fracture with regard to some pathological conditions

JÜRGEN THIELE

1. Introduction

The functional activity of the parathyroid gland in all terrestrial vertebrates is regulated by the concentration of the ionized calcium in the blood plasma. Lowering of the free or ionized fraction of the total plasma calcium acts as a feedback mechanism stimulating the parathyroid chief cell to synthesize and secrete more hormone. The delicate control of the calcium homoeostasis which is essential for normal neuromuscular reaction needs the vital function of the parathormone. Consequently functional and fine structural relationships are closely associated with changing aspects of the polypeptide-hormone producing organelles and alterations related to secretory events (1, 2, 3, 4, 5, 6). Additionally, experimental results after stimulation and suppression of parathyroid tissue (7, 8, 9, 10) and pathological conditions in animals (2, 4) and man with hyperplasias and adenomas of the parathyroid gland (1, 3, 4, 5) have to be considered. These findings obtained from experiments and naturally occurring pathological lesions in man (hyperparathyroidism) should be correlated with biochemical data of parathyroid hormone structure, synthesis and degradation (11, 12, 13, 14, 15, 16, 17).

By experimental studies it was demonstrated that the normal chief cell can be followed through a sequence of structural alterations corresponding to the functional cycle of hormone synthesis and regression (1, 18, 19). These different phases of cellular activity are summarized in Table 1. Interpretation of freeze-fracture replicas requires to realize that the cleavage plane may run across the cell thus exposing cytoplasmic features in a similar way than thin sections. Frequently however, the fracture follows the layers of the plasma membrane splitting it into two leaflets which are termed P-face (protoplasmic half) and E-face (external half) according to the commonly accepted nomenclature (20).

2. Cell types of the normal human parathyroid as revealed in freeze-fracture replicas

2.1. *Chief cell*

The parathyroid chief cell displays a diameter of 15–20 μm (3) and the most prominent feature of freeze-fracture replicas is the spherical nucleus containing many pores in random distribution (Fig. 1a). In stages of increased functional activity nuclear pore formation may be increased as calculated from other tissues (21) and nuclear diameter is found to be

Table 1. Semiquantitative evaluation of prominent ultrastructural alterations in the normal human chief cells due to its cyclic function [see also the schematic drawing of Shannon and Roth (19)].

Cyclic function of chief cell/ Ultrastructural features	Synthesis	Granulogenesis	Exocytosis	Regression
Rough endoplasmic reticulum	+++	++	++	+
Golgi apparatus	++	+++	++	+
Mitochondria	+++	++	++	+
Glycogen	++	+	++	+++
Lipid bodies	+	+	++	+++
Lysosomes	–	+	++	+++
Secretory granules	–	++	+++	+
Endocytotic vesicles	+	+	+++	–
Tortuosity of the plasma membrane	+	++	+++	–

Motta, PM (ed): Ultrastructure of endocrine cells and tissues. ISBN-13: 978-1-4613-3863-5

significantly greater in human adenomatous cells compared with the contralateral suppressed 'normal' glands (22). The cytoplasm contains organelles which are generally associated with protein synthesis (23): several stacks of cisternae of the rough endoplasmic reticulum (RER), perforated saccules and small Golgi vesicles and large spheroidal or roll shaped mitochondria (Fig. 1a). Further there are large lipid bodies with partial coalescence and which exhibit a conspicuous smooth surface on the cleavage planes. The plasma membrane follows a slightly undulating course and shows a few interdigitations with adjacent cells (5).

2.2. Oxyphil cell (Onlocyte)

The second cell type composing the normal parathyroid is the onkocyte or oxyphil cell. This derives from the chief cell and appears around puberty in the human gland (4). The number of oxyphils increases approximately from the fifth decade onwards (24) and functional activity is questionable since so called oxyphil adenomas associated with endocrine activity (primary hyperparathyroidism) always reveal an admixture of chief cells and transitional oxyphil cells (25, 26). The nuclei are generally round and smaller

Fig. 1. Survey of cell types of the normal human parathyroid gland. (a) *Chief cell* with nucleus (N) displaying many pores and cytoplasm containing several large lipid vacuoles in cross (L) and tangential cleavage planes (arrow) with a smooth surface, a small Golgi apparatus (G), several cisternae of the rough endoplasmic reticulum (ER) and numerous rod and spherical shaped bodies mostly corresponding to mitochondria; (b–c) *Oxyphil cells (onkocytes)* with abundance of mitochondria in predominantly cross (b) and tangential (c) fracture planes surrounding the nuclei (N) (a–c × 11,900). All illustrations of freeze-fracture replicas are mounted in such a way that shadowing is directed from below, the nomenclature of cleavage planes follows Branton et al. (20): E is E-face and P is P-face of the split plasma membrane.

than those of the chief cells and the cytoplasm is filled with numerous mitochondria often large and of a bizarre shape. These cells contain only few cisternae of the RER, have a very small Golgi region and infrequently secretory granules (1, 2, 3, 4). Freeze-fracture corroborates these thin section findings (5) by showing abundant and closely packed mitochondria in cross (Fig. 1b) and tangential cleavage planes (Fig. 1c).

3. Comparison of freeze-fracture with thin section ultrastructure

3.1. Rough endoplasmic reticulum (RER)

In all cells proteins for export i.e. secretory products are synthesized on the polysomes (polyribosomes) attached to the cisternal membranes of the RER (23). In the parathyroid chief cell one should therefore expect extensive stacks of cisternae during stimulated hormone synthesis comparable with the immunoglobulin producing plasma cells. This concurrence of measurable parathormone synthesis and development of the RER is obviously evident when human and bovine parathyroid tissues are cultured and treated with various stimulating and suppressive agents particularly changing calcium concentrations (7, 8, 9, 10). These *in vitro* experiments confirm and extend earlier ultrastructural studies on the secretory cycle of the chief cells (19) and parallel observations made on different pathological conditions particularly of the bovine (2, 4) and human parathyroid gland (3, 4, 5, 27, 28). After synthesis on the polysomes, a precursor compound of parathormone (so called pre-proparathormone) is channelled from the cisternae of the RER to the Golgi region (13, 15). The large precursor polypeptide or pre-proparathormone is cleaved by proteolysis probably intracisternally into pro-parathormone (13, 17) and translocated by an energy-dependent process in so called transition elements (vesicles) to the saccules of the Golgi apparatus. This important process is assumed to be mediated by microtubules (16, 29). Consequently the ultrastructure of the RER is most favourably encountered in actively synthesizing chief cells and consists of stacks of parallel arrays of flattened cisternae with studded rosette-like polysomes (Fig. 1a). These cisternae exhibit a few interruptions or pores which may be detectable in thin sections. Chief cells of adenomas or carcinomas with a high endocrine activity may not only show abundant cisternae arranged in extensive stacks, but also a number of annulate lamellae. These resemble piled up portions of the nuclear membrane and like it have frequently spaced pore complexes (1, 30, 31). In freeze-fracture replicas the cisternal or luminal aspect of the cleaved membrane (E-face) displays only a few scattered granules, whereas the cytoplasmic leaflet (P-face) has a finely particulated appearance and relative rarely pores are observed as described in other specimens (32, 33). Moreover, there are two additional features detectable on enface views of the split cisternal membrane besides those openings: small peak-like prominences or depressions are clearly distinguishable from the intramembrânous particles and probably indicate the irregular or slightly angulated outline of the cisternal membranes. Finally a striking patchwork pattern or so called mosaic of flat areas devoid of any particles is shown in both fracture faces (5) and this agrees e.g. with the findings of rodent liver cells (34).

3.2. Mitochondria

Number, size and shape of mitochondria may be quite variable and are probably related to endocrine activity of the parathyroid cells. Most striking are the cup or ring shaped giant mitochondria in chronically stimulated chief cells of secondary hyperparathyroidism (1, 5, 31, 35), in adenomas with high endocrine activity (5) and particularly in the oxyphil cells or onkocytes. This second cell type of the normal parathyroid gland is characterized by an excessive assembly of large mitochondria with well developed cristae (Figs. 1b, c), which cause their granulated and acidophilic appearance of light microscopy (4, 24). Besides the circular array of long cristae in the large cup shaped mitochondria, freeze-fracture may expose the inner surface (E-face) of the inner mitochondrial membrane and then shows small elevations possibly corresponding to the crossly cleaved bases of the cristae stalks (5). The outer mitochondrial membrane displays a smooth (P-face) or highly granulated (E-face) surface and consequently may be compared with the mitochondria of other tissues (34, 36). Another remarkable feature of the mitochondria in stimulated chief cells of adenomas or hyperplastic glands are dense round inclusions of the matrix. These occur exclusively in all cases with a hypercalcemic stage of primary hyperparathyroidism and are suggested to represent calcium deposits (1, 5).

3.3. Lipid and lysosomal bodies

Complex or isolated lipid bodies are a very conspicuous constituent of the normal and suppressed chief cell and seem to present residues of intracellular lysosomal hormone degradation (1, 2, 4, 5, 19). The relatively large amount of lipids in those residual or lipofuscin bodies particularly in the resting or atrophic 'normal' glands contralateral to parathyroid adenomas (in primary hyperparathyroidism), causes a positive reaction with Sudan II or IV stain and therefore allows a rapid identification (37). The lysosomal origin of these dense and vacuolated bodies and their generation from the Golgi apparatus was demonstrated in normal, adenomatous und hyperplastic human parathyroid glands (19). Thin sections show either relatively isolated electron dense or vacuolated bodies or large groupings of translucent, membrane bound vacuoles. Freeze-fracture reveals a smooth or slightly lamellated surface in tangential or cross cleavage planes (Fig. 1a), compatible with neutral fat and cholesterin compounds (38) and therefore corresponding with the sudanophilic droplets of light microscopy (37). Appearance and functional implications of these dense and vacuolated bodies of lysosomal nature is elucidated by several experiments *in vivo* and *in vitro* and pathological conditions as well. In acute hyperparathyroidism with a rapid extrusion of secretory granules from adenomatous chief cells and a maximum of coupled exo- and endocytosis (see section 4), lysosomal bodies are numerous following complete degranulation and possibly participate in the digestion of the retrieved granule membrane (6, 28). Ultrastructure of parathyroid glands (mostly human parathyroid adenomas) in organ culture with the possibility of an exact evaluation of hormonal activity, displays an increased number of those dense and vacuolated bodies when various suppressive agents are added to the culture medium (7, 8, 9, 10). In agreement with these *in vitro* experiments is the fine structure of surgically removed parathyroid adenomas in patients who received a suppressive therapy prior to operation (39). This remarkable new generation of lipid containing bodies confirms the speculation that in the majority, they present residual bodies due to autophagocytosis or autolysosomes i.e. reaction products of so called crinophagy or digestion of excess of secretory granules (7).

3.4. Glycogen

Large groupings of glycogen granules occupy extensive areas of the cytoplasm in the parathyroid chief cell especially in those which stay in the resting phase of their functional cycle (Table 1). However, the meaning of these extensive glycogen accumulations is poorly understood when considering the pathways of hormone synthesis and is generally regarded as energy supply or storage (7, 27). Since clusters of glycogen are present in both, active and so called atrophic chief cells, the quantitative alterations which occur during the different functional stages are not easily revealed. Animal experiments (40) support the assumption of an energy supply role, although chronically stimulated parathyroid glands in secondary hyperparathyroidism due to renal failure contain abundant glycogen as so called small water-clear cells of light microscopy (2, 3, 5, 31). The glycogen deposits consist of densely packed isolated spheroidal granules (β-particles, Fig 2a, b) as described in cells of various tissues (41, 42). This monoparticulated glycogen with a diameter of about 30 nm represents a complex structure since it is composed of many filamentous subunits, the so called γ-subunits or elements (41, 42, 43). These γ-subunits may be demonstrated by several methods i.e. negative staining (43) or the alkaline bismuth procedure (44) as shown in Figure 2a. In comparison with this appearance of the β-glycogen particle, freeze-fracture (Fig. 2b) shows larger cobble stone-like elements of about 50–60 nm diameter with a finely granulated suface. The granularity or roughness of the cleavage plane may result from the many γ-subunits. The enlarged size of the single particle is explained when considering the light halo around each individual β-particle which is not distinguished in freeze-fracture replicas. This translucent space should be consistent with the enzyme rich moiety bound to glycogen granules (45, 46) and is thought to be responsible for a rapid lytic or synthetic activity due to energy requirements of the parathyroid chief cell.

3.5. Golgi apparatus

The Golgi apparatus in the normal and pathologically altered parathyroid chief cell is greatly variable in shape and size. These alterations are closely related to secretory activity (Table 1) and are somewhat similar to the development of the RER (2, 4, 5, 9). These changes are unterstandable from the assumed three principal functions of the Golgi apparatus: 1) the conversion of the proparathyroid hormone into parathyroid hormone (13, 15), a process which may be inhibited by certain agents and then leads to a disruption of the Golgi saccules (47); 2) the encapsulation of the final hormone moiety by a carbohydrate rich limiting membrane as in other proteohormone secreting cells (23); 3) the production of lysosomal bodies (19) which may function as autolysosomes (i.e. crinophagy) or play a major role in the digestion of residues of the secretory granule membrane (endocytic invaginations and vesicles) following exocytosis (6). Serial thin sections and freeze-fracture replicas show a complex structure of the Golgi region which is in good agreement with the suggested pathways of parathormone synthesis and may be especially conspicuous in the actively secreting chief cell (Table 1) or hyperfunctioning gland of acute hyperparathyroidism (28). The ultrastructural aspect in transversely cut thin sections shows 4–6 saccules with several interruptions or pores arranged in a convex or forming face and a concave or maturing interior. A closer view on the saccules in freeze-fracture replicas exhibits a central almost non-perforated area (so called dictyosome) and a highly fenestrated periphery(Fig. 2c) obviously giving rise to tubules as in various animal and plant cells (48, 49, 50). These tubular structures may partially correspond to the small Golgi vesicles in the vicinity of the stack of saccules (50). Others are mostly representing transition elements which originate from the cisternal membranes of the RER and are speculated

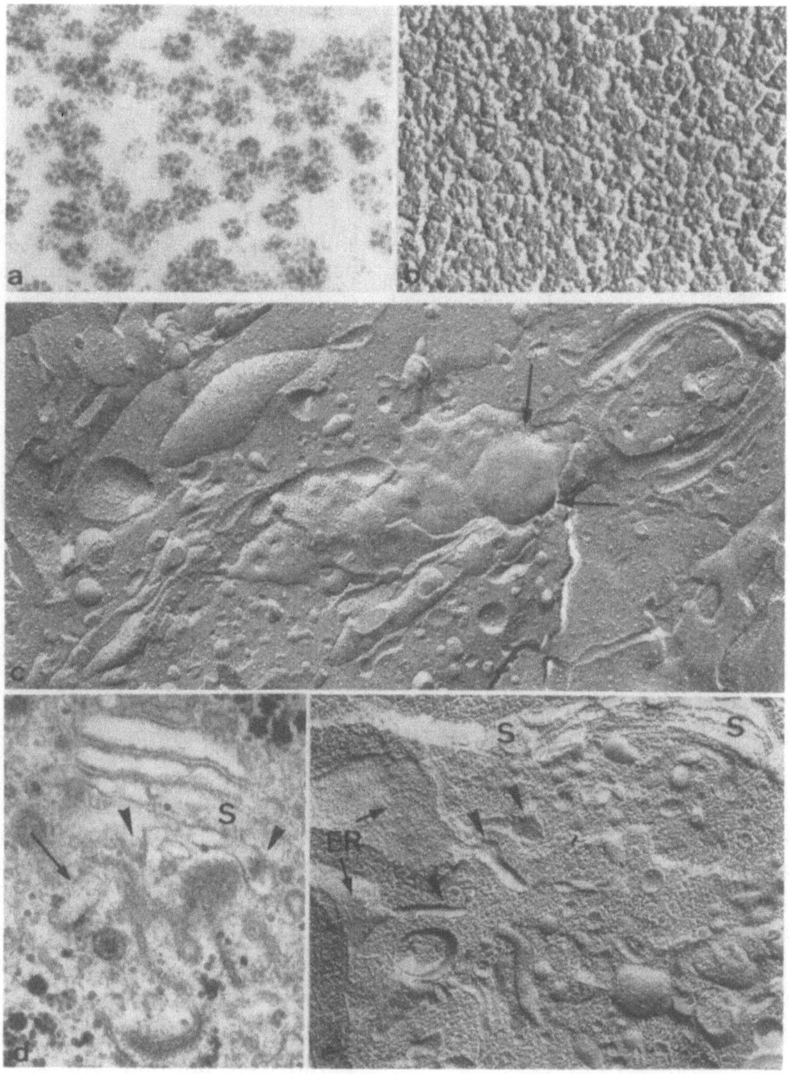

Fig. 2. Glycogen and Golgi apparatus. (a–b) Thin section of β-glycogen granules shows a loose arrangement with a translucent halo and many subunits (γ-subunits) following bismuth staining (a). Freeze-fracture reveals a large cobble stone-like surface without discrimination of the halo and a granulated aspect corresponding with the γ-subunits (b); (c) Freeze-fracture survey of the Golgi saccules shows a central non-perforated area or a so called dictyosome (arrows) and a periphery with many pores and tubular structures surrounded by small spheroidal bodies (vesicles); (d–e) Thin section and freeze-fracture of the interior of the Golgi region reveal a condensation in the inner saccules (S) possibly of secretory material (d, arrow heads). There is further a tubular network with connections to cisternal portions of the endoplasmic reticulum (arrow), besides production of dense bodies or pro-secretory granules (d). A corresponding replica demonstrates beneath the inner saccules (S) several tubular extensions (e, arrow heads), of the endoplasmic reticulum (ER) in close neighbourhood to spherical bodies or pro-secretory granules (e) (a, b × 140,000; c–e × 57,400).

to provide transport vesicles for the pro-parathormone. The vesicular and sometimes dense tubular structures may also be recognized in freeze-fracture replicas since showing extensions or processes of the cisternal membranes of the RER and small spheroidal structures (compare Fig. 2d with Fig. 2e).

The interior of the Golgi stack is occupied by a dense tubular reticulum which pursues an irregular course and therefore may be only partially demonstrable (Fig. 2d). Serial thin sections disclose that dense membrane limited bodies arise from these tubules which in the majority are consistent with so called pro-secretory granules (Fig. 2d). Therefore this electron dense reticulum is thought to be identical with the transtubular network of Rambourg (49) and the so called GERL of Novikoff et al. (51). This remarkable structure of the maturing face of the Golgi region is not only the source of lysosomal bodies, but the predominant site for secretory granule formation and may be clearly separated from the Golgi saccules by cytochemical methods (52, 53) and is obviously demonstrated by three-dimensional reconstruction (49, 53). In so called primary water-clear cell hyperplasia of the human parathyroid gland the cytoplasm of those tumor cells is filled by large almost translucent vacuoles (54). Ultrastructural studies reveal that these vacuoles probably originate from the Golgi apparatus (55) and are regarded as some kind of altered secretory granules due to a partial disturbance of the Golgi function (56).

3.6. Secretory granules and filaments

Secretory granules are spherical membrane surrounded bodies of about 200–300 nm diameter with a dense or finely particulated content in thin sections, but they exhibit no special features in freeze-fracture replicas (5). Density and uniform circular size distinguish them from the pleomorphous and frequently vacuolated lysosomal bodies (3). Frequently, dense bodies with a more flocculent content and various sizes are observed near the concave face of the Golgi apparatus and possibly present so called pro-granules (Fig. 2d), while the more mature forms have a trend to marginate along the plasma membrane (Fig. 3a). The amount of secretory granules is a reliable parameter for the secretory activity of the chief cell (Table 1): in fact, they are numerous preceding exocytosis in adenomatous and hyperplastic parathyroids (4, 6), sparce in adenomatous glands of acute hyperparathyroidism due to enforced extrusion (28) and show a variable accumulation in the normal human chief cell. Immunocytochemical localization of parathyroid hormone at the ultrastructural level reveals that the reaction product is largely confined to the secretory granules i.e. that those granules indeed contain the parathormone (57).

The secretory granules are characteristically associated with small bundles of microfilaments. These microfilaments are thought to be contractile in nature, present the so called cytoskeleton and probably function as a tool for the movement of the granules (2, 5). In addition to the small bundles of microfilaments, microtubules are only infrequently detected, often in close neighbourhood with the Golgi region. This location may support the experimental finding that microtubules mediate the transport of newly synthesized parathormone to the site of its intracellular cleavage (29).

3.7. Plasma membrane

The plasma membrane of the normal parathyroid chief cell exhibits an almost straight course with only a few finger-like processes at the junctions with several other cells. An increased tortuosity showing many interdigitations is observed starting at the secretory phase of the functional cycle and being excessively apparent in highly stimulated glands in hypocalcemic stages (2, 7, 8, 9, 18). This enlargement of surface area is easily recognized in freeze-fracture replicas as a very valuable indicator of the secretory condition, since it is suggested to present a sequel of enforced endo-exocytic activity during hormone release in highly activated parathyroid cells (28). Intercellular junctions between human chief cells consist of desmosomes with a regular ultrastructure in thin sections and assemblies of densely arranged granules on both fracture faces (P- and E-face) possibly compatible with the anchoring sites for the tonofilaments (16). No other specialized junctions comparable with the findings in the endothelium (see section 5.2) are described for human parathyroid tissue (5, 6). This is in contrast to the rat parathyroid gland where both tight and gap junctions have been shown between the chief cells (58). Pairs of cilia with well developed central filaments and basal rootlets are a common occurrence in normal, adenomatous and hyperplastic parathyroid cells (1, 3, 4, 5, 31), however, their function is not clear.

4. Exocytosis of secretory granules

Extrusion of the secretory granules is managed by a membrane fusion process of both the secretory granule membrane and the plasma membrane with following release of the parathormone containing interior or the core. This discharge obviously occurs very rapidly so that remnants of the core are not visible between the structures of the basement membrane or in the interstitial space. The different ultrastructural aspects are not well understood from thin sections since exocytosis is an infrequent event in normal chief cells and therefore it is only conspicuously expressed in highly stimulated cells as in adenomas associated with acute hyperparathyroidism (6). Using this cytological model for demonstration of exocytosis, a combination of freeze-fracture and thin section may help to clarify each stage of this important biological process. The very first phase consists of a margination of secretory granules which are lined up along the plasma membrane (Fig. 3a) and freeze-fracture reveals a bulging of each granule (Fig. 3b). However, the margination is not irregular, but displays a certain direction, because the limiting membranes of several granules seem to adhere to each other and thus lead to a

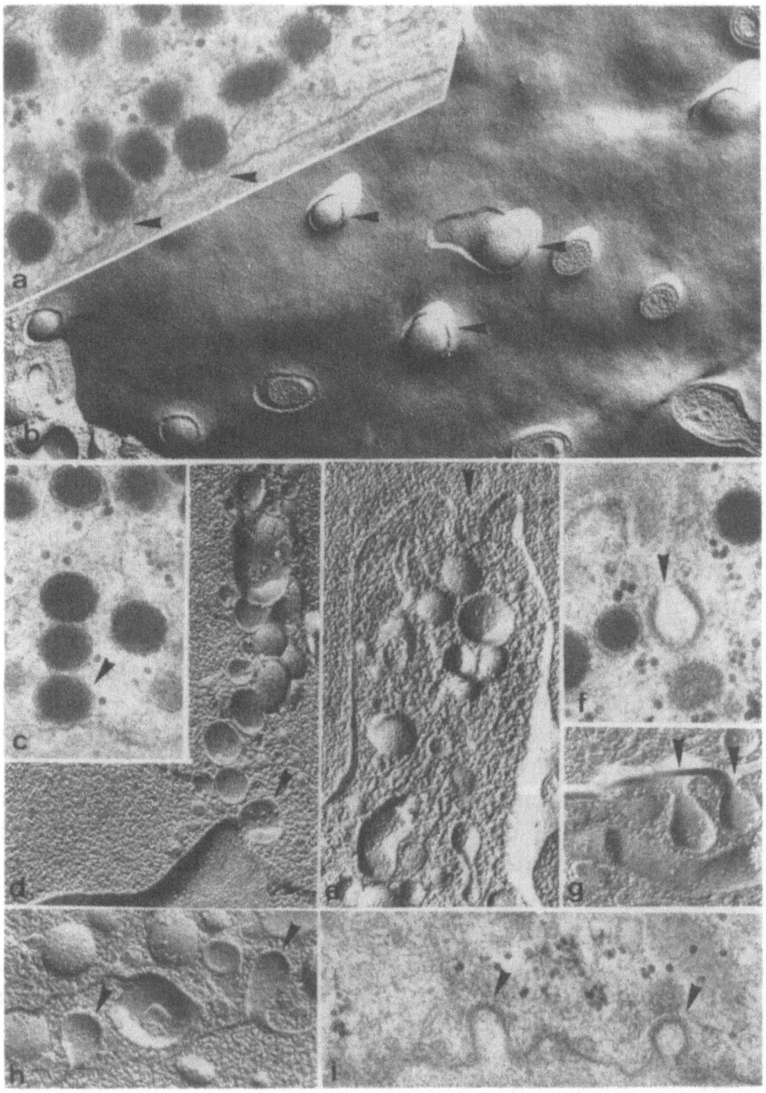

Fig. 3. Exo- and endocytosis in the highly stimulated chief cell of adenomas with the clinical picture of acute hyperparathyroidism. (a–b) Thin section shows a margination of secretory granules with attachments (arrow heads) of both granule and plasma membranes (a) followed by a bulging forward as demonstrated in tangential cleavage planes of freeze-fracture replicas (b, arrow heads); (c–d) Tandem-like arrangement of secretory granules during exocytosis with a row-like deployment behind the first fused granule (c, d arrow heads); (e–i) Endocytosis of granule membranes after extrusion of the core displays bristle coated invaginations of different shapes in thin sections (f, i, arrow heads) and corresponding bottle-like infoldings in freeze-fracture replicas (g, h, arrow heads) in close connection with vesicles and apparently intact secretory granules (f, arrow heads) or spherical bodies (h, arrow heads), respectively (a–i × 57,400).

tandem-like extrusion (Fig. 3c, d) comparable with that of the pancreatic beta-cells (59). The following events such as the fusion of granule and plasma membrane have been investigated in detail elsewhere [see review by Thiele (6)]. The final stage is characterized by an enormous endocytic activity with many bristle coated invaginations (Fig. 3f, i) or bottle shaped infoldings of various size (Fig. 3g, h) which bear evidence for a retrieval or recycling of the granule membrane in agreement with other endocrine glands (60, 61, 62). This assumption of a membrane shuttle secondary to discharge is further supported by *in vitro* experiments: here a significant increase of endocytic activity with many vesicles and a remarkable convolution of the cell membranes was noticed after enforced stimulation by low calcium treatment of parathyroid tissue in organ culture (8). In addition to this process of secretory granule extrusion, recent experimental studies lead to the conclusion that there may be also a so called by-pass secretion without package of the hormone molecules into membrane limited granules (10, 15).

5. Interstitium

5.1. Interstitial cells and nerves

The extracellular space is bordered by the basement membranes of both the parathyroid cells and the microvasculature and exhibits a flocculent, electron-dense material containing numerous bundles of collagen fibers. In between this felt-like reticulum of fibers a variety of cells may be noticed, but most frequently fibroblasts, small lymphocytes and occasionally mast cells.

Along the vessels many non-myelinated nerve fibers are observed. The latter non frequently parallel to arterioles and precapillary arterioles and exhibit vesicle-rich processes partly surrounded by Schwann cells. These findings indicate that parathyroid vessels have a rich vasomotor innervation predominantly adrenergic, which may play an important role in the control of the blood flow (63).

5.2. Vessels

The most prominent feature of the parathyroid interstitium are the vessels which may be properly designated as microvasculature since they are consistent with capillaries, small venules and arterioles. The endothelial cells are linked to one another by intercellular junctions of two different types: occluding (tight) junctions and communicating (gap) junctions. In thin sections tiny areas of a close or punctate attachment may be detected between the attenuated processes of the endothelial cells (Fig. 4a). Freeze-fracture following the intercellular plasma membranes demonstrate corresponding ridges (or grooves respectively) with a striking discontinuity pursuing an irregular course (Fig 4c–e). Although they show many interconnections, those strands or furrows are completely missing in some areas, therefore being more compatible with a fascia than a zonula occludens and consequently represent a very leaky type of interconnection between the endothelial cells. Gap junctions are relative rarely encountered and often exhibit a close relation to those strands of the fascia occludens (so called intercalated types, Fig. 4e). According to Simionescu et al. (64, 65) a segmental differentiation of cell junctions may be discriminated depending on vessel type. The capillary endothelium is formed by prominent profiles of more densely arranged strands with branchings or staggered deployment of particles which may also lay isolated (Fig. 4c). In contrast, venules show discontinuous low-profile ridges or furrows devoid of particles with a fascia-like appearance (Fig. 4d) and only infrequently small gap junctions. The cell junctions of arterioles have a more elaborate structure of several strands, many surrounding particles and extended often intercalated gap junctions (Fig. 4e).

The flattened processes of the endothelial cells of capillaries may be continuous or fenestrated. The latter type displays circular openings bridged by a diaphragm with a central knob (Fig. 4b) as described in a variety of specimens (66, 67). Freeze-fracture of these fenestrae (Fig. 4f) demonstrates a tiny central prominence (or knob) on the cytoplasmic face and a small disc-like elevation probably corresponding to this septum and its central knob (67, 68). Although there may be an appreciable variation within each capillary type, so that each vascular bed has a certain degree of structural specifity (69), those interruptions of the zonula (fascia) occludens could represent the morphological equivalent of the small pore or slit system, while the fenestrae may be compatible with the so called larger pore system as postulated by physiologists. The large areas of fenestrated endothelium may consequently allow an easy passage for the relatively large molecule of parathormone with its complex polypeptide chain of 84 amino acids (13, 14, 15).

A further conspicuous structure of small arteries are peculiar rod shaped dense bodies composed of several tubules (Fig. 4g) which are consistent with so called Weibel-Palade bodies (70). These bodies have a tubular substructure and are closely related to the

Golgi apparatus (71, 72), also in freeze-fracture replicas (Fig. 4h). However, they do not possess a lysosomal function (73), but are suggested to contain a procoagulative substance which may be expulsed towards the vessel lumen (71, 72).

6. Concluding remarks

The parathyroid chief cell may serve as an ideal model for a polypeptidehormone synthesizing cell with a relatively simple feed back mechanism of function. Thus it closely resembles e.g. the beta cell of the pancreatic islets also in regard to the similarities of ultrastructure. However, it may be speculated that the unique functional cycle of each normal parathyroid cell is probably responsible for the lack of any specialized i.e. communicating junction between adjacent cells which are observed in some other endocrine glands. There are striking dissimilarities between parathyroid cells of different species which may be related to differences in the calcium metabolism and development and maturation of the osseous skeleton (2, 4). In this context is should be

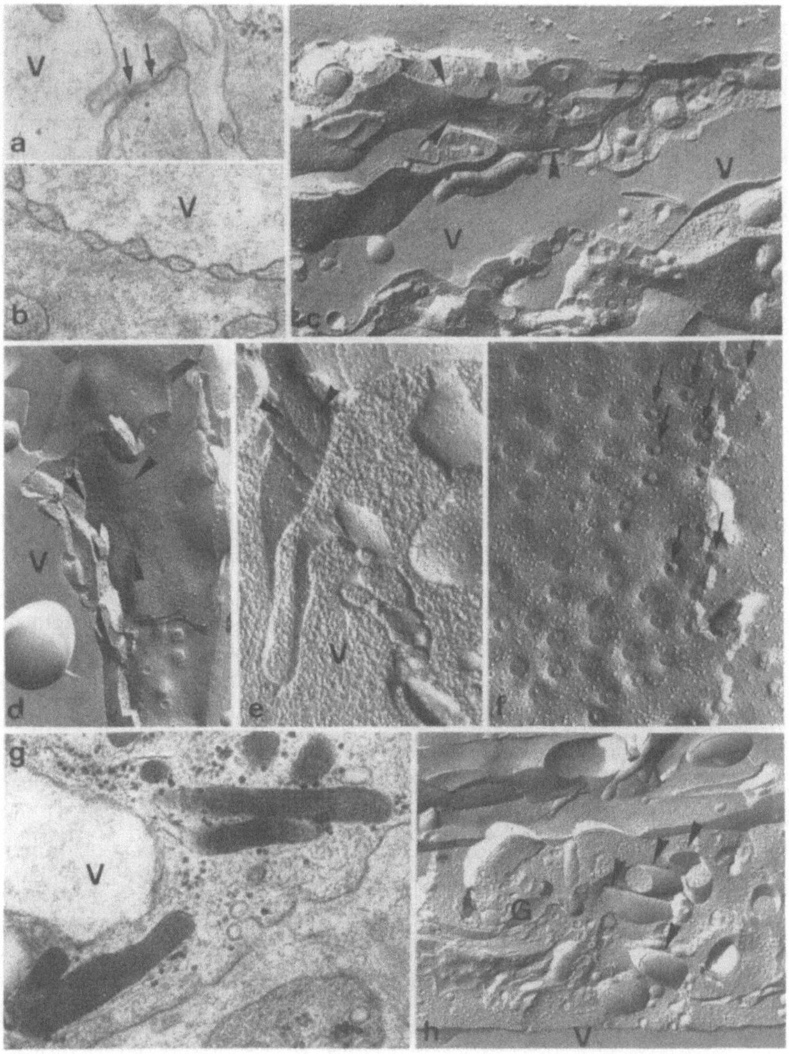

Fig. 4. Vessels of the parathyroid gland (V: vascular lumen). (a–b) Thin sections with flattened endothelial cells showing areas of close attachments (a, arrows) or fenestrations bridged by a thin diaphragm (b); (c–e) Freeze-fracture with a fascia occludens of a capillary endothelium displays prominent ridges (c, arrow heads) and low grooves possibly of a small venule (d, arrow heads); further an intercalated gap junction of an arteriole with assembly of particles and surrounding strands of a zonula occludends (e, arrow heads); (f) Endothelial fenestrations in en-face aspect demonstrate a diaphragm with a central prominence or knob (f, arrows); (g–h) Weibel-Palade granules with a rod shaped appearance in thin sections exhibit tubular inner structures (g). Corresponding freeze-fracture replicas (h) show cleaved elongated bodies (arrow heads) in close neighbourhood to the Golgi apparatus (G) (a–e, g–h × 29,400; f × 57,400).

mentioned that considerable progress has been made to keep tissue of human parathyroid glands (mostly surgically removed adenomas) in culture medium for some time. This offers the opportunity to study the effects of a changing calcium concentration and other agents very intensely by the alterations of the ultrastructure and in combination with direct measurements of parathormone production (7, 8, 9, 10). Comparable findings with certain functional implications may be also obtained by a careful selection of human parathyroid glands or tumors respectively. Patients with hyperparathyroidism are particularly useful for the investigation of some ultrastructural features which may be prominently expressed in this condition like endo-/exocytosis (6) or suppression of function in an atrophic gland (3, 4). An open question which currently needs future attention is the so called by-pass secretion (10, 15) with parathormone discharge preceding any package of secretory granules. In addition to the comprehensive study of parathormone biosynthesis, secretion and degradation of Morrissey and Cohn (17), these experiments have further to be extended to cultured human parathyroid tissue. Finally the vacuoles of the rarely occurring primary water-clear cell hyperplasia should by reviewed be freeze-fracture techniques to disclose possible relations with the membranes of the Golgi saccules (54, 55, 56). Future research certainly must include the fine structure also of the parafollicular cells in man in comparison with changes of the parathyroid chief cells, in order to better understand the various pathological conditions of these closely related endocrine tissues.

References

1. Altenähr E: Ultrastructural pathology of parathyroid glands. Curr Top Path 56: 1–54, 1972.
2. Capen CC: Functional and fine structural relationships of parathyroid glands. Adv Vet Sci Comp Med 19 (188) 249–286, 1975.
3. Nilsson O: Studies on the ultrastructure of the human parathyroid glands in various pathological conditions. Acta Path Microbiol Scand Section A 263: 5–88, 1977.
4. Roth SL, Capean CC: Ultrastructural and functional correlations of the parathyroid gland. Int Rev Exp Path 14: 161–221, 1974.
5. Thiele J: Human parathyroid gland: a freeze fracture and thin section study. Curr-Top Path 65: 31–80, 1977.
6. Thiele J: Changes in the plasma membrane associated with endocrine activity. A thin section and freeze-fracture study on the human parathyroid chief cell. Virchows Arch B Cell Path 34: 219–237, 1980.
7. Chertow BS, Buschmann RJ, Henderson WJ: Subcellular mechanism of parathyroid hormone secretion. Ultrastructural changes in response to calcium, vitamin A, vinblastine, and cytochalasin B. Lab Invest 32 (2): 190–200, 1975.
8. Chertow BS, Manke DJ, Williams GA, Baker GR, Hargis GK, Buschmann RJ: Secretory and ultrastructural responses of hyperfunctioning human parathyroid tissues to varying calcium concentration and vinblastine. Lab Invest 36 (2): 198–205, 1977.
9. Dietel M, Dorn G, Montz R, Altenähr E: Influence of vitamin D_3, 1,25-dihydroxyvitamin D_3 and 24,25-dihydroxyvitamin D_3 on parathyroid hormone secretion, adenosine 3′, 5′-monophosphate release, and ultrastructure of parathyroid glands in organ culture. Endocrinology 105 (1): 237–245, 1979.
10. Dietel M, Dorn-Quint G: By-pass secretion of human parathyroid adenomas. A particular intracellular form of rapid adaptation to external stimuli. Lab Invest 43 (2): 116–125, 1980.
11. Arnaud CD, Sizemore GW, Oldham SB, Fischer JA, Tsao HS, Littledike ET: Human parathyroid hormone: glandular and secreted molecular species. Amer J Med 50: 630–638, 1971.
12. Habener JF, Powell D, Murray TM, Mayer GP, Potts JT: Parathyroid hormone: secretion and metabolism *in vitro*. Proc Nat Acad Sci USA 68: 2986–2991, 1971.
13. Habener JF, Amherdt M, Ravazzola M, Orci L: Parathyroid hormone biosynthesis. Correlation of conversion of biosynthetic precursors with intracellular protein migration as determined by electron microscope autoradiography. J Cell Biol 80: 715–731, 1979.
14. O'Riordan JLH, Potts JT, Aurbach GD: Isolation of human parathyroid hormone. Endocrinology 89: 234–239, 1971.
15. Cohn DV, Hamilton JW: Newer aspects of parathyroid chemistry and physiology. Cornell Vet 66: 271–300, 1976.
16. Chu LLH, MacGregor RR, Cohn DV: Energy-dependent intracellular translocation of proparathormone. J Cell Biol 72: 1–10, 1977.
17. Morrissey JJ, Cohn DV: Secretion and degradation of parathormone as a function of intracellular maturation of hormone pools. Modulation by calcium and dibutyryl cyclic AMP. J Cell Biol 83: 521–528, 1979.
18. Roth SI, Raisz LG: The course and reversibility of the calcium effect on the ultrastructure of the rat parathyroid gland in organ culture. Lab Invest 15: 1187–1211, 1966.
19. Shannon WA, Roth SI: An ultrastructural study of acid phosphatase activity in normal, adenomatous and hyperplastic (chief cell type) human parathyroid glands. Am J Pathol 77: 493–506, 1974.
20. Branton D, Bullivant S, Gilula NB, Karnovsky MJ, Moor H, Mühlethaler K, Nothcote DH, Packer L, Satir B, Satir P, Speth V, Staehelin LA, Steere RL, Weinstein RS: Freeze-etching nomenclature. Science 190: 54–56, 1975.
21. Maul GG, Price JW, Lieberman MW: Formation and distribution of nuclear pore complexes in interphase. J Cell Biol 51: 405–418, 1971.
22. Lloyd HM, Jacobi JM, Cooke RA: Nuclear diameter in parathyroid adenomas. J Clin Pathol 32 (12): 1278–1281, 1979.
23. Palade G: Intracellular aspects of the process of protein synthesis. Science 189: 347–358, 1975.
24. Christie AC: The parathyroid oxyphil cells. J Clin Path 20: 591–602, 1967.
25. Heimann P, Hansson G, Nilsson O: Primary hyperparathyroidism in a case of oxyphilic adenoma. Acta Path Microbiol Scand Section A 79: 10–14, 1971.
26. McGregor DH, Lotuaco LG, Rao MS, Chu LLH: Functioning oxyphil adenoma of parathyroid gland. An ultrastructural and biochemical study. Am J Pathol 92 (3): 691–712, 1978.
27. Altenähr E, Arps H, Montz R, Dorn G: Quantitative ultrastructural and radioimmunologic assessment of parathyroid gland activity in primary hyperparathyroidism. Lab Invest 41 (4): 303–312, 1979.
28. Hehrmann R, Thiele J, Tidow G, Hesch R-D: Acute hyperparathyroidism. Clinical, laboratory and ultrastructural findings in a variant of primary hyperparathyroidism. Klin Wochenschr 58: 501–510, 1980.
29. Kemper B, Habener JF, Rich A, Potts JT: Microtubules and the intracellular conversion of proparathyroid hormone to parathyroid hormone. Endocrinology 96: 903–911, 1975.
30. Elliott RL, Arhelger RB: Fine structure of parathyroid adenomas; with special reference to annulate lamellae and septate desmosomes. Arch Path 81: 200–212, 1966.
31. Altenähr E, Seifert G: Ultrastruktureller Vergleich menschlicher Epithelkörperchen bei sekundärem Hyperparathyreoidismus und primärem Adenom. Virchows Arch Abt A Path Anat 353: 60–86, 1971.
32. Orci L, Matter A, Rouiller Ch: A comparative study of freeze-etch replicas and thin sections of rat liver. J Ultrastruct Res 35: 1–19, 1971.
33. Orci L, Perrelet A, Like AA: Fenestrae in the rough endoplasmic reticulum of the exocrine pancreatic cells. J Cell Biol 55: 245–249, 1972.

34. Breathnach AS, Stolinski C, Gross M: Freeze-fracture replication of rough endoplasmic reticulum of mouse liver cells. J Cell Sci 11: 477–489, 1972.
35. Hasleton PS, Ali HH: The parathyroid in chronic renal failure – a light and electron microscopical study. J Pathol 132: 307–323, 1980.
36. Wrigglesworth JM, Packer L, Branton D: Organization of mitochondrial structure as revealed by freeze-etching. Biochim Biophys Acta 205: 125–135, 1970.
37. Roth SI, Gallagher MJ: The rapid identification of 'normal' parathyroid glands by the presence of intracellular fat. Am J Pathol 84: 521–528, 1976.
38. Ruska H, Ruska C, Meyer-Delpho W: Das Gefrierätzbild von Lipidablagerungen in der Aortenintima cholesteringefütterter Kaninchen. Virchows Arch Abt B 11: 279–283, 1972.
39. Holmlund D, Boquist L, Larsson S-E, Lorentzon R: Influence of 1,25-dihydroxycholecalciferol on parathyroid activity in patients with primary hyperparathyroidism. Acta Chir Scand 145 (1): 27–33, 1979.
40. Nunez EA, Whalen JP, Krook L: An ultrastructural study of the natural secretory cycle of the parathyroid gland of the bat. Am J Anat 134: 459–480, 1972.
41. Drochmans P: Morphologie du glycogène. Etude au microscope électronique de colorations négatives du glycogène particulaire. J Ultrastruct Res 6: 141–163, 1962.
42. Biava C: Identification and structural forms of human particulate glycogen. Lab Invest 12 (12): 1179–1197, 1963.
43. Wanson J-C, Drochmans P: Rabbit skeletal muscle glycogen. A morphological and biochemical study of glycogen β-particles isolated by the precipitation-centrifugation method. J Cell Biol 38: 130–149, 1968.
44. Ainsworth SK, Ito S, Karnovsky MJ: Alkaline bismuth reagent for high resolution ultrastructural demonstration of periodate-reactives sites. J Histochem Cytochem 20: 995–1005, 1972.
45. Leloir LF, Goldemberg SH: Synthesis of glycogen from uridine diphosphate glucose in liver. J Biol Chem 235 (4): 919-923, 1960.
46. Luck DJL: Glycogen synthesis from uridine diphosphate glucose. The distribution of the enzyme in liver cell fractions. J Biophys Biochem Cytol 10: 195–209, 1961.
47. McGregor DH, Chu LLH, MacGregor RR, Cohn DV: Disruption of the Golgi zone and inhibition of the conversion of proparathyroid tissue by tris (hydroxymethyl) aminomethane. Am J Pathol 87 (3): 553–568, 1977.
48. Mollenhauer HH, Morré DJ, Bermann L: Homology of form in plant and animal Golgi apparatus. Anat Rec 158: 313–318, 1967.
49. Rambourg A, Clermont Y, Hermo L: Three-dimensional architecture of the Golgi apparatus in Sertoli cells of the rat. Am J Anat 154: 455–476, 1979.
50. Morré DJ, Ovtracht LA: Structure of rat liver Golgi apparatus: relationship to lipoprotein secretion. J Ultrastruc Res 74: 284–295, 1981.
51. Novikoff PM, Novikoff AB, Quintana N, Hauw J-J: Golgi apparatus, GERL, and lysosomes of neurons in rat dorsal root ganglia, studied by thick section and thin section cytochemistry. J Cell Biol 50: 859–886, 1971.
52. Novikoff AB, Novikoff PM: Cytochemical contributions to differentiating GERL from the Golgi apparatus. Histochem J 9: 525–551, 1977.
53. Broadwell RD, Oliver C: Golgi apparatus, GERL, and secretory granule formation within neurons of the hypothalamo-neurohypophysial system of control and hyperosmotically stressed mice. J Cell Biol 90: 474–484, 1981.
54. Sheldon H: On the water-clear cell in the human parathyroid gland. J Ultrastruct Res 10: 377–383, 1964.
55. Roth SI: The ultrastructure of primary water-clear cell hyperplasia of the parathyroid glands. Am J Pathol 61: 233–248, 1970.
56. Thiele J, Pichlmayr R: Der akute Hyperparathyreoidismus bei primärer Wasserheller-Zellen-Hyperplasie. Klin Wochenschr 52: 1063–1069, 1974.
57. Futrell JM, Roth SI, Su SPC, Habener JF, Segre GV, Potts JT: Immunocytochemical localization of parathyroid hormone in bovine parathyroid glands and human parathyroid adenomas. Am J Pathol 94 (3): 615–622, 1979.
58. Ravazzola M, Orci L: Intercellular junctions in the rat parathyroid gland: a freeze-fracture study. Biol Cell 28: 137–144, 1977.
59. Lacy PE, Howell SL, Young DA, Fink CJ: New hypothesis of insulin secretion. Nature 219: 1177–1179, 1968.
60. Farquhar M: Recovery of surface membrane in anterior pituitary cells. Variations in traffic detected with anionic and cationic ferritin. J Cell Biol 77: 35–42, 1978.
61. Theodosis DT, Dreifuss JJ, Orci L: A freeze-fracture study of membrane events during neurohypophysial secretion. J Cell Biol 78: 542–553, 1978.
62. Herzog V, Miller F: Membrane retrieval in epithelial cells of isolated thyroid follicles. Eur J Cell Biol 19: 203–215, 1979.
63. Yeghiayan E, Rojo-Ortega JM, Genest J: Parathyroid vessel innervation: an ultrastructural study. J Anat 112: 137–142, 1972.
64. Simionescu M, Simionescu N, Palade GE: Segmental differentiations of cell junctions in the vascular endothelium. The microvasculature. J Cell Biol 67: 863–885, 1975.
65. Simionescu M, Simionescu N, Palade GE: Characteristic endothelial junctions in different segments of the vascular system. Thromb Res (Suppl 2) 8: 247–256, 1976.
66. Rhodin JAG: The diaphragm of capillary endothelial fenestrations. J Ultrastruct Res 6: 171–185, 1962.
67. Maul GG: Structure and formation of pores in fenestrated capillaries. J Ultrastruct Res 36: 768–782, 1971.
68. Friederici HHR: On the diaphragm across fenestrae of capillary endothelium. J Ultrastruct Res 27: 373–375, 1969.
69. Simionescu M, Simionescu N, Palade GE: Morphometric data on the endothelium of blood capillaries. J Cell Biol 60: 128–152, 1974.
70. Weibel ER, Palade GE: New cytoplasmic components in arterial endothelia. J Cell Biol 23: 101–112, 1964.
71. Matsuda H, Sugiura S: Ultrastructure of 'tubular body' in the endothelial cells of the ocular blood vessels. Invest Ophthalmol 9: 919–925, 1970.
72. Sengel A, Stoebner P: Golgi origin of tubular inclusions in endothelial cells. J Cell Biol 44: 223–226, 1970.
73. Lemeunier A, Burri PH, Weibel ER: Absence of acid phosphatase activity in specific endothelial organelles. Histochemie 20: 143–149, 1969.

Author's address:
Institute of Pathology
Medical School, Hannover
Karl-Wiechert-Allee 9
3000 Hannover 61
F.R.G.

CHAPTER 27

Microvascularization of endocrine glands as studied by injection-replica SEM method

AKIO KIKUTA, AIJI OHTSUKA, OSAMU OHTANI, and TAKURO MURAKAMI

1. Introduction

Endocrine cells or glands secrete their products, hormones, into the blood capillaries through the surrounding connective tissue space. These hormones are transported by blood flow to their target cells or tissues. Routes of hormone transport to the targets are, generally, by way of the heart. However, since the discovery of the hypophyseal portal vascular system (1, 2), short-range vascular routes of hormone transport have received a great interest among endocrinologists. Most of the studies on microvascular architecture or microcirculation have been performed by light microscope methods mainly using injected sections. However, the limited depth of field with a light microscope does not provide sufficient visualization of blood vessels of tissues and organs.

Recently, a useful and convenient method for microvascular investigation has been developed (3). This method consists of vascular casting, microdissection and scanning electron microscopy (injection-replica SEM method), and allows wide- and long-range viewing of fine vascular distribution and connections. This chapter reviews the findings of the microvascular architecture of some endocrine glands as observed by the injection-replica SEM method.

2. Preparation and Scanning Electron Microscopy of vascular casts

Fresh organs or tissues are thoroughly irrigated through arteries with Ringer's or other physiological solutions to remove blood which interferes with resin injection into fine capillaries. Low viscosity casting media (3), including a commercially available methacrylate medium (Mercox; Japon Vilene Inc., Tokyo) (4), are injected through the arteries into the irrigated organs or tissues until the efferent veins are filled with the injected media. After the resin injection, the organs or tissues are immersed in a hot water bath (60° C) for 6 hours, macerated overnight in a 10 – 20% NaOH solution, and washed for 8 – 10 hours in running tap water. The blood vascular corrosion casts thus prepared are frozen in water, cut into blocks of appropriate sizes, and air-dried. The blocks are mounted on metal stubs and microdissected with needles and forceps to expose the point of interest. The specimens are then coated with gold and observed with a scanning electron microscope.

3. Microvascularization of endocrine glands

3.1. Hypophysis

The hypophysis consists of two lobes, the neurohypophysis and adenohypophysis. In accordance with this morphologic architecture, each lobe has its own capillary beds (Figs. 1, 2). The capillary beds of the neurohypophysis are continuous (2, 5, 6) and commonly supply the median eminence, the infundibular stem, and the pars nervosa (infundibular process) (Fig. 3). The capillary meshworks of the neurohypophysis are finer than those of the adenohypophysis. The capillaries of the adenohypophysis, including those of the pars tuberalis, are sinusoids and run along the long axis of the gland (Fig. 1, 2). The capillary meshwork of the pars intermedia is poorly developed. It is formed by blood vessels extending from the pars nervosa and drained into the pars distalis (7).

The capillaries of the neurohypophysis are directly supplied by three pairs of arteries, the superior hypophyseal arteries for the median eminence, the medial hypophyseal arteries for the infundibular stem, and the inferior hypophyseal arteries for the pars nervosa. The adenohypophysis receives the so-called hypophyseal portal vessels (see below), and has no direct arterial supply.

The capillaries of the neurohypophysis and aden-

Motta, PM (ed): Ultrastructure of endocrine cells and tissues. ISBN-13: 978-1-4613-3863-5

ohypophysis empty into the neurohypophyseal and adenohypophyseal veins, respectively. The former veins originate in the caudal edges of the pars nervosa, and the latter ones in the caudal edges of the pars distalis. They usually join (confluent veins) (8) and open into the cavernous sinus. The lateral veins, additional veins draining adenohypophyseal capillaries into the cavernous sinus (2), also occur, although some authors deny their presence (8).

Since the description by Popa and Fielding (1) it has been known that portal vessels (hypophyseal portal vessels) are intercalated between the neurohypophyseal and the adenohypophyseal capillary beds (Figs. 1, 2). Long portal vessels connect the median eminence and infundibular stem with the pars distalis of the adenohypophysis (8). Short portal vessels connect the infundibular stem and pars nervosa with the pars distalis (8). In man, monkey (9), and cat (10), the long portal vessels course down all sides of the infundibular stem. In the rat, the longest portal vessels arise near the rostral pole of the median eminence and usually run along the medial part of the infundibular stem. In this animal, the lateral and caudal parts of the median eminence are connected

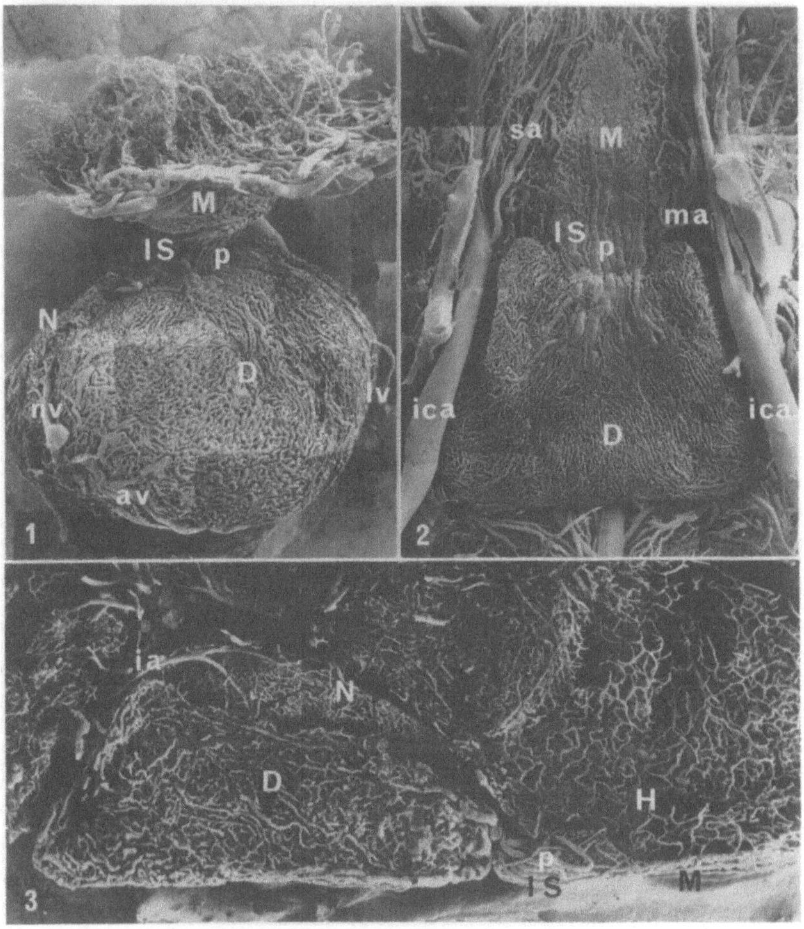

Figs. 1–3. (1) A scanning electron micrograph of an isolated vascular cast of the hypophysis of the rhesus monkey (right lateral view). M: median eminence; IS: infundibular stem; N: pars nervosa (infundibular process); D: pars distalis; p: portal vessels connecting the neurohypophysis and the pars distalis; av: adenohypophyseal veins; nv: neurohypophyseal veins; lv: lateral vein (× 13). (2) A ventral view of the rat hypophyseal vascular cast. The capillary plexus of the median eminence (M) of the neurohypophysis and the capillary plexus of the pars distalis (D) of the adenohypophysis are connected by portal vessels (p). The capillary plexus of the median eminence receives its arterial supply from the superior (sa) and medial (ma) hypophyseal arteries derived from the internal carotid artery (ica), while the pars distalis receives no direct arterial supply. Note the nearly parallel arrangement of the portal vessels. IS: infundibular stem (× 11). (3) A sagittal section of a cast of the rat hypophysis to show the vascular relationship within the gland. The capillary beds of the neurohypophysis are continuous throughout its three components, the median eminence (M), the infundibular stem (IS), and the pars nervosa (N). Note the long (p) and short (arrows) portal vessels connecting the neurohypophysis and the pars distalis (D). H: hypothalamus; ia: inferior hypophyseal arteries (× 25).

by relatively short portal vessels (Fig. 2).

Porter et al. (11) observed in their microinjection examination that, when the pressure was low, the spread of the perfusate in the pars distalis was often minimal. Based on this observation it has been hypothesized that a point-to-point circulation might exist. Thus, though the circulation has not yet been fully elucidated, it might be possible that specific regions of the adenohypophysis receive blood from specific portal vessels.

The capillary bed of the median eminence is composed of outer and inner plexuses. The outer plexus is often called the mantle plexus or external plexus, and the inner one is usually referred to as the subependymal plexus or internal plexus. In many species including the moneky (8, 10), cat (10) and rabbit (6), capillary loops of the outer plexus protrude towards the third ventricle to penetrate the innermost layer of the median eminence where the subependymal plexus is poorly developed. In the rat (7), the subependymal capillary plexus is well developed and the capillary loops are not conspicuous. The subependymal plexus has connections with the outer plexus and the capillary plexus of the hypothalamus. Although some authors regard the subependymal plexus as a portal route from the outer plexus to the hypothalamic plexus (10), our recent injection-replica SEM study (7) suggests that the subependymal plexus is directly derived from the arterial branches in the median eminence.

It is widely accepted that the blood in the hypophyseal portal vessels flows from the median eminence to the adenohypophysis, to convey neurosecretory products (hormone-releasing factors) to the adenohypophyseal cells. Injection-replica SEM studies also have confirmed this view (4, 5, 7, 8, 10). Recently, Oliver et al. (12) reported the presence of adenohypophyseal hormones in high concentrations within the portal blood collected from a single long portal vessel in the rat hypophysis. They considered that a part of the adenohypophyseal venous blood may return to the median eminence via the short portal vessels. This view has been supported by Bergland and Page (6) who further proposed that such blood may enter the brain through the external and internal capillary plexuses of the median eminence.

3.2. Pineal gland

The pineal gland (epiphysis) receives its blood supply from several arteries derived from the posterior cerebral arteries. These pineal arteries approach the epiphysis along the pineal stalk and give rise to a capillar network within the gland (Figs. 4a, b). The pineal capillaries run tortuously and freely anastomose to form an isotropical or homogeneous plexus. Hodde and Veltman (13) reported up to 12–16 veins in the rat pineal gland. The pineal veins are short and flow into an adjacent venous sytem, the great cerebral vein, the lateral veins and the confluent sinus or the sinus rectus.

Quay (14), by retrograde injection of dye in the cerebral veins of the rat proposed the presence of vascular routes from the pineal gland to the brain choroid plexus. However, our scanning electron microscopy of vascular casts has failed to confirm such shortrange vascular routes between the epiphysis and the choroid plexus.

3.3. Thyroid gland

Arteries and veins in the thyroid gland run through the interlobular connective tissues and divide into interfollicular arteries and veins, to supply the perifollicular capillary baskets. Scanning electron microscopy of vascular casts clearly demonstrates that the vascular beds of the thyroid gland are formed by numerous capillary baskets (Figs. 5–7). Each basket is composed of a single layer of capillary network (Figs. 6, 7) which encapsulates a thyroid follicle and receives one or more afferent and efferent vessels from the interfollicular arteries and veins.

The shape of the capillary baskets differs according to the species. In the monkey they are spherical or oval, but in the rat their shape is irregular (Fig. 6) (15). In the monkey (15) and rat, the capillary network of one basket is independent of the adjacent one (Fig. 7). In the dog, however, the wall of each basket consisting of a capillary network is often common with that of the adjacent basket (15). But regardless of these species differences, the meshworks of the baskets are fine and, the area of a single mesh space is usually smaller than that of the outer surface of a follicular epithelial cell. This indicates that all of the follicular epithelial cells directly face the perifollicular capillaries.

3.4. Parathyroid gland

The parathyroid gland receives its arterial supply from the thyroid arteries. The efferent veins of this gland drain into the thyroid veins (16). The capillaries of the parathyroid gland are distributed homogeneously throughout the gland (Figs. 5, 7). Although the parathyroid gland is embedded in the thyroid gland, there are no capillaries directly connecting the thyroid and parathyroid glands.

3.6. *Pancreatic islets*

The pancreas is a composite organ, consisting of exocrine acini and endocrine islets. The latter, the islets of Langerhans, are surrounded by the former exocrine acini. Such close contact between these two different tissues has been one of the insoluble enigmas. Recently, a hypothesis was proposed that exocrine tissues are directly perfused with the blood containing islet hormones (17, 18). The injection-replica SEM method has clearly confirmed this portal circulation from the pancreatic islet to exocrine acini (Fig. 8) (19, 20).

The pancreatic islet receives its afferent arteriole from the interlobular artery. In the monkey (19) and dog (18), the afferent arterioles penetrate deeply into the islet. In the rat and rabbit (20), the afferent arteriole divides in the cortical layer of the islet, the deeper portion of the islet being supplied by secondary branches of the cortical vessels. This difference

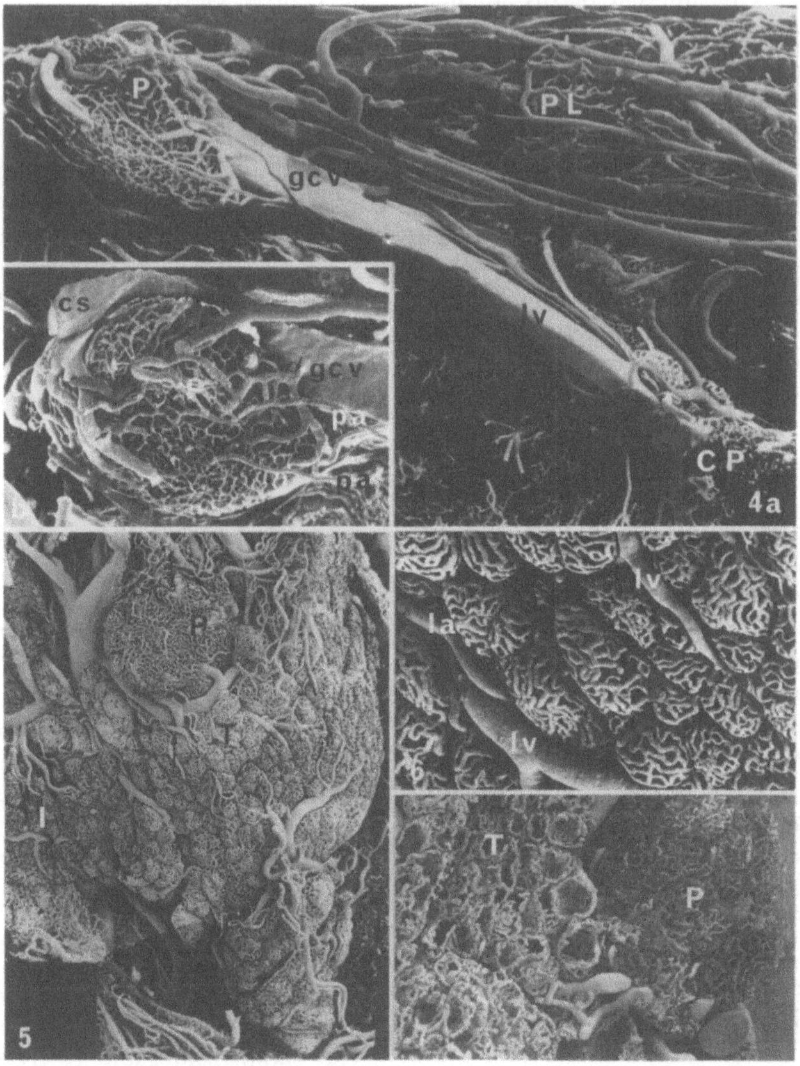

Figs. 4–7. (4) (a) A vascular cast of the rat pineal region in the sagittal plane (× 21); (b) The vascular cast of the rat pineal gland viewed from a slightly more dorsal side. Note the pineal veins (arrows) draining from the gland into the adjacent veins and sinuses. P: pineal gland; CP: choroid plexus of the third veintricle; PL: parietal lobe; pa: pineal arteries; gcv: great cerebral vein; cs: confluent sinus; lv: lateral vein (× 28). (5) A vascular cast of the right lobe of the thyroid gland (T) and the right parathyroid gland (P) of the rat. The vascular beds of the thyroid gland consist of round or oval capillary baskets. I: isthmus of the thyroid gland. Vertical arrows indicate the plane of section of (7) (× 21). (6) A closer view of the cast of the rat thyroid gland shown in (5). Note the round and oval capillary baskets.

between species in the arterial supply among the species is considered to be due to the difference in the distribution pattern of the islet cells. In the rat and rabbit, the cortical portion of the islet is mainly occupied by glucagon-secreting A cells and somatostatin-secreting D cells, while the core consists primarily of insulin-producing B cells (20). Fujita et al. (21) proposed that the blood in the islet generally flows in such a way that it takes up hormones from the A and D cells and transports them to B cells, to regulate release of insulin from the B cells which are stimulated by glucagon and inhibited by somatostatin.

Many efferent vessels radiate from the islet to the acinar capillary beds (Fig. 8). These efferent vessels, peri-insular portal (insulo-acinar) vessels, are considered to mediate the effect of the islet hormones on exocrine function (17, 18). A recent experimental study showed that insulin markedly potentiates the effect of pancreozymin upon the exocrine pancreas

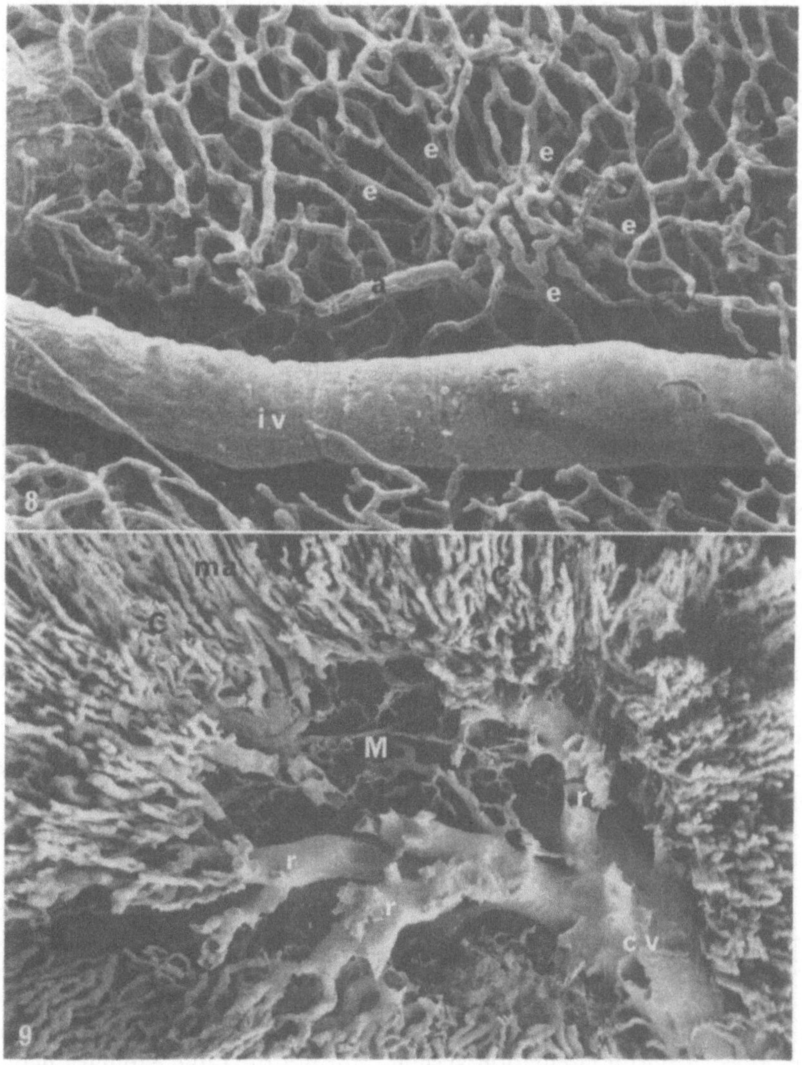

Figs. 8–9. (8) A vascular cast of the rhesus monkey pancreas. A vas afferens (a) penetrate deeply into the islet. From the periphery of the islet numerous vasa efferentia (e) radiate into the exocrine tissue to form a insulo-acinar portal system. iv: interlobular vein (× 175). (Micrograph by courtecy of T. Fujita and T. Murakami, from Archivum histologicum japonicum 35: 255–263, 1973). (9) A frontal section of a vascular cast of the rat adrenal gland. Cortical capillaries gather into larger vessels in the zona reticularis, then enter the medulla (M). These vessels continue into the peripheral radicles (r) of the central vein and finally drain into the central vein (cv) in a tree-like pattern. The medullary arteries (ma) penetrate the cortex and enter the medulla where they break up into capillaries to form a medullary capillary plexus. C: cortex (× 53).

(22). Furthermore, Fujita and Kobayashi (23) proposed that numerous nerve fibers in the islet may release their secretions into the capillaries, and these neurosecretions are conveyed in high concentrations to act upon the exocrine pancreas via the peri-insular portal vessels. Insulo-ductal portal vessels between the capillary plexus of the islet and pancreatic ducts are sometimes described, though their functional significance is unknown (20).

3.7. Adrenal gland

Adrenal arteries divide into ten or more arteries as they approach the gland. These arteries run through the capsule and form a subcapsular plexus. This plexus extends centripetally and constitutes the cortical capillary beds. The cortical capillary beds show characteristic arrangements reflecting the histological properties of the three zones (24). In the zona glomerulosa, the capillaries are arranged in a round basket pattern. In the zona fasciculata, the capillaries run centripetally in palisade fashion (Fig. 9). In the zona reticularis, the capillaries show tortuous arrangements. The capillaries of the zona reticularis receive the capillaries of the zonae glomerulosa and fasciculata and converge at the cortico-medullary junction into the peripheral radicles of the central vein and flow into the medulla.

Some branches of the adrenal arteries penetrate the cortex as the medullary arteries, and give rise to medullary capillaries which form a plexus among the radicles, including the peripheral ones, of the central vein in the medulla (Fig. 9). These medullary capillaries are thinner than those of the cortex and far thinner than central vein radicles. The medullary capillaries drain into the central vein via their venous radicles. The medullary capillaries adjacent to the cortex run into the peripheral radicles of the central vein.

Wurtman and Axelrod (25) showed that glucocorticoids in high concentrations activate N-methyl transferase, which converts noradrenaline to adrenaline. It is a widely accepted hypothesis that the cortical control of the synthesis of adrenaline in the medulla is mediated by blood vasculature (25, 26). Some authors described a portal circulation system from the cortical capillaries to medullary capillaries (27, 28, 29). However, our injection-replica SEM study (25) clearly showed that only a few connecting vessels occur between the cortical and medullary capillary plexuses. Our preliminary transmission electron microscopy of serial sections of the rat adrenal gland indicates that the vessels which allow penetration of cortical hormones into the medulla are probably peripheral radicles of the central vein.

3.8. Ovary

The ovary contains three principal functional subunits, follicles, corpora lutea, and stroma (16), each showing a characteristic vascular architecture (30, 31, 32) (Figs. 10–12). Primordial follicles are small in size and have no proper capillary bed. They are buried in the stroma which possesses a diffuse capillary plexus. Primary follicles have typical capillary baskets, each consisting of a capillary network of a single layer (Fig. 11). As the follicles develop, larger capillary baskets are formed. However, the wall of the capillary basket continues to be a single layer. No vessels penetrate the membrane propria of the follicle. The capillary beds of the large follicles are nearly independent of those of other follicles and stroma.

The follicular capillary beds are supplied by the follicular branches of the ovarian artery and converge into the follicular veins continuous with the ovarian vein.

The corpus luteum is highly vascularized. Vascular casts of the corpus luteum consist of highly packed capillaries with a sinusoidal appearance (Figs. 10, 12). The capillary beds of the corpora lutea are also independent of those of the surrounding tissues. In degenerative corpora lutea, however, their capillary beds are connected by many capillaries with the surrounding stroma vessels.

3.9. Testis

The testicular artery enters the testis and gives rise to testicular radiate arteries (33). The radiate arteries further divide and run as intertubular arterioles in the triangular interstices among the seminiferous tubules. These intertubular arterioles give off two types of capillaries, the intertubular and peritubular capillaries (Figs. 13, 14). The intertubular capillaries form the capillary plexus in the triangular interstices. The peritubular capillaries form the rope ladder-like plexus in the narrow connective tissue space between two adjacent seminiferous tubules (33, 34).

The intertubular and peritubular capillaries join to form a tube-like capillary network whose transverse section is hexagonal (Fig. 14). The longitudinal edges of the hexagonal tube are composed of the intertubular capillary plexuses. Their lateral surfaces are composed of the rope ladder-like peritubular capillary plexuses (Figs. 13, 14). Seminiferous tubules are surrounded by this hexagonal tube-like capillary network.

The intertubular and peritubular capillaries commonly drain into the intertubular venules. These venules consequently run on the surface of the testis. They drain into the testicular veins which empty into the venous pampiniform plexus.

4. Concluding remarks

Scanning electron microscopy of vascular casts has confirmed that the endocrine glands or tissues are richly vascularized. It has also confirmed that these glands or tissues possess their own specially differentiated microvascular patterns. This difference suggests that the functional significance of blood vasculature in secretory activities differs in each gland or tissue.

Hormone transport routes can be classified into two major types. One type may be called the long-range vascular route which is via the systemic circulation through the heart. The other type is the short-range vascular route, which directly transports hormones from an endocrine tissue to its target cells

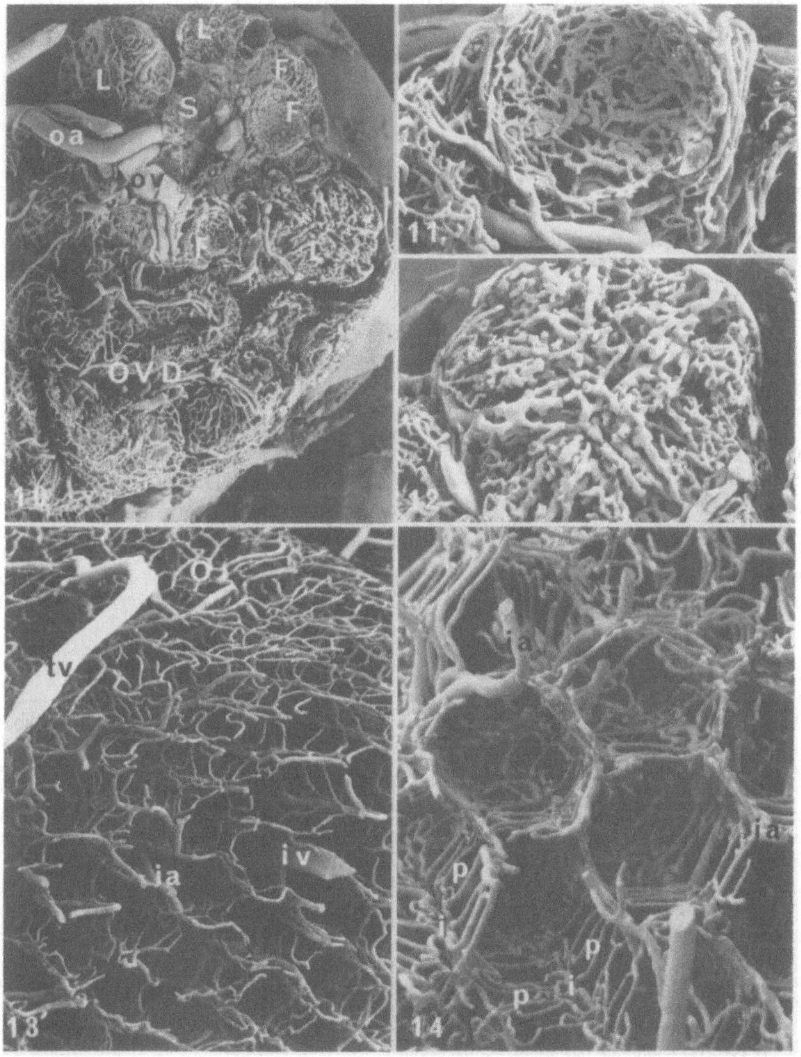

Figs. 10–14. (10) A vascular cast of the rat left ovary (frontal section). The ovarian vascular beds consist of three kinds of capillary plexuses, plexuses of developing follicles (F), those of corpora lutea (L) and those of the stroma (S). OVD: oviduct; oa: ovarian artery; OV: ovarian vein (× 18). (11) A transverse section of a developing follicle. Note the capillary network which constitutes the wall of the capillary basket (× 56). (12) A transverse section of the capillary plexus of a corpus luteum. Note the numerous anastomoses of sinusoidal vessels (× 45). (13) A vascular cast of the rat testis transversely sectioned. The outer surface (O) of the vascular cast of the testis is seen. tv: testicular vein; ia: intertubular arterioles; iv: intertubular venules (× 35). (14) A closer view of honeycomb-like testicular capillary plexuses. Each hexagonal tube-like capillary network consists of intertubular capillary plexuses (i) and rope ladder-like peritubular capillaries (p) which link the intertubular plexuses. ia: intertubular arterioles (× 56).

without passing through the heart. A typical example of the short-range vascular route is the hypophyseal portal vessels. The peri-insular portal system between the pancreatic islets and exocrine acini is another typical example of the short-range vascular system. Adrenocortical glucocorticoids are transported to the medullary chromaffin cells via the peripheral radicles of the central vein. This hormone transport system belongs to the category of the short-range route, though no portal vessels are intercalated.

As discussed above, the short-range vascular systems play an essential role in controlling or conditioning the functions of some endocrine glands. However, the details of these systems have not yet been elucidated. Various approaches, including the injection-replica SEM method, may be necessary for further understanding such functions.

References

1. Popa GI, Fielding U: A portal circulation from the pituitary to the hypothalamic region. J Anat 65: 88–91, 1930.
2. Wislocki GB, King LS: The permeability of the hypophysis and hypothalamus to vital dyes, with a study of the hypophyseal vascular supply. Am J Anat 58: 421–472, 1936.
3. Murakami T: Application of the scanning electron microscope to the study of the fine distribution of blood vessels. Arch Histol Jpn 32: 445–454, 1971.
4. Ohtani O, Murakami T: Peribiliary portal system in the rat liver as studied by the injection replica scanning electron microscope method. In: Scanning electron microscopy/1978/II, Becker RP, Johari O (eds), AMF O'Hare, SEM Inc., 1978, pp 241–244 and 20.
5. Page RB, Munger BL, Bergland RM: Scanning microscopy of pituitary vascular casts. Am J Anat 146: 273–302, 1976.
6. Bergland RM, Page RB: Pituitary-brain vascular relations: a new paradigm. Science 204: 18–24, 1979.
7. Murakami T: Pliable methacrylate casts of blood vessels: use in a scanning electron microscope study of the microcirculation in rat hypophysis. Arch Histol Jpn 38: 151–168, 1975.
8. Bergland RM, Page RB: Can the pituitary secrete directly to the brain? Endocrinol 102: 1325–1338, 1978.
9. Porter JC, Ondo JC, Cramer OM: Nervous and vascular supply of the pituitary gland. In: Handbook of physiology, Vol. IV. Knobel E, Sawyer H (eds), Section 7. Greep O, Eastwood EB (eds). Am Physiol Soc, Washington, DC, 1974, pp 33–43.
10. Page RB, Leure-duPree AE, Bergland RM: The neurohypophyseal capillary bed. II. Specialization within the median eminence. Am J Anat 153: 33–66, 1978.
11. Porter JC, Mical RS, Kamberi IA, Grazia YR: A procedure for the cannulation of a pituitary stalk portal vessel and perfusion of the pars distalis in the rat. Endocrinol 87: 197–201, 1970.
12. Oliver C, Mical RS, Porter JC: Hypothalamic-pituitary vasculature: evidence for retrograde blood flow in the pituitary stalk. Endocrinol 101: 598–604, 1977.
13. Hodde KC, Veltman AM: The vascularization of the pineal gland (Epiphysis cerebri) of the rat. In: Scanning electron microscopy/1978/III. Becker RP, Johari O (eds), SEM Inc., O'Hare AMF, 1979, pp 369–374.
14. Quay WB: Retrograde perfusions of the pineal region and the question of pineal vascular routes to brain and choroid plexuses. Am J Anat 137: 387–402, 1973.
15. Fujita H, Murakami T: Scanning electron microscopy on the distribution of the minute blood vessels in the thyroid gland of the dog, rat and rhesus monkey. Arch Histol Jpn 36: 181–188, 1974.
16. Turner CD, Bagnara JT: General endocrinology, Philadelphia, London, Toronto, WB Saunders Company, 1976.
17. Henderson JR: Why are the islet of Langerhans? Lancet 7618: 469–470, 1969.
18. Fujita T: Insulo-acinar portal system in the horse pancreas. Arch Histol Jpn 35: 161–171, 1973.
19. Fujita T, Murakami T: Microcirculation of monkey pancreas with special reference to the insulo-acinar portal system. A scanning electron microscope study of vascular casts. Arch Histol Jpn 35: 255–263, 1973.
20. Ohtani O, Fujita T: Microcirculation of the pancreas with special reference to periductular circulation. A scanning electron microscope study of vascular casts. Biomed Res 1: 130–140, 1980.
21. Fujita T, Yanatori Y, Murakami T: Insulo-acinar axis, its vascular basis and its functional and morphological changes caused by CCK-PZ and caerulein. In: Endocrine gut and pancreas. Fujita T (ed), Amsterdam, Elsevier, 1976, pp 347–357.
22. Kanno T, Saito A: The potentiating influences of insulin on pancreozymin-induced hyperpolarization and amylase release in the pancreatic acinar cell. J Physiol 261: 505–521, 1976.
23. Fujita T, Kobayashi S: Proposal of a neurosecretory system in the pancreas. An electron microscope study in the dog. Arch Histol Jpn 42: 277–295, 1979.
24. Kikuta A, Murakami T: Microcirculation of the rat adrenal gland. A scanning electron microscope study of vascular casts. Am J Anat 164: 19–28, 1982.
25. Wurtman RJ, Axelrod J: Control of enzymatic synthesis of adrenaline in the adrenal medulla by adrenal cortical steroid. J Biol Chem 241: 2301–2305, 1966.
26. Coupland RE, MacDougall JDB: Adrenaline formation in noradrenaline-storing chromaffin cells in vitro induced by corticosterone. J Endocrinol 36: 317–324, 1966.
27. Pohorecky LA, Wurtman RJ: Adrenocortical control of epinephrine synthesis. Pharmacol Rev 23: 1–35, 1971.
28. Henderson JR, Daniel PM: Portal circulation and their relation to countercurrent systems. Q J Exp Physiol 63: 355–369, 1978.
29. Motta PM, Muto M, Fujita T: Three-dimensional organization of mammalian adrenal cortex. A scanning electron microscopic study. Cell Tissue Res 196: 23–38, 1979.
30. Kardon RH, Kessel RG: SEM studies on vascular casts of the rat ovary. In: Scanning electron microscopy/1978/III. Becker RP, Johari O (eds), AMF O'Hare, SEM Inc., 1979, pp 743–750.
31. Van Blerkom J, Motta PM: A scanning electron microscopic study of the luteo-follicular complex. III. Formation of the corpus luteum and repair of the ovulated follicle. Cell Tissue Res 189: 131–154, 1978.
32. Motta PM, Makabe S: Morphodynamic changes of the mammalian ovary in normal and some pathological conditions. A scanning electron microscopic study. Biomed Res 2, Suppl 325–339, 1981.
33. Kormano M: An angiographic study of the testicular vasculature in the postnatal rat. Zeit Anat Entwick 126: 138–153, 1967.
34. Clark, RV: Three-dimensional organization of testicular interstitial tissue and lymphatic space in the rat. Anat Rec 184: 203–226, 1976.

Authors' address:
Department of Anatomy,
Okayama University Medical School,
2-5-1- Shikata-cho,
Okayama 700,
Japan

CHAPTER 28

Innervation of endocrine tissues

KLAUS UNSICKER

1. Introduction

Our understanding of neural regulation of endocrine glands is still fragmentary. Studies on neuroendocrine interactions have been hampered in the past by the fact that neuronal and endocrine systems have long been considered to operate independently and through opposite 'languages', the nervous system by highly localized release of rapidly inactivated transmitters, evoking a high-speed and short-duration response, the endocrine system, on the other hand, by secreting hormones, which are conveyed to their effector tissues by the blood and maintain their actions for appreciable lengths of time (1). Moreover, focusing at the examination of the link between the neural and the endocrine apparatus at the site of the hypothalamic neurosecretory centers and the lack of adequate methods for the demonstration of morphological and functional relationships between peripheral nerves and endocrine cells have, for a long time, prevented to successfully approach the problem of a direct nervous control of endocrine cells by conventional synaptic transmission.

With the advent of electron microscopy it became possible to look for ultramorphological evidence of endocrine cell innervation. As has been shown in earlier review articles covering this field (2, 3, 4, 5) neuroeffector junctions with endocrine cells demonstrated at an ultrastructural level are no longer a rarity and cover a variety of organs. These findings and correlative functional studies foster the view of a higher degree of functional interdependence between nervous and endocrine systems and their efferent and afferent pathways than has been previously thought. This article will not refer to two endocrine organs that have long been known to be regulated by peripheral nerves, the adrenal medulla and the pineal gland (See Chapter 15).

2. Adenohypophysis

The bulk of evidence supporting the concept of a hypothalamic control of the pituitary anterior and intermediate lobe by means of releasing and inhibiting hormones is overwhelming and a direct neuronal influence upon this organ might not be expected at all. However, electron microscopic studies have provided evidence for an innervation of adenohypophyseal cells in the *pars intermedia* of many vertebrate species (1, 7–22). In contrast, the innervation of the anterior lobe has been subjected to profound qualitative and quantitative changes during vertebrate phylogenesis. In teleost fishes nerve fibres transporting the hypothalamic releasing and inhibiting hormones do not terminate in the median eminence, but, instead, pass into the pars distalis making close contacts with endocrine cells (see 6, 23, for reviews). However, in the teleost species studied nerve fibres invade the anterior lobe to a different extent: in the eel, *Anguilla anguilla*, fibres are found in the dorsal portion of the organ only, whereas in the tench, *Tinca tinca*, the endocrine parenchyma is completely innervated. In amphibia, the innervation of the pars distalis is only scarce (24) or even lacking (25), while synapses at pars tuberalis cells have been demonstrated in *Rana temporaria* (25). Very few reports on nerve fibres in the *pars distalis* and *pars tuberalis* of the mammalian adenohypophysis have been published (26, 27, 28).

Although the presence of nerve fibres in both anterior and intermediate lobes of the adenohypophysis is no longer a matter of dispute, their precise nature and functional significance is largely enigmatic. Bargmann et al. (7) were the first to tentatively classify nerve terminals on endocrine cells of *pars intermedia* of cat pituitary into cholinergic, adrenergic and 'peptidergic' on the basis of synaptic vesicle ultramorphology and the distribution of vesicle populations in different nerve terminals. Adrenergic terminals were characterized by small dense core

Motta, PM (ed): Ultrastructure of endocrine cells and tissues. ISBN-13: 978-1-4613-3863-5

vesicles interspersed among electron-lucent ones. Another type of nerve ending, with only electron-lucent presynaptic vesicles was interpreted as a cholinergic synapse, but might well be aminergic, too, since the dense cores of adrenergic synaptic vesicles may be lost during the fixation procedure. Both adrenergic and cholinergic terminals induced the formation of membrane specializations at the postsynaptic site. A third type of nerve ending with large dense-core vesicles, 150–200 nm in diameter, resembled the neurosecretory terminals in the neurohypophysis and was therefore coined 'peptidergic'. The functional significance of these 'peptidergic' endings in the intermediate lobe has not been elucidated. Adrenergic terminals, which have been also clearly identified by their capacity to take up tritiated norepinephrine (teleosts, 14), may exert an inhibitory control of MSH release in the amphibian pars intermedia (29, 30).

In the rat pituitary anterior lobe bundles of unmyelinated nerve fibres and nerve terminals were found in the pericapillary connective tissue and intercellular spaces interspersed among glandular cells (20). Nerve endings containing mostly dense-core vesicles and tubular structures were seen to contact somatotrophs most frequently. In an earlier study Unsicker (28) had not observed any nerve profiles in the pituitary *pars distalis* of mouse, rat, hamster, cat and pig, although two types of axon terminals with light and neurosecretory type vesicles, respectively, were present in the *pars tuberalis*. In a careful investigation of the goldfish pituitary Kaul and Vollrath (23) were able to distinguish two types of neurosecretory fibre (Type A and Type B). Both types of fibres innervated gonadotrophic, STH and TSH_2 cells, mostly without an intervening basal lamina, whereas prolactin, ACTH and TSH_1 cells

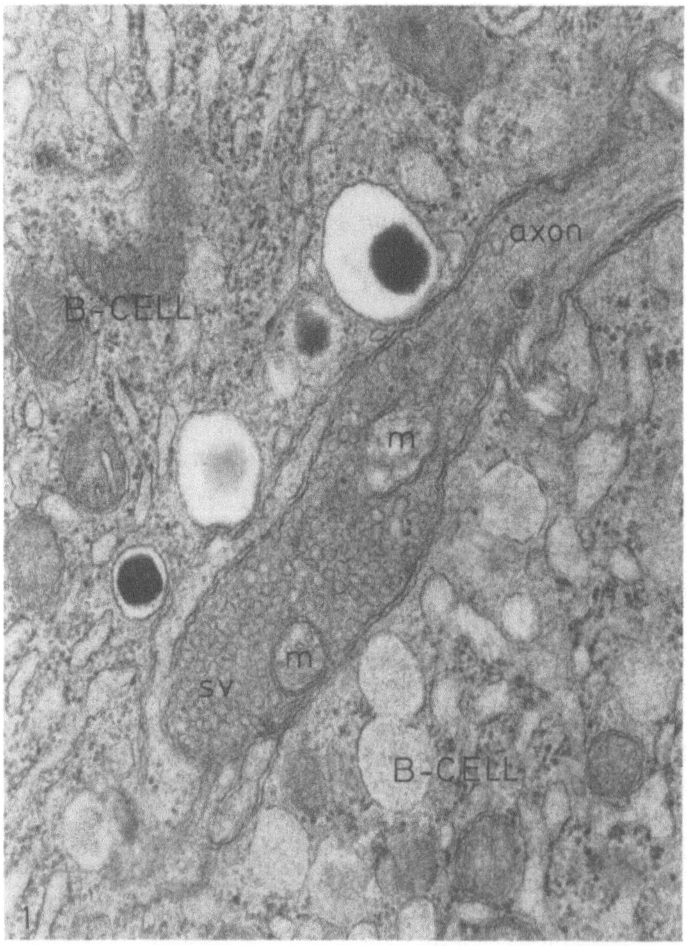

Fig. 1. Synapse at a B-cell of a rat pancreatic islet. An axon is deeply invaginated into the cell. Microtubules in parallel array and large dense core vesicles (arrows) are characteristic for the preterminal part of the synapse. The terminal portion of the axon contains an accumulation of small agranular vesicles. The synaptic cleft is 15 to 20 nm wide. Synaptic membrane specializations do not occur. Mitochondria (m) (× 35,910). Courtesy of Professor H.F. Kern, Marburg.

were innervated by Type B terminals. The amount of dense-core vesicles varied during the seasonal cycle suggesting a participation of nerve fibres in *pars distalis* control.

3. Thyroid gland

The innervation of thyroid follicle cells by unmyelinated nerves has first been demonstrated by Brettschneider (31) at an ultrastructural level. Nerve fibres using different types of transmitters appear to contribute to this innervation. In a series of studies Melander et al. (32–34) have provided evidence that sympathetic adrenergic fibres participate in the regulation of thyroid hormone secretion in a variety of mammalian species by releasing norepinephrine, which binds to adrenergic receptors in the follicle cells. These nerves can be easily visualized by fluorescence microscopy using the Falck-Hillarp technique. They form networks around blood vessels, and single fibres can be seen running between and around follicles. Their adrenergic character has also been revealed by using electron microscopic autoradiography for the demonstration of [^{3}H] norepinephrine uptake. Nerves taking up the cytotoxic false aminergic transmitters 6-hydroxydopamine and 5,7-dihydrotryptamine were seen in close apposition to the bases of thyroid epithelial cells (35).

Thyroid follicle cells also respond to cholinergic agents by an atropine-sensitive muscarinic mechanism (36, 39). Nerve fibres which can be stained by specific acetylcholine esterase and have a distribution very similar to that of adrenergic nerves may represent the morphological equivalent of the cholinergic inhibitory influence on hormone secretion (38, 40). Recently nerves containing vasoactive intestinal polypeptide (VIP) immunoreactivity have been shown to supply blood vessels and follicles in the mouse thyroid gland (41) and to regulate thyroid hormone secretion through a mechanism that is mediated by cyclic AMP (41). Cholinergic and VIP containing nerves have not been demonstrated as yet with respect to their ultramorphological apposition to the follicle cells.

3.1. Ultimobranchial body and thyroid C-cells

The ultimobranchial body originates from the neural crest and nerve fibres in relation to the secretory cells may therefore be expected. In the ultimobranchial body of *Rana pipiens* Robertson (42) has described axon terminals with the ultrastructural features of adrenergic nerves terminating on the basal portions of a few secretory cells. Synaptic granular and agranular vesicles accumulated at presynaptic membrane specializations and axons were found both deep within the secretory parenchyma and in the connective tissue matrix. From a study by Stoeckel and Porte (43) it is evident that nerve fibres are also present in the ultimobranchial body of birds. Terminal axon profiles which are associated with the calcitonin producing 'light' cells in the mammalian thyroid gland (44, 45) and located inside the follicular basal lamina might indicate that these cells, which are phylogenetically related to ultimobranchial secretory cells, are also under nervous control.

4. Parathyroid gland

Electron microscopic studies on the innervation of the parathyroid glands have confirmed earlier light microscopic findings as far as the presence of nerve fibres is concerned. Both myelinated and unmyelinated fibre bundles were described in the parathyroids of various mammalian species in relation to blood vessels (28, 46, 48), but only once direct contacts of terminal axon profiles at chief cells without intervening basal lamina or connective tissue have been reported (man:46). Ultramorphological alterations of secretory cells following vagotomy in the rabbit (49) and reports suggesting both an adrenergic and a vagal control in parathyroid hormone secretion (50–52) might lend support to speculations concerning a functional role of parenchymal and vascular nerve fibres in the parathyroid gland.

5. Pancreatic islets

Although secretion of the islet hormones insulin and glucagon may occur in the absence of an intact autonomic innervation of the pancreas, as has been shown in transplantation experiments (53–55), there is now abundant evidence indicating that stimulation of autonomic nerves modifies both insulin and glucagon secretion (56). The morphological substrate of the neuronal interference with islet cell function are nerve fibres in close connection to vascular walls and to the different types of islet cells using norepinephrine, VIP and probably also acetylcholine and serotonin as their transmitters.

Fluorescence microscopy employing the Falck-Hillarp technique has revealed a rich supply of catecholaminergic nerves in the pancreas of a variety of mammalian and submammalian species including man (57–62). Apparently the adrenergic nerve fibres enter the islets together with small blood vessels and mostly occupy the periphery of the islets. At an ultrastructural level adrenergic nerve terminals have been demonstrated upon endocrine A, B and D cells by employing the morphology of their synaptic vesicles, autoradiography and uptake of 6-hydroxydopamine as identification criteria (57, 63–68).

Physiological studies have established that stimulation of the sympathetic innervation via splanchnic nerves at frequencies within the physiological range as well as activation of the sympatho-adrenal system induce release of glucagon and inhibit that of insulin (56, 61, 69–74). Furthermore, norepinephrine administered *in vivo* inhibits the secretion of insulin (75, 76). Both α- and β-adrenergic mechanisms have been shown to play a role in insulin secretion: propranolol, a β-adrenergic blocking agent, reduces, and the α-adrenergic blocker phentolamine increases basal insulin secretion (57, 66, 77). However, species differences seem to exist with respect to regulation of basal insulin secretion by the sympatho-adrenal system (57). Sympathetic nerves appear to influence glucagon secretion to a much greater extent than insulin release (70). This is consistent with adrenergic fibres being mainly associated with A cells (66) and less frequently with B and D cells in mammals. In the domestic fowl Watanabe and Yasuda (68) have reported a rich innervation of D cells.

Most investigators agree that axons terminate on the surface of islet cells without any membrane specializations, being separated from the endocrine cells by a 20 nm wide intercellular cleft (64, 65–68, 78) (Fig. 1). However, by using lanthanum as an extracellular tracer Orci et al. (79) revealed narrowed zones between nerve endings and B cells, which they interpreted as gap junctions.

Positive acetylcholinesterase staining of nerve fibres has been described in many vertebrate species (64, 80–84). Ultrahistochemical studies have corroborated the light microscopic findings and demonstrated the axonal plasmalemma reaction product to be associated with axon profiles containing small clear vesicles (83, 84).

The concept of a cholinergic innervation of islet cells, in particular B and D cells, which is mediated by vagal nerves, has been substantiated by nerve transsection and stimulation experiments as well as by pharmacological studies. Electric stimulation of the vagal nerves elicits insulin release (85, 80, 87–89), while atropine inhibits the insulin-releasing actions of acetylcholine (90–93) and of electrical vagus stimulation (85–88). Atropine also causes long-term histophysiological changes in B, but not A cell morphology in terms of nuclear shrinkage (90). Vagotomy has also been shown to lead to profound ultrastructural changes of D cells in the domestic fowl including extensive depletion of secretory granules (94).

Axons in the pancreas of the bat are intensely labelled following in vitro incubation of the gland with ^{3}H-serotonin (95), some axons being closely related to A cells. Non-specific uptake of ^{3}H-serotonin by noradrenergic axons seemed unlikely under the experimental conditions employed.

Terminal axon profiles containing storage vesicles that might be attributed to hitherto unidentified peptide neuromodulators or – transmitters have been described to be closely apposed to endocrine islet cells in the domestic fowl (63, 68). VIP-immunoreactive fibres that possess large 'peptide'-type granular vesicles were shown at blood capillaries in the pancreatic islets of a snake (96). Several authors have described autonomic nerve cells in the pancreas of mammals (78 97–100) and birds (68, 101), which receive preganglionic axons making axo-dendritic synapses. The juxtaposition of pancreatic neurons and islets has led Fujita et al. (97, 98, 102) to propose the concept of the pancreatic neuro-insular complex thus re-emphasizing the earlier view of a specific relationship of nervous and endocrine elements in the pancreas ('complex neuro-insulaire', (103, 104). However, this concept still awaits its functional validation.

6. Endocrine cells in the gut

Few reports are available concerning an innervation of endocrine cells in the gut. Bundles of unmyelinated axons, which may be adrenergic, have been traced by fluorescence histochemistry and electron microscopy to the gastric and intestinal epithelium (105, 106). Terminal axons were found in close contact with enterochromaffin cells (107) and vagal nerve stimulation has been demonstrated to decrease 5-HT concentration in the enterochromaffin cells of the jejunum, midgut and ileum of the cat (108). Basal-granulated cells in the proventricular mucosa of the finch (101) and in the duodenum of the human fetus (110) were observed in intimate contact with vesicle-containing nerve profiles.

7. Juxtaglomerular apparatus

Investigations on the innervation of the juxtaglomerular apparatus in various mammalian and submammalian species have provided evidence for a possible involvement of catecholaminergic nerves in renin production and release (111–125).

Using serial section electron microscopy Barajas and Müller (113) analyzed axons for contact with the cells of the vascular and tubular components of the juxtaglomerular apparatus. They found single axons making 'en passant' synapses with both granular (renin-producing) and agranular (smooth muscle) cells of the afferent and efferent arterioles and with tubular cells. All the nerve endings contained small dense-core vesicles typical of adrenergic fibres along with agranular vesicles. Although acetylcholinesterase staining is frequently observed with nerves at the juxtaglomerular region (126), both the topographic distribution of positive fibres, which corresponds to that of catecholamine-containing fibres, and the failure of acetylcholine to modify renin secretion argue against a substantial cholinergic innervation of the renin-producing cells. The ultramorphology of the terminal axons and the relationships with their target cells are characteristic for peripheral autonomic neuroeffector connections (116). Adrenergic nerves also make contacts at a distance of 80 to 400 nm on juxtaglomerular epitheloid cells in the amphibia *Rana temporaria* and *Bufo bufo* (123). The adrenergic innervation of the juxtaglomerular apparatus in the primitive primate *Tūpaia belangeri* has been described to be particularly abundant (122). Recent physiological studies indicate that renin release is mediated by a β-adrenergic mechanism (111, 127). Dopamine as a transmitter in nerves associated with the glomerular vascular poles in canine kidney should also be discussed as a candidate to induce renin secretion (128).

8. Corpuscles of Stannius

Corpuscles of Stannius (CS), endocrine organs in teleosts, are likely to be involved in the control of bone metabolism, blood levels of ionic calcium and the sodium/potassium equilibrium in body fluids. Based on light microscopic studies many investigators have described nerve fibres within this organ and even in close proximity to endocrine cells (129). Fluorescence microscopy (Falck-Hillarp technique) has revealed the presence of intensely fluorescent cells and 'adrenergic' nerve fibres running in the proximity of blood vessels and endocrine cells of *Salmo irideus* CS. Fluorescent cells resembled chromaffin cells with respect to their ultrastructure. Nerve fibres contained small and large granular vesicles and few agranular vesicles. Synaptic contacts of these fibres on the surface of endocrine cells were not observed, the shortest distance between axon varicosities and presumed target cells being 120 nm (129).

9. Adrenal cortex

The issue of an efferent and afferent nerve supply to the adrenal cortex has been the subject of controversy during the light microscope era (130, 131, 132, 133). From 1969 onwards ultrastructural studies provided evidence for unmyelinated axons and terminals containing synaptic vesicles and abutting upon adrenocortical cells of a large variety of mammalian and submammalian species including man, monkey, pig, sheep, rat, mouse, hamster, guinea pig, birds and lizards (134–143) (Fig. 2). The axon terminals occasionally displayed accumulations of electron-dense material at the pre-synaptic site, but no typical pre- and postsynaptic membrane specializations; they were frequently located inside the basal lamina surrounding endocrine cells and separated from their target cells by an approximately 20 nm wide cleft. This innervation was usually described to be sparse as compared to the adrenal medulla and predominantly in the zona fasciculata and reticularis, although considerable variation concerning the density of innervation was reported even between closely related species (133, 139, 140, 141, 143). The nature of the transmitters in these nerve terminals has only been partly characterized. The endings described in the hamster could be cholinergic, since they contain large amounts of small clear vesicles (139). Adrenal cortical (interrenal) cells in the lizards *Lacerta dugesi* and *Lacerta pityusensis* were shown to receive terminals that remained unlabelled after administration of 6-hydroxydopamine and, hence, were tentatively classified to be cholinergic (143). In the sheep acetylcholinesterase-positive fibres forming truncs and distinct plexuses within all layers of the cortex, but only relatively sparse aminergic axons that were restricted to the outer zones, were described by Robinson et al. (137). In birds several types of nerve endings were found in close proximity of interrenal cells: adrenergic fibres did not penetrate the basal lamina and were separated by a distance of at least 80 nm from steroid-producing cells; in contrast, endings exhibiting reaction product for acetylcholinesterase at their surfaces and terminals characterized by their numerous large dense-core vesicles (peptidergic-type fibres) established intimate contacts with endocrine cells (133). In the guinea pig nerves originating from adrenomedullary neurons and small granule (SIF) cells could be traced back to adrenocortical cells (140). An intra- or periadrenal origin of the nerves innervating cortical cells may also be deduced from results by Robinson et al. (137), who reported remaining cortical fibres after sectioning the splanchnic nerves. Thus it appears that the efferent innervation to the vertebrate adrenocortical cells is subjected to considerable species-dependent variations as far as origin, transmitter type and density of nerve fibres are concerned.

There is also some morphological evidence for sensory nerve fibres in the adrenal cortex, although the occurrence of fibres near cortical cells that contain no vesicles is a very weak proof (140). Substantial experimental evidence supporting the concept of both an efferent and an afferent innervation of adrenocortical tissue comes from studies on compensatory adrenal growth. In 1959 Halász and Szentagothai (144) reported unilateral karyometric changes in the hypothalamic ventromedial nucleus following unilateral adrenalectomy. These findings were extended by Mary Dallman and her group, who showed that interruption of CNS pathways could inhibit the compensatory adrenal growth

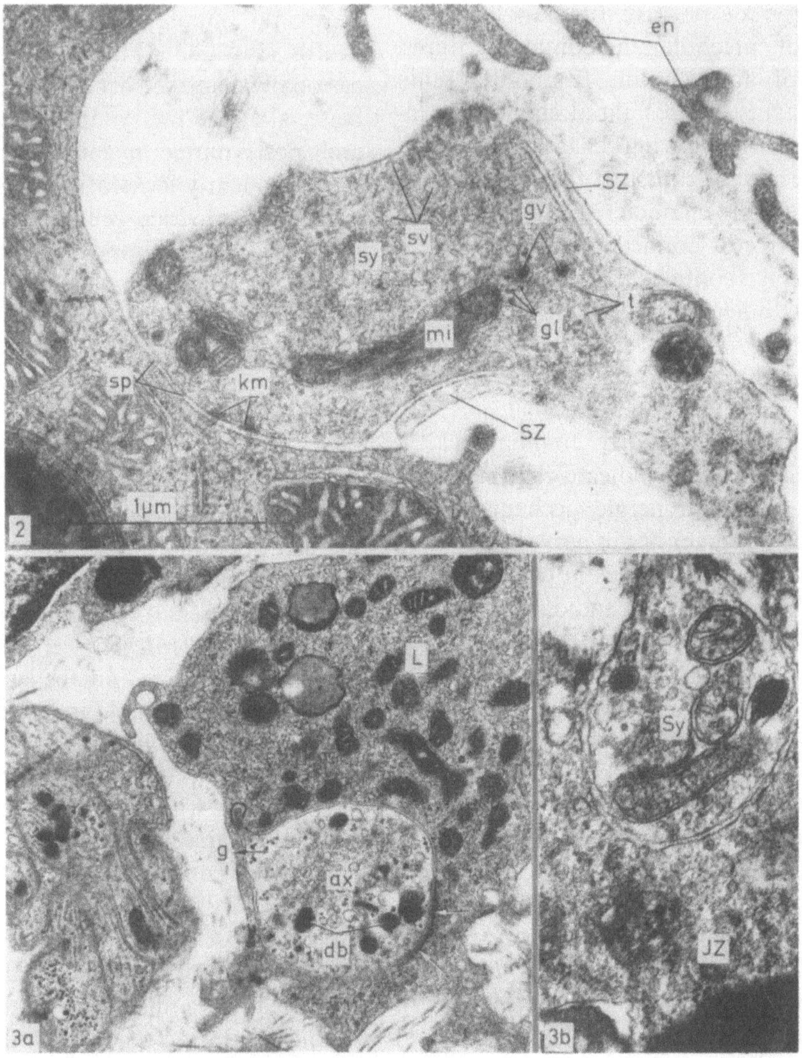

Figs. 2–3. (2) Synaptic terminal (sy) at an adrenocortical cell (hamster). The terminal 'bouton' contains large amounts of synaptic agranular vesicles (sv), few granular vesicles (gv), glycogen particles (gl), microtubules (t) and mitochondria (mi). Electron dense material (dm) is precipitated at the 'presynaptic' membrane. Synaptic cleft (sp), satellite cell (sz), endothelial cell (en) (× 29,400). From Unsicker (1969). (3) (a) Leydig cell (L) in the testis of a lizard, *Lacerta dugesi*, showing an axon terminal, which is deeply embedded into its cell body. The nerve fibre displays ultrastructural features typical of transmitter-containing portions of autonomic axons, such as granular and agranular vesicles. Glycogen particles (gp), dense bodies (db). White arrows point at a cell membrane specialization of the Leydig cell, which strongly resembles a postsynaptic membrane thickening (× 12,600). From Unsicker (1973); (b) Synaptic ending (sy) at an interstitial cell (iz) in the mouse ovary. The synaptic cleft is approximately 20 nm wide. (× 27,300). From Unsicker (1970).

response to unilateral adrenalectomy. Spinal cord hemisection between Th 2 and Th 3 inhibited compensatory adrenal growth when the contralateral but not the ipsilateral adrenal was removed (145). Growth inhibition was also observed after lesions of the premamillary or the ventromedial hypothalamic nucleus (146) of after squeezing the adrenal pedicle (147). Unilateral hemi-deafferentiations of the medial basal hypothalamus prevented compensatory adrenal growth on the side ipsilateral, but not contralateral to the removed adrenal (148). These results clearly suggest a neural reflex arc, whose efferent pathways have trophic effects upon the adrenal cortex.

More immediate secretory effects of autonomic nerves on adrenal cortical endocrine activities have also been reported. Vogt (149) using a biological assay technique showed that injection of adrenaline into the adrenal circulation caused an increase in steroid production. However, subsequent experiments using chemical assays failed to reproduce a direct action of catecholamines on steroid production (150). Infusion of acetylcholine into isolated calf adrenals caused a significant increase of steroid production (151). These findings, however, were not supported by observations reported by Blair-West et al. (152), who injected acetylcholine into the transplanted sheep adrenal. Moreover, incubation of slices of sheep adrenal cortex with acetylcholine did not produce consistent results (McDougall and Unsicker 1975, unpublished). Summing up, we have to state that conclusive experiments demonstrating a direct effect of any neuronal transmitter on cortical steroid synthesis and release are still lacking.

10. Testis

The reciprocal hormonal pituitary-gonad axis is the well-established main principle for the regulation of endocrine activities of the testis and little attention has been paid to the peripheral autonomic nervous system as an operating factor of testicular endocrine control. Suggestions, but no unequivocally reliable results have come from the early light microscopic literature regarding an innervation of Leydig cells (153).

Using the Falck-Hillarp method for the visualization of catecholamines Baumgarten and Holstein (154), Baumgarten et al. (155), Norberg et al. (156) and Dayan (157) found marked species differences in the adrenergic nerve supply of the mammalian testis. Few varicose nerve fibres were seen running through the interstitial tissue to approach Leydig cells (154;man); the bulk of adrenergic nerves, however, seemed to be exclusively of vasomotor nature (155, 157). At an ultrastructural level Baumgarten and Holstein (158) and Belt and Cavazos (159) described adrenergic varicosities in close proximity of Leydig cells in the human and boar testis, respectively. Catecholamines have been shown to increase testosterone synthesis and secretion in the isolated perfused testis of the dog (160). However, because of the interference of vascular and non-vascular nerves in the mammalian testis it is virtually impossible to determine the impact of adrenergic nerves in testicular hormone secretion (161). Elegant experiments performed by Nance (162) have provided evidence for a neural control of FSH release. Unilateral hypothalamic knife cuts prevent a rise in serum FSH following hemiorchidectomy of prepubertal rats on the ipsilateral, but not the contralateral side.

Histochemical and electron microscope studies on the innervation of avian and reptilian testes (Fig. 3a) have revealed adrenergic and acetylcholinesterase-positive nerves in the interstitial tissue with marked species variations (153, 158, 163, 164). The density of innervation fluctuated during the seasonal cycle. In *Natrix natrix* numerous adrenergic terminals were found deeply invaginated into Leydig cell perikarya (153).

Neuro-endocrine contacts on Leydig cells were not established in three amphibian species (*Xenopus laevis*, *Rana temporaria*, *Bufo bufo*; 123), but nerve terminals that contain clear vesicles ending on Leydig cells have been demonstrated in the teleost testis (165, 166).

11. Ovary

Although studies on neural control mechanisms of ovarian functions are by far less numerous than those on hormonal regulation, autonomic innervation of ovarian endocrine cells by peripheral nerves has received increased attention in recent years (167–169).

Histochemical studies employing the Falck-Hillarp-technique for the demonstration of catecholamines, acetylcholinesterase staining and electron microscopic investigations have clearly led to abandon the view based on former silver impregnation studies that virtually any type of ovarian cell including endocrine cells and even oocytes are contacted by nerves. As far as endocrine cells are concerned there is wide agreement that nerve fibres do not penetrate the basal lamina surrounding the granulosa layer of follicles and do not enter the corpus luteum (170,

171). Using a quantitative approach we have shown that a functional innervation of rat and pig corpora lutea is highly improbable (171). Thus, the interstitial gland of the ovary remains as the only endocrine target of ovarian nerves.

The density of adrenergic nerves in the ovary of *mammals* have been reported to vary greatly in different species. The fluorescent networks are dense in human (172, 173), feline (172–175) and guinea pig ovaries (170, 176–178), but sparse in the ovaries of rat and rabbit (170, 176–178). The density of ovarians aminergic nerves, however, does not generally reflect the amount of nerve fibres supplying the interstitial gland: in the rat the interstitial gland receives numerous adrenergic fibres (170, 179), whereas the murine ovary with its moderate to rich supply of adrenergic nerves shows very few fluorescent varicosities within interstitial cell complexes (181). Acetylcholinesterase-positive nerves in mammalian ovaries are mostly adrenergic or sensory, respectively, as has been demonstrated by ablating vagal and sacral inputs and by chemical sympathectomy with 6-hydroxydopamine (167). Electron microscopic proof for an innervation of interstitial cells in mouse, rat and guinea pig has been provided (179, 181, 182). (Fig. 3b). Terminal axons approach interstitial cells making close contacts at distances between 100 nm (rat; 179) and 20 nm (mouse; 181).

Bird ovaries receive a rich adrenergic innervation, which is up to an order of magnitude denser than that of mammalian ovaries, as judged by fluorescence microscopy and quantitative determinations of catecholamines (172, 183, 184). Nerve bundles form a varicose plexus in the follicle wall, particularly in the theca interna. Studies on the distribution of smooth muscle and smooth muscle-like elements using antibodies against chicken gizzard myosin and adrenergic nerves in the wall of the follicle have clearly documented that the endocrine cells are more densely innervated than the contractile elements (184). Electron microscopic observations by Dahl (185) and by us (184) have provided strong morphological evidence for an innervation of steroidogenic cells in the theca interna of the hen. Most terminal axons take up 5-hydroxydopamine and form synapses on the surface of endocrine cells (184). A cholinergic component in the innervation of theca interna cells has been suggested (185), but not sufficiently proven. It is worth mentioning that nerves in the hen ovary contain both norepinephrine and epinephrine (184), but yet it is unknown whether both amines are stored separately or concomitantly within individual axons. Physiological studies (167, 168) support the concept that peripheral autonomic nerves make an important contribution to the regulation of ovarian endocrine secretions. Thus, e.g., it has been demonstrated that norepinephrine significantly depresses progesterone secretion from pregnant rat ovarian interstitial gland slices. Afferent nerves in the ovary seem to be part of a mixed neuronal-hormonal reflex arch mediating alterations in gonadotrophin and steroid blood levels after interruption of the vagus nerve or mechanical stimuli of the ovarian pedicle.

In summary, studies on the efferent and afferent neural pathways in the ovary, which has previously been considered to be only under hormonal control, have greatly helped to provide us with a more complex view of the possibly modulatory role of ovarian nerves.

12. Conclusions

An increasing body of reports presenting morphological evidence for an innervation of endocrine glands reflects a growing interest in the neural directives to the endocrine apparatus. Neuroendocrinology is no longer restricted to investigating diencephalic-pituitary pineal relationships, but has gained a new dimension, which includes both efferent and afferent links between the central nervous system and endocrine organs. Certainly, search for secretomotor junctions between nerve profiles and endocrine cells by electron microscopy cannot be the only means for studying the innervation of endocrine glands; morphological information has to be substantiated by physiological data in an interdisciplinary effort. Further advances in the field may be expected from recombination studies of ganglia or isolated neurons and endocrine tissues in cell culture similar to those that have been performed with autonomic ganglia and pancreatic islets (186; ORCI personal communication), stimulation of isolated endocrine cells by various transmitters *in vitro* or characterization of transmitter receptor on endocrine cell membranes. Furthermore, more attention will have to be paid to the afferent neural pathways from endocrine organs to the central nervous system, e.g. by mapping the distribution of nerve fibres in these organs that contain the established primary afferent neurotransmitters substance P, somatostatin and VIP. Finally, a broad comparative approach may reveal the general validity of the concepts arising from increased knowledge of operational mechanisms in neuroendocrine interactions.

Acknowledgements

I am greatly indebted to Mrs. W. Börner, Mrs. I. Ganski and Mr. Ch. Fiebiger for editorial help. Figure 1 was kindly provided by Professor H.F. Kern, Department of Anatomy and Cell Biology, University of Marburg. Figures 2 and 3 were reproduced with permission from Springer-Verlag, Heidelberg. Work from our laboratory described in this article was supported by the Deutsche Forschungsgemeinschaft.

References

1. Gorbman A, Bern HA: A textbook of comparative endocrinology. New York: Wiley, 1962.
2. Bargmann W: On the innervation of vertebrate endocrine organs. Int Congr Ser No 273 Endocrin, Proc Fourth Int Congr Endocrin Washington, 18–24 June 1972, pp 220–223.
3. Bargmann W: Über die Innervation der endokrinen Organe der Wirbeltiere. Nova Acta Leopoldina nr 217, Bd 41: 25–35. Halle: Deutsche Akademie d Naturforsch Leopoldina, 1975.
4. Scharrer B: General principles of neuroendocrine communication. The Neurosciences: Second Study Program. Schmitt FO (ed.). New York, The Rockefeller Univ Press, 1970, pp 519–529.
5. Scharrer B: Neuroendocrine Communication (neurohormonal, neurohumoral, and intermediate. Progr Brain Res Vol. 38. Ariens Kappers J. Schade JP (eds). Amsterdam, Elsevier, 1972, pp 7–18.
6. Vollrath L: Zur Innervation endokriner Drüsen. Münch Med Wschr 111/27: 1464–1468, 1969.
7. Bargmann W, Lindner E, Andres KH: Über Synapsen an endokrinen Epithelzellen und die Definition sekretorischer Neurone. Z Zellforsch 77: 282–298, 1967.
8. Dent JN, Gupta BL: Ultrastructural observations on the developmental cytology of the pituitary gland in the spotted newt. Gen Comp Endocrinol 8: 273–288, 1967.
9. Doerr-Schott J, Follenius E: Localisation des fibres aminergiques dans l'hypophyse de Rana esculenta. Etude autoradiographique au microscope électronique. Compt Rend Acad Sci, sér D 269: 737–740, 1969.
10. Doerr-Schott J, Follenius E: Innervation de l'hypophyse intermédiaire de Rana esculenta et identification des fibres aminergiques par autoradiographie au microscope électronique. Z Zellforsch 106: 99–118, 1970.
11. Enemar A, Falck B, Iturriza FC: Adrenergic nerves in the pars intermedia of the pituitary in the toad, Bufo arenarum. Z Zellforsch mikr Anat 77: 325–330, 1967.
12. Follenius E: Bases structurales et ultrastructurales des corrélations diencéphalo-hypophysaires chez les sél ciens et les téléostéens. Arch Anat Micr 54: 195–216, 1965.
13. Follenius E: Innervation adrénergique de la méta-adénohypophyse de l'Epinoche (Gasterosteus aculeatus L.). Mise en évidence par autoradiographie au microscope éléctronique. CR Acad Sci (D) 267: 1208–1211, 1968.
14. Follenius E: Localisation fine des terminaisons nerveuses fixant la noradrénaline H^3 dans les différents lobes de l'adénohypophyse de l'Epinoche (Gasterosteus aculeatus L.). In: Aspects of neuroendocrinology Bargmann W, Scharrer B, (eds). Berlin, Springer, 1970, pp 232–244.
15. Howe A, Maxwell DS: Electron microscopy of the pars intermedia of the pituitary gland in the rat. Gen Comp Endocrinol 11: 169–185, 1968.
16. Jørgensen CB, Larsen LO: Neuroendocrine mechanisms in lower vertebrates. In: Neuroendocrinology, Vol. 2. Martini L, Ganong WF (eds). New York, Academic Press, 1967.
17. Knowles Sir F, Vollrath L: A dual neurosecretory innervation of pars distalis of the pituitary of the eel. Nature (Lond) 208: 1343–1344, 1965.
18. Knowles Sir F, Vollrath L: Neurosecretory innervation of the pituitary of the eels Anguilla and Conger. Phil Trans 250: 311–342, 1966.
19. Kobayashi Y: Functional morphology of the pars intermedia of the rat hypophysis as revealed with the electron microscope. II. Correlation of the pars intermedia with the hypophyseo-adrenal axis. Z Zellforsch 68: 155–171, 1965.
20. Pehlemann FW: Ultrastructure and innervation of the pars intermedia of the pituitary of Xenopus laevis. Proc Fourth Confer Europ Comp Endocr Carlsbad, Czechoslovakia, (abstr), 1967.
21. Saland LC: Ultrastructure of the frog pars intermedia in relation to hypothalamic control of hormone release. Neuroendocrinol 3: 72–88, 1968.
22. Ziegler B: Licht und elektronenmikroskopische Untersuchungen an Pars intermedia und Neurohypophyse der Ratte. Z Zellforsch 59: 486–506, 1963.
23. Kaul S, Vollrath L: The goldfish pituitary. II. Innervation. Cell Tiss Res 154: 231–249, 1974.
24. Mira-Moser F: L'ultrastructure de l'adénohypophyse du crapaud Bufo bufo L. III. Différenciation des cellules de la pars distalis au cours du dévelopment larvaire. Z Zellforsch 125: 88–107, 1972.
25. Dierickx K, Lombaerts-Vandenberghe MP, Druyts A: The structure and vascularization of the pars tuberalis of the hypophysis of Rana temporaria. Z Zellforsch 114: 135–150, 1971.
26. Kurosumi K, Kobayashi Y: Nerve fibers and terminals in the rat anterior pituitary gland as revealed by electron microscopy. Arch histol japon 43: 141–155, 1980.
27. Théret C, Tamboise E: Etude ultrastructurale des rapports expérimentaux entre des cellules alpha et des fibres neurovégétatives dans l'adénohypophyse du rat. Ann Endocrinol Paris 24: 421–440, 1963.
28. Unsicker K: On the innervation of mammalian endocrine glands (anterior pituitary and parathyroids). Z Zellforsch 121: 283–291, 1971b.
29. Iturrizza FC: Further evidences for the blocking effect of catecholamines on the secretion of melanocyte-stimulating hormone in toads. Gen Comp Endocrinol 12: 417–426, 1969.
30. Oshima K, Gorbman A: Pars intermedia: Unitary electrical activity regulated by light. Science 163: 195–197, 1969.
31. Brettschneider H: Elektronenmikroskopische Beobachtungen über die Innervation der Schilddrüse. Z mikr-anat Forsch 69: 630–649, 1963.
32. Melander A: Thyroid hormone secretion. Its regulation by intrathyroidal amines. Acta Physiol Scand Suppl 370: 7–31, 1971.
33. Melander A, Ericson LE, Sundler F, Westgren U: Intrathyroidal amines in the regulation of thyroid activity. Rev Physiol Biochem Pharmacol 73: 39, 1975a.
34. Melander A, Sundler F, Westgren U: Sympathetic innervation of the thyroid: variation with species and with age. Endocrin 96: 102, 1975b.
35. Tice LW, Creveling CR: Electron microscopic identification of adrenergic nerve endings on thyroid epithelial cells. Endocrinol 97: 1123–1129, 1975.
36. Decoster C, Van Sande J, Dumont J, Mockel J: Dissociation of cyclic 3′, 5′-guanosine monophosphate accumulation and secretory inhibition in the action of carbamylcholine on thyroid. FEBS Lett 66: 191, 1976.
37. Dumont JE, Boynaems JM, Decoster C, Erneux C, Lamy F, Lecocq R, Mockel J, Unger J, Van Sande J: Biochemical mechanisms in the control of thyroid function and growth. Adv Cyclic Nucleotide Res 9: 723, 1978.
38. Melander A, Sundler F: Presence and influence of cholinergic nerves in the mouse thyroid. Endocrin 105: 7–9, 1979.
39. Van Sande J, Erneux C, Dumont J: Negative control of TSH action by iodide and acetylcholine: mechanism of action in intact thyroid cells. J Cyclic Nucleotide Res 3: 335, 1977.
40. Amenta F, Caporuscio D, Ferrante F, Porcelli F, Zomparelli M: Cholinergic nerves in the thyroid gland. Cell Tiss Res 195: 367–370, 1978.
41. Ahrén B, Alumets J, Ericsson M, Fahrenkrug J, Fahrenkrug L, Håkanson R, Hedner P, Lorén I, Melander A, Rerup C, Sundler F: VIP occurs in intrathyroidal nerves and stimulates thyroid hormone secretion. Nature 287: 343–345, 1980.

42. Robertson DR: The ultimobranchial body in Rana pipiens. III. Sympathetic innervation of the secretory parenchyma. Z Zellforsch 78: 328–340, 1967.
43. Stoeckel ME, Porte A: Sur l'ultrastructure des corps ultimobranchiaux du poussin. CR Acad Sci D, 265: 2051–2053, 1967.
44. Young BA, Harrison RJ: Ultrastructure of light cells in the dolphin thyroid. Z Zellforsch 96: 222–228, 1969.
45. Welsch U: Zur histologischen und enzymhistochemischen Differenzierung der C-Zellen und des Follikelepithels der Säugerschilddrüse. Habilitationsschrift. Kiel, 1971.
46. Altenähr E: Electron microscopical evidence for innervation of chief cells in human parathyroid gland. Experientia 27: 1077, 1971.
47. Atwal OS: Myelinated nerve fibers in the parathyroid gland of the dog: A light and electron-microscopic study. Acta anat 109: 3–12, 1981.
48. Capen CC, Cole CR, Hibly JW: The ultrastructure, histopathology and histochemistry of the parathyroid glands of pregnant and non-pregnant cows fed a high level of vit. D. Lab Invest 14: 1809–1825, 1965.
49. Isono H, Shoumura S: Effects of vagotomy on the ultrastructure of the parathyroid gland of the rabbit. Acta Anat 108: 273–280, 1980.
50. Fischer JA, Blum JW, Binswanger U: Acute parathyroid hormone response to epinephrine in vivo. J clin Invest 52: 2434–2440, 1973.
51. Kukreja SC, Hargis GK, Bowser N, Henderson WJ, Fisherman EW, Williams GA: Role of adrenergic stimuli in parathyroid hormone secretion in man. J clin Endocr Metab 40: 478–481, 1975.
52. Morii H, Fujita T, Okinaka S: Effect of vagotomy and atropine on recovery from induced hypocalcemia. Endocrin 72: 173–179, 1963.
53. Gingerich R, Aronoff SL: Insulin and glucagon secretion from rat islets maintained in a tissue culture-perfusion system. Diabetes 28: 276–281, 1979.
54. Kemp CB, Knight MJ, Sharp DW, Lacy PE, Ballinger WF: Transplantation of isolated pancreatic islets into the portal vein of diabetic rats. Nature (Lond), 244: 447–448, 1973.
55. Ono J, Takaki R, Okano H, Fukuma M: Long-term culture of pancreatic islet cells with special reference to the β-cell function. In Vitro 15: 95–102, 1979.
56. Woods SC, Porte D: Neural control of the endocrine pancreas. Physiol Rev 54: 596–619, 1974.
57. Ahrén B, Ericson LE, Lundquist I, Lorén I, Sundler F: Adrenergic innervation of pancreatic islets and modulation of insulin secretion by the sympatho-adrenal system. Cell Tiss Res 216: 15–30, 1981a.
58. Cegrell L: Adrenergic nerves and monoamine-containing cells in the mammalian endocrine pancreas. A comparative study. Acta physiol scand, Suppl 314: 17–23, 1967.
59. Falck B, Hellman B: Evidence for the presence of biogenic amines in pancreatic islets. Experientia Basel, 19: 139–140, 1963.
60. Falck B, Hellman B: Monoaminergic mechanisms in the endocrine pancreas. In: The structure and metabolism of the pancreatic islets. Brolin SE, Hellman B, Knutson H (eds). p 429–35. New York, The MacMillan Comp, 1964, pp 429–435.
61. Järhult J, Holst J: The role of the adrenergic innervation to the pancreatic islets in the control of insulin release during exercise in man. Pflügers Arch 383: 41–45, 1979.
62. Trandaburu T: Comparative observations on adrenergic innervation and monoamine content in endocrine pancreas of some amphibians, reptiles and birds. Endokrinologi 59: 260–264, 1972b.
63. Dahl E: The fine structure of the pancreatic nerves of the domestic fowl. Z Zellforsch 136: 501–510, 1973.
64. Esterhuizen AC, Spriggs TLB, Lever LD: Nature of islet cell innervation in the cat pancreas. Diabetes 17: 33–36, 1968.
65. Legg PG: The fine structure and innervation of the beta and delta cells in the islets of Langerhans of the cat. Z Zellforsch 80: 307–321, 1967.
66. lundquist I, Ericson LE: β-Adrenergic insulin release and adrenergic innervation of mouse pancreatic islets. Cell Tiss Res 193: 73–85, 1978.
67. Shorr SS, Bloom FE: Fine structure of islet-cell innervation in the pancreas of normal and alloxan-treated rats. Z Zellforsch 103: 12–25, 1970.
68. Watanabe T, Yasuda M: Electron microscopic study on the innervation of the pancreas of the domestic fowl. Cell Tiss Res 180: 453–465, 1977.
69. Bloom SR, Edwards AV: The release of pancreatic glucagon and inhibition of insulin in response to stimulation of the sympathetic innervation. J Physiol 253: 157–173, 1975.
70. Girardier L, Seydoux J, Campfield LA: Control of A and B cells in vivo by sympathetic nervous input and selective hyper- or hypoglycemia in dog pancreas. J Physiol, Paris 72: 801–814, 1976.
71. Girardier L, Seydoux J, Berger M, Veicsteinas A: Selective pancreatic nerve section. An investigation of neural control of glucagon release in the conscious unrestrained dog. J Physiol, Paris 74: 731–735, 1978.
72. Miller RE: Neural inhibition of insulin secretion from the isolated canine pancreas. Amer J Physiol 229: 144–149, 1975.
73. Schalch DS: The influence of physical stress and exercise on growth hormone and insulin secretion in man. J Lab clin Med 69: 256–269, 1976.
74. Wright PH, Malaisse WJ: Effects of epinephrine, stress and exercise on insulin secretion by the rat. Amer J Physiol 214: 1031–1034, 1968.
75. Porte D Jr, Williams RH: Inhibition of insulin release by norepinephrine in man. Science 152: 1248–1250, 1966.
76. Porte D Jr: A receptor mechanism for the inhibition of insulin release by epinephrine in man. J clin Invest 46ij 86–94, 1967.
77. Ahrén B, Järhult J, Lundquist I: Influence of the symphathoadrenal system and somatostatin on the secretion of insulin in the rat. J Physiol 312: 563–575, 1981b.
78. Kern HF, Hofmann HV, Kern D: Licht- und elektronenmikroskopische Untersuchung der Langerhansschen Inseln von Nutria (Myocastor coypus), mit besonderer Berücksichtigung der neuroinsulären Komplexe. Z Zellforsch 113: 216–229, 1971.
79. Orci L, Perrelet A, Ravazzola M, Malaisse-Lagae F, Renold AE: A speciliazed membrane junction between nerve endings and B-cells in islets of Langerhans. Europ J clin Invest 3: 443–445, 1973.
80. Coupland RE: The innervation of pancreas of the rat, cat and rabbit as revealed by the cholinesterase technique. J Anat (Lond) 92: 143–149, 1958.
81. Libman LJ, Sutherland SD: An investigation into the intrinsic innervation of the pancreas (using cholinesterase and usual nervous tissue stains) monkeys, cats, rabbits, guinea pigs and rats. J Anat (Lond) 99: 420–421, 1965.
82. Trandaburu T: Comparative observations on AChE distribution in pancreas of some amphibians, reptiles and birds, with special reference to the islets of Langerhans. Histochemie 32: 271–279, 1972a.
83. Trandaburu T: The intrinsic innervation of the pancreas of the grass-snake (Natrix n. natrix L.), with particular reference to acetylcholinesterase activity in the islets of Langerhans. J Anat 117: 575–589, 1974a.
84. Trandaburu T: Ultrastructural and acetylcholinesterase investigations on the pancreas intrinsic innervation of two bird species (Columbia livia domestica Gm. and Euodice cantans Gm.). Gegenbaurs morph Jahrb, Leipzig 120: 888–904, 1974b.
85. Bergman N, Miller RE: Direct enhancement of insulin secretion by vagal stimulation of the isolated pancreas. Amer J Physiol 225: 481–486, 1973.
86. Daniel PM, Henderson JR: The effect of vagal stimulation on plasma insulin and glucose levels in the baboon. J Physiol (Lond) 192: 317–327, 1967.
87. Frohman LA, Ezdinli EZ, Javid R: Effect of vagotomy and vagal stimulation on insulin secretion. Diabetes 16: 443–448, 1967.
88. Kaneto A, Kosaka K, Nakao K: Effects of stimulation of the vagus nerve on insulin secretion. Endocrin 80: 530–536, 1967.
89. Porte D Jr, Girardier L, Seydoux J, Kanazawa Y, Posternak J: Neural regulation of insulin secretion in the dog. J clin Invest 52: 210–214, 1973.
90. Gagerman E, Idahl L-Å, Meissner HP, Täljedal I-B: Insulin release, cGMP, cAMP and membrane potential in acetylcholine-stimulated islets. Amer J Physiol 235: E493–500, 1978.
91. Gagerman E, Molin J, Täljedal I-B: Are pancreatic β-cells under vagal control? Medical Biology 57: 48–51, 1979.
92. Iversen J: Effect of acetylcholine on the secretion of glucagon and insulin from the isolated, perfused canine pancreas. Diabetes 22: 381–387, 1973.
93. Loubatières-Mariani MM, Chapal J, Alric R, Loubatières A: Studies of the cholinergic receptors involved in the secretion of insulin using isolated perfused rat pancreas. Diabetologia 9: 439–446, 1973.
94. Watanabe T, Fujioka T: Ultrastructural alterations of the pancreatic D cell in the domestic fowl following vagotomy. Cell Tiss Res 211: 171–174, 1980.
95. Nunez EA, Gershon P, Gershon MD: Serotonin and seasonal variation in the pancreatic structure of bats: possible presence of serotonergic axons in the gland. Amer J Anat 159: 347–360, 1980.

96. Fujii S, Kobayashi S, Fujita T, Yanaihara N: VIP-immunoreactive nerves in the pancreas of the snake, Elaphe quadrivirgata (Boie: Another model for insular neurosecretion. Biomed Res 1: 180–184,
97. Kobayashi S, Fujita T: Fine structure of mammalian and avian pancreatic islets with special reference to D cells and nervous elements. Z Zellforsch 100: 340–363, 1969.
98. Serizawa Y, Kobayashi S, Fujita T: Neuro-insular complex type I in the mouse. Re-evaluation of the pancreatic islet as a modified ganglion. Arch histol japon 42: 389–394, 1979.
99. Stahl M: Elektronenmikroskopische Untersuchungen über die vegetative Innervation der Bauchspeicheldrüse. Z mikr-anat Forsch 70: 62–102, 1963.
100. Watari N: Fine structure of nervous elements in the pancreas of some vertebrates. Z Zellforsch 85: 291–314, 1968.
101. Kudo S: Fine structure of autonomic ganglion in the chicken pancreas. Arch histol japon 32: 455–497, 1971.
102. Fujita T: Histological studies on the neuro-insular complex in the pancreas of some mammals. Z Zellforsch 50: 94–109, 1959.
103. Campenhout E van: Contribution á l'étude de l'histogénèse du pancréas chez quelques mammifères. Les complexes sympathicoinsulaires. Arch Biol (Liège) 37: 121–171, 1927.
104. Simard LC: Les complexes neuro-insulaires du pancréas humain. Arch Anat micr Morph exp 33: 49–64, 1937.
105. Jacobowitz D: Histochemical studies of the autonomic innervation of the gut. J Pharmocol exp Ther 149: 358–364, 1965.
106. Ratzenhofer M, Müller O, Becker H: Zur Innervation der Drüsen- und Stromazellen im Kaninchenmagen. Mikroskopie 25: 283–296, 1969.
107. Gasbarrini G, Melchionda N, Benfenati F, Mantovani BA, Aureli G: Studio del sistema enterocromaffine e del metabolismo triptaminico nell úomo. Boll Sci Med 141: 85–118, 1969.
108. Ahlman H, Dahlström A, Kewenter J, Lundberg J: Vagal influence on serotonin concentration in enterochromaffin cells in the cat. Acta physiol scand 97: 362–368, 1976.
109. Kataoka K: The neuro-endocrine complex in the gastro-entero-pancreatic endocrine system. Arch histol jap 40, Suppl: 119–127, 1977.
110. Osaka M, Kobayashi S: Duodenal basal-granulated cells in the human fetus with special reference to their relationship to nervous elements. In: Endocrine gut and pancreas Fujita T (ed) Amsterdam, Elsevier, 1976, pp 145–158.
111. Aoi W, Henry DP, Weinberger MH: Evidence for a physiological role of renal sympathetic nerves in adrenergic stimulation of renin release in the rat. Circ Res 38: 123–126, 1976.
112. Barajas L: Anatomical considerations in the control of renin secretion. In: Control of Renin Secretion. New York, Plenum Press, 1972, pp 1–16.
113. Barajas L, Müller J: The innervation of the juxtaglomerular apparatus and surrounding tubules: a quantitative analysis by serial section electron microscopy. J Ultrastruct Res 43: 107–132, 1973.
114. Coote JH, Johns EJ, MacLeod HV, Singer B: Effect of renal nerve stimulation, renal blood flow, and adrenergic blockade on plasma renin activity in the cat. J Physiol (Lond) 226: 15–36, 1972.
115. Ganong WF: Effects of sympathetic activity and ACTH on renin and aldosterone secretion. In: Hypertension. Genest J, Koiw E (eds). New York, Heidelberg, 1972, pp 2–14.
116. Gorgas K: Struktur und Innervation des juxtaglomerulären Apparates der Ratte. (Structure and innervation of the juxtaglomerular apparatus of the rat). Berlin-Heidelberg-New York: Springer, 1978.
117. Johnson JA, Davis JO, Witty RT: Effects of catecholamine and renal nerve stimulation on renin release in the nonfiltering kidney. Circ Res 29: 646–653, 1971.
118. La Grange RG, Sloop CH: Selective stimulation of renal nerves in the anesthetized dog. Effect on renin release during controlled changes in renal hemodynamics. Circ Res 33: 704–712, 1973.
119. Müller J, Barajas L: Electron microscopic and histochemical evidence for a tubular innervation in the renal cortex of the monkey. J Ultrastruct Res 41: 533–549, 1972.
120. Nilsson O: The adrenergic innervation of the kidney. Lab Invest 14: 1392–1395, 1965.
121. Silverman A-J, Barajas L: Effect of reserpin on the juxtaglomerular granular cells and renal nerves. Lab Invest 30: 723–731, 1974.
122. Taugner R, Forssmann WG, Ganten D, Schiller A: Studies on the juxtaglomerular apparatus VI. Sympathetic innervation, catecholamines and the renin-angiotensin-system in rats and tree-shrews (Tupaia belangeri). Cell Tiss Res 212: 375–382, 1980.
123. Unsicker K, Axelsson S, Owman Ch, Svensson K-G: Innervation of the male genital tract and kidney in the amphibia, Xenopus laevis Daudin, Rana temporaria L., and Bufo bufo L. Cell Tiss Res 160: 453–484, 1975.
124. Vander AJ: Effect of catecholamines and the renal nerves on renin secretion in anesthetized dogs. Amer J Physiol 209: 659–662, 1965.
125. Wagenmark J, Ungerstedt U, Ljungqvist A: Sympathetic innervation of the juxtaglomerular cells of the kidney. Circ Res 22: 149–153, 1968.
126. Barajas L, Silverman AJ, Muller J: Ultrastructural localization of acetylcholinesterase in the renal nerves. J Ultrastruct Res 49: 297–311, 1974.
127. Weinberger MH, Aoi W, Henry DP: The direct effect of β-adrenergic stimulation on renin release by rat kidney slice in vitro. Circ Res 37: 318–324, 1975.
128. Dinerstein RJ, Vannice J, Henderson RC, Roth LJ, Goldberg LI, Hoffmann PC: Histofluorescence techniques provide evidence for dopamine-containing neuronal elements in canine kidney. Science 205: 497–499, 1979.
129. Unsicker K, Polonius T: Catecholamines and 5-hydroxytryptamine in corpuscles of stannius of the salmonid, Salmo irideus L. A study correlating electron microscopical, histochemical and chemical findings. General comp Endocrin 31: 121–132, 1977.
130. Bachmann R: Die Nebenniere. In: Handbuch der mikroskopischen Anatomie des Menschen, Bd. VI/5. Berlin-Göttingen-Heidelberg: Springer 1954.
131. Chester Jones I: The adrenal cortex. Cambridge: University Press 1957.
132. Deane HW: The adrenocortical hormones. Their origin, chemistry, physiology, and pharmacology, part I. In: Handbuch der experimentellen Pharmakologie, Bd. XIV/1. Berlin, Springer, 1962.
133. Unsicker K: Fine structure and innervation of the avian adrenal gland. V. Innervation of interrenal cells. Z Zellforsch 146: 403–416, 1973b.
134. Garcia-Alvarez F: Estudio ultraestructural sobre la inervación de la corteza suprarenal. An Anat 19: 267–279, 1970.
135. Garcia-Alvarez F: Caracteristicas ultraestructurales y significación de las vesiculas sinapticas adrenergicas de la corteza suprarenal del cavia cobaya. An Anat 21: 301–310, 1972.
136. Migally N: The innervation of the mouse adrenal cortex. Anat Rec 194: 105–112, 1979.
137. Robinson PM, Perry RA, Hardy KJ, Coghlan JP, Scoggins BA: The innervation of the adrenal cortex in the sheep, Ovis ovis. J Anat 124: 117–129, 1977.
138. Uno H; Catecholaminergic terminals in the perisinusoidal spaces of the hepatic acini and adrenal cortex of macaques. Anat Rec 187: 735, 1977.
139. Unsicker K: Zur Innervation der Nebennierenrinde vom Goldhamster. Z Zellforsch 95: 608–619, 1969.
140. Unsicker K: On the innervation of the rat and pig adrenal cortex. Z Zellforsch 116: 151–156, 1971a.
141. Unsicker K: Synapsen an Interrenalzellen der Krähe. Naturwiss 59: 81, 1972b.
142. Unsicker K, Habura-Flüh O, Zwarg U: Different types of small granule-containing cells and neurons in the guinea-pig adrenal medulla. Cell Tiss Res 189: 109–130, 1978.
143. Unsicker K: Innervation of adrenal cells in the lizards Lacerta dugesi and Lacerta pityusensis. Gen comp Endocrin 24: 409–412, 1974b.
144. Halász B, Szentágothai J: Histologischer Beweis einer nervösen Signalübermittlung von der Nebennierenrinde zum Hypothalamus. Z Zellforsch 50: 297–306, 1959.
145. Engeland WC, Dallman MF: Neural mediation of compensatory adrenal growth. Endocrin 99: 1659–1662, 1976.
146. Engeland WC, Dallman MF: Compensatory adrenal growth in neurally mediated. Neuroendocrin 19: 352–362, 1975.
147. Dallman MF, Engeland WC, Shinsako J: Compensatory adrenal growth; a neurally mediated reflex. Amer J Physiol 231: 408–414, 1976.
148. Holzwarth MA, Dallman MF: The effect of hypothalamic hemi-islands on compensatory adrenal growth. Brain Res 162: 33–43, 1979.
149. Vogt M: Observations on some conditions affecting the rate of hormone output by the suprarenal cortex. J Physiol (Lond) 103: 317–332, 1944.
150. Cushman P, Alter S, Hilton JG: Cortisol secretion by the dog adrenal: effects of cyclic adenosine monophosphate, dichloroisoproterenol,

dihydroergotamine and adrenaline. J Endocrin 34: 271–272, 1966.
151. Rosenfeld G: Stimulative effect of acetylcholine on the adrenocortical function of isolated perfused calf adrenals. Amer J Physiol 183: 272–278, 1955.
152. Blair-West JR, Coghlan JP, Denton DA, Goding JR, Munro JA, Peterson RE, Wintour M: Humoral stimulation of adrenal cortical secretion. J clin Invest 41: 1606–1627, 1962.
153. Unsicker K: Innervation of the testicular interstitial tissue in reptiles. Z Zellforsch 146: 123–138, 1973a.
154. Baumgarten HG, Holstein AF: Catecholaminhaltige Nervenfasern im Hoden des Menschen. Z Zellforsch 79: 389–395, 1967.
155. Baumgarten HG, Falck B, Holstein AF, Owman Ch, Owman T: Adrenergic innervation of the human testis, epididymis, ductus deferens and prostate: a fluorescence microscopic and fluorimetric study. Z Zellforsch 90: 81–95, 1968a.
156. Norberg KA, Risley PL, Ungerstedt U: Adrenergic innervation of the male reproductive ducts in some mammals. I. The distribution of adrenergic nerves. Z Zellforsch 76: 278–286, 1967.
157. Dayan AD: Variation between species in the innervation of intratesticular blood vessels. Experientia (Basel) 26: 1359–1360, 1970.
158. Baumgarten HG, Holstein AF: Noradrenerge Nervenfasern im Hoden van Mammaliern und anderen Vertebraten. J neuro-visc Rel, Suppl 10: 563–572, 1971.
159. Belt WD, Cavazos LF: Personal communication. Quoted after Hudson N: The nerves of the testis, epididymis and scrotum. In: The testis, vol. I Johnson AD, Gomes WR, Vandemark NL (eds). New York, Academic Press, 1970, pp 47–99.
160. Eik-Nes KB: Production and secretion of testicular steroids. Rec Progr Horm Res 27: 517–535, 1971.
161. Hodson N: The nerves of the testis, epididymis and scrotum. In: The testis vol. I. Johnson AD, Gomes WR, Vandemark NL (eds). New York, Academic Press, 1970, pp 47–99.
162. Nance DM: Neural innervation and control of the testes: a role for the paraventricular nucleus? Soc Neurosci Abstr Vol 7, p XX, 1980.
163. Haase E: Histochemische und elektronenmikroskopische Untersuchungen über die Innervation des Hodens vom Bergfink (Fringilla montifringilla). Verh Dt Zool Ges 66: 106–110, 1973.
164. Unsicker K: Zur Innervation von Leydigzellen bei Reptilien. Verh Anat Ges 68: 273–276, 1974a.
165. Follenius E: Innervation des cellules interstitielles chez un poisson téléstéen Lebistes reticulatus R. Etude au microscope élétronique. CR Acad Sci 259: 228–230, 1964.
166. Gresik EW: Fine structural evidence for the presence of nerve terminals in the testis of the teleost, Oryzias latipes. Gen comp Endocrin 21: 210–213, 1973.
167. Burden HW: Neural modulation of ovarian function. TINS 1978a, p 85–86.
168. Burden HW: Ovarian Innervation. In: The Vertebrate Ovary. Comparative Biology and Evolution. Jones RE (ed). New York, Plenum Press, 1978b, pp 615–638.
169. Gerendai I, Halász B: Neural participation in ovarian control. TINS 1978, pp 87–88.
170. Burden HW: Adrenergic innervation in ovaries of the rat and guinea pig. Amer J Anat 133: 455–462, 1972.
171. Unsicker K: Qualitative and quantitative studies on the innervation of the corpus luteum of rat and pig. Cell Tiss Res 152: 513–523, 1974c.
172. Jacobowitz D, Wallach EE: Histochemical and chemical studies of the autonomic innervation of the ovary. Endocrin 81: 1132–1139, 1967.
173. Owman C, Rosengren E, Sjöberg N-O: Adrenergic innervation of the human female reproductive organs: A histochemical and chemical investigation. Obstet Gynec 30: 763–773, 1967.
174. Fink G, Schofield GC: Experimental studies on the innervation of the ovary in cats. J Anat 109: 115–126, 1971.
175. Rosengren E, Sjöberg N-O: The adrenergic nerve supply to the female reproductive tract of the cat. Amer J Anat 121: 271–284, 1967.
176. Burden HW: The distribution of smooth muscle in the cat ovary with a note on its adrenergic innervation. J Morphol 140: 467–476, 1973.
177. Jordan SM: Adrenergic and cholinergic innervations of the reproductive tract and ovary in the guinea pig and rabbit. J Physiol (Lond) 210: 115p–117p.
178. Kulkarni PS, Wakade AR, Kirpekar SM: Sympathetic innervation of guinea pig uterus and ovary. Amer J Physiol 230: 1400–1405, 1976.
179. Lawrence IE Jr, Burden HW: The autonomic innervation of the interstitial gland of the rat ovary during pregnancy. Amer J Anat 147: 81–94.
180. Owman C, Sjöberg N-O: Adrenergic nerves in the female genital tract of the rabbit: With remarks on cholinesterase-containing structures. Z Zellforsch mikr Anat 74: 182–197, 1966.
181. Unsicker K: Zur Innervation der interstitiellen Drüse im Ovar der Maus (Mus musculus L.). Z Zellforsch 109: 46–54, 1970.
182. Svensson K-G, Owman C, Sjöberg N-O, Sporrong B, Walles B: Ultrastructural evidence for adrenergic innervation of the interstitial gland in the guinea pig ovary. Neuroendocrinol 17: 40–47, 1975.
183. Bennett T, Malmfors T: The adrenergic nervous system of the domestic fowl. Z Zellforsch mikr Anat 106: 22–50, 1970.
184. Unsicker K, Seidel F, Gröschel-Stewart U, Lindmar R, Löffelholz K, Wolf U: Zur adrenergen Innervation des Hühnerovars. Verh Anat Ges 74: 443–445, 1980.
185. Dahl E: Studies on the fine structure of ovarian interstitial tissue. 3. The innervation of the thecal gland of the domestic fowl. Z Zellforsch mikr Anat 109: 212–226, 1970.
186. Brinn JE, Burden HW, Schweisthal MR: Innervation of the cultured fetal rat pancreas. Cell Tiss Res 182: 133–138, 1977.

Author's address:
Department of Anatomy and Cell Biology
Philipps-University
Robert-Koch-Str. 6
D-3550 Marburg
F.R.G.

Index